Shengxiang Yang, Yew-Soon Ong, Yaochu Jin (Eds.)

Evolutionary Computation in Dynamic and Uncertain Environments

Studies in Computational Intelligence, Volume 51

Editor-in-chief
Prof. Janusz Kacprzyk
Systems Research Institute
Polish Academy of Sciences
ul. Newelska 6
01-447 Warsaw
Poland
E-mail: kacprzyk@ibspan.waw.pl

Further volumes of this series
can be found on our homepage:
springer.com

Vol. 31. Ajith Abraham, Crina Grosan, Vitorino Ramos (Eds.)
Stigmergic Optimization, 2006
ISBN 978-3-540-34689-0

Vol. 32. Akira Hirose
Complex-Valued Neural Networks, 2006
ISBN 978-3-540-33456-9

Vol. 33. Martin Pelikan, Kumara Sastry, Erick Cantú-Paz (Eds.)
Scalable Optimization via Probabilistic Modeling, 2006
ISBN 978-3-540-34953-2

Vol. 34. Ajith Abraham, Crina Grosan, Vitorino Ramos (Eds.)
Swarm Intelligence in Data Mining, 2006
ISBN 978-3-540-34955-6

Vol. 35. Ke Chen, Lipo Wang (Eds.)
Trends in Neural Computation, 2007
ISBN 978-3-540-36121-3

Vol. 36. Ildar Batyrshin, Janusz Kacprzyk, Leonid Sheremetor, Lotfi A. Zadeh (Eds.)
Preception-based Data Mining and Decision Making in Economics and Finance, 2006
ISBN 978-3-540-36244-9

Vol. 37. Jie Lu, Da Ruan, Guangquan Zhang (Eds.)
E-Service Intelligence, 2007
ISBN 978-3-540-37015-4

Vol. 38. Art Lew, Holger Mauch
Dynamic Programming, 2007
ISBN 978-3-540-37013-0

Vol. 39. Gregory Levitin (Ed.)
Computational Intelligence in Reliability Engineering, 2007
ISBN 978-3-540-37367-4

Vol. 40. Gregory Levitin (Ed.)
Computational Intelligence in Reliability Engineering, 2007
ISBN 978-3-540-37371-1

Vol. 41. Mukesh Khare, S.M. Shiva Nagendra (Eds.)
Artificial Neural Networks in Vehicular Pollution Modelling, 2007
ISBN 978-3-540-37417-6

Vol. 42. Bernd J. Krämer, Wolfgang A. Halang (Eds.)
Contributions to Ubiquitous Computing, 2007
ISBN 978-3-540-44909-6

Vol. 43. Fabrice Guillet, Howard J. Hamilton (Eds.)
Quality Measures in Data Mining, 2007
ISBN 978-3-540-44911-9

Vol. 44. Nadia Nedjah, Luiza de Macedo Mourelle, Mario Neto Borges, Nival Nunes de Almeida (Eds.)
Intelligent Educational Machines, 2007
ISBN 978-3-540-44920-1

Vol. 45. Vladimir G. Ivancevic, Tijana T. Ivancevic
Neuro-Fuzzy Associative Machinery for Comprehensive Brain and Cognition Modeling, 2007
ISBN 978-3-540-47463-0

Vol. 46. Valentina Zharkova, Lakhmi C. Jain
Artificial Intelligence in Recognition and Classification of Astrophysical and Medical Images, 2007
ISBN 978-3-540-47511-8

Vol. 47. S. Sumathi, S. Esakkirajan
Fundamentals of Relational Database Management Systems, 2007
ISBN 978-3-540-48397-7

Vol. 48. H. Yoshida (Ed.)
Advanced Computational Intelligence Paradigms in Healthcare, 2007
ISBN 978-3-540-47523-1

Vol. 49. Keshav P. Dahal, Kay Chen Tan, Peter I. Cowling (Eds.)
Evolutionary Scheduling, 2007
ISBN 978-3-540-48582-7

Vol. 50. Nadia Nedjah, Leandro dos Santos Coelho, Luiza de Macedo Mourelle (Eds.)
Mobile Robots: The Evolutionary Approach, 2007
ISBN 978-3-540-49719-6

Vol. 51. Shengxiang Yang, Yew-Soon Ong, Yaochu Jin (Eds.)
Evolutionary Computation in Dynamic and Uncertain Environments, 2007
ISBN 978-3-540-49772-1

Shengxiang Yang
Yew-Soon Ong
Yaochu Jin
(Eds.)

Evolutionary Computation in Dynamic and Uncertain Environments

With 272 Figures and 89 Tables

Springer

Dr. Shengxiang Yang
Department of Computer Science
University of Leicester
University Road
Leicester LE1 7RH
United Kingdom
E-mail: s.yang@mcs.le.ac.uk

Dr. Yaochu Jin
Honda Research Institute Europe
Carl-Legien-Str. 30
63073 Offenbach
Germany
E-mail: Yaochu.Jin@honda-ri.de

Dr. Yew-Soon Ong
School of Computer Engineering
Nanyang Technological University
Block N4, Nanyang Avenue
Singapore 639798
E-mail: ASYSOng@ntu.edu.sg

Library of Congress Control Number: 2006939142

ISSN print edition: 1860-949X
ISSN electronic edition: 1860-9503
ISBN-10 3-540-49772-2 Springer Berlin Heidelberg New York
ISBN-13 978-3-540-49772-1 Springer Berlin Heidelberg New York

Springer is a part of Springer Science+Business Media
springer.com

Cover design: deblik, Berlin
Typesetting by the editors using a Springer LaTeX macro package
Printed on acid-free paper SPIN: 11431411 89/SPi 5 4 3 2 1 0

To our families

Preface

Evolutionary computation is a class of problem optimization methodology with the inspiration from the natural evolution of species. In nature, the population of a species evolves by means of selection and variation. These two principles of natural evolution form the fundamental of evolutionary algorithms (EAs). During the past several decades, EAs have been extensively studied by the computer science and artificial intelligence communities. As a class of stochastic optimization techniques, EAs can often outperform classical optimization techniques for difficult real world problems.

Due to the ease of use and robustness, EAs have been applied to a wide variety of optimization problems. Most of these optimization problems tackled are stationary and deterministic. However, many real-world optimization problems are subjected to dynamic and uncertain environments that are often impossible to avoid in practice. For example, the fitness function is uncertain or noisy as a result of simulation errors, measurement errors or approximation errors. In addition, the design variables or environmental conditions may also perturb or change over time. For these dynamic and uncertain optimization problems, the objective of the EA is no longer to simply locate the global optimum solution, but to continuously track the optimum in dynamic environments, or to find a robust solution that operates optimally in the presence of uncertainties. This poses serious challenges to classical optimization techniques and conventional EAs as well. However, conventional EAs with proper enhancements are still good tools of choice for optimization problems in dynamic and uncertain environments. This is because EAs are inspired by principles of natural evolution, which takes place in the ever-changing dynamic and uncertain environment in nature.

Handling dynamic and uncertain optimization problems has been a topic since the early days of evolutionary computation and has received increasing research interests over recent years due to its challenge and its importance in practice. Several events, e.g., journal special issues, workshops and conference special sessions, have taken place in recent years in the field of evolutionary computation in dynamic and uncertain environments. A variety of methods have been reported across a broad range of application backgrounds in recent years. This motivated the project of this book. This book aims to timely reflect the most recent advances, present sophisticated real-world applications, and explore future research directions in the field.

We have a total of 26 chapters in this book, which cover a broad range of topics relevant to evolutionary computation in dynamic and uncertain environments. Further, the chapters in this book are presented as the following four categories:

- Part I: Optimum Tracking in Dynamic Environments
- Part II: Approximation of Fitness Functions
- Part III: Handling Noisy Fitness Functions
- Part IV: Search for Robust Solutions

Part I: Optimum Tracking in Dynamic Environments

Most problems studied by the evolutionary computation community are stationary optimization problems where no change occurs over time. For stationary optimization problems, the goal is to design EAs that can quickly and precisely locate the optimal solution(s) to the problem at hand. However, for dynamic optimization problems (DOPs) where change occurs over time, the main task is not to find one optimal solution but to track the moving optimum as soon and narrow as possible. This poses a serious challenge to conventional EAs due to the convergence problem. For stationary optimization problems, convergence at a proper pace other than premature convergence is exactly what is expected for EAs to locate the optimal solution. However, convergence becomes a big problem for DOPs because once converged, it is difficult for conventional EAs to adapt to the changing environment. DOPs usually require EAs to maintain certain level of diversity in the population. In order to deal with this problem, several approaches have been developed in recent years to enhance the performance of EAs in dynamic environments. Part I of the book encapsulates nine chapters that reflect the state-of-the-art research on EAs for problem optimization in dynamic environments and their application to real world dynamic problems.

The first six chapters of Part I present advanced EA approaches for general DOPs. In Chapter 1, Yang investigates the application of two kinds of explicit memory schemes, direct memory and associative memory, for genetic algorithms (GAs) and univariate marginal distribution algorithms (UMDAs) for DOPs. Based on a series of systematically constructed dynamic test environments, experiments are carried out to compare the direct and associative memory schemes for GAs and UMDAs. Blackwell in Chapter 2 studies the use of charged swarms in the particle swarm optimization (PSO) algorithm for DOPs. A self-adapting multi-swarm approach with an exclusion operator that provides effective repulsion between swarms is advocated in this chapter.

A simple rule for swarm birth and death is proposed so that the multi-swarm may adjust its size dynamically and in relation to the number of peaks in the dynamic environments. Chapter 3 by Schönemann experimentally investigates evolution strategies (ESs) for dynamic numerical optimization problems. The results demonstrates that self-adaptive ESs are powerful methods for dynamic environments. To avoid the handicaps of existing performance measures, a new measurement, called average best function value (ABFV), is developed to compare EAs for DOPs. This chapter also discusses the choice for different strategy parameters, e.g., the optimal number of mutation step sizes, for ESs for practical application. An orthogonal dynamic hill-climbing algorithm (ODHC) is presented by Zeng et al. in Chapter 4 for continuous DOPs. In ODHC, the local peak climber is not a solution, but a "niche" (a small hyper-rectangle). An orthogonal design method is employed on the niche in order to seek a potential peak more quickly. An archive is also used to store the latest found higher peaks, so the ODHC algorithm can learn from the past search. Chapter 5 by Tinós and Yang presents a self-organizing random immigrants scheme for GAs to address DOPs. In this scheme, the worst individual and its neighbours are replaced by random immigrants, which are placed in a subpopulation to protect them from being replaced by fitter individuals in the main population. In this way, when the fitness of the individuals are close, one single replacement of an individual can affect a large number of individuals of the population in a chain reaction. This simple approach can take the system to a self-organization behaviour, which is useful for GAs in dynamic environments. Bosman in Chapter 6 investigates the use of learning and anticipation for EAs for online DOPs. The time–linkage property, i.e., decisions taken now may influence the score in the future, has been identified as an important source of problem–difficulty. A means to address time–linkage is to predict the future (i.e. anticipation) by learning from the past. This is formalized into an algorithmic framework. Experimental results show that in the presence of time–linkage EAs based on this algorithmic framework outperform conventional EAs.

The last three chapters of Part I present work on the application of EAs for real world dynamic problems. In Chapter 7, Dam et al. investigates XCS, a genetics-based learning classifier system, for online dynamic data mining problems with different degrees of concept changes. In order to reduce the recovery time of XCS after concept changes, three strategies are proposed to force the system to learn quickly after severe changes. The effect of noise on the recovery time after a concept change is also experimentally investigated. Chapter 8 by Michalewicz et al. discusses the prediction and optimization issues in dynamic environments and suggests a system architecture, called Adaptive Business Intelligence, to handle a kind of real world problems where the evaluation functions are based on the prediction of the future values of some variables. Three diverse case studies in dynamic environments: pollution control, ship navigation, and car distribution, are presented. All these problems require some level of prediction and optimization for recommending the best course of action. Quintão et al. in Chapter 9 present the application of EAs to the area coverage and node connectivity problems in wireless sensor networks (WSNs), a kind of ad-hoc networks with distributed communication, sensing, and processing capacities. EAs are provided to support the network manager with the concern of controlling the energy consumption in the network and the quality of service.

Part II: Approximation of Fitness Functions

A continuing trend in science and engineering is the use of increasingly accurate simulation codes in the design and analysis process so as to produce ever more reliable and high quality products. Such technologies now play a central role in aiding scientists validate crucial designs and to study the effects of altering key design parameters on product performance. Nonetheless, the use of accurate simulation methods can be very timing consuming, leading to possibly unrealistic design cycle. Further, it poses a serious impediment to the practical application of existing optimization methods for automatically establishing the critical design parameters present in real world problems in science and engineering. Particularly, EAs typically require many thousands of function calls to the simulation codes in order to locate a near optimal solution. One promising way to significantly reduce the computational cost of EAs by employing computationally cheap approximation models or surrogates in place of the original computationally expensive fitness functions during evolutionary optimization. The five chapters showcased in Part II of the book reflect the recent state-of-the-art research on single and multi-objective evolutionary frameworks for tackling problems with computationally expensive optimization functions in the context of real world applications.

To reduce the number of expensive fitness function evaluations in evolutionary optimization, Graning et al. in Chapter 10 present a study on several individual-based and generation-based adaptive strategies for neural network metamodel management. In their preliminary study, it was reported that some of adaptation mechanisms proposed do not perform well as expected. The individual-based meta-model management was found to be most promising among all and subsequently applied to real world 3D blade design optimization problem. Song in Chapter 11 considers the use of approximation models based on Gaussian Processes for structural shape optimization. Application examples of the proposed surrogate-assisted evolutionary approaches are given in areas of firtree shape optimization using finite element method and engine nacelle optimization using computational fluid dynamics.

The next three chapters contributed by Reyes-Sierra and Coello in Chapter 12, Deb and Nain in Chpater 13, and Mack et al. in Chapter 14 present three independent studies on using approximation models in the context of multi-objective optimization. In particular, Reyes-Sierra and Coello present an empirical study on using fitness inheritance over approximation models in the context of PSO and multi-objective optimization for enhancing evolutionary search. Deb and Nain, on the other hand, present a successive fitness landscape modelling for reducing the exact function evaluation calls while retaining the basic search capability of NSGA-II. Using a case study in space propulsion, Mack et al. show that besides obtaining substantial improvements in the efficiency of the evolutionary search, surrogate-based optimization is also useful for novel or exploratory design tasks by offering a global view of the characteristics of the design space, thus enabling one to define previously unknown feasible design space boundaries and to reveal important physics in the design.

Part III: Handling Noisy Fitness Functions

It rarely happens that the fitness of real-world problems can be calculated by a deterministic analytical function. In most cases, the quality of a candidate solution has either to be measured by sensors or estimated using a numerical method. The sensory measurements are usually contaminated with noise in the environment, and the estimations are often subject to randomness. Though EAs are more robust against noise compared to derivative-dependent optimization methods, special attention needs to be paid in many cases. This part of the book presents four interesting chapters describing various approaches to handling noise in fitness evaluations.

Chapter 15 by Neri and Mäkinen describes a hierarchical EA for optimal design of an electrical grounding grid and an elastic structure. In this hierarchical algorithm, the fitness of a population depends on the results from another population, which is therefore noisy. To achieve reliable results, counter measures including population sizing, sampling sizing and survivor selection are taken. Evolution of multi-rover systems in noisy environments has been discussed in Chapter 16 by Tumer and Agogino. Since it is unpractical to evaluate the fitness of rovers in collective, the authors presented different methods for designing the fitness function for individual rovers without degrading the performance. Noise introduced by sensors are also considered. A memetic algorithm combining a trust-region based local search with evolutionary global search is presented in Chapter 17 where a trust-region method is combined with an evolutionary search. It is shown that on the one hand, evolutionary algorithms are inherently more robust against noise due to their derivative-free characteristics, the quadratic model used for fitness estimation also contributes to reducing the influence of noise. Chapter 18 by Tezuka et al deals with a financial optimization problem where the fitness values are based on a Monte Carlo method. The explicit sampling method is adopted for reducing the influence of noise. To reduce the computational costs, a selection efficiency index is proposed and the sampling size is adapted in such a way that the selection efficiency is maximized.

Part IV: Search for Robust Solutions

Solving optimization problems using EAs has always been perceived as finding the optimal solution over the entire search space. However, the global optima may not always be the most desirable solution in many real world engineering design problems. In practice, if the global optimal solution is very sensitive to uncertainties, for example, small changes in design variables or operating conditions, then it may not be appropriate to use this highly sensitive solution. Part III showcases eight chapters primarily on new methodologies of EAs for robust search.

Lim et al. in Chapter 19 report a study on several single and multi-objective inverse robust evolutionary optimization schemes that make little assumption on the uncertainty structure. The inverse approach searches for solutions that guarantee a certain degree of maximum uncertainty and, at the same time, satisfy the desired nominal performance of the final design solution. A multi-objective algorithm is also proposed in Chapter 20 by Goh and Tan for robust optimization. Their method incorporates the features of micro-GA (as a local search) to locate a worst case scenario of the candidate solution, a memory-based feature of tabu restriction to guide the evolutionary process and periodic re-evaluation of archived solutions to reduce uncertainty of evolved solutions. In the context of real world robust design applications, Hu et al. in Chapter 21 describe a robust design approach that exploits the open-ended topological synthesis capability of genetic programming and bond graph modelling (GPBG) for evolving robust lowpass and highpass analog filters with respect to parameter perturbations. Handa et al. on the other hand, describes a novel route planning memetic optimization system for a fleet of salting trucks that remains robust under different road temperatures and different temperature distributions in a road network in Chapter 22. Fan et al. in Chapter 23 report a method for robust layout synthesis of micro-electromechanical resonators subjected to inherent geometric uncertainties such as the fabrication error on the sidewall of the structure. An alternative technique that hybridizes EAs and Interval Arithmetic is also described in Chapter 24 by Rocco et al. Barrico and Antunes in Chapter 25 present the concept of degree of robustness in a multi-objective evolutionary approach. The information on the degree of robustness of solutions can then be used to support the decision maker in the selection of a robust compromise solution. Finally, Ling et al. report a study on the effect of the sampling number of Monte-Carlo simulation method used in a standard crowding genetic algorithm for robust optimal design of varied-line-spacing holographic grating in recording optics in Chapter 26.

Generally speaking, this book fulfils the original aims quite well. The four parts represent a great variety of work in the area of evolutionary computation in dynamic and uncertain environments. We hope that the publication of this book will further promote this emerging research field.

Finally, we would like to thank Dr. Janusz Kacprzyk for inviting us to edit this book in the Springer book series "Studies in Computational Intelligence". We acknowledge the authors for their fine contributions and cooperation during the book preparation. We are grateful to Thomas Ditzinger and Heather King of Springer for their kind support for this book.

Shengxiang Yang
Yew-Soon Ong
Yaochu Jin
November 2006

Contents

Part IV Search for Robust Solutions

List of Contributors

Hussein A. Abbass
Artificial Life and Adaptive Robotics Laboratory
School of Information Technology and Electrical Engineering
The University of New South Wales
Australian Defence Force Academy
Canberra ACT 2600, Australia
abbass@itee.adfa.edu.au

Adrian Agogino
UC Santa Cruz
NASA Ames Research Center
Mailstop 269-3
Moffett Field, CA 94035, USA
adrian@email.arc.nasa.gov

Kiyoshi Akama
Information Initiative Center
Hokkaido University
Kita 11 Nishi 5
Sapporo, 060-0811, Japan
akama@iic.hokudai.ac.jp

Carlos Henggeler Antunes
Department of Electrical Engineering and Computers
University of Coimbra
3000-033 Coimbra, Portugal
ch@deec.uc.pt

Steven William Armfield
School of Aerospace, Mechanical and Mechatronic Engineering
University of Sydney
Sydney NSW 2006, Australia
armfield@aeromech.usyd.edu.au

Carlos Barrico
Department of Informatics
University of Beira Interior
6201-001 Covilhã, Portugal
cbarrico@inescc.pt

Tim Blackwell
Department of Computing
Goldsmiths College
University of London
New Cross, London SE14 6NW, U.K.
t.blackwell@gold.ac.uk

Peter A. N. Bosman
Centre for Mathematics and Computer Science (CWI)
P.O. Box 94079
1090 GB Amsterdam, Netherlands
Peter.Bosman@cwi.nl

Lee Chapman
School of Geography, Earth, and Environmental Science
The University of Birmingham
Edgbaston
Birmingham B15 2TT, U.K.
l.chapman@bham.ac.uk

Constantin Chiriac
SolveIT Software
PO Box 3161
Adelaide, SA 5000, Australia
cc@solveitsoftware.com

Carlos A. Coello Coello
CINVESTAV-IPN (Evolutionary Computation Group)
Departamento de Ingeniería Eléctrica, Sección Computación
Av. IPN No. 2508
Col. San Pedro Zacatenco
México D.F. 07360, México
ccoello@cs.cinvestav.mx

Hai H. Dam
Artificial Life and Adaptive Robotics Laboratory
School of Information Technology and Electrical Engineering
The University of New South Wales
Australian Defence Force Academy
Canberra ACT 2600, Australia
z3140959@itee.adfa.edu.au

Kalyanmoy Deb
Kanpur Genetic Algorithms Laboratory (KanGAL)
Dept. of Mechanical Engineering
Indian Institute of Technology Kanpur
Kanpur, PIN 208 016, India
deb@iitk.ac.in

Lixin Ding
State Key Laboratory of Software Engineering
Wuhan University
Wuhan 430072, Hubei, P. R. China.
lx_ding@263.net

Zhun Fan
Technical University of Denmark
Dept. of Mechanical Engineering
Lynby, 2800, Denmark
zf@mek.dtu.dk

Tushar Goel
231 MAE-A, P.O. Box 116250
Mechanical and Aerospace Engineering Department
University of Florida
Gainesville, FL 32611-6250, USA
tusharg@ufl.edu

Chi Keong Goh
Department of Electrical and Computer Engineering
National University of Singapore
4 Engineering Drive 3
Singapore 117576
ckgoh@nus.edu.sg

Erik Goodman
2120 Engineering Building
Michigan State University
East Lansing, MI 48824, USA
goodman@egr.msu.edu

Raphael Haftka
231 MAE-A, P.O. Box 116250
Mechanical and Aerospace Engineering Department
University of Florida
Gainesville, FL 32611-6250, USA
haftka@ufl.edu

Hisashi Handa
Graduate School of Natural Science and Technology
Okayama University,
Tsushima-Naka 3-1-1,
Okayama, 700-8530, JAPAN
handa@sdc.it.okayama-u.ac.jp

Jianjun Hu
MCB 403D
University of Southern California
Los Angeles, CA, 90089, USA
jianjunh@usc.edu

Yaochu Jin
Honda Research Institute Europe
Carl-Legien-Str 30
63073 Offenbach am Main, Germany
yaochu.jin@honda-ri.de

Lishan Kang
State Key Laboratory of Software Engineering
Wuhan University
Wuhan 430072, Hubei, P. R. China.
kang_whu@yahoo.com

Shaobo Li
CAD/CIMS Institute
Guizhou University
Guiyang 550003, Guizhou,
P. R. China
lishaobo@gzu.edu.cn

Dudy Lim
School of Computer Engineering
Nanyang Technological University
Nanyang Avenue, Singapore 639798
dlim@ntu.edu.sg

Meng-Hiot Lim
School of Electrical and Electronics Engineering
Nanyang Technological University
Nanyang Avenue, Singapore 639798
emhlim@ntu.edu.sg

Qing Ling
Department of Automation
University of Science and Technology of China
Hefei 230026, P. R. China
qingling@mail.ustc.edu.cn

Chris Lokan
Artificial Life and Adaptive Robotics Laboratory
School of Information Technology and Electrical Engineering
The University of New South Wales
Australian Defence Force Academy
Canberra ACT 2600, Australia
cjl@itee.adfa.edu.au

Yolanda Mack
231 MAE-A, P.O. Box 116250
Mechanical and Aerospace Engineering Department
University of Florida
Gainesville, FL 32611-6250, USA
tiki@ufl.edu

Raino A. E. Mäkinen
Dipartimento di Elettrotecnica ed Elettronica
Politecnico di Bari
Via E. Orabona 4, 70125
Bari, Italy
rainom@it.jyu.fi

Geraldo Robson Mateus
Computer Science Department
Universidade Federal de Minas Gerais (UFMG)
Av. Antônio Carlos, 6627
Belo Horizonte, MG, Brazil
mateus@dcc.ufmg.br

Matthew Michalewicz
SolveIT Software
P.O. Box 3161
Adelaide, SA 5000, Australia
mm@solveitsoftware.com

Zbigniew Michalewicz
School of Computer Science
University of Adelaide
Adelaide, SA 5005, Australia
zbyszek@cs.adelaide.edu.au

Masaharu Munetomo
Information Initiative Center
Hokkaido University
Kita 11 Nishi 5
Sapporo, 060-0811, Japan
munetomo@iic.hokudai.ac.jp

Pawan K. S. Nain
Kanpur Genetic Algorithms Laboratory (KanGAL)
Dept. of Mechanical Engineering
Indian Institute of Technology Kanpur
Kanpur, PIN 208 016, India
pksnain@iitk.ac.in

Fabíola Guerra Nakamura
Computer Science Department
Universidade Federal de Minas Gerais (UFMG)
Av. Antônio Carlos, 6627
Belo Horizonte, MG, Brazil
fgnaka@dcc.ufmg.br

Ferrante Neri
Department of Mathematical Information Technology,
P.O. Box 35 (Agora), FI-40014
University of Jyväskylä, Finland
neferran@cc.jyu.fi

Yew-Soon Ong
School of Computer Engineering
Nanyang Technological University
Blk N4, 2b-39
Nanyang Avenue, Singapore 639798
asysong@ntu.edu.sg

Frederico Paiva Quintão
Computer Science Department
Universidade Federal de Minas Gerais (UFMG)
Av. Antônio Carlos, 6627
Belo Horizonte, MG, Brazil
fred@dcc.ufmg.br

Margarita Reyes-Sierra
CINVESTAV-IPN (Evolutionary Computation Group)
Departamento de Ingeniería Eléctrica, Sección Computación
Av. IPN No. 2508
Col. San Pedro Zacatenco
México D.F. 07360, México
mreyes@computacion.cs.cinvestav.mx

Claudio M. Rocco S.
Facultad de Ingeniería
Universidad Central de Venezuela
Apartado Postal 47937
Los Chaguaramos
Caracas 1041A, Venezuela
crocco@reacciun.ve

Ronald Rosenberg
Department of Electrical and Computer Engineering
Michigan State University
East Lansing, MI 48823, USA
rosenben@egr.msu.edu

Daniel E. Salazar A.
Instituto de Sistemas Inteligentes y Aplicaciones Numéricas en Ingeniería (IUSIANI)
Universidad de Las Palmas de Gran Canaria
Edif. Central del Parque Científico y Tecnológico
2 planta, Campus de Tafira Baja
Las Palmas 35017, Spain
danielsalazaraponte@gmail.com

Martin Schmidt
SolveIT Software
P.O. Box 3161
Adelaide, SA 5000, Australia
ms@solveitsoftware.com

Lutz Schönemann
Department of Computer Science
University of Dortmund
D-44221 Dortmund, Germany
lutz.schoenemann@cs.uni-dortmund.de

Hui Shi
School of Computer Science
China University of GeoSciences
Wuhan 430074, Hubei, P. R. China.
shihui0205@163.com

Wei Shyy
Dept. of Aerospace Engineering
University of Michigan
1320 Beal Avenue
Ann Arbor, MI 48109-2140, USA
weishyy@umich.edu

Wenbin Song
School of Engineering Sciences
University of Southampton
University Road
Southampton SO17 1BJ, UK
w.song@soton.ac.uk

Kay Chen Tan
Department of Electrical
and Computer Engineering
National University of Singapore
4 Engineering Drive 3
Singapore 117576
eletankc@nus.edu.sg

Yoel Tenne
School of Aerospace, Mechanical
and Mechatronic Engineering
University of Sydney
Sydney NSW 2006, Australia
joel.tenne@aeromech.usyd.edu.au

Masaru Tezuka
Research and Development Section
Hitachi East Japan Solutions Ltd.
2-16-10, Honcho, Aoba
Senadi, 980-0014, Japan
tezuka@hitachi-to.co.jp

Renato Tinós
Departamento de Física e
Matemática
Universidade de São Paulo (USP)
Av. Bandeirantes 3900
Ribeirão Preto, SP, 14040-901, Brazil
rtinos@ffclrp.usp.br

Kagan Tumer
NASA Ames Research Center
Mailstop 269-4
Moffett Field, CA 94035, USA
ktumer@mail.arc.nasa.gov

Jiachuan Wang
Systems Department
United Technologies Research Center
East Hartford, 06128, USA
WangJ2@utrc.utc.com

Qiuping Wang
National Synchrotron Radiation
Laboratory
University of Science and Technology
of China
Hefei 230026, P. R. China
qiuping@ustc.edu.cn

Min Wen
Technical University of Denmark
Department of Informatics and
Mathematical Modelling
Lynby, 2800, Denmark
mw@imm.dtu.dk

Gang Wu
Department of Automation
University of Science and Technology
of China
Hefei 230026, P. R. China
wug@ustc.edu.cn

Shengxiang Yang
Department of Computer Science
University of Leicester
University Road
Leicester LE1 7RH, U.K.
s.yang@mcs.le.ac.uk

Xin Yao
CERCIA
School of Computer Science
The University of Birmingham
Edgbaston,
Birmingham B15 2TT, U.K.
x.yao@cs.bham.ac.uk

Sanyou Zeng
School of Computer Science
China University of GeoSciences
Wuhan 430074, Hubei, P. R. China.
sanyou-zeng@263.net

Part I

Optimum Tracking in Dynamic Environments

1

Explicit Memory Schemes for Evolutionary Algorithms in Dynamic Environments

Shengxiang Yang

Department of Computer Science, University of Leicester
University Road, Leicester LE1 7RH, United Kingdom
s.yang@mcs.le.ac.uk

Summary. Problem optimization in dynamic environments has atrracted a growing interest from the evolutionary computation community in reccent years due to its importance in real world optimization problems. Several approaches have been developed to enhance the performance of evolutionary algorithms for dynamic optimization problems, of which the memory scheme is a major one. This chapter investigates the application of explicit memory schemes for evolutionary algorithms in dynamic environments. Two kinds of explicit memory schemes: direct memory and associative memory, are studied within two classes of evolutionary algorithms: genetic algorithms and univariate marginal distribution algorithms for dynamic optimization problems. Based on a series of systematically constructed dynamic test environments, experiments are carried out to investigate these explicit memory schemes and the performance of direct and associative memory schemes are campared and analysed. The experimental results show the efficiency of the memory schemes for evolutionary algorithms in dynamic environments, especially when the environment changes cyclically. The experimental results also indicate that the effect of the memory schemes depends not only on the dynamic problems and dynamic environments but also on the evolutionary algorithm used.

1.1 Introduction

Evolutionary algorithms (EAs) have been widely applied to solve stationary optimization problems. However, many real world problems are actually dynamic optimization problems (DOPs) . For DOPs, the fitness function, design variables, and/or environmental conditions may change over time due to many reasons, e.g., machine breakdown and financial factors. Hence, for DOPs the aim of an optimization algorithm is no longer to locate an optimal solution but to track the moving optima with time. This challenges traditional EAs seriously since they cannot adapt well to the changing environment once converged. However, traditional EAs with proper enhancements are still good tools of choice for optimization problems in dynamic environments. This is

S. Yang: *Explicit Memory Schemes for Evolutionary Algorithms in Dynamic Environments*, Studies in Computational Intelligence (SCI) **51**, 3–28 (2007)
www.springerlink.com

because EAs are basically inspired by principles of natural evolution, which has been taking place in the ever-changing dynamic environments in nature.

In recent years, there has been a growing interest in investigating EAs for DOPs. This trend reflects the importance of the practical application of EAs for real world optimization problems, many of which are DOPs [4]. Several approaches have been developed into EAs to address DOPs, such as maintaining diversity during the run via random immigrants [8, 23, 25], increasing diversity after a change [6, 7], using memory schemes to store and reuse useful information [3, 26], and multi-population approaches [12].

Among the approaches developed for EAs in dynamic environments, memory schemes have proved to be beneficial for many DOPs. Memory schemes work by storing useful information from the current environment and reusing it later in new environments. The useful information may be stored in two mechanisms: by implicit memory or by explicit memory. For implicit memory schemes, EAs use genotype representations that contain redundant information to store good (partial) solutions to be reused later. Typical examples are genetic algorithms (GAs) based on multiploidy representations [9, 9, 10, 12], structured encoding [7], or dualism mechanisms [24, 29]. Explicit memory schemes use precise representations but split an extra memory space to explicitly store useful information, e.g., good solutions [2, 3, 13, 22] and/or environmental information [21, 25], from the current generation for reuse in later generations or environments.

In this chapter, we focus on studying explicit memory schemes for EAs in dynamic environments. Two kinds of explicit memory schemes, *direct memroy* and *associative memory*, are investigated within two classes of EAs, GAs and univariate marginal distribution algorithms (UMDAs), for DOPs. For the direct memory scheme good solutions are stored in the memory and reused in new environments. For the associative memory scheme, the environmental information as well as good solutoins are stored and associated in the memory. When a change occurs, the stored environmental information associated with the best re-evaluated memory solution is used to create new individuals into the population. Using the dynamic problem generator proposed in [24, 29, 30], a series of dynamic test problems are constructed from a set of stationary functions and experiments are carried out to compare the performance of investigated GAs and UMDAs, with and without explicit memory schemes. The experimental results validates the efficiency of the memory scheme for GAs and UMDAs in dynamic environments.

The outline of this chapter is given as follows. The next section briefly reviews explicit memory schemes developed for EAs in dynamic environments. Section 1.3 describes the memory enhanced GAs investigated in this study while Section 1.4 describes the memory enhanced UMDAs investigated in this study. Section 1.5 presents the dynamic test environments for this study. The experimental results and relevant analysis are presented in Section 1.6. Section 1.7 concludes this paper with discussions on relevant future work.

1.2 Explicit Memory for EAs in Dynamic Environments

The application of memory schemes has proved to be able to enhance EA's performance in dynamic environments, especially when the environment changes cyclically in the search space[1]. In these environments, with time going an old environment will reappear exactly and the associated solution in the memory, which exactly remembers the old environment, will instantaneously move EAs to the reappeared environment.

As mentioned before, the basic principle of memory schemes is to, implicitly or explicitly, store useful information from the current environment and reuse it later in new environments. Implicit memory schemes for EAs in dynamic environments depend on redundant representations to store useful information for EAs to exploit during the run. On the contrast, explicit memory schemes make use of precise representation but split an extra storage space where useful information from the current generation can be explicitly stored and reused in later generations or environments. For explicit memory schemes there are three major technical considerations: what to store in the memory, how to update the memory, and how to retrieve the memory.

For the first aspect, a natural choice is to store good solutions and reuse them when the environment change is detected. This is called *direct memory scheme*. For example, Louis and Xu [13] studied the open shop re-scheduling problem. Whenever a change (in a known pattern) occurs, the GA is restarted from a population with partial (5-10%) individuals inherited from the old run while the rest are randomly initialized. The authors reported a significant improvement of their GA over the GA with totally random restart scheme. Instead of storing good solutions only, the environmental information can also be stored and associated with good solutions in the memory. When the environment changes, the stored environemntal information can be used to associate with certain stored good solutions and reuse them more efficiently or used to create new individuals into the population. This memory scheme is called *associative memory scheme*. For example, Ramsey and Greffenstette [21] studied a GA for robot control problem, where good candidate solutions are stored in a permanent memory together with information about the robot current environment. When the robot incurs a new environment that is similar to a stored environment instance, the associated stored controller solution is re-activated. This scheme was reported to yield significant improvements. In Yang [26] and Yang and Yao [30], an associative memory scheme was introduced into population-based incremental learning (PBIL) algorithms [1]. In this memory scheme, the best sample in the population together with the propability vector, which represents the current environment, is stored in the memory.

[1] For the convenience of description, we differentiate the environmental changing periodicality in time and space by wording *periodical* and *cyclic* respectively. The environment is said to be *periodically* changing if it changes in a fixed time interval, e.g., every certain EA generations, and is said to be *cyclically* changing if it visits several fixed states in the search space in a certain order repeatedly.

When a change is detected, the probability vector associated with the best re-evaluated memory sample is used to create new samples. The associative memory greatly improves PBIL's performance in dynamic environments.

The memory space is usually limited (and fixed) for the efficiency of computation and searching. This leads to the second consideration of explicit memory schemes: memory organization and updating mechanisms. As to the memory organization, there exist two mechanisms: *local mechanism* where the memory is individual-oriented and *global mechanism* where the memory is population-oriented. Trojanowski and Michalewicz [22] introduced a local memory approach, where for each individual the memory stores a number of its ancestors. When the environment changes, the current individual and its ancestors are re-evaluated and compete together with the best becoming the active individual while the others stored in the memory. The global memory mechanism is more natural and popular. In the global memory mechanism, the best individual of the population is replaced into the memory every certain or random generations according to a certain replacement policy, see [3, 3].

As to the memory updating mechanism, a general principle is to select one memory individual to be removed for or updated by the best individual from the population in order to make the stored individuals to be of above average fitness, not too old, and distributed across several promising areas of the search space. Branke [3] has discussed several memory replacement strategies: 1). replacing the least important one with the importance value of individuals being the linear combination of age, contribution to diversity, and fitness; 2). replacing the one with least contribution to memory variance; 3). replacing the most similar one if the new individual is better; and 4). replacing the less fit of a pair of memory individuals that has the minimum distance among all pairs. The third strategy seems the most practical one due to its simplicity and will be applied in the memory enhanced EAs studied in this chapter. Bendtsen and Krink [2] proposed a different memory updating scheme where the memory individual closest to the best population individual is moved toward the best population individual, instead of being replaced from the memory by the best population individual.

For the third concern regarding how to retrieve the memory, a natural idea is to retrieve the best memory individual(s) to replace the least fit individual(s) in the population. This can be done every generation or only when the environment changes. The memory retrieval is sort of coupled with the above two concerns. For example, for the direct memory scheme the whole memory individuals may enter the new population as in [13] or compete with the population individuals for the new population as in [3], while for the associative memory scheme only the associated memory individual(s) [21] or new individuals created by the associated environmental information [26, 30] may enter the new population. And for the local memory organization scheme the best ancestor of an active individual competes with it to become active in the population [22], while for the global memory scheme the best memory individual(s) may compete with all individuals in the population.

```
t := 0 and initialize population P(0) randomly
repeat
  evaluate(P(t))
  replace the worst individual in P(t) by elite from P(t − 1)
  P'(t) := selectForReproduction(P(t))
  crossover(P'(t), p_c)
  mutate(P'(t), p_m)
until termination condition holds        // e.g., t > t_max
```

Fig. 1.1. Pseudo-code for the standard GA (SGA) where elitism of size 1 is used

In the following two sections we will respectively describe the GAs and UMDAs with direct and associative memory schemes, which are investigated in this chapter.

1.3 Description of Investigated GAs

1.3.1 The Standard GA

The standard GA maintains and evolves a population of candidate solutions through selection and variation. New populations are generated by first probabilistically selecting relatively fitter individuals from the current population and then performing crossover and mutation on them to create new off-spring. This process continues until some termination condition becomes true, e.g., the maximum allowable number of generations t_{max} is reached. The pseudo-code for the standard GA (SGA) investigated in this chapter is shown in Fig. 1.1, where p_c and p_m are the crossover and mutation probabilities respectively and the elitism is used.

Usually, with the iteration of SGA, individuals in the population will eventually converge to the optimam or near optimum solution(s) in stationary environments due to the pressure of selection. Convergence at a proper pace, instead of pre-mature, may be beneficial and, in fact, is expected in many optimization problems for GAs to locate expected solutions in stationary environments. However, convergence becomes a big problem for GAs in dynamic environments. In fact, it is the main reason why traditional GAs do not perform well in dynamic environments. Convergence deprives the population of genetic diversity. Consequently, when a change occurs, it is hard for GAs to adpat to the new environment. Hence, in dynamic environments additional approaches are required to maintain the population diversity by random immigrants or adapt the GA directly to the new environment by memory schemes. The next two sub-sections describe respectively GAs with direct memory and associative memory enhancements, which are the main concern of this chapter.

```
t := 0 and initialize population P(0) randomly
t_M := rand(5, 10) and initialize memory M(0) randomly
repeat
  evaluate(P(t), M(t))
  replace the worst individual in P(t) by elite from P(t − 1)

  if environmental change detected then
    if DMGA then P'(t) := retrieveBestMembers(P(t), M(t))
    else              // for AMGA and HMGA
      denote the best memory point <B_M(t), D_M(t)>
      I(t) := create α ∗ (n − m) individuals from D_M(t)
      P'(t) := replace the worst individuals in P(t) by ones in I(t)
      if HMGA then P'(t) := retrieveBestMembers(P'(t), M(t))
  else P'(t) := P(t)

  if t = t_M then t_M := t + rand(5, 10)    // time to update memory
    denote the best individual in P'(t) by B_P(t)
    if not DMGA then D_P(t) := allele distribution vector in P'(t)
    if still any random point in M(t) then
      if DMGA then replace a random memory point by B_P(t)
      else replace a random memory point by <B_P(t), D_P(t)>
    else              // memory is full
      if DMGA then S^c_M(t) := the memory point closest to B_P(t)
        if f(B_P(t)) ≥ f(S^c_M(t)) then S^c_M(t) := B_P(t)
      else <S^c_M(t), D^c_M(t)> := the memory point closest to <B_P(t), D_P(t)>
        if f(B_P(t)) ≥ f(S^c_M(t)) then <S^c_M(t), D^c_M(t)> := <B_P(t), D_P(t)>

  // standard genetic operations
  P'(t) := selectForReproduction(P'(t))
  crossover(P'(t), p_c)
  mutate(P'(t), p_m)
until termination condition holds        // e.g., t > t_max
```

Fig. 1.2. Pseudo-code for the memory enhanced GAs: GA with direct memory (DMGA), GA with associative memory (AMGA), and GA with hybrid memory (HMGA)

1.3.2 GA with Direct Memory

The pseudo-code for the GA with direct memory, denoted *DMGA* in this chapter, is shown in Fig. 1.2, where $f(\cdot)$ is the fitness function. DMGA (and other memory enhanced EAs in this study) uses a memory of size $m = 0.1 * n$, which is randomly initialized. When the memory is due to update, if any of the randomly initialized points still exists in the memory, the best individual of the population will replace one of them randomly; otherwise, it will replace the closest memory point if it is better (the most similar memory updating strategy). Instead of updating the memory in a fixed time interval, the memory in DMGA is updated in a stochastic time pattern. After a memory updating at generation t, the next memory updating time t_M is given by: $t_M = t + rand(5, 10)$. This way, the potential effect that the environmental change period coincides with the memory updating period (e.g., the memory is updated whenever the environment changes) can be smoothed away.

The memory in DMGA is re-evaluated every generation to detect environmental changes. The environment is detected as changed if at least one individual in the memory is detected having changed its fitness. If a change is detected, the memory is merged with the current population and the best $n - m$ individuals are selected as an interim population to undergo standard genetic operations for a new population while the memory remains unchanged.

1.3.3 GA with Associative Memory

In [26, 30], an associative memory has been developed for PBILs in dynamic environments. The idea can be extended to GAs for DOPs [28]. That is, we can store the environmental information together with good solutions in the memory for later reuses. Here, the key thing is how to represent the current environment. As mentioned before, given a problem in a certain environment the population of a GA will eventually converge toward the optimum or near optimum of the environment when the GA progresses its searching. The convergence information, i.e., the *allele distribution* in the population, can be taken as the natural representation of the current environment. Each time when the best individual of the population is stored in the memory, the statistics information on the allele distribution for each locus, called the *allele distribution vector*, can also be stored in the memory and associated with the best individual.

The pseudo-code for the GA with the associative memory, denoted *AMGA*, is also shown in Fig. 1.2. Within AMGA, the memory is used to store solutions and associated environmental information. That is, each memory point consists of a pair $< S, \mathbf{D} >$, where S is the stored solution and $\mathbf{D}$ is the associated allele distribution vector. For binary encoding (as per this study), the frequency of ones over the population in a gene locus can be taken as the allele distribution for that locus.

As in DMGA, the memory in AMGA is re-evaluated every generation. If an environmental change is detected, the allele distribution vector of the best memory point $< B_M(t), \mathbf{D}_M(t) >$, i.e., the memory point with its solution $B_M(t)$ having the highest re-evaluated fitness, is extracted. And a set of $\alpha * (n - m)$ new individuals are created from this allele distribution vector $\mathbf{D}_M(t)$ and swapped into the population by replacing the worst individuals. Here, the parameter $\alpha \in [0.0, 1.0]$, called *associative factor*, determines the number of new individuals to be generated and hence the impact of the associative memory to the current population. A new individual $S = \{s_1, \cdots, s_l\}$ is created by $\mathbf{D}_M(t) = \{d_1, \cdots, d_l\}$ (l is the encoding length) as follows:

$$s_i = \begin{cases} 1, & \text{if } rand(0.0, 1.0) < d_i \\ 0, & otherwise \end{cases} \tag{1.1}$$

The memory replacement strategy in AMGA is similar to that in DMGA. When the memory is due to update, if there are still any randomly initialized memory points in the memory, a random one will be replaced by $< B_P(t),\ \mathbf{D}_P(t) >$, where $B_P(t)$ and $\mathbf{D}_P(t)$ are the best individual and allele distribution vector of the current population respectively; otherwise, we first find the memory point $< S_M^c(t), \mathbf{D}_M^c >$ with its solution $S_M^c(t)$ closest to $B_P(t)$. If $B_P(t)$ is fitter than $S_M^c(t)$, i.e., $f(B_P(t)) > f(S_M^c(t))$, the memory point is replaced by $< B_P(t), \mathbf{D}_P(t) >$.

The aforementioned direct and associative memory can be combined into GAs. The resulted GA is called the *hybrid memory based GA* (HMGA in short). The pseudo-code of HMGA is also shown in Fig. 1.2. HMGA differs from AMGA only as follows. When a change is detected, new individuals are created from the allele distribution vector of the best memory point and swapped into the population. Then, the original memory solutions $M(t)$ are merged with the main population to select $n - m$ best ones as the interim population to go though standard genetic operations.

1.4 Description of Investigated UMDAs

1.4.1 The Standard UMDA

Mühlenbein [19] introduced the UMDA as the simplest version of estimation of distribution algorithms (EDAs) [18]. Thereafter, there have been several modifications of UMDAs [14] and UMDAs have been applied to many optimization problems [11]. In the binary search space, UMDAs evolve a probability vector $\mathbf{p}(t) = (p(1,t), \ldots, p(l,t))$ where all the variables are assumed to be independent of each other. The pseudo-code for the standard UMDA (SUMDA) studied in this chapter is shown in Fig. 1.3, where the mechanisms of mutation and elitism are used.

SUMDA starts from the *central probability vector* that has a value of 0.5 for each locus and falls in the central point of the search space. Sampling this probability vector creates random solutions because the probability of creating a 1 or 0 on each locus is equal[2]. At iteration t, a population $S(t)$ of n individuals are sampled from the probability vector $\mathbf{p}(t)$. The samples are evaluated and an interim population $D(t)$ is formed by selecting μ ($\mu < n$) best individuals, denoted $\mathbf{x}_1(t), \cdots, \mathbf{x}_\mu(t)$, from $S(t)$. Then, the probability vector is updated by extracting statistics information from $D(t)$ as follows:

[2] Without loss of generality, a binary-encoded solution $\mathbf{x} = (x_1, \ldots, x_l) \in \{0,1\}^l$ is sampled from a probability vector $\mathbf{p}(t)$ as follows: for each locus i, if a randomly created number $r = rand(0.0, 1.0) < p(i,t)$, its allele x_i is set to 1; otherwise, x_i is set to 0.

```
t := 0 and initialize the probability vector p(0) := 0.5
repeat
   sample a population S(t) of individuals by p(t)
   evaluate(S(t))
   replace the worst individual in S(t) by elite from S(t − 1)
   select the best μ individuals from S(t) to form D(t)
   build p'(t) according to D(t) by Eqn. (1.2)
   mutate p'(t) by Eqn. (1.3)
until termination condition holds        // e.g., t > t_max
```

Fig. 1.3. Pseudo-code of the standard UMDA (SUMDA) with mutation and elitism

$$\mathbf{p}'(t) := \frac{1}{\mu} \sum_{k=1}^{k=\mu} \mathbf{x}_k(t) \tag{1.2}$$

After the probability vector is updated according to $D(t)$, in order to keep the diversity of generated samples in dynamic environments, a bitwise mutation is applied in SUMDA. The mutation operation always changes the probability vector toward the central probability vector as follows. For each locus $i = \{1, \ldots, l\}$, if a random number $r = rand(0.0, 1.0) < p_m$ (p_m is the mutation probability), then mutate $p(i,t)$ using the following formula:

$$p'(i,t) = \begin{cases} p(i,t) * (1.0 - \delta_m), & p(i,t) > 0.5 \\ p(i,t), & p(i,t) = 0.5 \\ p(i,t) * (1.0 - \delta_m) + \delta_m, & p(i,t) < 0.5, \end{cases} \tag{1.3}$$

where δ_m is the mutation shift that controls the amount a mutation operation alters the value in each bit position. After the mutation operation, a new set of samples is generated by the new probability vector and this cycle is repeated.

As the search progresses, the elements in the probability vector move away from their initial settings of 0.5 towards either 0.0 or 1.0, representing smaples of high fitness. The search stops when some termination condition holds, e.g., the maximum allowable number of iterations t_{max} is reached.

1.4.2 UMDA with Direct Memory

The direct memory scheme for GAs can be easily extended to UMDAs for DOPs. The pseudo-code for the investigated UMDA with the direct memory, denoted *DMUMDA*, is shown in Fig. 1.4. In Fig. 1.4, n is the number of evaluations per iteration including the memory samples and $f(\mathbf{x})$ denotes the fitness of individual $\mathbf{x}$.

As in DMGA, DMUMDA uses a memory to store best samples from the population. And the memory in DMUMDA is updated using the same stochastic time pattern as in DMGA: after a memory update at time t, the next

```
t := 0 and initialize p(0) := 0.5
t_M := rand(5,10) and initialize memory M(0) randomly
repeat
  sample a population S(t) of individuals by p(t)
  evaluate(S(t), M(t))
  replace the worst individual in S(t) by elite from S(t − 1)

  if environmental change detected then
    if DMUMDA then S'(t) := retrieveBestMembers(S(t), M(t))
    else              // for AMUMDA and HMUMDA
      denote the best memory point <B_M(t), p_M(t)>
      I(t) := create α * (n − m) individuals from p_M(t)
      S'(t) := replace the worst individuals in S(t) by ones in I(t)
      if HMUMDA then S'(t) := retrieveBestMembers(S'(t), M(t))
  else S'(t) := S(t)

  if t = t_M then t_M := t + rand(5,10)    // time to update memory
    denote the best individual in S'(t) by B_S(t)
    if still any random point in M(t) then
      if DMUMDA then replace a random memory point by B_S(t)
      else replace a random memory point by <B_S(t), p(t)>
    else              // memory is full
      if DMUMDA then S^c_M(t) := the memory point closest to B_S(t)
        if f(B_S(t)) ≥ f(S^c_M(t)) then S^c_M(t) := B_S(t)
      else <S^c_M(t), p^c_M(t)> := the memory point closest to <B_S(t), p(t)>
        if f(B_S(t)) ≥ f(S^c_M(t)) then <S^c_M(t), p^c_M(t)> := <B_S(t), p(t)>

  select the best μ individuals from S'(t) to form D(t)
  build p'(t) according to D(t) by Eqn. (1.2)
  mutate p'(t) by Eqn. (1.3)
until termination condition holds      // e.g., t > t_max
```

Fig. 1.4. Pseudo-code for the memory enhanced UMDAs: UMDA with direct memory (DMUMDA), UMDA with associative memory (AMUMDA), and UMDA with hybrid memory (HMUMDA)

memory updating time is $t_M = t + rand(5, 10)$. When the memory is due to update, we first find the memory point closest to the best population sample in terms of Hamming distance. If the best population sample has higher fitness than this memory sample, it is replaced by the best population sample; otherwise, the memory stays unchanged.

The memory in DMUMDA is re-evaluated every iteration. If any memory sample has its fitness changed, the environment is detected to be changed. Then, the memory will be merged with the current population to form an intermit population. If no environmental change is detected, DMUMDA progresses just as the standard UMDA does.

1.4.3 UMDA with Associative Memory

Using associative memory for UMDAs is more straightforward than for GAs because the probability vector that is evolved within UMDAs can be directly taken as the environmental information without any cost of further calculation. Each time when the best sample of the population is stored in the memory, the probability vector is also stored in the memory and associated with the sample. The pseudo-code of the UMDA with the associative memory, denoted *AMUMDA*, is also shown in Fig. 1.4.

The memory in AMUMDA has $m = 0.1 * n$ points, each consisting of a pair $<S, \mathbf{p}>$, where S is a stored sample and $\mathbf{p}$ is the associated probability vector. The memory is re-evaluated every generation. If an environmental change is detected, the probability vector of the best memory point $<B_M(t), \mathbf{p}_M(t)>$ is extracted to create a set of $\alpha * (n - m)$ new samples to replace the worst ones in the population. Here, the parameter $\alpha \in [0.0, 1.0]$ is the associative factor. The memory in AMUDMA is updated similarly as in AMGA. When the memory is due to update, if there are still any randomly initialized memory points in the memory, a random one is replaced by $<B_S(t), \mathbf{p}(t)>$, where $B_S(t)$ is the best sample in the population; otherwise, the memory point $<S_M^c(t), \mathbf{p}_M^c(t)>$ closest to $B_S(t)$ is replaced by $<B_S(t), \mathbf{p}(t)>$ if $B_S(t)$ is fitter than $S_M^c(t)$.

Similarly, the above direct and associative memory can be combined into UMDAs. The pseudo-code of the UMDA with a hybrid direct and associative memory, denoted *HMUMDA*, is also shown in Fig. 1.4. In HMUMDA, when a change is detected, after integrating the individuals that are sampled from the best memory probability vector into the population, the memory samples $M(t)$ are also merged with the population to select $n - m$ best ones as the interim population to build a new model.

1.5 Dynamic Test Environments

The dynamic problem generator proposed in [24, 29] can construct *random dynamic environments* from any binary-encoded stationary function $f(\mathbf{x})$ ($\mathbf{x} \in \{0,1\}^l$) by a bitwise exclusive-or (XOR) operator. Suppose the environment changes every τ generations. For each environmental period k, an XORing mask $\mathbf{M}(k)$ is incrementally generated as follows:

$$\mathbf{M}(k) = \mathbf{M}(k-1) \oplus \mathbf{T}(k), \tag{1.4}$$

where "$\oplus$" is the XOR operator (i.e., $1 \oplus 1 = 0$, $1 \oplus 0 = 1$, $0 \oplus 0 = 0$) and $\mathbf{T}(k)$ is an intermediate binary template randomly created with $\rho \times l$ ones for environmental period k. For the first period $k = 1$, $\mathbf{M}(1)$ is set to a zero vector. Then, the population at generation t is evaluated as below:

$$f(\mathbf{x}, t) = f(\mathbf{x} \oplus \mathbf{M}(k)), \tag{1.5}$$

where $k = \lceil t/\tau \rceil$ is the environmental period index. With this generator, the parameter τ controls the change speed while $\rho \in (0.0, 1.0)$ controls the severity of environmental changes. Bigger ρ means severer environmental change.

Recently, the XOR dynamic problem generator has been extended to construct *cyclic dynamic environments* in [27] and *cyclic dynamic environments with noise* further in [30]. With the XOR generator, cyclic dynamic environments are constructed as follows. First, we can generate $2K$ XORing masks $\mathbf{M}(0), \mathbf{M}(1), \cdots, \mathbf{M}(2K-1)$ as the *base states* in the search space randomly. Then, the environment can cycle among these base states in a fixed logical ring. Suppose the environment changes every τ generations, then the individuals at generation t is evaluated as follows:

$$f(\mathbf{x}, t) = f(\mathbf{x} \oplus \mathbf{M}(I_t)) = f(\mathbf{x} \oplus \mathbf{M}(k\%(2K))), \tag{1.6}$$

where $k = \lfloor t/\tau \rfloor$ is the index of the current environmental period and $I_t = k\%(2K)$ is the index of the base state the environment is in at generation t.

The $2K$ XORing masks can be generated in the following way. First, we construct K binary templates $\mathbf{T}(0), \cdots, \mathbf{T}(K-1)$ that form a random partition of the search space with each template containing $\rho \times l = l/K$ bits of ones[3]. Let $\mathbf{M}(0) = \mathbf{0}$ denote the initial state. Then, the other XORing masks are generated iteratively as follows:

$$\mathbf{M}(i+1) = \mathbf{M}(i) \oplus \mathbf{T}(i\%K), i = 0, \cdots, 2K-1 \tag{1.7}$$

The templates $\mathbf{T}(0), \cdots, \mathbf{T}(K-1)$ are first used to create K masks till $\mathbf{M}(K) = \mathbf{1}$ and then orderly reused to construct another K XORing masks till $\mathbf{M}(2K) = \mathbf{M}(0) = \mathbf{0}$. The Hamming distance between two neighbour XORing masks is the same and equals $\rho \times l$. Here, $\rho \in [1/l, 1.0]$ is the distance factor, determining the number of base states.

From the XOR generator, we can further construct cyclic dynamic environments with noise as follows. We can construct a set of base states and let the environment cycles among the base states just as above. However, each time the environment is about to move to a next base state $\mathbf{M}(i)$, $\mathbf{M}(i)$ is bitwise flipped with a small probability, denoted p_n in this chapter.

In this experimmental study, three 100-bit binary functions, denoted *OneMax*, *Royal Road* and *Deceptive* respectively, are selected as the base stationary functions to construct dynamic test environments. They all consist of 25 contiguous 4-bit building blocks and have an optimum fitness of 100. As shown in Fig. 1.5, the building block for each function is defined based on the unitation function, i.e., the number of ones inside the building block. The building block for *OneMax* is just a *OneMax* sub-function, which aims to maximize the number of ones in a chromosome. The building block for *Royal Road* contributes 4 to the total fitness if its unitation is 4; otherwise, it contributes 0. The building block for *Deceptive* is fully deceptive. These three stationary functions have increasing difficulty for EAs in the order from *OneMax* to *Royal Road* to *Deceptive*.

---

[3] In the partition each template $\mathbf{T}(i)$ $(i = 0, \cdots, K-1)$ has randomly but exclusively selected $\rho \times l$ bits set to 1 while other bits to 0. For example, $\mathbf{T}(0) = 0101$ and $\mathbf{T}(1) = 1010$ form a partition of the 4-bit search space.

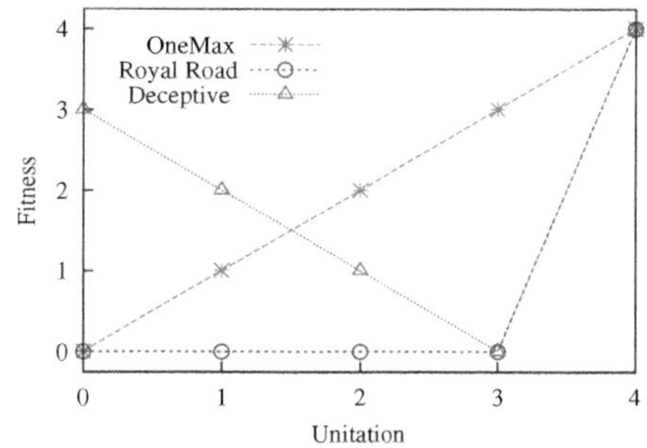

Fig. 1.5. Building block of the three stationary functions

Three kinds of dynamic environments, cyclic, cyclic with noise and random, are constructed from each base function using the XOR DOP generator. For each kind of dynamic environments, the landscape is periodically changed every τ generations during the run of an EA. In order to compare the performance of EAs in different dynamic environments, the parameters τ is set to 10 and 25 and ρ is set to 0.1, 0.2, 0.5, and 1.0 respectively. For cyclic dynamic problems with noise, the noise probability p_n is set to 0.05. Totally, a series of 24 DOPs, 2 values of τ combined with 4 values of ρ under three kinds of dynamic environments, are constructed from each stationary function.

1.6 Experimental Study

1.6.1 Experimental Design

Experiments were carried out to compare the performance of investigated EAs on the dynamic test environments. For all EAs, the parameters are set as follows: the total population size is set to $n = 100$, including memory size $m = 0.1*n = 10$ if used, and the elitism size is set to 1. For all GAs, parameters are set as: standard uniform crossover with the crossover probability $p_c = 0.6$, bit flip mutation with the mutation probability $p_m = 0.01$. For all UMDAs, the mutation probability $p_m = 0.02$ with the mutation shift $\delta_m = 0.05$, μ is set to $0.5 * n$ for SUMDA or $0.5 * (n - m)$ for memory enhanced UMDAs. For AMGAs and AMUMDAs, in order to test the effect of the associative factor α on their performance, α is set to 0.1, 0.5, and 1.0 respectively. For HMGAs and HMUMDAs, the associative factor α is set to 0.5. And an EA with associative memory will be reported as α-AMGA, α-HMGA, α-AMUMDA, or α-HMUMDA respectively in the experimental results.

For each experiment of an EA on a dynamic test problem, 50 independent runs were executed with the same set of random seeds. For each run 5000 generations were allowed, which are equivalent to 500 and 200 environmental changes for $\tau = 10$ and 25 respectively. For each run the best-of-generation fitness was recorded every generation. The overall offline performance of an algorithm on a problem is defined as:

$$\overline{F}_{BOG} = \frac{1}{G}\sum_{i=1}^{G}(\frac{1}{N}\sum_{j=1}^{N} F_{BOG_{ij}}), \tag{1.8}$$

where $G = 5000$ is the total number of generations for a run, $N = 50$ is the total number of runs, and $F_{BOG_{ij}}$ is the best-of-generation fitness of generation i of run j. The offline performance $\overline{F}_{BOG}$ is the best-of-generation fitness averaged over 50 runs and then averaged over the data gathering period.

1.6.2 Experimental Results and Analysis of GAs on DOPs

The experimental results of GAs on the dynamic test problems under cyclic, cyclic with noise, and random dynamic environments are plotted in Fig. 1.6 to Fig. 1.8 respectively. The corresponding statistical results of comparing GAs by one-tailed t-test with 98 degrees of freedom at a 0.05 level of significance are given in Table 1.1 to Table 1.3 respectively. In Table 1.1 to Table 1.3, the t-test result regarding *Alg.* 1 − *Alg.* 2 is shown as "=", "+", "−", "$s+$" or "$s-$" if *Alg.* 1 is statistically equivalent to, insignificantly better than, insignificantly worse than, significantly better than, or significantly worse than *Alg.* 2 respectively. From the figures and tables several results can be observed.

First, both DMGA and AMGAs perform significantly better than SGA on most dynamic problems, especially in cyclic environments. This result validates the efficiency of introducing memory schemes, either direct or associative, into GAs in dynamic environments. Viewing across Fig. 1.6 to Fig. 1.8, it can be seen that both DMGA and AMGAs achieve the largest performance improvement over SGA in cyclic environments. For example, when $\tau = 10$ and $\rho = 0.5$, the performance difference of DMGA over SGA, $\overline{F}_{BOG}(DMGA) - \overline{F}_{BOG}(SGA)$, is $94.1 - 58.9 = 35.2$, $67.2 - 59.8 = 7.4$, and $67.2 - 65.6 = 1.6$ under cyclic, cyclic with noise, and random environments respectively. This result indicates that the effect of memory schemes depends on the cyclicity of dynamic environments. When the environment changes randomly and slightly (i.e., ρ is small), both DMGA and AMGAs are beaten by SGA. This is because under these conditions, the environment is unlikely to return to a previous state that is memorized by the memory scheme. And hence inserting stored solutions or creating new ones according to the stored allele distribution vector may mislead or slow down the progress of the GAs.

Second, comparing AMGAs over DMGA, it can be seen that AMGAs outperform DMGA on many DOPs, especially under cyclic environments. This happens because the extracted memory allele distribution vector is much stronger than the stored memory solutions in adapting the GA to the new environment. However, when ρ is small and the environment changes randomly, AMGAs are beaten by DMGA for most cases, see the t-test results regarding α-AMGA − DMGA. This is because under these environments the negative effect of the associative memory in AMGAs may weigh over the direct memory in DMGA.

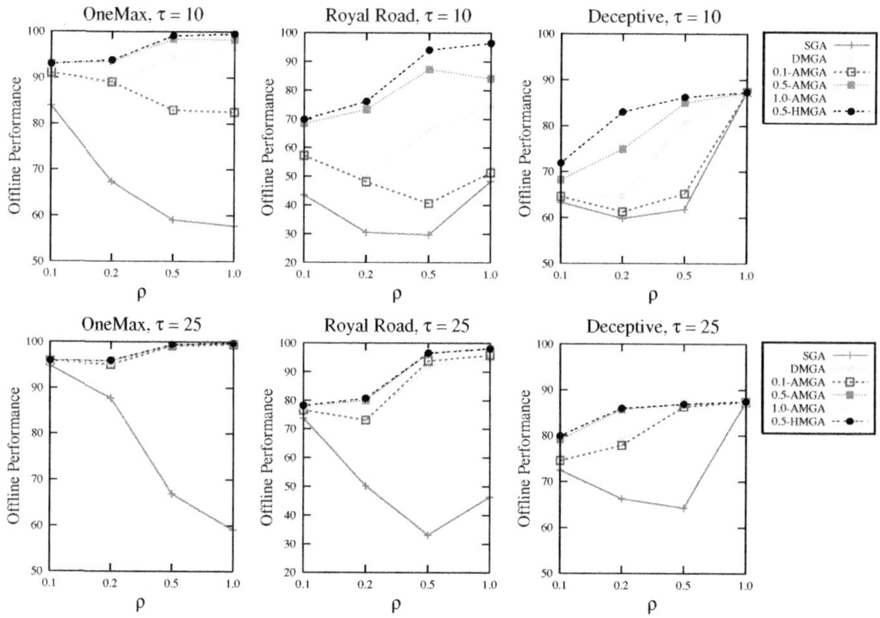

Fig. 1.6. Experimental results of GAs on cyclic DOPs

Table 1.1. The t-test results of comparing GAs on cyclic DOPs

t-test Result	*OneMax*				*Royal Road*				*Deceptive*			
$\tau = 10$, $\rho \Rightarrow$	0.1	0.2	0.5	1.0	0.1	0.2	0.5	1.0	0.1	0.2	0.5	1.0
DMGA − SGA	$s+$	$s+$	$s+$	$s+$	$s+$	$s+$	$s+$	$s+$	$s+$	$s+$	$s+$	$s+$
0.5-AMGA − SGA	$s+$	$s+$	$s+$	$s+$	$s+$	$s+$	$s+$	$s+$	$s+$	$s+$	$s+$	$-$
0.1-AMGA − DMGA	$s+$	$s+$	$s-$	$s-$	$s+$	$s-$	$s-$	$s-$	$+$	$s-$	$s-$	$s-$
0.5-AMGA − DMGA	$s+$	$s+$	$s+$	$s+$	$s+$	$s+$	$s+$	$s+$	$s+$	$s+$	$s+$	$s-$
1.0-AMGA − DMGA	$s+$	$s+$	$s+$	$s+$	$s+$	$s+$	$s+$	$s+$	$s+$	$s+$	$s+$	$s-$
0.5-AMGA − 0.1-AMGA	$s+$	$s+$	$s+$	$s+$	$s+$	$s+$	$s+$	$s+$	$s+$	$s+$	$s+$	$s-$
1.0-AMGA − 0.5-AMGA	$s-$	$s-$	$s+$	$s+$	$s+$	$s+$	$s+$	$s+$	$s+$	$s+$	$s+$	$s-$
0.5-HMGA − 0.5-AMGA	$-$	$s+$	$s+$	$s+$	$s+$	$s+$	$s+$	$s+$	$s+$	$s+$	$s+$	$+$
$\tau = 25$, $\rho \Rightarrow$	0.1	0.2	0.5	1.0	0.1	0.2	0.5	1.0	0.1	0.2	0.5	1.0
DMGA − SGA	$s+$	$s+$	$s+$	$s+$	$s+$	$s+$	$s+$	$s+$	$s+$	$s+$	$s+$	$s+$
0.5-AMGA − SGA	$s+$	$s+$	$s+$	$s+$	$s+$	$s+$	$s+$	$s+$	$s+$	$s+$	$s+$	$s+$
0.1-AMGA − DMGA	$s-$	$s-$	$s+$	$-$	$-$	$+$	$s+$	$s+$	$s-$	$-$	$-$	$-$
0.5-AMGA − DMGA	$-$	$+$	$s+$	$s+$	$s+$	$s+$	$s+$	$s+$	$s+$	$s+$	$s+$	$s-$
1.0-AMGA − DMGA	$s-$	$s-$	$s+$	$s+$	$s+$	$s+$	$s+$	$s+$	$s+$	$s+$	$s+$	$-$
0.5-AMGA − 0.1-AMGA	$+$	$s+$	$s+$	$s+$	$s+$	$s+$	$s+$	$s+$	$s+$	$s+$	$s+$	$-$
1.0-AMGA − 0.5-AMGA	$s-$	$s-$	$s+$	$s+$	$-$	$s+$	$s+$	$s+$	$s+$	$s+$	$s+$	$-$
0.5-HMGA − 0.5-AMGA	$s+$	$+$	$s+$	$s+$	$+$	$s+$	$s+$	$+$	$s+$	$s+$	$+$	$s+$

In order to better understand the performance of GAs, the dynamic performance of GAs regarding best-of-generation fitness against generations on dynamic *OneMax* functions with $\tau = 10$ and $\rho = 0.5$ under different cyclicity of dynamic environments is plotted in Fig. 1.9. In Fig. 1.9, the first and last 10 environmental changes (i.e., 100 generations) are shown and the data were

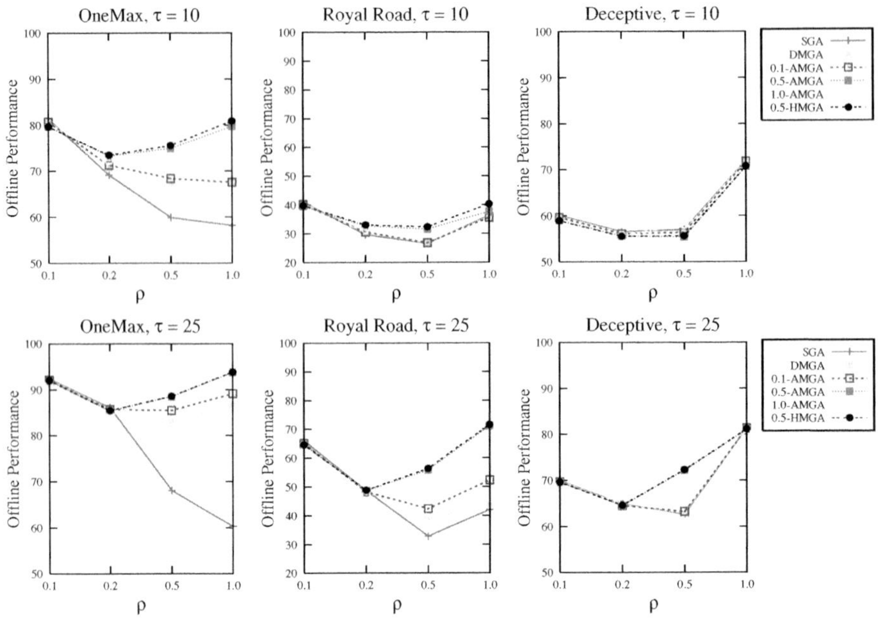

Fig. 1.7. Experimental results of GAs on cyclic DOPs with noise

Table 1.2. The t-test results of comparing GAs on cyclic DOPs with noise

t-test Result	*OneMax*				*Royal Road*				*Deceptive*			
$\tau = 10$, $\rho \Rightarrow$	0.1	0.2	0.5	1.0	0.1	0.2	0.5	1.0	0.1	0.2	0.5	1.0
DMGA − SGA	$s-$	$s+$	$s+$	$s+$	$s-$	$s+$	$s+$	$s-$	$-$	$-$	$-$	$-$
0.5-AMGA − SGA	$s-$	$s+$	$s+$	$s+$	$s-$	$s+$	$s+$	$s+$	$s-$	$s-$	$s-$	$s-$
0.1-AMGA − DMGA	$s-$	$s+$	$s+$	$-$	$s-$	$s+$	$+$	$s-$	$s-$	$-$	$-$	$s-$
0.5-AMGA − DMGA	$s-$	$s+$	$s+$	$s+$	$s-$	$s+$	$s+$	$s+$	$s-$	$s-$	$s-$	$s-$
1.0-AMGA − DMGA	$s-$	$s+$	$s+$	$s+$	$s-$	$s+$	$s+$	$s+$	$s-$	$s-$	$s+$	$s-$
0.5-AMGA − 0.1-AMGA	$s-$	$s+$	$s+$	$s+$	$s-$	$s+$	$s+$	$s+$	$s-$	$s-$	$s-$	$s-$
1.0-AMGA − 0.5-AMGA	$s-$	$s-$	$s+$	$s+$	$s-$	$s-$	$s+$	$s+$	$s-$	$=$	$s+$	$s-$
0.5-HMGA − 0.5-AMGA	$-$	$s+$	$s+$	$s+$	$s+$	$s+$	$s+$	$s+$	$+$	$+$	$s+$	$+$
$\tau = 25$, $\rho \Rightarrow$	0.1	0.2	0.5	1.0	0.1	0.2	0.5	1.0	0.1	0.2	0.5	1.0
DMGA − SGA	$s-$	$s-$	$s+$	$s+$	$s-$	$s-$	$s+$	$s+$	$-$	$-$	$+$	$+$
0.5-AMGA − SGA	$s-$	$s-$	$s+$	$s+$	$s-$	$+$	$s+$	$s+$	$s-$	$s-$	$s+$	$s-$
0.1-AMGA − DMGA	$-$	$s-$	$s+$	$s+$	$-$	$-$	$s+$	$s+$	$-$	$-$	$s+$	$s-$
0.5-AMGA − DMGA	$s-$	$s-$	$s+$	$s+$	$s-$	$s+$	$s+$	$s+$	$s-$	$-$	$s+$	$s-$
1.0-AMGA − DMGA	$s-$	$s-$	$s+$	$s+$	$s-$	$s-$	$s+$	$s+$	$s-$	$s+$	$s+$	$s-$
0.5-AMGA − 0.1-AMGA	$s-$	$s-$	$s+$	$s+$	$s-$	$s+$	$s+$	$s+$	$s-$	$-$	$s+$	$s-$
1.0-AMGA − 0.5-AMGA	$s-$	$s-$	$s+$	$s+$	$s-$	$s-$	$s+$	$s+$	$s-$	$s+$	$s+$	$s-$
0.5-HMGA − 0.5-AMGA	$+$	$+$	$+$	$s+$	$+$	$s+$	$s+$	$s+$	$+$	$+$	$s+$	$+$

averaged over 50 runs. From Fig. 1.9, it can be seen that, under cyclic and cyclic with noise environments, after several early stage environmental changes, the memory schemes start to take effect to maintain the performance of DMGA and AMGAs at a much higher fitness level than SGA. And the associative memory in AMGAs works better than the direct memory in DMGA, which

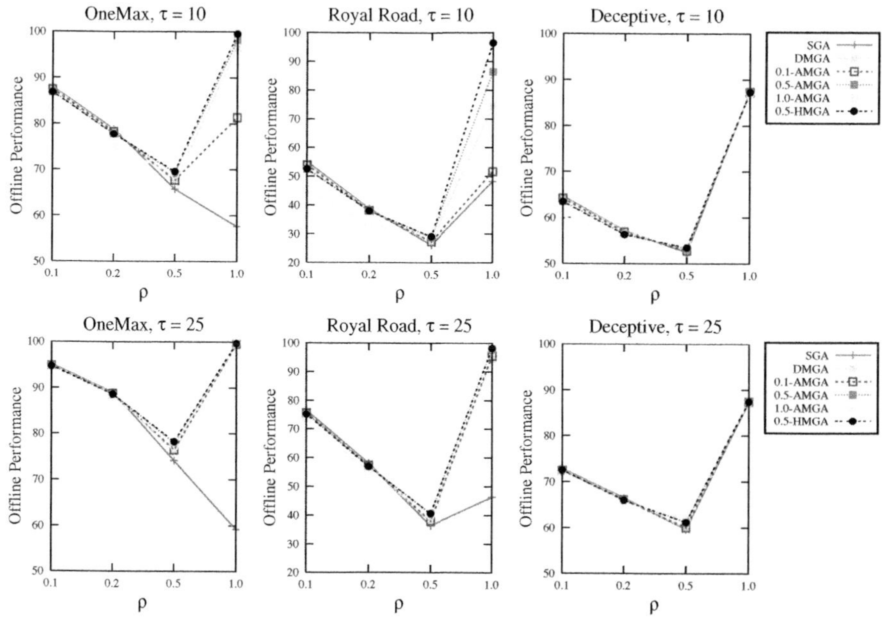

Fig. 1.8. Experimental results of GAs on random DOPs

Table 1.3. The t-test results of comparing GAs on random DOPs

t-test Result	*OneMax*				*Royal Road*				*Deceptive*			
$\tau = 10$, $\rho \Rightarrow$	0.1	0.2	0.5	1.0	0.1	0.2	0.5	1.0	0.1	0.2	0.5	1.0
DMGA − SGA	$s-$	$s-$	$s+$	$s+$	$s-$	$s-$	$s+$	$s+$	$s-$	$s-$	$s+$	$s+$
0.5-AMGA − SGA	$s-$	$s-$	$s+$	$s+$	$s-$	$s-$	$s+$	$s+$	$s-$	$s-$	$s+$	$-$
0.1-AMGA − DMGA	$s-$	$s-$	$s+$	$s-$	$s-$	$s-$	$s+$	$s-$	$s-$	$s-$	$s+$	$s-$
0.5-AMGA − DMGA	$s-$	$s-$	$s+$	$s+$	$s-$	$s-$	$s+$	$s+$	$s-$	$s-$	$s+$	$s-$
1.0-AMGA − DMGA	$s-$	$s-$	$s+$	$s+$	$s-$	$s-$	$s+$	$s+$	$s-$	$s-$	$s+$	$s-$
0.5-AMGA − 0.1-AMGA	$s-$	$s-$	$s+$	$s+$	$s-$	$s-$	$s+$	$s+$	$s-$	$s-$	$s+$	$s-$
1.0-AMGA − 0.5-AMGA	$s-$	$s-$	$s-$	$s+$	$s-$	$s-$	$s+$	$s+$	$s-$	$s-$	$s+$	$s-$
0.5-HMGA − 0.5-AMGA	$+$	$+$	$s+$	$s+$	$+$	$s+$	$s+$	$s+$	$+$	$+$	$s+$	$s+$
$\tau = 25$, $\rho \Rightarrow$	0.1	0.2	0.5	1.0	0.1	0.2	0.5	1.0	0.1	0.2	0.5	1.0
DMGA − SGA	$s-$	$s-$	$s+$	$s+$	$s-$	$s-$	$s+$	$s+$	$-$	$s-$	$s+$	$s+$
0.5-AMGA − SGA	$s-$	$s-$	$s+$	$s+$	$s-$	$s-$	$s+$	$s+$	$s-$	$s-$	$s+$	$s+$
0.1-AMGA − DMGA	$-$	$s-$	$s+$	$+$	$-$	$-$	$s+$	$s+$	$-$	$s-$	$s+$	$s-$
0.5-AMGA − DMGA	$s-$	$s-$	$s+$	$s+$	$s-$	$s-$	$s+$	$s+$	$-$	$s-$	$s+$	$s-$
1.0-AMGA − DMGA	$s-$	$s-$	$s+$	$s+$	$s-$	$s-$	$s+$	$s+$	$s-$	$s-$	$s+$	$s-$
0.5-AMGA − 0.1-AMGA	$s-$	$s-$	$s+$	$s+$	$s-$	$s-$	$s+$	$s+$	$+$	$s-$	$s+$	$+$
1.0-AMGA − 0.5-AMGA	$s-$	$s-$	$s-$	$s+$	$s-$	$s-$	$s-$	$s+$	$s-$	$s-$	$s+$	$-$
0.5-HMGA − 0.5-AMGA	$-$	$-$	$s+$	$s+$	$-$	$+$	$s+$	$+$	$+$	$+$	$+$	$s+$

can be seen in the late stage behaviour of GAs. Under random environments the effect of memory schemes is greatly deduced where all GAs behave almost the same and there is no clear vision regarding the effect of the memory schemes on the performance of DMGA and AMGAs.

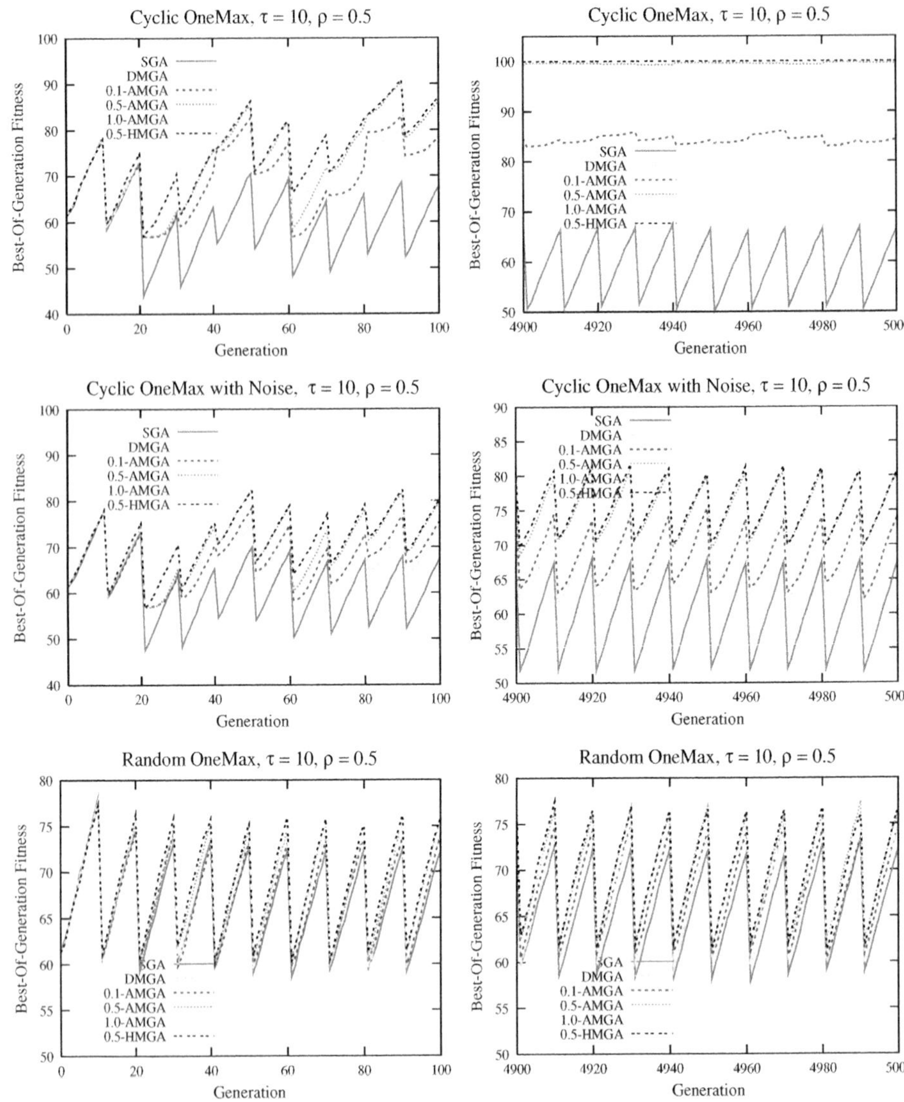

Fig. 1.9. Dynamic performance of GAs on the dynamic *OneMax* problems

Third, when examining the effect of α on AMGA's performance, it can be seen that 0.5-AMGA outperforms 0.1-AMGA on most dynamic problems, see the t-test results regarding 0.5-AMGA – 0.1-AMGA. This is because increasing the value of α enhances the effect of associative memory for AMGA. However, 1.0-AMGA is beaten by 0.5-AMGA on many cases, especially when ρ is small, see the t-test results regarding 1.0-AMGA – 0.5-AMGA. When

$\alpha = 1.0$, all individuals in the population are replaced by the new individuals created by the re-activated memory allele distribution vector when a change occurs. This may be disadvantageous. Especially, when ρ is small, the environment changes slightly and good solutions of the previous environment are likely also good for the new one. It is better to keep some of them instead of discarding them all.

Finally, comparing the performance of HMGA over AMGAs for DOPs, it can be seen that HMGA outperforms AMGAs for most dynamic problems, see the t-test results regarding 0.5-HMGA – 0.5-AMGA. For example, on the cyclic dynamic *Royal Road* function with $\tau = 10$ and $\rho = 0.5$, the performance of 0.5-HMGA is $\overline{F}_{BOG}(0.5\text{-}HMGA) = 94.0$, which is significantly better than the performance of 0.5-AMGA with $\overline{F}_{BOG}(0.5\text{-}AMGA) = 87.2$. However, the performance improvement of α-HMGA over α-AMGA is relatively small in comparison with the performance improvement of α-AMGA over SGA.

1.6.3 Experimental Results and Analysis of UMDAs on DOPs

The experimental results of UMDAs on the dynamic test problems under cyclic, cyclic with noise, and random dynamic environments are plotted in Fig. 1.10 to Fig. 1.12 respectively. The corresponding statistical results of comparing UMDAs by one-tailed t-test with 98 degrees of freedom at a 0.05 level of significance are given in Table 1.4 to Table 1.6 respectively. And the dynamic performance of UMDAs with respect to best-of-generation fitness against generations on the dynamic *OneMax* problems with $\tau = 10$ and $\rho = 0.5$ is plotted in Fig. 1.13, where the first and last 10 environmental changes are shown and the data were averaged over 50 runs. From the tables and figures, several results can be observed. The observations are similar to the previous observations regarding the experimental results of GAs and will be briefly recapped below. The main concern will be focused on the differences between the performance of UMDAs and GAs.

First, both direct and associative memory schemes significantly improve the performance of UMDAs on most DOPs, see the t-test results regarding DMUMDA – SUMDA and 0.5-AMUMDA – SUMDA. And it seems the memory schemes have a more consistent positive effect on the performance of UMDAs on all DOPs than on the performance of GAs. For example, 0.5-AMUMDA significantly outperforms SUMDA not only on all cyclic DOPs but also on almost all noisy and random DOPs. For example, from the dynamic behaviour of UMDAs shown in Fig. 1.13, it can be see that memory enhanced UMDAs maintain a much higher fitness level than SUMDA not only on cyclic *OneMax* problem but also on the random *OneMax* problem. This result indicates that the effect of memory schemes depends not only on the dynamic problems and environments but also on the EA used.

Second, the associative factor α has the similar effect on the performance of AMUMDAs as on the performance of AMGAs. Increasing the value of α

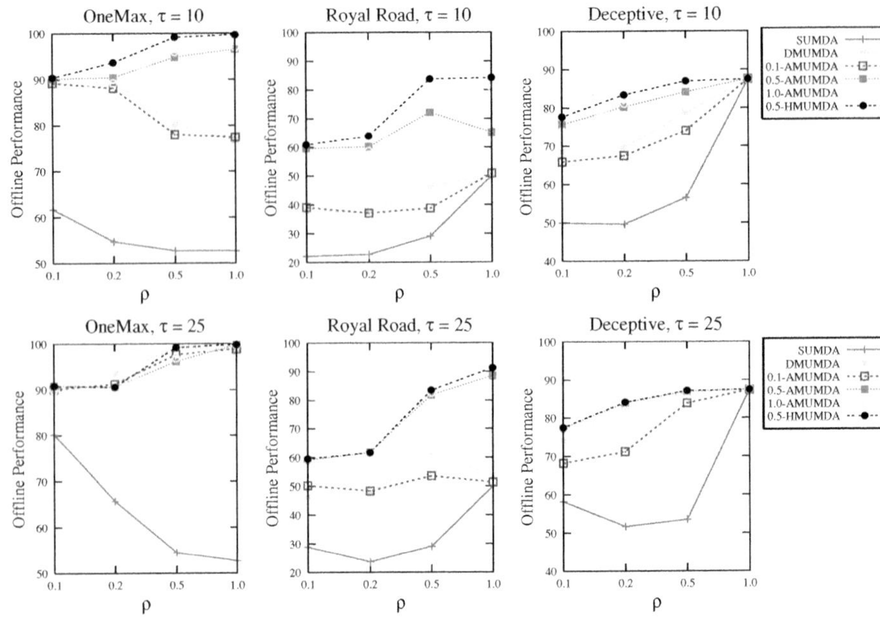

Fig. 1.10. Experimental results of UMDAs on cyclic DOPs

Table 1.4. The t-test results of comparing UMDAs on cyclic DOPs

t-test Result	*OneMax*				*Royal Road*				*Deceptive*			
$\tau = 10$, $\rho \Rightarrow$	0.1	0.2	0.5	1.0	0.1	0.2	0.5	1.0	0.1	0.2	0.5	1.0
DMUMDA – SUMDA	$s+$	$s+$	$s+$	$s+$	$s+$	$s+$	$s+$	$s+$	$s+$	$s+$	$s+$	$s+$
0.5-AMUMDA – SUMDA	$s+$	$s+$	$s+$	$s+$	$s+$	$s+$	$s+$	$s+$	$s+$	$s+$	$s+$	$s+$
0.1-AMUMDA – DMUMDA	$-$	$s-$	$s-$	$+$	$s+$	$-$	$s-$	$s+$	$s-$	$s-$	$s-$	$=$
0.5-AMUMDA – DMUMDA	$s+$	$s+$	$s+$	$s+$	$s+$	$s+$	$s+$	$s+$	$s+$	$s+$	$s+$	$+$
1.0-AMUMDA – DMUMDA	$s-$	$-$	$s+$	$s+$	$s+$	$s+$	$s+$	$s+$	$s+$	$s+$	$s+$	$-$
0.5-AMUMDA – 0.1-AMUMDA	$s+$	$s+$	$s+$	$s+$	$s+$	$s+$	$s+$	$s+$	$s+$	$s+$	$s+$	$+$
1.0-AMUMDA – 0.5-AMUMDA	$s-$	$s-$	$s+$	$+$	$s-$	$s-$	$s+$	$s+$	$s-$	$+$	$s+$	$-$
0.5-HMUMDA – 0.5-AMUMDA	$s+$	$s+$	$s+$	$s+$	$s+$	$s+$	$s+$	$s+$	$s+$	$s+$	$s+$	$-$
$\tau = 25$, $\rho \Rightarrow$	0.1	0.2	0.5	1.0	0.1	0.2	0.5	1.0	0.1	0.2	0.5	1.0
DMUMDA – SUMDA	$s+$	$s+$	$s+$	$s+$	$s+$	$s+$	$s+$	$s+$	$s+$	$s+$	$s+$	$s+$
0.5-AMUMDA – SUMDA	$s+$	$s+$	$s+$	$s+$	$s+$	$s+$	$s+$	$s+$	$s+$	$s+$	$s+$	$s+$
0.1-AMUMDA – DMUMDA	$s+$	$s-$	$s-$	$s+$	$s+$	$+$	$s-$	$s-$	$s-$	$s-$	$s-$	$s-$
0.5-AMUMDA – DMUMDA	$s+$	$s-$	$s-$	$s+$	$s+$	$s+$	$s+$	$s+$	$s+$	$s+$	$s+$	$-$
1.0-AMUMDA – DMUMDA	$s+$	$s-$	$s-$	$s+$	$s+$	$s+$	$s+$	$s+$	$s+$	$s+$	$s+$	$+$
0.5-AMUMDA – 0.1-AMUMDA	$s+$	$-$	$s-$	$s+$	$s+$	$s+$	$s+$	$s+$	$s+$	$s+$	$s+$	$+$
1.0-AMUMDA – 0.5-AMUMDA	$s-$	$-$	$+$	$+$	$s-$	$-$	$+$	$s+$	$-$	$+$	$s+$	$+$
0.5-HMUMDA – 0.5-AMUMDA	$+$	$+$	$s+$	$s+$	$+$	$+$	$s+$	$s+$	$+$	$+$	$s+$	$+$

from 0.1 to 0.5 improves the performance of AMUMDA while further raising the value of α to 1.0 degrades the performance of AMUMDA, see the t-test results regarding 0.5-AMUMDA – 0.1-AMUMDA and 1.0-AMUMDA – 0.5-AMUMDA in Table1.4 to Table1.6. This result can also be seen in their dynamic performance shown in Fig. 1.13, where 1.0-AMUMDA achieves a

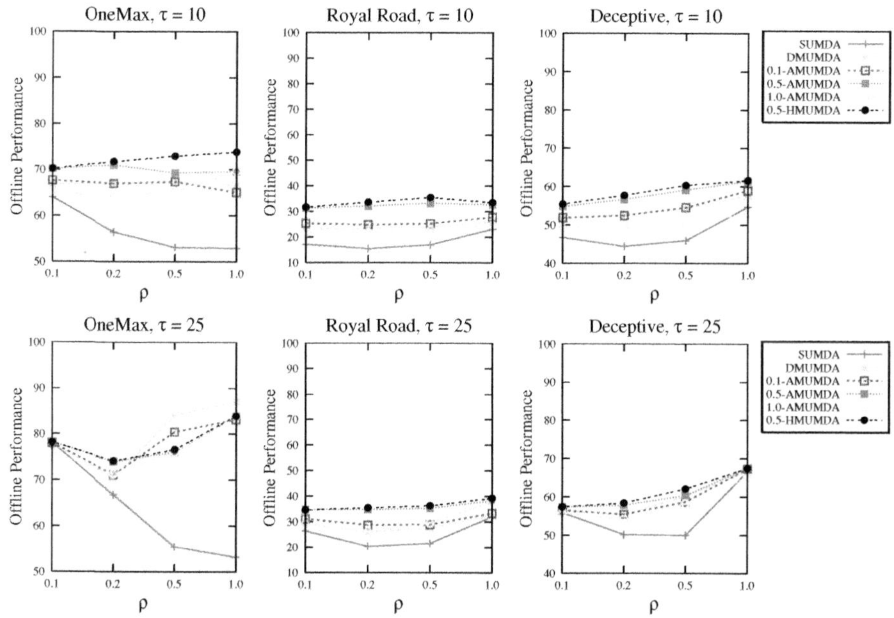

Fig. 1.11. Experimental results of UMDAs on cyclic DOPs with noise

Table 1.5. The t-test results of comparing UMDAs on cyclic DOPs with noise

t-test Result	*OneMax*				*Royal Road*				*Deceptive*			
$\tau = 10$, $\rho \Rightarrow$	0.1	0.2	0.5	1.0	0.1	0.2	0.5	1.0	0.1	0.2	0.5	1.0
DMUMDA – SUMDA	$s+$	$s+$	$s+$	$s+$	$s+$	$s+$	$s+$	$s+$	$s+$	$s+$	$s+$	$s+$
0.5-AMUMDA – SUMDA	$s+$	$s+$	$s+$	$s+$	$s+$	$s+$	$s+$	$s+$	$s+$	$s+$	$s+$	$s+$
0.1-AMUMDA – DMUMDA	$s+$	$+$	$-$	$s-$	$s+$	$s+$	$s+$	$s+$	$s+$	$s-$	$s+$	$s+$
0.5-AMUMDA – DMUMDA	$s+$	$s+$	$s+$	$s+$	$s+$	$s+$	$s+$	$s+$	$s+$	$s+$	$s+$	$s+$
1.0-AMUMDA – DMUMDA	$s-$	$s-$	$s-$	$s+$	$s-$	$s-$	$s-$	$s-$	$s-$	$s-$	$s-$	$s-$
0.5-AMUMDA – 0.1-AMUMDA	$s+$	$s+$	$s+$	$s+$	$s+$	$s+$	$s+$	$s+$	$s+$	$s+$	$s+$	$s+$
1.0-AMUMDA – 0.5-AMUMDA	$s-$	$s-$	$s-$	$+$	$s-$	$s-$	$s-$	$s-$	$s-$	$s-$	$s-$	$s-$
0.5-HMUMDA – 0.5-AMUMDA	$+$	$s+$	$s+$	$s+$	$s+$	$s+$	$s+$	$s+$	$s+$	$s+$	$s+$	$s+$
$\tau = 25$, $\rho \Rightarrow$	0.1	0.2	0.5	1.0	0.1	0.2	0.5	1.0	0.1	0.2	0.5	1.0
DMUMDA – SUMDA	$s-$	$s+$	$s+$	$s+$	$s+$	$s+$	$s+$	$s+$	$s+$	$s+$	$s+$	$+$
0.5-AMUMDA – SUMDA	$s+$	$s+$	$s+$	$s+$	$s+$	$s+$	$s+$	$s+$	$s+$	$s+$	$s+$	$s+$
0.1-AMUMDA – DMUMDA	$s+$	$s+$	$s-$	$s-$	$s+$	$s+$	$s+$	$s+$	$s+$	$+$	$s-$	$s+$
0.5-AMUMDA – DMUMDA	$s+$	$s+$	$s-$	$s-$	$s+$	$s+$	$s+$	$s+$	$s+$	$s+$	$+$	$s+$
1.0-AMUMDA – DMUMDA	$s-$	$s+$	$s-$	$s-$	$-$	$+$	$s+$	$s+$	$s-$	$s-$	$s-$	$s-$
0.5-AMUMDA – 0.1-AMUMDA	$s+$	$s+$	$s-$	$s+$	$s+$	$s+$	$s+$	$s+$	$s+$	$s+$	$s+$	$+$
1.0-AMUMDA – 0.5-AMUMDA	$s-$	$s-$	$-$	$+$	$s-$	$s-$	$s-$	$-$	$s-$	$s-$	$s-$	$s-$
0.5-HMUMDA – 0.5-AMUMDA	$-$	$s+$	$s+$	$+$	$-$	$s+$	$s+$	$s+$	$+$	$s+$	$s+$	$+$

much lower fitness level than 0.5-AMUMDA during each environmental period, especially under cyclic with noise and random environments.

Third, combining the direct memory with the associative memory further improves the performance of AMUMDAs, see the t-test results with respect to 0.5-HMUMDA – 0.5-HMUMDA. This result can be further seen in the

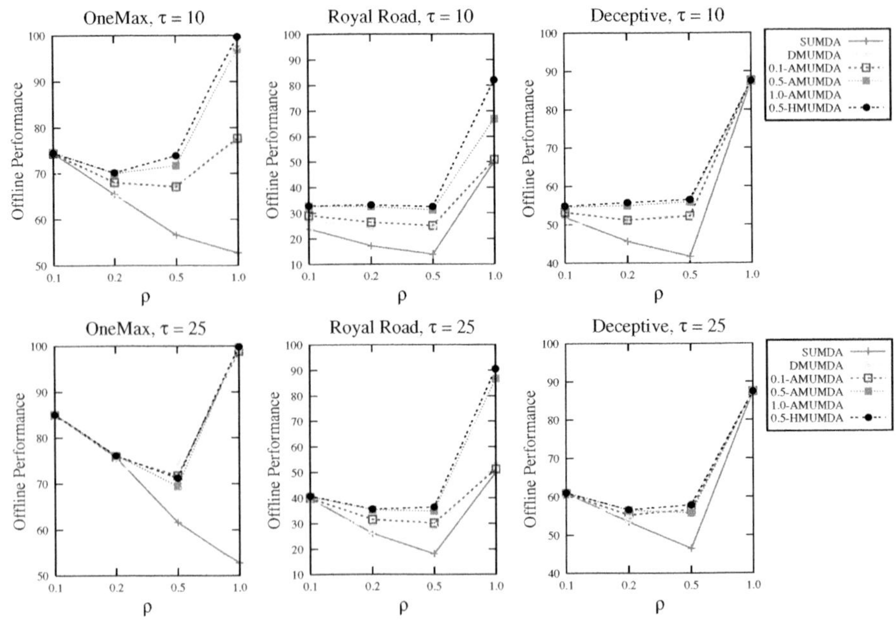

Fig. 1.12. Experimental results of UMDAs on random DOPs

Table 1.6. The t-test results of comparing UMDAs on random DOPs

t-test Result	*OneMax*				*Royal Road*				*Deceptive*			
$\tau = 10$, $\rho \Rightarrow$	0.1	0.2	0.5	1.0	0.1	0.2	0.5	1.0	0.1	0.2	0.5	1.0
DMUMDA − SUMDA	−	s+	s+	s+	s+	s+	s+	s+	s+	s+	s+	s+
0.5-AMUMDA − SUMDA	s+	s+	s+	s+	s+	s+	s+	s+	s+	s+	s+	s+
0.1-AMUMDA − DMUMDA	s+	s+	s−	+	s+	s+	s+	+	s+	s+	s−	−
0.5-AMUMDA − DMUMDA	s+	s+	s+	s+	s+	s+	s+	s+	s+	s+	s+	+
1.0-AMUMDA − DMUMDA	s−	s−	s−	s+	s−	s−	s−	s+	s−	s−	s−	−
0.5-AMUMDA − 0.1-AMUMDA	+	s+	s+	s+	s+	s+	s+	s+	s+	s+	s+	+
1.0-AMUMDA − 0.5-AMUMDA	s−	s−	s−	+	s−	s−	s−	s+	s−	s−	s−	s−
0.5-HMUMDA − 0.5-AMUMDA	−	s+	s+	s+	+	s+	s+	s+	s+	s+	s+	−
$\tau = 25$, $\rho \Rightarrow$	0.1	0.2	0.5	1.0	0.1	0.2	0.5	1.0	0.1	0.2	0.5	1.0
DMUMDA − SUMDA	s−	−	s+	s+	−	s+	s+	s+	s−	s+	s+	s+
0.5-AMUMDA − SUMDA	+	s+	s+	s+	s+	s+	s+	s+	s+	s+	s+	s+
0.1-AMUMDA − DMUMDA	+	s+	s−	s+	s+	s+	s+	s−	s+	s+	s−	+
0.5-AMUMDA − DMUMDA	s+	s+	s−	s+	s+	s+	s+	s+	s+	s+	s−	−
1.0-AMUMDA − DMUMDA	s−	s−	s−	s+	s−	s−	s−	s+	s−	s−	s−	+
0.5-AMUMDA − 0.1-AMUMDA	+	s+	s−	s+	s+	s+	s+	s+	s+	s+	s−	−
1.0-AMUMDA − 0.5-AMUMDA	s−	s−	s−	+	s−	s−	s−	s+	s−	s−	s−	s+
0.5-HMUMDA − 0.5-AMUMDA	=	+	s+	s+	+	s+	s+	s+	+	s+	s+	−

dynamic performance of 0.5-HMUMDA shown in Fig. 1.13, where 0.5-HMUMDA maintains the highest level of fitness during each environmental period under cyclic, noisy and random environments.

Finally, let's compare the performance of investigated UMDAs and GAs on DOPs. The t-test results of comparing UMDAs and GAs on the dynamic

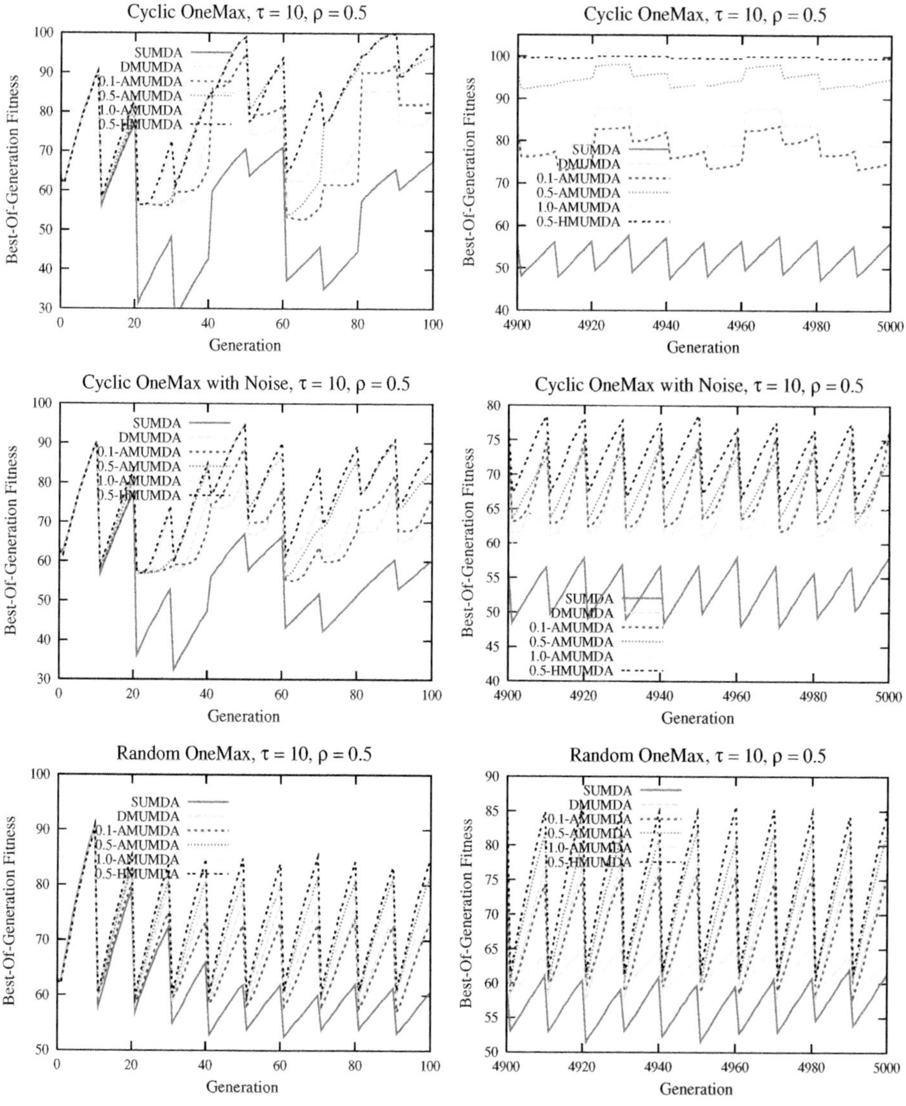

Fig. 1.13. Dynamic performance of UMDAs on the dynamic *OneMax* problems

test problems with $\tau = 10$ are given in Table 1.7. From Table 1.7, it can be seen that GAs outperform corresponding UMDAs on most dynamic test problems.

Table 1.7. The t-test results of comparing UMDAs and GAs on DOPs with $\tau = 10$

t-test Result	*OneMax*				*Royal Road*				*Deceptive*			
Cyclic, $\rho \Rightarrow$	0.1	0.2	0.5	1.0	0.1	0.2	0.5	1.0	0.1	0.2	0.5	1.0
SUMDA − SGA	$s-$	$s-$	$s-$	$s-$	$s-$	$s-$	$s-$	$s+$	$s-$	$s-$	$s-$	$s+$
DMUMDA − DMGA	$s-$	$s+$	$s-$	$s-$	$s-$	$s-$	$s-$	$s-$	$s+$	$s+$	$s-$	$s+$
0.1-AMUMDA − 0.1-AMGA	$s-$	$s-$	$s-$	$s-$	$s-$	$s-$	$s-$	$-$	$s+$	$s+$	$s+$	$s+$
0.5-AMUMDA − 0.5-AMGA	$s-$	$s-$	$s-$	$s-$	$s-$	$s-$	$s-$	$s-$	$s+$	$s+$	$s-$	$s+$
1.0-AMUMDA − 1.0-AMGA	$s-$	$s-$	$s-$	$s-$	$s-$	$s-$	$s-$	$s-$	$s-$	$s-$	$-$	$s+$
0.5-HMUMDA − 0.5-HMGA	$s-$	$s+$	$s-$	$s-$	$s-$	$s-$	$s-$	$s-$	$s-$	$s-$	$s-$	$s+$
Cyclic with Noise, $\rho \Rightarrow$	0.1	0.2	0.5	1.0	0.1	0.2	0.5	1.0	0.1	0.2	0.5	1.0
SUMDA − SGA	$s-$	$s-$	$s-$	$s-$	$s-$	$s-$	$s-$	$s-$	$s-$	$s-$	$s-$	$s-$
DMUMDA − DMGA	$s-$	$s-$	$+$	$s-$	$s-$	$s-$	$s-$	$s-$	$s-$	$s-$	$s-$	$s-$
0.1-AMUMDA − 0.1-AMGA	$s-$	$s-$	$s-$	$s-$	$s-$	$s-$	$s-$	$s-$	$s-$	$s-$	$s-$	$s-$
0.5-AMUMDA − 0.5-AMGA	$s-$	$s-$	$s-$	$s-$	$s-$	$s-$	$s+$	$s-$	$s-$	$s+$	$s+$	$s-$
1.0-AMUMDA − 1.0-AMGA	$s-$	$s-$	$s-$	$s-$	$s-$	$s-$	$s-$	$s-$	$s-$	$s-$	$s-$	$s-$
0.5-HMUMDA − 0.5-HMGA	$s-$	$s-$	$s-$	$s-$	$s-$	$s-$	$s-$	$s-$	$s-$	$s-$	$s+$	$s-$
Random, $\rho \Rightarrow$	0.1	0.2	0.5	1.0	0.1	0.2	0.5	1.0	0.1	0.2	0.5	1.0
SUMDA − SGA	$s-$	$s-$	$s-$	$s-$	$s-$	$s-$	$s-$	$s+$	$s-$	$s-$	$s-$	$s+$
DMUMDA − DMGA	$s-$	$s-$	$s+$	$s-$	$s-$	$s-$	$s-$	$s-$	$s-$	$s-$	$+$	$s+$
0.1-AMUMDA − 0.1-AMGA	$s-$	$s-$	$s-$	$s-$	$s-$	$s-$	$s-$	$s-$	$s-$	$s-$	$s-$	$s+$
0.5-AMUMDA − 0.5-AMGA	$s-$	$s-$	$s+$	$s-$	$s-$	$s-$	$s+$	$s-$	$s-$	$s-$	$s+$	$s+$
1.0-AMUMDA − 1.0-AMGA	$s-$	$s-$	$s-$	$s-$	$s-$	$s-$	$s-$	$s-$	$s-$	$s-$	$s-$	$s+$
0.5-HMUMDA − 0.5-HMGA	$s-$	$s-$	$s+$	$s-$	$s-$	$s-$	$s-$	$s-$	$s-$	$s-$	$s+$	$s+$

1.7 Conclusions

This chapter investigates the application of explicit memory schemes for EAs in dynamic environments. Two kinds of explicit memory schemes, i.e., direct memory and associative memory, are applied into two kinds of EAs, i.e., GAs and UMDAs, to address dynamic optimization problems. The direct memory scheme just stores and reuses best solutions in the memory. In the contrast, in the associative memory scheme, best solutions together with the current environmental information, (the allele distribution vector for GAs or working probability vector for UMDAs) are stored in the memory. When an environmental change is detected, the stored allele distribution vector (for GAs) or probability vector (for UMDAs) that is associated with the best re-evaluated memory solution is extracted to create new individuals into the population.

Based on the XOR dynamic problem generator, a series of dynamic test problems were systematically constructed, featuring three kinds of dynamic environments: cyclic, cyclic with noise, and random. Based on this dynamic test problems, experimental study was carried out to test the memory schemes for GAs and UMDAs. From the experimental results, the following conclusions can be drawn on the dynamic test environments. First, memory schemes are efficient to improve the performance of GAs and UMDAs in dynamic environments and the cyclicity of dynamic environments greatly affect the performance of memory schemes for GAs and UMDAs in dynamic environments. Second, generally speaking the associative memory scheme outperforms traditional direct memory scheme for GAs and UMDAs in dynamic environments. Third, the associative factor has an important impact on the performance of AMGAs and AMUMDAs. Setting α to 0.5 seems a good choice for AMGAs and AMUMDAs. Fourth, combining direct memory with associative memory may further improve the performance of GAs and UMDAs in dynamic environments. The hybrid memory scheme is a good approach for EAs for DOPs.

The work studied in this chapter can be extended in several ways. Developing other memory management and retrieval mechanisms would be an interesting future work for memory-based UMDAs and other estimation of distribution algorithms [1, 18] in dynamic environments. Comparing the investigated explicit memory schemes with implicit memory schemes is another future work. And it is also an interesting work to further investigate the integration of the memory schemes with other approaches, such as multi-population, diversity approaches, and adaptive operators, for EAs in dynamic environments.

References

1. S. Baluja (1994). Population-based incremental learning: A method for integrating genetic search based function optimization and competitive learning. *Technical Report CMU-CS-94-163*, Carnegie Mellon University, USA.
2. C. N. Bendtsen and T. Krink (2002). Dynamic memory model for non-stationary optimization. *Proc. of the 2002 Congress on Evol. Comput.*, pp. 145-150.
3. J. Branke (1999). Memory enhanced evolutionary algorithms for changing optimization problems. *Proc. of the 1999 Congress on Evolutionary Computation*, vol. 3, pp. 1875-1882.
4. J. Branke, T. Kaußler, C. Schmidth, and H. Schmeck (2000). A multi-population approach to dynamic optimization problems. *Proc. of the 4th Int. Conf. on Adaptive Computing in Design and Manufacturing*, pp. 299-308.
5. J. Branke (2002). *Evolutionary Optimization in Dynamic Environments.* Kluwer Academic Publishers.
6. H. G. Cobb and J. J. Grefenstette (1993). Genetic algorithms for tracking changing environments. *Proc. of the 5th Int. Conf. on Genetic Algorithms*, pp. 523-530.
7. D. Dasgupta and D. McGregor (1992). Nonstationary function optimization using the structured genetic algorithm. *PPSN II*, pp. 145-154.
8. D. E. Goldberg and R. E. Smith (1987). Nonstationary function optimization using genetic algorithms with dominance and diploidy. *Proc. of the 2nd Int. Conf. on Genetic Algorithms*, pp. 59-68.
9. D. E. Goldberg (1989). *Genetic Algorithms in Search, Optimization, and Machine Learning.* Reading, MA: Addison-Wesley.
10. J. J. Grefenstette (1992). Genetic algorithms for changing environments. *Parallel Problem Solving from Nature II*, pp. 137-144.
11. P. Larrañaga and J. A. Lozano (2002). *Estimation of Distribution Algorithms: A New Tool for Evolutionary Computation.* Kluwer Academic Publishers.
12. E. H. J. Lewis and G. Ritchie (1998). A comparison of dominance mechanisms and simple mutation on non-stationary problems. *Proc. of the 5th Int. Conf. on Parallel Problem Solving from Nature*, pp. 139-148.
13. S. J. Louis and Z. Xu (1996). Genetic algorithms for open shop scheduling and re-scheduling. *Proc. of the 11th ISCA Int. Conf. on Computers and their Applications*, pp. 99-102.

14. T. Mahnig and H. Mühlenbein (2000). Mathematical analysis of optimization methods using search distributions. *Proc. of the 2000 Genetic and Evolutionary Computation Conference Workshop Program*, pp. 205-208.
15. N. Mori, H. Kita and Y. Nishikawa (1997). Adaptation to changing environments by means of the memory based thermodynamical genetic algorithm. *Proc. of the 7th Int. Conf. on Genetic Algorithms*, pp. 299-306.
16. R. W. Morrison and K. A. De Jong (1999). A test problem generator for non-stationary environments. In *Proc. of the 1999 Congress on Evolutionary Computation*, vol. 3, pp. 2047-2053.
17. R. W. Morrison and K. A. De Jong (2000). Triggered hypermutation revisited. In *Proc. of the 2000 Congress on Evolutionary Computation*, pp. 1025-1032.
18. H. Mühlenbein and G. Paaß(1996). From recombination of genes to the estimation of distributions I. Binary parameters. *Proc. of the 4th Int. Conf. on Parallel Problem Solving from Nature*, pp. 178-187.
19. H. Mühlenbein (1998). The equation for response to selection and its use for prediction. *Evolutionary Computation*, 5: 303–346.
20. K. P. Ng and K. C. Wong (1997). A new diploid scheme and dominance change mechanism for non-stationary function optimisation. *Proc. of the 6th Int. Conf. on Genetic Algorithms.*
21. C. L. Ramsey and J. J. Greffenstette (1993). Case-based initializtion of genetic algorithms. *Proc. of the 5th Int. Conf. on Genetic Algorithms.*
22. K. Trojanowski and Z. Michalewicz (1999). Searching for optima in non-stationary environments. *Proc. of the 1999 Congress on Evolutionary Computation*, pp. 1843-1850.
23. F. Vavak and T. C. Fogarty (1996). A comparative study of steady state and generational genetic algorithms for use in nonstationary environments. *AISB Workshop on Evolutionary Computing, LNCS 1143*, pp. 297-304.
24. S. Yang (2003). Non-stationary problem optimization using the primal-dual genetic algorithm. *Proc. of the 2003 Congress on Evolutionary Computation*, vol. 3, pp. 2246-2253.
25. S. Yang (2005). Memory-based immigrants for genetic algorithms in dynamic environments. *Proc. of the 2005 Genetic and Evolutionary Computation Conference*, vol. 2, pp. 1115-1122.
26. S. Yang (2005). Population-based incremental learning with memory scheme for changing environments. *Proc. of the 2005 Genetic and Evolutionary Computation Conference*, vol. 1, pp. 711-718.
27. S. Yang (2005). Memory-enhanced univariate marginal distribution algorithms for dynamic optimization problems. *Proc. of the 2005 Congress on Evolutionary Computation*, vol. 3, pp. 2560-2567.
28. S. Yang (2006). Associative memory scheme for genetic algorithms in dynamic environments. *Applications of Evolutionary Computing, LNCS 3907*, pp. 788-799.
29. S. Yang and X. Yao (2005). Experimental study on population-based incremental learning algorithms for dynamic optimization problems, *Soft Computing*, 9(11): 815–834.
30. S. Yang and X. Yao (2006). Population-based incremental learning with associative memory for dynamic environments, submitted to *IEEE Transactions on Evolutionary Computation.*

2

Particle Swarm Optimization in Dynamic Environments

Tim Blackwell

Department of Computing, Goldsmiths College, University of London
New Cross, London SE14 6NW, UK
t.blackwell@gold.ac.uk

Summary. This chapter reviews the application of particle swarms to dynamic optimization and explains why the canonical particle swarm optimization (PSO) algorithm must be modified for good performance in environments such as the moving peaks benchmark. The chief obstacle to good performance, namely diversity loss, can be overcome in various ways; this article focusses on the use of charged swarms to provide particle repulsion between members of a sub-swarm. Although diversity enhancing mechanisms can help a population track a single peak, they cannot help the population to watch other peaks that may become optimal as the landscape changes. For this reason, a multi-swarm approach is advocated here; an exclusion operator provides effective repulsion between swarms so that each may settle on a peak. New results on self-adaptation are also presented; a simple rule for swarm birth and death is proposed so that the multi-swarm may adjust its size dynamically and in relation to the number of peaks. The self-adapting multi-swarm is demonstrated to be competitive with the best results for a hand-tuned multi-swarm.

2.1 Introduction

Particle Swarm Optimization (PSO) is a versatile population-based optimization technique, in many respects similar to evolutionary algorithms (EAs). PSO has been shown to perform well for many static problems [30]. However, many real-world problems are dynamic in the sense that the global optimum location and value may change with time. The task for the optimization algorithm is to track this shifting optimum. It has been argued [14] that EAs are potentially well-suited to such tasks, and a review of EA variants tested in the dynamic problem is given in [13, 15]. It might be wondered, therefore, what promise PSO holds for dynamic problems.

Optimization with particle swarms has two major ingredients, the particle dynamics and the particle information network. The particle dynamics are derived from swarm simulations in computer graphics [21], and the information sharing component is inspired by social networks [25, 32]. These ingredients combine to make PSO a robust and efficient optimizer of real-valued objective

T. Blackwell: *Particle Swarm Optimization in Dynamic Environments*, Studies in Computational Intelligence (SCI) **51**, 29–49 (2007)
www.springerlink.com

functions (although PSO has also been successfully applied to combinatorial and discrete problems too). PSO is an accepted computational intelligence technique, sharing some qualities with Evolutionary Computation [1].

The application of PSO to dynamic problems has been explored by various authors [6, 9, 17, 23, 24, 30]. The overall consequence of this work is that PSO, just like EAs, must be modified for optimal results on dynamic environments typified by the moving peaks benchmark (MPB). (Moving peaks, arguably representative of real world problems, consist of a number of peaks of changing with and height and in lateral motion [7, 12].) The origin of the difficulty lies in the dual problems of *outdated memory* due to environment dynamism, and *diversity loss*, due to convergence.

Of these two problems, diversity loss is by far the more serious; it has been demonstrated that the time taken for a partially converged swarm to re-diversify, find the shifted peak, and then re-converge is quite deleterious to performance [3]. Clearly, either a re-diversification mechanism must be employed at (or before) function change, and/or a measure of diversity can be maintained throughout the run. There are four principle mechanisms for either re-diversification or diversity maintenance: randomization [23], repulsion [5], dynamic networks [24, 36] and multi-populations [6, 29].

Multi-swarms combine repulsion with multi-populations [6, 7]. Interestingly, the repulsion occurs between particles, and between swarms. The multi-population in this case is an interacting super-swarm of charged swarms. A charged swarm is inspired by models of the atom: a conventional PSO nucleus is surrounded by a cloud of 'charged' particles. The charged particles are responsible for maintaining the diversity of the swarm. Furthermore, and in analogy to the exclusion principle in atomic physics, each swarm is subject to an exclusion pressure that operates when the swarms collide. This prohibits two or more swarms from surrounding a single peak, thereby enabling swarms to watch secondary peaks in the eventuality that these peaks might become optimal. This strategy has proven to be very effective for MPB environments.

This chapter starts with a description of the canonical PSO algorithm and then, in Section 3, explains why dynamic environments pose particular problems for unmodified PSO. The MPB framework is also introduced in this section. The following section describes some PSO variants that have been proposed to deal with diversity loss. Section 5 outlines the multi-swarm approach and the subsequent section presents new results for a self-adapting multi-swarm, a multi-population with swarm birth and death.

2.2 Canonical PSO

In PSO, population members (particles) possess a memory of the best (with respect to an objective function) location that they have visited in the past, *pbest*, and of its fitness. In addition, particles have access to the best location of any other particle in their own network. These two locations (which will coincide for the best particle in any network) become attractors in the search space of the swarm. Each particle will be repeatedly drawn back to spatial neighborhoods close to these two attractors, which themselves will be updated if the global best and/or particle best is bettered at each particle update. Several network topologies have been tried, with the star or fully connected network remaining a popular choice for unimodal functions. In this network, every particle will share information with every other particle in the swarm so that there is a single *gbest* global best attractor representing the best location found by the entire swarm.

Particles possess a velocity which influences position updates according to a simple discretization of particle motion

$$\mathbf{v}(t+1) = \mathbf{v}(t) + \mathbf{a}(t+1) \tag{2.1}$$

$$\mathbf{x}(t+1) = \mathbf{x}(t) + \mathbf{v}(t+1) \tag{2.2}$$

where $\mathbf{a}$, $\mathbf{v}$, $\mathbf{x}$ and t are acceleration, velocity, position and time (iteration counter) respectively. Eqs. 2.1, 2.2 are similar to particle dynamics in swarm simulations, but PSO particles do not follow a smooth trajectory, instead moving in jumps, in a motion known as a flight [28] (notice that the time increment dt is missing from these rules). The particles experience a linear or spring-like attraction, weighted by a random number, (particle mass is set to unity) towards each attractor. Convergence towards a good solution will not follow from these dynamics alone; the particle flight must progressively contract. This contraction is implemented by Clerc and Kennedy with a constriction factor χ, $\chi < 1$, [20]. For our purposes here, the Clerc-Kennedy PSO will be taken as the canonical swarm; χ replaces other energy draining factors extant in the literature such as a decreasing 'inertial weight' and velocity clamping. Moreover the constricted swarm is replete with a convergence proof, albeit about a static attractor (although there is some experimental and theoretical support for convergence in the fully interacting swarm where particles can move attractors [10]).

Explicitly, the acceleration of particle i in Eq.2.1 is given by

$$\mathbf{a}_i = \chi[c\boldsymbol{\epsilon} \cdot (\mathbf{p}_g - \mathbf{x}_i) + c\boldsymbol{\epsilon} \cdot (\mathbf{p}_i - \mathbf{x}_i)] - (1-\chi)\mathbf{v}_i \tag{2.3}$$

where $\boldsymbol{\epsilon}$ are vectors of random numbers drawn from the uniform distribution $U[0,1]$, $c > 2$ is the spring constant and $\mathbf{p}_i$, $\mathbf{p}_g$ are particle and global attractors. This formulation of the particle dynamics has been chosen to demonstrate explicitly constriction as a frictional force, opposite in direction, and proportional to, velocity. Clerc and Kennedy derive a relation for $\chi(c)$: standard values are $c = 2.05$ and $\chi = 0.729843788$. The complete PSO algorithm for maximizing an objective function f is summarized as Algorithm 1.

Algorithm 1 Canonical PSO

FOR EACH particle i
    Randomly initialize $\mathbf{v}_i, \mathbf{x}_i = \mathbf{p}_i$
    Evaluate $f(\mathbf{p}_i)$
    $g = \arg\max f(\mathbf{p}_i)$
REPEAT
    FOR EACH particle i
        Update particle position $\mathbf{x}_i$ according to eqs.. 2.1, 2.2 and 2.3
        Evaluate $f(\mathbf{x}_i)$
        //Update personal best
        IF $f(\mathbf{x}_i) > f(\mathbf{p}_i)$ THEN
            $\mathbf{p}_i = \mathbf{x}_i$
        //Update global best
        IF $f(\mathbf{x}_i) > f(\mathbf{p}_g)$ THEN
            $\mathbf{p}_g = \arg\max f(\mathbf{p}_i)$
UNTIL termination criterion reached

2.3 PSO Problems with Moving Peaks

As has been mentioned in Sect 22.1, PSO must be modified for optimal results on dynamic environments typified by the moving peaks benchmark (MPB). These modifications must solve the problems of outdated memory, and of lost diversity. This explains the origins of these problems in the context of MPB, and shows how memory loss is easily addressed. The following section then considers the second, more severe, problem.

2.3.1 Moving Peaks

The dynamic objective function of MPB, $f(\mathbf{x}, t)$, is optimized at 'peak' locations $\mathbf{x}^*$ and has a global optimum at $\mathbf{x}^{**} = \arg\max\{f(\mathbf{x}^*)\}$ (once more, assuming optimization means maximizing). Dynamism entails a small movement of magnitude s, and in a random direction, of each $\mathbf{x}^*$. This happens every K evaluations and is accompanied by small changes of peak height and width. There are p peaks in total, although some peaks may become obscured. The peaks are constrained to move in a search space of extent X in each of the d dimensions, $[0, X]^d$.

This scenario, which is not the most general, nevertheless has been put forward as representative of real world dynamic problems [12] and a benchmark function is publicly available for download from [11]. Note that small changes in $f(\mathbf{x}^*)$ can still invoke large changes in $\mathbf{x}^{**}$ due to peak promotion, so the many peaks model is far from trivial.

2.3.2 The Problem of Outdated Memory

Outdated memory happens at environment change when the optima may shift in location and/or value. Particle memory (namely the best location visited in the past, and its corresponding fitness) may no longer be true at change, with potentially disastrous effects on the search.

The problem of outdated memory is typically solved by either assuming that the algorithm knows just when the environment change occurs, or that it can detect change. In either case, the algorithm must invoke an appropriate response. One method of detecting change is a re-evaluation of f at one or more of the personal bests $\mathbf{p}_i$ [17, 23]. A simple and effective response is to re-set all particle memories to the current particle position and f value at this position, and ensuring that $\mathbf{p}_g = \arg\max f(\mathbf{p}_i)$. One possible drawback is that the function has not changed at the chosen $\mathbf{p}_i$, but *has* changed elsewhere. This can be remedied by re-evaluating f at all personal bests, at the expense of doubling the total number of function evaluations per iteration.

2.3.3 The Problem of Lost Diversity

Equally troubling as outdated memory is insufficient diversity at change. The population takes time to re-diversify and re-converge, effectively unable to track a moving optimum.

It is helpful at this stage to introduce the swarm diameter $|S|$, defined as the largest distance, along any axis, between any two particles [2], as a measure of swarm diversity (Fig. 2.1). Loss of diversity arises when a swarm is converging on a peak. There are two possibilities: when change occurs, the new optimum location may either be within or outside the collapsing swarm. In the former case, there is a good chance that a particle will find itself close to the new optimum within a few iterations and the swarm will successfully track the moving target. The swarm as a whole has sufficient diversity. However, if the optimum shift is significantly far from the swarm, the low velocities of the particles (which are of order $|S|$) will inhibit re-diversification and tracking, and the swarm can even oscillate about a false attractor and along a line perpendicular to the true optimum, in a phenomenon known as linear collapse [5]. This effect is illustrated in Fig. 2.2.

These considerations can be quantified with the help of a prediction for the rate of diversity loss [2, 3, 10]. In general, the swarm shrinks at a rate determined by the constriction factor and by the local environment at the optimum. For static functions with spherical symmetric basins of attraction, the theoretical and empirical analysis of the above references suggest that the rate of shrinkage (and hence diversity loss) is scale invariant and is given by a scaling law

$$|S(t)| = C\alpha^t \tag{2.4}$$

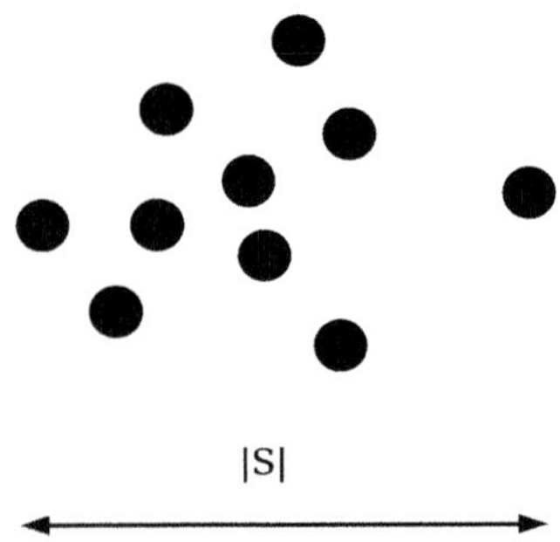

Fig. 2.1. The swarm diameter.

for constants C and $\alpha < 1$, where $\alpha \approx 0.92$ and C is the swarm diameter at iteration $t = 0$. The number of function evaluations between change, K, can be converted into a period measured in iterations, L by considering the total number of function evaluations per iteration including, where necessary, the extra test-for-change evaluations. We might expect that, for peak shift distance s and box size $X = |S(0)|$, if $s >> S_L = X\alpha^L$, tracking will be very hard since the swarm has already converged to a very small ball at the first change. The experiments reported in [3] show that canonical PSO fails to track a single peaked dynamic environment defined by $s = 8.7, L = 100, X = 10$. Since S_L computes to 0.0024, this is hardly surprising.

2.4 Diversity Lost and Diversity Regained

There are two solutions in the literature to the problem of insufficient diversity. Either a diversity increasing mechanism can be invoked at change (or at pre-determined intervals), or some permanent means can be put in place to ensure there is sufficient diversity at all times [7]. These modifications are the subject of this section.

2.4.1 Re-diversification

Hu and Eberhart [23] study a number of re-diversification mechanisms. These all involve randomization of the entire, or part of, the swarm. This happens when re-evaluation of the objective function at one or several of the attractors detects change, or at a pre-set interval. Clearly the problem with this approach is the arbitrariness of the extra parameters. Since randomization implies information loss, there is a danger of erasing too much information and effectively re-starting the swarm. On the other hand, too little randomization might not introduce enough diversity to cope with the change. And, of course, if tests for change happen at pre-determined intervals, there is a danger of missing a shift. The arbitrariness of the extra parameters can only be solved if much

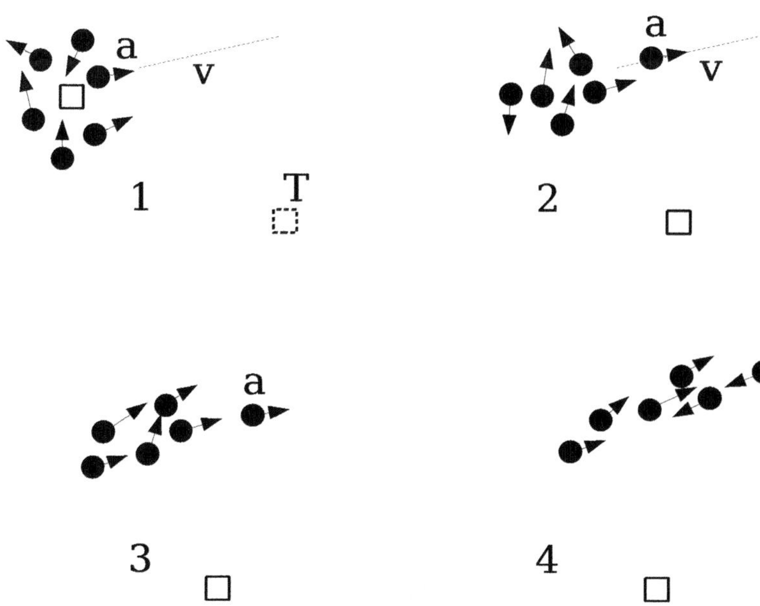

Fig. 2.2. Sequence of frames showing possible behavior when optimum shift is greater that swarm diversity. When the attractor T (square box, frame 1) shifts, particle a is at the global best, p_g. a continues along trajectory v since it is not accelerated in this update (frame 2). Particle a continues to move along v, repositioning p_g at each update, becoming in effect the swarm leader (frame 3). After a while, the swarm oscillates along v, about a point perpendicular to T (frame 4). Eventually random fluctuations will cause another particle to deviate from v and move closer towards the attractor. The swarm soon follows and converges on T.

prior knowledge about f's dynamism is available, or some other higher-level modification mechanism scheme is implemented. Such a scheme could infer details about f's dynamism during the run, making appropriate adjustments to the re-diversification parameters. So far, though, higher level modifications such as these have not been studied.

2.4.2 Maintaining Diversity by Repulsion

A constant, and hopefully good enough, degree of swarm diversity can be maintained at all times either through some type of repulsive mechanism, or by adjustments to the information sharing neighborhood. Repulsion can either be between particles, or from an already detected optimum. For example, Krink et al [34] study finite-size particles as a means of preventing premature convergence. The hard sphere collisions produce a constant diversification pressure. Alternatively, Parsopoulos and Vrahatis [30] place a repeller at an already detected optima, in an attempt to divert the swarm and find new optima. Neither technique, however, has been applied to the dynamic scenario.

An example of repulsion that has been tested in a dynamic context is the atom analogy [4–6, 9]. In this model, a swarm is comprised of a 'charged' and a 'neutral' sub-swarm. The model can be depicted as a cloud of charged particles orbiting a contracting, neutral, PSO nucleus, Fig. 2.3. The charged particles can be either classical or quantum particles; either type are discussed in some depth in references [6, 7] and in the following section. Charge enhances diversity in the vicinity of the converging PSO sub-swarm, so that optimum shifts within this cloud should be trackable. Good tracking (outperforming canonical PSO) has been demonstrated for unimodal dynamic environments of varying severities [3].

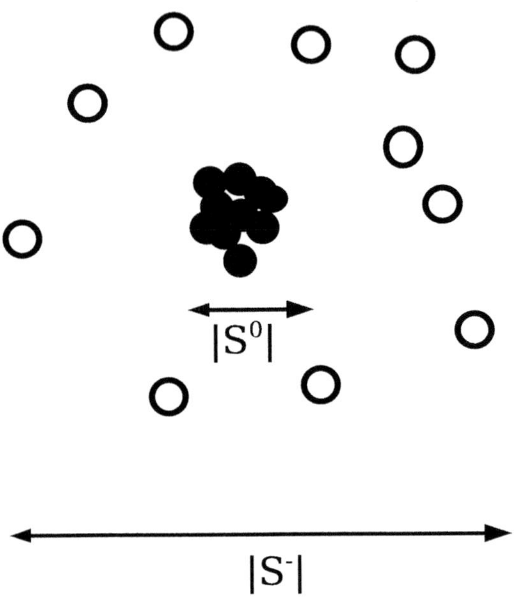

Fig. 2.3. The Atom Analogy. The situation depicted here shows a PSO sub-swarm of neutral particles (filled circles), converging at an optimum. The neutral swarm diameter, $|S^0|$, is shrinking by a factor of 0.92 at each iteration. This sub-swarm is surrounded by a number of charged particles with constant diversity $|S^-|$. Both sub-swarms share the same global attractor p_g. Optimum moves to locations within the charged sub-swarm will be rapidly re-optimized by the swarm as a whole.

2.4.3 Maintaining Diversity with Dynamic Network Topology

Adjustments to the information sharing topology can be made with the intention of reducing, maybe temporarily, the desire to move towards the global best position, thereby enhancing population diversity. Li and Dam use a grid-like neighborhood structure, and Jansen and Middendorf test a hierarchical structure, reporting improvements over unmodified PSO for unimodal dynamic environments [24, 36].

2.4.4 Maintaining Diversity with Multi-populations

A number of research groups have considered multi-populations as a means of enhancing diversity. The multi-population idea is particularly helpful in multi-modal environments such as many peaks. The aim here is to allow each population to converge on a promising peak. Then, if any secondary peak becomes the global optimum as a result of change, a population is close at hand. Multi-population techniques include niching, speciation and multi-swarms.

In the static context, the niching PSO of Brits et al [16] can successfully optimize some static benchmark problems. In *nichePSO*, if a particle's fitness changes very little (the variance in fitness is less than a threshold) over a small number of iterations, a two particle sub-swarm is created from this particle and its nearest spatial neighbor. This technique, as the authors point out, would fail in a dynamic environment because niching depends on a homogeneous distribution of particles in the search space, and on a training phase.

A speciation PSO variation, known as clearing [31] has been adopted by Li in the static context and generalized by Parrot and Li to dynamic functions [27, 29]. Under clearing, the number and size of swarms is adjusted dynamically by constructed an ordered list of particles, ranked according to their fitness, with spatially close particles joining a particular species. This method relies on a speciation radius and has no further diversity mechanism. Other related work includes using different swarms in cooperation to optimize different parts of a solution [35], a two swarm min-max optimization algorithm [33] and iteration by iteration clustering of particles into sub-swarms [26]. Apart form Parrot and Li's speciation, none of these multi-population techniques have been generalized as dynamic optimizers. The multi-swarm approach of Blackwell and Branke is described in detail in the next section.

2.5 Multi-swarms

A combined approach might be to incorporate the virtues of the multi-population approach and of swarm diversity enhancing mechanisms such as repulsion. Such an optimizer would be well suited to the many peaks environment. Multi-swarms, first proposed in a non-optimization context [8] would seem to do just this. The extension of multi-swarms to a dynamic optimizer was made by Blackwell and Branke [6], and is inspired by Branke's own self-organizing scouts (SOS) [13]. The scouts have been shown to give excellent results on the many peaks benchmark.

A multi-swarm is a colony of charged swarms interacting locally via *exclusion* and globally by *anti-convergence*. The motivation for these operators is that a mechanism must be found to prevent two or more swarms from trying to optimize the same peak (exclusion) and also to maintain multi-swarm diversity, that is to say the diversity amongst the population of swarms as a whole (anti-convergence). Multi-swarms have been compared very favorably to both hierarchical swarms and to self-organizing scouts.

We consider below the main ingredients of the multi-swarm algorithm in more depth. In particular we will assess the values of parameters with relation to the many peaks benchmark. A complete discussion of parameter choices is given in [7]. The multi-swarm algorithm is presented in Algorithm 2.

2.5.1 Atom Analogy

In the atom analogy, each swarm is pictured as an atom with a contracting nucleus of neutral PSO particles, and an enveloping cloud of charged particles. All particles are in fact members of the same information network so that they all (in the star topology) have access to $\mathbf{p}_g$. The mutual repulsions between the charged particles may follow a deterministic, classical, rule (Coulomb repulsion, parameterized by particle charge Q). Alternatively, in a quantum atom, the particles are positioned within a hypersphere of radius r_{cloud} centered on $\mathbf{p}_g$ according to a probability distribution. So far, two uniform distributions have been tested. References [6, 7] consider a uniform shell distribution, $p(r, dr) = \rho(r)dr = const$, where ρ is a probability density, r is a shell radius, dv is a volume element and p is a probability. Recent work has investigated a uniform volume distribution, $p(\mathbf{x}, dv) = \rho(\mathbf{x})dv = const$. In both cases, the distributions are normalized so that $p(r > r_{cloud}) = 0$. Other, non-uniform and possibly dynamic distributions might be favorable for the quantum swarm in some cases, but such distributions remain unexplored.

The quantum atom has the advantages of lower complexity and an easily controllable distribution: the Coulomb repulsion has quadratic complexity and highly fluctuating electron orbits [2, 3]. An order of magnitude estimation for the parameters r_{cloud} (for quantum swarms), or Q, for classically charged clouds, can be made by supposing that good tracking will occur if the mean charged particle separation $< |\mathbf{x}^- - \mathbf{p}_g| >$ is comparable to s. This separation is easy to compute for the quantum swarm: only empirical data is available for classical charged particles.

2.5.2 Exclusion

In order to demonstrate the necessity for exclusion, first consider an assembly of non-interacting swarms, also known as a many-swarm. A many-swarm has M swarms, and each swarm, for symmetrical configurations has N^0 neutral and N^- charged particles. Such a many-swarm is written $M * (N^0 + N^-)$. Since the swarms do not interact - either dynamically through the particle velocity and positions updates, or by sharing information - the M swarms are completely independent, and any number of them may try to optimize the

Algorithm 2 Multi-Swarm

```
//Initialization
FOR EACH particle ni
  Randomly initialize v_ni, x_ni = p_ni
  Evaluate f(p_ni)
FOR EACH swarm n
  p_ng := argmax{f(p_ni)}
REPEAT
  // Anti-Convergence
  IF all swarms have converged THEN
    Re-initialize worst swarm.
  FOR EACH swarm n
    // Test for Change
    Evaluate f(p_ng).
    IF new value is different from last iteration THEN
      Re-evaluate each particle attractor.
      Update swarm attractor.
    FOR EACH particle i of swarm n
      // Update Particle
      Apply equations (3) - (9) depending on particle type.
      // Update Attractor
      Evaluate f(x_ni).
      IF f(x_ni) > f(p_ni) THEN
        p_ni := x_ni.
      IF f(x_ni) > f(p_ng) THEN
        p_ng := x_ni
    // Exclusion.
    FOR EACH swarm m ≠ n
      IF swarm attractor p_ng is within r_excl of p_mg THEN
        IF f(p_ng) ≤ f(p_mg) THEN
          Re-initialize swarm n
        ELSE
          Re-initialize swarm m
        FOR EACH particle in re-initialized swarm
          Re-evaluate function value.
          Update swarm attractor.
UNTIL number of function evaluations performed > max
```

same peak. This is undesirable because it is clearly inefficient to have two or more swarms on the same peak, and in any case, we wish to distribute the swarms throughout the search space for peak watching. Hence the swarms must interact in some way. One possibility is to allow the swarms to interact topologically and share a single information network. The multi-swarm approach is to seek a *spatial* interaction between swarms. Such an interaction might repel entire swarms from already occupied peaks. However, Coulomb repulsion, or some such similar physics-inspired repulsion would not be satisfactory, because the attractive pull towards the peak might be balanced by the repulsive force away from other nearby swarms. In such an equilibrium, no swarm would be able to optimize the peak.

Exclusion is inspired by the exclusion principle in atomic and molecular physics. This principle states that no two electrons may occupy the same state. The exclusion principle provides an effective repulsive force between two gas molecules with overlapping electron clouds [22]. However the effective force does not arise from any deterministic equation that governs the electron motion, but is a rule imposed on the probability distributions of the electron positions. A version of this principle for interacting swarms is a rule that forbids two swarms moving to within r_{excl} of each other, where the distance between swarms is defined as the distance between their $\mathbf{p}_g$'s. The exclusion operator simple randomizes, in the entire search space, the worse swarm in any collision, as judged by the current best value determined by the swarm, $f(\mathbf{p}_g)$. The configuration of the interacting multi-swarm is written $M(N^0 + N^-)$.

An order of magnitude estimation for r_{excl} can be made by assuming that all p peaks are evenly distributed in X^d. The linear diameter of the basin of attraction of a peak is then, on average, $d_{boa} = X/p^{1/d}$. It is reasonable to assume that swarms that are closer than this distance should experience exclusion, since the overall strategy is to place one swarm on each peak.

2.5.3 Anti-Convergence

Anti-convergence is a simple operator that is designed to ensure there is at least one free swarm in the multi-swarm at all times. A free swarm is one that is patrolling the search space rather than converging on a peak. A swarm is assumed to be converging when the neutral swarm diameter is less than a convergence diameter, $2r_{conv}$. The idea is that if the number of swarms is less than the number of peaks, all swarms may converge, leaving some peaks unwatched. One of these unwatched peaks may later become optimal. The presence of free swarms maintains multi-swarm diversity and encourages response to peak promotion.

Estimations of r_{conv} are difficult, but some progress can be made by considering the rate of convergence of the neutral swarm, as given by Equation 2.4. A lower bound on r_{conv} can be estimated from the ideal case that a swarm immediately tracks a shifted peak. This means that the swarm size at the shift is about s and the swarm has K function evaluations worth of time to contract around the peak. On the other hand, r_{conv} should certainly be less than r_{excl} because exclusion occurs before convergence.

Note that there are two levels of diversity. Diversity at the swarm level, as enforced by exclusion, enables a single swarm to track a single moving peak and diversity at the multi-swarm level enables the multi-swarm as a whole to find new peaks.

2.5.4 Multi-swarm Cardinality

The multi-swarm cardinality M can be estimated from p. If possible we would expect that $M > p$ is undesirable since free swarms absorb valuable function evaluations and there is no need to have many more swarms than peaks. Anti-convergence is expected to be beneficial for $M < p$. Optimally, we suppose that $M = p$, and in this case anti-convergence can be switched off, since the multi-swarm has just the right number of swarms.

2.5.5 Results

A exhaustive series of multi-swarm experiments has been conducted for the many peaks benchmark. The standard settings for MPB are number of peaks, $p = 10$, change period in function evaluations, $K = 5000$, peak shift severity, $s = 1.0$, dimensionality, $d = 5$ and search space range $X = 100$. The peak heights and widths vary randomly but are constrained to $[30, 70]$ and $[1, 12]$ respectively. Each experiment actually consists of 50 runs for a given set of multi-swarm parameters. Each run uses a different random number generator seed for initialization of the swarms, the swarm update algorithm and the MPB generator. Other non-standard MPB's were also tested for comparisons. Multi-swarm performance is quantified by the offline error which is the average, at any point in time, of the error of the best solution found since the last environment change. This measure is commonly used for scenarios such as the MPB and is zero for perfect tracking.

Various values of multi-swarm cardinality M were tested for fixed total number of particles as given by the expression $N_{total} = M(N^0 + N^-)$. As expected, $M = p$ was found to be optimal. The multi-swarm coped well with shift severities between 1.0 and 6.0. The multi-swarm offline errors for MPB's with different numbers of peaks were lass than 3.0 for $5 \leq p \leq 200$. Anti-convergence was found to bring a significant improvement when $M < p$. The robustness of the algorithm, and its generalizability into higher dimensions, was also tested by taking $d = 10$ and varying the predicted optimal parameter settings (i.e. r_{excl}, r_{conv}, r_{cloud} and Q) by 20%. The multi-swarm is most sensitive to r_{excl}, but even here offline errors varied by less than 10%.

In all cases, the multi-swarms with charge outperform many-swarms, PSO and multi-swarms without charge. Furthermore, quantum swarms perform better than classical charged swarms. The offline error, as compared to hierarchical swarms and self-organizing scouts, is cut by a half, and the improvement over a randomization scheme is about an order of magnitude. It seems, therefore, that the multi-swarm is a very promising approach for problems of the MPB class.

2.6 Self-adapting Multi-swarms

The multi-swarm model of the previous section introduced a number of new parameters. Although recommendations can be made for these settings, this analysis depends on prior knowledge of the environment and on many test runs. A laudable goal for any optimization technique is the reduction of hand-tunable parameters. Although parameter adjustments might improve performance for any particular problem, a general purpose method that might perform reasonably well across a spectrum of problems is certainly attractive. We therefore wonder to what extent PSO and PSO-variants can find their own best parameter settings during a single run. Such self-adapting algorithms might not return the best performance on any particular problem instance. However to make fair comparisons, the total number of function evaluations (or iterations) involved in all the trials of the hand-tuned method must be taken into account. This point is emphasized by Clerc in his explorations of 'Tribes', a self-adapting PSO [18, 19].

Here we will describe self-adaptations at the level of the multi-swarm. Future work will seek to incorporate adaptations of individual swarms into this scheme. The following will assume $(5^0 + 5^-)$ swarms with canonical PSO neutral particles and quantum charged particles and with the swarm diversity parameter, r_{cloud}, determined by the peak shift severity. This recipe gave the best results for the environments studied in the previous section. The parameters at the multi-swarm level are the number of swarms M and the exclusion and convergence radii r_{excl} and r_{conv}. It is assumed in the following that the multi-swarm has access to the dimensionality d and the extent of the search space X, but not the MPB parameters p or K. (Previously, r_{excl}, r_{conv} and M were determined with knowledge of p and K.)

2.6.1 Swarm Birth and Death

The basic idea is to allow the multi-swarm to regulate its size by bringing new swarms into existence, or by removing redundant swarms. The aim, as before, is to place a swarm on each peak, and to maintain multi-swarm diversity with (at least one) patrolling swarm. The multi-swarm therefore needs a new swarm if all swarms are currently converging. Alternatively, if there are too many free swarms (i.e. those that fail the convergence criterion), a free swarm should be removed. If there is more than one free swarm, the choice for removal is arbitrary and a simple rule might be to remove the worst of the free swarms, as judged by $f(\mathbf{p}_g)$.

This simple birth/death mechanism removes the need for the anti-convergence operator, and for specifying the multi-swarm cardinality. The self-adapting version of Algorithm 2 is given below in Algorithm 3, where M_{free} is the number of free swarms at iteration t, and identical steps in Algorithm 2 have been abbreviated.

Algorithm 3 Self-adapting multi-swarm

Begin with a single free swarm, randomized in X^d
At each iteration t:
  IF $M_{free} = 0$, generate a new free swarm
  ELSE If $M_{free} > n_{excess}$, remove worst free swarm
  FOR EACH swarm n
    test for change
    IF swarm n is excluded, randomize
    ELSE update particle velocities and positions
    update attractors
    test for convergence
  apply exclusion
REPEAT

For generality, Algorithm 3 also specifies a redundancy parameter n_{excess} which is set to the desired number of free swarms. A simple choice is to suppose that $n_{excess} = 1$, but this may not give sufficient diversity if there are many peaks. Alternatively, $n_{excess} = \infty$ means that no swarm can ever be removed. Intermediate values control the amount of multi-swarm diversity. Part of the purpose of the experiments reported below is to assess the algorithm for robustness in n_{excess}.

The multi-swarm size $M(t)$ is dynamic and at any iteration t is given by

$$M(0) = 1$$
$$M(t) = \begin{cases} M(t-1)+1, & M_{free} = 0 \\ M(t-1)-1, & M_{free} > n_{excess} \end{cases} \tag{2.5}$$

Swarm convergence and exclusion are now determined by a dynamic convergence radius $r(t)$ defined by

$$r(t) = \frac{X}{2M^{1/d}} \tag{2.6}$$

which has been chosen to ensure a mean volume per swarm of $(2r)^d = \frac{X^d}{M}$, a condition which might be expected to be if the peaks are, on average, uniformly distributed in X^d. The number of free swarms at any time is the difference between the multi-swarm size and the number of converging swarms. A swarm is defined as 'converging' if its diameter is less than $2r$. The exclusion radius is replaced by $r(t)$. Hence two parameters, M and r_{excl} and one operator, anti-convergence, have been removed from the multi-swarm algorithm, at the expense of introducing a new parameter, n_{excess}.

2.6.2 Results

A number of experiments using the MPB of Section 2.5.5 with 10 and 200 peaks were conducted to test the efficacy of the self-adapting multi-swarm for various values of n_{excess}. The uniform volume distribution described in Section 2.5.1 was used. An empirical investigation revealed that $r_{cloud} = 0.5s$ for shift length s yields optimum tracking for these MPB's, and this was the value used here.

Table 2.1 shows the raw and rounded offline errors for $1 \leq n_{excess} \leq 7$ and for $n_{excess} = \infty$. Only the rounded errors are significant, but the pre-rounded values have been reported in order to examine algorithm functionality. For comparison, the best performance of an unadapted $10(10^0+10^-)$ multi-swarm for the $p = 10$ and $p = 200$ environments is, respectively, 1.75(0.06) and 2.26(0.03) [7]. The best self-adapting multi-swarm errors are 1.77(0.05) and 2.37(0.03), only slightly higher than the hand-tuned values. The constancy of the raw offline error for $n_{excess} \geq 5$ shows that the algorithm never tries to generate 6 or more swarms: $n_{excess} = 5$ is equivalent to setting n_{excess} to ∞.

Table 2.1. Variation of offline error with n_{excess} for 10 and 200 dynamic peaks. The raw data demonstrates identical algorithm behavior for $n_{excess} \geq 5$.

	Raw		Rounded (standard error)	
n_{excess}	$p = 10$	$p = 200$	$p = 10$	$p = 200$
1	1.9729028458116666	2.5383187958098343	1.97(0.09)	2.54(0.04)
2	1.879641879674056	2.398361911062971	1.88(0.07)	2.40(0.02)
3	1.7699076299648027	2.386396596554201	1.77(0.05)	2.39(0.03)
4	1.8033988974332964	2.372590853208213	1.80(0.06)	2.37(0.03)
5	1.8013758537004643	2.365026825401844	1.80(0.06)	2.37(0.03)
6	1.8010120393728533	2.3651325663361167	1.80(0.06)	2.37(0.03)
7	1.8010120393728533	2.3651325663361167	1.80(0.06)	2.37(0.03)
infinity	1.8010120393728533	2.3651325663361167	1.80(0.06)	2.37(0.03)

2.6.3 Discussion

Theoretically, $n_{excess} = 1$ would appear to be an ideal setting, allowing the multi-swarm to adapt to the number of peaks, whilst always maintaining a single free swarm. However, inspection of the numbers of free and converged swarms for a single function instance revealed that the self-adaptation mechanism at $n_{excess} = 1$ frequently adds a swarm, only to remove one at the subsequent iteration. The explanation is believed to be the following: suppose a swarm (swarm A) has started to converging on a peak. A new free swarm, swarm B, will be created. Since A is just at the edge of convergence, fluctuations in the swarm size may cause the swarm to appear to be free at the next iteration. Hence there will be two free swarms, and one must be removed - almost certainly swarm B, since this has had little chance to improve its $\mathbf{p}_g$. Swarm A will again start to converge (according to the criterion), causing the creation of a new free swarm at the next iteration. Such repeated creations and annihilations of the free swarm waste valuable function evaluations.

Another possible setting is $n_{excess} = \infty$. This effectively turns swarm removal off. Although there is no check to the number of swarms, it is not unreasonable to suppose that given enough time, the multi-swarm would stabilize at $M_{conv} = p$ and $M_{free} = 1$. The convergence criterion is rather naive and may mark a free swarm as converging even though it is not associated with a peak. This will cause the generation of another free swarm, with no means of removal. This will only be a problem when $M_{conv} > p$, a situation that might not even happen within the time-scale of the run. For example, Figures 2.4 and 2.5 show convergence data for a single run at $p = 10$ and $p = 200$. For $p = 10$, the eleventh swarm is generated at function evaluation, $n_{eval}, = 254054$. There are 500000 evaluations in a run, and in the remaining evaluations the multi-swarm size is, at most 13, and remains fairly steady after the 400000th evaluation. The $p = 200$ trial shows a steadily growing multi-swarm, which never attains complete coverage of all peaks, ending with 47 converging swarms and 1 free swarm.

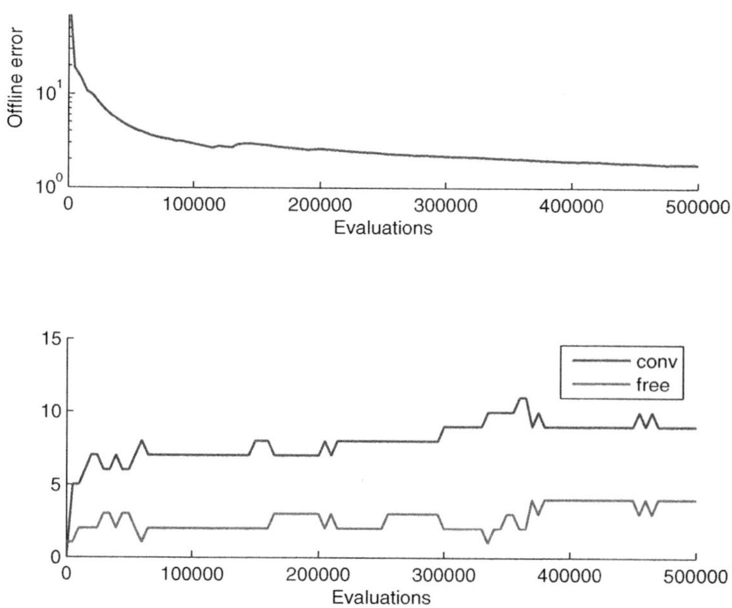

Fig. 2.4. Convergence of the self-adapting $n_{excess} = \infty$ multi-swarm for a single instance of the 10 peak MPB environment. Upper plot shows offline error, lower plot shows number of converged and free swarms.

The flexibility of a swarm removal mechanism is desirable for a number of reasons. For example, two peaks might move within r_{excl} of each other, causing

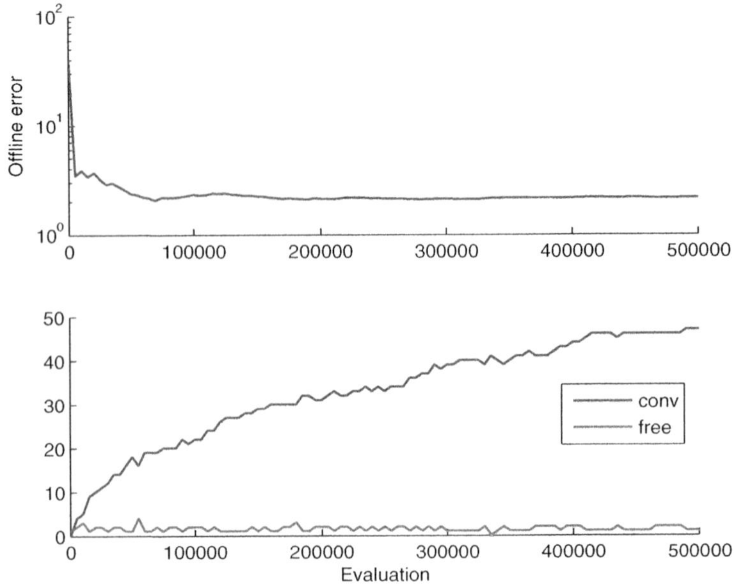

Fig. 2.5. Convergence of the self-adapting $M_{excess} = \infty$ multi-swarm for a single instance of the 200 peak MPB environment. Upper plot shows offline error, lower plot shows number of converged and free swarms.

a previously converged swarm to vaporize through exclusion (the better swarm remains). Or maybe a peak i becomes invisible if its height is smaller than other peak heights at its optimizer i.e. if $f(\mathbf{x_i^*}) < f(\mathbf{x_j})$ for some $j \neq i$. In either case, superfluous free swarms will consume function evaluations.

The redundancy n_{excess} can be set at any value between the two extremes of $n_{excess} = 1$ and $n_{excess} = \infty$. ($n_{excess} = 0$ gives very bad performance, no swarms at all may be added and the single swarm converges, and remains on, the first peak it finds.) The results for $p = 10$ and $p = 200$ indicate that tuning of n_{excess} can improve performance for runs where the multi-swarm has found all the peaks. Tuning can prevent the multi-swarm from generating too many free swarms - for example 3 free swarms are optimal for $p = 10$. However, setting n_{excess} at either extreme still produces good performance, and better than the comparison algorithms cited in Section 2.5.5. Perhaps the multi-swarm itself could tune n_{excess} during a run. A more sophisticated convergence criterion would also have to be devised. For example, the convergence criterion could take into account both the swarm diameter and the rate of improvement of $f(\mathbf{p}_g)$.

2.7 Conclusions

This chapter has reviewed the application of particle swarms to dynamic optimization. The canonical PSO algorithm must be modified for good performance in environments such as many peaks. In particular the problem of diversity loss must be addressed. The most promising PSO variant to date is the multi-swarm; a multi-swarm is a colony of swarms, where each swarm, drawing from an atomic analogy, consists of a canonical PSO surrounded by a cloud of charged particles. The underlying philosophy behind multi-swarms is to place a separate PSO on the best peaks, and to maintain a population of patrolling particles for the purposes of identifying new peaks. Movement of any peak that is being watched by a swarm is tracked by the charged particles. An exclusion operator ensures that only one swarm can watch any one peak, and anti-convergence seeks to maintain a free, patrolling swarm.

New work on self-adaptation has also been presented here. Self-adaptation aims at reducing the number of tunable parameters and operators. Some progress has been made at the multi-swarm level, where a mechanism for swarm birth and death has been suggested; this scheme eliminates one operator and allows the number of swarms and an exclusion parameter to adjust dynamically. One free parameter, the number of patrolling swarms, still exists, but results suggest that the algorithm is not overly sensitive to this number. Self-adaptations at the level of each swarm, in particular allowing particles to be born and to die, and self-regulation of the charged cloud radius remain unexplored.

References

1. A. Engelbrecht. *Computational Intelligence*. John Wiley and sons, 2002.
2. T. M. Blackwell. Particle swarms and population diversity I: Analysis. In J. Branke, editor, *GECCO Workshop on Evolutionary Algorithms for Dynamic Optimization Problems*, pages 9–13, 2003. http://www.ubka.uni-karlsruhe.de/cgi-bin/psview?document=2003%2Fwiwi%2F1.
3. T. M. Blackwell. Particle swarms and population diversity II: Experiments. In J. Branke, editor, *GECCO Workshop on Evolutionary Algorithms for Dynamic Optimization Problems*, pages 14–18, 2003. http://www.ubka.uni-karlsruhe.de/cgi-bin/psview?document=2003%2Fwiwi%2F1.
4. T. M. Blackwell and P. Bentley. Don't push me! collision avoiding swarms. In *Congress on Evolutionary Computation*, pages 1691–1696, 2002.
5. T. M. Blackwell and P. J. Bentley. Dynamic search with charged swarms. In W. B. Langdon et al., editors, *Genetic and Evolutionary Computation Conference*, pages 19–26. Morgan Kaufmann, 2002.
6. T. M. Blackwell and J. Branke. Multi-swarm optimization in dynamic environments. In G. R. Raidl, editor, *Applications of Evolutionary Computing*, volume 3005 of *LNCS*, pages 489–500. Springer, 2004.

7. T. M. Blackwell and J. Branke. Multi-swarms, exclusion and anti-convergence in dynamic environments. *IEEE transactions on Evolutionary Computation*, 10(4): 459–472, 2006.
8. T. M. Blackwell. Swarm music: Improvised music with multi-swarms. In *Proc AISB'03 Symposium on artificial intelligence and creativity in arts and science*, pages 41–49, 2003.
9. T. M. Blackwell. Swarms in dynamic environments. In E. Cantu-Paz, editor, *Genetic and Evolutionary Computation Conference*, volume 2723 of *LNCS*, pages 1–12. Springer, 2003.
10. T. M. Blackwell. Particle swarms and population diversity. *Soft Computing*, 9(11): 793–802, 2005.
11. J. Branke. The moving peaks benchmark website. *http://www.aifb.uni-karlsruhe.de/jbr/movpeaks.*
12. J. Branke. Memory enhanced evolutionary algorithms for changing optimization problems. In *Congress on Evolutionary Computation CEC99*, volume 3, pages 1875–1882. IEEE, 199. ftp://ftp.aifb.uni-karlsruhe.de/pub/jbr/branke_cec1999.ps.gz.
13. J. Branke. Evolutionary approaches to dynamic environments - updated survey. In *GECCO Workshop on Evolutionary Algorithms for Dynamic Optimization Problems*, pages 27–30, 2001. http://www.aifb.uni-karlsruhe.de/jbr/EvoDOP/Papers/gecco-dyn2001.pdf.
14. J. Branke. *Evolutionary Optimization in Dynamic Environments.* Kluwer, 2001. http://www.aifb.uni-karlsruhe.de/jbr/book.html.
15. J. Branke and H. Schmeck. Designing evolutionary algorithms for dynamic optimization problems. *Theory and application of evolutionary computation: recent trends*, pages 239–262, 2002. S. Tsutsui and A. Ghosh, editors.
16. R. Brits, A. P. Engelbrecht, and F. van den Bergh. A niching particle swarm optimizer. In *Fourth Asia-Pacific conference on simulated evolution and learning*, pages 692–696, 2002.
17. A. Carlisle and G. Dozier. Adapting particle swarm optimisationto dynamic environments. In *Proc of int conference on artificial intelligence*, pages 429–434, 2000.
18. M. Clerc. Think locally act locally - a framework for adaptive particle swarm optimizers. Technical report, 2002. http://clerc.maurice.free.fr/pso/ (accessed June 29, 2006).
19. M. Clerc. *Particle Swarm Optimization.* ISTE publishing company, 2006.
20. M. Clerc and J. Kennedy. The particle swarm: explosion, stability and convergence in a multi-dimensional space. *IEEE transactions on Evolutionary Computation*, 6:158–73, 2000.
21. C. Reynolds. Flocks, herds and schools: a distributed behavioral model. *Computer Graphics*, 21:25–34, 1987.
22. A. P. French and E. F. Taylor. *An introduction to quantum physics.* W. W. Norton and Company, 1978.
23. X. Hu and R. C. Eberhart. Adaptive particle swarm optimisation: detection and response to dynamic systems. In *Proc Congress on Evolutionary Computation*, pages 1666–1670, 2002.
24. S. Janson and M. Middendorf. A hierachical particle swarm optimizer for dynamc optimization problems. In G. R. Raidl, editor, *Applications of evolutionary computing*, volume 3005 of *LNCS*, pages 513–524. Springer, 2004.

25. J. Kennedy and R. C. Eberhart. Particle swarm optimization. In *Proceedings of the 1995 IEEE International Conference on neural networks*, pages 1942–1948, 1995.
26. J. Kennnedy. Stereotyping: improving particle swarm performance with cluster analysis. In *Congress on Evolutionary Computation*, pages 1507–12, 2000.
27. X. Li. Adaptively choosing neighborhood bests in a particle swarm optimizer for multimodal function optimization. In K. Deb et al, editor, *Proceedings of the Genetic and Evolutionary Copmutation Conference, GECCO-2004*, volume 3102 of *LNCS*, pages 105–116. Springer, 2004.
28. B. Mandelbrot. *The Fractal Geometry of Nature.* W. H. Freeman and Company, 1983.
29. D. Parrott and X. Li. A particle swarm model for tracking multiple peaks in a dynamic environment using speciation. In *Congress on Evolutionary Computation*, pages 98–103, 2004.
30. K. E. Parsopoulos and M. N. Vrahatis. Recent approaches to global optimization problems through particle swarm optimization. *Natural Computing*, pages 235–306, 2002.
31. A. Petrowski. A clearing procedure as a niching method for genetic algorithms. In J. Grefenstette, editor, *Int'l Conference on Evolutionary Computation*, pages 798–803. IEEE, 2003.
32. R. Eberhart and Y. Shi. *Swarm intelligence.* Morgan Kaufmann, 2001.
33. Y. Shi and A. Khrohling. Co-evolutionary particle swarm optimization to solve min-max problems. In *Congress on Evolutionary Computation*, pages 1682–1687, 2002.
34. J. Vesterstrom T. Krink and J. Riget. Particle swarm optimisation with spatial particle extension. In *Congress on Evolutionary Computation*, page 14741479, 2002.
35. F. van den bergh and A. P. Englebrecht. A cooperative approach to particle swarm optimization. *IEEE transactions on Evolutionary Computation*, pages 225–239, 2004.
36. X. Li and K. H. Dam. Comparing particle swarms for tracking extrema in dynamic environments. In *Congress on Evolutionary Computation*, pages 1772–1779, 2003.

3

Evolution Strategies in Dynamic Environments

Lutz Schönemann

Chair of Algorithm Engineering and Computational Intelligence
Department of Computer Science, University of Dortmund
D-44221 Dortmund, Germany
lutz.schoenemann@cs.uni-dortmund.de

Summary. Numerical parameter optimization is an often needed task. In many times, it is not necessary to find the exact optimum but a good solution in an appropriate time. Especially in dynamic environments the main task is not to find one nearly optimal solution but to track the moving optimum as narrow as possible. For this type of problems it is necessary for an optimization algorithm to own mechanisms for adaptation to the problem at hand. Evolutionary algorithms with self-adaptive features are state-of-the-art and known as good problem solvers for this type of optimization tasks. In this chapter, we present a detailed description of one main variant of this class of problem solvers, namely evolution strategies.

3.1 Introduction

Evolutionary algorithms (EA) are a class of nature inspired problem solvers. They use mechanisms known from natural evolution and have in common the transfer of their biological background into optimization. EA have proven their potentials in many real world applications. It is known that in static environments evolutionary algorithms find good or nearly optimal solutions even for difficult problems in a short time. In addition, they own a high robustness against changes of the problem instance over a certain range. Moreover, state-of-the-art EA like evolution strategies (ES) can adapt to different situations during the run. Such a feature makes them interesting for application in dynamic environments . These types of problems seem to be the more interesting ones since most real world problems are of non-stationary type.

In dynamic optimization we can divide two distinct phases. At first, the algorithm needs some time to search for the optimum. The period needed for this task is called the searching phase. Once the algorithm has found the optimum within a certain accuracy the algorithm must follow the moving optimum with a distance as small as possible. This phase is called the tracking phase and lasts potentially eternally.

Dynamic optimization problems are harder to optimize than static problems. The more difficulty of optimization in non-stationary environments is

L. Schönemann: *Evolution Strategies in Dynamic Environments*, Studies in Computational Intelligence (SCI) **51**, 51–77 (2007)
www.springerlink.com

reflected by the circumstance that only a few theoretical analyses were performed on problems in dynamic environments. In contrast to this, many investigations concentrate on the case of static environments. Consequentially, for this type of problems many theoretical results are known. But for the dynamic case the situation is different. Here, the dependencies between the algorithm and the changing environments are more difficult to analyze. Therefore, the results are not transferable to the dynamic case in a straight forward manner. Hence, experimental investigations are necessary in the foreseeable future.

ES benefit from their self-adaptive mechanisms. Algorithms with this feature are able to adjust their endogenous parameters to the problem at hand during the evolutionary run. But the ability to let the adaptation process work depends on good settings of exogenous parameters which are not varied during the evolutionary run. One exogenous parameter is the number of step sizes used for changing the object variables. With some experiments we determine the influence of different choices for this parameter.

To get a deeper insight into the behavior of EA in dynamic environments it is helpful to start with a thorough investigation of easier problems. This is done in the remaining text, which is structured as follows. Section 3.2 illustrates in general the underlying dynamic problems. Evolution strategies are introduced in detail in Sect. 3.3. Beside the main algorithm some specialized variants are mentioned. Section 3.4 gives an introduction into the topic of comparing two or more strategies. A detailed specification of the test scenarios investigated here is described in Section 3.5. Experiments concerning the optimal number of mutation step sizes are then performed in Sect. 3.6. Finally, Sect. 12.9 offers concluding remarks.

3.2 Problem Definition

Although evolution strategies were applied successfully to combinatorial as well as discrete optimization problems, for ES the main field of application is continuous numerical parameter optimization. Therefore, such problems are lying in the main focus of this study. The reader who is interested in ES working on other search spaces is referred to [2].

Real-valued parameter optimization deals with optimization problems with which a user is often confronted. In the static case it is characterized by a minimization problem of the type

$$\min f_s : \mathbb{R}^n \to \mathbb{R} \ .$$

A maximization could be ascribed to a minimization by $\max f_s = -\min -f_s$. In the static case, the problem to be optimized remains the same during the whole run. In contrast to this, dynamic problems are non-stationary. This circumstance causes that a good solution found once, get obsolete during the shorter or longer next time. Hence, whereas for a static function it is sufficient to find one good solution, in dynamic environments the algorithm must track the moving target within a distance being as small as possible.

Depending on the type of changes the problem could be more or less difficult to optimize. In one type the underlying function does not change, but the optimum moves through the search space. Such problems could be ascribed by

$$\min_t f_d(x; x^*(t)) := f_s(x - x^*(t))$$

to the static case where the optimum $x^*(t)$ is included as an additional parameter. The optimum $x^*(t)$ moves during the optimization process and depends on t. Be aware, that despite the chance to wait some time, t could not be influenced by the user.

The exact dynamic test scenarios used for our investigations are derived from this general definitions and described in Sect. 3.5.

3.3 Evolution Strategies

Before we introduce evolution strategies in detail, we start with a brief history of evolutionary algorithms. For a more detailed description of the historical development of evolutionary algorithms the reader is referred to [8].

3.3.1 Evolution Strategies in the Historical Context of Evolutionary Algorithms

In many scientific areas new theoretical ideas were developed before their practical usefulness could be demonstrated. A similar development could be seen in the field of evolutionary algorithms. These algorithms were inspired by Darwins theory of natural selection [7]. Indeed, such a procedure suggested itself since the nature is a complex search space and had developed well-adapted life-forms.

The origins of nature inspired methods could be followed back to the last 1950s [4, 6, 11, 12]. But at these times computational power was less developed to support their practical importance. This changes in the early 1960s when the demand for numerical optimization methods grew more and more. Several authors started independently from each other to search for nature inspired techniques. From these, three main variants were established and exist today. Namely, these are genetic algorithms (GA) developed by Holland [15, 16], evolutionary programming (EP) schemed by Fogel [9, 10], and evolution strategies designed by Rechenberg and Schwefel [19, 25]. Nowadays all methods firm under the term of evolutionary algorithms.

Whereas the former two methods started more or less as design studies, evolution strategies have shown their practical power in several applications in early days. Evolution strategies were not developed for numerical optimization problems. Instead, they were used to assist a user in adjusting a discrete system to find an optimal shaping. The most famous example is the development of a two-phase nozzle, which results in a so-far optimal shape of strange appearance. These early successes helped ES to become known to a greater circle of researchers. Afterwards, evolution strategies left the application area of system design and entered the field of numerical (continuous) optimization problems. In numerical optimization the task is to find an optimal setting for real-valued parameters resulting in a minimum or maximum of a given function. Today, this is the main application area for evolution strategies.

The development of evolution strategies started with the simplest possible strategy. One given solution called the parental individual is slightly varied by applying variation operators resulting in a new solution called offspring individual. In this so-called $(1+1)$-ES the new solution is tested against its parent. If the offspring has a better (more fitter) function value the new solution becomes the next parent. Otherwise the offspring is discarded. The execution of the complete loop is called one generation and repeated for a given number of generations.

Whereas this is an easy to use and resource sparing variant the increasing power of computer systems allowed more sophisticated implementations. Consequently, the next step was the introduction of the $(1+\lambda)$-ES in which the parent generates λ offspring in one generation. Then the best solution from this offspring population substitutes the parent if it has a better fitness. Otherwise the parent is transferred to the next generation.

Although very unlikely it is possible for the parent to survive for the complete evolutionary run. This is the case if all generated offspring have a worse function value. Now, the introduction of more than one offspring admits a different proceeding. Since the offspring compete against each other a natural selection exists independently from the competition with the parent. Therefore it is possible to prevent the parent to be transferred to the next generation regardless of its fitness. This idea results in the $(1,\lambda)$-ES in which the best offspring becomes the parent of the next generation. It was observed that due to this proceeding in many applications the EA was able to find a solution of adequate quality in a shorter time. This means that the so-called mechanism of adaptation often works better than in an $(1+\lambda)$-ES. If the exact selection method does not matter both methods are summarized to the $(1 \overset{+}{,} \lambda)$-ES.

A last step was the introduction of more than one parent. In the $(1 \overset{+}{,} \lambda)$-ES every offspring is derived slightly modified from the only parent. In the multimembered $(\mu \overset{+}{,} \lambda)$ strategy exists a population of parents. If an offspring is generated directly from one parent, every parent generates in the mean λ/μ offspring. But the multimembered ES allows more. Here, several parents could be combined to one recombinant. This solution is then mutated to form a new offspring. The next parental population may be the best individuals from the offspring or the union of offspring and parent population.

3.3.2 State-of-the-art: the $(\mu, \kappa, \lambda, \rho)$-ES

In this section we present the most general form of evolution strategies in detail.

General evolutionary loop

Every evolutionary algorithm has so-called strategy parameters that influence the working of the variation operators. An improper setting of these parameters may hamper the EA from finding the optimum. We distinguish exogenuous and endogenuous parameters. On the one hand, exogenuous parameters are those parameters which are set at the start of the ES and are not changed by the ES during the run. E.g., such parameters are the population sizes which usually remain constant during the evolutionary run. On the other hand, endogenuous parameters could change during the run. An ES in which strategy parameters could change is called an adaptive ES. This change could be done from time to time by a routine which use the current progress for the parameter control. Alternatively, the strategy parameters may be coded in the individuals. In this case the strategy parameters are part of the variation process. The selection of good strategy parameters is done indirectly through the selection of the whole individuals. As seen above the selection of an individual is based on the fitness of the individual which reflects the quality of the object variables. An ES with this feature is called self-adaptive. In static environments self-adaptation is known as a helpful mechanism to locate the optimum with a high accuracy. But it is even more useful in dynamic environments than in static ones since the problem changes from time to time and therefore the strategy parameters are frequently obsolete.

Today there exists several variants of ES designed to met the efforts of different users. The most popular one is a generalization of the ones mentioned above and was introduced by Schwefel and Rudolph [26]. Fig. 3.1 presents the basic evolutionary loop of their $(\mu, \kappa, \lambda, \rho)$-ES. In the $(\mu, \kappa, \lambda, \rho)$-ES the algorithms work on a set of μ parental solutions. With the help of variation operators a set of λ additional solutions are generated. The variation is split into the recombination of ρ parents to one new solution and subsequently mutating this solution resulting in one offspring.

Because it is supposed that we optimize in dynamic environments the optimization process never ends. Hence, in real scenarios the following procedure, called generational loop, is repeated for infinity: In every generation t a multiset O_t of λ offspring are generated by variation operators. After evaluation, the μ best elements from the union of offspring and old parents are selected for the next parent population. In contrast to this plus strategy the comma ES takes only the offspring into account. This is regarded as helpful for the adaptation process, especially if the environment changes with a high frequency (every generation).

```
t ← 0
P_t ← init()
evaluate(P_t)
while TRUE do
    move of the optimum
    O_t ← mutate(recombine(P_t))
    evaluate(O_t)
    P_{t+1} ← select(O_t ∪ P_t)
    t ← t + 1
end while
```

Fig. 3.1. Basic $(\mu, \kappa, \lambda, \rho)$-ES

The selection operator choses the new parents from the parental and the offspring population depending on the fitness and the age of each individual. Every individual could survive for up to $\kappa \in \{1, \dots, \infty\}$ generations. A whole cycle of the loop of performing the variation operators and the selection operator is called a generation.

After selecting the new population, the age of every parent and the generation counter are increased and the loop starts anew.

Representation

The evolutionary loop starts with initializing and evaluating the initial population P_0 of potential solutions. A single solution (individual) comprises of n object variables x_i, one for every problem dimension, and a set of up to $n_\sigma + n_\alpha$ strategy parameters which affects the working of the variation operators. The n_σ mutation step sizes control the mean amount of changes affected by the mutation operator. Due to their values new solutions are found nearby old solutions or they are lying more far away. The optional n_α rotation angles facilitates the ES to walk along directions independently of the coordinate axes. Although it is possible to use $1 < n_\sigma < n$ step sizes usual choices are $n_\sigma \in \{1, n\}$. The usage of rotation angles is very rare since otherwise the adaptation process lasts often very long.

Initialization

The first population is mostly initialized equally distributed over the complete or a wide range of the domain. If a good solution is known the whole population could be initialized with this solution. The same holds for constrained problems. In such a case it could be difficult to find a valid solution. Then the algorithm could be initialized with one or more known valid solution(s). The step sizes are initialized depending on the size of the domain.

Reproduction and Variation

The variation of solutions is done in two steps. In the first one, the recombination operator often combines two individuals to produce one offspring. Frequently, the object variables of every coordinate of the child are chosen independently from each other randomly from one of the parents (discrete recombination). Whereas the strategy parameters are set to the mean of the corresponding strategy parameters of both parents (intermediate recombination).

In the second step, the mutation operator changes the values (x, σ) of a formally generated offspring to the new values (x', σ') using the following operations:

- In case of one mutation step size:

$$\sigma' = \sigma \cdot \exp(\tau_0 \cdot N(0,1)),$$

  where $\tau_0 \propto \frac{1}{\sqrt{n}}$.
- In case of $1 < n_\sigma \leq n$ mutation step sizes:

$$\sigma_i' = \sigma_i \cdot \exp(\tau' \cdot N(0,1) + \tau \cdot N_i(0,1)),$$

  where $\tau \propto \frac{1}{\sqrt{2\sqrt{n}}}$ and $\tau' \propto \frac{1}{\sqrt{2n}}$.

In both variants the mutation of the object variables is usually done by

$$x_i' = x_i + N_i(0, \sigma_i') \ \ \forall i \in \{1, \ldots, n\}$$

Here, $N(\cdot)$ is a normally distributed random variable common for all i. $N_i(\cdot)$ is a normally distributed random variable chosen independently from each other for every coordinate. $\tau_0, \tau, \tau' > 0$ are so-called learning rates, that follow the rules given above. From theory it is known that in static environments the optimal settings for the learning rates depend on the used population sizes [1, S. 302]. Since the underlying function as well as the optimal population sizes are unknown in advance, in most cases this holds for the optimal settings for the learning rates, too. Hence, in static environments the proportionality factor of the learning rates are often set to one. In dynamic environments no theoretical analyses regarding the optimal settings for the learning rates are known. Because the function to optimize changes during the evolutionary run, a higher setting for the learning rates seems to be favorable. Therefore, in the experimental investigations in this study the proportionality factors of the learning rates are set to three.

The mutation of the object variables is done using a normally distributed random variable. It is a nearby idea to try other random variables. Several attempts in this direction were taken. Some theoretical analyses proved that other distributions guarantee that an EA could leave a local optimum in the static case [21, 22]. But since the results are only valid for infinity long runs, none of the distributions have etablished as a standard for some problem classes.

Principally, the number of mutation step sizes can be varied between the two extremes of one and n. If $1 < n_\sigma < n$ step sizes are used, $n_\sigma - 1$ coordinates get their own step size and the remaining $n - n_\sigma + 1$ coordinates share one step size. The number of used strategy parameters has an influence on the obtainable results. First, they affect the time needed for adaptation. The higher the number of strategy parameters the higher is the time needed for adaptation. Second, depending on the underlying moving type, with a less number of strategy parameters grows the distance in which the algorithm follows the optimum. In Sect. 3.6, our main focus is the effect of different choices for the number of mutation step sizes.

Selection

The algorithm uses the κ selection scheme in which the best μ individuals of the set of parents and offspring are chosen for the next reproduction cycle as long as their age are less than or equal to the maximal life span κ. The κ selection is the generalization of the comma and plus selection. In the comma strategy the new parents are chosen as the μ best individuals from the set of offspring. This means that the total population is exchanged after every generation. Hence, every individual survives for exactly one generation. Here, the quotient λ/μ is called the selection pressure of the strategy. In the plus selection scheme the new parents are selected from the union of sets of parents and offspring. In this case, every individual could survive potentially for eternity. It is ensured that the best individual found so far is transferred to the next generation. Such a strategy is called elitist.

The κ selection was introduced in the 1990s and it is assumed that in many situations the best choice for κ is different from one and infinity. But since then, no theoretical analysis were done in this research area and the knowledge of an optimal maximal life span is still outstanding. Indeed, it seems to be a difficult task to analyze an ES with a maximal life span $1 < \kappa < \infty$. To brighten this field the author made a first fundamental experimental investigation [24]. In that study the influence of different κ values on the obtainable results were tested. Although this study only looked at static environments the results give hints for the dynamic case. It was observed that on the relative easy test functions used after a short period the adaptation process works well. This is also reflected by the fact that almost all individuals do not exceed a life span of three. Mostly, the individuals died with an age of one or two. Hence, a limitation of the life span to a value greater than three showed no effect on the obtained results. In dynamic environments the situation is threefold. If the changes occur very seldom the environment remains constant for longer periods. In this case the situation is similar to the static case and different settings for κ do not promise better results. If the environment changes very often a permanent adaptation is necessary and almost all individuals are replaced in the next generation automatically. Again, a limitation of the life span seems not promising. If the changes occur in moderate intervals Δg the limitation of the life span to exact this value ($\kappa = \Delta g$) is a close-by idea. But first investigations in this direction were not successful.

Additionally, in [24] it was shown that a single individual of high fitness and with bad settings for the strategy parameters could hamper a proper adaptation to the problem at hand. Hence, it should be guaranteed that an individual could not survive for too long. So, in practical applications a small value like $\kappa = 5$ is chosen. Hence, a single individual can bequeath their genes, but dies after a few generations.

Parameter Settings

Theoretical analyses on the behavior of evolutionary algorithms in dynamic environments are very rare. Often they look at simple strategies or test problems. These analyses failure if harder problems or more sophisticated EA are regarded. Such EA uses self-adaptation and complex variation mechanisms like a whole set of n mutation step sizes. With more complicated dynamics the analysis gets even harder. Due to this, at the moment experimental investigations are necessary to get a deeper insight into the functioning of evolutionary algorithms.

Beside setting the initial values of the endogenuous parameters as the mutation step sizes, it is necessary to choose good values for the exogenuous parameters. The main ones of these parameters are the population sizes. As optimal values are unknown, researchers use often an $(15, 100)$ strategy because this have proven as a good compromise. In our experimental investigations in Sect. 3.6 we try other settings for these parameters.

The learning rates τ, τ', τ_0 described above are chosen based on theoretical analyses of some test functions. On other functions these heuristics are not optimal. Kursawe examined the best settings for the learning rates on some test functions empirically [17]. The resulting optimal settings follow no obvious rule. Hence, most authors fall back on the heuristic settings.

3.3.3 Other ES variants

In addition to the standard forms of EA, several variants were developed for particular purposes. Especially for the case of dynamic environments some sophisticated evolutionary algorithms exist [5, 18]. Mostly, they have in common that they use several subpopulations to search and follow moving local optima of different quality. In these cases, it is hoped that a formally local optimum which changed to be the global one during the run, is already occupied by a subpopulation. Such mechanisms are even known from optimization in static environments [27].

An open question is the usefulness of derandomized mutation schemes. Two of those strategies are the *cumulative step size adaptation (CSA)* and the *covariance matrix adaptation (CMA)* developed by Hansen and Ostermeier [14]. The underlying idea of both strategies is to adapt the mutation based on the information about the length and direction realized so far. The second variant is able to rebuild correlated mutations by adapting a covariance matrix. The exact working mechanism of both strategies lies beyond the scope of this introductory article. For a detailed discussion the reader is referred to the original literature cited above.

Of course, there exists other EA variants which may be an interesting approach for our problem definition [2]. But none of them has reached the status of a standard method. We use the type of ES described above because of two reasons. First, a user with only a small knowledge of EA would prefer an easy to use algorithm. Often, specialized EA have some additional parameters which must be set by the user in advance. A wrong setting of these parameters may hamper the EA to follow the optimum. Thus, many users fall back on standard variants for which only some parameters must be set by the user. Second, the used algorithm must behave well in static environments, too. Evolution strategies follow these demands.

3.4 Performance Measures and Comparison of Strategies

Although it is interesting to investigate the time needed to search for the optimum, we are only interested in the behavior of the algorithm during the tracking phase.

An investigation concerning the best strategy to use needs a ranking of the tested strategies. Such a ranking is done by comparing the strategies. A comparison of two strategies needs an objective criteria. Usually, such a criteria is based on an unary performance measure for a single strategy. Measuring the performance of an EA is not easy even in the static case. It is a matter of opinion whether a local or global performance measure is taken. In addition, the performance could be measured in the fitness space or in the search space. For a detailed discussion of this topic see [3].

Especially in the case of dynamic environments the comparison of different strategies is a difficult task. To define a helpful measurement it is necessary to have an imagination of the exact object of investigation. An EA in a dynamic environment passes two phases. In the first one (searching phase) the EA starts from an initial point and moves into the direction of the optimum. If it falls below a particular distance the tracking phase begins. During the tracking phase the EA follows the optimum with more or less equal distance. The mean distance to the optimum depends on the abilities of adaptation of the EA. If the changes are not too large, this period lasts for the rest of the run. Otherwise a new searching phase starts, because the optimum had departed too far from its old position. If the EA diverges from the optimum it is not able to reach the tracking period.

Several attempts were done to define an appropriate measurement [28]. But all of them have more or less deficiencies. First, often the user is only interested in one of the two phases. But most measures do not distinguish between them. Hence, an algorithm may be handicapped if its searching phase lasts longer, but follows the optimum within a smaller distance. Second, a necessity is to abstract from one single run and to regard the general behavior observed in multiple runs. All existing measurements calculate the arithmetic mean of the observed function values. This has one disadvantage. The arithmetic mean is vulnerable against large outliers. In extreme, a single outlier of large intense could distort the result in an improper manner. This is almost undesirable.

To get rid of these deficiencies, in [23] we defined a new measurement for comparing two or more strategies in dynamic environments. It is called **average best function value (ABFV)**. Abridging, the ABFV is a robust measure for the average fitness of a mean run. In the case of minimization, a smaller ABFV means a better behavior. In the following we present this measurement in a nutshell. Fig. 3.2 (top) shows a common single run of an ES in a dynamic environment. After a few generations (searching period) the EA has found a solution with a certain accuracy. Due to statistical fluctuations in the next generations (tracking period) the function values oscillate around this value.

In the following we concentrate our investigations on the tracking period. The period to find the optimum for the first time is not considered any more. Although in dynamic environments the process of optimization is infinite we have to restrict our investigations to a limited time horizon. Depending on the objective function, the problem dimension, the moving frequency and severity we run every EA for an equal number of function evaluations. This guarantees that the results are comparable within the same optimization problem.

As a first measurement we could calculate the mean function value of a single run during the tracking period. But a single run of one strategy is not sufficient to get meaningful results. We reduce the random fluctuations by taking the median (0.5-quantile) of every generation of repeated runs. In contrast to the mean the median is more resistant against statistical outliers. To get an impression of the statistical fluctuations fig. 3.2 (bottom) shows in addition to the median the 0.05- and 0.95-quantiles of 50 runs of the same experiment. Adding these quantiles leads to an approximative 90% confidence interval for the function value of every generation of a single run.

For a simple comparison of two strategies we need a single measurement for every strategy. We calculate such a measurement in the following way. In every single run and every generation of the tracking period we print the best function value of the current population. To reduce inaccuracies due to random fluctuations every strategy is performed m times. Repeating it m times we get m function values for every generation. For every generation we take the median of these m function values and get a *median run*. The mean of this median run during the tracking period serves as the tracking measurement $M_{(\mu,\lambda)}$, which we call the *average best function value*. Fig. 3.3 demonstrates our

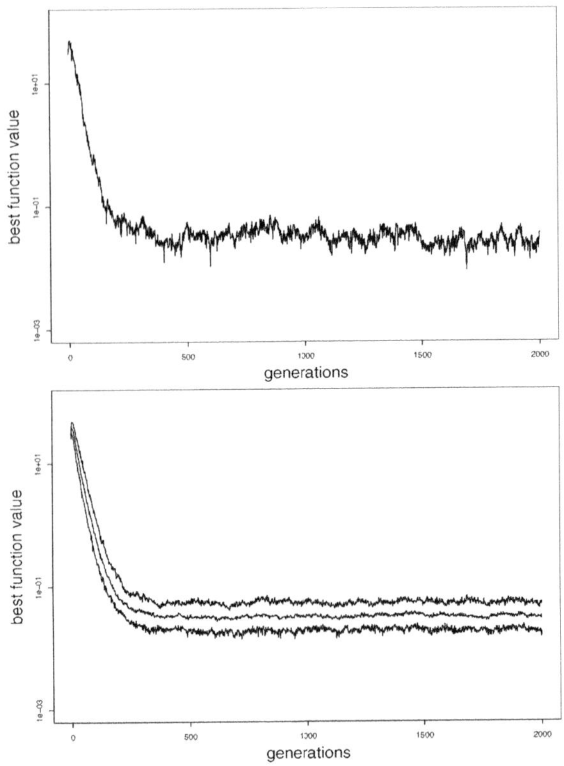

Fig. 3.2. Results of a (15, 100)-ES on the dynamic sphere with $n = 30$. The optimum moved every generation in one dimension with a constant $s = 0.1$. Single run (top) and 0.05-, 0.5- and 0.95-quantiles of the best function values of every generation of 50 runs of the same ES (bottom).

approach. In this example the average best function value of the generations from 1000 to 2000 is approximately $M_{(15,100)} = 0.03$. Keep in mind that if the EA is already in the tracking period and we increase the number of function evaluations the resulting value will not change substantially. This is because we calculate the mean function value of the tracking period and the expected value will not change if we calculate it for more generations.

Although the ABFV was developed for dynamic environments we can define it for static environments, too. In static environments we do not have a tracking period. In most cases, the ES reaches the global optimum with a certain distance or it gets stuck in a local optimum. In both situations we define the tracking period of the ES as the last generation of a single run. This definition has advantages as well as disadvantages. But it includes the situation that in static environments the ES may improve till the last generation. Using a certain number of runs will make sure that this value is not a statistical outlier.

We return to the former experiment. As an alternative to the $(15, 100)$-ES we use a $(10, 100)$-ES on the same problem. Fig. 3.4 (top) shows the median runs for both strategies. In this figure we see a noticeable better performance of the variant with $\mu = 10$. The respective tracking measurement is $M_{(10,100)} = 0.01$. On this problem, the ABFV of the $(10, 100)$-ES is better than the ABFV of the $(15, 100)$-ES.

The ABFV gives a first hint which strategy performs better on a given problem instance. But it is not sufficient to decide if two strategies are significant different. To get statistical certainty we need additional calculations. For every generation we sort the belonging m values of the m runs. After this we join the best value of every generation to a new best virtual run, the second best value of every generation to a new second best virtual run, ..., and the worst value of every generation to a new worst virtual run. Totally, we get m virtual runs which we call the α-quantile runs ($\alpha \in \{0, 1/(m-1), 2/(m-1), \ldots, (m-2)/(m-1), 1)$. The 0.5-quantile run forms the median run from above. This median run has a counterpart in statistical time series analyses. In this area, the mean value function is calculated. This is nearly exact our median run, abstained from the fact that instead of the median the mean of all values of a generation is regarded. But as noticed before, the mean is not robust against already one single large outlier.

Be aware that every α-quantile run may be compounded of values of different real runs and must not reflect a real single run. Imagine two single runs. For generations 101-200 the first one has the function value 1 and the other one has value 10. For generations 201-300 the first one has the value 10 and the other one has value 1. So both runs have an equal mean function value. The resulting virtual runs have the function value 1 for the best virtual run

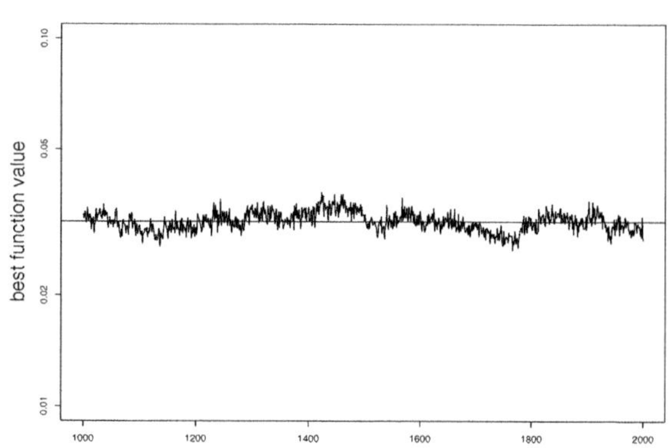

Fig. 3.3. The median function values of generations 1000 – 2000 of 50 runs of a $(15, 100)$-ES on the dynamic sphere with $n = 30$. The optimum moved every generation in one dimension with a constant $s = 0.1$. Additionally, the horizontal line shows the mean of the 1001 plotted values, which we call the *average best function value*

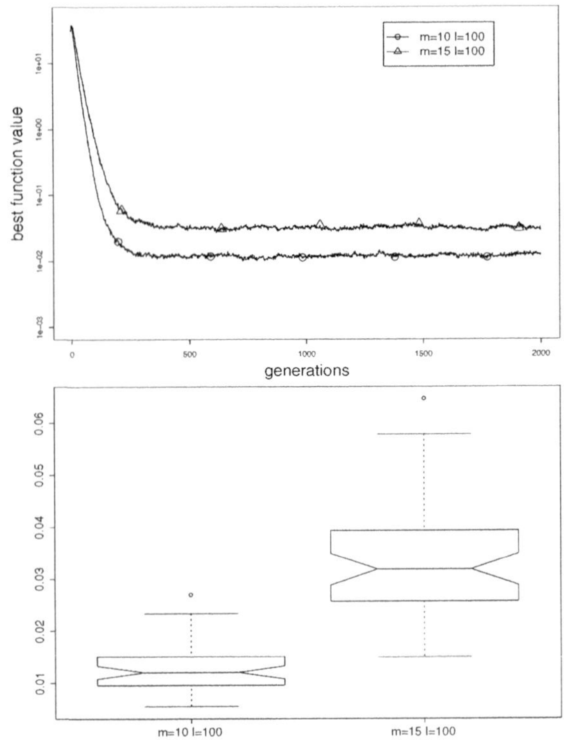

Fig. 3.4. Comparison of 50 runs of a (10, 100)- and a (15, 100)-ES on the dynamic sphere with $n = 30$. The optimum moved every generation in one dimension with a constant $s = 0.1$. Median runs (top) and boxplot of the mean function values of the α-quantile runs (bottom).

and 10 for the worst one. But no one of them reflect the real situation in which every single run has a mean function value of 5.5. This is an extreme example and we can assume that the median run will reflect the average situation regarding the median function value of a single run.

If we calculate the mean value of every α-quantile run we get m values for the mean best function value. These m values can serve for a statistical test, which tests the median of two distributions on equality. To do this, the strategies m mean values are compared with the ones of another strategy. We do not know the exact underlying distribution of these values. So it is the best to use a non-parametric test. Due to space limitation, in this study we are not presenting an example for this statistical test. Instead, we demonstrate a graphical method to show the differences between two strategies. For the example given above, fig. 3.4 (bottom) shows the boxplots of the mean values of the α-quantile runs. The non-overlapping notches of the boxplots show that the median of the function values reached by the (10, 100)-ES is significant (at the 5%-level) better than the one of the (15, 100)-ES.

In this study we want to investigate the performance of several ES. Comparing two strategies with disparate number of offspring is somewhat problematic. Think of strategy A as a (15, 100)-ES and strategy B as a (15, 200)-ES and a total number of 100,000 function evaluations. Taking the fitness after every generation, in the first case one has 1,000 values (one after every 100 function evaluations) and in the second case one has only 500 values (one after every 200 function evaluations). The ABFV calculates the mean of these values. Regard the situation when we have a true continuous movement, that means, that the optimum moves during every function evaluation. Hence, the function values probably grow higher between two complete generations. But the new value is taken after calculating the whole population. And this behavior increases with an increasing number of offspring. Fig. 3.5 makes this connection clearer. It shows a possible run of a $(\mu, 100)$-ES and a $(\mu, 200)$-ES

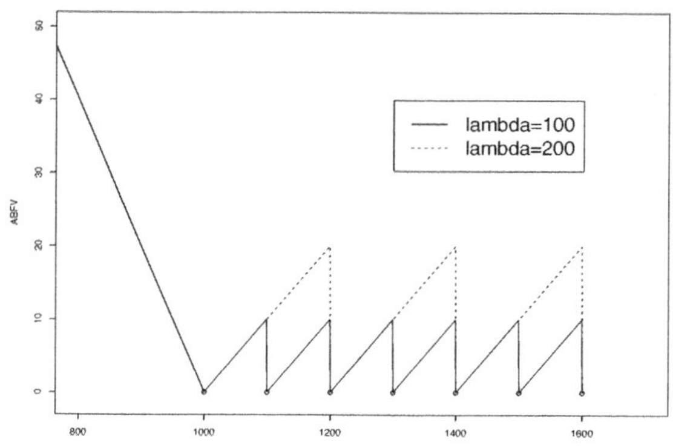

Fig. 3.5. The process of dynamic, the point of function evaluations and a probable fitness development (see text)

on a fictitious dynamic function. Every strategy calculates a whole generation and prints the ABFV of this population. In the case of $\lambda = 100$ this is done at function evaluations $100, 200, 300, \ldots$ (indicated by circles) and for $\lambda = 200$ this is done at $200, 400, 600, \ldots$ Assume that both strategies are able to find the minimum with function value 0 at every generation. Due to this, for both strategies the according ABFV has the same value 0. But between the generations the optimum moves. And this is not noticed by the strategies. So, between two complete generations the real function values of the current population may increase with increasing time. Because for the $(\mu, 200)$-ES the time between two generations is longer than for the $(\mu, 100)$-ES the effects may be greater there. In the test cases below we disregard these difficulties and assume that the environment changes after every generation.

3.5 Test Scenarios

In the next section we investigate some aspects of evolution strategies with a series of test experiments. It is the topic of this section to define the underlying test scenarios.

The type of dynamic optimization problems with which a user is faced is very manifold. One is the moving direction which could be linear, cyclic or random. We assume that the number of changes is high enough to speak of a dynamic problem. If the problem changes only once we will be confronted with two static problems (one before the change and one after it). Therefore, the moving frequency is a second one and could vary between a change after every generation ($\Delta g = 1$) and a high number ($\Delta g \gg 1$). The moving severity s is another criteria. It is obvious that an algorithm can not follow a moving optimum if the moving severity is extremely high. In addition it is likely that it is a difference if the severity is constant or not.

Our experimental investigations are conducted on three test functions. Every test function is a representative of a class of similar optimization problems. We use

- the sphere model as a representative of the class of unimodal functions,
- the Ackley function which is a multimodal function of moderate difficulty, and
- the Rastrigin function since it is considered to be more multimodal and therefore harder to optimize.

All test functions have their optimum in the origin. This makes it easy to handle the position of the moving optimum. The first test function used here is the well-known sphere model

$$f_{\text{Sphere}}(x) = \sum_{i=1}^{n} x_i^2 \ .$$

The sphere model as well as the following two other functions is almost always included in a catalog of standard test problems. This is a simple function with one local optimum, which is therefore also global. In the static case the consideration of this function serves for measuring the convergence velocity of an algorithm.

The Ackley function

$$f_{\text{Ackley}}(x) = -20 \exp\left(-0.2 \times \sqrt{\frac{1}{n}\sum_{i=1}^{n} x_i^2}\right) - \exp\left(\frac{1}{n}\sum_{i=1}^{n} \cos\left(2\pi x_i\right)\right) + 20 + e$$

is an often used function with a moderate multimodal structure. The Ackley function should cause no complications to an optimization algorithm.

The Rastrigin function

$$f_{\text{Rastrigin}}(x) = \sum_{i=1}^{n} \left(x_i^2 - 3 \times \cos\left(2\pi \cdot x_i\right)\right) + 3n$$

used here is regarded to have a much higher multimodality. Assumingly, the results on this function should be significant different from the ones on the other two functions.

Many different choices for the type of dynamics are possible. For a demonstration of the effects of different parameter settings we concentrate on selected types of dynamics. For every test the optimum is relocated in one of three ways:

I: The optimum moves only in one (the first) coordinate, all other coordinates remain constant.
II: The optimum moves in all n coordinates. In every coordinate the moving strength is equal.
III: Here, the optimum moves in all coordinates, too. But the covered distance is different in every coordinate. The covered distance in every coordinate increases from the first coordinate to the last one. For an n-dimensional problem this means that the optimum moves in the first coordinate with an amount a, in the second one with $\sqrt{2} \cdot a$, in the third one with $\sqrt{3} \cdot a$, ..., and in the last coordinate with an amount of $\sqrt{n} \cdot a$. To result in a total severity s the factor must be set to $a = s \cdot \sqrt{2/(n^2 + n)}$.

Figure 3.6 demonstrates an example for all moving types in case of two dimensions.

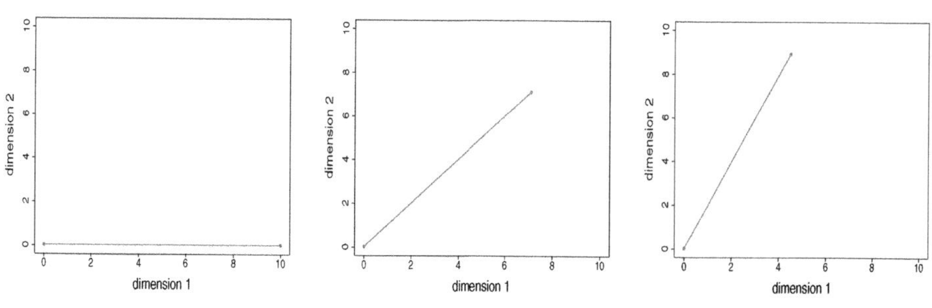

Fig. 3.6. Moving types I–III (from left to right). In every case the optimum starts in the origin and moves with a total covered distance of 10

All moving types give us freedom in choosing the moving frequency and severity. In our study the moving frequency stays constant over a whole run. It will mostly be $\Delta g = 1$ meaning that the optimum moves every generation. We call this *pseudo-continuous movement*. It is obvious that an algorithm can not follow a moving optimum if the moving severity s is too high. Therefore, the moving intensity will vary over a moderate range. Additionally, the changes must not be constant. We use a moving deviation d specifying the standard deviation of the normal distributed random variable with mean s. The results for the different moving types are comparable because d is chosen small in relation to s and the mean total severity remains constant for all runs. It is hoped that the algorithm could adapt to the mean severity and follows the optimum with a small distance.

3.6 Experimental Investigations

In this study the main focus of our experiments lies on the choice of the number of step sizes. Doing this, we start with a short consideration about the total number of combinatorial possibilities to choose the number of step sizes and to allocate them to the single coordinates. But before this, a description of the experimental setup is given.

In [23] we presented a study concerning the choices for population sizes. It was observed that good settings depend on the characteristics of the given problem. The characteristics are namely the problem dimension and the fitness landscape. For an unknown problem often standard settings are used. The population sizes are frequently set to $\mu = 15, \lambda = 100$ resulting in an $(15, 100)$-ES. These settings have revealed to guarantee a minimal diversity in the parental population and a sufficient selection pressure of $\lambda/\mu \approx 7$ ([2, 13, 20]). Thus, we follow these suggestions and use a $(15, 100)$-ES.

Since we are only interested in ES behavior during the tracking phase the first population is initialized in the optimum and the mutation step sizes are set near to the optimal values. This means, that the step sizes match the expected distance of the moving optimum in the next step. This procedure reduces the time the ES needs to adapt to the problem. Experimental investigations showed that other initial step sizes only have the effect that the ES needs a longer period for adaptation. After adaptation the behavior is still the same. Optimization in dynamic environments is an endless task. Obviously, for testing a strategy the user has to stop the optimization process at a certain time. Because the ES was initialized nearly optimal, no time is needed to reach the tracking phase in which the ES follows the optimum with a certain mean distance. Therefore, the used run time of 1000 generations is sufficient for informative results. To compare the results for different strategies and moving types we use the performance measure ABFV described above.

The main task in dynamic environments is to track the moving optimum. We assume that the movement is continuously meaning that the optimum changes every generation. As described in Sect. 3.5 the movement itself is linear. The only unknowns are the moving severity and the direction. It may occur that the optimum moves only in one coordinate or in all coordinates. But other variants may be appear as well. E.g., the optimum may move in coordinates $2, 3, 5, 7$ – each with a different amount –, whereas the other coordinates remain constant. It is obvious that two coordinates should share a common step size only in the case when the necessary changes are nearly the same. Otherwise the coordinates need separate step sizes.

This example shows that there exist numerous possibilities for choosing the number of step sizes and the assignment to the coordinates. We can differ three cases:

- $n_\sigma = 1$
- $n_\sigma = n$
- $1 < n_\sigma < n$

In the first case everything is easy. The one step size is responsible for changing all coordinates. Of course, if the coordinates must change with different amounts an appropriate adaptation may be impossible. This is because the single step size has an average amount which is too high for the coordinates which must change slow and too low for the coordinates which must change fast.

This problem could be avoided if we use n step sizes. In this case every coordinate has its own step size. Under the condition that the step sizes have their optimal settings the optimum could be reached fast (static case) or the optimum could be followed with a small distance (dynamic case). After initialization the step sizes are presumably not optimal. It is known as a difficulty that a growing number of step sizes increases the time needed to find the best settings for every step sizes. To reduce this time, it seems to be a good idea to regard other choices.

When choosing $1 < n_\sigma < n$ step sizes some coordinates share a common step size. Then two main problems arise. First, the exact number of step sizes must be chosen. Second, the step sizes must be assigned to the coordinates. Assume the easiest case, meaning that $n_\sigma = 2$. Then it is possible that one coordinate has an own step size and the remaining $n - 1$ coordinates share the second step size. Becaus the movement could appear in every of the n coordinates, we have now n possibilities to assign the single step size to one coordinate. As another alternative, we may decide to assign the two step sizes to one half of the coordinates each. In this situation we have $\binom{n}{n/2}$ possibilities to assign the first step size to $n/2$ coordinates and the second step size to the other $n/2$ coordinates. Other distributions are possible. E.g., the first step size could be shared among two coordinates and the second step size could be shared by the remaining $n - 2$ coordinates. Then we have $\binom{n}{2}$ possibilities to select the two coordinates for the first step size.

These few examples should point out that learning the right number of step sizes and the correct assignment to the coordinates is an almost impossible task even in dynamic environments. Hence, it is necessary to choose a seemingly optimal number of step sizes and an appropriate assignment to the coordinates. In the lack of problem dependent knowledge one decides to concentrate on the two extreme cases of $n_\sigma = 1$ and $n_\sigma = n$. This is exactly what we do in the next paragraphs. We investigate the influence of the used number of different step sizes on the obtainable results. For this, the three types of dynamics described above are used in conjunction with several severities. Moreover, in some experiments the severity underlies a stochastically variation during the run.

The first experiments are conducted on the sphere model f_S. If the optimum moves following moving type I the usage of one mutation step size is an inappropriate choice, because the ES must move in the first coordinate whereas the other coordinates remain constant. Hence, the results of such a strategy are very poor (Fig. 3.7). Notice, that on the y-axis the square root

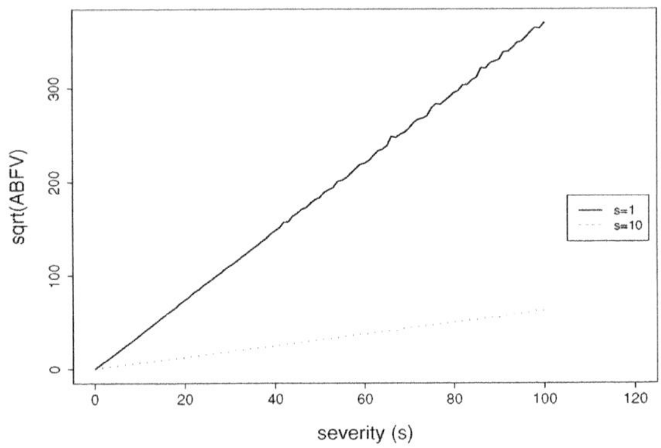

Fig. 3.7. ABFV of an (15, 100)-ES with one and ten mutation step size/s on the dynamic 10-dimensional sphere. The optimum moves every generation in one coordinate with a total severity s (moving type I).

of the ABFV is plotted. The linear run of the curve proofs that the results perfectly depend quadratically on the moving severity s. The ES is able to follow the moving optimum for the moving severities tested here. This behavior was recognized for all moving types. Random samples assists the assumption that this assertion is true even for much higher severities. In contrast to the worse results with one step size, an ES with n strategy parameters is able to follow the moving optimum with a smaller distance. This result holds for higher problem dimensions, too.

If the optimum moves in all dimensions with an equal severity only one step size is necessary. Because an ES with n step sizes must adapt n strategy parameters instead of only one, the reader could assume that an ES with one mutation step size would perform better than an ES with n mutation step sizes. Figure 3.8 compares the behavior of both strategies for moving type II. The curves are nearly the same. This means that the ES with n step sizes behaves not worse than the ES with only one step size. Figure 3.9 shows the

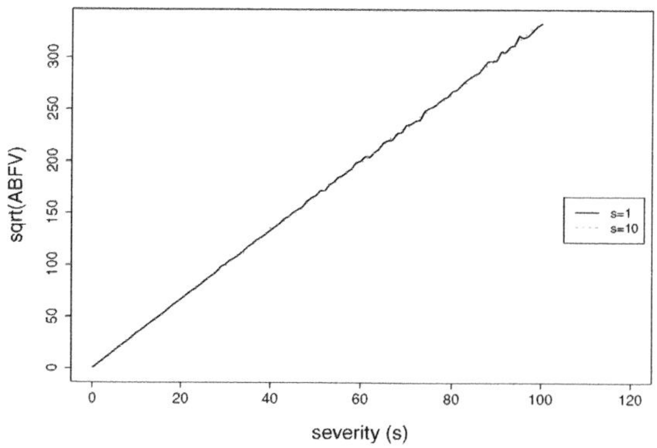

Fig. 3.8. ABFV of an (15, 100)-ES with one and ten mutation step size/s on the dynamic 10-dimensional sphere. The optimum moves every generation in all coordinates with moving type II and a total severity s.

results for moving type III. Here, the ES with n step sizes performs better. But the differences are smaller than expected. One reason may be that in this situation the trade off to adapt n step sizes is so high that the benefits of the different step sizes is compensated. If this holds we expect that for much higher problem dimensions the situation changes meaning that the ES with one step size performs better. For $n = 30$ the observed results are depicted in Fig. 3.10. Indeed, the results for the ES with n mutation step sizes are worse.

Moreover, the results are similar if we use stochastically distributed severities with a moderate standard deviation (less than 10% of the severity). The

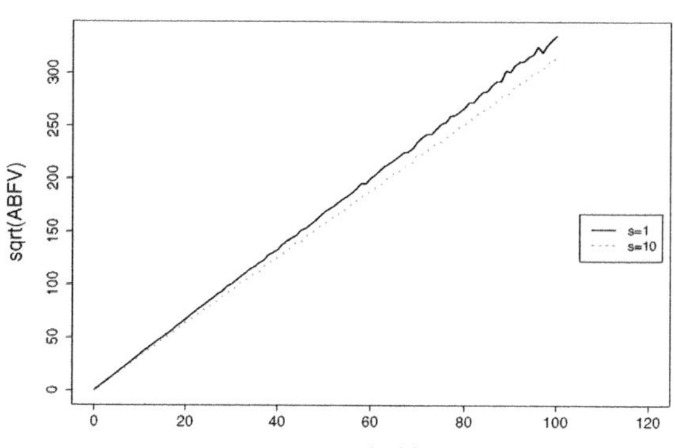

Fig. 3.9. ABFV of an (15, 100)-ES with one and n mutation step size/s on the dynamic 30-dimensional sphere. The optimum moves every generation in all coordinates with moving type III and a total severity s.

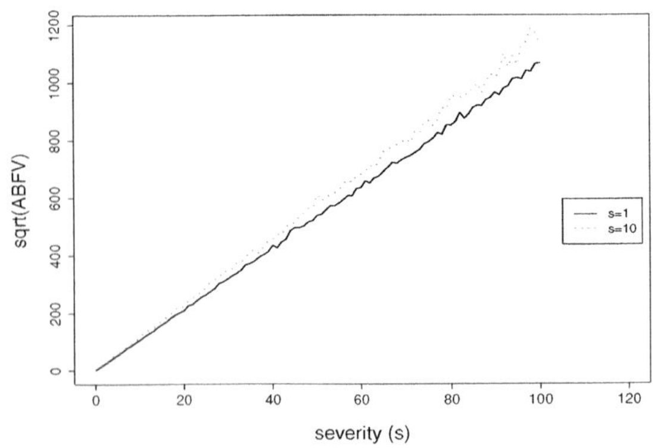

Fig. 3.10. ABFV of an (15, 100)-ES with one and 30 mutation step size/s on the dynamic 30-dimensional sphere. The optimum moves every generation in all coordinates with moving type III and a total severity s.

influence of this stochastically term is negligible. Going ahead we remark that the same holds for the other test functions used in this study.

An interesting question concerns the optimal number of step sizes which should be used if we know in advance in how many coordinates the optimum moves. The number of step sizes could then be chosen suitable for the number of moving coordinates. In the case of moving type I the best choice are two step sizes. One for the moving coordinate and the other for the $n-1$ constant coordinates. Using n_σ step sizes, the first $n_\sigma - 1$ coordinates have their own step size. The last step size is then shared between the remaining $n - n_\sigma + 1$ coordinates. Figure 3.11 shows the different ABFV if the number of step sizes is varied. The best results are obtained as expected with two step sizes. The worst performance is observed for one step size. Be aware that we still hang on to a logarithmic y-axis. For an increasing number of step sizes the ABFV grows more and more worse. But even for n step sizes it is relatively small.

In another experiment the number of different step sizes is varied, too. The assignment of the n_σ different step sizes are chosen as before, meaning that the first $n_\sigma - 1$ coordinates have their own step size and that the last step size is shared by the last $n - n_\sigma + 1$ coordinates. But now the optimum moves in the last coordinate. This means that the moving coordinate shares its step size with some other (constant) coordinates. Figure 3.12 depicts the results for such a setting. Somewhat surprising, the worst results are not gotten for one step size but with $n_\sigma = 8$ and $n_\sigma = 9$ step sizes. On the one hand it is better if the last step size is used for as few as possible other coordinates. On the other hand the more step sizes are used the more difficult is the adaptation.

The next experiments are fulfilled on the Ackley function f_{Ackley}. The results are very similar to the ones observed for the sphere model. Therefore,

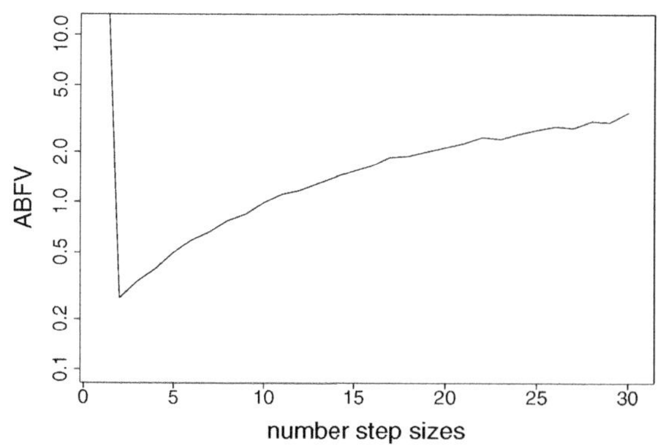

Fig. 3.11. ABFV of an (15, 100)-ES with different number mutation step sizes on the dynamic 30-dimensional sphere. The optimum moves every generation in one (the first) coordinate (moving type I) and a severity $s = 1$.

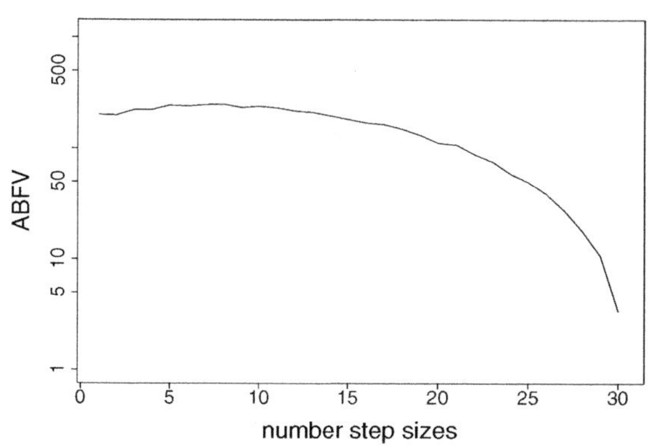

Fig. 3.12. ABFV of an (15, 100)-ES with different number mutation step sizes on the dynamic 30-dimensional sphere. The optimum moves every generation in one (the last) coordinate (moving type I) and a severity $s = 1$.

we do not want to show the exact results for this function. Instead we go directly to the consideration of the Rastrigin function.

The problem difficulty of the Rastrigin function $f_{\text{Rastrigin}}$ should be demonstrated by a simple experiment on the static Rastrigin. Figure 3.13 shows a representative run of two ES variants. The ES with one mutation step size converges very fast. Whereas in the first generations the convergence velocity of the ES with n mutation step sizes is identical, in the following generations it is much slower. Moreover, at generation ≈ 150 the ES stagnates for about 25 generations. We observe a slight regression before the ES starts to converge again. At the end, the ES runs in a local optimum, but has the ability to leave

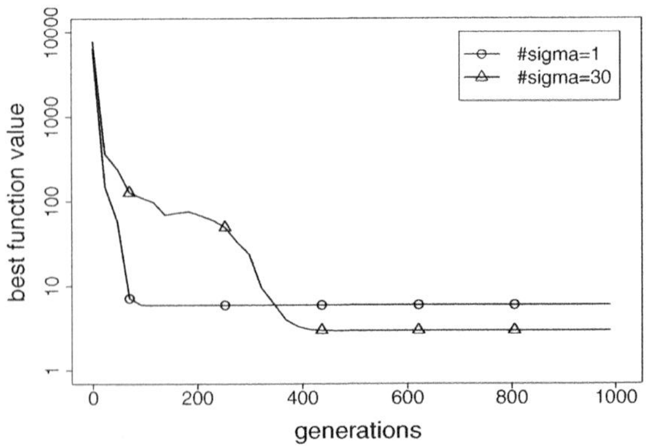

Fig. 3.13. Exemplary run of an (15, 100)-ES with one and n mutation step size/s on the static 30-dimensional Rastrigin

it. The quality of the overall reached optimum is better than the one reached by the ES with one step size.

After this short excursion to the static case we look at the dynamic case. From the point of characteristic of the curves the results with moving type I are comparable to the ones on the sphere function. For both ES variants Fig. 3.14 reveals that the ABFV increases quadratically with the moving severity. Again, the ES with n step sizes is superior to the ES with only one step size. Due to the problem hardness the ABFV for the Rastrigin function are somewhat worse than the obtained values on the sphere. But the differences are not very high. For the reason of completeness we mention, that there exist

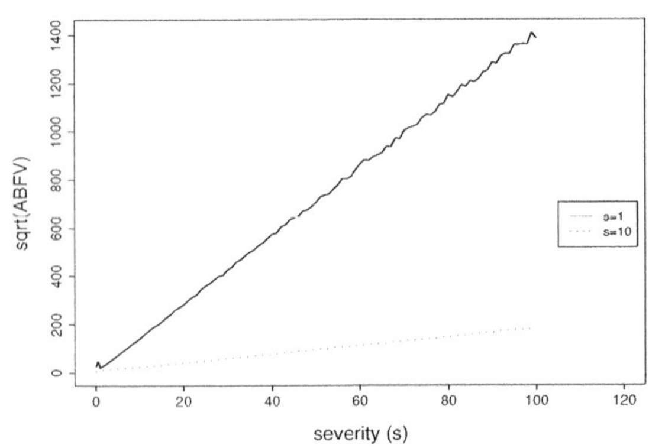

Fig. 3.14. ABFV of an (15, 100)-ES with one and n mutation step size/s on the dynamic 30-dimensional Rastrigin. The optimum moves every generation in one coordinate with a total severity s (moving type I).

no differences between the results of the two ES variants when the optimum moves following type II and III. For the last case Fig. 3.15 shows the exact

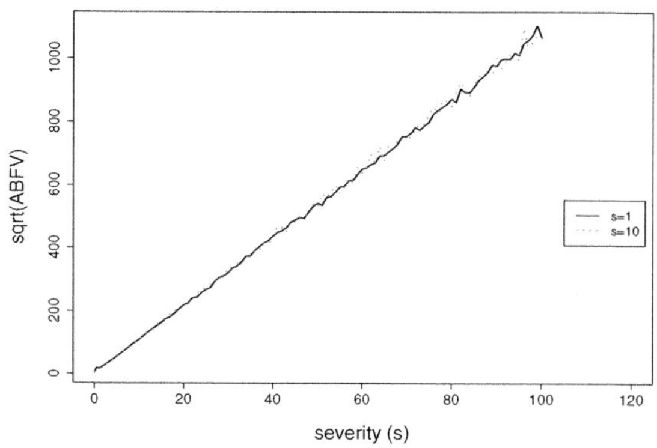

Fig. 3.15. ABFV of an (15, 100)-ES with one and n mutation step size/s on the dynamic 30-dimensional Rastrigin. The optimum moves every generation in all coordinates with a total severity s (moving type III).

values. A similar result was already observed for the two other functions. A direct comparison of the results with the according ones on the sphere model (Fig. 3.10) shows no significant differences in the obtained function values. This is an observation of high importance. Remember that the Rastrigin is a function of high multimodality. In the static case, this resulted in a premature convergence to a local optimum. But for the dynamic case, the last figure shows that the difficulties are dominated by the dynamics. If higher problem dimensions are used the basic results are the same. But the obtainable function values increase with a growing problem dimension.

3.7 Conclusions

In this chapter we gave an introduction to current evolution strategies. The main focus lies on dynamic numerical optimization problems. After an exact description of the algorithm, the general working mechanisms of ES were examplarily shown by selected experimental investigations. The results demonstrated that contemporary self-adaptive evolution strategies are powerful optimization methods for dynamic environments. They are able to follow a moving optimum of even high moving severity.

A new algorithm must compete against the older ones. The comparison of two or more algorithms is based on the quality of every strategy. The quality of an algorithm is usually specified as a performance measure based on the fitness function. The main existing performance measures were introduced with their advantages and disadvantages. To avoid the handicaps the ABFV was developed as a new measurement.

The reader was then assisted in the choice for different strategy parameters for practical application. The main investigated aspect was the choice for the optimal number of mutation step sizes. Since in most cases it is not known in how many and in which coordinates the optimum changes, we saw that it is beneficial to choose n mutation step sizes, one for every problem dimension.

Acknowledgments

This work was supported by the Deutsche Forschungsgemeinschaft (DFG) as part of the *Collaborative Research Center "Computational Intelligence" (SFB 531).*

References

1. H.-G. Beyer. *The Theory of Evolution Strategies.* Springer, Berlin, 2001.
2. H.-G. Beyer and H.-P. Schwefel. Evolution strategies: A comprehensive introduction. *Natural Computing*, 1(1):3–52, 2002.
3. H.-G. Beyer, H.-P. Schwefel, and I. Wegener. How to analyse evolutionary algorithms. *Theoretical Computer Science*, (287):101–130, 2002.
4. G. E. P. Box. Evolutionary operation: A method for increasing industrial productivity. *Applied Statistics*, VI(2):81–101, 1957.
5. J. Branke. *Evolutionary Optimization in Dynamic Environments.* Kluwer, 2002.
6. H. J. Bremermann. Optimization Through Evolution and Recombination. In M. C. Yovits, G. T. Jacobi, and G. D. Goldstein, editors, *Proceedings of the Conference on Self-Organizing Systems*, pages 93 – 106. Spartan Books, Washington, D.C., 1962.
7. C. Darwin. *The origin of species by means of natural selection.* The Modern Library — Random House, New York, 1872.
8. K. De Jong, D. B. Fogel, and H.-P. Schwefel. *A history of evolutionary computation.* Oxford University Press, New York, and Institute of Physics Publishing, Bristol, 1997.
9. L. J. Fogel. Autonomous automata. *Industrial Research*, (4):14–19, 1962.
10. L. J. Fogel, A. J. Owens, and M. J. Walsh. *Artificial Intelligence through Simulated Evolution.* Wiley, New York, 1966.
11. R. M. Friedberg. A learning machine: Part I. *IBM Journal*, 2(1):2–13, 1958.
12. R. M. Friedberg, B. Dunham, and J. H. North. A learning machine: Part II. *IBM Journal*, 3(7):282–287, 1959.
13. J. Grefenstette. Rank-based selection. In T. Bäck, D. B. Fogel, and Z. Michalewicz, editors, *Handbook of Evolutionary Computation*, pages C2.4:1–6. Oxford University Press, New York, and Institute of Physics Publishing, Bristol, 1997. Release 97/1.
14. N. Hansen and A. Ostermeier. Completely derandomized self-adaptation in evolution strategies. *Evolutionary Computation*, 2(9):159–195, 2001.

15. J. H. Holland. Outline for a logical theory of adaptive systems. *Journal of the ACM*, 9(3):297–314, 1962.
16. J. H. Holland. *Adaptation in natural and artificial systems*. Ann Arbor: The University of Michigan Press, 1975.
17. F. Kursawe. *Grundlegende empirische Untersuchungen der Parameter von Evolutionsstrategien — Metastrategien*. PhD thesis, University of Dortmund, Germany, 1999.
18. R. W. Morrison. *Designing Evolutionary Algorithms for Dynamic Environments*. Springer, Berlin, 2004.
19. I. Rechenberg. *Cybernetic solution path of an experimental problem*. 1965. Royal Aircraft Establishment, Farnborough, Library Translation no. 1122.
20. G. Rudolph. Evolution strategies. In T. Bäck, D. B. Fogel, and Z. Michalewicz, editors, *Handbook of Evolutionary Computation*, pages B1.3:1–6. Oxford University Press, New York, and Institute of Physics Publishing, Bristol, 1997. Release 97/1.
21. G. Rudolph. Local convergence rates of simple evolutionary algorithms with cauchy mutations. *IEEE Trans. Evolutionary Computation*, 4(1):249–258, 1997.
22. G. Rudolph. Asymptotical convergence rates of simple evolutionary algorithms under factorizing mutation distributions. In J.-K. Hao, E. Lutton, E. Ronald, M. Schoenauer, and D. Snyers, editors, *Artificial Evolution: Third European Conf. (AE'97)*, pages 223–233, Berlin, 1998. Springer.
23. L. Schönemann. On the influence of population sizes in evolution strategies in dynamic environments. In J. Branke, editor, *GECCO Workshop on Evolutionary Algorithms for Dynamic Optimization Problems (EvoDOP)*, pages 123–127, 2003.
24. L. Schönemann. Maximal life span in evolutionary algorithms. In B. Filipič and J. Šilc, editors, *Proc. Int'l Conf. Bioinspired Optimization Methods and Their Applications (BIOMA'04)*, pages 21–30. Jožef Stefan Institute, Ljubljana, Slowenien, 2004.
25. H.-P. Schwefel. *Kybernetische Evolution als Strategie der experimentellen Forschung in der Strömungstechnik*. PhD thesis, Technical University of Berlin, Hermann Fttinger-Institute for Fluid Dynamics, 1965.
26. H.-P. Schwefel and G. Rudolph. Contemporary evolution strategies. In F. Morán, A. Moreno, J. J. Merelo, and P. Chacón, editors, *Advances in Artificial Life, Proc. of the Third European Conference on Artificial Life*, pages 893–907. Springer, Berlin, 1995.
27. J. Sprave. A unified model of non-panmictic population structures in evolutionary algorithms. In P. J. Angeline and V. W. Porto, editors, *Proc. 1999 Congress on Evolutionary Computation (CEC'99), Washington D.C.*, volume 2, pages 1384–1391. IEEE Press, Piscataway NJ, 1999.
28. K. Weicker. An analysis of dynamic severity and population size. In M. Schoenauer, K. Deb, G. Rudolph, X. Yao, E. Lutton, J. J. Merelo, and H.-P. Schwefel, editors, *Proc. Parallel Problem Solving from Nature (PPSN VI)*, pages 159–168. Springer, Berlin, 2000.

4

Orthogonal Dynamic Hill Climbing Algorithm: ODHC

Sanyou Zeng[1,2], Hui Shi[1], Lishan Kang[3], and Lixin Ding[3]

[1] School of Computer Science, China University of GeoSciences
Wuhan 430072, Hubei, P. R. China
sanyou-zeng@263.net, shihui0205@163.com

[2] Department of Computer Science, Hunan University of Technology
Zhuzhou 412008, Hunan, P. R. China

[3] State Key Laboratory of Software Engineering, Wuhan University
Wuhan 430072, Hubei, P. R. China
kang_whu@yahoo.com, lx_ding@263.net

Summary. An orthogonal hill-climbing algorithm for dynamic optimization problems with continuous variables, labeled *ODHC*, is proposed in this chapter. The local peak climber is not a solution, but a "niche" (a small hyperrectangle). An orthogonal design method is employed on the niche, in order to seek a potential peak more quickly. An archive is used to store the latest found higher peaks, so the ODHC algorithm can learn from the past search, this can also enhance the performance of the ODHC algorithm. The randomly created niches implement the global search. Numerical experiments show that the ODHC algorithm performs much better than the Self Organizing Scouts algorithm.

4.1 Introduction

Most research in evolutionary computation focuses on the optimization of static problems. Many real world optimization problems, however, are actually dynamic. Hence, optimization methods, capable of continuously adapting to a changing environment, are needed. The solution to dynamic optimization problems (DOPs) using genetic algorithms (GAs) was first introduced by Goldberg and Smith [1] and has attracted growing interest from the evolutionary algorithm (EA) community in recent years [2, 3]. Researchers have used many GA based approaches to address this problem, as surveyed in [4, 18], such as the hypermutation scheme [6, 7], the random immigrants scheme [8], memory-based methods [9–11], and multi-population approaches [12].

We restrict our attention in this chapter to those problems whose fitness values (usually objective function values) vary only a little over a small region of the search space and whose fitness landscapes display some exploitable similarities before and after a change. The approaches mentioned above, all try to

S. Zeng et al.: *Orthogonal Dynamic Hill Climbing Algorithm: ODHC*, Studies in Computational Intelligence (SCI) **51**, 79–104 (2007)
www.springerlink.com

exploit landscape similarities before and after a change, but do not fully employ those similarities in small regions. The "orthogonal" design [13] however does make use of such similarities. Zeng et al [14, 15] designed an orthogonal multi-objective evolutionary algorithm (OMOEA) for multi-objective optimization problems. The OMOEA algorithm employed an orthogonal design to find Pareto-optima statistically. By the way, another excellent characteristic of the "orthogonal" design is its uniform exploration of the experiment space, which was used in the literatures [14, 15] and [16, 17]. One of the characteristic features of the ODHC algorithm is its fast and robust exploration of small regions of the fitness landscape.

4.2 Background on Dynamic Optimization Problems

4.2.1 Definition of Dynamic Optimization

Whenever a change in a dynamic optimization problem occurs, i.e., when the optimization goal, the problem instance, or some restrictions change, the optimum to that problem might change as well. If this is the case, an adaptation of the old solution is necessary.

The goal of dynamic optimization is to find the optimal control profile of one or more control variables or control parameters of a system. Optimality is defined as the minimization or maximization of a objective function without violating given constraints.

A definition of dynamic optimization problems, which the Orthogonal Dynamic Hill-Climbing (ODHC) algorithm manages to solve in this chapter, is described as follows:

Definition 1. *Dynamic Optimization Problem:*

$$\vec{x}_{max}(t) = arg \underset{\vec{x} \in S}{Maximize} \quad f(\vec{x}, t) \tag{4.1}$$

where $\vec{x} = (x_1, x_2, ..., x_N)$ *is continuous variables and* S *is the search space.*

In addition to depending on the variables $\vec{x}$, the function values of f are time dependable in Definition 1. But the dimension N of $\vec{x}$ and the search space S are fixed with time. $\vec{x}_{max}(t)$ is the optimal solution at time t. We know it is easy to turn minimization into maximization. Hence, only maximization is discussed here.

It is impractical and even impossible to find the precise time-dependent optimal solutions $\vec{x}_{max}(t)$ for an optimization approach. The practical way is to track the moving optimum. The optimization approaches must spend time on finding the optimal solution at any time point. We must, therefore, suppose that the environment has a sudden change with a comparative long static stage. That is, the function values of f are only dependent on variables $\vec{x}$ without time-dependent in the static stage where we hope the algorithm approaches the optimum as much as possible in addition to adapt to dynamic environments.

4.2.2 Detecting Changes in the Environment

In order to be able to react to changes explicitly, one first has to detect that a change in the environment actually takes place. For many applications, changes can be made explicitly known to the system, e.g., when a new job arrives in the queue to be scheduled. However, in other applications, e.g., when the quality of raw material varies over time, such changes have to be detected by the system.

A commonly used indicator for changes in the environment is deterioration of the performance or the time-averaged best performance of the algorithms. However, this assumes more or less that a change in the environment will actually decreases the quality of the old solution, which is not necessarily the case. The fitness values are also used to decide when the new optimum is reached, which of course implies that the new optimum has at least the quality of the old one.

Slightly less restrictive is the approach in [18], where several individuals are reevaluated every generation and a change in the environment is detected if the fitness of at least one individual has changed.

Fogarty *et al.* [18] briefly mentioned the idea of using a "validation module" to evaluate the performance of EAs and restart EAs when a certain limit value is exceeded, i.e., when the environment has changed so much that the EA's performance is no longer satisfactory. However, no clues were given as to how this validation module might work.

Some other approaches [19–21] explicitly maintain a module of the environment and constantly monitor whether the module is still consistent with the real environment. If the response predicted by the module and the actual response obtained from the environment differ too much, it is concluded that the environment has changed. The model is then updated and the EA can be restarted.

4.2.3 Benchmark Problems

Common Characteristics of Benchmark Problems

So far, a number of different types of dynamic optimization problems have been used to test algorithms, ranging from simple mathematical functions over all kinds of scheduling problems to applications in artificial life. We restrict our attention here to a small subset of problems that might eventually form a common benchmark suite and thus will be discussed in more detail here. The following aspects have been considered to be relevant for a suite of benchmark problems:

- It should be possible to vary many of the environmental variables.
- There should be benchmarks for at least binary and real-valued encoding.
- They should be simple to implement and to describe.
- They should be computationally efficient.
- They should allow conjectures to real world problems.

The Moving Peaks Function

Branke [4] suggested a problem called "the moving peaks function" that consisted of a multi-dimensional landscape with several peaks. Independently, Morrison and De Jong [22] and Yang [23] suggested other benchmarks.

With the aim to bridge the gap between very complex, hard to understand real-world problems and all too simple toy problems, in [4], Branke suggested a problem with a multidimensional landscape consisting of several peaks, where the height, the width and the position of each peak is altered slightly every time a change in the environment occurs.

Note that a small change in the landscape may have two consequences: sometimes it may be sufficient to adapt the current solution to reach the new optimum, and sometimes it may be necessary to switch to another, previously slightly inferior but now better solution. The former happens when the optimum shifts slightly, the latter happens when the height of the peaks changes such that a different peak becomes the maximum peak. In these cases, the optimization algorithm basically has to "jump", or cross a valley, to reach the new maximum peak.

A test function with N dimensions and m peaks can be formulated as:

$$
\begin{array}{l}
F(\vec{x},t) = max(B(\vec{x}), \max_{i=1\ldots m} P(\vec{x}, h_i(t), w_i(t), \vec{p_i}(t))), \\
\begin{cases}
h_i(t) = h_i(t-1) + height\_severity \cdot \sigma \\
w_i(t) = w_i(t-1) + width\_severity \cdot \sigma \\
\vec{p_i}(t) = \vec{p_i}(t-1) + \vec{v_i}(t) \\
\vec{v_i}(t) = \frac{s}{|(1-\lambda)\vec{r}+\lambda\vec{v_i}(t-1)|}((1-\lambda)\vec{r} + \lambda\vec{v_i}(t-1))
\end{cases} \\
where \\
\begin{cases}
B(\vec{x}): & variant\ with\ \vec{x} \\
P(\vec{x}, h_i(t), w_i(t), \vec{p_i}(t))): & variant\ with\ both\ \vec{x}\ and\ t \\
h_i(t), w_i(t), \vec{p_i}(t): & variant\ with\ t \\
m_h \le h_i(t) \le M_h, & m_w \le w_i(t) \le M_w \\
m_{\vec{x}} \le x_j \le M_{\vec{x}}, & j = 1, 2, ..., N
\end{cases} \\
and\ where \\
\begin{cases}
height\_severity,\ width\_severity,\ s,\ \lambda: & constant \\
m_h,\ M_h;\ m_w,\ M_w; m_{\vec{x}},\ M_{\vec{x}}: & constant \\
\sigma, \quad \vec{r}: & random\ constant \\
0 \le \lambda \le 1, \quad \sigma \in N(0,1), & \| \vec{r} \| = s
\end{cases}
\end{array}
\tag{4.2}
$$

Where $B(\vec{x})$ is a time-invariant "basis" landscape, and P is the function defining a peak shape, where each of the m peaks has its own time-varying parameters: height (h), width (w), and location ($\vec{p}$).

For the shape of function P, two examples are given here, one with cone shape of peaks, the other with hilly peaks. They are defined as follows respectively:

Cone shape function:

$$P(\vec{x}, h_i(t), w_i(t), \vec{p_i}(t)) = h_i(t) - w_i(t) * \| \vec{x} - \vec{p_i}(t) \|, \tag{4.3}$$

Hilly shape function:

$$P(\overrightarrow{x}, h_i(t), w_i(t), \overrightarrow{p_i}(t)) = h_i(t) - w_i(t) * \parallel \overrightarrow{x} - \overrightarrow{p_i}(t) \parallel^2 \\ -0.1 * \sin(20 * \parallel \overrightarrow{x} - \overrightarrow{p_i}(t) \parallel^2), \tag{4.4}$$

where $\parallel x - \overrightarrow{p_i}(t) \parallel = \sqrt{\sum_{j=1}^{N} (x_j - p_j(t))^2}$, $\overrightarrow{p_i}(t) = (p_{i,1}(t), p_{i,2}(t), ..., p_{i,N}(t))$ and $\overrightarrow{x} = (x_1, x_2, ..., x_N)$.

The location, the height and the width of each peak are initialized according to an in-built random number generator[4]. Then, every Δe evaluations, the height and width of every peak are changed by adding a random gaussian variable. The location of every peak i is moved by a vector $\overrightarrow{v_i}$ of fixed length s in a random direction (for $\lambda = 0$) or a direction depending on the previous direction (for $\lambda > 0$).

Overall, the parameter s allows to control the severity of a change, Δe will determine the frequency of change, λ allows to control whether the changes exhibit a trend. More formally, a change of environment can be described as $t - 1 \Rightarrow t$, see Equation(4.2). The shift vector $\overrightarrow{v_i}$ is a linear combination of a random vector $\overrightarrow{r}$ and the previous shift vector $\overrightarrow{v_i}(t-1)$, and normalized to length s. The random vector $\overrightarrow{r}$ is created by drawing random numbers for each dimension and normalizing its length to s.

The function's complexity may easily be scaled by increasing the number of dimensions or the number of peaks, or by using complex peak- and base-functions. Furthermore, the benchmark provides a number of different peak functions P and allows to vary the step size s over time in many ways. For more details of the moving peaks function, please visit the website: http://www.aifb.uni-karlsruhe.de/~jbr/MovPeaks/.

4.2.4 Measuring the Performance

Since for dynamic optimization problems a single, time-invariant optimal solution does not exist, the goal is not to find the extreme but to track their progression through the space as closely as possible. To demonstrate superiority of one approach over another, in the current literature it is quite common to just display the convergence plots of the different approaches and to compare them visually. For an accurate comparison, however, numeric measures are definitely preferable.

Offline performance and offline error have been used for the dynamic case in this chapter. They are defined as follows: let $f_t^{close-to-otp}$ be the best value found, and f_t^{otp} the optimum, during the time between two consecutive environmental changes $[t-1, t]$, denote $e_t = |f_t^{otp} - f_t^{close-to-otp}|$, and let T be the number of environmental changes considered. Then,

[4] By seeding this random number generator differently, numerous test problems with the same fundamental characteristics can be generated.

- *Offline performance*

$$f^* = \frac{1}{T}\sum_{t=1}^{T} f_t^{close-to-otp} \tag{4.5}$$

  is the average of the best values found so far at each time step, i.e. with and being the last time step<t at which a change in the environment occurred. This assumes that optimization is done in a simulated environment and only the best solutions are actually transferred into the real world. Of course this requires that the environmental changes are known to the observer.
- *Offline error*

$$e^* = \frac{1}{T}\sum_{t=1}^{T} e_t \tag{4.6}$$

  is the average of all current errors, i.e. the average deviation of the currently best individual from the optimum since the last change.

Since in general the EA's tracking ability is the interesting aspect, it may be reasonable for either measurement to ignore the first few generations, i.e. the adaptation process to the first solution, and only measure the performance after the system has passed the initial startup. If the initial startup phase is considered important, one might report on the startup phase performance and the long-run performance separately.

4.3 Techniques Relevant to ODHC

4.3.1 Brief Introduction to the Hill-Climbing Algorithm

Generally speaking, there are two kinds of search strategies. The first kind of search strategies does not have inspiration information as the guidance, so the search is blind and the efficiency is low. The second kind is the heuristic search strategy that uses the inspiration information as the guidance and hence the search is toward the goal direction, avoiding detouring and enhancing the solution efficiency.

Hill climbing is one of the heuristic search techniques. Hill climbing strategies expand the current state in the search of the neighbors of the current states and evaluate its children. The best child is selected for further expansion and neither its siblings nor its parent are retained. Search halts when it reaches a state that is better than any of its children. Hill climbing is named for the strategy that might be used by an eager, but blind mountain climber that goes uphill along the steepest possible path until he can go no further.

A major problem of hill climbing strategies is their tendency to become stuck at foothills, a plateau or a ridge. If the algorithm reaches any of the above mentioned states, then the algorithm fails to find a solution.

- Foothills or local maxima is a state that is better than all its neighbors but is not better than some other states farther away. At a local maximum, all moves appear to make things worse. Foothills are potential traps for the algorithm.
- A plateau is a flat area of the search space in which a whole set of neighboring states have the same value. On a plateau, it is not possible to determine the best direction in which to move by making local comparisons.
- A ridge is a special kind of local maximum. It is an area of the search space that is higher that the surrounding areas and that itself has a slope. But the orientation of the high region, compared to the set of available moves and the directions in which they move, makes it impossible to traverse a ridge by single moves. Any point on a ridge can look like peak because movement in all probe directions is downward.

To solve the above problem, group hill climbing is used or the size of the search neighbors of the current state at each climbing step is expanded. The basic operation of a hill-climbing algorithm, denoted *Algorithm 1* in this chapter, is described in Fig. 4.1.

Step 1: Pick a random point in the search space
Step 2: Create children in the neighbor of the current state
Step 3: Choose the child with the best quality and move to that state
Step 4: Repeat 2 to 4 until all the children states are of lower quality
Step 5: Return the current state as the solution state

Fig. 4.1. The basic operation of a hill-climbing algorithm

The neighbor is called *niche* in the hill-climbing algorithm in this chapter. The children are created by using the orthogonal design method.

4.3.2 Orthogonal Design Method

An Example to Introduce the Orthogonal Design Method

We use a concrete example in this section to introduce the basic concept of an "orthogonal design method". For further details, see [13]. The example is concerned with the yield of vegetable growth. The yield of a vegetable depends on at least the following three factors:

1) the temperature
2) the amount of fertilizer used
3) the pH value of the soil

Table 4.1. Experimental design with three factors and three levels per factor

Level	Factor		
	Temperature($^{\circ}C$)	Fertilizer(g/m^2)	pH
Level 1	20	100	6
Level 2	25	150	7
Level 3	30	200	8

In this example, each factor has three possible values, as shown in Table 4.1. We say that each factor has three "levels".

To find the best combination of levels for a maximum yield, we can perform an experiment for each combination, and then select the combination with the highest yield. In the above example, there are $3 \times 3 \times 3 = 27$ combinations, and hence there are 27 experiments. In general, when there are N factors, each with Q levels, there are Q^N possible combinations. When N and Q are large, it may not be possible to perform all Q^N experiments. Therefore, it is desirable to sample a small, but representative set of combinations, for the experimentation. The "orthogonal design method" was developed for this purpose [13], where an orthogonal array is constructed to represent the sampled set of combinations which evenly distribute over the experimentation space. An orthogonal array $L_M(Q^N)$ is a $M \times N$ array with Q levels for each column, denoted by $\{1, 2, ..., Q\}$. We select M combinations to be tested, where M may be much smaller than Q^N. Equation (4.7) is an example of an orthogonal array where $M = 9, N = 3, Q = 3$.

$$L_9(3^3) = \begin{bmatrix} 1 & 1 & 1 \\ 1 & 2 & 2 \\ 1 & 3 & 3 \\ 2 & 1 & 2 \\ 2 & 2 & 3 \\ 2 & 3 & 1 \\ 3 & 1 & 3 \\ 3 & 2 & 1 \\ 3 & 3 & 2 \end{bmatrix} \tag{4.7}$$

The $L_9(3^3)$ has three factors, three levels per factor, and nine combinations of levels. The three factors have respective levels 1, 1, 1 in the first combination, 1, 2, 2 in the second combination, etc. We apply the orthogonal matrix $L_9(3^3)$ to select nine combinations to be tested. Assume the corresponding yields of the M combinations are denoted by $[y_i]_{M\times1}$, where the ith combination (experiment) has a yield y_i. The nine combinations and their yields in the above example are shown in Table 4.2.

From the yield of the selected combinations, a promising solution can be obtained by the following statistical method.

Table 4.2. The yield of nine representative combinations, based on the orthogonal matrix $L_9(3^3)$

Combination	Factor			
	Temperature	Fertilizer	pH	yield
1	$1(20^oC)$	$1(100g/m^2)$	1(6)	2.75
2	$1(20^oC)$	$2(150g/m^2)$	2(7)	4.52
3	$1(20^oC)$	$3(200g/m^2)$	3(8)	4.65
4	$2(25^oC)$	$1(100g/m^2)$	2(7)	4.60
5	$2(25^oC)$	$2(150g/m^2)$	3(8)	5.58
6	$2(25^oC)$	$3(200g/m^2)$	1(6)	4.10
7	$3(30^oC)$	$1(100g/m^2)$	3(8)	5.32
8	$3(30^oC)$	$2(150g/m^2)$	1(6)	4.10
9	$3(30^oC)$	$3(200g/m^2)$	2(7)	4.37

1) Calculate the mean value of the yield for each factor at each level, where each factor has a level with best mean value (cf. Fig. 4.4).
The mean yields of the temperature are:
   - $\Gamma_{1,1} = (2.75 + 4.52 + 4.65)/3 = 3.97$ at level $1(20^oC)$,
   - $\Gamma_{2,1} = (4.60 + 5.58 + 4.10)/3 = 4.76$ at level $2(25^oC)$,
   - $\Gamma_{3,1} = (5.32 + 4.10 + 4.37)/3 = 4.60$ at level $3(30^oC)$.

   The mean yields of the fertilizer are:
   - $\Gamma_{1,2} = (2.75 + 4.60 + 5.32)/3 = 4.22$ at level $1(100g/m^2)$,
   - $\Gamma_{2,2} = (4.52 + 5.58 + 4.10)/3 = 4.73$ at level $2(150g/m^2)$,
   - $\Gamma_{3,2} = (4.65 + 4.10 + 4.37)/3 = 4.37$ at level $3(200g/m^2)$.

   The mean yields of the PH value are:
   - $\Gamma_{1,3} = (2.75 + 4.10 + 4.10)/3 = 3.65$ at level 1(6),
   - $\Gamma_{2,3} = (4.52 + 4.60 + 4.37)/3 = 4.50$ at level 2(7),
   - $\Gamma_{3,3} = (4.65 + 5.58 + 5.32)/3 = 5.18$ at level 3(8).

   These mean yields are shown in Table 4.3.

Table 4.3. The mean yield for each factor at different levels

Level	Mean yield		
	Temperature	Fertilizer	pH
Level 1	3.97	4.22	3.65
Level 2	4.76	4.73	4.50
Level 3	4.60	4.37	5.18

2) Choose the combination of the best levels as a promising solution (cf. Fig. 4.5).
The temperature has the best mean yield, 4.76, at level 2 (i.e., 25^oC). The fertilizer has the best yield, 4.73, at level 2 (i.e., $150g/m^2$). The pH value has the best yield, 5.18, at level 3 (i.e., 8). We therefore consider

$(25^oC, 150g/m^2, 8)$ to be a promising and robust solution. The solution may not be optimal when used with an orthogonal design. But for additive and quadratic models, it is provably optimal.

Definition of an Orthogonal Array

Definition 2. *An orthogonal array $L_M(Q^N)$ is a $M \times N$ array with Q levels for each column, denoted by $\{1, 2, ..., Q\}$. Denote the orthogonal array $L_M(Q^N)$ by $[a_{i,j}]_{M\times N}$, such that*

1) In any column $\{a_{1,j}, a_{2,j}, ..., a_{M,j}\}$, each of the Q symbols $1, 2, ..., Q$ occurs the same number of times, i.e. $(M/Q), j = 1, 2, ..., N$.
2) In any two different columns $\{(a_{1,j}, a_{1,k}), (a_{2,j}, a_{2,k}), ..., (a_{M,j}, a_{M,k})\}$, each of the Q^2 possible pairs $\{(1,1), (1,2), ..., (1,Q), (2,1), (2,2), ..., (2,Q), ..., (Q,1), (Q,2), ..., (Q,Q)\}$ occurs the same number of times $(M/Q^2), j, k = 1, ..., N, j \neq k$.

Every row of $L_M(Q^N) = [a_{i,j}]_{M\times N}$ represents a different combination of levels, where $a_{i,j}$ means that the jth factor in the ith combination has a level value $a_{i,j}$, and $a_{i,j}$ takes a value from the set $\{1, 2, ..., Q\}$.

Steps of Orthogonal Design Method

As we will explain shortly, the technique proposed in this chapter may require different orthogonal matrices for different optimization problems. The construction of an orthogonal matrix is not a trivial task, since we do not know whether an orthogonal matrix of a given size exists. Many orthogonal matrices have been presented in the literature. It is impossible, however, to tabulate them all. To construct a class of orthogonal matrices $L_M(Q^P)$, we introduce a simple permutation method that is derived from the mathematical theory of Galois fields (cf. [24]). The M, P, Q in $L_M(Q^P)$ fulfill the following:

$$\begin{cases} M = Q^J \\ P = (Q^J - 1)/(Q - 1), \end{cases} \tag{4.8}$$

where Q is prime and J is a positive integer.

Denote the *jth* column of the orthogonal matrix $[a_{i,j}]_{M\times P}$ by $\vec{a}_j$. Column $\vec{a}_j$ for $j = 1, 2, (Q^2 - 1)/(Q - 1) + 1, (Q^3 - 1)/(Q - 1) + 1, ..., (Q^{J-1} - 1)/(Q-1) + 1$ are called the basic columns. The others are called the non-basic columns. The algorithm first constructs the basic columns and then generates the non-basic columns. The details of the algorithm, denoted *Algorithm 2*, are given in Fig. 4.2.

$L_M(Q^P)$, constructed by Algorithm 2, has P columns. For a problem with N decision variables, we discard the last $P - N$ columns of $L_M(Q^P)$ and obtain a matrix $L_M(Q^N)$ which is still orthogonal, according to Definition 2. For example, let $Q = 3, J = 2$, we get an orthogonal array $L_9(3^4)$ with 4

```
// Construct the basic columns as follows
for k = 1 to J do
   j = (Q^(k-1) - 1)/(Q - 1) + 1
   for i = 1 to Q^J do
      a_{i,j} = floor((i - 1)/(Q^(J-k))) mod Q
   endfor
endfor
// Construct the non-basic columns as follows
for k = 2 to J do
   j = (Q^(k-1) - 1)/(Q - 1) + 1;
   for s = 1 to j - 1, t = 1 to Q - 1 do
      a→_{j+(s-1)(Q-1)+t} = (a→_s × t + a→_j) mod Q;
   endfor
endfor
```

Fig. 4.2. Constructing the orthogonal matrix $L_M(Q^P)$

columns ($P = 4$) by Algorithm 2. For the problem of vegetable yield, there are only 3 factors ($N = 3$). Therefore, we discard the last column of $L_9(3^4)$ for yielding $L_9(3^3)$. Fig. 4.3 shows that the last column of $L_9(3^4)$ is discarded for getting $L_9(3^3)$.

$$L_9(3^4)=\begin{bmatrix}1&1&1&1\\1&2&2&2\\1&3&3&3\\2&1&2&3\\2&2&3&1\\2&3&1&2\\3&1&3&2\\3&2&1&3\\3&3&2&1\end{bmatrix}\quad \overset{\textit{Discard the}}{\underset{\textit{last column}}{\Rightarrow}}\quad L_9(3^3)=\begin{bmatrix}1&1&1\\1&2&2\\1&3&3\\2&1&2\\2&2&3\\2&3&1\\3&1&3\\3&2&1\\3&3&2\end{bmatrix}$$

Fig. 4.3. Discard the last column of the $L_9(3^4)$ for yielding $L_9(3^3)$

The orthogonal design method uses the mean value of the objective at each level of each factor. We denote the objective values of the orthogonal experiments by $[y_i]_{M\times 1}$ where the objective has the value y_i at the ith combination, where the mean values are denoted by $[\Gamma_{k,j}]_{Q\times N}$, where the objective has the mean value $\Gamma_{k,j}$ at the kth level of the jth factor, and

$$\Gamma_{k,j} = \frac{Q}{M} \sum_{a_{i,j}=k} y_i, \tag{4.9}$$

where the orthogonal matrix $L_M(Q^N)$ has the value $a_{i,j}$ at the ith row and jth column, i.e., the jth factor has level $a_{i,j}$ in the ith combination (experiment). The objective has value y_i at the ith combination, and $\sum_{a_{i,j}=k} y_i$ implies the sum of y_i where any i satisfies $a_{i,j} = k$. The details of the algorithm, called *Algorithm 3*, are shown in Fig. 4.4.

// Sum the objective results for each factor at each level
set $\Gamma_{i,j} = 0$ for $i = 1, 2, ..., Q$; $j = 1, 2, ..., N$
for $i = 1$ to $M, j = 1$ to N **do**
  $q = a_{i,j}; \Gamma_{q,j} = \Gamma_{q,j} + y_i$
endfor
// Average the results for each factor at each level
$[\Gamma_{k,j}]_{Q\times N} = [\Gamma_{k,j}]_{Q\times N} \times Q/M$

Fig. 4.4. Calculation of the mean value $[\Gamma_{k,j}]_{Q\times N}$

Each factor has its best level, found from the mean value matrix $[\Gamma_{k,j}]_{Q\times N}$. Usually, the combination of the best levels is a good solution, and for additive or quadratic models, it is optimal. The details of calculating the combination of the best levels (*Algorithm 4*) are given in Fig. 4.5.

---

for $j = 1$ to N **do**
  $b_j = arg \max_{i\in\{1,2,\ldots,Q\}} \Gamma_{i,j}$
endfor
return Potentially good combination $\overrightarrow{b} = (b_1, b_2, ..., b_N)$

Fig. 4.5. Calculation of potentially good combination $[b_j]_{1\times N}$

4.3.3 The Concept of Niche

Niche is a function of a particular species in an ecological community; all aspects of an organisms existence that enable it to survive and reproduce. For example, an osprey primarily preys upon fish therefore its niche is near water, and it could not survive in the desert.

The niche technology in this chapter simulates the principle of niche in ecology. A niche W is a hyper-rectangle in the search space S here. Each niche evolves (moves or shrinks). It has its fitness by which it is selected or deleted. Each niche selects its best solution found so far as its representative, and the fitness value of the representative is regarded as that of the niche.

4.4 The ODHC Algorithm

4.4.1 Brief Introduction to the ODHC Algorithm

In the ODHC algorithm, the climber is a niche. Each niche has its fitness. Each one selects its best solution found so far as its representative, and the fitness value of the representative is regarded as that of the niche.

Since the orthogonal design method works well for niches, it is used to find a potentially good solution that probably is situated near the optimal solution in the niche. The potentially good solution is probably the representative of the niche. Because the representative is likely to be close to the optimal solution in its niche, we determine whether or not the niche covers a peak of the whole search space according to the position of the representative. The representative must stay inside its niche, or at least at its boundary. If the representative stays inside the niche, we say that the niche covers a peak. If it stays at the boundary of the niche, we say the niche does not yet cover a peak.

The operator of climbing to a peak for a niche in the ODHC algorithm consists of two stages: At the first stage, the niche does not cover a peak. ODHC repeats the moving operator to approach the niche toward a potential peak until it covers a peak. Then, comes the second stage. At the second stage, ODHC repeats the shrinking operator to shrink the niche to get a "close-to-peak" with a higher precision until the niche size becomes less than a threshold.

An efficient dynamic optimization method must approach the optimal as much as possible in the static stage before the next environment change. We suppose that a peak shifts to a position near its previous position without changing much of its height after a sudden environment change. Group climbing technique is used in the ODHC algorithm by using an archive mechanism. The archive is to store the current found high peaks. Once the environment changes, the ODHC algorithm searches peaks in the following two ways: One is that the ODHC algorithm constructs niches (named exploiting niche), the centers of which are the past peaks in the archive. These niches climb to get back the known peaks. And the other is that it creates niches (named exploring niche) randomly to explore new peaks. orthogonal design method is used to speed up the climbing in the both ways. The former climbers learn from the past search results. The later climbers implement the global search. By the way, both searches can be implemented in parallel.

We implement the ODHC algorithm in the following way. Suppose the archive can store K peaks. The representative of each niche is denoted by $\overrightarrow{s}$. We use a flag to signal whether or not a niche covers a peak. This flag is called $PeakIsInside$, with one such flag per niche. $PeakIsInside == false$ means that the niche does not cover a peak, while $PeakIsInside == true$ means that the niche covers a peak. Let $PeakIsInside = false$ be the initial value. The design of the ODHC algorithm is as follows.

4.4.2 Design of the ODHC Algorithm

The Framework of the ODHC Algorithm

The ODHC algorithm finds peaks in the following ways: It constructs exploiting niches, the centers of which are the past peaks in the archive. These niches climb to get back the known peaks. And the algorithm creates exploring niches randomly to explore new peaks. The operations depicted above can be run parallel. The exploiting climbers learn from the past search results. The exploring climbers implement the global search. The framework of the ODHC algorithm, denoted *Algorithm 5*, is given in Fig. 4.6

```
// Initiation
Empty the archive;
// Begin climbing
repeat
  // Learn from past search
  for each past peak in the archive do
    Construct a niche W with the past peak being its center
    Calculate the representative of the niche W
    The niche W climbs to a peak; (cf. Algorithm 6)
  endfor
  repeat
    // Explore new peaks
    Randomly create an initial niche E (cf. Algorithm 7)
    The niche E climbs to peak (cf. Algorithm 6)
    Insert E into the archive
    if the archive is full then
      Delete an inferior peak, respecting the need for diversity
    endif
  until FALSE
until the termination criterion is satisfied
```

Fig. 4.6. The framework of the ODHC algorithm

NOTE: The "deletion of a peak, respecting the need for diversity" means that if there are two peaks that have a distance to each other less than a given threshold ε_1 in the archive then delete any one of them, otherwise delete the worst niche.

Climb to Peak

The operator of climbing to a peak for a niche in the ODHC algorithm consists of two stages: At the first stage, the niche does not cover a peak. ODHC repeats the moving operator to approach the niche toward a potential peak until it covers a peak. Then, comes the second stage. At the second stage, ODHC repeats the shrinking operator to shrink the niche to get a "close-to-peak" with a higher precision until the niche size becomes less than a threshold. This peak climbing algorithm, *Algorithm 6*, is shown in Fig. 4.7.

```
repeat
  if PeakIsInside == false then
    // W covers no peak
    Store current representative s⃗ to pres⃗
    Move W (cf. Algorithm 9)
    Calculate a new representative s⃗ for W (cf. Algorithm 8)
    if s⃗ stays inside W or s⃗ == pres⃗ then
      PeakIsInside = true
    endif
    if having detected an environmental change then
      Reset the search pointer
    endif
  else
    // W covers a peak
    if the solution's precision is inferior to the requirement then
      Shrink W (cf. Algorithm 10);
      Calculate a representative s⃗ for W (cf. Algorithm 8)
      if having detected an environmental change then
        Reset the search pointer
      endif
    endif
  endif
until the niche size is less than the threshold ε2
```

Fig. 4.7. Climb to peak for niche W

NOTE: Resetting the search pointer means updating the niches in the archive. The centers of those niches are the past peaks, while their sizes are initially given. To create the initial size of a niche, each side of the search space is divided into several equal portions. Hence the search space is equally divided into units. Assume each unit covers no more than a single peak, which conserves the availability of the orthogonal design. We let the size of the unit be the size of the initial niche. The initial side lengths of each niche are denoted by $d_1, d_2, ..., d_N$.

Randomly Create a Niche

The operator that creates a niche randomly is shown as *Algorithm 7* in Fig. 4.8.

Randomly pick a unit from the search space as an initial niche W
Calculate a representative $\vec{s}$ for W (cf. Algorithm 8)
$PeakIsInside = false$

Fig. 4.8. Randomly create a niche

Calculate a Representative for a Niche

The algorithm for calculating a representative uses the orthogonal design method which can speed up the search by statistical calculation. Suppose the dynamic function has N factors, and the range of each factor is quantized into Q levels. By executing Algorithm 2 (see Fig. 4.2), we obtain the orthogonal matrix $L_M(Q^P)$. P must be larger than or equal to N here. The details of calculating a representative, *Algorithm 8*, are given in Fig. 4.9.

Step 1: Execute Algorithm 2 to construct the orthogonal matrix $L_M(Q^P)$
Step 2: Delete the last $P-N$ columns of $L_M(Q^P)$ to obtain $L_M(Q^N)$
Step 3: Using $L_M(Q^N)$, execute Algorithm 3 to construct the matrix $[\Gamma_{q,j}]_{Q\times N}$
Step 4: Execute Algorithm 4 to obtain a promising solution $\vec{b}$
Step 5: The best of the M combinations corresponding to $L_M(Q^N)$ is assigned to $\vec{b}_1$
Step 6: The superior value among $\vec{b}$, $\vec{b}_1$ and the previous representative $\overrightarrow{pres}$ is assigned to current representative $\vec{s}$.

Fig. 4.9. Calculate a representative for a niche

NOTE: Algorithm 8 ensures that the representative is the best solution found so far in the niche.

Adjust a Niche

Given the following niche W

$$W = \{\vec{x} = (x_1, x_2, ..., x_N) | l_j \leqslant x_j \leqslant u_j, j = 1, 2, ..., N\} \tag{4.10}$$

and its representative $\vec{s} = (s_1, s_2, ..., s_N)$. And niche W' is

$$W' = \{\vec{x} = (x_1, x_2, ..., x_N) | l'_j \leqslant x_j \leqslant u'_j, j = 1, 2, ..., N\} \tag{4.11}$$

The details of moving niche W to W', called *Algorithm 9*, are shown in Fig. 4.10. Fig. 4.11 shows the process of moving a niche. The details of the shrinking operator, denoted *Algorithm 10*, are given in Fig 4.12.

for $j = 1$ to N **do**
  if $(l_j(t) \leq s_j \leq u_j(t))$ **then**
    $l'_j = s_j - (u_j - l_j)/2;$
    $u'_j = s_j + (u_j - l_j)/2;$
  endif
  if $(l_j(t) == s_j)$ **then**
    $l'_j = s_j - (Q-2)(u_j - l_j)/(Q-1);$
    $u'_j = s_j + (u_j - l_j)/(Q-1);$
  endif
  if $(u_j(t) == s_j)$ **then**
    $l'_j = s_j - (u_j - l_j)/(Q-1);$
    $u'_j = s_j + (Q-2)(u_j - l_j)/(Q-1);$
  endif
endfor
if W' lies outside of the problem's search space **then**
  move it appropriately into the problem's search space
endif

Fig. 4.10. Move the niche W to W'

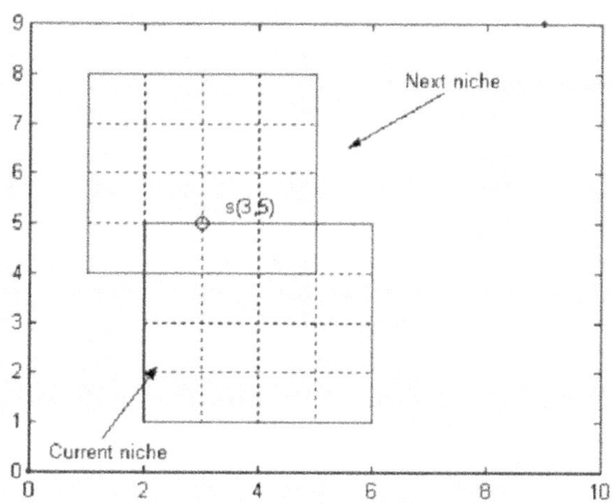

Fig. 4.11. An illustration of a moving niche where $N = 2$ and $Q = 5$

for $j = 1$ to N **do**
  $l'_j = s_j - (u_j - l_j)/(Q-1);$
  $u'_j = s_j + (u_j - l_j)/(Q-1);$
endfor

Fig. 4.12. Shrink the niche W to W'

4.5 Numerical Experiments

4.5.1 Parameter Setting for the ODHC Algorithm

Five parameters need to be set in the ODHC algorithm.

1) The initial size of the niche: $d_1, d_2, ..., d_N$
2) The orthogonal array size for computing niche representative: $M = Q^J$
3) The number of peaks the archive can store: K
4) The threshold for conserving peak diversity in the archive: ε_1
5) The threshold for stopping niche shrinking operator: ε_2

NOTES: Suppose that the side lengths of the search space have values $D_1, D_2, ..., D_N$ and that they are divided respectively into $Q_1, Q_2, ..., Q_N$ equal portions. The side lengths of the initial niche are then $d_1 = D_1/Q_1, d_2 = D_2/Q_2, ..., d_N = D_N/Q_N$. A niche should cover no more than a single peak, to ensure that the ODHC algorithm can work efficiently. If the niche size is too large, it may cover more than one peak, hence, the representative calculated by Algorithm 4.9 would be poor, and the ODHC algorithm would fail to track the moving best. On the other hand, if the niche size is too small, then exploring the peak would be time consuming, because moving the niche would be very slow. However, the ODHC algorithm is insensitive to the initial size of the niches to some extent.

In the experiments mentioned below, each side of the search space is divided into 30 equal portions, i.e., $Q_1 = Q_2 = ... = Q_N = 30$. This makes it likely that a niche covers no more than a single peak for the test problems below. For the orthogonal array size, we choose a small number of levels Q, because a large Q would increase the number of experiments Q^J, so the number of computations would increase. The default value for Q is 5. We let parameter J be determined by both Q and the number of dimensions N. The orthogonal matrix $L_M(Q^N)$ consists of the former N columns of the orthogonal matrix $L_M(Q^P)$ constructed by Algorithm 4.2. According to Equation(4.8), we have $P = (Q^J - 1)/(Q - 1)$. P, N must therefore satisfy

$$P = (Q^J - 1)/(Q - 1) \geq N \tag{4.12}$$

Given N and Q, as J increases, the number of combinations Q^J increases exponentially. Therefore we choose the smallest integer J that satisfies Equation (4.12). Thus J is a parameter determined internally, and need not be set outside the ODHC algorithm. The default setting is to have $K = 10$, $\varepsilon_1 = d/(2 * (Q - 1))$, and $\varepsilon_2 = 0.001$.

Parameter Setting for Test Function

We choose the moving peaks function which has been mentioned in the Equation (4.2) of sec 4.2.3 segment of this chapter as our test problem. Unless stated otherwise, the default settings, defining the benchmark, are employed in our experiments. These settings are shown in Table 4.4, where the change frequency means the number of evaluations between two environmental changes. The term "evaluation" refers to the creation and fitness measurement of an individual. The term "shift length s" means that a peak will move randomly with a length s, at the next environmental change.

Table 4.4. Default settings for the moving peaks benchmark used in this paper

Parameter	
Number of peaks m	10
Change frequency	5000
Height_severity	7.0
Width_-severity	1.0
λ	0
Peak shap e	cone(cf. Equation (4.3))
Basic function	no
Shift length s	1.0
Number of dimension N	5
The range of decision vector $[m_{\vec{x}}, M_{\vec{x}}]$	[0.0,100.0]
The range of peak height changing $[m_h, M_h]$	[30.0,70.0]
The range of peak width parameter $[m_w, M_w]$	[1,12]
Initial peak height for all peaks	50.0

Brief Introduction to the Self-Organizing Scouts Algorithm

The Self-Organizing Scouts (SOS) algorithm has been published in [25]. The basic idea of SOS is that once a peak has been found (i.e. the population converged to a high-performance region), the population should split: a small fraction, called the "scout population", should "watch" over that peak, while the remainder of the population ("base population") should spread out and continue searching for new peaks. When a watched peak moves, the scout population may follow it through the space, and even request reinforcement. Since the population size is limited, individuals are continuously redistributed to those populations where they seem to be needed most, and peaks that seem too unpromising may be abandoned. In general, the SOS algorithm is described as in Fig. 4.13.

4.5.2 Results of Experiments

In order to be able to derive conclusions independent of the algorithm's random seed and for a class of problems, instead of only a particular problem instance, we need more than a single test run for comparison. There are two

---

repeat
  Compute next generation of base population and scout population
  Adjust search space of scout populations
  if forking generation **then**
    Create new scout population if suitable cluster is found
    Adjust number of individuals in base and scout populations
  endif
until termination criterion

Fig. 4.13. The basic SOS algorithm

principal possibilities to reduce the stochastic influence of random seed and particular test instance: (1) Run both algorithms multiple times with different random seeds on a single problem instance. (2) Run both algorithms multiple times on different instances of a class of problems with similar characteristics.

For most results reported, we will be using an average over 50 runs of the ODHC algorithm, each run with a specific random seed and a specific instance of the problem, such that both SOS algorithm and ODHC algorithm face the same 50 environments and have exactly the same starting position. Assuming that the problem solutions fall into the same value range, or at least, that the performance of one algorithm over the other is independent of the average solution quality for a specific problem, this seems to be a viable compromise.

Since for dynamic fitness functions it is not useful to report the best solution achieved, we will here mainly report on the average offline error (cf. Equation(4.6)). The offline errors after 500,000 evaluations in the ODHC algorithm are compared with those of the SOS algorithm.

Tables 4.5, 4.6, 4.7, and 4.8 show the effects of peak movements, frequencies of change, changing the number of peaks and higher dimensionality in both the ODHC and SOS algorithms respectively.

The Effect of Peaks Movements

In the Moving Peaks Benchmark, the distance by which a peak shifts can be set explicitly. For shift length $s = 0$, the peaks still stay at the same place, but the optimum now switches irregularly between 10 peaks which are stored in the archive. By increasing s slowly, we can examine the effect of different algorithms on the performance.

Tables 4.5 displays the offline error after 500,000 evaluations over a range of values for s. When the length s of the shift vector is increased, the performance of both SOS algorithm and ODHC algorithm decreases. That is to be expected, since in addition to the ability to jump from one location to the other when the optimum peak changes, the algorithms now have to deal with the additional challenge to trace the peaks as they move through the search space.

The ODHC algorithm is less affected than SOS algorithm. The scout niches in the ODHC algorithm seem to be able to successfully and simultaneously trace the movements of "their" peaks, thus maintaining valid information before and after the change of the environment.

Table 4.5. The offline errors of both SOS and ODHC for different shift lengths

Shift Length	SOS	ODHC
1.0	4.01	1.78
2.0	5.12	2.16
3.0	6.54	2.47

The Effect of Changing Frequency

So far, the environment gradually changed every 5000 evaluations. When the change frequency increase (the numbers of individual evaluations between changes decrease), the performance of both SOS algorithm and ODHC algorithm suffers significantly (the offline error increase). In part, this may be due to the nature of the offline error measure : When the environment changes too fast, new peaks may hardly be explored, even in the worse situation, the past known peaks in the archive may not be exploited before the next environment change.

Tables 4.6 reveals that the ODHC algorithm performs better than the SOS algorithm. The ODHC algorithm performs better because an archive is used to store peaks. And the created exploit niches are able to find back the old optimums which have been found before the change of environment. Also, the ODHC algorithm creates new explore niches to find new optimums after the change of environment. The explore niches implement the global search. The niche technique might be capable of tracking the quickly moving target. The orthogonal design method used in ODHC algorithm speeds up the search. Thus, there is more time for convergence and a jump from one peak to another is required less often.

The Effect of Changing the Number of Peaks

We now examine the number of peaks on the performance of both the SOS algorithm and the ODHC algorithm. For the experiments reported here, we varied the number of peaks from 1 to 200, always using a shift lengths $s = 1.0$.

The results are summarized in Table 4.7. Increasing the number of peaks up to 20 reduce the performance of both the SOS algorithm and the ODHC algorithm. Because with more than one peak it is no longer sufficient to follow a peak, but it becomes more and more necessary to jump from one peak to the other as the peaks change in height.

Table 4.6. The offline errors of both SOS and ODHC, for different numbers of individual evaluations between changes

Evaluations between Environment Changes	SOS	ODHC
10000	3.62	1.71
5000	4.01	1.78
2500	4.93	2.19
1000	6.01	2.43
500	8.59	5.70

When the number of peaks is increased above 20, some of the offline error becomes actually smaller. This may be due to two factors: first, the more peaks there are, the easier it becomes to jump from one peak onto another, simply because there will very likely be another peak nearby. And second, due to the characteristics of the landscape always evaluating to the maximum of all peaks, the average fitness of the landscape $F(\overrightarrow{x}, t)$ (cf. Equation(4.2)) increases with the number of peaks, and lower peaks may actually be hidden by high, broad peaks, thus bounding the maximum possible error.

As can be seen, the ODHC algorithm again clearly outperforms the SOS algorithm for any number of peaks in the landscape. The experiment results indicate that the ODHC algorithm can make full use of the neighbor information while the peaks move.

Table 4.7. The offline errors of both SOS and ODHC for different numbers of peaks

Peak numbers	SOS	ODHC
1	2.06	0.86
10	4.01	1.78
20	4.43	1.93
30	4.20	1.90
40	4.06	1.88
50	4.12	1.78
100	3.75	1.29
200	3.62	1.10

The Effect of Higher Dimensionality

To reduce the necessary computation time, most of the considered test problems are relatively simple. In our experiments, we look at a problem with 200 peaks and 20 dimensions. The offline error for the SOS algorithm and the ODHC algorithm on that problem is depicted in Table 4.8.

Note that except for dimensionality and the number of peaks, all other parameters of the Moving Peaks Benchmark are kept at their default value.

Thus, while the search space is much larger now, the shape and width parameter of the peaks is kept, leading to an environment with a sparse distribution of peaks, and a much lower average height of the landscape. Also, due to the dimensionality, tracking of a peak becomes significantly more difficult.

Table 4.8 indicates that the SOS algorithm is not very useful in such an environment, because the SOS algorithm may not be able to switch peaks due to the diversity in the memory .

The ODHC algorithm performs reasonably well. Firstly, the ODHC algorithm uses the hill climbing strategy. Hill climbing strategy is one of the heuristic search techniques. The heuristic search strategy uses the inspiration information as the guidance and hence the search is toward the goal direction, avoiding detouring and enhancing the solution efficiency. And secondly, the orthogonal design method is employed on niches. The orthogonal design methods selects a small number of evenly distributed samples from a space with huge number of individuals and then finds a potentially good solution by statistical calculation. This potentially good solution probably is situated near the optimal solution in the niche. Therefore, the niche climbs a potentially peak fast.

The good performance of the ODHC algorithm indicates that the ODHC algorithm is a new sound approach to solve dynamic optimization problems which have a higher dimensionality.

Table 4.8. The offline errors of both SOS and ODHC on a moving peaks problem with 200 peaks and 20 dimensions

SOS	ODHC
12.67	5.17

Overall, all of these results displayed in Tables 4.5, 4.6, 4.7, and 4.8 show that the ODHC algorithm can track the moving best solution a lot better than can the SOS algorithm. We therefore conclude that the ODHC algorithm is an improvement in algorithms of its type.

4.6 Conclusions

4.6.1 Advantages of ODHC Algorithm

The peak climber in the ODHC algorithm is not a solution $\overrightarrow{x}$ but rather a "niche" which has a representative representing the best solution so far found. An orthogonal design method is employed on niche. Therefore, the niche climbs a potentially peak fast. The climbing operator for a niche to climb a peak in the ODHC algorithm consists of two stages: At the first stage, the niche does not cover a peak. It repeats moving operator to approach a potential peak until covering a peak and then comes the second stage. At the second stage, it repeats shrinking operator to obtain a "close-to-peak" with a higher precision until the niche size less than threshold.

We suppose that a peak shifts to a position near its previous position without changing much of its height after a sudden environment change. The ODHC algorithm keeps an archive to store the current found higher peaks. Once the environment change occurs, the ODHC algorithm searches peaks in two ways: One is that it constructs exploiting niches, the centers of which are the past peaks in the archive, and search from those niches to get back the higher known peaks. The other is to create exploring niches randomly to explore new peaks. The former search learns from the past search results. The later implements the global search.

4.6.2 Limitations of ODHC Algorithm

Each niche should only cover a peak, such that the ODHC algorithm may work well. If the distance of two peaks is too near, a niche may cover more than two peaks, which will decrease the effectiveness of the orthogonal design, then the ODHC algorithm may work comparatively poor. But, the robustness of the orthogonal design conserves that it can statistically find potentially good solutions no mater how many peaks the niche has. Therefore, the ODHC algorithm should not work too poor.

Acknowledgments

This work was supported by the National Natural Science Foundation of China (No. 60473037). We thank Dr. Shengxiang Yang for his assistance with the expression of this chapter and helpful suggestions.

References

1. D. E. Goldberg and R. E. Smith. Nonstationary function optimization using genetic algorithms with dominance and diploidy. In J. J. Grefenstette, editor, Second International Conference on Genetic Algorithms, pages 59-68. Lawrence Erlbaum Associates, 1987.
2. T. Back (1998). On the behavior of evolutionary algorithms in dynamic fitness landscapes. In Proc. of the 1998 IEEE Int. Conf. on Evolutionary Computation, pp 446-451. IEEE Press.
3. N. Mori, S. Imanishi, H. Kita, and Y. Nishikawa. Adaptation to changing environments by means of the memory based thermodynamical genetic algorithm. In T. Bäck, editor. Seventh International Conference on Genetic Algorithms, pp 299-306. Morgan Kaufmann, 1997.
4. J. Branke. Evolutionary Optimization in Dynamic Environments. Kluwer Academic Publishers, 2002.

5. J. Branke. Evolutionary approaches to dynamic optimization problems - Instruction and recent trends-. in J. Branke, editor, Proceedings of the Workshop on Evolutionary Algorithms for Dynamic Optimization Problems, pp 1-3. Chicago, USA, 2003.
6. H. G. Cobb. An investigation into the use of hypermutation as adaptive operator in genetic algorithms having continuous, time-dependent nonstationary environments. Technical Report AIC-90-001, Naval Research Laboratory, Washington, USA, 1990.
7. R. W. Morrison and K. A. De Jong (2000). Triggered hypermutation revisited. Proc. of the 2000 Congress on Evolutionary Computation, pp 1025-1032.
8. J. J. Grefenstette. Genetic algorithms for changing environments. In R. Maenner and B. Manderick, editors, Parallel Problem Solving from Nature 2, pp 137-144. North-Holland, 1992.
9. J. Lewis, E. Hart and G. Ritchie (1998). A comparison of dominance mechanisms and simple mutation on non-stationary problems. Proc. of the 5th Int. Conf. on Parallel Problem Solving from Nature, pp 139-148.
10. K. P. Ng and K. C. Wong (1995). A new diploids scheme and dominance change mechanism for non-stationary function optimisation. In L. J. Eshelman (ed.), Proc. of the 6th Int. Conf. on Genetic Algorithms.
11. S. Yang. Memory-based immigrants for genetic algorithms in dynamic environments. Proceedings of the 2005 Genetic and Evolutionary Computation Conference, Vol. 2, pp 1115-1122, 2005.
12. J. Branke, T. Kaufler, C. Schmidt, and H. Schmeck. A multipopulation approach to dynamic optimization problems. Adaptive Computing in Design and Manufacturing. Springer, 2000.
13. D. C. Montgomery, Design and Analysis of Experiments, 3rd ed. New York: Wiley, 1991.
14. S. Zeng, L. Kang, L. Ding. An Orthogonal Multi-objective Evolutionary Algorithm for Multi-objective Optimization Problems with Constraints. Evolutionary Computation. Vol.12, No.1, pp 77-98, MIT Press, 2004.
15. S. Zeng, S. Yao, L. Kang, and Y. Liu. An Efficient Multi-objective Evolutionary Algorithm: OMOEA-II. In C. A. Coello Coello et al. (Eds.), proceedings of the Third International Conference on Evolutionary Multi-Criterion Optimization, LNCS series, Springer-Verlag, pp 108-119, 2005.
16. Y. W. Leung and Q. Zhang (1997). Evolutionary algorithms + experimental design methods: A hybrid approach for hard optimization and search problems, Res. Grant Proposal, Hong Kong Baptist Univ.
17. Y. W. Leung and Y. Wang (2001). An orthogonal genetic algorithm with quantization for global numerical optimization. IEEE Trans. Evol. Comput. vol.5, No.1, pp. 40-53.
18. T. C. Fogarty, F. Vavak, and P. Cheng. Use of the genetic algotithm for load balancing of sugar beet presses. Proceedings of the 6th Int. Conf. on Genetic Algorithms, pages 617-624. Morgan Kanufmann, 1995.
19. C. L. Karr. Genetic algorithms and fuzzy logic for adaptive process control. In S. Goonatilake and S. Khebbal, editors, Intelligeht Hybrid Systems, chapter 4, pages 63-64, John Wiley, 1995.
20. C. L. Karr. Adaptive process control using biologic paradigms, In L. C. Jain, editor, Proceedings of Electronic Technology Directions to the Year 2000, volume 1, pages 128-136, 1995.

21. C. L. Ramsey and J. J. Grefenstette. Case-based initialization of genetic algorithms, In S. Forrest, editor, Proceedings of the 5th International Conference on Genetic Algorithms, pages 84-91. Morgan Kaufmann, 1993.
22. R. W. Morrison and K. A. DeJong. A test problem generator for nonstationary environments. Proceedings of the 1999 Congress on Evolutionary Computation, volume 3, pages 2047-2053, 1999.
23. S. Yang. Constructing dynamic test environments for genetic algorithms based on problem difficulty. Proceedings of the 2004 Congress on Evolutionary Computation, Vol. 2, pages 1262-1269, 2004.
24. A. S. Hedayat, N. J. A. Sloane and J. Stufken. Orthogonal Arrays: Theory and Applications. New York: Springer-Verlag, 1999.
25. J. Branke, T. Kaubler, and H. Schmeck. Guiding multiobjective evolutionary algorithms towards interesting regions. Technical Report No. 399, Institute AIFB, University of Karlsruhe, Germany, 2000

5

Genetic Algorithms with Self-Organizing Behaviour in Dynamic Environments

Renato Tinós[1] and Shengxiang Yang[2]

[1] Departamento de Física e Matemática, FFCLRP, Universidade de São Paulo (USP), Av. Bandeirantes 3900, Ribeirão Preto, SP, 14040-901, Brazil
rtinos@ffclrp.usp.br

[2] Department of Computer Science, University of Leicester
University Road, Leicester LE1 7RH, United Kingdom
s.yang@mcs.le.ac.uk

Summary. In recent years, researchers from the genetic algorithm (GA) community have developed several approaches to enhance the performance of traditional GAs for dynamic optimization problems (DOPs). Among these approaches, one technique is to maintain the diversity of the population by inserting random immigrants into the population. This chapter investigates a self-organizing random immigrants scheme for GAs to address DOPs, where the worst individual and its next neighbours are replaced by random immigrants. In order to protect the newly introduced immigrants from being replaced by fitter individuals, they are placed in a subpopulation. In this way, individuals start to interact between themselves and, when the fitness of the individuals are close, one single replacement of an individual can affect a large number of individuals of the population in a chain reaction. The individuals in a subpopulation are not allowed to be replaced by individuals of the main population during the current chain reaction. The number of individuals in the subpopulation is given by the number of individuals created in the current chain reaction. It is important to observe that this simple approach can take the system to a self-organization behaviour, which can be useful for GAs in dynamic environments.

5.1 Introduction

A significant part of optimization problems in real world is dynamic optimization problems (DOPs), where the evaluation function and the constraints of the problem are not fixed [24]. When changes occur in the problem, the solution given by the optimization procedure may be no longer effective, and a new solution should be found [4]. The optimization problem may change for several factors, like faults, machine degradation, environmental or climatic modifications, and economic factors. In fact, the natural evolution, which is the inspiration for genetic algorithms (GAs), is always dynamic. The occurrence of natural cataclysms, geological modifications, competition for natural

R. Tinós and S. Yang: *Genetic Algorithms with Self-Organizing Behaviour in Dynamic Environments*, Studies in Computational Intelligence (SCI) **51**, 105–127 (2007)
www.springerlink.com

resources, coevolution between species, and climatic modifications are only a few examples of changes related to natural evolution.

The simplest approach to deal with DOPs is to start a new optimization process whenever a change in the problem is noticed. However, the optimization process generally requires time and substantial computational effort. If the new solution after the change in the problem is, in some sense, related to the previous solution, the knowledge obtained during the search for the old solution can be utilized to find the new solution [16]. In this case, the search for new solutions based on the old solutions can save substantial processing time. Evolutionary algorithms are particularly attractive to such problems as individuals representing solutions of the problem before the changes can be transferred into the new optimization process.

However, in GAs, the population of solutions generally converges in the fitness landscape to points close to the best individual of the population. If the fitness landscape abruptly changes, the actual population can be trapped in local optima located close to the old solution. In fact, the premature convergence of the solution to a local optima is not a problem exclusive to DOPs, but it can be a serious problem in stationary optimization problems too [20]. In order to avoid the premature convergence, several approaches where the diversity level is re-introduced or maintained throughout the run have appeared in literature over the past years (see the surveys [4, 16, 24]). Typical examples of such approaches are the *random immigrants GA* (RIGA) [14], the sharing or crowding mechanisms [5], the variable local search [25], the thermodynamical genetic algorithm [21], and the use of hypermutation [6].

RIGA, which is inspired by the flux of immigrants that wander in and out of a population between two generations in nature, is very interesting and simple [7, 14]. In RIGA, some individuals of the current population are replaced by randomly generated individuals in each generation of the run. A replacement strategy, like replacing random or worst individuals of the population, defines which individuals are replaced by the immigrants. The RIGA tries to maintain the diversity level of the population, which can be very useful to prepare the population for possible fitness landscape changes [7].

However, in some cases, when the number of genes in the individual is high and the local optimum where the population is found has fitness much higher than the mean fitness of all possible solutions of the search space, the survival probability of the new random individuals is generally very small. This occurs because the selection methods employed in GAs preserve, directly or indirectly, the best individuals of the population, and the probability that the fitness of the new random individuals is higher than (or close to) the fitness of the current individuals is generally small.

In this work, instead of substituting the worst individuals or the random individuals in each generation like in the standard RIGA, the worst individual and its next neighbours are replaced. In order to protect the newly introduced immigrants from being replaced by fitter individuals, they are placed in a subpopulation. In this way, individuals start to interact between themselves and, when the fitness of the individuals are close, as in the case where the diversity level is low, one single replacement of an individual can affect a large number of individuals of the population in a chain reaction. The individuals in the subpopulation are not allowed to be replaced by individuals of the main population during the current chain reaction. The number of individuals in the subpopulation is not defined by the programmer, but is given by the number of individuals created in the current chain reaction. It is important to observe that this simple approach can take the system to a self-organization behaviour, which can be useful in DOPs.

The experimental results suggest that the proposed GA presents a kind of self-organizing behaviour, known as self-organized criticality (SOC) [1], which is described in Section 5.2. The proposed GA is presented in Section 5.3, and the experimental results are presented in Section 16.4. In Section 5.4.3, the proposed GA and the experimental results are analyzed. Finally, Section 5.5 concludes the work with discussions on relevant future work.

5.2 Self-Organized Criticality

Systems consisting of several interacting constituents may present an interesting kind of self-organizing behaviour known as SOC [2], [15]. Researchers have suggested that several phenomena exhibit SOC, like sand piles, earthquakes, forest fires, electric breakdowns, and growing interfaces [1].

An interesting behaviour appears in systems exhibiting SOC: they self-organize into a particular critical state without the need of any significant tuning action from outside. The critical state is described by the response of a system to external perturbation. In a system exhibiting noncritical behaviour, the distribution of responses to perturbation at different positions and at different times is narrow and well described by an averaged value. In a system exhibiting critical behaviour, no single characteristic response exists, i.e., the system exhibits scale invariance. A small perturbation in one given location of the system may generate a small effect on its neighbourhood or a chain reaction that affects all the constituents of the system.

The statistical distributions describing the response of the system exhibiting SOC are given by power laws in the form

$$P(s) \sim s^{-\tau} \tag{5.1}$$

and

$$P(d) \sim d^{-\alpha}, \tag{5.2}$$

where s is the number of constituents of the system affected by the perturbation, d is the duration of the chain reaction (lifetime), and τ and α are constants. The sand pile model described in [2], where a single grain is added at a random position in every interval of time Δt is an example of a system exhibiting SOC. In order to characterize the response of the sand pile model, one can measure the number of sand grains (s) involved in each avalanche induced by the addition of a single grain and the duration (d) of each avalanche. In the critical state, the statistical distributions describing the response of the system to the addition of a single grain are given by Eqs. 5.1 and 5.2, and the addition of a single grain can affect only a grain in its neighbourhood or can affect the whole sand pile.

Bak has suggested that SOC occurs in natural evolution too [1]. An evidence of SOC in evolution would be the fact that it does take place through bursts of activity intercalated by calm periods, instead of gradually at a slow and constant pace [13]. There are many more small extinction events than large events, such as the Cretaceous extinction of dinosaurs and many other species, and extinction events occur on a large variety of length scales [23]. Bak has suggested that extinctions propagate through ecosystems, such as avalanches in a sand pile, and perturbations of the same size can unleash extinction events of a large variety of sizes [1]. In such hypothesis, species coevolve to a critical state [17].

A very simple simulation model to study the connection between evolution and SOC was proposed by Bak and Sneppen [1]. In the one-dimensional version of the model, the individuals (or species in the authors' terminology) are placed in a circle, and a random value of fitness is assigned to each one of them. In each generation of the simulation, the values of fitness of the individual with the lowest fitness in the current population, one individual located in its right position, and one located in its left position are replaced by new random values. An analogy of the connection between neighbours in this simple model is the interaction between species in nature: if a prey is extinct, the fitness of its predators will change. The Bak-Sneppen Model is summarized in Fig. 5.1.

---

begin
  Find the index j of the individual with the lowest fitness
  Replace the fitness of the individuals with index j, $j-1$, and $j+1$ by random values drawn with uniform density
end

Fig. 5.1. The Bak-Sneppen model

The Bak-Sneppen model presents an interesting behaviour. In the beginning of the simulation, the mean fitness of the population is low, but, as the number of generation increases, the mean fitness increases too. Eventually, the mean fitness ceases to increase, and the critical state is reached. In the Bak-Sneppen Model, a replacement of the fitness of the worst individual causes the replacement of its two next neighbours. In the critical state, the values of fitness of the neighbours are very often replaced by random numbers with smaller values. The new worst individual can be then one of these two neighbours, which are replaced with its two next neighbours, originating a chain reaction, called replacement event in this work, that can affect all the individuals of the population. The replacement events exhibit scale invariance and their statistical distributions are given by power laws in the form of Eqs. 5.1 and 5.2. Large replacement events generally occur when almost all individuals of the population have similar high values of fitness.

It is important to observe that SOC avoids the situation where the species get trapped in local optima in the fitness landscape in the Bak-Sneppen evolution model. The idea is interesting and relatively simple, and soon researchers proposed the use of SOC in optimization processes. Boettcher and Percus [3] proposed the optimization with extremal dynamics, a local-search heuristic for finding solutions in problems where constituents of the system are connected, e.g., the spin glass optimization problem. Løvbjerg and Krink [19] extended Particle Swarm Optimization with SOC in order to improve the optimization process and to maintain the diversity level.

In GAs, Krink and Thomsen [18] proposed the use of the sand pile model previously discussed to generate power laws to be utilized to control the size of spatial replacement zones in a diffusion model. When an individual is extinct, a mutated version of the best individual of the population is created in its place. It is important to observe that, in the algorithm proposed in [18], SOC appears in the sand pile model utilized to control the size of the replacements, and not as a result of the self-organization of the constituents of the system (individuals of the GA).

5.3 Random Immigrants Genetic Algorithm with Self-Organizing Behaviour

In the standard RIGA, randomly chosen individuals of the current population $\mathbf{P_t}$ are replaced by randomly generated individuals. A replacement rate specifies the number of individuals replaced in each generation. The standard RIGA can be summarized in Fig. 5.2, which differs from the generational *standard GA* (SGA) only by the inclusion of the procedure "replace($\mathbf{P_t}$)", where randomly chosen individuals of the current population are replaced by randomly generated individuals.

In this work, we propose the replacement of the individual with the lowest fitness of the current population and its two next neighbours for new randomly generated individuals in RIGA. The indices of the individuals are used to determine the neighboring relations. In each generation of the algorithm, the individual with the lowest fitness in the current population (index j), one individual located in its right position (index $j+1$), and one located in its left position (index $j-1$) are replaced by new random individuals. One can observe that, as the proposed GA is not spatially distributed, the neighbouring relations are random.

Require: N: population size; p_c: crossover rate; p_m: mutation rate
begin
  $t \leftarrow 1$
  initialize($\mathbf{P_t}$,N)
  evaluate($\mathbf{P_t}$)
  while (stop criteria are not satisfied) **do**
    $\mathbf{P_t} \leftarrow$ replace($\mathbf{P_t}$)
    for $i = 1$ to N **do**
      $P_{t+1}(i) \leftarrow$ selection($\mathbf{P_t}$,i)
    end for
    crossover($\mathbf{P_{t+1}}$,p_c)
    mutation($\mathbf{P_{t+1}}$,p_m)
    evaluate($\mathbf{P_{t+1}}$)
    $t \leftarrow t + 1$
  end while
end

Fig. 5.2. The random immigrants genetic algorithm (RIGA)

A second strategy is still adopted, where the new immigrants created during the current chain reaction (called replacement event in this work), which occurs along the generations, are preserved in a subpopulation. The size of this subpopulation is not defined by the programmer, but is given by the number of individuals created in the current replacement event. The individuals in the current population that do not belong to the subpopulation are not allowed to replace individuals present in the subpopulation. The individuals that belong to the subpopulation are allowed to evolve, i.e., they are submitted to selection, crossover, and mutation. It is important to observe that selection and crossover are allowed only among individuals that belong to the subpopulation.

We hope that, with such strategies, the system can exhibit SOC in order to increase the diversity level of the population in a self-organized way and, then, to avoid the situation where the individuals get trapped in local optima in the fitness landscape when the problem changes.

In the proposed *self-organizing random immigrants GA* (SORIGA), there are two major modifications from the standard RIGA. In the first modification, the procedure "replace($\mathbf{P_t}$)" is modified as presented in Fig. 5.3. The current size (or duration) of each replacement event, i.e., the number of times that we replace the worst individual and its neighbours in the current replacement event, is recorded and denoted by d, and the minimum and maximum index values of the replaced individuals ($i_{min} - 1$ and $i_{max} + 1$) are employed to compute the number of individuals affected by the current replacement event. The number of individuals in the subpopulation, i.e., the size of subpopulation, equals to $(i_{max}+1)-(i_{min}-1) = i_{max}-i_{min}+2$. When the chain reaction ceases, i.e., the individual with the lowest fitness does not belong to the subpopulation, the size of the replacement is set to 1.

Procedure replace($\mathbf{P_t}$)
begin
  Find the index j of the individual with the lowest fitness
  Replace the individuals of $\mathbf{P_t}$ with indices j, $j-1$, and $j+1$ by randomly generated individuals
  if $(i_{min} - 1 \le j \le i_{max} + 1)$ **then**
    $d \leftarrow d + 1$
    if $(j = i_{min} - 1)$ **then**
      $i_{min} \leftarrow j$
    end if
    if $(j = i_{max} + 1)$ **then**
      $i_{max} \leftarrow j$
    end if
    else
      $d \leftarrow 1$
      $i_{min} \leftarrow j$
      $i_{max} \leftarrow j$
  end if
end

Fig. 5.3. The replace approach

Procedure selection($\mathbf{P_t}, i$)
begin
  if $(i < i_{min} - 1)$ or $(i > i_{max} + 1)$ **then**
    Select an individual from $\mathbf{P_t}$
  else
    Select an individual from the subset of individuals of $\mathbf{P_t}$ with index in $[i_{min} - 1, i_{max} + 1]$
  end if
end

Fig. 5.4. The selection approach

The second modification, which is presented in Fig. 5.4, lies in the selection approach for each individual in the population. Two cases can occur. If the index of the new individual was not affected by the current replacement event, the new individual is selected according to the standard approach. Otherwise, i.e., if the index was affected by the current replacement, the new individual

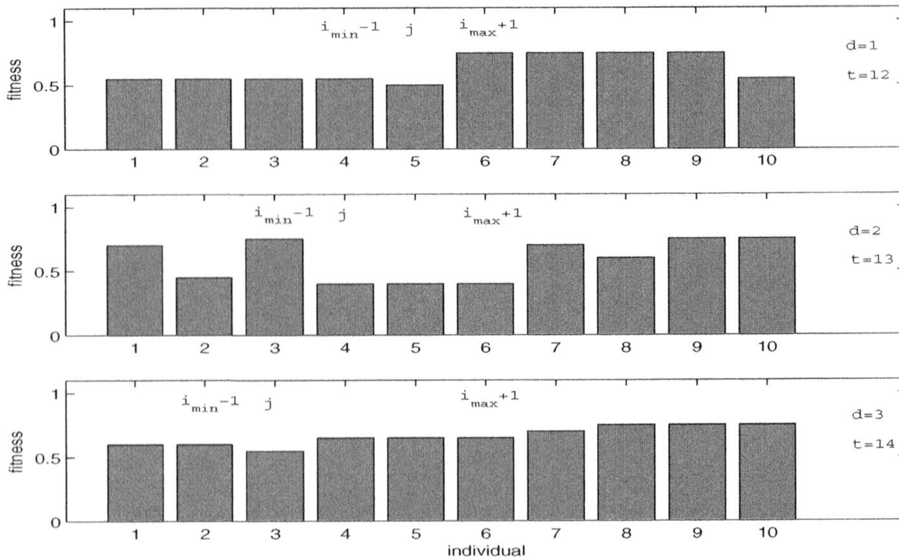

Fig. 5.5. Fitness of the individuals of the current population at generations 12, 13, and 14 in a run of the SORIGA on the example problem

is selected from the subpopulation that consists of the individuals replaced in the current replacement event (individuals with index values from $i_{min} - 1$ to $i_{max} + 1$).

In order to illustrate its working, SORIGA is applied to a simple problem where the fitness function is defined as

$$f(\mathbf{x}) = \frac{u(\mathbf{x})}{l}, \tag{5.3}$$

where $u(\mathbf{x})$ is the unitation function of a binary vector (individual) $\mathbf{x}$ of length l, which returns the number of ones in vector $\mathbf{x}$. The individuals, which are randomly generated in the first generation, are selected according to elitism and the roulette wheel method. Mutation with rate $p_m = 0.01$ and two-point crossover with rate $p_c = 0.7$ are employed. The number of individuals in the population is equal to 10 and $l = 20$. Fig. 5.5 presents the first three steps of an replacement event in a run of SORIGA on this example. The figure shows the fitness of all individuals in the current population in generations 12, 13, and 14 respectively. In generation 12, the individual with index 5 (index j in Fig. 5.3) has the lowest fitness in the population. In this way, individual with index 5 and its two next neighbours (individuals with indices 4 and 6) are replaced by randomly generated individuals. In the next generation, the individual with index $j = 4$ has now the lowest fitness, and it together with its two next neighbours (individuals with indices 3 and 5) are then replaced. In generation 14, the individual with index $j = 3$ has the lowest fitness. It can be observed that the chain reaction is propagated because the remaining individuals have fitness values higher than the individuals in the subpopulation defined by the limits $i_{min} - 1$ and $i_{max} + 1$ (see Fig. 5.3 and Fig. 5.4). The individuals that do not belong to this subpopulation are not allowed to replace an individual of this subpopulation.

5.4 Experimental Study

In order to evaluate the performance of proposed SORIGA, two sets of experiments are carried out. In the first set of experiments, the dynamic test environment for GAs proposed by Yang [26] is employed (Subsection 5.4.1). In the second set of experiments, evolutionary robots are simulated in dynamic environments (Subsection 5.4.2). In the experiments, SORIGA is compared to SGA, and two versions of RIGA. In the first version, denoted *RIGA1*, three individuals randomly chosen from the current population are replaced by randomly generated individuals in each generation. In the second version, denoted *RIGA2*, the three worst individuals, i.e., the individuals with the lowest fitness, are replaced by randomly generated individuals. The analysis of the results is presented in Subsection 5.4.3.

5.4.1 Dynamic Test Environment

In order to evaluate the performance of different GAs in DOPs, Yang [26] proposed a dynamic environment generator based on unitation and trap functions. A trap function is defined as follows

$$f(\mathbf{x}) = \begin{cases} \frac{a}{z}(z - u(\mathbf{x})), & \text{if } u(\mathbf{x}) \leq z \\ \frac{b}{l-z}(u(\mathbf{x}) - z), & \text{otherwise,} \end{cases} \tag{5.4}$$

where $u(\mathbf{x})$ is the unitation function of a binary vector $\mathbf{x}$ of length l, a is the local and possibly deceptive optimum, b is the global optimum, and z is the slope-change location which separates the attraction basin sizes of the two optima. A trap function can be a deceptive function for GAs, i.e., a function where there exist low-order schemata that, instead of combining to form high-order schemata, forms schemata resulting in a deceptive solution that is sub-optimal [11]. A trap function is deceptive on average if the ratio of the fitness of the local optimum to that of the global optimum is constrained by the following relation [9]

$$r = \frac{a}{b} \geq \frac{2 - 1/(l - z)}{2 - 1/z} \tag{5.5}$$

Deception is not the only element that can generate difficulty to a GA. The problem difficulty can also be caused by exogenous noise and scaling. The scaling problem arises in functions that consist of several schemata with different worth to the solution [12]. A scaling problem can be simulated using additively decomposable functions as follows

$$f(\mathbf{x}) = \sum_{i=1}^{m} c_i \; f_i(\mathbf{x}_{I_i}), \tag{5.6}$$

where m is the number of schemata that are juxtaposed and summed together, I_i is the set of the fixed bit positions that form schema i, and c_i is the scaling factor for each sub-function f_i.

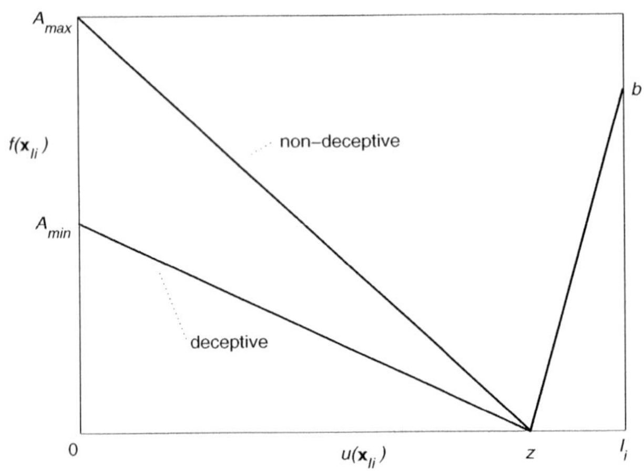

Fig. 5.6. *Illustration of the trap function $f(\mathbf{x})$. The global optimum changes between b and A_{max} in every δ_t generations.*

Employing Eqs. 5.4 and 5.6, it is possible to create different dynamic environments where the problem difficulty can be adjusted. In this work, dynamic environments where the deception difficulty is modified by changing the peak heights of optima are employed [26]. In these dynamic environments, the fitness of an individual $\mathbf{x}$ is given by additively decomposable trap functions defined as follows

$$f(\mathbf{x}) = \sum_{i=1}^{m} c_i \; f_i(\mathbf{x}_{I_i}, t) \tag{5.7}$$

$$f(\mathbf{x}_{I_i}, t) = \begin{cases} \frac{a_i(t)}{z_i}(z_i - u(\mathbf{x}_{I_i})), & \text{if } u(\mathbf{x}_{I_i}) \leq z_i \\ \frac{b_i}{l_i - z_i}(u(\mathbf{x}_{I_i}) - z_i), & \text{otherwise,} \end{cases} \tag{5.8}$$

where $i = 1, \ldots, m$, $\mathbf{x}^{\mathrm{T}} = [\mathbf{x}_{I_1}^{\mathrm{T}} \ \ldots \ \mathbf{x}_{I_m}^{\mathrm{T}}]$, $\mathbf{x}_{I_i} = [x_{(i-1)l_i+1} \cdots x_{(i-1)l_i+l_i}]^{\mathrm{T}}$, $b_i = b$=1.0, $z_i = z$, the scaling is given by $c_i = 2^{i-1}$, and a_i switches between $A_{min} > 0$ and $A_{max} > b_i$ in every δ_t generations. The parameter A_{min} is constrained by Eq. 5.5. That is, it is chosen in order that the trap functions are deceptive on average. In this way, in every δ_t generations, the global optimum changes between b and $a_i = A_{max}$, and the problem changes between deceptive and non-deceptive (Fig. 5.6).

Three dynamic environments are generated as the test bed for all GAs in this work. In the first (Environment 1), $l = 36$, $z = 5$, $m = 6$, l_i=6, $A_{min} = 0.6$, and $A_{max} = 1.4$. In the second (Environment 2), $l = 36$, $z = 4$, $m = 6$, l_i=6, $A_{min} = 0.9$, and $A_{max} = 1.9$. In the third (Environment 3), $l = 45$, $z = 4$, $m = 9$, l_i=5, $A_{min} = 0.6$, and $A_{max} = 1.4$.

Experimental Design

For each run of an algorithm in a dynamic environment, the individuals of the initial population are randomly chosen. The individuals are selected in each generation according to elitism and the roulette wheel method. Mutation with rate p_m and two-point crossover with rate p_c are utilized. Nine experiments with different parameters are presented in this section. Table 5.1 presents the parameters utilized in each experiment.

Table 5.1. Experimental Setttings - Dynamic Test Environments

Experiment	Environment	Population Size	p_m	p_c	δ_t
1a	1	100	0.01	0.7	5000
1b	1	20	0.01	0.7	5000
1c	1	300	0.01	0.7	5000
1d	1	100	0.001	0.7	5000
1e	1	100	0.05	0.7	5000
1f	1	100	0.01	0.2	5000
1g	1	300	0.01	0.7	10000
2	2	100	0.01	0.7	5000
3	3	100	0.01	0.7	5000

The comparison of the results obtained by different algorithms on DOPs is more complex than the same comparison on stationary problems [24]. For DOPs, it is necessary to evaluate not the final result, but rather the optimization process itself. Here, the measure *adaptability*, proposed in [24] and based on a measure proposed by De Jong [8], is utilized to evaluate the GAs. Adaptability is computed as the difference, averaged over the entire run, between the fitness of the current best individual of each generation and the corresponding optimum value. The best results for the adaptability measure are those with the smallest values.

Table 5.2. Adaptability - Dynamic Test Environments

Experiment	SGA	RIGA1	RIGA2	SORIGA
1a	0.1882 (15.68%)	0.0177 (1.47%)	0.0217 (1.81%)	0.0086 (0.72%)
1b	0.1996 (16.63%)	0.0267 (2.22%)	0.0325 (2.71%)	0.0173 (1.44%)
1c	0.1622 (13.52%)	0.0137 (1.15%)	0.0132 (1.10%)	0.0068 (0.57%)
1d	0.2001 (16.67%)	0.0198 (1.65%)	0.0263 (2.19%)	0.0106 (0.88%)
1e	0.0368 (3.07%)	0.0071 (0.59%)	0.0076 (0.63%)	0.0064 (0.54%)
1f	0.1924 (16.03%)	0.0370 (3.08%)	0.0414 (3.46%)	0.0200 (1.67%)
1g	0.1491 (12.42%)	0.0071 (0.59%)	0.0073 (0.61%)	0.0036 (0.30%)
2	0.1956 (13.49%)	0.0143 (0.98%)	0.0166 (1.15%)	0.0084 (0.58%)
3	0.1655 (13.79%)	0.0090 (0.75%)	0.0104 (0.87%)	0.0052 (0.43%)

Table 5.3. Mean Fitness of the Population - Dynamic Test Environments

Experiment	SGA	RIGA1	RIGA2	SORIGA
1a	0.9286	1.0559	1.0755	1.0174
1b	0.9610	1.0340	1.0796	0.8730
1c	0.9198	1.0480	1.0569	1.0368
1d	0.9891	1.1356	1.1534	1.0865
1e	0.8590	0.8614	0.8729	0.8342
1f	0.9263	1.0566	1.0694	1.0330
1g	0.9323	1.0544	1.0631	1.0405
2	1.1519	1.2914	1.3199	1.2360
3	0.9503	1.0726	1.0930	1.0321

Experimental Results

Tables 5.2 and 5.3 respectively present the experimental results regarding the adaptability and the mean fitness of all individuals of the population averaged over 20 trials, each one with a different random seed. In Table 5.2, the percentage inside the parentheses indicates the adaptability over the optimum fitness values.

Hypothesis tests, considering the Student's t-distribution, indicate that the measure adaptability is smaller for SORIGA with a level of significance equal to 0.01 in all experiments except Experiment 1e, where the level of significance equals 0.095.

Fig. 5.7 shows the fitness of the best individuals averaged over the 20 trials for SGA and SORIGA in Experiment 1a.

5.4.2 Evolutionary Robotics

Robots in which artificial evolution is used as a fundamental form of adaptation or design are known as evolutionary robots [22]. In the experiments presented in this section, mobile robots are simulated in DOPs using a modified version of the Evorobot simulator developed by S. Nolfi [22]. In the simulator

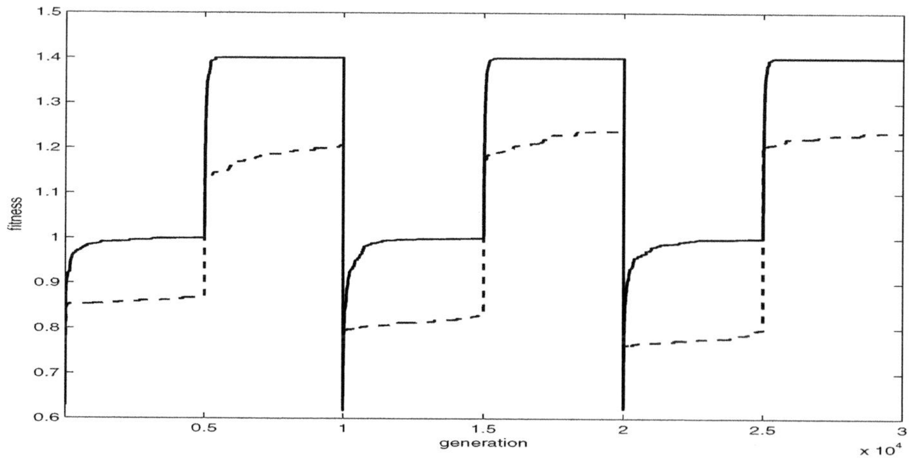

Fig. 5.7. *Averaged fitness of the best individual in Experiment 1a (dynamic test environment). The solid and dashed lines represents the results for SORIGA and SGA respectively.*

utilized in the experiments presented in this section, the robots are controlled by a recurrent artificial neural network (Elman Network). GAs have been employed to perform several tasks in artificial neural networks (ANNs), such as architecture design, synaptic weight adjustment, learning rule adaptation, and synaptic weight initialization [21]. In the experiments presented here, the synaptic weights of the ANN used to control the robot are adjusted by GAs.

Two evolutionary robot experiments are presented in this section. The two experiments are inspired by the experiment proposed by Floreano and Mondada [10], where a Khepera robot with eight infrared distance sensors (six sensors in one side and two in another side of the robot), two ambient light sensors, and one floor brightness sensor navigates in an arena. The robot has a measurable limited energy, which is recharged every time the robot crosses a battery recharge area. The battery recharge area is indicated by a different color of the floor and by a light source mounted in a tower inside the area.

Experimental Design

In the experiments presented in this section, the fitness function is given by the accumulated averaged rotation speed of the two wheels of the robot during its life time, i.e., while the battery has energy and while the robot does not crash into a wall or an obstacle, considering a maximum limit of 60 seconds. A fully charged battery allows the robot to move for 20 seconds. The fitness is not computed while the robot remains in the battery recharge area. Although the fitness function does not specify that the robot should return to the battery recharge area, the individuals that develop the ability to find it and periodically return to it while exploring the arena without hitting the obstacles accumulate more fitness. The neural network utilized to control the robots has 17 inputs (8 infrared sensors, 2 ambient light sensors, 1 floor brightness sensor, 1 sensor for the battery energy, and 5 recurrent units), 5 hidden neurons, and 2 outputs (2 motors).

Two experiments with 2000 generations each are presented in this section. In the first experiment (Experiment 4), the environment where the robot is evolving is changed after 1000 generations. Environment changes frequently occur in real problems, where some aspects of the environment are frequently modified. Besides, robots are frequently evolved in simulations to avoid damage, and, when a satisfactory behaviour is reached, the neural networks employed to control the simulated robot are transferred to the real ones. In the experiments, the robot evolves in an arena of 40cm $\times$ 45cm free of obstacles during the first 1000 generations, and in an arena of 60cm $\times$ 35cm with four cylindrical obstacles during the last 1000 generations.

In the second experiment (Experiment 5), the robot is affected by a failure after 1000 generations. The responses of the six infrared sensors located in one side of the robot are set to zero when it is affected by this failure. We are interested in investigating the reconfiguration of the robot after an abrupt failure. The robot should evolve in an arena of 60cm $\times$ 35 cm with three cylindrical obstacles.

In the runs, the individuals of the initial population are randomly chosen. The evolving robot always starts in a fixed position on the environment, but with a random initial orientation. The individuals are represented by a vector of real values corresponding to the synaptic weights of the ANN. In each generation of the GAs, the 20 best individuals are selected and each one generates 5 children ($N = 100$). In both experiments, $p_m = 0.01$ and crossover is not utilized.

Experimental Results

Table 5.4 presents the experimental results with respect to the adaptability (supposing a maximum fitness equal to 1.0), the mean fitness of all individuals of the population, and the fitness of the best individual after 2000 generations averaged over 20 trials, each one with a diffente random seed, which indicates the performance of the robot after the change in the problem. For most of the times, a new solution is found, which allows the robot to navigate in the environment and to return to the battery recharge area only when the battery level is low. When the problem changes, the fitness values of the robot become small, and a new solution is searched.

Hypothesis tests, considering the Student's t-distribution, indicate that the fitness of the best individual after 2000 generations is higher for SORIGA in Experiment 4 with the level of significance equal to 0.04, 0.1, and 0.28 when compared to SGA, RIGA1, and RIGA2 respectively. In Experiment 5,

Table 5.4. Experimental Results - Evolutionary Robotics

Measure	Algorithm	Experiment 4 (environment changing)	Experiment 5 (failure reconfiguration)
Adaptability	SGA	0.3614 (36.14%)	0.6002 (60.02%)
	RIGA1	0.3508 (35.08%)	0.5201 (52.01%)
	RIGA2	0.3129 (31.29%)	0.6268 (62.68%)
	SORIGA	0.3022 (30.22%)	0.5191 (51.91%)
Mean Fitness	SGA	0.2991	0.1861
	RIGA1	0.3406	0.2317
	RIGA2	0.3540	0.1837
	SORIGA	0.3484	0.2301
Fitness of the best individual (end of the simulation)	SGA	0.6380	0.2840
	RIGA1	0.6690	0.3880
	RIGA2	0.7270	0.3590
	SORIGA	0.7550	0.4410

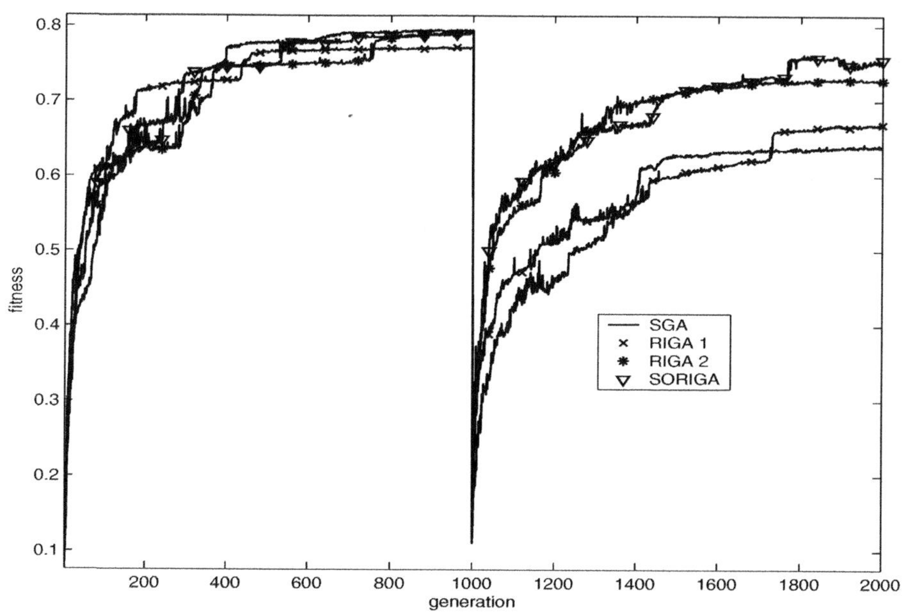

Fig. 5.8. *Averaged fitness of the best individual in Experiment 4 (evolutionary robots)*

the fitness of the best individual after 2000 generations is higher for SORIGA with the level of significance equal to 0.1, 0.34, and 0.25 when respectively compared to SGA, RIGA1, and RIGA2.

Figure 5.8 and Figure 5.9 show the averaged fitness for all GAs in Experiment 4 and Experiment 5 respectively.

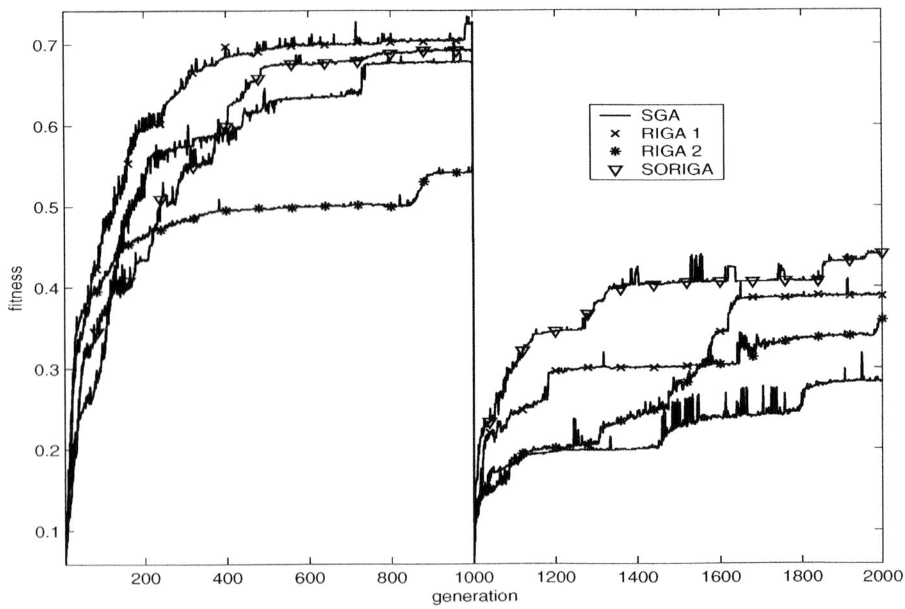

Fig. 5.9. *Averaged fitness of the best individual in Experiment 5 (evolutionary robots)*

5.4.3 Analysis of the Results

In the experiments presented in Section 5.4.1, the values of adaptability for the three GAs with random immigrants are smaller than the values for SGA, i.e., the averaged fitness of the best individuals is higher for the GAs with random immigrants. These results can be explained by the fact that it is difficult for the SGA to escape from the local optima induced by the deceptive problem and by changing the global optima. However, random immigrants inserted in every generation provide diversity to the populations of the last three GAs, which explains their better results.

Let us now analyze the results of the three GAs with random immigrants. First, let us investigate how the proposed SORIGA works. In the beginning of the experiments, the individuals of the initial populations generally have low fitness. In SORIGA, the new individuals that replace the individual with the lowest fitness and its neighbours generally have low fitness too. Since several individuals in the population have low fitness, the probability that one of the neighbours of the current worst individual becomes the new worst is small. As a consequence, a single replacement of an individual generally does not generate large chain reactions of replacements. That is, the distribution of the duration of replacement events is narrow and well described by a small average value. As the number of generations increases, the mean fitness increases too. In this situation, several individuals of the current population have fitness values higher than the average fitness of the new random individuals. Then, the probability that one of the two neighbours of the old worst individual, which were replaced in the last generation, becomes the new worst individual increases. When this new worst individual is replaced with its two next neighbours, a chain reaction can be developed and the replacement events can have, then, a large variety of sizes. In this case, the replacement events can not be characterized by a narrow distribution.

The better results of SORIGA over other RIGA in the experiments presented here can be explained by two major factors. First, the number of different individuals that are replaced in a fixed period of generations is generally large for SORIGA. In the RIGA where the worst individuals are replaced, it is common that new individuals replace individuals with the same index in next generation, because the new individuals usually have small fitness values. In this way, the number of different individuals that are replaced in a fixed period of generations is usually smaller in comparison with SORIGA, and hence the diversity becomes smaller too. This fact can be observed by analyzing the results presented in Table 5.2. SORIGA presents the smallest values of the mean fitness of the population, even though its fitness values of the best individuals are the highest (i.e., its adaptability values are the smallest).

The second major fact that explains the better results of SORIGA is that the survival probability of a new random individual, which can be evolved to become a solution of the problem, is generally smaller in the standard GAs with random immigrants. This happens because the fitness values for the current individuals, whose locations are generally located in (or close to) local maxima after several generations, are generally much higher than the mean fitness of the search space (i.e., the mean fitness of all possible individuals). This occurs because the selection methods employed in GAs preserves, directly or indirectly, the best individuals of the population. An immigrant usually survives during the evolution only if its fitness is close to the mean fitness of the population. This is a rare event when the number of parameters in the solution is high or when the local optimum where the population is found has fitness values much higher than the mean fitness of the search space. On the other hand, SORIGA preserves a new potential solution in a subpopulation and allows it to evolve while the current replacement event is in progress. When the replacement event ends, evolved versions of possible new solutions given by fair immigrants are generally present in the current population and can be combined with the individuals of the main population to generate new solutions.

Figures 5.10 and 5.11 show the mean fitness and the duration of the replacement events (d) in the first trial of Experiments 1a and 1e respectively. It can be seen that when the global optimum changes from a smaller to a higher value, the mean fitness of the population increases. This leads to higher mean duration values of the replacement events and hence increases the diversity of the population. Such interesting behaviour is reached by self-organization, and not by a rule imposed by the programmer. The size of the subpopulation

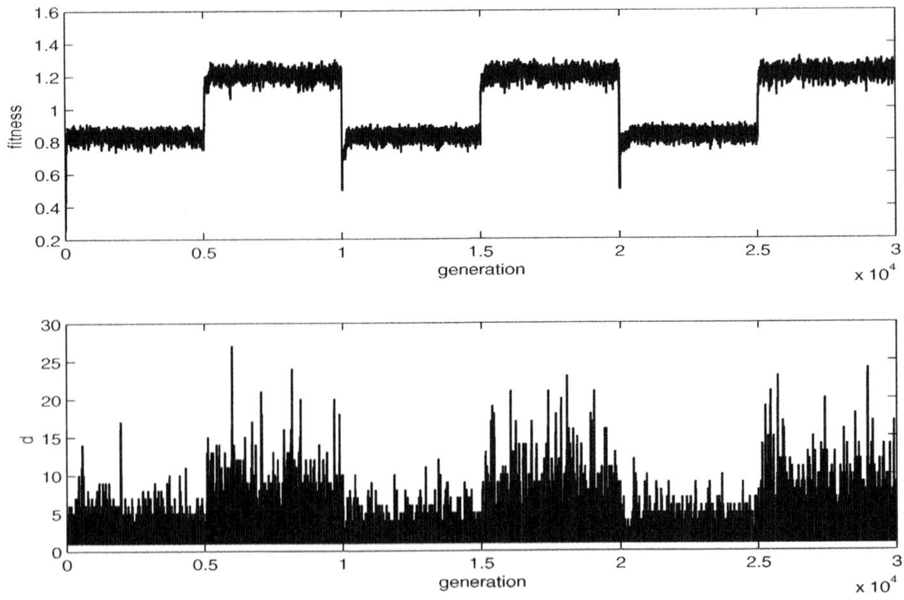

Fig. 5.10. *Mean fitness and duration of replacement events (in generations) in the first trial of Experiment 1a*

self-organizes according to the population diversity level. This fact can be seen by comparing Figures 5.10 and 5.11. In Experiment 1e the mutation rate is higher than in Experiment 1a (see Table 5.1), which results in a higher diversity level. In this way, it is not necessary to generate large replacement events to increase the diversity level. It can be observed that the duration of the larger replacement events is higher in Experiment 1a. This fact explains the worse results of SORIGA when compared to other GAs in Experiment 1e (Table 5.2).

The same analysis can be done to the evolutionary robot experiments (Table 5.4). In experiments 4 and 5, the changes in the problems are so strong that new solutions completely different from the old ones should be found. One can consider, as an example, the experiment where a failure is introduced in the robots (Experiment 5). In the experiments with evolutionary robots, navigation strategies where the robot always moves in the same direction are initially developed. Most of the times, the developed direction of moving is that one that provides more sensing capabilities to the robot, i.e. that one where the front of the robot is the side with more infrared sensors (6). When a failure in the 6 infrared sensors is introduced, the current navigation strategy is no longer interesting. Moving the robot in the developed former direction generally causes collision since it can not detect walls and obstacles in its front side. In this case, a new navigation strategy should be developed, where

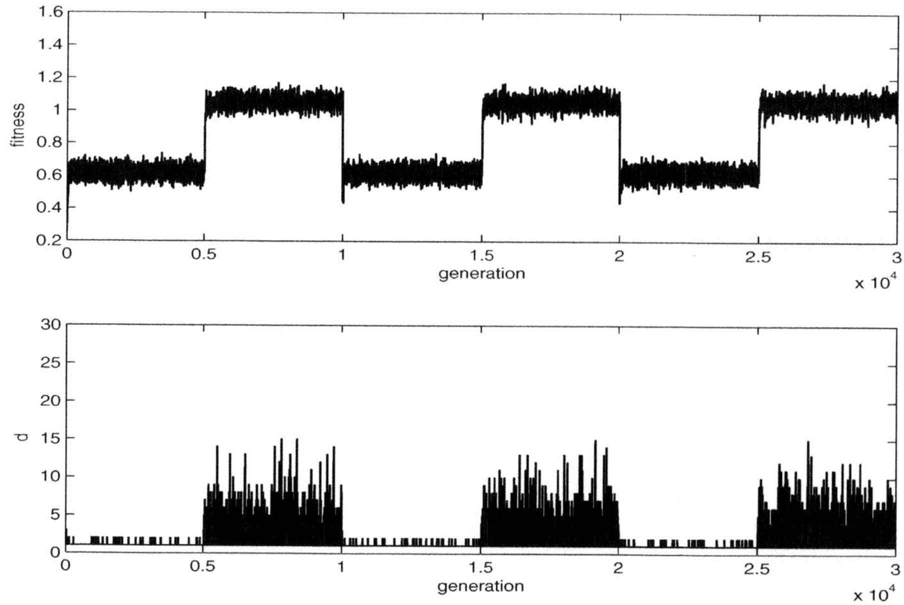

Fig. 5.11. *Mean fitness and duration of replacement events (in generations) in the first trial of Experiment 1e*

the side originally chosen as the rear of the robot, i.e., the side with the two infrared sensors that are working, is now in the front of the robot. Such changes causes a drastic modification to the ANN responsible for the robot control. In the SGA, the probalibity to find a new solution with such characteristics after the introduction of the failure is generally smaller, as the SGA can became trapped in local optima given by the old solution. The old solution is generally better than most new solutions, where the robot generally does not know how to navigate in a straight way.

However, SORIGA can eventually generate new solutions far from the local optima and develop them in the subpopulation. These facts can explain the better results of SORIGA regarding the final fitness of the best individuals presented in Table 5.4. Fig. 5.12 presents the fitness of the best individual and the duration of the replacement events for SORIGA in the tenth trial of this experiment. One can observe that, after the introduction of the failure at generation 1000, the fitness of the best individual becomes low. After some replacement events, a new solution is found (after generation 1800). This solution allows the robot to navigate in the environment and return to the battery recharge area only when the battery level is low, even with the presence of 6 faulty infrared sensors. One can still observe that, like in the results shown in Fig. 5.10, higher mean values for the duration of the replacement events occur for higher fitness values.

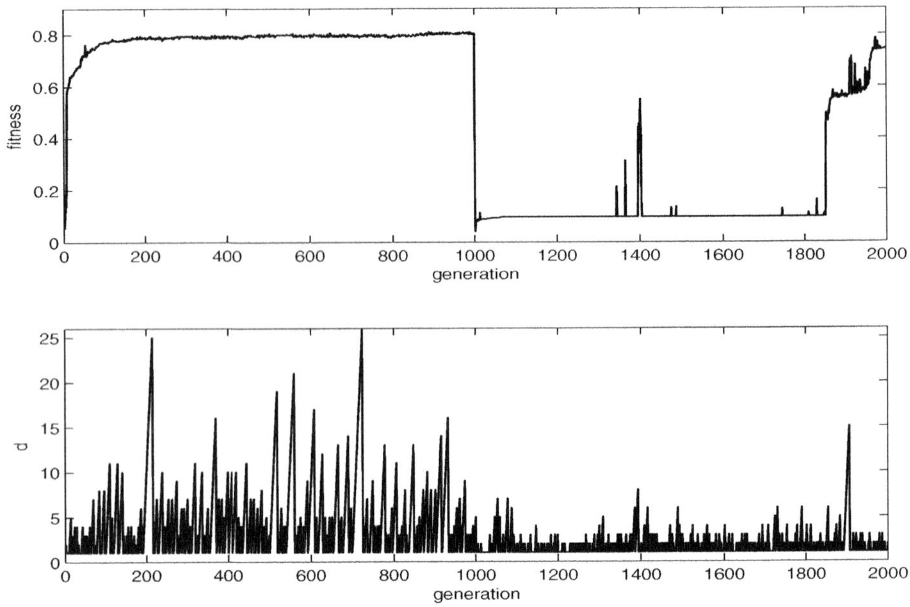

Fig. 5.12. *Fitness of the best individual and duration of replacement events in the tenth trial of the evolutionary robots Experiment 5 (failure reconfiguration)*

In the experiments presented in the last section, like in the fossil recorded data for the replacement events in nature [23], there are more small than large replacement events, and the replacement events occur on a large variety of length scales. In Figures 5.13 and 5.14, the distribution of the number of replacement events against each size is plotted in a log-log scale for trials of the dynamic test environment Experiment 1g and evolutionary robots Experiment 5 respectively. One can observe that the results exhibit power laws (see Section 5.2) even without any apparent tuning, indicating the presence of SOC. This kind of self-organization behaviour arises in systems where many degrees of freedom are interacting and the dynamics of the system is dominated by the interaction between these degrees of freedom, rather than by the intrinsic dynamics of the individual degrees of freedom [15]. In SORIGA, the population self-organizes in order to allow the occurrence of replacement events with a large variety of length scales. Large replacement events generally occur when the mean fitness of the population is high and the diversity level of the population is low. The population diversity is controlled by self-organization, allowing the GA to escape from local optima when the problem changes.

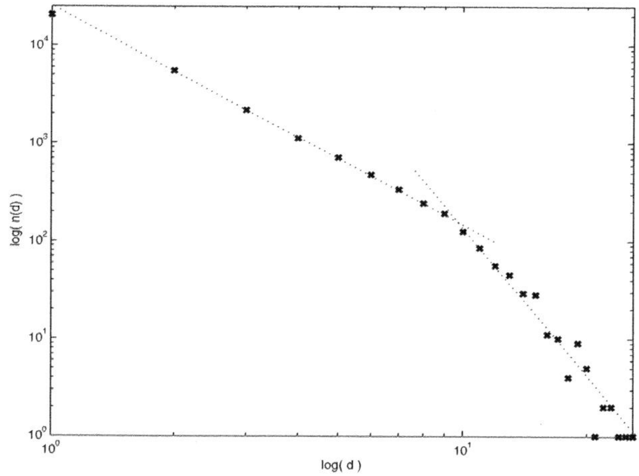

Fig. 5.13. *Number of occurrences for each size of the replacement events in the first trial of Experiment 1g*

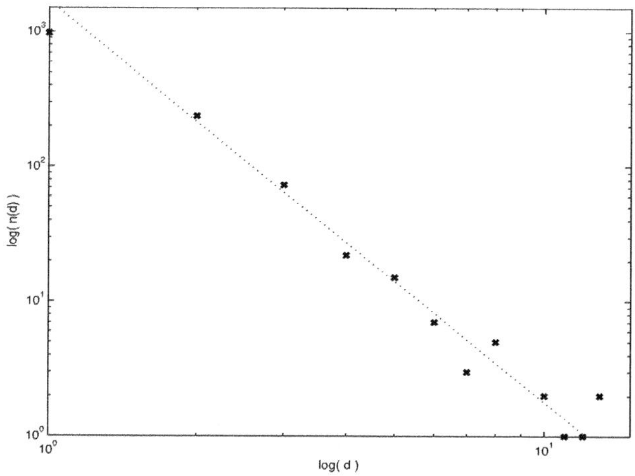

Fig. 5.14. *Number of occurrences for each size of the replacement events in the fourth trial of Experiment 5*

5.5 Conclusions

In this work, a GA with random immigrants where the worst individual and its next neighbors are replaced in every generation is proposed. The new individuals are preserved in a subpopulation, which size is not defined by the programmer, but is given by the number of individuals created in the current replacement event. In SORIGA, the individual starts to interact between themselves and, when the fitness of the individuals are close, as in the case where the diversity level is low, one single replacement can affect a large number of individuals in an replacement event. It is important to observe that this simple approach can take the system to a self-organization behaviour, which can be useful for DOPs to maintain the diversity of the solutions and, then, to allow the GA to escape from local optima when the problem changes. In this way, the proposed GA is interesting for DOPs where the new solution is located in a peak that is hardly reached from the location of the old solution by traditional GA operators.

Studying and combining self-organizing behaviours, such as the self-organized criticality studied in this work, into GAs have shown to be beneficial for their performance under dynamic environments. Further work can be done in this area. A relevant future work is to compare the self-organizing property with other properties, such as the speciation schemes, for GAs under more comprehensive dynamic environments. Another future work is to investigate the use of other neighbouring relations in the proposed algorithm.

Acknowledgments

The authors would like to thank FAPESP (Proc. 04/04289-6) for the financial support.

References

1. P. Bak. *How nature works: the science of self-organized criticality.* Oxford University Press, 1997.
2. P. Bak, C. Tang, and K. Wiesenfeld. Self-organized criticality. an explanation of 1/f noise. *Physical Review Letters*, 59(4):381–384, 1987.
3. S. Boettcher and A. G. Percus. Optimization with extremal dynamics. *Complexity*, 8(2):57–62, 2003.
4. J. Branke. Evolutionary approaches to dynamic optimization problems - introduction and recent trends. In J. Branke, editor, *GECCO Workshop on Evol. Alg. for Dynamic Optimization Problems*, pages 2–4, 2003.
5. W. Cedeno and V. R. Vemuri. On the use of niching for dynamic landscapes. In *Proc. of the 1997 IEEE Int. Conf. on Evolutionary Computation*, pages 361–366, 1997.
6. H. G. Cobb. An investigation into the use of hypermutation as an adaptive operator in genetic algorithms having continuouis, time-dependent nonstationary environments. Technical Report AIC-90-001, Naval Research Laboratory, Washington, USA, 1990.
7. H. G. Cobb and J. J. Grefenstette. Genetic algorithms for tracking changing environments. In S. Forrest, editor, *Proc. of the 5th Int. Conf. on Genetic Algorithms*, pages 523–530, 1993.
8. K. A. De Jong. *An analysis of the behavior of a class of genetic adaptive systems.* PhD Dissertation, University of Michigan, 1975.

9. K. Deb and D. E. Goldberg. Analyzing deception in trap functions. In *Foundation of Genetic Algorithms 2*, pages 93–108, 1993.
10. D. Floreano and F. Mondada. Evolution of homing navigation in a real mobile robot. *IEEE Trans. on Systems, Man, and Cybernetics - Part B: Cybernetics*, 26(3):396–407, 1996.
11. D. A. Goldberg. *Genetic Algorithms in Search, Optimization, and Machine Learning.* Addison-Wesley Publishing Company, Inc., 1989.
12. D. A. Goldberg. *The Design of Innovation: Lessons from and for Competent Genetic Algorithms.* Boston, MA: Kluwer Academic Publishers., 2002.
13. S. J. Gould. *Wonderful Life: The Burgess Shale and the Nature of History.* W. W. Norton and Company, 1989.
14. J. J. Grefenstette. Genetic algorithms for changing environments. In R. Maenner and B. Manderick, editors, *Parallel Problem Solving from Nature 2*, pages 137–144. North Holland, 1992.
15. H. J. Jensen. *Self-organized criticality: emergent complex behavior in physical and biological systems.* Cambridge University Press, 1998.
16. Y. Jin and J. Branke. Evolutionary optimization in uncertain environments - a survey. *IEEE Trans. on Evol. Computation*, 9(3):303–317, 2005.
17. S. A. Kauffman. *The origins of order: self-organization and selection in evolution.* Oxford University Press, 1993.
18. T. Krink and R. Thomsen. Self-organized criticality and mass extinction in evolutionary algorithms. In *Proc. of the 2001 Congress on Evolutionary Computation*, volume 2, pages 1155–1161, 2001.
19. M. Løvbjerg and T. Krink. Extending particle swarm optimisers with self-organized criticality. In *Proc. of the 2002 Congress on Evolutionary Computation*, volume 2, pages 1588–1593, 2002.
20. M. Mitchell. *An Introduction to Genetic Algorithms.* MIT Press, 1996.
21. N. Mori, H. Kita, and Y. Nishikawa. Adaptation to a changing environment by means of the feedback thermodynamical genetic algorithm. In A. E. Eiben, T. Bäck, M. Schoenauer, and H.-P. Schwefel, editors, *Parallel Problem Solving from Nature*, number 1498 in LNCS, pages 149–158. Springer, 1998.
22. S. Nolfi and D. Floreano. *Evolutionary robotics: the biology, intelligence, and technology of self-organizing machines.* MIT Press/Bradford Books: Cambridge, USA, 2000.
23. D. M. Raup. Biological extinction in earth history. *Science*, 231:1528–1533, 1986.
24. K. Trojanowski and Z. Michalewicz. Evolutionary algorithms for non-stationary environments. In M. A. Klopotek and M. Michalewicz, editors, *Intelligent Inf. Systems, Proc. of the 8th Int. Workshop on Intelligent Inf. Syst.*, pages 229–240, 1999.
25. F. Vavak, T. C. Fogarty, and K. Jukes. A genetic algorithm with variable range of local search for tracking changing environments. In H.-M. Voigt, editor, *Parallel Problem Solving from Nature*, number 1141 in LNCS. Springer Verlag Berlin, 1996.
26. S. Yang. Constructing dynamic test environments for genetic algorithms based on problem difficulty. In *Proc. of the 2004 Congress on Evolutionary Computation*, volume 2, pages 1262–1269, 2004.
27. X. Yao. Evolving artificial neural networks. *Proc. of the IEEE*, 87(9):1423–1447, 1999.

6

Learning and Anticipation in Online Dynamic Optimization

Peter A.N. Bosman

Centre for Mathematics and Computer Science (CWI), P.O. Box 94079, 1090 GB Amsterdam, The Netherlands, Peter.Bosman@cwi.nl

Summary. In this chapter we focus on the importance of the use of learning and anticipation in (online) dynamic optimization. To this end we point out an important source of problem–difficulty that has so far received significantly less attention than the traditional shifting of optima. Intuitively put, decisions taken now (i.e. setting the problem variables to certain values) may influence the score that can be obtained in the future. We indicate how such time–linkage can deceive an optimizer and cause it to find a suboptimal solution trajectory. We then propose a means to address time–linkage: predict the future (i.e. anticipation) by learning from the past. We formalize this means in an algorithmic framework and indicate why evolutionary algorithms (EAs) are specifically of interest in this framework. We have performed experiments with two benchmark problems that feature time–linkage. The results show, as a proof of principle, that in the presence of time–linkage EAs based on this framework can obtain better results than classic EAs that do not predict the future.

6.1 Introduction

The majority of the literature on dynamic optimization [11] involves the tracking of optima as the search space transforms over time. If evolutionary algorithms (EAs) [14] are used to achieve this goal, issues such as maintaining diversity around (sub)optima and continuously searching for new regions of interest that may appear over time are the most important [1–4, 7–9, 13, 15, 16, 19, 23, 25, 29]. The shifting of optima in dynamic optimization problems is important to study and to (re)design EAs for. However, there is another feature of dynamic optimization problems that is common in real–world problems such as scheduling [10] and vehicle routing [21, 22, 27] that has received less attention in the literature. We will call this feature *time–linkage*.

Intuitively put, the presence of time–linkage in a dynamic optimization problem causes decisions that are made now, which are often made on the basis of optimizing a certain score right now, to influence the optimal score that can be obtained in the future. This in turn decreases the overall score obtained in the long run. A typical and illustrative example is the case of

Peter A.N. Bosman: *Learning and Anticipation in Online Dynamic Optimization*, Studies in Computational Intelligence (SCI) **51**, 129–152 (2007)
www.springerlink.com

dynamic vehicle routing where the locations to visit are announced over time. If locations are clustered, but the clusters themselves are far apart, routing on the basis of the currently available locations will likely lead to oscillatory behavior of the vehicles if the announced locations oscillate between the clusters. More efficient routes could be formed by keeping vehicles inside clusters and only occasionally letting them move to another cluster. In addition, quality of service (e.g. being on time) as determined by the routing influences future customer demand. Poor performance for a specific customer will likely not result in repeated orders from that customer. Hence, the revenue of a company over time is determined by the current performance, but also by the impact the current way of routing has on future events.

The most important contribution made in this chapter is that we will show that dependencies between decisions over time requires their explicit processing during optimization to ensure the best performance in the long run. Any approach that does not explicitly process these dependencies and instead only solves the problem for the current time will never obtain an optimal result.

In this chapter we also present an algorithmic framework for solving dynamic optimization problems. The algorithmic framework is specifically equipped with the possibility of processing time–linkage. To this end, we propose the incorporation of learning (e.g. statistical [28] or machine [20]) with the explicit task of predicting the future to prevent being deceived over time. An evolutionary approach in which the future is predicted for dynamic optimization has been proposed before [26]. However, the cited approach only predicts the future for a single discrete time step. As a result, the algorithm cannot process longer, arbitrarily sized, time–linkage intervals. Moreover, the approach was only tested on a problem that doesn't contain time–linkage. As a result, no significant difference was observed in using either a good predictor or a bad predictor. In this chapter, we present two new benchmark problems that contain time–linkage and show, as a proof of principle, how they can be solved using an instance of our proposed framework.

It should be noted that it is not our goal in this chapter to propose a new state–of–the–art EA for dynamic optimization. Instead, we want to point out the influence that time–linkage can have and how, in a general manner, EAs can be equipped with tools to cope with time–linkage.

The remainder of this chapter is organized as follows. In Section 6.2 we characterize online dynamic optimization problems. Next, in Section 6.3 we discuss solving these problems. In Section 6.4 we describe our algorithmic framework and in Section 6.5 we present results of running experiments with EAs based on this framework. Finally, possible directions for future research as well as conclusions are presented in Section 6.6.

6.2 Defining Online Dynamic Optimization Problems

In order to formally define the class of online dynamic optimization problems, we first give a general definition of optimization problems. Without loss of generality we assume that the goal is maximization.

Definition 1 (Optimization problem). *An optimization problem is defined as follows:*

$$\max_{\zeta \in \mathbb{P}} \{\mathfrak{F}_\gamma(\zeta)\} \quad (6.1)$$
$$\textbf{s.t. } \mathfrak{C}_\gamma(\zeta) = \textit{feasible}$$

where $\mathfrak{F}_\gamma : \mathbb{P} \to \mathbb{O}$ *is the optimization function,* $\mathbb{P}$ *is the parameter space,* $\mathbb{O} = \mathbb{R}^{n_o}$ *is the* n_o*–dimensional objective space,* $\mathfrak{C}_\gamma : \mathbb{P} \to \{\textit{feasible}, \textit{infeasible}\}$ *is the constraint function and* $\gamma \in \mathbb{G}$ *are problem–specific parameters.*

6.2.1 Static versus Dynamic

An optimization problem can either be dynamic or not. If it is not dynamic, it is said to be static (instead of non–dynamic).

Definition 2 (Static optimization problem). *An optimization problem is said to be static if it cannot be written as a dynamic optimization problem.*

Definition 3 (Dynamic optimization problem). *An optimization problem is said to be dynamic if the optimization function has the specific form of a functional. Moreover, the function space to optimize over consists of functions of a single variable* $t \in \mathbb{T} = [0, t^{\text{end}}]$, $t^{\text{end}} > 0$. *Variable* t *is commonly referred to as time. An optimization problem is said to be dynamic if the optimization function and the constraint function can be written as below in Equation 6.2 and there are no equivalent definitions of* $\mathfrak{F}_\gamma$ *and* $\mathfrak{C}_\gamma$ *that do not involve the time variable.*

$$\mathfrak{F}_\gamma(\zeta) = \int_0^{t^{\text{end}}} \mathfrak{F}^{\text{dyn}}_{\gamma^{\text{dyn}}(t, Z(t,\zeta))} \left(\zeta^{\text{dyn}}_\zeta(t)\right) dt \quad (6.2)$$

$$\mathfrak{C}_\gamma(\zeta) = \begin{cases} \textit{feasible} & \text{if } \forall t \in [0, t_{\text{end}}] : \mathfrak{C}_{\gamma^{\text{dyn}}(t, Z(t,\zeta))} \left(\zeta^{\text{dyn}}_\zeta(t)\right) = \textit{feasible} \\ \textit{infeasible} & \text{otherwise} \end{cases}$$

where $\mathfrak{F}^{\text{dyn}}_{\gamma^{\text{dyn}}(t, Z(t,\zeta))} : \mathbb{P}^{\text{dyn}} \to \mathbb{O}^{\text{dyn}}$ *is the dynamic optimization function,* $\mathbb{P}^{\text{dyn}}$ *is the dynamic parameter space,* $\mathbb{O}^{\text{dyn}} = \mathbb{O} = \mathbb{R}^{n_o}$ *is the* n_o*–dimensional dynamic objective space,* $\zeta^{\text{dyn}}_\zeta : \mathbb{T} \to \mathbb{P}^{\text{dyn}}$ *is the dynamic variable function that for any time* $t \in \mathbb{T}$ *returns the settings for the variables of function* $\mathfrak{F}^{\text{dyn}}$, $\zeta \in \mathbb{P}$ *are the parameters of function* ζ^{dyn}, $Z : \mathbb{T} \times \mathbb{P} \to$

$\{\mathbb{T} \times \mathbb{P}^{\text{dyn}}\}$ *is a function that for any time* $t \in \mathbb{T}$ *and any setting* $\zeta \in \mathbb{P}$ *of the parameters of function* ζ^{dyn} *returns a set that contains the settings of the parameters of function* ζ^{dyn} *for all times before* t*, i.e.* $Z(t,\zeta) = \bigcup_{0 \leq t' < t} \left\{ \left(t', \zeta_\zeta^{\text{dyn}}(t') \right) \right\}$, $\gamma^{\text{dyn}} : (\mathbb{T}, \{(\mathbb{T} \times \mathbb{P}^{\text{dyn}})\}) \to \mathbb{G}^{\text{dyn}}$ *is a function that for any time* $t \in \mathbb{T}$ *and the dynamic variable function restricted to all earlier times* $t' < t$ *returns the problem–specific parameters,* $\mathfrak{C}_{\gamma^{\text{dyn}}(t,Z(t,\zeta))} : \mathbb{P}^{\text{dyn}} \to \{feasible, infeasible\}$ *is the dynamic constraint function and* $t^{\text{end}} > 0$ *is the horizon of the dynamic optimization problem. Moreover, we use the convention that* $\int_a^b (f_0(x), \ldots, f_{n_o-1}(x))\, dx = \left(\int_a^b f_0(x)dx, \ldots, \int_a^b f_{n_o-1}(x)dx \right)$ *Concordantly, the optimization problem to solve can now be written as:*

$$\max_{\zeta \in \mathbb{P}} \left\{ \int_0^{t^{\text{end}}} \mathfrak{F}^{\text{dyn}}_{\gamma^{\text{dyn}}(t,Z(t,\zeta))} \left(\zeta_\zeta^{\text{dyn}}(t) \right) dt \right\} \tag{6.3}$$

$$\textbf{s.t. } \forall t \in [0, t^{\text{end}}] : \mathfrak{C}_{\gamma^{\text{dyn}}(t,Z(t,\zeta))} \left(\zeta_\zeta^{\text{dyn}}(t) \right) = \textit{feasible}$$

In the dynamic optimization problem as defined above it is assumed that time is continuous. However, this does not always have to be the case. If time is not continuous, the integral can be written as a discrete sum and the dynamic optimization problem is said to be discrete.

6.2.2 Offline versus Online

A dynamic optimization problem can either be offline or not. If it is not offline it is said to be online (instead of non–offline).

Definition 4 (Offline dynamic optimization problem). *A dynamic optimization problem is said to be offline if the dynamic optimization function can be evaluated completely.*

From definition 4 we have that in offline dynamic optimization problems the optimal dynamic variable function can be found by constructing complete dynamic variable functions and subsequently evaluating their corresponding optimization values by integrating over all time. The advantage of solving offline dynamic optimization problems is that with the exception of a practical one, there is no time–limit in solving the problem. In a sense, this type of problem is very close to the traditional definition of optimization problems in that we have a problem that must be optimized and we can evaluate the optimization function. The only thing that we know additionally here is that the optimization problem has the specific form of a dynamic optimization problem.

Definition 5 (Online dynamic optimization problem). *A dynamic optimization problem is said to be online if the dynamic optimization function* *cannot be evaluated for all future times $t > t^{now}$ and the dynamic optimization problem must therefore be solved as time goes by.*

Online dynamic optimization is by far the most practical variant of dynamic optimization. It must continually be decided what values to use for the dynamic variables from one moment to the next. The only thing that can be evaluated is how well the algorithm has done so far, which we call the history function.

Definition 6 (History function). *The history function is defined as follows:*

$$\mathfrak{H}^{\text{dyn}}(t^{\text{now}}, \zeta) = \int_{0}^{t^{\text{now}}} \mathfrak{F}^{\text{dyn}}_{\gamma^{\text{dyn}}(t,Z(t,\zeta))}\left(\zeta^{\text{dyn}}_{\zeta}(t)\right) dt \tag{6.4}$$

Like many other optimization problems, online dynamic optimization problems can be constructed artificially. Since the problem is then fully known, there is in design no difference with the offline case. The difference only resides in the fact that the problem is *treated* as being online. Online optimization problems are however typically real–world problems in which the problem–specific parameters actually change over time in an unknown fashion.

6.3 Solving Online Dynamic Optimization Problems

First we identify two important and commonly distinguished sources of problem–difficulty in dynamic optimization problems in Section 6.3.1. In Section 6.3.2 we then discuss solving online dynamic optimization problems by only taking into account the current situation. Finally, in Section 6.3.3 we describe the advantages of solving online dynamic optimization problems by also taking into account future implications of decisions.

6.3.1 System Influence and Control Influence

We distinguish between two types of influence that cause the dynamic optimization problem to change with time: system influence and control influence.

Definition 7 (System influence). *System influence is the type of influence that the problem solver has no control over. It is the part of the dynamic system that changes over time, regardless of choices made for the problem variables. It is the inherent reason why the optimization problem is dynamic and hence why the optimization function parameters $\gamma^{\text{dyn}}(t)$ are a function of time.*

Definition 8 (Control influence). *Control influence is the response of the dynamic system at time t^{now} to the choices for the problem variables made in the past by the problem solver, i.e. the trajectory $\zeta^{\text{dyn}}_{\zeta}(t)$ with $t \in [0, t^{\text{now}})$.*

Most EAs designed for dynamic optimization problems and studied in the literature are specifically designed to cope with system influence. Without taking into account the (possible) presence of control influence however, the online dynamic optimizer risks falling victim to time–deception.

6.3.2 Optimizing the Present: Falling Victim to Time–Deception

The Approach

An often–used approach to solving online dynamic optimization problems is to optimize $\mathfrak{H}^{\text{dyn}}(t^{\text{now}}, \boldsymbol{\zeta})$ continuously or whenever an event resulting from system influence takes place. Since we cannot change the past, we can only vary the settings of the variables at t^{now}. Hence, the optimization problem to solve at time t^{now} using this approach is actually the optimization of the value of the dynamic optimization function at time t^{now}. To cope with a variety of system influences when using EAs, diversity preserving mechanisms are often used to prevent convergence as are other techniques such as detecting (major) changes in the landscape to trigger a restart or forking off multiple sub–populations from a general optimizer to search various parts of the search space more closely as they become more interesting over time.

How Bad Can It Be?

Unfortunately, the answer is *arbitrarily bad*. The most important reason for this is the presence of control influence. Of course system influence could make the problem change in a random way, clearly already making the problem arbitrarily difficult. However, even if system influence is smooth and the landscape is not complex, optimizing only the current situation can lead to arbitrarily bad results due to the presence of time–linkage.

Definition 9 (Time–linkage). *A dynamic optimization problem is said to contain time–linkage if and only if there exists at least one time* $0 \leq t \leq t^{\text{end}}$ *for which the dynamic optimization value at time* t *is dependent on at least one earlier solution* $\boldsymbol{\zeta}_{\boldsymbol{\zeta}}^{\text{dyn}}(t')$, $0 \leq t' < t$.

As an example, consider the following unconstrained l–dimensional dynamic optimization problem; a simple adaptation of the sphere problem that shifts with time:

$$\max_{\boldsymbol{\zeta}_{\boldsymbol{\zeta}}^{\text{dyn}}(t)} \left\{ \int_0^{t^{\text{end}}} \varphi\left(\boldsymbol{\zeta}_{\boldsymbol{\zeta}}^{\text{dyn}}(t), t\right) dt \right\} \tag{6.5}$$

where

$$\varphi\left(\zeta_\zeta^{\text{dyn}}(t),t\right)=\begin{cases}-\sum_{i=0}^{l-1}\left(\zeta_\zeta^{\text{dyn}}(t)_i-t\right)^2 & \text{if } 0\leq t<1\\ -\sum_{i=0}^{l-1}\left(\left(\zeta_\zeta^{\text{dyn}}(t)_i-t\right)^2+\psi\left(\left|\zeta_\zeta^{\text{dyn}}(t-1)_i\right|\right)\right) & \text{otherwise}\end{cases}$$

Now, when optimizing only the present in an online setting, a value for $\zeta_\zeta^{\text{dyn}}(t^{\text{now}})$ is chosen by maximizing $\varphi(\zeta_\zeta^{\text{dyn}},t^{\text{now}})$. For any t, $\varphi(\zeta_\zeta^{\text{dyn}},t)$ is just a hyperparabola with a unique maximum for $\zeta_\zeta^{\text{dyn}}(t)_i = t$. For $0 \leq t < 1$ the associated optimization value is 0 and for $t \geq 1$ this value is $-\sum_{i=0}^{l-1}\psi\left(\left|\zeta_\zeta^{\text{dyn}}(t-1)_i\right|\right)$. It is this construction that deceives an approach in which only the present is optimized because then the actual value of function $\psi(\cdot)$ is not taken into account although it may decrease at an arbitrary rate, depending on its form. However, if the simple choice of $\zeta_\zeta^{\text{dyn}}(t)_i = 0$ is always made, then, assuming that $\psi(0) = 0$, the optimization value that is reached is $l\int_0^{t^{\text{end}}}-t^2dt = -\frac{l}{3}\left(t^{\text{end}}\right)^3$, regardless of function $\psi(\cdot)$.

Now if for instance $\psi(x) = x^2$, the result is better if only the present is optimized than if just $\zeta_\zeta^{\text{dyn}}(t)_i = 0$ is chosen. Although a better result can still be obtained because $\zeta_\zeta^{\text{dyn}}(t)_i = 0$ is not the optimal solution, the penalty of disregarding time–linkage is only small. But if $\psi(\cdot)$ is a higher–order increasing function, such as $\psi(x) = e^x - 1$, a (much) worse optimization value will be obtained and the price to pay for not taking into account time–linkage is (much) higher. A graphical illustration of the difference in obtained optimization values for the case of $l = 1$ is given in Figure 6.1.

Since in the online case the behavior of the optimization function in the future is not known, optimizing only the present can thus significantly reduce overall solution quality. Hence, optimizing only the present is not a good approach unless the problem provably does not contain time–linkage. Otherwise, the problem is time–deceptive for this approach to solving online dynamic optimization problems.

Definition 10 (Time–deception). *A dynamic optimization problem is said to be time–deceptive for an optimizer if the problem contains time–linkage and the optimizer has no means to efficiently take this time–linkage into account during optimization and therefore cannot find the optimal solution trajectory.*

6.3.3 Optimizing the Present and the Future: Learning to Avoid Time–Deception

The Approach

The approach of optimizing only the present is deceived over time because the true problem definition (i.e. Equation 6.2) is not used. Future changes that occur as a result of decisions made earlier are neglected. To remedy this, optimization over future choices is required. In the online case however, an evaluable future is absent. Hence, the only option is to predict the future.

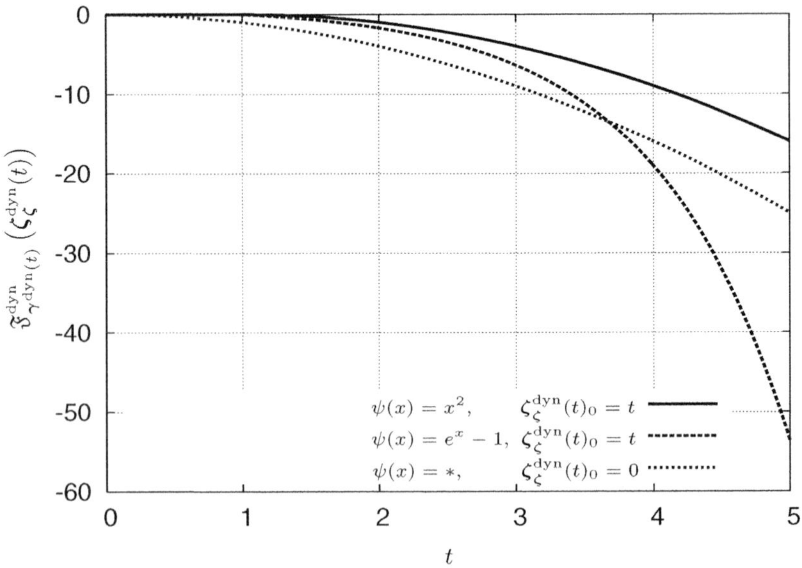

Fig. 6.1. Illustration of the optimization values obtained for different variable trajectories and different forms of $\psi(\cdot)$ in the case of $l = 1$ and $t^{\text{end}} = 5$. $\psi(x) = *$ is any function such that $\psi(0) = 0$.

The better the prediction, the closer the algorithm can get to optimality. The available information to base the prediction upon besides problem–specific information is information that was collected in the past.

The optimization problem to be solved using this approach at time t^{now} is based on an approximation of the value of the dynamic optimization function over a future time span of length t^{plen}:

$$\max_{\zeta^{\text{dyn}}_{\zeta}(t)} \left\{ \int_{t^{\text{now}}}^{\min\{t^{\text{now}}+t^{\text{plen}},t^{\text{end}}\}} \hat{\mathfrak{F}}^{\text{dyn}}_{\boldsymbol{\alpha}}\left(t, \zeta^{\text{dyn}}_{\zeta}(t)\right) dt \right\} \tag{6.6}$$

$$\text{s.t. } \forall t \in \left[t^{\text{now}}, \min\left\{t^{\text{now}}+t^{\text{plen}}, t^{\text{end}}\right\}\right] : \hat{\mathfrak{C}}^{\text{dyn}}_{\boldsymbol{\alpha}}\left(t, \zeta^{\text{dyn}}_{\zeta}(t)\right) = \textit{feasible}$$

where

$$\begin{cases} \hat{\mathfrak{F}}^{\text{dyn}}_{\boldsymbol{\alpha}}\left(t^{\text{now}}, \zeta^{\text{dyn}}_{\zeta}(t^{\text{now}})\right) = \mathfrak{F}^{\text{dyn}}_{\gamma^{\text{dyn}}(t^{\text{now}}, Z(t^{\text{now}}, \zeta))}\left(\zeta^{\text{dyn}}_{\zeta}(t^{\text{now}})\right) \\ \hat{\mathfrak{C}}^{\text{dyn}}_{\boldsymbol{\alpha}}\left(t^{\text{now}}, \zeta^{\text{dyn}}_{\zeta}(t^{\text{now}})\right) = \mathfrak{C}^{\text{dyn}}_{\gamma^{\text{dyn}}(t^{\text{now}}, Z(t^{\text{now}}, \zeta))}\left(\zeta^{\text{dyn}}_{\zeta}(t^{\text{now}})\right) \end{cases}$$

Prediction in the complete BBO case

The complete BBO (Black–Box Optimization) case is the most general case. No prior knowledge is assumed on the problem to be solved other than the number of variables and their types. Additional knowledge can only be gained by evaluating solutions. Since nothing is known about the optimization function, only a very general form of induction can be performed to predict future function values.

We assume that the number of variables and their semantics do not change. To predict the (expected) value of the dynamic optimization function, an approximation based on previously evaluated solutions can be used. Computing this approximation is a (statistical) learning problem. The available data in the learning problem is:

$$\bigcup_{i=0}^{n_{\text{data}}-1} \left\{ \left(\left(t^i, \boldsymbol{\zeta}_{\boldsymbol{\zeta}}^{\text{dyn}}(t^i), Z(t^i, \boldsymbol{\zeta}) \right), \boldsymbol{y}^i \right) \right\} \tag{6.7}$$

where on the input–side of the pattern we have the time t^i, $0 < t^i \leq t^{\text{now}}$, the value of the variables $\boldsymbol{\zeta}_{\boldsymbol{\zeta}}^{\text{dyn}}(t^i)$ at time t^i and the history of the variable–value–trajectory $Z(t^i, \boldsymbol{\zeta})$ up to time t^i and on the output–side of the pattern we have the value of the dynamic optimization function $\boldsymbol{y}^i$ for solution $\boldsymbol{\zeta}^{\text{dyn},i}$ at time t^i. Note that the use of an EA can greatly add to the availability of data and can hence increase the accuracy of the predictions because a (diverse) population is used. Each population member can serve as a pattern. Note that the integration of the history of the solution–trajectory into the data set is essential for the processing of time–linkage.

The goal of learning is to estimate the value of the dynamic optimization function for future times (assuming that the constraint function does not need to be estimated) by minimizing the generalization error over the time span that contains the data to learn from. In the single–objective case the learning problem can be formalized as follows:

$$\min_{\boldsymbol{\alpha} \in \mathbb{A}} \left\{ \int_{t^{\min}}^{t^{\max}} \left(\mathfrak{F}_{\boldsymbol{\gamma}^{\text{dyn}}(t, Z(t,\boldsymbol{\zeta}))}^{\text{dyn}} \left(\boldsymbol{\zeta}_{\boldsymbol{\zeta}}^{\text{dyn}}(t) \right) - \hat{\mathfrak{F}}_{\boldsymbol{\alpha}}^{\text{dyn}} \left(\boldsymbol{\zeta}_{\boldsymbol{\zeta}}^{\text{dyn}}(t) \right) \right)^2 dt \right\} \tag{6.8}$$

where

$$\begin{cases} t^{\min} &= \min_{i \in \{0,1,\ldots,n_{\text{train}}-1\}} \{t^i\} \\ t^{\max} &= \max_{i \in \{0,1,\ldots,n_{\text{train}}-1\}} \{t^i\} \end{cases}$$

and $\boldsymbol{\alpha} \in \mathbb{A}$ are the parameters of the function class from which to choose the approximated optimization function.

Prediction in the partial BBO case

In the presence of problem–specific information, the learning task may be less involved which may improve the reliability of the predictions. A typical case is when the function can be evaluated for any $0 \leq t < t^{\text{end}}$, as long as the required parameters are set. Then, if we are able to predict the values for the parameters accurately, we automatically get an accurate function evaluation. The less parameters to be estimated, the better the hope is of obtaining good approximations.

Prediction length, prediction base and history length

In addition to the choice of approximation class and learning mechanism, there are two key issues of importance in using predictions in dynamic optimization:

1. **How far into the future should predictions be made?**
   (we call this prediction length, denoted t^{plen})
2. **From how far into the past should information be used to base the prediction upon?**
   This question is actually twofold:
   a) **How far into the past should t^i go in Equation 6.7?**
   (we call this prediction base, denoted t^{pbase})
   b) **For each pattern, how far into the past, starting from the time of that pattern, should we take into account the history of the variable trajectory?**
   (we call this history length, denoted t^{hlen})

A proper choice for the prediction length and the history length depends on the time–linkage time span, i.e. how far into the future do current choices have a significant influence? Certainly this is also the minimal choice for the prediction base. However, larger values for prediction length, prediction base and history length may be required to look beyond the linkage and observe the general dynamics of the optimization problem.

Another issue that influences the proper choice for the prediction length is the reliability of predictions. As predictions are made further into the future, they are bound to become less reliable, giving a trade–off between the required prediction length as a result of time–linkage and the feasible prediction length as a result of reliability issues.

How Good Can It Be?

Fortunately, the answer is *arbitrarily good*. However, although it is intuitively clear that the optimum is attainable, this does require *perfect* predictions. Then, the problem can be solved to optimality by optimizing at any time the integral over the predictions with $t^{\text{plen}} = t^{\text{end}} - t^{\text{now}}$.

The strength of the optimization method (with respect to the problem at hand) is still a key component to success. However, the success of the approach now also heavily depends on the strength of the prediction method. Bad predictions may even lead to worse results than are obtained by optimizing only the present. Hence, careful design and performance assessment of methods that predict the future are certainly called for. In the following section we present a general framework for solving dynamic optimization problems by incorporating learning techniques as described above.

6.4 An Algorithmic Framework

6.4.1 Components

Solver

The solver, denoted S, is an optimization algorithm, possibly equipped with tools to allow for adaptability as time goes by. The function to be optimized is provided by the function component.

Predictor

The predictor, denoted P, is a learning algorithm that approximates either the optimization function directly or several of its parameters. The data set from which to estimate a function is provided by the database component. When called upon, the predictor returns either the predicted function value directly or predicted values for parameters.

Function

The function, denoted F, is the optimization function to be maximized by the solver. If the future is not to be taken into account, this function is just the dynamic optimization function. Otherwise, the dynamic optimization function is used to compute the optimization value that pertains to the current time and the predictor is used to predict future optimization values. The trajectory of the variables in the predicted future is to be set by the solver. To specify this trajectory a dynamic variable function is needed that can supply a solution for each possible time between t^{now} and $\min\{t^{\text{now}} + t^{\text{plen}}, t^{\text{end}}\}$. It is computationally convenient to divide the trajectory–future interval as well as the trajectory–history interval into non–zero sub–intervals of length t^{pint} and t^{hint} respectively. The optimization value then is a discrete approximation of the integral over the future interval where future predictions are in addition to other data based on a discretized past trajectory. If the dynamic optimization function is not stochastic, a list of solutions can be used such that there is one solution for each sub–interval. If the dynamic optimization function is stochastic however, a predefined future list of actions may not be optimal. The reason for this is that different actions may be optimal in different situations. Because the future is stochastic, the actual future situation is unknown. Hence, a responsive strategy is then required instead to be able to obtain optimality. The solution to optimize by the solver then is thus not a list of assignments to the problem variables, but a strategy in the form of a function of time that returns a solution conditioned on the actual future situation. Note that the list of assignments in the non–stochastic case is a specific form of the strategy.

Database

The database, denoted D, is a collection of patterns upon which the predictor bases its predictions. Patterns are added either by the function component (i.e. in the complete BBO case whenever a new solution is evaluated) or by the system whenever an event occurs that is related to the parameters of interest (i.e. in the partial BBO case). All patterns are time–stamped. The database only contains patterns with a time stamp t such that $t^{\text{now}} - t^{\text{pbase}} \leq t \leq t^{\text{now}}$.

Timer

The timer, denoted T, can provide the current time t^{now}.

6.4.2 Dividing Resources

Clearly, optimization becomes more involved if we also want to take into account predictions of the future. Not only does the number of variables to optimize over increase (at least if we regard the complete BBO case), but also additional computation time is required for learning to make predictions.

It is important to note that there is a trade-off between how much time should be spent on running the solver and how much time should be spent on running the predictor. We propose to implement the solver and the predictor components as threads. This allows both for a scheme in which the solution component and the predictor component run simultaneously as well as a scheme where the predictor and solver are run sequentially by synchronization using signals. An example of the second scheme is given by the scenario in which the solver sends a signal when a certain number of generations have passed and then awaits completion of the learning task before continuing.

6.4.3 Definition

To complete the framework, we provide an algorithmic description of how the components are used together to solve online dynamic optimization problems. First, the trajectory is made empty and all the components are initialized. Then, the solver and the predictor are started and the actual optimization begins. Although the solver may store a solution into the trajectory at any time (e.g. at the end of each generation for an EA), we want to ensure that at least a few solutions are stored in the trajectory. To this end, the solver is requested for a solution at regular intervals of length t^{sint}. These requests are issued until t^{end} is reached. Then, the solver and the predictor are halted and the resulting trajectory is returned. Pseudo–code for the framework is given in Figure 6.2.

An important part of the framework that is of a specific form is the way in which a solution in the form of a future trajectory is evaluated. It is here that the prediction component can influence the way in which the solver searches

FRAMEWORK$(S, P, F, D, T, t^{\text{end}}, t^{\text{pbase}}, t^{\text{plen}}, t^{\text{sint}}, t^{\text{pint}})$

1 $Z \leftarrow ()$
2 S.INITIALIZE$(S, P, F, \ldots, t^{\text{pint}})$
3 P.INITIALIZE$(S, P, F, \ldots, t^{\text{pint}})$
4 F.INITIALIZE$(S, P, F, \ldots, t^{\text{pint}})$
5 D.INITIALIZE$(S, P, F, \ldots, t^{\text{pint}})$
6 T.INITIALIZE$(S, P, F, \ldots, t^{\text{pint}})$
7 S.START()
8 P.START()
9 **do**
  9.1 $t^{\text{now}} \leftarrow T$.GETTIME()
  9.2 $\zeta \leftarrow S$.REQUESTSOLUTION()
  9.3 $\boldsymbol{Z} \leftarrow \boldsymbol{Z} \sqcup ((\zeta, t^{\text{now}}))$
  9.4 $t^{\text{next}} \leftarrow \min\{t^{\text{next}} + t^{\text{sint}}, t^{\text{end}}\}$
  9.5 AWAITTIME(t^{next})
  while $t^{\text{now}} \leq t^{\text{end}}$
10 S.STOP()
11 P.STOP()
12 **return**$(\boldsymbol{Z})$

Fig. 6.2. Pseudo–code for the algorithmic framework

for a solution at time t^{now} because the predictor is used to evaluate all parts of the trajectory that pertain to future times. Pseudo–code for this specific part of the framework is given in Figure 6.3.

F.EVALUATE$(\zeta_\zeta^{\text{dyn}})$

1 $t^{\text{now}} \leftarrow T$.GETTIME()
2 $\boldsymbol{y} \leftarrow t^{\text{pint}} \mathfrak{F}^{\text{dyn}}_{\gamma^{\text{dyn}}(t^{\text{now}}, Z(t^{\text{now}}, \zeta))}(\zeta_\zeta^{\text{dyn}}(t^{\text{now}}))$
3 **if** COMPLETEBBOCASE() **then**
  3.1 D.ADDPATTERN$(((\zeta_\zeta^{\text{dyn}}(t^{\text{now}}), t^{\text{now}}), \boldsymbol{y}))$
  3.2 **for** $i \leftarrow 1$ **to** $\lceil t^{\text{plen}} / t^{\text{pint}} \rceil - 1$ **do**
    3.2.1 $\boldsymbol{y} \leftarrow \boldsymbol{y} + t^{\text{pint}} P$.PREDICT$(\zeta_\zeta^{\text{dyn}}, t^{\text{now}} + i \cdot t^{\text{pint}})$
4 **else**
  4.1 **for** $i \leftarrow 1$ **to** $\lceil t^{\text{plen}} / t^{\text{pint}} \rceil - 1$ **do**
    4.1.1 $\gamma^{\text{predicted}} \leftarrow P$.PREDICT$(\zeta_\zeta^{\text{dyn}}, t^{\text{now}} + i \cdot t^{\text{pint}})$
    4.1.2 $\boldsymbol{y} \leftarrow \boldsymbol{y} + t^{\text{pint}} \mathfrak{F}^{\text{dyn}}_{\gamma^{\text{predicted}}}(\zeta_\zeta^{\text{dyn}}(t^{\text{now}} + i \cdot t^{\text{pint}}))$
5 **return**$(\boldsymbol{y})$

Fig. 6.3. Pseudo–code for the evaluation of future trajectories

6.5 Experiments

6.5.1 EA

The optimization problems that we perform experiments with are real–valued, but at any point in time not very daunting because the most important thing we focus on in this chapter is time–linkage. Therefore, we opt for a simple and fast real–valued EA. We use an EDA (Estimation–of–Distribution Algorithm) for real–valued optimization [5, 18] without learning dependencies between problem variables. The main difference with traditional EAs is that in EDAs a probabilistic model is learned using the selected solutions. The probabilistic model can capture various properties of the optimization problem. By drawing new solutions from the probabilistic model these properties can be exploited to obtain more efficient optimization.

In this chapter we performed experiments with a real–valued EDA based on the normal distribution in which each variable is taken to be independent of all the other variables. Such an EDA is also known as the naive I𝔻𝔼A (Iterated Density–Estimation Evolutionary Algorithm) [6]. In the naive variant the mean and standard deviation of a one–dimensional normal distribution are estimated from the selected solutions for each variable separately. A new solution is constructed by sampling one value per variable from the associated one–dimensional normal distribution. Since the optimization problem is dynamic, we prevented total premature convergence by bounding the estimated variance for each variable to a minimum of 0.1. Finally, all results were averaged over 100 independent runs.

6.5.2 BBO: Time–Linkage Numerical Problem

The Problem

We first investigate the real–valued time–linkage problem introduced in Section 6.3.2, Equation 6.5. We regard two variants by setting $\psi(x) = x^2$ and $\psi(x) = e^x - 1$. Moreover, we have used a dimensionality of $l = 1$.

Instantiating the Framework

In this problem the goal of the predictor is to predict the value of the optimization function directly. For clarity we point out that the predicted functions are estimated from a set of patterns with timestamps at most t^{pbase} time ago. The patterns in the data set contain the actually chosen trajectory in the past up to time $t^i - t^{\text{hlen}}$ in steps of t^{hint}, i.e. if $t^{\text{hlen}} = t^{\text{hint}} = 1$, a pattern is given by $((t^i, \boldsymbol{\zeta}_{\zeta}^{\text{dyn}}(t^i), \boldsymbol{\zeta}_{\zeta}^{\text{dyn}}(t^i - 1)), \boldsymbol{y}^i)$. We used three different predictor instances.

1. **Optimal**
   Returns the true value of the dynamic optimization function.
2. **Linear estimator**
   Returns the value of a linear function that was estimated using the least–squares technique.
3. **Quadratic estimator**
   Returns the value of a quadratic function that was estimated using the least–squares technique.

For the case of $\psi(x) = x^2$ only the function class used by the quadratic estimator contains the target function. Hence, effective future predictions are possible with proper estimations in this case, which should ensure the prevention of time–deception. For the case of $\psi(x) = e^x - 1$, neither of the estimators can represent the target function. However, the quadratic estimator should be capable of far better approximations than the linear estimator.

Since we present a proof of principle, we do not investigate the selection of t^{plen} during optimization. Instead, we fix it to either 0, corresponding to the traditional approach of not looking into the future, or to the optimal value of 1. Moreover, we set $t^{\text{hlen}} = t^{\text{hint}} = t^{\text{pbase}} = t^{\text{pint}} = t^{\text{plen}}$.

Results

A population size of 25 was experimentally found to be adequate for solving the optimization problem in each time step. We set $t^{\text{end}} = 10$ and advanced time by a time step of 0.001 every generation. Since the database contains all patterns over a time span of length 1 and the time steps are of size 0.001, the size of the database can become quite large. Although this allows for a higher precision of estimations, it also results in large time requirements for the learning task. Learning was performed after a predefined number of generations. To investigate the impact on the overall quality of optimization, we performed experiments with various values for the number of generations between learning phases: $1, 10, 100$ and 1000.

The average trajectories obtained for the quadratic and the exponential time–linkage numerical problem are shown in Figures 6.4 and 6.5 respectively. The overall results (i.e. the integral over $t \in [0, 10]$) are tabulated in Table 6.1.

Theoretically, under the assumption that the length of the time–linkage is known, the optimal trajectory can be obtained if the target function is in the function class used by the learner and the learner is competent in that it will indeed find that target when learning. In the case of the quadratic time–linkage numerical problem this is experimentally verified by the results. The use of the quadratic estimator leads to results that are very close to optimality (i.e. when the future is known). The discrepancy is explained by the startup time the learner needs before being able to construct a model based upon previously encountered data. Moreover, results improve if learning is performed more frequently because the model is then constructed earlier.

In the case of the exponential time–linkage numerical problem, neither the linear nor the quadratic estimator have a function class that contains the target exponential function. However, for the time–linkage in this problem that depends only on a single point in the past over a distance of 1, a quadratic function can quite closely approximate an exponential function. For this reason the use of the quadratic estimator leads to good results here as well, albeit not optimal. Small deviations from the optimal trajectory as a result of a small learner error can indeed be seen in Figure 6.5. The linear estimator

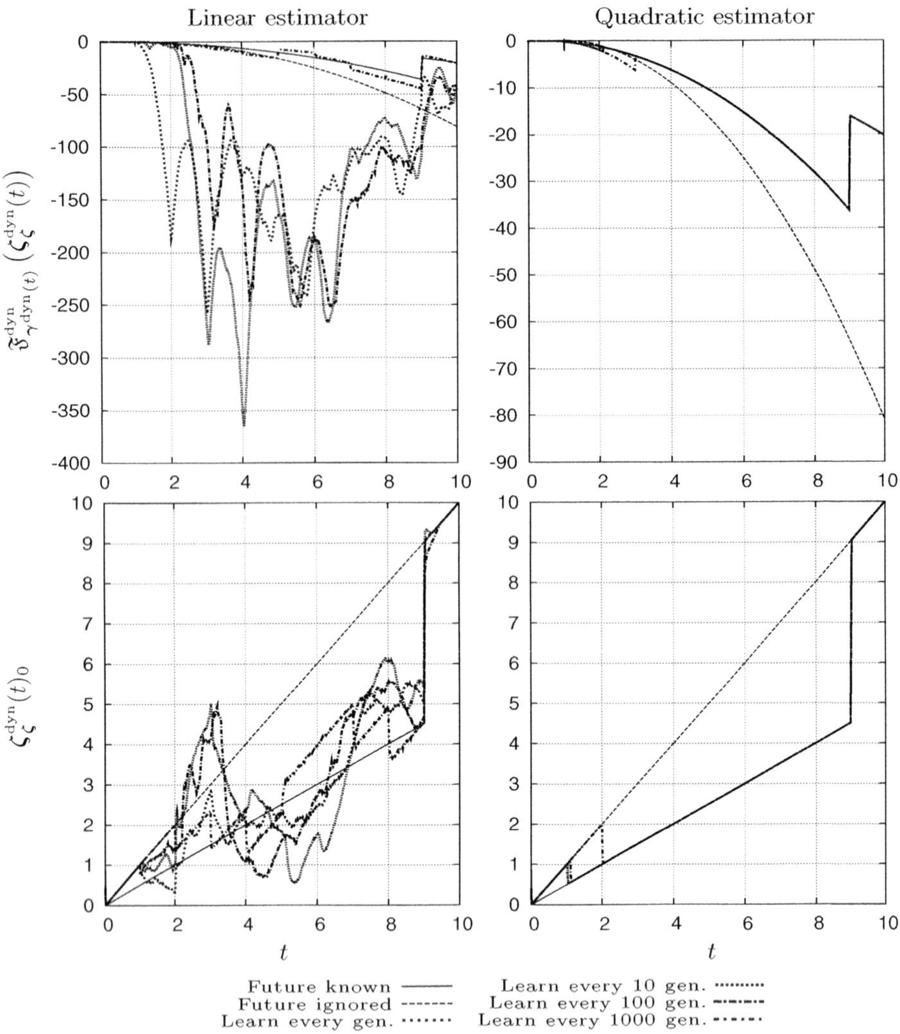

Fig. 6.4. Results averaged over 100 runs on the time–linkage numerical problem with $\psi(x) = x^2$

is not capable of approximating a quadratic function well. For the exponential function, linear estimation is even worse. The use of the linear estimator therefore leads to far worse results. An even more important point to note is that the results using the linear estimator can be even worse than when prediction is not used because of the large errors in the predictions. Hence, another important issue in using learning for online dynamic optimization is the assessment of the reliability of predictions and the use of predictions only if this reliability is large enough.

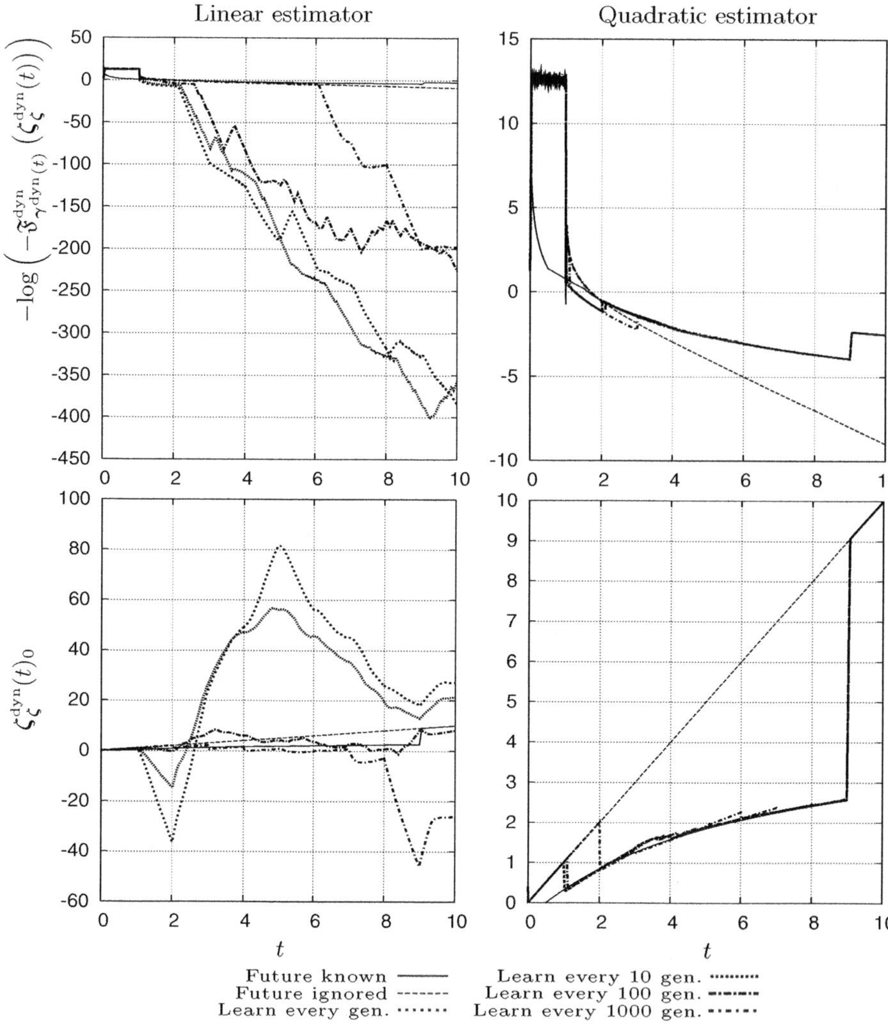

Fig. 6.5. Results averaged over 100 runs on the time–linkage numerical problem with $\psi(x) = e^x - 1$

The results lead to the expected conclusion that competent learners are called for and that reliability of predictions is a major issue. The competence of the learner in the BBO case depends on general/overall competence which is very hard to obtain. In the problem–specific case however, achieving learner competence may be easier because the shape of the model to be learned (i.e. parametric learning) is known from domain knowledge, ensuring that the target function is in the function class used by the learner.

		$\psi(x) = x^2$	$\psi(x) = e^x - 1$
	Future known	$-1.21846 \cdot 10^2$	$-1.55430 \cdot 10^2$
	Future ignored	$-2.42940 \cdot 10^2$	$-8.08692 \cdot 10^3$
Linear	**Learn every gen.**	$-1.09434 \cdot 10^3$	$-2.04922 \cdot 10^{65}$
Linear	**Learn every 10 gen.**	$-1.18946 \cdot 10^3$	$-7.93853 \cdot 10^{172}$
Linear	**Learn every 100 gen.**	$-9.98665 \cdot 10^2$	$-3.02553 \cdot 10^{96}$
Linear	**Learn every 1000 gen.**	$-1.38907 \cdot 10^2$	$-1.22634 \cdot 10^{86}$
Quadratic	**Learn every gen.**	$\mathbf{-1.22010 \cdot 10^2}$	$\mathbf{-1.55966 \cdot 10^2}$
Quadratic	**Learn every 10 gen.**	$-1.22013 \cdot 10^2$	$-1.55969 \cdot 10^2$
Quadratic	**Learn every 100 gen.**	$-1.22062 \cdot 10^2$	$-1.56069 \cdot 10^2$
Quadratic	**Learn every 1000 gen.**	$-1.23178 \cdot 10^2$	$-1.58092 \cdot 10^2$

Table 6.1. Overall results (i.e. $\int_0^{t^{\text{end}}} \mathfrak{F}^{\text{dyn}}_{\gamma^{\text{dyn}}(t,Z(t,\zeta))}(\zeta^{\text{dyn}}_{\zeta}(t)))$ on the time–linkage numerical problem

6.5.3 Partial BBO: Dynamic Pick–up Problem

The Problem

The second problem that we investigate is a discrete partial BBO problem. Although it is based on a simple model, the time–linkage in the problem is large: any decision made now influences the result of the dynamic optimization function for all future time steps. The intuitive description is that at time step t a truck is located at $\boldsymbol{x}^{\text{truck}}(t)$ and a package appears at location $\boldsymbol{x}^{\text{package}}(t)$. It must now be decided whether to send the truck to go and pick up the package or to drive elsewhere. If the package is not picked up, it disappears. Picking up the package pays a value of 1, but driving costs a value equal to the Euclidean distance traveled. The number of packages is $n_{\text{packages}} = t^{\text{end}} + 1$, i.e. the time steps are of size 1. A solution at time t now is a tuple $\boldsymbol{\zeta}^{\text{dyn}}_{\zeta}(t) = (b(t), \boldsymbol{x}^{\text{alternative}}(t))$ where $b \in \{0, 1\}$ indicates whether the package at time t should be picked up ($b(t) = 1$) and $\boldsymbol{x}^{\text{alternative}}(t)$ is the location to drive to if the package is not to be picked up ($b(t) = 0$). Mathematically:

$$\mathfrak{F}^{\text{dyn}}_{\gamma^{\text{dyn}}(t)}\left(\left(b(t), \boldsymbol{x}^{\text{alternative}}(t)\right)\right) = \begin{cases} 1 - \| \boldsymbol{x}^{\text{package}}(t) - \boldsymbol{x}^{\text{truck}}(t) \| & \text{if } b(t) = 1 \\ 0 - \| \boldsymbol{x}^{\text{alternative}}(t) - \boldsymbol{x}^{\text{truck}}(t) \| & \text{otherwise} \end{cases} \tag{6.9}$$

where

$$\boldsymbol{x}^{\text{truck}}(t) = \begin{cases} \sim \prod_{i=0}^{l-1} \mathcal{N}(0,1) & \text{if } t = 0 \\ \boldsymbol{x}^{\text{package}}(t-1) & \text{if } t = 1 \text{ and } b(t-1) = 1 \\ \boldsymbol{x}^{\text{alternative}}(t-1) & \text{otherwise} \end{cases}$$

For simplicity, the model used to generate new package locations is a univariately factorized normal distribution with zero mean and unit variance, i.e. $\boldsymbol{x}^{\text{package}}(t) \sim \prod_{i=0}^{l-1} \mathcal{N}(0,1)$.

Instantiating the Framework

A simple strategy is given by a hillclimber. The decision taken at each time step is to move only to pick up a package and moreover only to do so if the distance to the package is less than 1. In other words, a negative score is never accepted.

We have compared the hillclimber with an EA instance of the general framework. Since the optimization function is completely known with the exception of $\boldsymbol{x}^{\text{package}}(t)$, the problem is only partially a BBO problem. Hence, we can restrict the prediction task to predicting future values for $\boldsymbol{x}^{\text{package}}(t)$. We have performed experiments where we assumed the distribution of $\boldsymbol{x}^{\text{package}}(t)$ to be known and where we estimated this distribution from data, assuming only that the data is indeed normally distributed.

In theory, the influence of any decision at time t influences the outcome of the dynamic optimization function at any time $t' > t$. However, the larger $t' - t$, the smaller the remaining impact on the situation at time t'. Although in theory it would be optimal to set t^{plen} to t^{end} with $t^{\text{pint}} = 1$, such a choice gives rise to two practical problems. First, a large t^{plen} gives extremely large trajectories to optimize. This drastically increases the resources required by the EA to solve the problem. Second, since the future is stochastic in this problem, a proper estimation of the expected future profits requires averaging evaluations over multiple calls. Moreover, the variability of these outcomes increases as t^{plen} increases because more uncertainty is introduced. Hence, unless an infinite number of calls is used, a smaller value for t^{tplen} is expected to be optimal in practice.

Because the problem is stochastic, we require the solution in the solver component to be a dynamic variable function. For the problem at hand this means that we need to evolve a decision strategy for where to move the truck, given a certain (predicted) situation. The strategy we choose here is a simple one. For the current situation, a decision is directly subject to evolution. For future, predicted, situations up to a time span of t^{plen}, the hillclimber strategy is used. Hence, the EA only provides a solution for the current time. Certainly, better results may be obtained by allowing the EA to evolve a more elaborate strategy. However, using the hillclimber already gives an impression of the quality of a certain starting point. This information, albeit an approximation of what can truly be achieved, can therefore still give additional insights into the quality of a decision for the current situation, i.e. where to move the truck to right now. Since the dynamic optimization function is stochastic, we point out again that multiple calls are required to estimate the expected future payoff even when using the hillclimber to evaluate the future. To reduce the number of statistical errors, the best evolved decision is compared to the default choice of doing nothing, i.e. $b(t^{\text{now}}) = 0$ and $\boldsymbol{x}^{\text{alternative}}(t^{\text{now}}) = \boldsymbol{x}^{\text{truck}}(t^{\text{now}})$. Only if the mean fitness of the best evolved decision averaged over 100 calls to the dynamic optimization function is statistically significantly larger than the mean fitness of the default decision, the evolved decision is used. The statistical hypothesis test used to this end is the Aspin–Welch–Satterthwaite (AWS) T–tests at a significance level of $\alpha = 0.05$. The AWS T–test is a statistical hypothesis test for the equality of means in which the equality of variances is not assumed [17].

Results

A population size of 100 was experimentally found to be adequate for solving the optimization problem in each time step. We set $t^{\text{end}} = 100$ and advanced time by a time step of 1 every 50 generations. Since only one pattern was added to the data set each time step, and only a normal distribution is estimated from data, learning can be done very fast for this problem. Therefore learning was performed whenever time was advanced. The final scores (i.e. the integral of the dynamic optimization function over $[0, 100]$) are shown in Figure 6.6 as a function of the prediction length t^{plen}. Indeed as expected and motivated earlier in the previous subsection, the best value for t^{plen} is not the maximum length of t^{end}, but a smaller length, even if the model is fully known.

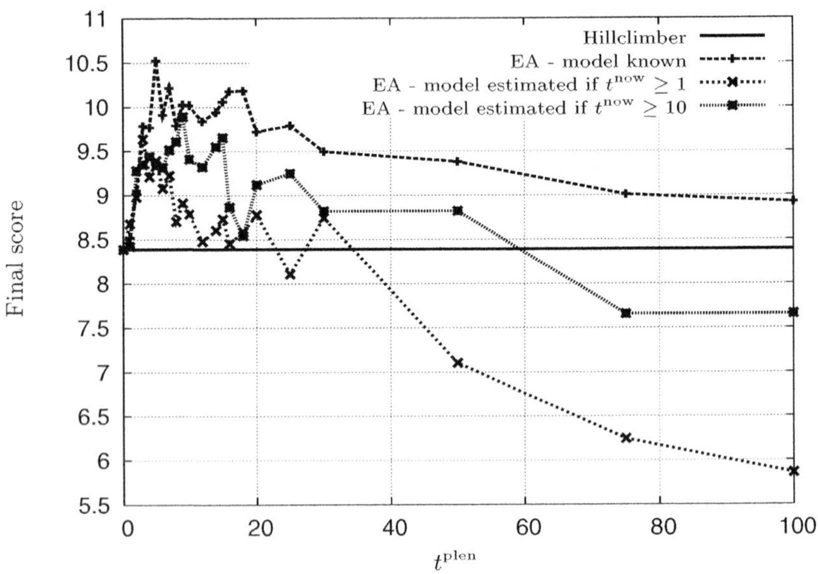

Fig. 6.6. Final score (i.e. $\int_0^{t^{\text{end}}} \mathfrak{F}^{\text{dyn}}_{\gamma^{\text{dyn}}(t,Z(t,\zeta))}(\zeta^{\text{dyn}}_{\zeta}(t))$) averaged over 100 runs on the dynamic pick–up problem as a function of the prediction length

The trajectory of the cumulative fitness for the best value and maximum value of t^{plen} are shown in Figure 6.7. This figure also reveals why the use of information about the future leads to a better result in the end. All algorithms other than the hillclimber are willing to accept negative scores in a single turn if the prospect on future gains is larger. This happens if the truck moves more towards the origin as the density of the normal distribution is the highest there. The better strategy adopted by the system is thus to initially move towards the region close to the origin and never move too far away from it, even if doing so means making a profitable pick–up.

Finally, it is again interesting to note that postponing the use of learning until a higher reliability is obtained leads to better results. This indicates again the importance of reliable predictions in the proposed approach.

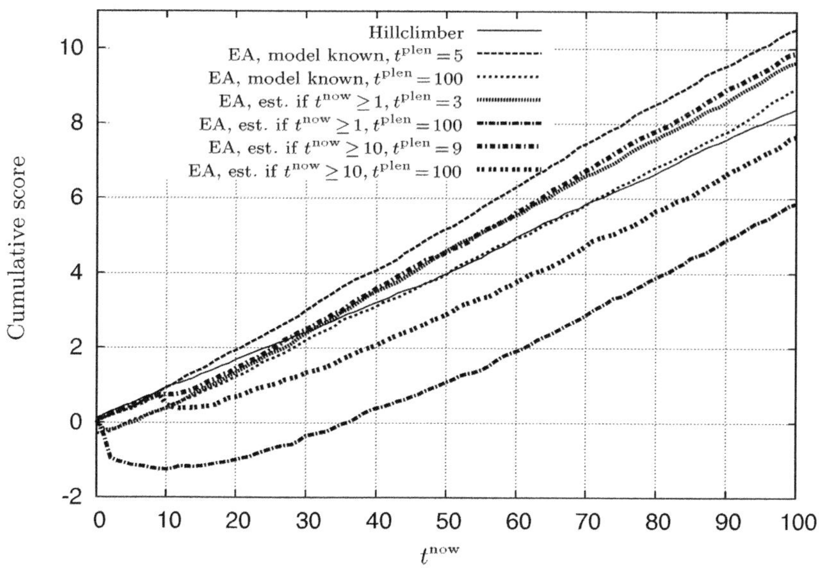

Fig. 6.7. Cumulative score (i.e. $\mathfrak{H}^{\text{dyn}}(t^{\text{now}}, \zeta)$) averaged over 100 runs on the dynamic pick–up problem as a function of time

6.6 Discussion and Conclusions

In this chapter we have highlighted a specific source of difficulty in online dynamic optimization problems. We have labeled the difficulty time–linkage. In the worst case time–linkage can lead to time–deception. In that case any optimization algorithm is mislead and finds suboptimal results unless future implications of current decisions are taken into account. To tackle problems exhibiting this type of problem difficulty, we have proposed a framework that learns to predict the future and optimizes not only the current situation but also future predicted situations. We have proposed and used two new benchmark problems, but a larger suite of problems containing time–linkage is called for and should become a standard in dynamic optimization research.

In our experiments, we have fixed the future prediction time span as well as the history data time span. An interesting question is whether the time spans required to prevent deception can be measured during optimization.

This calls for techniques for time–linkage identification in a similar sense as gene–linkage identification techniques are required in standard GAs to prevent deception as a result of dependencies between a problem's variables [12, 24].

Another important and related issue is how quickly the reliability of prediction degrades into the future. Even if we know how far into the future we must predict, these predictions are hardly of any use if they are unreliable. The prediction reliability is influenced mostly by the difficulty of the function to predict (i.e. relatively steady or heavily fluctuating) and by the availability of data.

Ultimately, the expansion of dynamic EAs to process time–linkage information should be integrated with current state–of–the–art dynamic EAs that are capable of tackling other important problem difficulties that arise in dynamic optimization such as the overtaking of the optima by other local optima as time goes by. An EA that is capable of efficiently tackling both sources of problem difficulty is likely to be robust and well–suited to be used in practice and hence to be tested in real–world scenario's. To that end however, a further expansion that makes the approach well–suited for the multi–objective case is also likely to be crucial.

References

1. M. Andrews and A. Tuson. Diversity does not necessarily imply adaptability. In J. Branke, editor, *Proceedings of the Workshop on Evolutionary Algorithms for Dynamic Optimization Problems at the Genetic and Evolutionary Comp. Conference – GECCO 2003*, pages 24–28, 2003.
2. P. J. Angeline. Tracking extrema in dynamic environments. In P. J. Angeline et al., editors, *Sixth Int. Conf. on Evol. Programming*, pages 335–345, Berlin, 1997. Springer Verlag.
3. D. V. Arnold and H.-G. Beyer. Random dynamics optimum tracking with evolution strategies. In J.J. Merelo et al., editors, *Parallel Problem Solving from Nature – PPSN VII*, pages 3–12, Berlin, 2002. Springer Verlag.
4. T. M. Blackwell. Particle swarms and population diversity II: Experiments. In J. Branke, editor, *Proceedings of the Workshop on Evolutionary Algorithms for Dynamic Optimization Problems at the Genetic and Evolutionary Comp. Conference – GECCO 2003*, pages 14–18, 2003.
5. P. A. N. Bosman and D. Thierens. Advancing continuous ideas with mixture distributions and factorization selection metrics. In M. Pelikan and K. Sastry, editors, *Proc. of the Optimization by Building and Using Probabilistic Models OBUPM Workshop at the Genetic and Evolutionary Comp. Conference – GECCO 2001*, pages 208–212, 2001.
6. P. A. N. Bosman and D. Thierens. The naive MIDEA: a baseline multi–objective EA. In C. A. Coello Coello et al., editors, *Evolutionary Multi–Criterion Optimization – EMO'05*, pages 428–442, Berlin, 2005. Springer–Verlag.
7. J. Branke. Memory enhanced evolutionary algorithms for changing optimization problems. In *Proceedings of the 99 Congress on Evolutionary Computation – CEC 99*, pages 1875–1882, Piscataway, New Jersey, 1999. IEEE Press.

8. J. Branke. *Evolutionary Optimization in Dynamic Environments.* Kluwer, Norwell, Massachusetts, 2001.
9. J. Branke, T. Kaußler, C. Schmidt, and H. Schmeck. A multi–population approach to dynamic optimization problems. In I. C. Parmee, editor, *Adaptive Computing in Design and Manufacture – ACDM 2000*, pages 299–308, Berlin, 2000. Springer Verlag.
10. J. Branke and D. Mattfeld. Anticipation in dynamic optimization: The scheduling case. In M. Schoenauer et al., editors, *Parallel Prob. Solving from Nature – PPSN VI*, pages 253–262, Berlin, 2000. Springer Verlag.
11. M. R. Caputo. *Foundations of Dynamic Economic Analysis.* Cambridge University Press, Cambridge, 2005.
12. K. Deb and D. E. Goldberg. Sufficient conditions for deception in arbitrary binary functions. *Annals of Mathematics and Artificial Intelligence*, 10:385–408, 1994.
13. S. M. Garrett and J. H. Walker. Genetic algorithms: Combining evolutionary and 'non'–evolutionary methods in tracking dynamic global optima. In W. B. Langdon et al., editors, *Proceedings of the Genetic and Evolutionary Computation Conference – GECCO 2002*, pages 359–366. Morgan Kaufmann, 2002.
14. D. E. Goldberg. *Genetic Algorithms in Search, Optimization and Machine Learning.* Addison Wesley, Reading, Massachusetts, 1989.
15. J. Grefenstette. Evolvability in dynamic fitness landscapes: a genetic algorithm approach. In *Proceedings of the 99 Congress on Evolutionary Computation – CEC 99*, pages 2031–2038, Piscataway, New Jersey, 1999. IEEE Press.
16. K. De Jong. Evolving in a changing world. In Z. W. Ras and A. Skowron, editors, *Foundations of Intelligent Systems*, pages 512–519, Berlin, 1999. Springer Verlag.
17. M. G. Kendall and A. Stuart. *The Advanced Theory Of Statistics, Volume 2, Inference and Relationship.* Charles Griffin & Company Limited, 1967.
18. P. Larrañaga, R. Etxeberria, J. A. Lozano, and J. M. Peña. Optimization in continuous domains by learning and simulation of Gaussian networks. In M. Pelikan et al., editors, *Proceedings of the Optimization by Building and Using Probabilistic Models OBUPM Workshop at the Genetic and Evolutionary Computation Conference – GECCO 2000*, pages 201–204, 2000.
19. A. M. L. Liekens, H. M. M. ten Eikelder, and P. A. J. Hilbers. Finite population models of dynamic optimization with alternating fitness functions. In J. Branke, editor, *Proc. of the Workshop on Evolutionary Algorithms for Dynamic Optimization Problems at the Genetic and Evol. Comp. Conference – GECCO 2003*, pages 19–23, 2003.
20. T. M. Mitchell. *Machine Learning.* McGraw-Hill, New York, New York, 1997.
21. W. B. Powell. Algorithms for the dynamic vehicle allocation problem. In B. L. Golden and A. A. Assad, editors, *Vehicle Routing: Methods and Studies*, pages 249–292. Elsevier Science, Amsterdam, 1988.
22. H. N. Psaraftis. Dynamic vehicle routing problems. In B. L. Golden and A. A. Assad, editors, *Vehicle Routing: Methods and Studies*, pages 223–248. Elsevier Sc., Amsterdam, 1988.
23. L. Schöneman. On the influence of population sizes in evolution strategies in dynamic environments. In J. Branke, editor, *Proceedings of the Workshop on Evolutionary Algorithms for Dynamic Optimization Problems at the Genetic and Evolutionary Computation Conference – GECCO 2003*, pages 29–33, 2003.

24. D. Thierens. Scalability problems of simple genetic algorithms. *Evolutionary computation*, 7:331–352, 1999.
25. R. K. Ursem. Multinational gas: Multimodal optimization techniques in dynamic environments. In D. Whitley et al., editors, *Proceedings of the Genetic and Evolutionary Computation Conference – GECCO 2000*, pages 19–26. Morgan Kaufmann, 2000.
26. J. I. van Hemert, C. Van Hoyweghen, E. Lukschandl, and K. Verbeeck. A "futurist" approach to dynamic environments. In J. Branke and T. Bäck, editors, *Proceedings of the Workshop on Evolutionary Algorithms for Dynamic Optimization Problems at the Genetic and Evolutionary Computation Conference – GECCO 2001*, pages 35–38, 2001.
27. J. I. van Hemert and J. A. La Poutré. Dynamic routing problems with fruitful regions: models and evolutionary computation. In X. Yao et al., editors, *Parallel Problem Solving from Nature – PPSN VIII*, pages 692–701, Berlin, 2004. Springer Verlag.
28. V. Vapnik. *Statistical learning theory.* Wiley, New York, New York, 1998.
29. M. Wineberg and F. Oppacher. Enhancing the ga's ability to cope with dynamic environments. In D. Whitley et al., editors, *Proceedings of the Genetic and Evolutionary Computation Conference – GECCO 2000*, pages 3–10. Morgan Kaufmann, 2000.

7

Evolutionary Online Data Mining: An Investigation in a Dynamic Environment

Hai H. Dam, Chris Lokan, and Hussein A. Abbass

Artificial Life and Adaptive Robotics Laboratory
School of Information Technology and Electrical Engineering
The University of New South Wales
Australian Defence Force Academy
Canberra ACT 2600, Australia
{z3140959,cjl,abbass}@itee.adfa.edu.au
http://www.itee.adfa.edu.au/~alar

Summary. Recently, traditional data mining algorithms are challenged by two main problems: streaming data, and changes in the hidden context. These challenges emerged from real-world applications such as network intrusion detection, credit card fraud detection, etc. Online or incremental learning becomes more important than ever for dealing with these problems. This chapter investigates XCS, a genetics-based learning classifier system, which offers an incremental learning ability and also is able to handle an infinite amount of continuously arriving data. XCS has been tested on many data mining problems and demonstrated as a potential online data mining approach. Most experiments with XCS assume a static environment. Since environments are more likely to be dynamic in real life, noise and environmental factors need to be taken into account in a good data mining approach. This chapter investigates XCS in dynamic environments, in the presence of noise in the training data. An essential requirement of an algorithm in dynamic environments is to be able to recover quickly from hidden changes, while reusing previous knowledge. Our results show that XCS is capable of recovering quickly from small changes in the underlying concepts. However, it requires significant time to re-learn a model after severe changes. We propose several strategies to force the system to learn quickly after severe changes. These are adaptive learning rate; re-initializing the parameters; and re-initializing the population. Experiments show improvement in the predictive performance of XCS, when compared to the traditional XCS, in both noisy and noise-free environments.

7.1 Introduction

Data mining is a sophisticated process to discover novel and potentially useful knowledge from massive databases [14]. Data mining employs machine learning techniques to accomplish different tasks such as classification, regression, prediction, and clustering. The focus of this chapter is on classification, which

Hai H. Dam et al.: *Evolutionary Online Data Mining: An Investigation in a Dynamic Environment*, Studies in Computational Intelligence (SCI) **51**, 153–178 (2007)
www.springerlink.com

attempts to develop a model from previous observed data for future prediction of values in a finite set.

Genetics-based learning classifier systems (GBLCS) [18] are a modern heuristic method of data mining in which a genetic algorithm (GA) and reinforcement learning (RL) are employed. In this framework, a population of classifiers (or so-called rules) is evolved over time based on the concept of natural selection. GA is involved to discover new classifiers and RL is used to evaluate classifiers with guidance from the environment. The advantages of using GA are robustness to tolerate noise and an ability to track changes over time [16].

The first version of GBLCS was implemented by Holland in 1978 [20], to simulate animal behaviors. The simulation showed that animal behaviors were predicted successfully. Wilson [33] confirmed later that GBLCS can simulate knowledge growth in animals. Goldberg [15] applied GBLCS to a real life application for controlling a pipeline system. Bernado [3] compared the predictive performance of GBLCS with six other well known learning algorithms on fifteen data sets, and claimed that GBLCS reach and even exceed the performance of other traditional learning schemes.

Work on GBLCS can be grouped into two main strategies: the Pittsburgh [28] and Michigan [19] approaches. The key difference is that an individual in the Pittsburgh approach is a set of classifiers representing a complete solution to the learning problem; while an individual in the Michigan approach is a single classifier that represents a partial solution to the overall learning task. Thus, the Michigan and Pittsburgh systems are quite different approaches. Which is preferred depends on the nature of the application.

Many studies have compared the performance of the two approaches, on several data mining problems, in order to determine circumstances in which one approach would perform better than the other and to understand the behavior of each system [1, 2, 24]. In some cases the Pittsburgh approach is more robust, but it is computationally very expensive compared to the Michigan approach (since the Pittsburgh approach maintains a population of complete solutions, the cost of evaluating the population is high).

XCS [7, 34, 35] is widely accepted as one of the most reliable Michigan-style GBLCS for data mining [2, 3, 5, 12]. The framework was introduced by Wilson in 1995 as an enhanced version of the traditional GBLCS. The two major changes in XCS are: fitness is based on the accuracy of the reward prediction, instead of the strength (or reward directly received from an environment); and a niche genetic algorithm. Many studies showed that XCS performs at least as well as other traditional machine learning techniques on several data mining problems [2, 26]. The advantages of XCS are its rule-based representation, which can easily recognize the mining patterns; its online learning, which is able to deal with an infinite stream of one-pass instances; and its incremental learning, which continuously refines and revises the current knowledge.

Many experiments on XCS, as well as many data mining applications, assume that an environment is static: once a model is learned, it can be used indefinitely. However, in real life the environment is never static. Noise may affect a data item; noise levels may vary over time; the underlying data model may change. Being able to cope with noise and changes in the data is an essential requirement for a good application in a dynamic environment. Branke stated that a good system should be capable of continuously adapting its solution to any changes in the environment, while reusing information gained in the past [4]. If the problem changes completely, without any reference to the history, it would be regarded as much simpler dynamic environment because the system has to learn from scratch.

In this chapter, we investigate XCS in the dynamic environment with and without noise in the training data. We also propose several strategies to help XCS recover quickly in the face of changes in the dynamic environment. A justification of each strategy is provided in terms of the recovery time.

The chapter is structured as follows. The next section explains about the dynamic environment in data mining, and also discusses related work in the field. Section 7.3 provides a brief review of XCS. Section 7.4 describes our experiment setup and a test problem. Section 7.5 investigates XCS in noise-free and noisy environments with different levels of concept change. Section 7.6 proposes several strategies for XCS to cope with dynamic environments. Section 7.7 compares the performance of those strategies in noise-free and noisy dynamic environments. Finally, conclusions and future work are presented in the last section.

7.2 A Review of Dynamic Environments in Data Mining

7.2.1 Concept Change

In the literature of data mining, a dynamic environment is mainly concerned with *concept change* (or *concept drift*), in which a target learning concept changes over time [32]. In the real world, concept change occurs so frequently that any online data mining application needs to take it into account seriously. For example, a company's policy may change every day to reflect the consumer's market.

The target learning concept can change in many ways under many aspects. Abbass et al. [1] determined six common kinds of changes:

- Change in the model. A learned model may become incorrect after a period of time. For example, customers' preferences for shopping change due to changes in seasons, fashion, etc. Hence a model of customers' preferences in summer is quite different to the one in winter.
- Change in the number of underlying concepts. The number of concepts may not stay constant forever in real-life applications. Some new concepts may be introduced and some old ones may become obsolete. For example, a school may introduce a new class on network security, due to high demand from students, and may stop teaching programming in Pascal because not

many students take the unit. A model to classify a unit based on students' preferences needs to be modified to capture new concepts and eliminate obsolete ones.

- Change in the number of features. The number of available features may vary over time.
- Change in the level of noise. Noise is an unwanted factor but it occurs very often in real world data. The noise level may change reflecting the condition of an environment. For instance, voice data may have a high level of noise at a supermarket during daytime, but may have a low level of noise during nighttime.
- Change in the class distribution.
- Change in the sample bias.

Abbass et al. grouped these changes into two main areas: model boundary changes and sample changes. The first three changes belong to the first category and the last three changes belong to the latter category. This chapter focuses mainly on changes in the model boundary, or the concepts of the model, and on noise.

The target concepts may change at different rates depending on the nature of the application. Some concepts may change slowly, resulting in ambiguity and uncertainty in between periods of stability. Other concepts might change suddenly; new instances become no longer consistent with the current concepts of the learned model. In this chapter, we investigate the second case; we leave the first one for future work.

7.2.2 Data Mining in the Presence of Concept Change

An effective learning algorithm for tracking concept change is one that can identify changes in the target concept without being explicitly informed about them; can recover quickly from changes by adjusting its knowledge; and can use previous knowledge in the case that old concepts reappear [32].

The first system designed for concept change is STAGGER [27], using probabilistic concept description. The system responds to concept change by adjusting weights in the model, and discarding any concepts that fell below a threshold accuracy.

Since then, researchers have proposed several rule-based algorithms for dealing with concept change. A family of FLORA algorithms [32] is one of them, which learns rules from an explicit window of training instances. A window means that oldest examples are replaced by newly arrived ones so that they fit into the window. The learner trusts only the latest examples. FLORA2 allows dynamic size of the window in response to the system performance. FLORA3 stores concepts for future use and reassesses their utilities when context changes are perceived. FLORA4 is designed to deal with noise in the input data.

Klinkenberg and Thorsten [21] developed an alternative method to detect concept change using the support vector machine. They also use a window of recent instances to detect changes.

A drawback of the windowing technique is that the system becomes sensitive to changes in the distribution of instances.

Classifier ensembles are an alternative approach in data mining for dealing with concept change. A set of experts is maintained. The prediction is made by combining the experts' knowledge, using voting and/or weighted voting.

Street and Kim [30] suggest that building separate classifiers on sequential chunks of training data is also effective to handle concept change. These classifiers are combined into a fixed size ensemble using a heuristic replacement strategy. Kolter and Maloof [22] present an approach based on the weighted majority algorithm to create and remove base learners in response to changes in performance.

In general, a disadvantage of the ensemble approach is that the system needs to have a strategy to create new experts and eliminate old ones to adapt to the environment.

In this chapter, we will explain and explore XCS in the presence of noise and concept change. XCS is an incremental learner. By its nature it is able to handle a stream of data. After returning a prediction to the environment, XCS updates its current knowledge of the instance based on the feedback in terms of reward. The use of a genetic algorithm helps the system to explore the search space and escape from a local optimum. The genetic algorithm is based on natural selection, which aims at deleting weak rules and evolving good rules over time. Therefore XCS is able to handle concept change effectively.

7.3 XCS

This section provides a brief description of the XCS system. XCS is a rule-based evolutionary learning classifier system, in which each classifier represents a partial solution to the target problem. A typical goal of XCS is to evolve a population of classifiers $[P]$ to represent a complete solution to the target problem. XCS relies on RL for evaluating classifiers in the population, and GA for exploring a search space and introducing new classifiers in the population.

During the learning process, the system receives inputs from the environment and returns its prediction. Feedback from the environment is given in terms of a reward reflecting how good the prediction was.

XCS is designed for both single-step (receiving immediate reward from the environment) or multi-step (receiving delayed reward from the environment) environments. In classification, a single-step environment is mainly used.

Each classifier consists of the *Condition* (the body of the rule), the *Action* (the prediction of the classifier) and some parameters. The *Condition* refers to several environment states, to which the classifier may match. The *Action* will be chosen as a system prediction if the classifier is fired. There are three main parameters associated with a classifier that are used to determine how good it is: reward prediction P, prediction error ε, and fitness F. The reward prediction P refers to an amount of reward, which the classifier predicts to be received from the environment if its prediction is chosen. The prediction error ε is the absolute difference between the reward prediction P and the actual reward received. The fitness is an inverse function of the reward prediction error. Another two important parameters are *numerosity* and *experience*. *Numerosity* records the number of copies the classifier has in $[P]$. *Experience* indicates how often the classifier is chosen for prediction making; in other words, how general the classifier is.

A classifier in XCS is a macro–classifier, which represents a distinct rule (a pair of *Condition:Action*) in the population. Whenever a new classifier is introduced, the population is scanned through to check if its copy already exists. If it does not, the classifier is added to the population. Otherwise, the *numerosity* of its copy is incremented by one.

Upon receiving an input from the environment, a *match set* $[M]$ is formed. It contains all classifiers in $[P]$ whose *condition* matches the input. Classifiers in $[M]$ will participate in a system decision making to decide on the system prediction. In exploitation phase, a *prediction array* $[PA]$ is formed for estimating the value of each possible action in $[M]$. An action having the highest value in $[PA]$ will be selected to export to the environment. The exploration phase, on the other hand, chooses a random action in $[M]$ so that it will give a chance for every classifier to be evaluated. An *action set* $[A]$ is then formed, containing those classifiers in $[M]$ that have the chosen action.

In exploration phase, the parameters of classifiers in $[A]$ are updated incrementally and GA might be involved in $[A]$. The *fitness, reward prediction* and *prediction error* parameters of the classifiers in $[A]$ are updated, based on the difference between the predicted and actual reward.

GA is activated only when the average *experience* of classifiers in $[A]$ is higher than a threshold defined by the user. Two parents are selected from $[A]$ with probability proportional to their fitness. Two offspring are generated by reproducing, crossing–over, and mutating the parents with certain probabilities. Offspring are inserted in $[P]$. The parents also remain in $[P]$ to compete with the offspring. If the population size exceeds the predefined limit, some inaccurate classifiers are removed from $[P]$.

The algorithm of XCS is presented in Fig. 7.1. The parameters it uses are listed in Table 7.1.

7.4 Experiments

7.4.1 Test Problems

Some concept change handling systems have been tested on real-world data such as spam filtering data [11], US Census Bureau data [30], and credit

Table 7.1. Parameters of XCS

Parameter	Meaning
N	maximum population size
$P_{\#}$	don't care probability
β	learning rate
α, ϵ_0, v	accuracy determination parameters
θ_{GA}	threshold to activate GA
χ	probability of applying crossover
μ	probability of applying mutation
θ_{del}	minimum experience to be considered during deletion
θ_{sub}	minimum experience to be considered for subsumption
p_I, ε_I, f_I	initial parameter values for each classifier
θ_{nma}	minimum number of different actions in $[M]$ for covering

```
Initialize XCS parameters and initialize [P] to empty
repeat
  for each training instance I_trn from env do
    Form a match set [M] of those classifiers in [P] that satisfy I_trn
    while the number of different actions in [M] < θ_nma do
      Generate a random classifier cl that covers I_trn
      Add classifier cl to set [P] and set [M]
    end while
    Select an action randomly from those in [M]
    Generate an action set [A] of classifiers in [M] that have the chosen action
    Return the action to env
    Receive a reward R from env estimating how good the action was w.r.t. I_trn
    Adjust parameters of those classifiers in [A]
    if an average experience of classifiers in [A] is higher than θ_GA then
      Select and reproduce (mutation and crossover) two classifiers in [A]
      Offspring inherit fitness, reward prediction, prediction error from parents
      Insert offspring in [P]
      while the current population size is larger than N do
        Remove a classifier from [P] wrt the fitness and action set size
      end while
    end if
  end for
  for each testing instance I_tst from env do
    Form a match set [M]
    Form a prediction array [PA] estimating the value of each possible action
    in [M] based on the fitness and reward prediction
    Select the best action that has the highest value in [PA]
    Return the action to the environment
  end for
until the termination conditions are met
```

Fig. 7.1. Algorithm of XCS

card fraud data [31]. A repository of such data sets [13] is maintained by the University of California at Irvine (UCI).

The main disadvantage of those data is that they do not contain much concept change. Thus concept changes are added artificially, and therefore it becomes an artificial problem.

Artificial problems tend to be favored by researchers, however. They have the advantage that the researcher can control the type and rate of concept change, context recurrence, presence of noise, irrelevant features, etc.

The most popular artificial testing benchmark for concept change in data mining is the STAGGER concept [17, 22, 27, 32]. The problem contains 3 simple Boolean concepts of 3 features with a total of 120 instances and 3 cycles of change. However, this problem can not be extended to a large scale of data. Scalability is very important as concept change mostly occurs in a stream of data.

In this chapter, we experiment on the multiplexer problem. A binary string of length $k + 2^k$ is used as an input. The first k bits determine the position of an output bit in the last 2^k bits. The multiplexer problem is one of the most popular testing benchmarks in learning classifier system research in general and XCS in particular [6, 10, 34].

Wilson [36] extended the multiplexer problem to a continuous domain by introducing a real-threshold to convert a real number to a binary number. For example, assume a real number is in the range $[0, 1)$ and the real-threshold is 0.5. The real input number r will be converted to 0 in binary if $r < 0.5$, otherwise it is set to 1 in binary. The real multiplexer problem then becomes a traditional binary multiplexer problem.

The real multiplexer problem is also considered a challenging artificial problem. It can extend to large scale data. The magnitude and rate of changes, as well as noise in the system, can be controlled easily by the researcher.

7.4.2 Experimental Setup

We conducted two experiments to study the real-6-multiplexer problem in a dynamic environment with/without noise.

To simulate the dynamic environment for the real-6-multiplexer problem, we change the underlying concept of data by altering the real-threshold. This influences the mapping from a real number to a binary number. The concept at the continuous level is changed, but the concept at the binary level remains unchanged. Figure 7.2 illustrates the effect of concept change to an actual label of an instance. In this case, a label of an instance $I =< 0.6, 0.2, 0.05, 0.7, 0.3, 0.4 >$ is changed from 1 to 0 when the real-threshold is shifted from 0.1 to 0.5.

The first experiment aims to learn the impact of different magnitudes of concept change (MoC) on the system's predictive performance. A single run consists of two cycles. The real-threshold of the first cycle is set to 0.1. After 100,000 time steps (each step in our experiment includes one training instance and one testing instance) the threshold is changed. We choose the $100,000^{th}$ time step because it gives any learning algorithm enough time to accumulate full knowledge in this problem. Hence, we can evaluate accurately its adaptive ability after the concept change. The real-threshold is altered with different levels of MoC, such as $MoC = \{0.1, 0.2, 0.3, 0.4, 0.5, 0.6, 0.7, 0.8\}$. For example, if $MoC = 0.4$, the real-threshold of the second cycle will change from 0.1 to 0.5.

In essence, a low level of MoC will cause a small change to the underlying model. A high level of MoC, in contrast, will affect a larger amount of data and therefore present a bigger challenge for a learning system. For example, if MoC is 0.1, approximately 10% of the data might be affected. If MoC is 0.4, 40% of the data might be changed.

The experiment is performed initially in a noise-free environment, and then under different levels of noise. Noise is incorporated only in the training instances by flipping their classes with a certain probability. For example, a noise level of 0.05 means that the flipping probability is 5% (or approximately 5 noisy instances occur in each 100 training inputs).

The results of the first experiment are presented in Section 7.5.

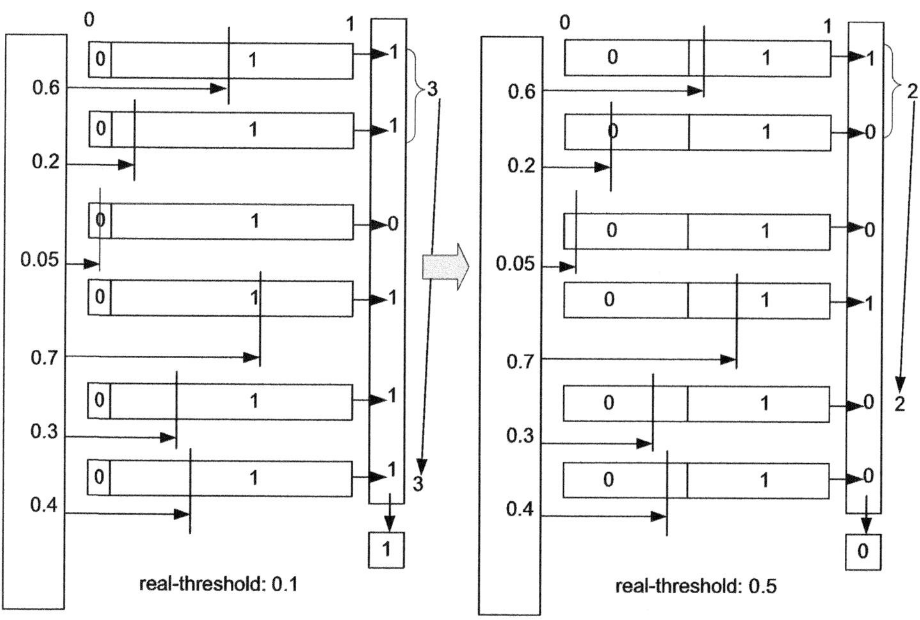

Fig. 7.2. The real-threshold influences the class of an input instance. The instance $I =< 0.6, 0.2, 0.05, 0.7, 0.3, 0.4 >$ belongs to class 1 when the real-threshold is 0.1, and class 0 when the real-threshold is 0.5.

The second experiment aims to compare the predictive performance of XCS, using three different proposed strategies for dealing with concept change.

The strategies are presented in Section 7.6. The second experiment is also performed initially in a noise-free environment, and then under different levels of noise. The results of the experiment are presented in Section 7.7.

In [9], we found out that imbalance of class distribution has an effect on the system performance. Experiments in this chapter avoid this problem, by assigning equal distribution of each class.

All experiments presented in this paper are averaged over 30 independent runs.

7.4.3 System Setup

XCS is set up with the same parameter values[1] used by Wilson [36], Stone and Bull [29], and Dam et al [8] as follows: $N = 1,000, \beta = 0.2, \alpha = 0.1, \epsilon_0 = 10, v = 5, \theta_{GA} = 12, \chi = 0.8, \mu = 0.04, \theta_{del} = 20, \delta = 0.1, \theta sub = 20, p_I = 10, \varepsilon_I = 0, f_I = 0.01, \theta_{nma} = 2$.

7.5 Performance with Different Magnitudes of Change

7.5.1 Performance in a Noise-Free Environment

In the real multiplexer problem, concept change occurs when the real-threshold for mapping a real number to a binary number is changed. The magnitude of change (MoC) in this problem is the absolute difference in the real-threshold before and after the change. For example, if the real-threshold is changed from 0.4 to 0.5, the MoC is 0.1.

Figure 7.3 shows the predictive performance of XCS over 200,000 time steps, with the change occurring at the $100,000^{th}$ time step. The magnitudes of change are $MoC = \{0.1, 0.2, 0.3, 0.4, 0.5, 0.6, 0.7, 0.8\}$ respectively. Observing the graphs, a sharp decrease in the predictive accuracy is obvious at the $100,000^{th}$ time step, where the concept change takes place.

When MoC is small (e.g. 0.1), the accuracy at the beginning of the second cycle does not drop as low as the starting point of the first cycle. Also, XCS adjusts very quickly to the change and soon returns to nearly 100% accuracy. The time needed for XCS to learn in the second cycle is shorter than in the first cycle. This is because XCS starts without any prior knowledge in the first cycle. The system must rely on GA to explore new classifiers and reinforcement component to evaluate these classifiers. XCS needs a sufficient time to explore completely the environment. In the second cycle, on the other hand, the system already has a complete knowledge of the environment. After the change, only small part of the knowledge of the system becomes inaccurate. Therefore the system has to discover only a few more concepts for completion.

Increasing MoC results in a more significant change to the underlying model. Therefore, a longer time is required for the system to recover. When

[1] The parameters are defined in Table 7.1

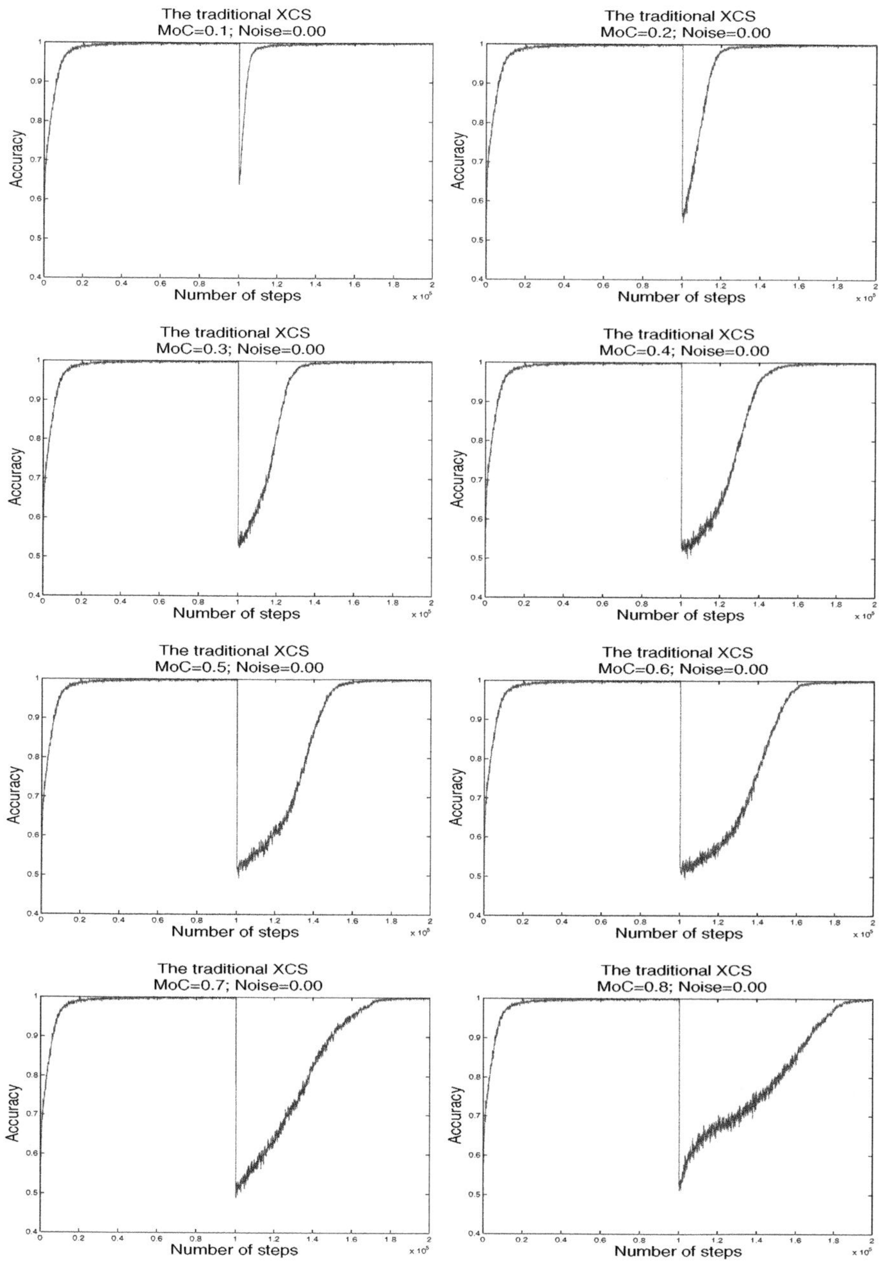

Fig. 7.3. The real-6-multiplexer problem; Performance with different MoC values in noise-free environments; The real threshold value is changed from 0.1 to 0.2,0.3,0.4,0.5,0.6,0.7,0.8,0.9 respectively; The curve plots the predictive accuracy of the testing data over time.

MoC is 0.3 or more, the learning time in the second cycle is much longer than in the first cycle.

The population in XCS contains a set of classifiers, which represent its current knowledge of the environment. If a severe change happens to the underlying model, a large number of classifiers that were accurate in the previous cycle may become inaccurate under new concepts. These classifiers normally acquired high fitness in the previous cycle. They are unlikely to be eliminated quickly from the population. XCS needs time to re-evaluate these classifiers, gradually reducing their fitness until it is small enough for them to be deleted from the population as a result of natural selection. In other words, it takes time for the system to unlearn the old concepts as well as to learn the new ones. This is the key reason for the increased recovery time and decreased performance when MoC is high.

Table 7.2 presents XCS's performance right after a concept is changed, and the number of time steps required for XCS to recover to 99% accuracy, at each level of MoC.

Table 7.2. System performance with different magnitudes of change at the $100,000^{th}$ time step, when the real thresholding is changed from 0.1 to 0.2, 0.3, 0.4, 0.5, 0.6, 0.7, 0.8, and 0.9 respectively. The result is averaged over 30 runs.

MoC	Performance after the change	Time to recover to 99% accuracy
0.1	0.6433 ± 0.0495	11,500
0.2	0.5610 ± 0.0379	22,600
0.3	0.5353 ± 0.0581	34,000
0.4	0.5270 ± 0.0501	49,400
0.5	0.5227 ± 0.0715	54,600
0.6	0.5210 ± 0.0512	61,100
0.7	0.5193 ± 0.0449	70,400
0.8	0.5347 ± 0.0431	84,600

The table reveals that XCS can recover quickly from a small magnitude of change. If MoC is large (more than 0.2), XCS requires a much longer time to learn and adapt to the new concepts. If MoC>0.4, it is quicker to start learning from scratch rather than continuing with the current knowledge.

7.5.2 Effect of Noise in Dynamic Environments

There are several studies of XCS in noisy static environments. Lanzi et. al [25] found that XCS can only converge to an optimal solution when the level of uncertainty is limited. They argued that XCS is unable to distinguish inaccurate classifiers caused by over-generalization and inaccurate classifiers caused by uncertainty. To cope with uncertainty they introduced a new parameter, μ, in all classifiers for estimating its minimum prediction error overtime.

They explained that μ will help XCS to separate the inaccuracy due to overgeneralization from that due to uncertainty. Like other parameters in XCS, μ is updated whenever a classifier appears in $[A]$ with a step size depending on a learning rate.

To the best of our knowledge, the literature does not include studies of XCS in dynamic environments with noise effects. This section will explore this area by investigating the performance and population of XCS in dynamic environments with several levels of noise.

Figure 7.4 shows the population and performance of XCS before and after an underlying concept is changed severely (MoC=0.8), under several noise levels.

Observing the graph, we notice three regions of noise that affects the recovery time of XCS from concept changes. These are: low noise (Figure 7.4, the first row), which results in a slow recovery; medium noise (Figure 7.4, the second row), which helps the system to recover almost as fast as the original learning; and high noise (Figure 7.4, the last row), which increases the recovery time dramatically. The levels of noise in each region are $[0.00, 0.03)$,$[0.03, 0.15)$, and $[0.15, 1.00)$ for low, medium, and high respectively.

In the first cycle, XCS starts without prior knowledge. The population curve increases dramatically after a few steps at the beginning for all three regions. At this stage, GA is working hard to introduce new classifiers into the system. Those classifiers might be either good or bad. The learning component is responsible for updating their parameters accurately to reflect their goodness.

When noise is small, XCS is able to differentiate effectively good classifiers from the bad ones. As the result, the population curve starts to decrease gradually after the peak. By the time the concept change happens the system has achieved a compact population, which contains only accurate and maximally general classifiers [23]. Since the concept change is severe ($MoC = 0.8$), most of these classifiers become inaccurate after the change. GA starts working hard again as a large number of classifiers are inserted in the population. In order to recover completely, the incorrect classifiers need to be re-evaluated and removed from the population. In order to remove those classifiers, it requires a sufficient time to update their parameters from good to bad so that they can be eliminated. This explains why the performance in the second cycle improves much more slowly than in the first cycle.

Medium noise, on the other hand, affects the learning process as Lanzi [25] has suggested. XCS is not able to compact the population as much as it could in low noise cases. The population curve decreases insignificantly in the first cycle. Before a concept change, the population contains both good and bad classifiers. As a result, the performance can never achieve 100% accuracy. Observing the decrease in the population suggests that XCS has achieved a certain level of generalization. After the concept change, only a small number of classifiers are inserted in the population. Some inaccurate classifiers in the previous cycle might become accurate under new conditions. XCS does not

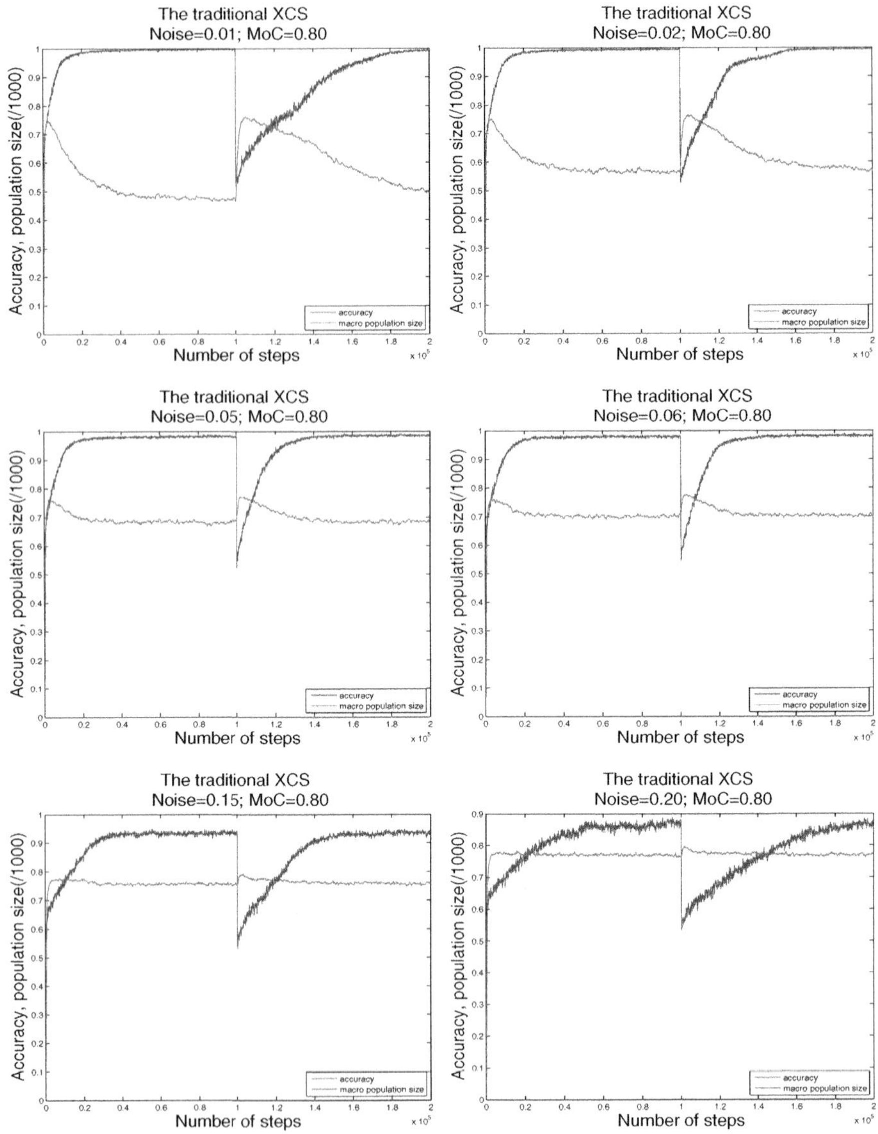

Fig. 7.4. Performance and population of XCS on several noise levels under a severe concept change (MoC=0.8)

have to remove and discover these classifiers. Hence the recovering time is reduced significantly.

When the noise level is high, XCS is unable to eliminate bad classifiers completely. The population always evolves both bad and good classifiers. The last row of Figure 7.4 shows that the population curves don't decrease at all after the initial peak. A large population implies that the system was not able to generalize. It seems that the population contains many bad and specific classifiers. Unlike the previous cases, some inaccurate classifiers appear to be still inaccurate after the concept change because they are over-specific. Hence, the recovery time starts increasing as the noise level increases in this region.

7.6 Strategies for Recovering from Concept Change

In Section 7.5.1 we saw that it can be quicker to start learning from scratch rather than continuing with current knowledge after severe changes. Figure 7.4 shows that this is also true in noisy environments.

This conclusion goes against the requirement of a good learning algorithm in dynamic environments, where previous knowledge should be reused to reduce the learning time. This motivates us to propose some strategies to improve XCS, where the system is able to reuse previous knowledge and thus recover quickly from severe changes in the underlying concept.

We propose three strategies. The first involves adjusting the learning rate dynamically. The other two involve first recognizing that a concept change has occurred, and then changing the knowledge represented in the population [P]. They differ in how the knowledge is changed.

7.6.1 An Adaptive Learning Strategy

The reinforcement learning component of XCS is responsible for updating parameters of the classifiers in the action set [A], based on feedback (reward) from the environment. The Widrow-Hoff technique is used to adjust the reward prediction P toward the actual reward received from the environment; and also the prediction error ε toward the absolute difference between the actual and predicted reward.

The following equations are used to update the parameters at time step i:

$$\varepsilon_i \leftarrow \varepsilon_{i-1} + \beta(|R - P_{i-1}| - \varepsilon_{i-1}) \tag{7.1}$$

$$P_i \leftarrow P_{i-1} + \beta(R - P_{i-1}) \tag{7.2}$$

where P is the reward prediction, ε is the prediction error, R is the actual reward received from the environment, and β is a learning rate ($0 < \beta \leq 1$). The learning rate β determines how much the *reward prediction* and *prediction error* parameters are adjusted at each time step.

Butz [5] has shown that XCS can perform better if the learning rate β is lowered in a noisy environment. The lower learning rate effectively decreases the noise in the parameter estimations of the classifiers. On the other hand, increasing the learning rate is useful when the environment changes. XCS then needs to accelerate the learning process in order to re-adjust quickly the parameters of the affected classifiers. Increasing the learning rate makes these parameters convert faster to the correct values. Thus in a noisy and dynamic environment, we want the learning rate to be low when the concept is stable, and raised when the concept changes.

Therefore, we propose an adaptive learning strategy that adjusts the learning rate according to the prediction performance of the system. The learning rate become an adaptive variable and can be expressed in a formula as follows:

$$\Delta E = \frac{current\_error - previous\_error}{R} \tag{7.3}$$

$$\beta' = \begin{cases} \beta_{min} + (\beta_{max} - \beta_{min})e^{\frac{\Delta E - 1}{\Delta E}\alpha} & \text{if } \Delta E > 0, \\ \beta_{min} & \text{if } \Delta E \leq 0 \end{cases} \tag{7.4}$$

$$\beta_t = \frac{\beta_{t-1} + \beta'}{2} \tag{7.5}$$

where β_{min}, β_{max} are minimum and maximum bounds of the learning rate defined by the user; β_t, β_{t-1} are the learning rate at the current and previous time steps; ΔE is the change in the prediction error between the current and previous time steps. α is a constant. R is the maximum reward given by the environment.

β' is a function of system error. A positive ΔE implies performance degradation, a negative ΔE implies performance improvement. The learning rate is increased (though not beyond a specified maximum value) when performance degrades. However, when performance is stable or improving, the learning rate is reduced (though not below a specified minimum value).

7.6.2 Recognizing a Concept Change

The previous subsection proposed an adaptive method for accelerating or decelerating the learning process according to the system performance. Another approach is to revise the knowledge maintained in the population when a concept change occurs. This means it is necessary to be able to recognize when the concept change occurs.

XCS can recognize a concept change by observing two main factors: the reward prediction error and the number of covering instances within a window of training instances.

The reward prediction error of XCS is the absolute difference between the reward prediction made by the system and the actual reward received from the environment upon a prediction. In an exploitation phase, XCS attempts to export an action, which maximizes the reward prediction within its current knowledge. If the difference between the reward prediction and the actual reward is low, XCS is perfectly adequate for classification. Otherwise, it indicates that XCS is unstable and is exploring the search space.

If a concept is changed, a large increase in the prediction error is likely. Turning this around, if the system experiences a massive increase in the prediction error after a period of stability, a concept change is a likely explanation.

There is a second possible explanation for this phenomenon: the covering. The covering process of XCS is activated when a previously unseen input arrives. Normally, XCS forms a match set $[M]$ for each input and an action is chosen from the classifiers in the set. If the match set $[M]$ is empty, no existing classifiers satisfy the input, so the covering technique will generate random classifiers to match the input. Those classifiers are initialized with some random values and inserted into the population. It is then the reinforcement component's responsibility to update these parameters appropriately to make the classifiers more accurate. This takes some time; until there has been a sufficient number of evaluations they are not accurate. If they are used before they become accurate, it may cause a decrease in the predictive performance of the system.

Thus another explanation for an increase in prediction error after a period of stability is that XCS is exploring a new area in the search space that it has not experienced before.

To recognize a possible concept change, the user can specify an error threshold and a window size. If the average prediction error within the window exceeds the threshold, it indicates that a concept change may have occurred.

To ensure the phenomenon is caused by a concept change, and not by a covering process, a counter is used in the window. It counts the number of times covering has been used within the window of training instances. A covering threshold is specified in advance by the user. The system will not conclude a concept change when the prediction error increases markedly, if the covering threshold is reached.

7.6.3 Revising Knowledge

Figure 7.3 reveals that XCS needs a long time to recover from severe changes. Observing the performance of XCS before and after the change, we notice that the second cycle requires much longer to reach 100% accuracy than the first cycle. In the first cycle, the system starts from scratch, and knowledge accumulates with experience. In the second cycle, many classifiers of XCS suddenly become inaccurate after the severe change. The system has to fulfill several tasks to recover, such as re-evaluating inaccurate classifiers, exploring new classifiers, etc.

We seek to revise the system's knowledge when a concept change is recognized, to avoid the long time needed to recover from the concept change. We propose two strategies:

1. **Re-initializing the population**: This strategy is for the current knowledge from the previous cycle to be entirely deleted after detecting a concept change. The system then starts from scratch without any knowledge

about the environment. It becomes similar to the learning process in the first cycle, and therefore it will not take much longer than the first cycle to achieve a stable performance.

2. **Re-initializing classifier parameters**: XCS maintains a complete action set, which contains all possible combination pairs of rules and actions in the population that are maximally general and accurate.
   The criterion for an accurate classifier in XCS is not only that it has a correct pair of rule and action, but that it can predict a future reward as closely as possible to the actual reward. For example, suppose an environment gives two reward levels: 0 and 1000 for an incorrect and correct prediction respectively. A classifier having a correct rule (a correct pair of condition and action) is considered accurate if it can predict a reward of 1000 for its action. An incorrect classifier (a wrong pair of condition and action) is also considered accurate if its prediction is close to 0. However, incorrect but accurate classifiers always have low reward predictions compared to correct and accurate ones. Therefore, they are not usually chosen from the match set $[M]$.
   In essence, the population of XCS always maintains a set of incorrect but accurate classifiers. These classifiers may become correct after the environment changes. Hence it is not efficient to delete them completely, as in the previous strategy, because XCS will need time to discover and re-introduce them in the population. However, it is also not efficient to keep the population as it is, because XCS needs to re-evaluate all existing classifiers in order to reduce the fitness of classifiers that have become inaccurate after the change.
   Therefore, our third strategy is to keep the whole population, but re-initialize their parameters so that XCS can start the second cycle with some prior knowledge.

7.6.4 Parameter Setup

To detect the concept change, we set the thresholds as follows:

- Error threshold: $E = 0.4$
- Covering threshold: $C = 1$
- Window size: $w = 50$

In the third strategy, the parameters of classifiers in $[P]$ will be reset after detecting the change. The parameters are re-initialized as follows:

- Reward prediction $P = 10$
- Reward prediction error $\varepsilon = 0$
- Fitness $F = 0.01$
- Numerosity $n = 1$
- Experience $e = 0$
- The action set size $a = 1$

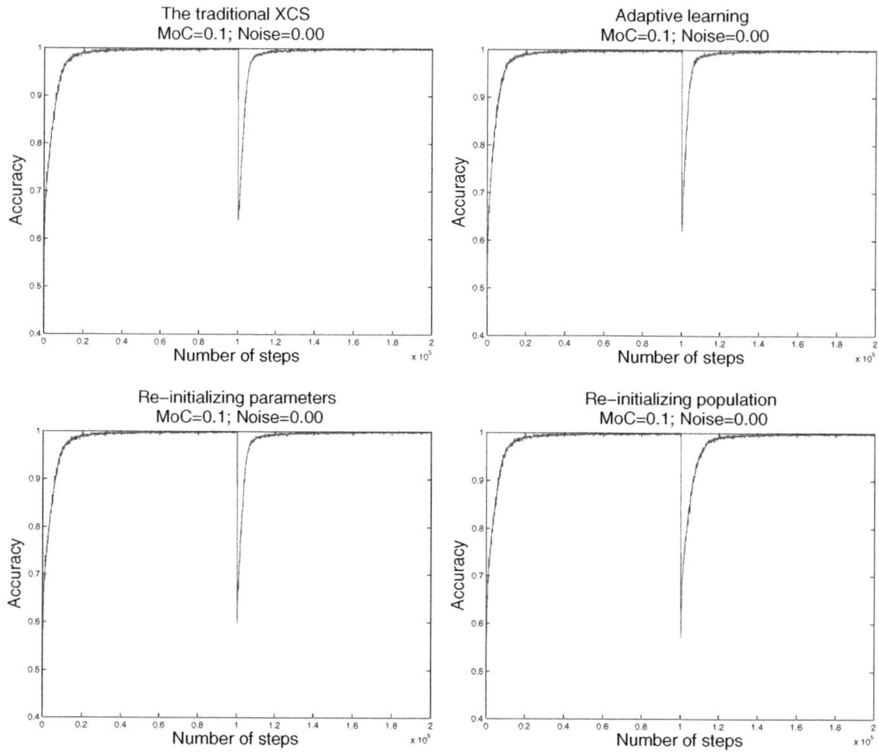

Fig. 7.5. Performance of different strategies when magnitude of change is 0.1. The curve plots the predictive accuracy of the system.

7.7 Comparing the Performance of Different Strategies

7.7.1 Noise-Free Environment

Figure 7.5 and Figure 7.6 show the performance of XCS with our proposed strategies, at two levels of MoC: small and large or 0.1 and 0.8 respectively.

When MoC is small (Figure 7.5), the performance of XCS is quite similar under each strategy, after the change at the $100,000^{th}$ time step. The re-initializing population strategy seems to require a little more time than other strategies to recover from changes. A small MoC results in small change in the model, which makes only a small portion of classifiers become inaccurate. By re-initializing the population, XCS needs to learn from scratch. Other strategies take advantage of reusing previous knowledge and therefore can recover more quickly. Hence, in the case of small MoC, the traditional XCS, or the adaptive learning strategy, or the re-initializing classifier parameters strategy, all seem to be better than the re-initializing population strategy.

When MoC is large (Figure 7.6), the graphs show that XCS with the adaptive learning strategy improves enormously over the traditional XCS in

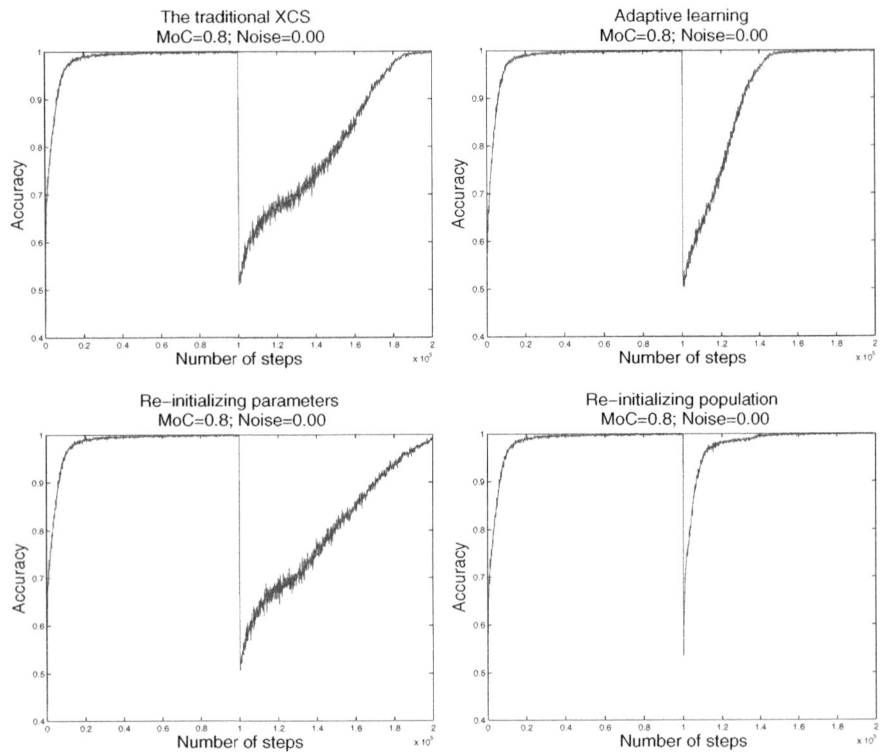

Fig. 7.6. Performance of different strategies when magnitude of change is 0.8. The curve plots the predictive accuracy of the system.

terms of recovery time. XCS with this strategy requires as little as half of the normal time needed to approach 100% accuracy. Because MoC is large, a large portion of classifiers in the population becomes inaccurate. The traditional XCS requires a long time to explore new rules and delete inappropriate rules from the population. The increased learning rate in the adaptive learning strategy speeds up the learning process of the system. Therefore, it helps the system recover more quickly from changes in the underlying concepts.

Even with a faster learning process, XCS needs some time to re-evaluate a number of inaccurate but high fitness classifiers. When MoC is large, the number of affected classifiers is high, this re-evaluation time is big enough that it is more effective to re-learn from scratch. In this case, the re-initializing population strategy can be seen as the best approach.

The main disadvantage of the re-initializing population strategy is that the system needs to be capable of detecting the concept change accurately before the population is re-initialized. For recognizing the concept change, the system depends on the user's specified error threshold. An important question is how to set the error threshold correctly. If the error threshold is too high, the system may ignore small changes. If the threshold is low and the data is noisy, it might cause false alarms, so the system may re-initialize its population unnecessarily when there has not been a true concept change. So noisy data might cause the population to be re-initialized frequently, and hence knowledge is not able to accumulate during the learning.

In our experiments, it seems the system was able to detect concept changes accurately with the proposed thresholds. However, depending on the nature of the target problem, these thresholds need to be changed accordingly. In the future, we plan to investigate an adaptive error threshold, which depends on the performance degradation.

The re-initializing parameters strategy does not appear to work. It performs no better than the other strategies when MoC is low, and performs worse than the others when MoC is high. Re-initializing the parameters makes all classifiers equally important in the population. Initially, we hoped that good and bad classifiers will be identified faster, therefore speeding up the recovery process. However, it might be harder to XCS to detect inaccurate classifiers because the evaluation needs to start from scratch.

The next section will look into more details of noise and how these strategies behave with different levels of noise.

7.7.2 The Effect of Noise

Figure 7.7 shows the performance of XCS with different strategies, when noise is medium and the concept changes significantly. The performance of XCS in all experiments could not reach 100% accuracy due the noise in the training data. XCS requires a similar period of time to recover from the change, under all strategies.

Unlike the previous subsection, when noise is incorporated in the training data the re-initialized population strategy does not show better performance than other strategies. This is because the error threshold was set to 0.4. With noiseless data the error difference between before and after the concept change is nearly 0.5, which is greater than the threshold. Thus, the system could recognize the concept change. The noisy environment makes the error difference drop below 0.4. Therefore, the system cannot recognize the change, and it behaves like the traditional XCS.

Figure 7.8 shows the performance of XCS with the different strategies, when the noise level is 0.15 and a severe change occurs.

The adaptive learning strategy performs better than all other strategies. The predictive curve of this learning strategy is steeper than the curves for the other strategies. XCS becomes more flexible in adapting to a new environment. By having an adaptive learning rate, XCS is able to detect useless classifiers faster so that they can be eliminated as quickly as possible. Also, the useful classifiers will be re-evaluated and become accurate more quickly, so they can participate correctly in the decision making process.

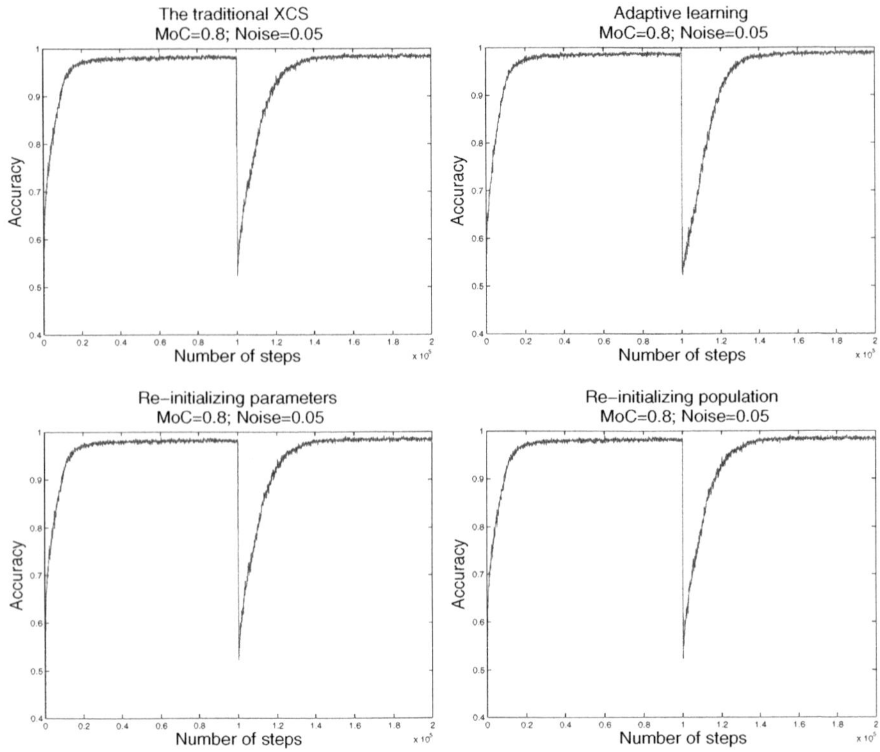

Fig. 7.7. Magnitude of change is 0.8. Noise is 0.05. The curve plots the predictive accuracy of the system.

In conclusion, the re-initializing population strategy outperforms all other strategies in a noise free environment. The strategy depends on the error threshold to recognize the concept change explicitly in order to response to the change. The adaptive learning strategy seems to be better than other strategies, when the noise level is high.

7.8 Conclusions

This chapter explored XCS in dynamic environments with different degrees of concept change. We found that the conventional XCS is capable of recovering quickly when dealing with small magnitudes of change. However, when the magnitude of change is high, a long recovery time is required for the system to achieve a stable performance.

We proposed three strategies, which aim to reduce the recovery time of XCS after concept changes. We found that the re-initializing population

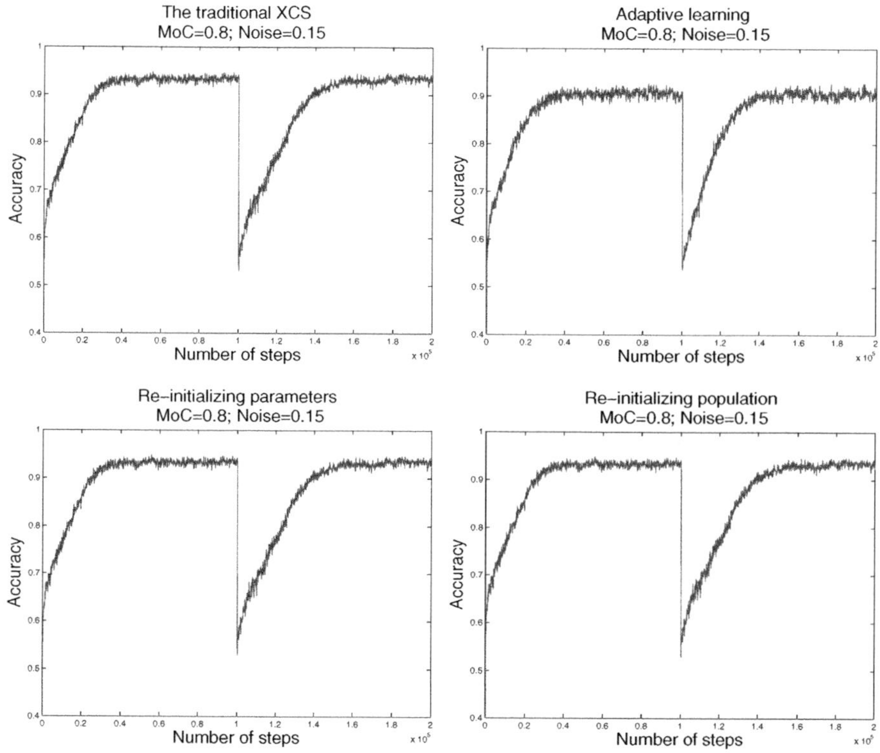

Fig. 7.8. Magnitude of change: 0.8, Noise: 0.15. The curve plots the predictive accuracy of XCS.

strategy dramatically reduces recovery time in a noise free environment. The adaptive learning approach is the next best.

We also investigated the effect of noise on recovery time after a concept change. We have found out that adding small noise to the conventional XCS requires a longer recovery time in comparison to medium noise. Also a very noisy environment is a big challenge for XCS to perform accurately and recover from a concept change. The adaptive learning strategy achieves a better predictive performance when compared to other strategies.

In the future, we are interested in testing further our strategies on different data sets from the UCI repository [13].

Acknowledgments

We would like to thank Stewart Wilson and Martin Butz for their valuable comments. The research reported in this paper was funded by the Australian Research Council Linkage grant number LP0453657.

References

1. H. A. Abbass, J. Bacardit, M. V. Butz, and X. Llora. *Online Adaptation in Learning Classifier Systems: Stream Data Mining.* Illinois Genetic Algorithms Laboratory, University of Illinois at Urbana-Champaign, June 2004. IlliGAL Report No. 2004031.
2. J. Bacardit and M. V. Butz. *Data Mining in Learning Classifier Systems: Comparing XCS with GAssist.* Illinois Genetic Algorithms Laboratory, University of Illinois at Urbana-Champaign, June 2004. IlliGAL Report No. 2004030.
3. E. Bernadó, X. Llorà, and J. M. Garrell. XCS and GALE: a comparative study of two learning classifier systems with six other learning algorithms on classification tasks. In *Proceedings of the 4th International Workshop on Learning Classifier Systems (IWLCS-2001)*, pages 337–341, 2001. Short version published in Genetic and Evolutionary Compution Conference (GECCO2001).
4. J. Branke. *Evolutionary Optimization in Dynamic Environments*, volume 3 of *Genetic Algorithms and Evolutionary Computation.* Kluwer Academic Publishers, 2002.
5. M. V. Butz. *Rule-based Evolutionary Online Learning Systems: Learning Bounds, Classification, and Prediction.* PhD thesis, University of Illinois at Urbana-Champaign, 2004.
6. M. V. Butz, T. Kovacs, P. L. Lanze, and S. W. Wilson. Toward a theory of generalization and learning in XCS. *IEEE Tranactions on Evolutionary Computation*, 7(6), 2003.
7. M. V. Butz and S. W. Wilson. An algorithmic description of XCS. In *IWLCS '00: Revised Papers from the Third International Workshop on Advances in Learning Classifier Systems*, pages 253–272. Springer-Verlag, 2001.
8. H. H. Dam, H. A. Abbass, and C. Lokan. DXCS: an XCS system for distributed data mining. In *Proceedings of the Genetic and Evolutionary Computation Conference, GECCO-2005*, Washington D.C., USA, 2005.
9. H. H. Dam, H. A. Abbass, and C. Lokan. Investigation on DXCS: An XCS system for distribution data mining, with continuous-valued inputs in static and dynamic environments. In *Proceedings of IEEE Cogress on Evolutionary Computation*, Edinburgh, Scotland, 2005.
10. K. A. De Jong, W. M. Spears, and D. F. Gordon. Using genetic algorithms for concept learning. *Machine Learning*, 13(2-3):161–188, 1993.
11. S. Delany, P. Cunningham, A. Tsymbal, and L. Coyle. A case-based technique for tracking concept drift in spam filtering. *Journal of Knowledge Based Systems*, 18(4–5):187–195, 2005.
12. P. W. Dixon, D. Corne, and M. J. Oates. A preliminary investigation of modified XCS as a generic data mining tool. In *Advances in Learning Classifier Systems: 4th International Workshop, IWLCS*, pages 133–150. Berlin Heidelberg: Springer-Verlag, 2001.
13. C. B. D. J. Newman, S. Hettich and C. Merz. UCI repository of machine learning databases. University of California, Irvine, Department of Information and Computer Sciences. http://www.ics.uci.edu/~mlearn/MLRepository.html, 1998.
14. U. Fayyad, G. Piatetsky-Shapiro, P. Smyth, and R. Uthurusamy. From data mining to knowledge discovery: An overview. In *Advances in Knowledge Discovery and Data Mining*, pages 1–36. The MIT Press, 1996.
15. D. E. Goldberg. *Computer-Aided Gas Pipeline Operation using Genetic Algorithms and Rule Learning.* PhD thesis, The University of Michigan, 1983.

16. D. E. Goldberg. *Genetic Algorithms in Search, Optimization, and Machine Learning.* Addision-Wesley Publishing Company, INC., 1989.
17. M. B. Harries, C. Sammut, and K. Horn. Extracting hidden context. *Machine Learning*, 32(2):101–126, 1998.
18. J. H. Holland. *Adaptation in Natural and Artificial Systems.* University of Michigan Press, Ann Arbor, 1975. Republished by the MIT press, 1992.
19. J. H. Holland. Escaping Brittleness: The Possibilities of General-Purpose Learning Algorithms Applied to Parallel Rule-Based Systems. In Mitchell, Michalski, and Carbonell, editors, *Machine Learning, an Artificial Intelligence Approach. Volume II*, chapter 20, pages 593–623. Morgan Kaufmann, 1986.
20. J. H. Holland and J. S. Reitman. Cognitive systems based on adaptive algorithms. In D. A. Waterman and F. Hayes-Roth, editors, *Pattern-directed Inference Systems.* New York: Academic Press, 1978. Reprinted in: Evolutionary Computation. The Fossil Record. David B. Fogel (Ed.) IEEE Press, 1998. isbn: 0-7803-3481-7.
21. R. Klinkenberg and T. Joachims. Detecting concept drift with support vector machines. In *ICML '00: Proceedings of the Seventeenth International Conference on Machine Learning*, pages 487–494, San Francisco, CA, USA, 2000. Morgan Kaufmann Publishers Inc.
22. J. Kolter and M. Maloof. Dynamic weighted majority: A new ensemble method for tracking concept drift. In *Proceedings of the Third IEEE International Conference on Data Mining*, pages 123–130, Los Alamitos, CA, 2003. IEEE Press.
23. T. Kovacs. XCS classifier system reliably evolves accurate, complete, and minimal representations for boolean functions. In C. Roy and Pant, editors, *Soft Computing in Engineering Design and Manufacturing (WSC2)*, pages 59–68. Springer-Verlag, 1997.
24. T. Kovacs. Two views of classifier systems. In *Fourth International Workshop on Learning Classifier Systems - IWLCS-2001*, pages 367–371, San Francisco, California, USA, 7 2001.
25. P. L. Lanzi and M. Colombetti. An extension to the XCS classifier system for stochastic environments, 1999.
26. S. Saxon and A. Barry. XCS and the Monk's problems. In *Learning Classifier Systems, From Foundations to Applications*, pages 223–242, London, UK, 2000. Springer-Verlag.
27. J. C. Schlimmer and D.H. Fisher. A case study of incremental concept induction. In *Proceedings of the Fifth National Conference on Artificial Intelligence*, pages 496–501, Philadelpha, PA, 1986. Morgan Kaufmann.
28. S. F. Smith. *A Learning System Based on Genetic Adaptive Algorithms.* PhD thesis, University of Pittsburgh, 1980.
29. C. Stone and L. Bull. For real! XCS with continuous-valued inputs. *Evolutionary Computation*, 11(3):299–336, 2003.
30. W. N. Street and Y. Kim. A streaming ensemble algorithm (sea) for large-scale classification. In *KDD '01: Proceedings of the seventh ACM SIGKDD international conference on Knowledge discovery and data mining*, pages 377–382, New York, NY, USA, 2001. ACM Press.
31. H. Wang, W. Fan, P. S. Yu, and J. Han. Mining concept-drifting data streams using ensemble classifiers. In *KDD '03: Proceedings of the ninth ACM SIGKDD international conference on Knowledge discovery and data mining*, pages 226–235, New York, NY, USA, 2003. ACM Press.

32. G. Widmer and M. Kubat. Learning in the presence of concept drift and hidden contexts. *Machine Learning*, 23(1):69–101, 1996.
33. S. W. Wilson. Knowledge Growth in an Artificial Animal. In *Proceedings of the First International Conference on Genetic Algorithms and their Applications*, pages 16–23. Lawrence Erlbaum Associates, 1985.
34. S. W. Wilson. Classifier fitness based on accuracy. *Evolutionary Computation*, 3(2):149–175, 1995.
35. S. W. Wilson. Generalization in the XCS classifier system. In J. R. Koza, W. Banzhaf, K. Chellapilla, K. Deb, M. Dorigo, D. B. Fogel, M. H. Garzon, D. E. Goldberg, H. Iba, and R. Riolo, editors, *Genetic Programming 1998: Proceedings of the Third Annual Conference*, pages 665–674, University of Wisconsin, Madison, Wisconsin, USA, 1998. Morgan Kaufmann.
36. S. W. Wilson. Get real! XCS with continuous-valued inputs. In P. Lanzi, W. Stolzmann, and S. Wilson, editors, *Learning Classifier Systems, From Foundations to Applications, LNAI-1813*, pages 209–219, Berlin, 2000. Springer-Verlag.

8

Adaptive Business Intelligence: Three Case Studies

Zbigniew Michalewicz[1], Martin Schmidt[2], Matthew Michalewicz[2], and Constantin Chiriac[2]

[1] School of Computer Science, University of Adelaide, Adelaide, SA 5005, Australia, also with Institute of Computer Science, Polish Academy of Sciences and Polish-Japanese School of Information Technology, Warsaw, Poland
zbyszek@cs.adelaide.edu.au

[2] SolveIT Software, PO Box 3161, Adelaide, SA 5000, Australia
{ms,mm,cc}@solveitsoftware.com

Summary. This chapter contains a general discussion on the prediction and optimization issues present in dynamic environments, and explains the ideas behind Adaptive Business Intelligence. The chapter also presents three diverse case studies. The first deals with pollution control, the second one with ship navigation, and the third one with car distribution. All these problems are set in dynamic environments; all three problems require some level of prediction (prediction of weather for pollution optimization, prediction of paths for unidentified ships at sea, and prediction of prices for cars sold at different auction sites). All these problems also require optimization for recommending the best course of action.

8.1 Introduction

Every problem has an objective. Usually, this is a general statement describing what we are looking for. The objective defines the goal (or set of goals[3]) for a particular problem. These goals are translated into evaluation functions, which provide mappings from the solution space to a set of numbers. Thus, evaluation functions assign numeric values for each solution for each specified goal.

Evaluation functions (for single-objective problems) or a set of evaluation functions (for multi-objective problems) are key components of any heuristic method (whether genetic algorithms, tabu search, simulated annealing, ant system, or even simple hill-climbers), as they define the connection between the method and the problem. By assigning a numeric quality measure to

[3] The objective may consist of several goals, thus making the problem multi-objective. Note, some confusing terminology: a problem, where the objective consists of a few goals, is called a multi-objective problem.

Z. Michalewicz et al.: *Adaptive Business Intelligence: Three Case Studies*, Studies in Computational Intelligence (SCI) **51**, 179–196 (2007)
www.springerlink.com

each solution, evaluation functions allow comparison between the quality of various candidate solutions. Note that evaluation functions may return just the rank of a candidate solution among a set of solutions, a precise number (when the evaluation function is defined as a closed formula), or they may include various components (as penalty expressions for cases when a candidate solution violates some problem-specific constraints).

Many real world problems are set in uncertain (possibly changing) environments. There is a general agreement [2] that such uncertainties can be categorized into four classes: (1) noise, (2) robustness, (3) approximation, and (4) time-varying environments. Consequently, evaluation functions should be modified accordingly to deal with each particular case. However, it seems that the above classification misses the most important (and probably most frequent) real world scenario: namely, where the evaluation functions are based on predictions of the future values of some variables. Using case studies, we illustrate three such scenarios, expose the similarities between them, and we suggest a system architecture (called Adaptive Business Intelligence) to deal with such problems.

This chapter is organized as follows: The next section provides a brief overview of four categories of uncertainties and introduces a new category, where the evaluation functions are based on predictions of some variables. The next three sections present three case studies: optimisation of pollution in Poland, path planning (ship navigation), and car distribution. We then discuss the common characteristics of these case studies, and propose an Adaptive Business Intelligence architecture to deal with such problems. The last section concludes this chapter.

8.2 Uncertain Environments

As mentioned in the Introduction, uncertain (and possibly changing) environments are usually categorized into four classes: (1) noise, (2) robustness, (3) approximation, and (4) time-varying environments. Before we present and discuss the fifth category, and argue that this fifth category is the most common in real word situations, let's first discuss the main features of these four categories.

Noise. Sometimes evaluation functions are subject to noise. This happens when evaluation functions return sensory measurements or results of randomised simulations. In other words, the evaluation procedure for the same solution (i.e., the solution defined as a vector of some design variables) may return different values. The common approach in such scenarios is to approximate a noisy evaluation function eval by an averaged sum of several evaluations:

$$eval(\boldsymbol{x}) = 1/n \sum_{i=1}^{n} (f(\boldsymbol{x}) + z_i), \tag{8.1}$$

where $\boldsymbol{x}$ is a vector of design variables (i.e., variables controlled by a method), $f(\boldsymbol{x})$ is the evaluation function, z_i represents additive noise, and n is the sample size. Note that the only measurable (returned) values are $f(\boldsymbol{x}) + z$.

Robustness. Sometimes design variables, other variables, or constraints of the problem are subject to perturbations after the solution is determined. The general idea is that such (slightly modified) solutions should have quality evaluations (thus making the original solution robust). This is important in scenarios involving manufacturing tolerances, or when it is necessary to modify the original solution because of employee illness or machine failure. The common approach to such scenarios is to use evaluation function eval based on the probability distribution of possible disturbances δ, which is approximated by Monte Carlo integration:

$$eval(\boldsymbol{x}) = 1/n \sum_{i=1}^{n} f(\boldsymbol{x} + \delta_i). \tag{8.2}$$

Note that $eval(\boldsymbol{x})$ depends on the shape of $f(\boldsymbol{x})$ at point $\boldsymbol{x}$; in other words, the neighbourhood of $\boldsymbol{x}$ determines the value of $eval(\boldsymbol{x})$.

Approximation. Sometimes it is too expensive to evaluate a candidate solution. In such scenarios, evaluation functions are often approximated based on experimental or simulation data (the approximated evaluation function is often called the meta-model). In such cases, evaluation function eval becomes:

$$eval(\boldsymbol{x}) = f(\boldsymbol{x}) + E(\boldsymbol{x}), \tag{8.3}$$

where $E(\boldsymbol{x})$ is the approximation error of the meta-model. Note that the approximation error is quite different than noise, as it is usually deterministic and systematic.

Time-varying environments. Sometimes evaluation functions depend on an additional variable: time. In such cases, evaluation function eval becomes:

$$eval(\boldsymbol{x}) = f(\boldsymbol{x}, t), \tag{8.4}$$

where t represents the time variable. Clearly, the landscape defined by the function f changes over time; consequently, the best solution may change its location over time. There are two main approaches for handling such scenarios: (1) to restart the method after a change, or (2) require that the method is capable of chasing the changing optimum.

However, it seems the largest class of real world problems is not included in the above four categories. From our business/industry experience of the last decade, it is clear that in many real world problems the evaluation functions are based on the predicted future values of some variables. In other words, evaluation function eval is expressed as:

$$eval(\boldsymbol{x}) = f(\boldsymbol{x}, P(\boldsymbol{x}, \boldsymbol{y}, t)), \tag{8.5}$$

where $P(\boldsymbol{x}, \boldsymbol{y}, t)$ represents an outcome of some prediction for solution vector $\boldsymbol{x}$ and additional (environmental, beyond our control) variables $\boldsymbol{y}$ at time t. Let's compare this category with the four categories defined earlier to see the differences between them.

First of all, noise may or may not be involved. If the prediction model is deterministic, then there is no noise in the scenario: every solution vector $\boldsymbol{x}$ is evaluated to the same value. On the other hand, if the prediction model involves simulations, noise might be present. Second, the meaning of robustness is quite different. Unexpected disturbances (e.g., delays) influence the outcomes of the prediction model, and should be handled accordingly. Third, the concept of approximation is different. Note, that we can evaluate a candidate solution precisely (i.e., the evaluation function is not expensive), however, approximation is connected with uncertainties of the predictions. Finally, the time-changing environment also has a different meaning. As the real world changes, the prediction model needs constant updates and/or parameter adjustments, thus changing the problem landscape in an implicit way.

In the following three sections we illustrate this particular category of real world problems by providing three case studies. It is important, however, to keep in mind that *many* other business problems clearly fall into this category: from inventory control problems (where it is necessary to forecast demand), through marketing problems (where it is necessary to predict the ratings of some programs several weeks ahead of time), to portfolio management problems (where it is necessary to predict some economic variables). All of these examples share the common characteristic: optimization of the evaluation function is based on some prediction model. This also means that the quality of the results provided by the optimizer depends on the quality of the prediction model: there is no point in optimising anything if the predictions are bad to begin with! These three case studies briefly describe three diverse problems and illustrate (in general terms) solutions based on Adaptive Business Intelligence (we will return to these concepts towards the end of this chapter).

8.3 Case Study #1: Pollution Control

This case study is based on a "pollution control" research project [5] completed during late 1990s. The main objective of this project was to reduce the ecological damage caused by 132 power stations in Poland. To solve this challenging problem, a system was developed to control the production and distribution of energy. The system included data on emission sources, atmospheric pollution dispersion, deposition models, as well as the ecosystem's capacity to sustain high levels of pollution.

To create a precise model of Poland (which has an approximate land area of 900 km by 750 km), a computational grid with a spatial resolution of 30 km by 30 km was used. This model considered three groups of emission sources: 1) sources from abroad (mainly from Germany and the Czech Republic); 2) Polish sources from the private sector; and 3) Polish sources from the public sector. The first two groups of SO_2 concentrations were defined as *background concentrations*, as the system did not control these sources (it only controlled the third group). However, the resulting concentration of SO_2 in each square of the computational grid was the sum of both of the background concentration and the concentration caused by the third group.

To find the optimal production level for each power station, an evolutionary algorithm was used. This optimisation technique was extended with memory structures, which were particularly useful for "remembering" past patterns. The system based its recommendations (i.e., the optimal energy production level for each power station) on weather forecasts for the following five days. When new weather reports and forecasts became available (every four hours), the system incorporated this new information thus changing the problem landscape. However, "chasing" the optimum was not straightforward because of problem-specific constraints: drastic changes in energy production levels are cost prohibitive.

Several experiments with real data (meteorological data and weather forecasts, and emission data for selected summer and winter periods) were conducted using this new system. Actual pollution levels were calculated for the preceding year, and then the optimization process was applied to this historical data and the results were compared with the actual, non-optimized production schedules. From these experiments, the following conclusions were drawn: The amount of produced energy was equal in both cases, the aggregate operating cost was also equal in both cases, but the ecological damage in Poland was substantially less (12-15%) when the optimization process was applied. A "savings" of 12-15% in ecological damage is a tremendous result, especially when one considers that human lives are at risk in this particular problem. As indicated at the beginning of this section, similar "savings" can be realized in other problem areas, even if the problem particulars are substantially different.

The simplest way to present the results of this system is through a graphical interfaces that compares the optimised solutions with non-optimized solutions (see Figure 8.1). The left-hand-side map shows the non-optimized results for ecological damage in Poland, while the right-hand-side map shows the optimized results for the same time period. Clearly, there are fewer dark squares on the right-hand-side (optimized) map. This is reflected in the concentration and emission numbers shown at the bottom of the screen. It is important to note that the system achieved these lower pollution levels while maintaining the same power production output of approximately 23,849 megawatts per hour.

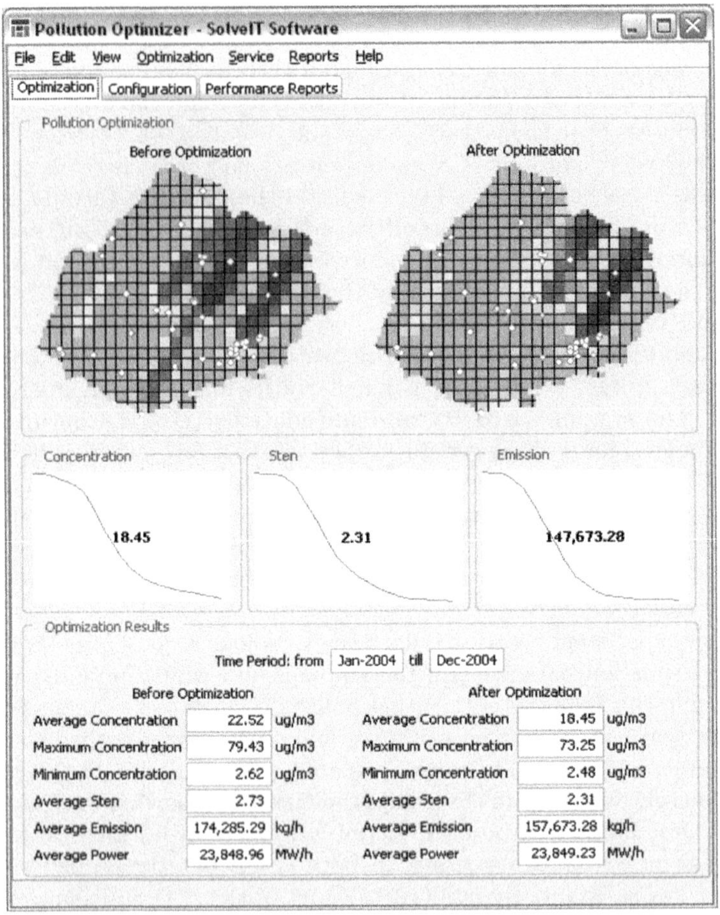

Fig. 8.1. Graphical interface for pollution control system

As the weather forecast provided precise values of temperatures, humidity, pressures, wind velocity and directions, cloud coverage and levels, etc., the problem was not noisy. Robustness of the solution has a different flavour however, as it was connected with uncertainties in the forecasting; thus the evaluation function was:

$$eval(\boldsymbol{x}) = 1/n \sum_{i=1}^{n} f(\boldsymbol{x}, P(\boldsymbol{x}, \boldsymbol{y} + \delta_i, t)), \tag{8.6}$$

where δ_i represents possible disturbances in accuracy of weather forecasts. Approximation was not the issue, as no meta-model was built. A time-changing environment was implied by the forecasting model, which received new information at regular intervals. However, the algorithm did not follow "the moving peak" of a dynamic landscape (which is the focus of the majority of research papers in this area, see [1, 8, 12]), but rather, it tried to find a production pattern that satisfied problem-specific constraints, kept production cost at the same level, and minimized ecological damage over the next few days.

8.4 Case Study #2: Path Planning

This case study is based on an ongoing project that started in the mid-1990s for the Gdynia Maritime University in Poland [9]. The problem can be described as follows: A ship sails in an environment with natural constraints (e.g., lands, canals, shallow waters), and other constraints resulting from formal regulations (e.g., traffic restricted zones, fairways, etc.). These constraints are assumed to be stationary and are defined by polygons-in a similar manner to those used in creating electronic maps.

When sailing in a stationary environment, a ship (called "own ship") meets other sailing ships (called "strange ships" or "targets"), some of which constitute a collision threat. The degree of the collision threat with dangerous targets is not constant and depends on the approach parameters: *Distance at Closest Point of Approach* and *Time of Closest Point of Approach*, as well as the speed ratio of both ships and the distance and bearing of the target. It is assumed that the dangerous target has appeared in the area of observation and can cross the predicted course of the own ship at a dangerous distance. Actual values of this distance depend on the assumed time horizon. The ranges of five to eight nautical miles in front of the bow, and two to four nautical miles behind the stern of the ship are assumed. The threatening targets are interpreted as having dangerous paths and speeds as determined by the Automatic Radar Plotting Aids (ARPA) system. The moving constraints represent the approaching ships, and the shape of each constraint depends on the safety conditions (on an assumed value of the safe approach distance, assumed safe distance, speed ratio, and bearing of the moving target).

The operator selected a safe distance on the basis of weather conditions, sailing area, and speed of the ship. When planning the safe trajectory, the system should take into account both the fixed constraints, and the areas of danger represented by the moving targets, which dynamically change their locations. Figure 8.2 displays model of the environment where:

- fixed navigation constraints are modelled using convex and concave polygons,
- moving targets are modelled as moving hexagons,
- the dimensions of the own ship are neglected due to the small length of the own ship with respect to the maximum length of the areas representing the moving targets.

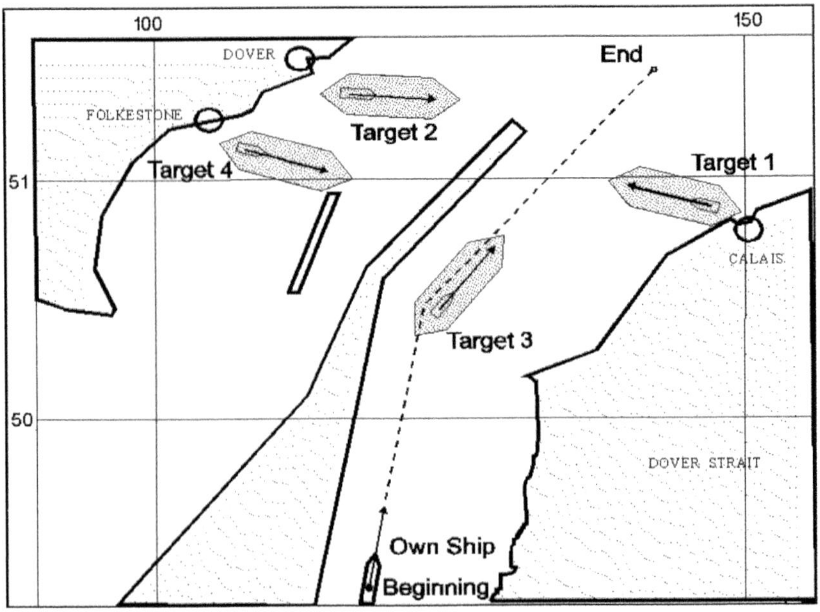

Fig. 8.2. A sample environment

According to transport plans, the own ship should cover a given route in some assumed time. On the other hand, it has to move along a given trajectory safely (i.e., it must avoid navigation obstacles and cannot come too close to other moving targets). Estimating a ship's trajectory in a collision situation represents a difficult trade-off between a necessary deviation from a given course and the safety of sailing. Hence, this is a multi-criterion planning problem, which takes into account the safety and economy of the ship's motion.

Many simulation experiments were conducted for this problem. For example, figure 8.3 represents the navigational situation in which the own ship passes around three islands and four moving targets coming from different directions and at different speeds.

The speed of the own ship was defined as equal to 3.6, 8.6, or 13.6 knots. The progress of trajectory adaptations is shown above after 200, 500, and 1,000 generations, respectively. As with the pollution optimiser, evolutionary algorithms were used to evolve the optimal trajectory for the own ship. Moving along the determined trajectory with changing speed, the own ship can sail in between the islands and pass the targets in front of their bows or behind their sterns. The ship speed is changed along subsequent trajectory sections. Initially, the ship reduces the speed to pass the first two targets and then, after sailing in between the islands, it increases the speed to pass Target 3 and 4, at the same time making it possible to reach the final destination in the shortest time. The execution of the proposed solution gives the optimum trajectory

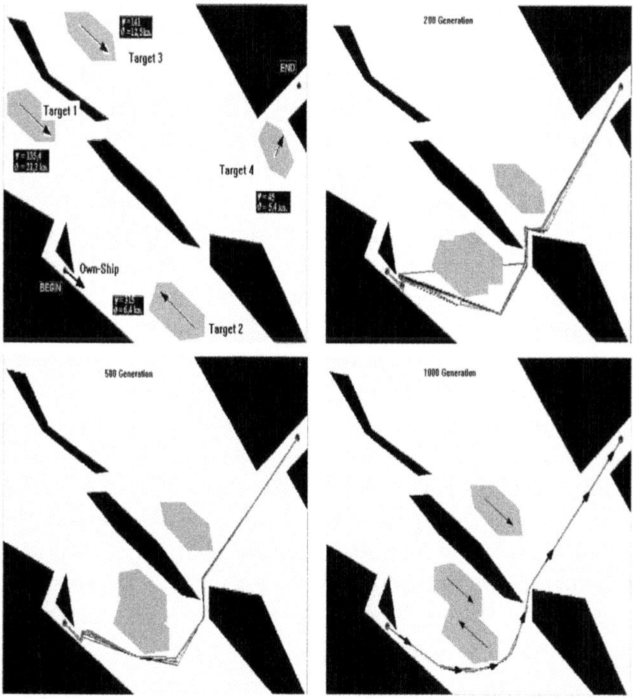

Fig. 8.3. Evolution of paths for particular environment with four moving targets

with respect to safety and economic criteria. Many additional experiments were reported in [9] and [10].

In this case study, the prediction model assumed that the target ships would continue on their courses at their current speeds. Because the model returned precise values (directions and speeds of target ships), the problem was not noisy. Robustness of the solution had a different flavour once again, as it was connected with uncertainties of external events (e.g., small changes of directions and speeds of target ships). Thus:

$$eval(\boldsymbol{x}) = 1/n \sum_{i=1}^{n} f(\boldsymbol{x}, P(\boldsymbol{x}, \boldsymbol{y} + \delta_i, t)), \tag{8.7}$$

where δ_i represented possible disturbances in the accuracy of external variables. Approximation was not an issue, as no meta-model was built. A time-changing environment was implied by the changes in speed and direction of target ships.

8.5 Case Study #3: Car Distribution System

This case study is based on a software project completed in 2002 for a major car leasing company in United States [6]. The problem was to recommend the best distribution of "off-lease" cars (which are returned to the company after the lease term is over) to many available auction sites. By "best distribution," we mean a distribution that maximizes the net total proceeds from all these cars. Many issues have to be considered in the process of finding the optimal solution to this problem, including price predictions for the various types of cars at different auction sites, price depreciation, volume effects, and transportation costs.

A leased car is owned by a finance company, which gets the car back at the end of the lease term. Each car is different, and these differences include make and model, mileage, model year, colour, type of transmission, body style and options, wear and tear (i.e., the damage level), etc. There are hundreds of auction sites around the United States, and each off-lease car has to be sold at one of them. The characteristics listed above (plus some others) influence the sale price of a car at each particular location. The central question is, Where should each car be sent to maximize the total proceeds from all these auctions sales?

Fig. 8.4. Cars to be distributed (darker circles) and 50 auction sites (lighter circles)

Figure 8.4 illustrates the case (for a particular day of operation). The darker circles illustrate areas where the leased cars were returned on a particular day.

The larger the circle, the more cars were returned in that area (clearly, most cars were returned in the eastern side of the United States). Note that the distribution of cars would look different each day, as people and organizations will return their cars at different locations. The lighter circles, on the other hand, illustrate 50 auction sites that are available for sending the off-lease cars to. The locations of these auction sites are fixed: they are the same every day for any number of returned cars.[4] At the first glance, the problem may look easy. One might be tempted just take one car at a time, consult some report[5] to find the average sale price for this particular type of car (remember to adjust the price for mileage, options, etc.) at each particular auction, and decide to send the car to the auction with the highest average sale price. Of course, the transportation cost should also be taken into account (usually, the longer distance, the higher cost), but all these calculations are manageable. However, there are other issues that must be taken into account:

- *Transportation.* When a whole truckload of cars is shipped from one location to another, the company is charged a cheaper rate. Therefore, it is important to take into account the number of cars that are transported on each truck.
- *Risk factors.* Cars can be damaged, stolen, or the transportation truck might be involved in an accident. Longer trips also increase the probability of a delay, and the solution must take this into account.
- *Volume sensitivity effect.* If many cars of the same type are sent to the same auction site, then the volume effect applies: oversupply of a particular type of a car will decrease its price.
- *Size of the search space.* The distribution of 4,000 cars to 50 auction sites gives us 50^{4000} possible distributions, which is much larger than the estimated numbers of atoms in the Universe.
- *Price depreciation.* Every auction site has a typical sale day (e.g., every Wednesday at 10 am, or every second Thursday at 11am). If some cars arrive one day after the sale date, then they will have sit at the auction site for one or two weeks (until the next sale date) and the price depreciation is often around $10 per day, per average car.
- *Recent history.* When making a recommendation, all decisions made during the past few weeks must be taken into account. Many of these cars

[4] Although the locations of the 50 auction sites are fixed, the company may change the sites it does business from time to time by dropping some sites and adding new ones (thereby changing the layout of the 50 yellow circles). This may happen if the cars are routinely damaged at some sites, auction fees go up, or some other reason. However, these decisions raise several additional questions, such as: How do we evaluate the monetary impact of dropping some sites and adding others? and, Can we increase profits by replacing some auction sites with others? These are important considerations and we will address them later in the chapter.

[5] There are plenty of such reports, including *Black Book*, *Kelly Blue Book*, *Manheim Auction Report*, etc.

may still be in transit, and if they are going to the same auction site, then they might be sold on the same day.

- *Inventory.* It is important to monitor the inventory level of cars at all auction sites, as each site has a particular throughput: If an auction can handle 250 cars per sale day, and the current inventory is larger than 250, then additional time should be added to the estimated sale date.
- *Dynamic market changes.* Market prices for cars change quite frequently; sometimes slowly and sometimes very quickly. Leasing companies (like most businesses) operate in a non-stationary environment that is influenced by many external factors, such as: (1) seasonality (e.g., it is not easy to sell convertibles in New York during the wintertime), (2) the arrival of new models (e.g., new models enter the marketplace in August, influencing the price of older models), and (3) weather (which influences the number of dealers present at an auction, which in turn influences the sale price). Business rules. It is essential to accommodate various business rules that can be added or dropped at any time (e.g., "do not send any red cars to South-East auctions"). This feature is important for analysing what-if scenarios.
- *Business rules.* It is essential to accommodate various business rules that can be added or dropped at any time (e.g., "do not send any red cars to south-east auctions"). This feature is important for analysing what-if scenarios.

Because the market price for various cars changes quite frequently, it is important to stay up-to-date with the current auction prices (which may change from one day to the next). This makes the decision process more difficult. Furthermore, car prices may change in different regions in different ways. For example, it is not easy to sell a convertible Corvette in Boston in the fall, but it might be an excellent idea to sell the car in Florida since the temperature is becoming just right for a convertible (this is referred to as the *seasonality effect*). It is also necessary to deal with new models entering the market each year (typically in August), and every few years a new body style is introduced for particular make/models, which can cause an even bigger price drop for the older body styles. Also note that it takes time to transport a specific car to a specific auction site. The truck has to get there, pick up the car, pick up additional cars (possibly somewhere close by), and then, finally, drive the cars to the assigned auction site. This can take two weeks or more, during which time the auction prices might have already changed. Hence, the sale price for all cars should be estimated a couple of weeks ahead of time (for some auction sites), while taking seasonality and market changes into account. Clearly, the quality of the solution found by the optimiser (again, evolutionary algorithms were used for this purpose) depends on quality of the sale price predictions for the off-lease cars.

Figure 8.5 displays a report showing the distribution of cars grouped by auction site. It shows all the cars distributed by the software, specifying the distribution centre, recommended auction site, predicted sale price, transportation cost, net price lift, and other data. The lift is calculated as the difference between the predicted net price of the vehicle if sent to the recommended auction site, and the net price and the net price for the standard solution (which is based on expert rules that were developed by business managers over the years).

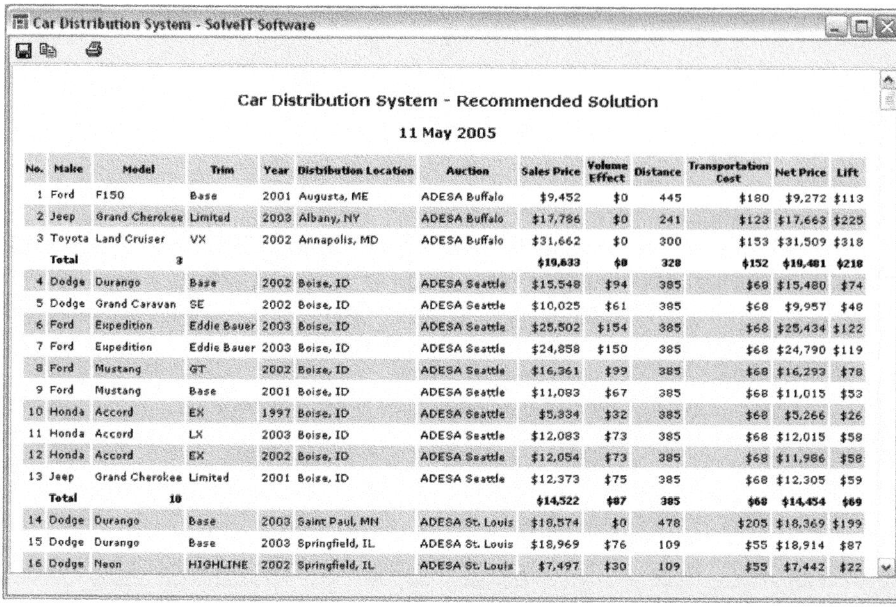
Car Distribution System - SolveIT Software

Car Distribution System - Recommended Solution

11 May 2005

No.	Make	Model	Trim	Year	Distribution Location	Auction	Sales Price	Volume Effect	Distance	Transportation Cost	Net Price	Lift
1	Ford	F150	Base	2001	Augusta, ME	ADESA Buffalo	$9,452	$0	445	$180	$9,272	$113
2	Jeep	Grand Cherokee	Limited	2003	Albany, NY	ADESA Buffalo	$17,786	$0	241	$123	$17,663	$225
3	Toyota	Land Cruiser	VX	2002	Annapolis, MD	ADESA Buffalo	$31,662	$0	300	$153	$31,509	$318
	Total	**3**					**$19,633**	**$0**	**328**	**$152**	**$19,481**	**$218**
4	Dodge	Durango	Base	2002	Boise, ID	ADESA Seattle	$15,548	$94	385	$68	$15,480	$74
5	Dodge	Grand Caravan	SE	2002	Boise, ID	ADESA Seattle	$10,025	$61	385	$68	$9,957	$48
6	Ford	Expedition	Eddie Bauer	2003	Boise, ID	ADESA Seattle	$25,502	$154	385	$68	$25,434	$122
7	Ford	Expedition	Eddie Bauer	2003	Boise, ID	ADESA Seattle	$24,858	$150	385	$68	$24,790	$119
8	Ford	Mustang	GT	2002	Boise, ID	ADESA Seattle	$16,361	$99	385	$68	$16,293	$78
9	Ford	Mustang	Base	2001	Boise, ID	ADESA Seattle	$11,083	$67	385	$68	$11,015	$53
10	Honda	Accord	EX	1997	Boise, ID	ADESA Seattle	$5,334	$32	385	$68	$5,266	$26
11	Honda	Accord	LX	2003	Boise, ID	ADESA Seattle	$12,083	$73	385	$68	$12,015	$58
12	Honda	Accord	EX	2002	Boise, ID	ADESA Seattle	$12,054	$73	385	$68	$11,986	$58
13	Jeep	Grand Cherokee	Limited	2001	Boise, ID	ADESA Seattle	$12,373	$75	385	$68	$12,305	$59
	Total	**10**					**$14,522**	**$87**	**385**	**$68**	**$14,454**	**$69**
14	Dodge	Durango	Base	2003	Saint Paul, MN	ADESA St. Louis	$18,574	$0	478	$205	$18,369	$199
15	Dodge	Durango	Base	2003	Springfield, IL	ADESA St. Louis	$18,969	$76	109	$55	$18,914	$87
16	Dodge	Neon	HIGHLINE	2002	Springfield, IL	ADESA St. Louis	$7,497	$30	109	$55	$7,442	$22

Fig. 8.5. The recommended distribution of cars; report screen

This Adaptive Business Intelligence system helps the remarketing team with decisions on the daily, near-optimal distribution of cars. As discussed earlier, the problem is extremely complex and the implemented system addresses the issues of transportation, volume effect, price depreciation, recent history, current inventory, risk factors, and dynamic market changes. As mentioned earlier in the chapter, this software was successfully installed as a production system in 2002. It has been used since then on a daily basis to recommend the best distribution of cars and generates a significant sales lift (multi-million dollars per year).

As with the pollution control problem, the prediction model returned precise values (sale prices of cars for various distributions of these cars over many auction sites), thus the problem was not noisy. Robustness of the solution also had a different flavour, as it was connected with uncertainties of external events (e.g., delay of a truck). Thus:

$$eval(\boldsymbol{x}) = 1/n \sum_{i=1}^{n} f(\boldsymbol{x}, P(\boldsymbol{x}, \boldsymbol{y} + \delta_i, t)), \tag{8.8}$$

where δ_i represents possible disturbances in accuracy external variables (e.g., the constant time to deliver between points A and B can be disturbed by unexpected weather patterns and/or truck failures). Approximation was not an issue, as no meta-model was built. A time-chaining environment was implied again by the prediction model, which received new information (new sale records of cars) at regular intervals.

8.6 Adaptive Business Intelligence

Adaptive Business Intelligence is the discipline of using prediction and optimisation techniques to build self-learning "decisioning" systems. Adaptive Business Intelligence systems include elements of data mining, predictive modelling, forecasting, optimisation, and adaptability, and are used by business managers to make better decisions. The implementation of Adaptive Business Intelligence assumes the existence of a few key modules that interact with each other other. We discuss them in turn.

The prediction module, in general, generates a predicted output based upon some input (see Figure 8.6).

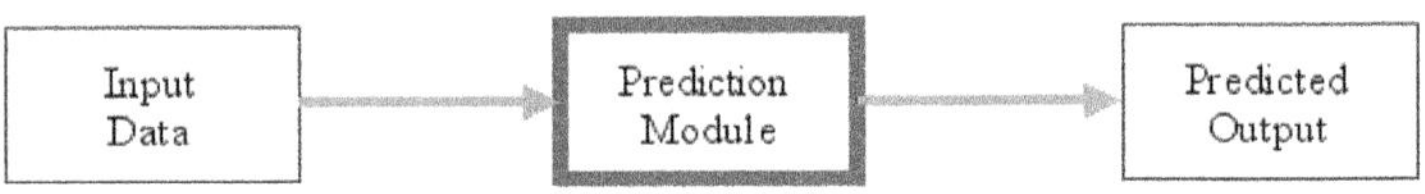

Fig. 8.6. Prediction module

There are many techniques for constructing a prediction model, from classic forecasting methods [3]) to modern heuristic methods, such as neural networks, evolutionary programming, and genetic programming [4, 13]. A recent study [11] investigated the development of a new dynamic genetic programming model specifically tailored for prediction in time-changing environments. The model incorporated methods to automatically adapt to changes in the environment, as well as retain knowledge learned from previously encountered environments. A similar approach was proved to be effective in the price prediction module of the car distribution system.

Next, the optimisation module has to be capable of recommending the best answer. Note, that the optimisation module's recommendation is based on the prediction module's output, so there is a strong relationship between the prediction and optimization modules. The overall concept is displayed in Fig. 8.7.

What happens here is that the optimization module generates some possible solutions to the problem at hand (e.g., creates a possible distribution of

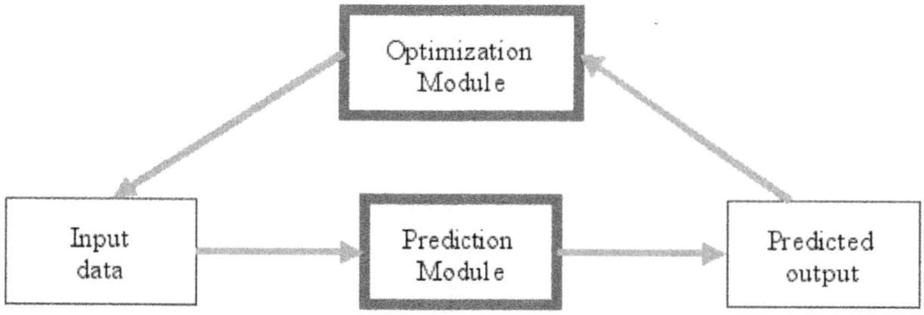

Fig. 8.7. Prediction and optimization modules

cars to the auction sites), which serves as *input data* for the prediction module. The predicted *output data* is then used evaluate the generated answers. In other words, the optimisation module tries different *input data* combinations in order to find the best *predicted output.*

Developing effective prediction and optimization modules is a great start, but by themselves, they are not sufficient in today's constantly changing environment. Because today's accurate prediction might be inaccurate tomorrow, the prediction module must be capable of "learning from" and "adapting to" changes in the environment. Adaptability can be accomplished by slightly changing the learned relationship between input and output each time it is needed. This could be every minute, hour, day, week, month, or for any other time period. The update frequency depends on how fast the environment changes. Some classic forecasting methods (e.g., exponential smoothing methods) [3] approach this problem by putting more emphasis on more recent data. However, a really good adaptive solution can make its own decision on the frequency of update: it will continuously measure its own prediction errors and adjust its parameters. Hence, a *really* good adaptive system would adapt its own speed of adaptation!

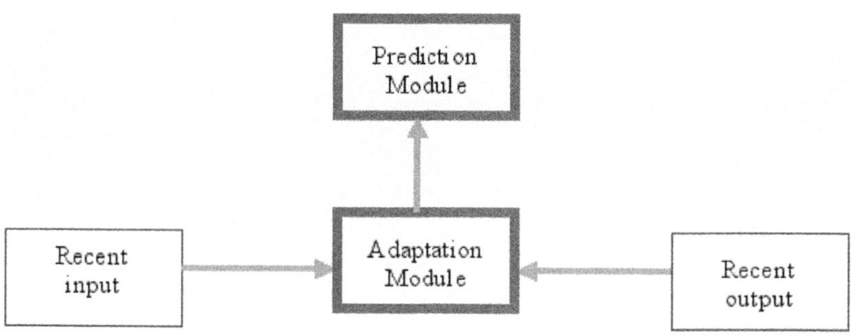

Fig. 8.8. Adaptation and prediction modules

Figure 8.8 illustrates the adaptation process. The *recent input* and *recent output* are taken from very recent history, and historic data is used to construct and train the prediction module. The adaptability module would take the recent input and output and, if necessary, adapt the parameters of the prediction module to decrease the prediction error-in other words, to adapt the prediction module to changes in the environment so it can make better predictions in the future.

The prediction, optimization, and adaptability modules are the core components of an Adaptive Business Intelligence system. However, this does not mean that other components are not important (e.g., an easy-to-use graphical user interface, a database for storing information). Thus, the overall structure of an Adaptive Business Intelligence system resembles the diagram of Fig. 8.9.

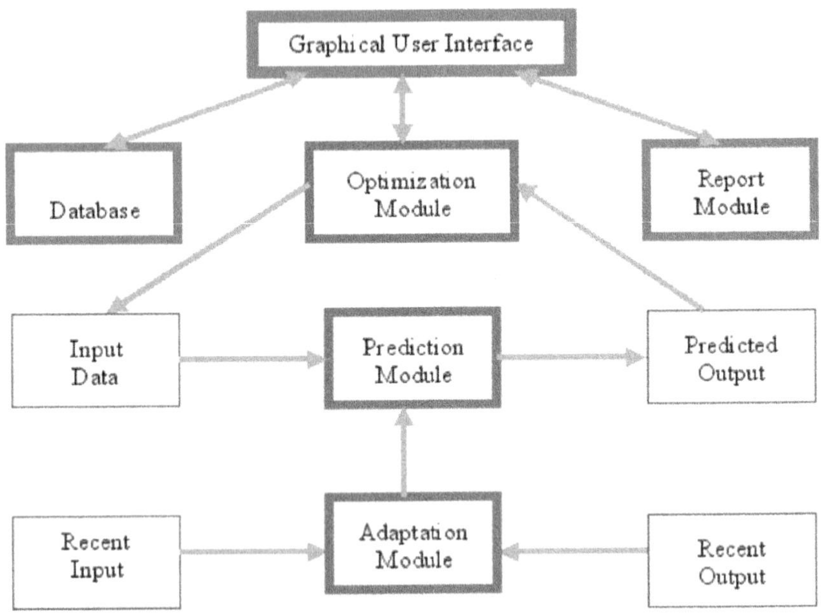

Fig. 8.9. Architecture of Adaptive Business Intelligence system

In all three case studies, we have seen examples of these modules at work. The prediction module was responsible for predicting car prices (cars distribution system) and the trajectory of target ships (path planning). The optimization module was responsible for recommending the production levels of 132 power stations (pollution control), the distribution of cars from collection points to auction sites (car distribution system), and optimal ship trajectories (path planning). Finally, the adaptability module was present in car distributionsystem, where it was responsible for adapting the various parameters of the prediction model to tune its performance. Furthermore, in all three cases, there were reporting and visualization modules incorporated into the final software, which interfaced with historical databases.

8.7 Conclusions

Adaptive Business Intelligence addresses the two fundamental questions that are of the utmost importance for all people, businesses, and government agencies. These are:

- What is likely to happen in the future?
- What is the best course of action?

Whether we realize it or not, these two questions pervade our everyday lives - both on a personal and professional level. When driving to work, for instance, we have to make a traffic prediction before we can choose the quickest driving route. At work, we need to predict the demand for our product before we can decide how much to produce. And before investing in a foreign market, we need to predict future exchange rates and economic variables. It seems that regardless of the decision being made or its complexity, we first need to make a prediction of what is likely to happen in the future, and then choose the best course of action based on that prediction. This fundamental process underpins the basic premise of Adaptive Business Intelligence.

It is clear that effective solutions for complex business problems require software that is capable of operating in time-changing environments and detecting current trends, and which can recommend near-optimum solution and Through the concept of adaptability, a software system can also learn and adapt by evaluating the actual outcome of each decision made. This gives enterprises the ability to monitor business trends, evolve and adapt quickly as situations change, and make intelligent decisions on uncertain and incomplete information [7].

The main research issue is connected with investigating the properties of an Adaptive Business Intelligence system, where the evaluation function is based on predictions of the future values of some variables. In other words, evaluation function eval is expressed as:

$$eval(\boldsymbol{x}) = f(\boldsymbol{x}, P(\boldsymbol{x}, \boldsymbol{y}, t)), \tag{8.9}$$

where $P(\boldsymbol{x}, \boldsymbol{y}, t)$ represents an outcome of some prediction for solution vector $\boldsymbol{x}$ and additional (environmental, beyond our control) variables $\boldsymbol{y}$ at time t. This was the case to various extents-for pollution control, which was based on weather prediction, path planning, which was based on the prediction of target movements, car distribution, which was based on the prediction of sale prices, as well as various other problems: inventory control problems (which are based on demand forecasts), media problems (which are based on

TV rating predictions), portfolio management problems (which are based on economic predictions), and so forth. Furthermore, in the formulation of the problem, we should also take into account dynamic constraints and multiple objectives.

References

1. J. Branke. *Evolutionary Optimization in Dynamic Environments*, Kluwer, 2001.
2. Y. Jin, and J. Branke, Evolutionary Optimization in Uncertain Environments A Survey, *IEEE Transactions on Evolutionary Computation*, Vol.9, No.3, June 2005, pp. 303-317.
3. S. Makridakis, S.C. Wheelwright, and R.J. Hyndman, *Forecasting. Methods and Applications*, Wiley, 1998.
4. Z. Michalewicz and D.B. Fogel, *How to Solve It: Modern Heuristics*, 2nd edition, Springer, Berlin, 2004.
5. M. Michalewicz, Juda-Rezler, K., Trojanowski, K., Matuszewski, A., Michalewicz, Z., Trojanowski, M., Evolutionary Real-Time Optimization System for Ecological Power Control, Proceedings of the Nineth International Symposium on Intelligent Information Systems IIS'2000), Bystra, Poland, June 13 – 17, 2000, in Advances in Soft Computing, Physica-Verlag, pp. 227-242, 2000.
6. Z. Michalewicz, M. Schmidt, M. Michalewicz, C. Chiriac, A Decision-Support System based on Computational Intelligence: A Case Study, *IEEE Intelligent Systems*, Vol.20, No.4, July-August 2005, pp. 4449.
7. Z. Michalewicz, M. Schmidt, M. Michalewicz, C. Chiriac, *Adaptive Business Intelligence*, Springer, Berlin, 2006.
8. R.W. Morrison, *Designing Evolutionary Algorithm for Dynamic Environments*, Springer-Verlag, Berlin, 2004.
9. R. Smierzchalski and Michalewicz, Z., Modeling of Ship Trajectory in Collision Situations by an Evolutionary Algorithm, *IEEE Transactions on Evolutionary Computation*, Vol.4, No.3, pp. 227-241, 2000.
10. R. Smierzchalski and Z. Michalewicz, Path Planning in Dynamic Environments, chapter in *Innovations in Machine Intelligence and Robot Perception*, S. Patnaik, L.C. Jain, S.G. Tzafestas, and V. Banoore (Eds), Springer, 2005.
11. N. Wagner, Z. Michalewicz, M. Khouja, and R.R. McGregor, Time Series Forecasting for Non-static Environments: the DyFor Genetic Program Model, to appear in *IEEE Transactions on Evolutionary Computation*, 2006.
12. K. Weicker, *Evolutionary Algorithms and Dynamic Optimization Problems*, Der Andrere Verlad, Berlin, 2001.
13. S.M. Weiss and N. Indurkhya. *Predictive Data Mining*, Morgan Kaufmann, San Francisco, 1998.

9

Evolutionary Algorithms for Combinatorial Problems in the Uncertain Environment of the Wireless Sensor Networks

Frederico Paiva Quintão, Fabíola Guerra Nakamura, and Geraldo Robson Mateus

Computer Science Department, Universidade Federal de Minas Gerais (UFMG)
Av. Antônio Carlos, 6627, Belo Horizonte, MG, Brazil
{fred,fgnaka,mateus}@dcc.ufmg.br

Summary. This chapter presents basic concepts on the recent technology of the Wireless Sensor Networks (WSNs) and discusses some problems that arise in this new kind of ad-hoc network. Mathematical formulations and evolutionary algorithms are provided to support the network manager . Our approaches are concerned in controlling the energy consumption in the network and the Quality of Service aspects, such as area coverage and nodes connectivity, through the use of fast and efficient algorithms.

9.1 Introduction

A Wireless Sensor Network (WSN) is a kind of ad-hoc network, with distributed communication, sensing and processing capacities. A WSN can be composed by tens or even hundreds of small battery-powered devices, called sensor nodes. They can be used in a large number of applications, such as:

- indoor environments control, like the maintenance of complex equipments in factories and assembly lines [1];
- air pollution level and animal life monitoring [2];
- military spies, providing information about enemy movements in a battle field.

Figure 9.1 shows a common architecture: many sensor nodes, also called source nodes, monitoring an area and reporting data of a phenomenon to the sink node, which is a special node that works as a network access point sending the collected data to an outside observer. These collected data are processed and can be combined with other data, such as the topography of the monitoring area, providing more accurate information.

F. P. Quintão et al.: *Evolutionary Algorithms for Combinatorial Problems in the Uncertain Environment of the Wireless Sensor Networks*, Studies in Computational Intelligence (SCI) **51**, 197–222 (2007)
www.springerlink.com

Data collected by the source nodes can be transmitted to the sink node by two kinds of routes, using single or multi-hop communication. In the single-hop style, each source node transmits its data straightly to the sink node. In the other hand, the source nodes can use the intermediary nodes to relay their data: this is the multi-hop route, illustrated by Fig. 9.2. Multi-Hop is considered more interesting because it is more scalable and energy-efficient, although the delay may increase with the number of hops. Moreover, in many situations source nodes cannot reach the sink node directly, so multi-hop should be used.

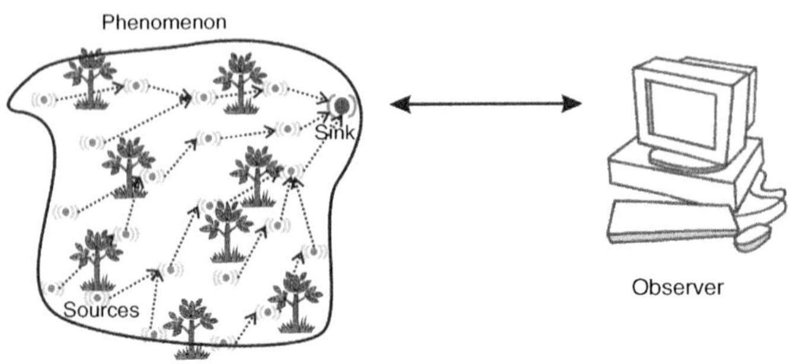

Fig. 9.1. A common architecture of a WSN

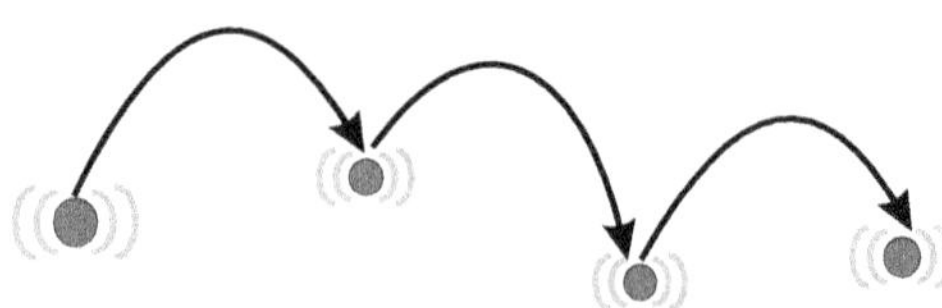

Fig. 9.2. Multi-Hop communication in Wireless Sensor Networks

Figure 9.3 shows the sensor node MicaZ [3] from Mica Motes Family, which can perform activities like sensing, communication and processing.

To perform all activities expected for a WSN, the hardware of each sensor node generally includes:

- a sensor board, containing at least one kind of sensor device;
- a limited quantity of memory;
- a processor, with limited power of processing;
- a radio to perform wireless communication; and
- a battery, which provides energy to all components.

For instance, the Mica Motes Family has the following basic configuration: a memory of 128 KB, an 8 MHz processor, IEEE 802.15 communication protocol for node MicaZ and two AA batteries as the energy provider.

Fig. 9.3. Sensor node MicaZ from Mica Motes Family

There are several challenges regarding WSNs once these networks present several unique features when compared to traditional ad-hoc networks, therefore existing ad-hoc solutions must be extended and adapted to be used in WSNs. Between these features can be stressed application dependency, energy restrictions, node redundancy, high node density, limited bandwidth, and dynamic topology, just to point out a few.

WSNs can be classified according to their composition. Regarding composition, WSNs are static when the nodes positions are the same during all the network lifetime or mobile when the position of the sensor nodes changes. In the mobile case, the sensor nodes can be part of a mobile equipment such as a robot. WSNs are homogenous when all sensor nodes have the same features and are heterogenous when the nodes are different.

The topology of a WSN can be dynamic basically because the sensor nodes are susceptible to failures during the network lifetime, which means that some nodes can leave the network at any time, and because some natural phenomenon can change the nodes position, the wind for instance.

These unique features have allowed a wide variety of research in areas such as energy-efficient network protocols and low-power hardware design. They also have encouraged proposals of management architectures for WSNs, which aim to increase the network resources productivity and maintain the Quality of Service (QoS) provided. A good and common example of management functions are node scheduling schemes, whose efforts are to make the networks more energy-efficient, while trying to maintain application requirements, such as coverage and connectivity. Furthermore, once WSNs can be deployed into a hostile area (such as a volcano crater), and the number of nodes can be high, recharging or replacing nodes' battery may be inconvenient or even impossible.

The development of energy-efficient protocols for WSNs organization can help extending their lifetime. However, given their features, these protocols should work as fast as possible. For example, when the nodes are deployed in an area susceptible to a lot of changes, the network manager will have to evaluate and define the best set of nodes to maintain the application requirements several times during the network lifetime. Thus a fast algorithm for these evaluations is ideal and necessary.

There are a lots of different ways of organizing WSNs. The most common are:

- Hierarchical WSNs: in these networks, the sensor nodes are grouped into clusters. Each cluster has a leader (also called cluster header), which is, in general, the one that sends the collected data to outside the cluster. In this scenario the algorithms should provide solutions considering the area coverage, nodes connectivity, and also the clustering process. Approaches for this kind of network are provided by Heinzelman *et al* [11] and Chiaresserini *et al* [16].
- Flat WSNs: in these networks, sensor nodes are not organized into clusters, which means that all nodes have the same hierarchical level. Coverage and connectivity are the main issues concerning flat networks. The works of Cerpa and Estrin [17], Ye *et al* [18] and Zhang and Hou [19] deal with these problems.

Different kinds of WSNs need different algorithms to address their problems. In section 9.3 of this chapter are presented two combinatorial problems in flat WSNs and evolutionary algorithms to solve them. The algorithms work in a centralized way and were developed after studies of Integer Linear Programming (ILP) and Mixed Integer Linear Programming (MILP) mathematical formulations of the problems. The mathematical formulations provide metrics to evaluate the solutions of the algorithms.

Two work approaches are presented: in the first one, given a set of sensor nodes, it is evaluated the best subset of nodes that should be active to assure the monitoring area coverage and the nodes connectivity in a certain time t of the network lifetime. The second approach works in a multi-period way providing a solution that foresees the subset of active sensor nodes to maintain the area coverage for each time period and not taking in account connectivity issues. Each model presents advantages and disadvantages. The first one can be useful for very dynamic networks, like those ones exposed to a lot of natural phenomenons and under less controlled environments. The second approach is useful for steadier networks, and, for some classes of networks, it can achieve the best node scheduling possible.

The remainder of chapter is organized as follows: Section 9.2 presents important concepts to the problems addressed. The combinatorial problems definition, models and evolutionary algorithms appear in Section 9.3.

9.2 Basic Concepts

This section brings some important concepts and assumptions when working with node scheduling schemes in WSNs.

9.2.1 Coverage in Wireless Sensor Networks

Megerian *et al* define coverage as a measure of the ability of the network to detect and observe a phenomenon in the sensor field [8] and relate the coverage to the WSN quality of service requirements once it can indicate the network observability levels [7].

In order to quantify the coverage area of a WSN, one can define the node sensing area as the region around the node where a phenomenon can be detected and define this region as a circle of range R, where R is the sensing range as showed on Figure 9.4(a). The coverage area of a WSN consists of sensing areas union of all active nodes in the network, as showed on Fig. 9.4(b). If there is any kind of obstacle in the monitoring area, the nodes sensing area can decrease as showed on Figure 9.4(c). In this work, the presence of obstacles is not addressed. See Menezes [15] for a further discussion on this topic.

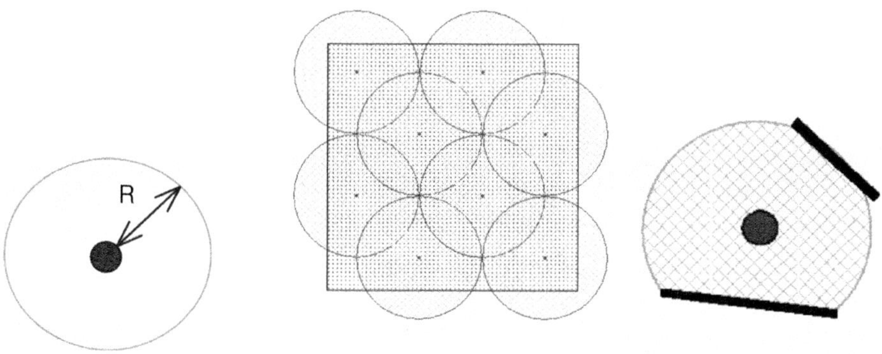

(a) Node sensing area

(b) Coverage area

(c) Effect of obstacles on the node sensing area

Fig. 9.4. Coverage in Wireless Sensor Networks

The coverage area can be modeled through the use of demand points, which represent the center of a small square area in the sensor field. This concept allows to evaluate the coverage in a discrete space and it is very useful for modeling purposes. To guarantee the coverage, each demand point must be covered by at least one active sensor, otherwise the coverage fails.

An approach to calculate the coverage area is to verify the percentage of demand points reached by at least a sensor node. Likewise, coverage fail is defined as the percentage of demand points not covered.

9.2.2 Connectivity in Wireless Sensor Networks

A WSN is said to be connected when there is a communication infra-structure (at least one path) that allows all the collected data to reach the observer, which means that all active nodes can reach the sink node, directly or in a multi-hop way.

The node connectivity in WSNs is an important issue because based on the physic characteristics of radio waves and environmental features, such as the presence or absence of obstacles in the monitoring area, every sensor node will have a maximum communication range. Obstacles can attenuate the signal decreasing the communication range. For the Mica Motes Family, this range can (in theory) reach the value of $100m$ ($300fts$) [3] in the maximum transmission power and given the absence of obstacles in the area.

Vieira *et al* [6] discuss three possibilities of network modeling concerning the sensing and communication ranges: sensing range greater than, less than or equal to communication range. However, it has been accepted that the communication range (R_c) is many times greater than the sensing range (R_s).

The relationship between the nodes sensing and communication ranges is discussed by Wang *et al* [20]. Using geometric analysis, the authors prove the following theorem:

Theorem 1. *For a set of sensors that at least 1-cover a convex region A*[1]*, the communication graph is connected if* $R_c \geq 2R_s$*, where* R_c *is the node communication range and* R_s *is the sensing range.*

This theorem means that when the sensing range R_s is less than half of the radio range R_c, if one assures the area coverage he also assures the network connectivity. Therefore, when working in scenarios where this assumption is true, one does not need to consider the connectivity between the sensor nodes a problem.

9.2.3 Energy Consumption Model

As mentioned before, one of the main features of WSNs is a high energy restriction, which is due to the limited battery of the sensor node, and in many cases due to the impossibility of battery recharge/replacement. The definition of a node energy consumption model can allow WSN researches to focus the studies on topics that have higher impacts on the network lifetime.

To define this model one should consider the node basic operations: transmission, reception, sensing, and processing. The operations energy consumption depends on the current necessary to perform the task and time period to execute the task. The energy consumption can be estimated by the following equation:

---

[1] *1-cover* means that each demand point is covered by at least one sensor node

$$E = \alpha \times \Delta t, \tag{9.1}$$

where:

- E is the total energy consumed in mAh,
- α is the current consumed in mA,
- Δt is the period of time in h.

Different battery discharge models can be used to calculated the node residual energy but, for the sake of simplicity, many works adopt the linear model of battery discharge where the node residual energy is the difference between the battery capacity and the total node consumed energy.

The next section explores some of the combinatorial problems that can appear in the uncertain environment of WSNs.

9.3 Combinatorial Problems in Wireless Sensor Networks

The WSN application dependency makes really important the definition of a work scenario. The next subsections define two node scheduling problems in WSNs, the Coverage and Connectivity Problem in Flat Wireless Sensor Networks and the Multi-Period Coverage Problem in Flat Wireless Sensor Networks. Each one of these problems is modeled through a mathematical formulation based on Integer Linear Programming (ILP) and solved by optimization packages and evolutionary algorithms. The issues addressed in these problems are really important because according to Vieira *et al* [6] and the results of Tilak *et al* [5] a node scheduling control in WSNs, besides minimizing the network energy consumption, can reduce problems such as radio interference between neighbors nodes, collision of packets and media congestion.

On the development of the mathematical formulations the following assumptions were made: each sensor node knows its position and has an unique identification and the battery discharge follows a linear model. The traffic in the network is generated only by source nodes. It is also assumed that the nodes consume an amount of energy to stay active called maintenance energy (ME) that represents the consumption with activation, sensing and processing. The energy consumption with transmission (TE) and energy reception (RE) are treated separately. The monitoring area is obstacles free. Regarding the composition the network modeled can be either homogenous or heterogeneous and is organized in an flat way.

9.3.1 The Coverage and Connectivity Problem in Wireless Sensor Networks

This problem can be stated as: *Given a monitoring area A, a set of demand points D, a set of sensor nodes S and a sink node m, the Coverage and*

Connectivity Problem in Wireless Sensor Networks (CCP-WSN) consists of assuring that at least one sensor node from S is covering each demand point $j \in D$ in the monitoring area A and there is a path between these nodes and the sink node m, minimizing the consumption of energy by the sensor nodes.

CCP-WSN Mathematical Model

The CCP-WSN is formulated as a Mixed Integer Linear Programming (MILP) problem.

The following parameters are used in the formulation.

S: set of sensor nodes.
D: set of demand points.
A^d: set of arcs connecting sensor nodes to demand points.
A^s: set of arcs connecting sensor nodes.
A^m: set of arcs connecting sensor nodes to the sink node.
I^d: set of arcs (i, j) incoming on the demand point $j \in D$.
I^s: set of arcs (i, j) incoming on the sensor node $j \in S$.
O^s: set of arcs (i, j) outgoing the sensor node $i \in S$.
ME_i: maintenance energy for node $i \in S$.
TE_{ij}: transmission energy between nodes i and j, $\{i, j\} \in \{A^s \cup A^m\}$.
RE_i: reception energy for node $i \in S$.
NC_j: coverage penalty, cost of no coverage of a demand point $j \in D$.

The model variables are:

x_{ij}: variable that has value 1 if node i covers demand point j, and 0 otherwise.
z_{lij}: decision variable that has value 1 if arc (i, j) is in the path between sensor node l and the sink node m, and 0 otherwise.
y_i: decision variable that has value 1 if node i is active, and 0 otherwise.
h_j: has value 1 if demand point j is not covered and 0 otherwise.
e_i: variable to indicate the energy consumed by node i.

The formulation proposed is presented below.

The objective function (9.2) minimizes the network energy consumption and penalize the not covered demand points. Since it minimizes the network energy consumption, the mathematical formulation reduces the number of active nodes.

$$min \sum_{i \in S} e_i + \sum_{j \in D} NC_j h_j \tag{9.2}$$

The set of constraints (9.3), (9.4), (9.5), and (9.6) deals with the coverage problem. Constraints (9.3) specifically assure that, if possible, at least one sensor node will cover each demand point, otherwise the variables h_j will have the value 1. Constraints (9.4) indicate that a node can only cover a point if it is active. Constraints (9.5) and (9.6) set limits for variables x and h respectively.

$$\sum_{ij \in I^d} x_{ij} + h_j \geq 1, \forall j \in D \tag{9.3}$$

$$x_{ij} \leq y_i, \forall i \in S, \forall ij \in A^d \tag{9.4}$$

$$0 \leq x_{ij} \leq 1, \forall ij \in A^d \tag{9.5}$$

$$h_j \geq 0, \forall j \in D \tag{9.6}$$

The set of constraints (9.7), (9.8), (9.9) and (9.10) are related to the connectivity problem. Constraints (9.7) and (9.8) assure a path between each active sensor node $l \in S$ and the sink node m and constraints (9.9) and (9.10) only allow active nodes to be part of these paths.

$$\sum_{ij \in I^s} z_{lij} - \sum_{jk \in O^s} z_{ljk} = 0, \forall j \in (S \cup m - l), \forall l \in S \tag{9.7}$$

$$-\sum_{jk \in O_j^s} z_{ljk} = -y_l, j = l, \forall l \in S \tag{9.8}$$

$$z_{lij} \leq y_i, \forall i \in S, \forall l \in (S - j), \forall ij \in (A^s \cup A^m) \tag{9.9}$$

$$z_{lij} \leq y_j, \forall j \in S, \forall l \in (S - j), \forall ij \in (A^s \cup A^m) \tag{9.10}$$

The energy constraints (9.11) and (9.12) define the energy bound values. The lower bound is zero and the upper bound is the node battery capacity. The constraints (9.11) also define that a node spends its energy with maintenance, packets transmission and reception.

$$ME_i \times y_i + \sum_{l \in (S-i)} \sum_{ki \in I_i^s} RE_i \times z_{lki} +$$
$$\sum_{l \in S} \sum_{ij \in O_i^s} TE_{ij} \times z_{lij} \leq e_i, \forall i \in S \tag{9.11}$$

$$e_i \geq 0, \forall i \in S \tag{9.12}$$

Constraints (9.13) define the decision variables as boolean.

$$h, y, z \in \{0, 1\} \tag{9.13}$$

The model solution consists of a subset of active nodes, represented by the variables y_i with value 1, and the demand points not covered, represented by the variables h_j with value 1, that assure the best possible coverage. The variables x_{ij} with value 1 indicate that the active sensor node i covers the demand point j. The solution also provides a path between the active nodes and the sink node assuring the network connectivity and that is given by the variables z_{lij} with value 1. The solution also estimates the network energy consumption in the variables e_i.

CCP-WSN Evolutionary Algorithm

The CCP-WSN mathematical model requires hard computational effort to reach optimal solutions. Thus, it is proposed to solve the CCP-WSN problem with a hybrid heuristic that decomposes it into two simpler sub-problems. The strategy used to solve the problem consists of:

1. Solving a coverage problem in order to find the minimal number of nodes needed to cover all the monitoring area or achieve the best possible coverage. For this sub-problem a genetic algorithm is used;
2. Applying, to the best solution found in the previous step, a local search in order to ensure the connectivity between the nodes. In this step Prim's Minimum Spanning Tree and Dijkstra's Shortest Path algorithms are used.

A heuristic based on genetic search can solve the coverage problem and it is a good option because it usually provides more than one good and feasible solution, and this redundance can be interesting to the network manager. Otherwise, to solve the CCP-WSN and its many constraints with a genetic algorithm is not advisable because one would have to use very special operators to avoid the generation of non-feasible solutions, which could involve even hard computational tasks.

Given this first look over the strategy, the sub-problems are detailed below, starting with the coverage problem.

Coverage Problem

The problem can be stated as: *Given a monitoring area A, a set of sensor nodes S and a set of demand points D, the Coverage Problem in Wireless Sensor Networks (CP-WSN) consists of assuring that at least one sensor node $s \in S$ will cover each demand point $j \in D$, and minimizing the energy consumption to active sensor nodes.* One formulation for the Coverage Problem in WSN is found in the CCP-WSN mathematical model [14].

The formulation for the CP-WSN uses the same parameters and the variables x, h, y of the CCP-WSN model. Thus, the mathematical model for the CP-WSN can be formulated as follows.

The objective function minimizes the number of active nodes and penalize the not covered demand points.

$$min \sum_{i \in S} ME_i \times y_i + \sum_{j \in D} NC_j \times h_j \tag{9.14}$$

The model is subject to a set of coverage constraints and a set of variable types constraints. Constraints (9.15) assure that each demand point may be covered by a sensor node or not be covered. Constraints (9.16) impose that a node only can cover a demand point if it is active. The constraints (9.17) and (9.18) present the variables' limits and the constraints (9.19) define the variables types.

$$\sum_{ij} x_{ij} + h_j \geq 1, \forall j \in D \text{ and } \forall ij \in A^d \tag{9.15}$$

$$x_{ij} \leq y_i, \forall i \in S \text{ and } \forall ij \in A^d \tag{9.16}$$

$$0 \leq x_{ij} \leq 1, \forall ij \in A^d \tag{9.17}$$

$$h_j \geq 0, \forall j \in D \tag{9.18}$$

$$y \in \{0, 1\} \tag{9.19}$$

To improve the whole algorithm efficiency, the coverage problem objective function can be modified as follows:

$$min \sum_{i \in S} (ME_i + PC_i) \times y_i + \sum_{j \in D} NC_j \times h_j \tag{9.20}$$

where PC_i is a parameter that contains the cost of the shortest path from each node $i \in S$ to the sink node m. This cost is computed by the Dijkstra's shortest path algorithm applied to all nodes of the network, during a pre-processing phase that considers the transmission energy between nodes as edges costs. The variable PC_i is used as a penalty for those nodes whose path to the sink node is expensive, and the results show that this change is quite interesting, improving the algorithm. In fact, this is a way to consider the connectivity problem during the genetic algorithm search.

To solve the Coverage Problem, using the function (9.20) as a fitness function, a genetic algorithm is proposed, based on binary encoding, described as follows:

Encoding

Each chromosome uses binary encoding and has size $|S|$ equal to the number of sensor nodes in the network. Each position of the chromosome represents a gene: if a chromosome position is set to 1, this implies that the node corresponding to this position is turned on in this chromosome. For example, suppose a network containing 10 nodes and one of the chromosomes has the following nodes activated: (1, 3, 8, 10); so, its binary representation would be as follows:

1	0	1	0	0	0	0	1	0	1

Given a set of active nodes from one chromosome, the coverage is evaluated through a binary coverage matrix, which reports, for an input (i, j), whether a sensor node i covers demand point j.

Genetic algorithm general specification

The following operators are implemented:

- Selection uses Cost Weighting Pairing (CWP) [21], a kind of roulette-wheel algorithm that gives bigger probability of matching for the chromosomes with the best values of the fitness function. When two chromosomes are selected, they are always combined;
- Recombination (crossover) implements the simplest one-point crossover, in which a random number c (where $1 \leq c \leq |S|$) is generated and in this position the crossover takes place. Each couple of recombined chromosomes generates two offsprings, which replace the two worst chromosomes of the population;
- Mutation occurs with a small probability μ. All chromosomes (with the exception of the best one) are visited, and with a probability μ one of their genes undergoes mutation.

Ensuring connectivity to CCP-WSN

The solution generated by the genetic algorithm random search may assure a good monitoring area coverage, but some active nodes can be disconnected, which means that the data collected by them cannot reach the observer. To assure connectivity Prim's minimum spanning tree algorithm is used to connect the nodes at a minimum cost and identify if there are nodes disconnected from the network. Then Dijkstra's shortest path algorithm is used to turn the disconnected network into a connected one.

The solution strategy is detailed as follows:

1. Initially the Prim's minimum spanning tree (MST) algorithm is applied over a graph G_1 containing the network active nodes (set during the last step). The condition for an edge to belong to G_1 is the following: *An edge (u, v) can belong to G_1 only if the distance between nodes u and $v \in G_1$ is shorter than the maximum communication range of the nodes* (as discussed in Section 9.2.2). The result of the MST algorithm is a tree connecting the nodes. This tree can even be used as a routing tree. However, given the condition above, some of the active nodes may be disconnected from the tree. When this happens, the next step is applied.
2. A graph G_2 is created, containing all the nodes of the network. The same condition described above is applied: *An edge (u, v) can belong to G_2 only if the distance between nodes u and $v \in G_2$ is shorter than the maximum*

communication range of the nodes. So, the Dijkstra's shortest path algorithm is applied from each one of the disconnected active nodes to the sink node. This path surely goes across non-active nodes, otherwise all active nodes would be connected during the previous step. Thus, this step turns on all the sensor nodes that appear on the path and that are not on the genetic algorithm solution.

Fig. 9.5 summarizes the hybrid approach used to solve the CCP-WSN.

```
input: set S of sensor nodes, sink node, set D of demand points, sensing range,
       communication range
begin
  for all i ∈ S do
    PC_i ← Compute Dijkstra(i, sink);
  end for
  Compute coverage matrix; /*reports whether a sensor node i covers a demand
  point j*/
  Compute connectivity matrix; /*reports the Euclidian distance between sensor
  nodes i and l*/
  /*Genetic algorithm*/
  Create random initial population, with size |PopInitial|;
  Evaluate initial population using equation (9.20);
  ShellSort(Initial population); /*Sort the population in ascending order */
  Natural selection(Initial Population); /*removes the worst chromosomes from
  the initial population*/
  while NOT stop condition then
    Select chromosomes for matching using Cost Weighting Pairing;
    Proceed matching/crossover;
    Execute Mutation with probability μ;
    Evaluate new population using equation (9.20);
    Shellsort(new Population);
  end while
  /*end of Genetic Algorithm*/
  /*Local search for connectivity ensuring*/
  Compute Prim(active nodes);
  while there is a disconnected node i do
    Compute Dijkstra(i, sink);
  end while
  /*end of local search*/
  Compute objective function - equation (9.2);
  Return the best solution found;
  /*end of Hybrid Algorithm*/
end
```

Fig. 9.5. The hybrid algorithm for CCP

Computational Results

The computational results compare the values obtained by solving the mathematical model with the commercial optimization package CPLEX 9.0 [22] to the values obtained by the hybrid algorithm for the CCP-WSN. The tests consider scenarios with 32 sensor nodes and one sink node and the size of the monitoring area vary from $30m$ x $30m$, to $40m$ x $40m$, $50m$ x $50m$ and $60m$ x $60m$. For each scenario there are four different configurations: cfg1, cfg2, cfg3 and cfg4 where the communication range assumes values $15m$, $20m$, $30m$ and $40m$, respectively. The sensing range is kept constant and equals to $15m$.

The first test considers a scenario with a square area of size $30m$ x $30m$. The evolutionary algorithm runs for 15 generations. Results follow in Table 9.1. Each result of the evolutionary algorithm represents an average value of 33 runs. The column $Active_c$ reports the number of sensor nodes activated by CPLEX, $Time_c$ reports the computational time spent by CPLEX and $Objective_c$ reports for the value of the objective function obtained by CPLEX. The subscription $_h$ represents the results of the hybrid algorithm.

Configuration	**Active$_h$**	**Active$_c$**	**Time$_h$**	**Time$_c$**	**Objective$_h$**	**Objective$_c$**
`cfg1`	4.51	4	2.96	115.78	98.87	80.82
`cfg2`	4.00	4	2.96	106.81	84.74	80.82
`cfg3`	4.00	4	2.92	61.41	84.87	80.82
`cfg4`	4.00	4	2.99	68.16	85.12	80.82

Table 9.1. *Comparison of hybrid algorithm and CPLEX - 32 nodes -* $30m$ x $30m$

In this scenario, the hybrid algorithm has a great advantage compared to CPLEX, although its values of objective function are worse than the ones from CPLEX (23% in the worst case and 5% in the best case). Regarding the execution time, in CPLEX best case the hybrid algorithm runs 20 times faster and in CPLEX worst case of comparison, it is almost 40 times faster.

Figure 9.6 reports the ratios between objective function values, number of active nodes and processing time of the hybrid algorithm and CPLEX for the results in Table 9.1. It is clear that the hybrid algorithm results are close to the optimal for the objective function values and for the number of active nodes. Moreover, the ratios between the processing times of both algorithm are almost zero, showing that the time spent by the hybrid algorithm is despicable.

The next scenario has 32 nodes in an area of size $40m$ x $40m$. Results are reported in Table 9.2.

The hybrid algorithm is again faster than CPLEX (about 15 times for CPLEX best case), but in its best case regarding the value of the objective function, it is 15% worse.

Figure 9.7 shows the processing time of each solver according to the communication range, for the last set of tests. The hybrid algorithm has few

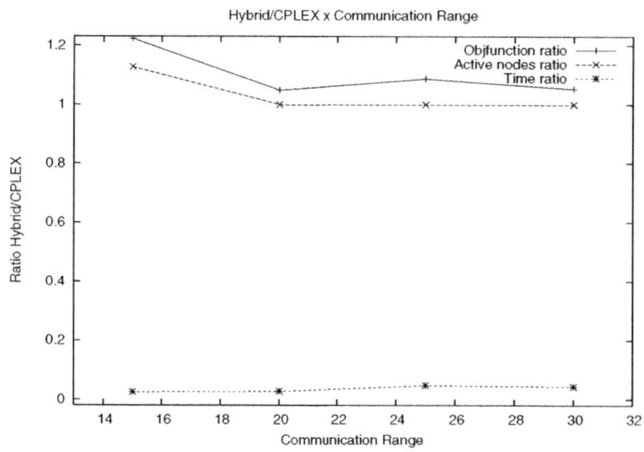

Fig. 9.6. *Ratios between hybrid algorithm and CPLEX*

Configuration	Active$_h$	Active$_c$	Time$_h$	Time$_c$	Objective$_h$	Objective$_c$
cfg1	7.63	5	6.06	105.74	172.12	110.00
cfg2	5.63	5	5.84	80.38	123.54	101.52
cfg3	5.27	5	5.76	117.11	116.78	101.52
cfg4	5.48	5	6.13	92.06	120.76	101.52

Table 9.2. *Comparison of hybrid algorithm and CPLEX - 32 nodes -* 40m x 40m

variations in the processing time, but CPLEX presents great variations. Actually, the complexity of the hybrid algorithm depends of the area size, because the "hard" computational part of the algorithm deals only with the coverage. CPLEX is more sensible to the size of the communication range, and the area as well, so its behavior is less predictable.

Following the process comparison, the tests consider a square area of 50m x 50m. The results are summarized in Table 9.3. For this scenario, `cfg1` is not considered and the hybrid algorithm runs for 20 generations.

Configuration	Active$_h$	Active$_c$	Time$_h$	Time$_c$	Objective$_h$	Objective$_c$
cfg2	11.18	9	8.41	46.63	276.92	200.02
cfg3	9.00	9	8.52	42.73	230.61	191.92
cfg4	9.00	9	8.52	31.62	229.97	187.82

Table 9.3. *Comparison of hybrid algorithm and CPLEX - 32 nodes -* 50m x 50m

In this test, the hybrid algorithm is 5 times faster than CPLEX and the objective function values are about 25% worse than the ones reached by CPLEX.

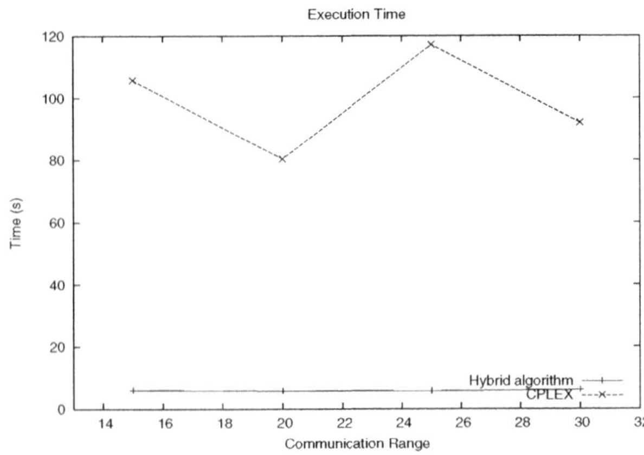

Fig. 9.7. *Processing time of each solver, according to communication range*

The results obtained in instances with 32 nodes in a square area of size $60m$ x $60m$ are showed on Table 9.4. Configuration `cfg1` is not considered again.

Configuration	**Active**$_h$	**Active**$_c$	**Time**$_h$	**Time**$_c$	**Objective**$_h$	**Objective**$_c$
`cfg2`	14.90	14	12.66	41.33	386.66	331.92
`cfg3`	13.63	13	12.80	60.82	379.74	300.30
`cfg4`	13.54	13	12.49	56.28	373.86	278.62

Table 9.4. *Comparison of hybrid algorithm and CPLEX - 32 nodes -* $60m$ x $60m$

The hybrid algorithm is about 4 times faster than CPLEX, and regarding the objective function is 34% worse in the worst case. Again, regarding the number of active nodes, the hybrid algorithm performs well. It should be said that the original topology presented 3 demand points not covered, and for all configurations the hybrid algorithm reached the best possible coverage.

Figure 9.8 shows the hybrid algorithm improvement as the communication range increases. One can conclude that the hybrid algorithm should work better on scenarios where the communication range is many times larger than the sensing range. Actually, in these scenarios, one should solve only the Coverage Problem, as proposed by Wang *et al* [20] and discussed on Section 9.2.2. The hybrid algorithm proposed solves the Coverage Problem in a very efficient way, so one can conclude that this algorithm will be useful in situations when this assumption is considered, which probably will be very common in practice.

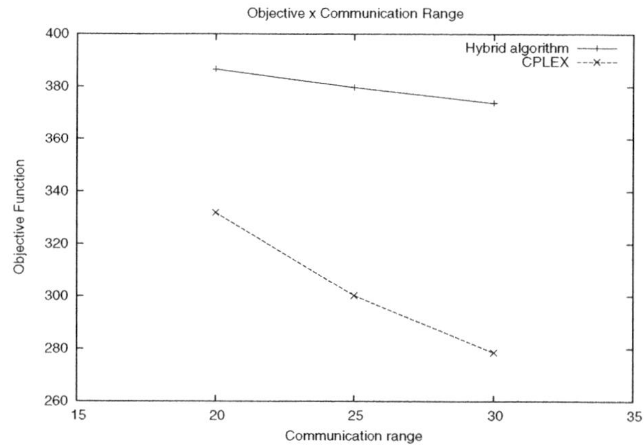

Fig. 9.8. *Hybrid algorithm improves as increasing the communication range*

9.3.2 The Multi-Period Coverage Problem in Wireless Sensor Networks

This problem can be stated as: *Given a monitoring area A, a set of sensor nodes S, a set of demand points D and t periods of time, the Multi-Period Coverage Problem in Wireless Sensor Networks (MCP-WSN) consists of assuring that at least one sensor node from S is covering each demand point $j \in D$ in each period t.*

MCP-WSN Mathematical Formulation

MCP-WSN is formulated as an Integer Linear Programming (ILP) problem. The following parameters are used in the formulation.

S: set of sensor nodes.
D: set of demand points.
T: set of periods of time.
A^d: set of arcs. connecting sensor nodes to demand points.
ME_i: maintenance energy for node $i \in S$.
NC_j: coverage penalty of no coverage of a demand point $j \in D$.
n: maximal number of periods that a sensor node can be active.

The model variables are:

x_{ij}^t: variable that has value 1 if node i covers demand point j during period $t \in T$, and 0 otherwise,
y_i^t: decision variable that has value 1 if node i is active in period $t \in T$, and 0 otherwise,

h_j^t: variable that has value 1 if demand point j is not covered in period $t \in T$, and 0 otherwise.

The formulation proposed is presented below. The objective function (9.21) minimizes the number of active nodes and penalize the not covered demand points in each period.

$$min \sum_{i \in S} \sum_{t \in T} ME_i \times y_i^t + \sum_{j \in D} \sum_{t \in T} NC_j \times h_j^t \tag{9.21}$$

The objective function is subject to a set of coverage and activation constraints.

Constraints (9.22), (9.23), (9.24), and (9.25) deal with the multi-period coverage problem. They assure that active nodes cover the demand points and that just active nodes can sense the environment. Constraints (9.22) also guarantee that a demand point can be uncovered, if none sensor node can reach it.

$$\sum_{ij \in I^d} x_{ij}^t + h_j^t \geq m, \forall j \in D \text{ and } \forall t \in T \tag{9.22}$$

$$x_{ij}^t \leq y_i^t, \forall i \in S, \forall ij \in A^d \text{ and } \forall t \in T \tag{9.23}$$

$$0 \leq x_{ij}^t \leq 1, \forall ij \in A^d \tag{9.24}$$

$$h_j^t \geq 0, \forall j \in D \tag{9.25}$$

The activation constraints (9.26) indicate the maximal number of periods in which a sensor node can be active. In practical situations, $n < T$, since the sensor nodes are energy-constrained and probably will not survive during all the network lifetime. When $n > T$, it means that the nodes can remain turned on during all the network lifetime (T), and in this case the solution will be the same in all time periods and these constraints can even be removed.

$$\sum_{t \in T} y_i^t \leq n, \forall i \in S \tag{9.26}$$

Constraints (9.27) define the decision variables as boolean.

$$h, y \in \{0, 1\} \tag{9.27}$$

The model solution indicates, for each period, the set of active nodes (variables y), which demand points they cover (variables x) and also the set of demand points(variables h) not covered. Figure 9.9 shows the input parameters of the algorithm: the monitoring area, the demand points and the sensor nodes. Figures 9(d) and 9(e) show a possible MCP-WSN solution, considering two periods and $n = 1$.

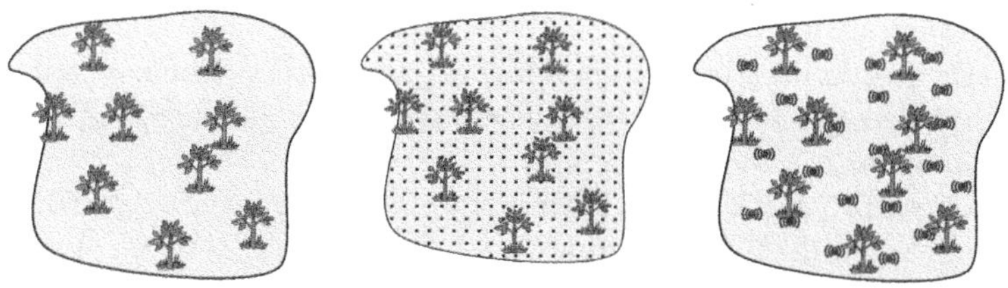

(a) Monitoring area. (b) Demand points. (c) Sensor nodes deployed in the area.

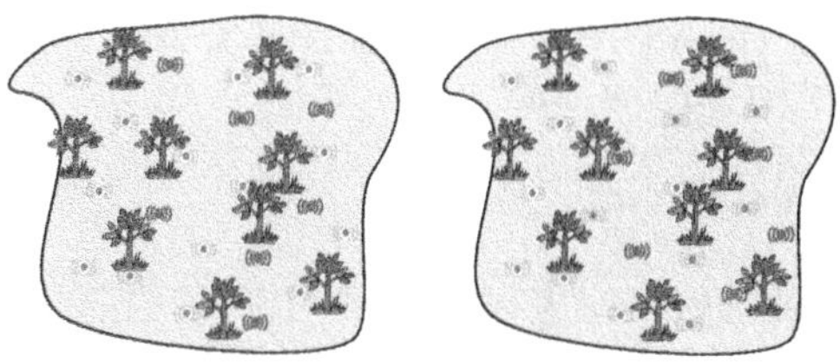

(d) Active nodes in period 1. (e) Active nodes in period 2.

Fig. 9.9. Example of WSN and MCP solution

The previous discussion and the example of Figure 9.9 make clear that there is a relationship among the number of sensor nodes $|S|$, the number of periods that a sensor can remain active n and the network lifetime T. For instance, if the number of sensor nodes is small and $n << T$, it is expected that the solutions for the last periods of time will not contain active sensor nodes. In this work, the parameters values are chosen to difficult the algorithm search, avoiding trivial solutions. In general, the algorithms will try to balance the number of active sensor nodes and demand points not covered in each period of time $t \in T$.

MCP-WSN Evolutionary Algorithm

The evolutionary solution for the MCP-WSN is based on a genetic search and a local search. Both search mechanisms are integrated in order to find good and feasible solutions. The genetic search looks for good solutions for the MCP-WSN, but many times these solutions can be unfeasible. So, it is applied a local search that turns unfeasible solutions into feasible ones, using an incremental greedy algorithm. Both mechanisms are detailed in the next section.

Genetic search

The genetic search uses binary encoding of parameters. The concept of multi-period chromosome is introduced, each one of them carrying information about all periods $t \in T$.

The representation of the chromosomes in the genetic algorithm for the MPC-WSN is similar to the one used for the CCP-WSN, presented in Section 9.3.1. Each chromosome can be represented by a binary matrix, with dimension θ:

$$\theta = [|T|][|S|] \tag{9.28}$$

where $|T|$ represents the total number of periods (or the network expected lifetime). Thus, each chromosome have, in each line t, $t \in T$, the sensor nodes which should be active in the time period t.

Operations over chromosomes

The multi-period chromosomes features allow a lot of possible interesting bio-inspired operations over them. Some of them are briefly specified next.

Initial population The generation of the initial population can be implemented as follows:

- For each member of the initial population, a matrix of size $[|T|][|S|]$ is generated. Each gene is generated using a random process with uniform distribution. It is generated a random number z, and whether z is smaller than a pre-defined parameter κ, than the gene is set to 1, otherwise it is set to 0.

The initial population generally can have a high number of chromosomes. However, it can be computationally hard to work with a lot of chromosomes. So, in a population of size Ξ, just Π chromosomes are selected to be used during the iterations of the algorithm, where $\Pi \leq \Xi$.

Random mutation (RM): In this operation, each chromosome of the population, except the best one, is visited. With a probability μ, a chromosome is selected for mutation, and a period t and a gene i are randomly chosen. The position $[t][i]$ get its value inverted.

Greedy Period mutation (GPM): This operation performs like the previous one. But, when a period is selected, each one of its genes is visited. For each gene with value 1 a random number r is generated. If r is smaller than a pre-defined probability β, then the gene is set to 0. This operation intends to decrease the number of active nodes in the chromosome.

Random Matching (RMa): This operation chooses which chromosomes will be combined for generating new chromosomes in the population. The random process chooses two chromosomes using an uniform distribution.

Cost Weighting Pairing (CWP): The CWP process, defined in Section 9.3.1, chooses two chromosomes for matching. The chromosomes with better values of objective function have high probability of matching. So, it is intended to get a good solution through the combination of good solutions. This algorithm presents a problem due to the precision of the float point representation. During the algorithm execution, this problem is monitored, and when it is about to happen, the mode of matching is changed to RMa.

Crossover When two chromosomes are selected (using RMa or CWP), they are combined as follows:

- The first child of the couple receives the first $T/2$ periods of the father, and the remainder from the mother.
- The second child receives the first $T/2$ periods from the mother, and the remainder from the father.

These two new chromosomes go to the place of the two worst chromosomes in the population. This can be seen as a kind of natural selection.

Local Search Process

During the process of mutation, matching and crossover one can get unfeasible solutions as the best ones. This occurs because the genetic search just considers constraints (9.22), (9.23), (9.24) and (9.25), not worrying about constraints (9.26). The goal of the local search proposed is to work just on these constraints turning unfeasible solutions into feasible ones.

The local search works as follows. Over all chromosomes of the population, it starts computing an array γ that keeps the number of periods that each gene is active in the chromosome. So, for each gene i and for each period t, it is verified if i is active. If during this process it is verified that i is active during more periods than possible (equation (9.26)) the current gene i in the period t is changed to 0 (*zero*) and its status in the array γ is updated. Next, the algorithm searches the gene j closest to i (the sensor node j closest to i), and, if j can be activated for one more period, its corresponding gene is turned to 1 in the same period t, and γ is updated. The algorithm performance depends on the parameter T, and if T is very large, the Local Search process spends a lot of time, once this algorithm worst case is $O(|S|^3|T|)$.

This local search always guarantees that unfeasible solutions will be turned into feasible ones by its interactive and greedy process. If some sensor node should be turned off, the values of variables x and h are updated to fit the remainder constraints (from (9.22) to (9.25)).

Since the local search process changes the solutions found by the genetic search, they can be worse than the previous ones regarding the fitness function, but they are feasible. Fig. 9.10 summarizes the main loop of the evolutionary algorithm developed.

input: set S of sensor nodes, size of area, number of demand points
begin
  while NOT stop condition **do**
    Select chromosomes for matching using RMa or CWP;
    Proceed matching/crossover;
    Mutation using RM or GPM;
    Local Search;
    Evaluate new population using equation (9.21);
    Shellsort(new Population);
  end while
end

Fig. 9.10. The evolutionary algorithm for MCP

Computational Results

For this problem, an instance that contains 16 nodes with sensor ranges of $15m$, deployed into a square area of size $50m$ x $50m$ is used. Two configurations are considered; the first one uses RMa + RM and the second uses CWP + RM. Table 9.5 summarizes the results[2]. In this network, the original coverage of

CFG	Active	Δ(Active)	Unc. area (%)	Δ(Unc. area)	Time (s)	Δ(Time)
CPLEX	4.00	0.82	24.05	9.44	1052.18	-
CF1	3.96	0.09	28.21	1.34	13.10	0.52
CF2	4.00	0.00	27.57	1.53	15.15	4.09

Table 9.5. *Comparisons of evolutionary algorithm and CPLEX - 16 nodes -* $50m$ x $50m$ *- Sensing range* = $15m$

the monitoring area is 99.48%. CF2 reaches a very interesting result, reaching the best number of active nodes during all simulations, and a good coverage when compared to CPLEX results. Figure 9.11 shows the behavior of both configurations for each generation. CF2 converges faster, and the population in the end is almost equals to the best one, showing that the best solution has spread its genes for the population through the use of CWP. CF1 presents a light convergence, but the best solution finds a result similar to CF2's. Considering only the objective function, the best solution of all tests obtained by CF2 is about 4% worse than the result obtained by CPLEX, showing that this configuration reached a very interesting result for this instance.

In the second set of tests, each sensor node has enough battery to remain active for only two periods. The first instance has 16 nodes in the square area of size $60m$ x $60m$ and sensor range of $15m$, and it is used a $\kappa = 40\%$. The first

[2] Again the algorithm runs for 33 times to get average values.

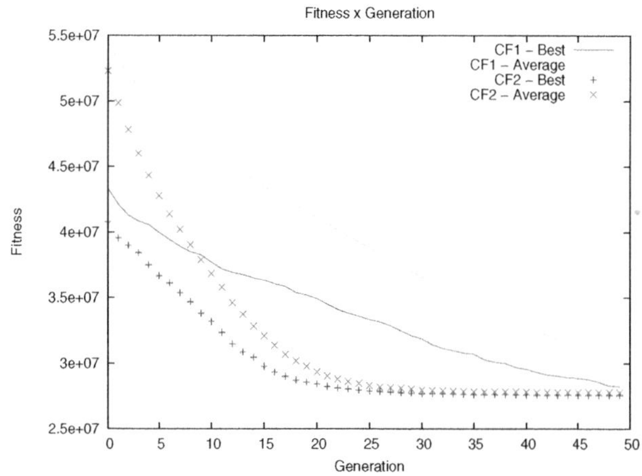

Fig. 9.11. Algorithm evolutions

configuration uses RMa + RM, and the second uses RMa + GPM. Table 9.6 reports the results.

CFG	Active	Δ(Active)	Unc. area (%)	Δ(Unc. area)	Time (s)	Δ(Time)
CPLEX	7.00	0.82.	32.19	6.21	159.39	-
CF1	7.34	0.25	34.08	0.83	27.84	9.70
CF2	6.75	0.40	35.67	0.74	25.00	7.17

Table 9.6. *Comparisons of evolutionary algorithm and CPLEX - 16 nodes - 60m x 60m - Sensing range* = 15m

It is important to say that this network presents 77.78% of original coverage. CF1 presents the best result, being close to CPLEX regarding the number of active nodes and coverage. Figure 9.12 reports the configurations behavior showing that CF1 is better. Regarding the value of the objective function, the best solution found by CF1 is just 1% worse than CPLEX optimal solution.

9.4 Conclusions

A wireless sensor network (WSN) is a kind of ad-hoc network, with distributed communication, sensing and processing capacities. A WSN can be composed by tens or even hundreds of small battery-powered devices, called sensor nodes. There are several challenges regarding WSNs once these networks present several unique features when compared to traditional ad-hoc

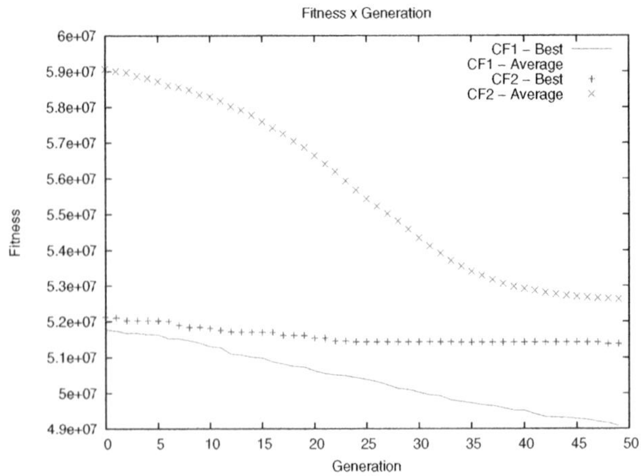

Fig. 9.12. Algorithm evolutions

networks, therefore existing ad-hoc solutions must be extended and adapted to be used in WSNs. So it is a area with a great variety of combinatorial problems where the evolutionary approaches can be applied.

This chapter discusses two combinatorial problems in the area of wireless sensor networks and presents formal definitions of these problems, a mathematical formulation and evolutionary algorithms for each one of them. The solutions of the mathematical formulations are optimal, but to reach these solutions it is spent a lot of computational time and demanded hard computational effort and dedicated and generally not cheap software. The evolutionary algorithms are fast, and in many situations can find good solutions in a feasible time. Moreover, no kind of commercial software is necessary. The optimal and evolutionary approaches can run together with a WSN management architecture, like the one proposed in [4].

References

1. Park, S., Savvides, A., Srivastava, M.B.: Simulating Networks of Wireless Sensors. In: 2001 Winter Simulation Conference, Informs Simulation Society (2001)
2. Mainwaring, A., Polastre, J., Szewczyk, R., Culler, D., Anderson, J.: Wireless Sensor Networks for Habitat Monitoring. In: ACM 1st International Workshop on Sensor Networks and Applications (WSNA'02), ACM Press (2002)
3. Crossbow Technology, I.: MPR - Mote Processor Radio Board MIB - Mote Interface / Programming Board User Manual. Crossbow Technology, Inc (2003)
4. Ruiz, L.B.: Manna: A Management Architecture for Wireless Sensor Networks. PhD thesis, Universidade Federal de Minas Gerais (2003)

5. Tilak, S., Abu-Ghazaleh, N., Heinzelman, W.: Infrastructure Tradeoffs for Sensor Networks. In: ACM 1st International Workshop on Sensor Networks and Applications (WSNA'02), ACM (2002) 49–58
6. Vieira, M.A.M., Vieira, L.F.M., Ruiz, L.B., Loureiro, A.A.F., Fernandes, A.O., Nogueira, J.M.S.: Scheduling Nodes in Wireless Sensor Networks: A Voronoi Approach. In: LCN '03: Proceedings of the 28th Annual IEEE International Conference on Local Computer Networks, IEEE (2003) 423
7. Meguerdichian, S., Koushanfar, F., Potkonjak, M., Srivastava, M.B.: Coverage Problems in Wireless Ad-Hoc Sensor Networks. In: INFOCOM' 01, IEEE (2001) Volume 3, 1380 – 1387
8. Megerian, S., Potkonjak, M.: Low Power 0/1 Coverage and Scheduling Techniques in Sensor Networks. UCLA. Technical report (2003)
9. Slijepcevic, S., Potkonjak, M.: Power Efficient Organization of Wireless Sensor Networks. In: IEEE International Conference on Communications (ICC) 2001, IEEE (2001) 1260–1265
10. Siqueira, I., Ruiz, L., Loureiro, A., Nogueira, J.M.: A Management Service for Density Control in Wireless Sensor Networks. In: 22nd Brazilian Simposium on Computer Networks (SBRC), Brazilian Computer Society (in portuguese) (2003) 249 – 262
11. Heinzelman, W., Chandrakasan, A., Balakrishnan, H.: An Application-Specific Protocol Architeture for Wireless Microsensor Networks. IEEE Transactions on Wireless Communications (2002) 660–670
12. Quintão, F.P., Mateus, G.R., Nakamura, F.G.: An Evolutive Approach for the Coverage Problem in Wireless Sensor Networks. In: Proceedings of 24th Brazilian Computer Society Congress, Brazilian Computer Society (in portuguese) (2004)
13. Quintão, F.P., Nakamura, F.G., Mateus, G.R.: A Hybrid Approach to Solve the Coverage and Connectivity Problem in Wireless Sensor Networks. In: IV European Workshop on Meta-heuristics, EUME (2004)
14. Nakamura, F.G. Quintão F.P., Menezes, G.C., Mateus, G.R.: An Optimal Node Scheduling for flat Wireless Sensor Networks In: ICN 2005 - International Conference on Networking, IEEE (2005)
15. Menezes, G.C.: Model and Algorithms for the definition of Density and Position of nodes in a Wireless Sensor Network. Master's thesis, Federal University of Minas Gerais (in portuguese) (2004)
16. Chiasserini, C.F., Chlamtac, I., Monti, P., Nucci, A.: An Energy-efficient Method for Nodes Assignment in Cluster-Based Ad-Hoc Networks. Wireless Networks Vol. 10 (2004)
17. Cerpa, A., Estrin, D.: ASCENT: Adaptive Self-Configuring sEnsor Networks Topologies. In: Proceedings of the 21st International Annual Joint Conference of the IEEE Computer and Communications Society (INFOCOM02) Vol. 3 (2002).
18. Ye, F., Zhong, G., Cheng, J., Zhang, L.: PEAS: A Robust Energy Conserving Protocol for long-lived Sensor Networks. In: Proceedings of the 23rd International Conference on Distributed Computing Systems (ICDCS03) pages 28–37 (2003).
19. Zhang, H., Hou, J.: Mantaining Sensing Coverage and Connectivity in Large Sensor Networks. Wireless ad hoc and Sensor Networks, Vol. 1, pages 89–123 (2005).

20. Wang, X., Xing, G., Zhang, Y., Lu, C., Pless, R., Gill, C.: Integrated Coverage and Connectivity Configuration in Wireless Sensor Networks. In: SenSys '03: Proceedings of the 1st International Conference on Embedded Networked Sensor Systems, ACM Press (2003) 28–39
21. Haupt, R.L., Haupt, S.E.: Practical Genetic Algorithms. John Wiley & Sons, Inc (1998)
22. ILOG, Inc.: High-Performance Software for Mathematical Programming and Optimization. http://www.ilog.com/products/cplex/ (2005)
23. Tarjan, R.E.: Data Structures and Network Algorithms. Society for Industrial and applied mathematics (1983)
24. Tanembaum, A.S.: Computer Networks, 3rd Edition. Prentice Hall PTR (1996)

Part II

Approximation of Fitness Functions

10

Individual-based Management of Meta-models for Evolutionary Optimization with Application to Three-Dimensional Blade Optimization

Lars Gräning, Yaochu Jin, and Bernhard Sendhoff

Honda Research Institute Europe
Carl-Legien-Str. 30, 63073 Offenbach am Main, Germany
{lars.graening,yaochu.jin,bernhard.sendhoff}@honda-ri.de

Summary. To reduce the number of expensive fitness function evaluations in evolutionary optimization, individual-based and generation-based strategies for meta-model management (evolution control) have been proposed. In this work, four individual-based frameworks for meta-model management are investigated. A feed-forward neural network is employed to construct an approximation model of the fitness function. Structure optimization of the neural network is used to reduce the approximation error. In an attempt to adapt the number of controlled individuals, adaptation mechanisms are suggested based on the model error, selection error, rank correlation, and fitness correlation. Preliminary results indicated that the adaptation mechanisms do not work well as expected.

Two of the frameworks are implemented in 3D blade design optimization. The results showed that individual-based meta-model management is promising, though further efforts are still needed to improve the performance of the evolutionary algorithms with meta-models for fitness estimation.

10.1 Introduction

It has been shown that evolutionary algorithms are very powerful in solving many real-world optimization tasks such as 3D turbine blade aerodynamic design optimization of a jet engine [5, 13, 14], of micro heat exchanger [2] or transonic wing design [17]. The advantage of evolutionary algorithms is that they stochastically search the fitness landscape for the optimal solution without the need of any gradient information. However, this advantage is at the cost of a large number of fitness evaluations. In 3D blade optimization, one evaluation of the fitness will take huge computational time because computational fluid dynamics (CFD) simulations have to be performed to evaluate the performance of the blade.

L. Gräning et al.: *Individual-based Management of Meta-models for Evolutionary Optimization with Application to Three-Dimensional Blade Optimization*, Studies in Computational Intelligence (SCI) **51**, 225–250 (2007)
www.springerlink.com

To reduce the number of fitness evaluations, one idea is to estimate the fitness using computationally efficient meta-models, see [8] for an overview of existing methods. One problem to deal with in real-world optimization problems is that it is difficult to acquire enough data so that the meta-model can sufficiently approximate the original fitness landscape, which could result in false convergence [11, 15]. Therefore it is not advisable to use meta-models only as a surrogate for the original fitness function. To avoid false convergence, the neural network model should be used in conjunction with the original fitness function. This is termed evolution control or model management [11, 15]. If evolution control is used, new data become available during optimization, which can then be used for on-line update of the meta-model. Meta-models can also be employed in the local search embedded in evolutionary optimization [19].

In this work, four individual-based evolution control methods are compared on three benchmark problems. The two most promising methods are adopted for 3D blade design optimization. In an attempt to improve the performance of the methods, adaptation mechanisms are suggested to adjust the impact of the meta-model on the optimization process during optimization.

10.2 Evolutionary Optimization with Neural Network Based Fitness Estimation

10.2.1 Evolutionary Optimization

The evolution strategy with covariance matrix adaptation (ES-CMA) [4] is adopted for blade optimization in this work. No recombination has been used since negative influence has been observed in blade optimization. The mutation operator adds normally distributed random values to the design parameters of the individual in order to search the design space. Adaptation of the parameters of the normal distribution in each generation plays an essential role for the performance of the search algorithm. ES-CMA uses a derandomized self-adaptation mechanism where the whole covariance matrix is adapted to adjust the parameters of the normal mutation distribution. These parameters, called strategy parameters, are also encoded in the chromosome of the individuals.

One major problem in evolutionary design optimization process is the high cost of computation resources for evaluating the quality of the designs. For example, a 3D design optimization run takes upto 3 months for 200 generations of evolution on high performance computers. In this work, we employ neural networks as the meta-model to partially substitute the computationally expensive fitness evaluations.

10.2.2 Artificial Neural Networks

Up to now, polynomials, kriging model, radial-basis-function networks, and multi-layer perceptrons (MLP) have been used as meta-models in evolutionary optimization [8, 16]. We decided to use MLPs in this work because it has been shown that MLPs are very powerful in function approximation and classification. The neural network adapts its parameters and structure to learn the functional mapping between the design parameters and the performance with the help of a number of training data obtained from previous optimization. After the network is trained, it can be used to predict the fitness of new designs, given the design parameters.

Multi-layer Perceptros (MLPs)

In [1, 6] it is shown that one hidden layer is sufficient to approximate any continuous functions, provided that a sufficient number of hidden layer neurons is used. The number of hidden neurons depends strongly on the characteristics of the target function [20]. Mostly, the characteristics of the function is unknown. In this case, it is suggested that structure optimization of the MLPs should be considered [20].

In this work, the algorithm introduced by Hüsken et al [7] is employed. The architecture of the neural network is encoded in a connection matrix and a weight matrix. The values in the connection matrix determine which nodes in the network are connected and the values in the weight matrix determine the strength of connections. Fig. 10.1 illustrates the mapping between a neural network and the connection and weight matrices.

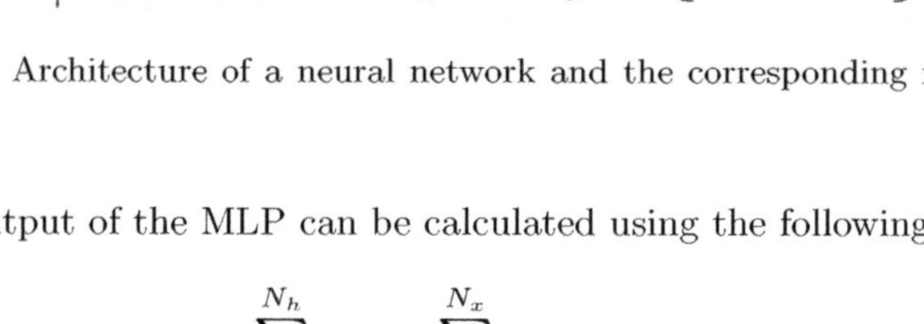

Fig. 10.1. Architecture of a neural network and the corresponding matrix representation

The output of the MLP can be calculated using the following equations:

$$y_l = \sum_{k=1}^{N_h} w_{lk} \cdot g(\sum_{i=1}^{N_x} w_{ki} x_i + t_h) + t_y, \tag{10.1}$$

where N_x is the number of input neurons and N_h is the number of hidden neurons. In the neural network the following activation function $g(z)$ is used, whose characteristic is similar to the sigmoidal function.

$$g(z) = \frac{z}{1 + |z|}. \tag{10.2}$$

The weights of the neural network are trained online during the optimization when new training samples are available. For neural network training, Rprop [18], an improved version of the back propagation algorithm is used. The main difference between Rprop and the back propagation algorithm is that the learning rate adjustments and weight changes do not depend on the magnitudes of the gradient, rather on the signs of the gradient terms.

Structure Optimization of Neural Networks

To improve the approximation quality of the neural network, one way is to optimize the architecture of the neural network during optimization. Yao [21] provided a comprehensive review of the optimization of neural networks using evolutionary algorithms. In this work, a genetic algorithm has been used for this purpose. The connections a_i and the value of the weights w_i of the neural network are encoded into the genotype of an individual. This means that each individual encodes a neural network with a different architecture and different weights. To generate offspring representing different neural networks, specific mutation methods are used. The mutation methods allow to insert or delete a single connection or neuron and the weights are mutated by adding a normally distributed random number. After mutation, the Lamarckian mechanism is used for lifetime learning of the weights. Finally, the weights are coded back into the individuals. EP-tournament-selection is used to select the individuals representing the neural networks with the lowest mean square error with respect to the training data.

10.3 Individual-Based Evolution Control Methods

It is found that if a meta-model such as a neural network is used to estimate the fitness of the individuals, the evolutionary algorithm probably will converge to a false optimum [15], which is not one of the original fitness function. In these cases, it is essential that the model be used in conjunction with the original fitness function. How often the model should be used instead of the original fitness evaluation is the task of *evolution control* or *model management* methods.

As illustrated in Fig. 10.2, evolution control methods can be divided into two basic approaches, namely individual-based and generation-based [11].

In the generation-based approach, one has to decide generation by generation for all individuals whether the fitness will be determined using the meta-model or using the original fitness function. If the individual-based approach is used, one has to decide for each individual in every generation whether the meta-model or the time-consuming fitness function should be used. In this

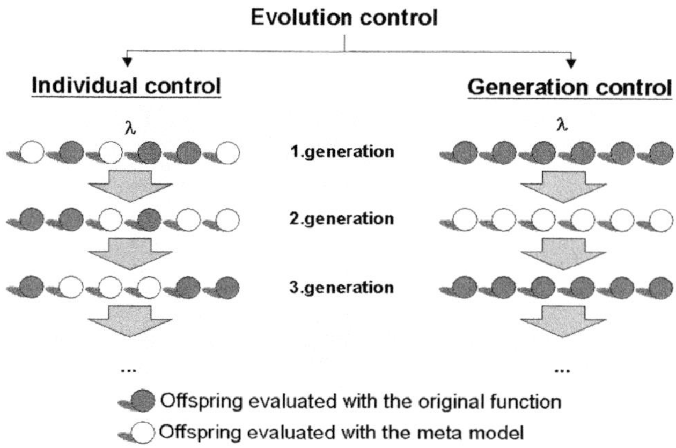

Fig. 10.2. Principle of individual-based and generation-based evolution control methods

work we concentrate on the individual-based approaches and will introduce several methods in detail.

As can be seen in Fig. 10.3, the individual-based evolution control method can be described by a common evolutionary optimization process. In each iteration, λ' offspring are generated out of the μ parents by mutation. After that, the individual-based control method decides which λ^* offspring are evaluated by the real fitness function. The results are used to train the neural network before the fitness of the remaining $\lambda' - \lambda^*$ offspring will be estimated by the neural network. In the end μ parents will be selected out of the λ individuals according to their fitness.

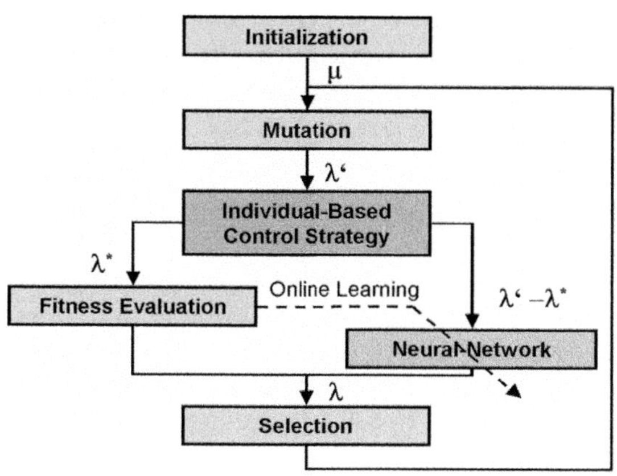

Fig. 10.3. Evolutionary optimization process including individual-based evolution control methods

10.3.1 Best Selection (BS)

In the best selection strategy [15], all λ' offspring are pre-evaluated by the neural network at first to find the most promising (best) λ^* individuals. These λ^* individuals are evaluated by the original fitness function and the results are used for training. After training the neural network, the remaining $\lambda' - \lambda^*$ individuals are again evaluated by the neural network to get a better estimation. At the end of each generation, the μ best individuals out of all $\lambda = \lambda'$ individuals become parents for the next generation.

10.3.2 Pre-Selection (PreS)

A pre-selection has been introduced in [19], in which the Gaussian processes are used as meta-model instead of neural networks. The idea is as follows. $\lambda' > \lambda$ offspring are generated by mutation, and the neural network is used to estimate the fitness of these offspring. The λ^* most promising individuals are pre-selected out of the λ' offspring. Like in the best selection strategy, the λ^* individuals are evaluated using the original fitness function. The main difference to the best selection strategy is that the μ parents are selected only out of the λ^* individuals, which are all evaluated with the original fitness function.

10.3.3 Clustering Technique (CT)

In [12] a different approach is described to find out which individuals have to be evaluated with the original fitness function. Using the k-means clustering technique, all λ' individuals of a generation are grouped into λ^* clusters. Now the λ^* individuals closest to the cluster center are evaluated using the original fitness function. The results of the fitness evaluations, as in all other methods, are used to train the neural network during the optimization. The fitness of the remaining $\lambda' - \lambda^*$ individuals is estimated using the neural network. Last but not least the μ parents are selected out of all $\lambda = \lambda'$ individuals.

10.3.4 Clustering Technique with Best Strategy (CTBS)

The idea of the clustering technique with best strategy is the same as in clustering technique. The offspring are also grouped into a number of clusters. Now the neural network is used to predict the fitness of each offspring. Instead of evaluating the individuals closest to the cluster center, the λ^* individuals with the best predicted fitness of each cluster will be evaluated by the original fitness function.

10.3.5 Simulation Results

The main reason for using individual-based evolution control is to reduce computational costs. In real-world optimization problems like the blade optimization, the calculation of the fitness needs a large amount computational time. It is impossible to test all the algorithms on the real-world design optimization problem. So all methods are tested first on three widely used benchmark functions. The test functions are the Sphere, Rosenbrock and Ackley functions. To get an impression of how the functions looks like, the equation and the two dimensional plotting of the functions are illustrated in Fig. 10.4, where n is the dimension of the test functions. In the following simulations, the dimension is set to 10. Part of the following results has been reported in [3].

(a) (b)

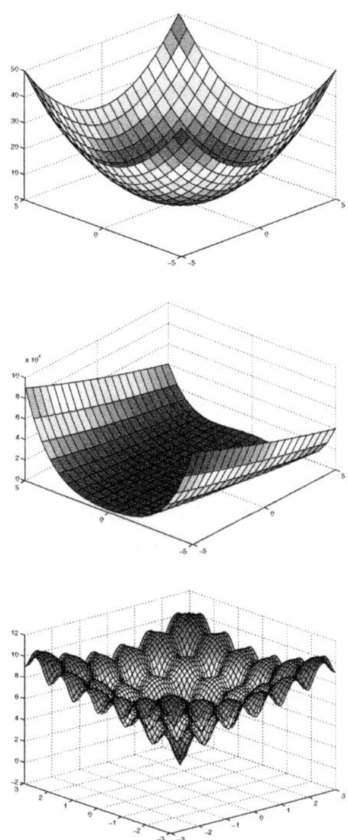

Sphere

$$f_{sphere}(x) = \sum_{i=1}^{n} x_i^2$$

Rosenbrock

$$f_{rosen}(x) = \sum_{i=1}^{n-1} \left(100\left(x_{i+1} - x_i^2\right)^2 + \left(x_i - 1\right)^2\right)$$

Ackley

$$f_{ackley}(x) = 20 + e - 20\exp\left(-0.2 \cdot \sqrt{\frac{1}{n} \cdot \sum_{i=1}^{n} x_i^2}\right) - \exp\left(\frac{1}{n}\sum_{i=1}^{n} \cos\left(2\pi x_i\right)\right)$$

Fig. 10.4. Overview of the used test-functions with an (a) 2D illustration and (b) the equation of the functions

Simulation Setup

In all simulations, a (μ, λ) ES-CMA without recombination is adopted. The strategy parameters of the covariance matrix are randomly initialized between $\sigma_{min} = 0.05$ and $\sigma_{max} = 4$.

We compare the model management frameworks in both serial and parallel computing environments. The parameters for the evolutionary optimization process are set according to the different requirements of the used computational environment.

When the optimization is conducted in a serial environment, the performance depends only on the number of fitness evaluations needed to reach a near optimum. So in each generation μ parents are selected out of the same amount of λ offspring. The remaining parameters are adjusted according to the values of μ and λ as combined in Table 10.1. For clarity the following notation is used: $(\mu, [\lambda']\lambda[\lambda^*])$. μ parents are always selected out of λ offspring. λ' defines the number of pre-selected and λ^* the number of controlled individuals. The ratio according to [4] is set to $\mu \leq \frac{\lambda}{3}$. The ratio of λ to λ' and λ^* are based on recommendations or findings in [12] and [19].

PlainES	PreS	BS, CT, CTBS
(3, [12]12[12])	(3, [24]12[12])	(3, [12]12[6])

Table 10.1. Settings of the strategy parameters $(\mu, [\lambda']\lambda[\lambda^*])$ to compare the performance of the individual-based control methods for optimization in a serial computational environment

If it can be assumed that in a parallel computational environment enough computers are available to evaluate all individuals in parallel, the number of fitness evaluations itself is less important. In that case the number of generations needed to reach a near-optimal solution is the main concern. We also assume that the number of used machines equals the number of fitness evaluations λ^*, which is held constant for all methods. The remaining parameters are adjusted with respect to λ^*. The entire setup is listed in Table 10.2.

PlainES	PreS	BS, CT, CTBS
(2, [6]6[6])	(2, [12]6[6])	(2, [12]12[6])

Table 10.2. Settings of the parameters μ, λ' and λ^* to compare the performance of the individual-based control methods for optimization in a parallel computational environment

The neural network used in the simulations consists of 10 input nodes, one hidden layer with four hidden neurons, and one output node. If structure optimization is carried out, the number of hidden neurons is not fixed.

To achieve a good local approximation of the original fitness landscape, only data of the most recent evaluations are used for training.

Serial Optimization

In a serial computational environment, only the number of expensive fitness evaluations is of importance to reach a near-optimum. 20 independent runs are performed for each optimization to reduce the randomness. The median fitness value of the best offspring in each generation is plotted versus the number of fitness evaluations to compare the performance of the introduced methods. The results from the Sphere, Rosenbrock and Ackley functions are presented in Fig. 10.5 and Fig. 10.6. The left column presents the results with and the right column without structure optimization of the neural network.

To show the statistical significance between the evolution control methods and the plain evolution strategy, the boxplot of the results are given in Fig. 10.6. The boxplot illustrates the median and the variance of the fitness values of the best individual in the final generation over 20 runs. The notches of the boxes in the plot are the graphical equivalence to the student t-test. If the notches of two boxes do not overlap, there is a significant difference between the medians of the two strategies at a significance level of $p = 0.05$.

From Fig. 10.5 and Fig. 10.6, we can see that all evolution control methods except the best strategy improve the performance of the plain evolution strategy significantly on the 10D Sphere function. But there are no statistically significant differences between the model-assisted strategies themselves, no matter whether structure optimization of the neural networks are performed or not.

It turns out that the clustering technique with best strategy outperforms other algorithms on the 10D Rosenbrock function, when no structure optimization of the neural network is carried out. However, all algorithms fail to improve the performance of the plain evolution strategy significantly. This result may be attributed to the fact that the number of hidden neurons is not sufficiently large to approximate the Rosenbrock function. Meanwhile, the result indicates that with structure optimization, the neural networks perform locally very well on the Rosenbrock function.

From Fig. 10.5, it can be seen that the individual-based evolution control methods perform well on the Ackley function. But as we can see in the boxplots in Fig. 10.6 the variance of the strategies is very high except the pre-selection strategy. The pre-selection strategy outperforms the plain evolution strategy in almost all of the 20 runs.

In summary, it turned out that the pre-selection method shows the most stable and promising results and only fails to improve the evolution strategy on the Rosenbrock function. The reason for the stability of the pre-selection methods might be that the parents of the next generation are only selected from the individuals that evaluated with the original fitness function.

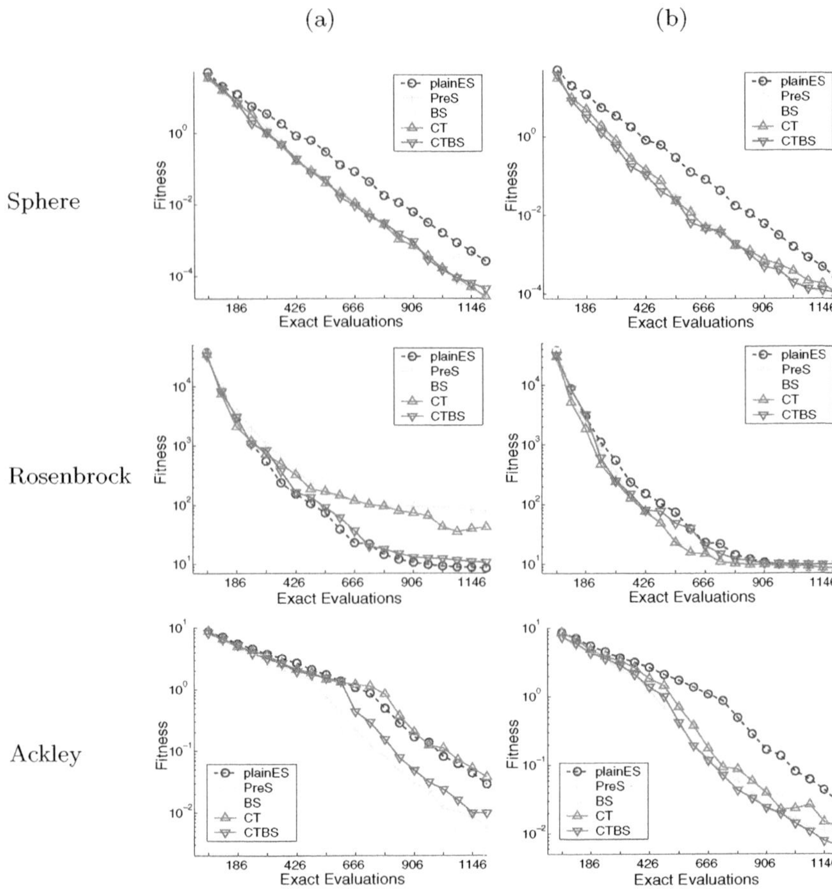

Fig. 10.5. Performance comparison of the individual-based methods in a serial computational environment: (a) without structure optimization and (b) with structure optimization of the neural network

In the following, we only show the results with structure optimization of the neural networks because the above results show that using structure optimization mostly improves the performance of the neural network and the optimization process.

Parallel Optimization

As mentioned before, to compare the performance of the individual-based evolution control methods in a parallel computational environment, the number of exact fitness evaluations is not as important as the number of generations. Therefore the fitness values are plotted versus the number of generations. To make sure that the comparison is fair, the number of real fitness evaluations

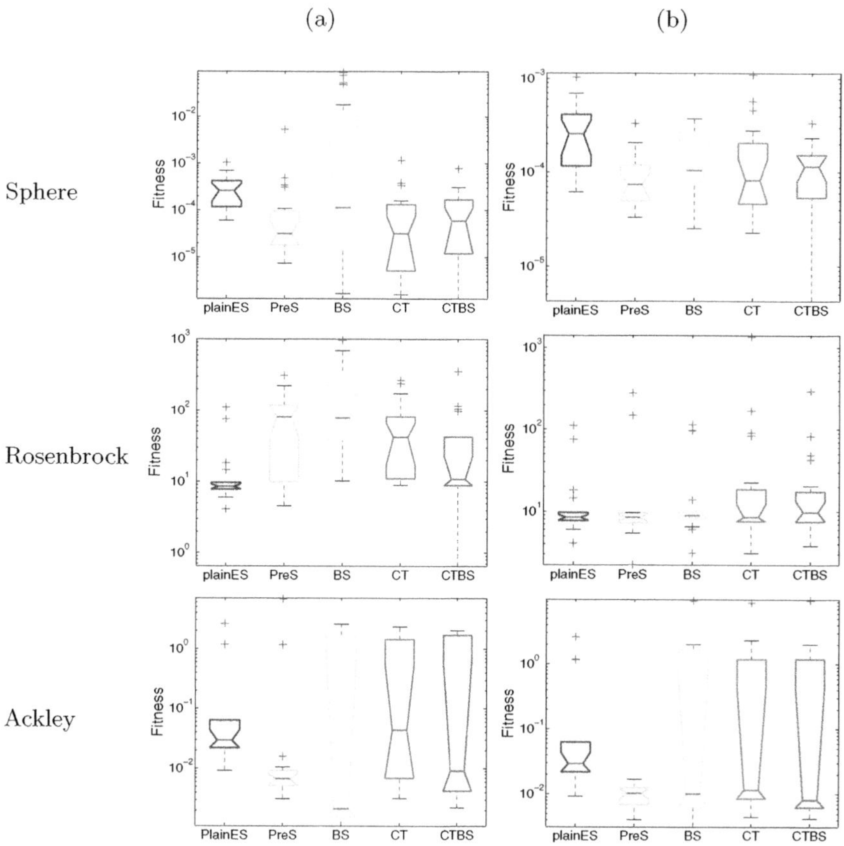

Fig. 10.6. Boxplot of the best fitness in the final generation over 20 runs after 1200 exact fitness evaluations are done: (a) without structure optimization, and (b) with structure optimization

each generation in all methods is the same. Fig. 10.7(a) shows the characteristics of the median best fitness value over the generations and in Fig. 10.7(b) the boxplot for statistical analysis is illustrated using structure optimization of the neural network.

As can be seen in Fig. 10.7, the pre-selection strategy improves the plain evolution strategy on all used test functions. So it might improve the optimization process better if the parents are only selected out of the λ^* offspring evaluated with the original fitness function. Using BS, CT or CTBS, the parents are selected out of all λ individuals, whose fitness has either been evaluated with the real fitness function or estimated by the neural network. The possible reason is that if some individuals are selected according to the estimated fitness, the optimization algorithm might be misled, which degrades the performance.

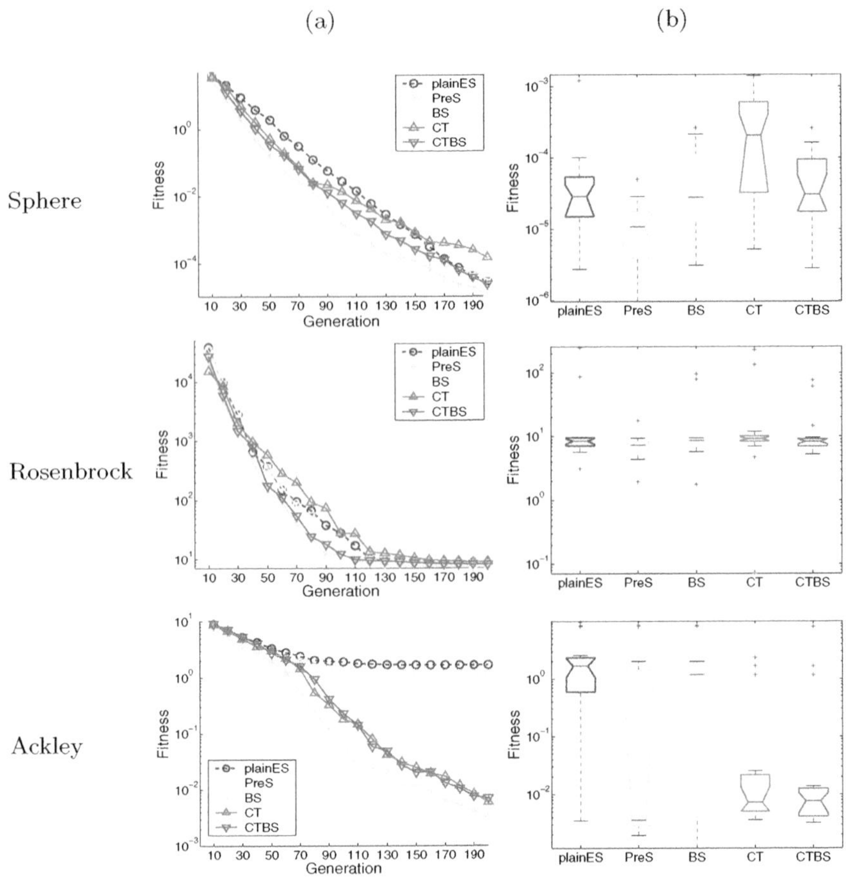

Fig. 10.7. Results of the individual-based methods in a parallel computational environment: (a) performance comparison and (b) boxplot over 20 runs after 200 generations

We can see that clustering the design space on bumpy fitness landscapes like the Ackley function gives a benefit to the algorithm. The clustering technique with best strategy performs also well on all other test functions, especially if the fitness function becomes more complex. However, it can not outperform the pre-selection strategy.

In both serial and parallel computational environments, the pre-selection and clustering technique with best strategy delivered the most promising results.

10.4 Adaptation in Individual-based Evolution Control

In the above comparisons, the number of controlled individuals is fixed during optimization. In this section, we consider adapting the number of controlled individuals. Three different adaptation frameworks are introduced. The idea of adaptation is that if the performance of the neural network increases during optimization, it should be used more often to substitute the original fitness function. There are two parameters that can be taken into account for adaptation. One is the number of fitness evaluations λ^*, and the other is the number of pre-selected individuals λ'. Adjustment of λ^* only makes sense of on-line scheduling of computational resources is possible.

10.4.1 Normalized Squared Error Driven Adaptation Mechanism (NERD)

The first quality measure we considered here to adapt the number of individuals is the approximation quality of the neural network. The approximation quality of the neural network can often be measured using the squared error between the individual's original fitness $\phi_i^{(Orig.)}$ and the estimated fitness of the neural network $\phi_i^{(MLP)}$:

$$E_i^{(SE)} = \left(\phi_i^{(MLP)} - \phi_i^{(Orig.)}\right)^2. \tag{10.3}$$

The error can be determined for each offspring i which has been evaluated with the original fitness function. The main idea of this adaptation method is that if the error of the neural network becomes smaller in the next generation $t+1$, the neural network should be used more often. But the error in the generation $t+1$ is unknown. It is only possible to compare the error in the current generation t with the error in the last generation $t-1$. There are two problems in comparing these two errors. In general, the fitness values by itself decline during optimization and so probably the value of the squared error will also decline. Therefore the squared error should be normalized by the use of the mean squared error:

$$E_i^{(NSE)} = \frac{E_i^{(SE)}}{\frac{1}{\lambda^*}\sum_{i=1}^{\lambda^*} E_i^{(SE)}}. \tag{10.4}$$

Comparing the mean of the normalized squared errors of all offspring will lead to the second problem. If there is one individual with a large error while the error of the rest of the individuals is small the comparison might be misleading. Instead of comparing the mean values, the ranks of the individuals' normalized errors are compared. An example how the rank of the error in generation $t-1$ and generation t can be calculated is illustrated in Fig. 10.8. First the values of the two sets of errors are combined and sorted. After that, each element is given its corresponding rank. If some samples carry the same error value, the rank will be averaged. Then, the ranks of the sets from generation $t-1$ and generation t are accumulated.

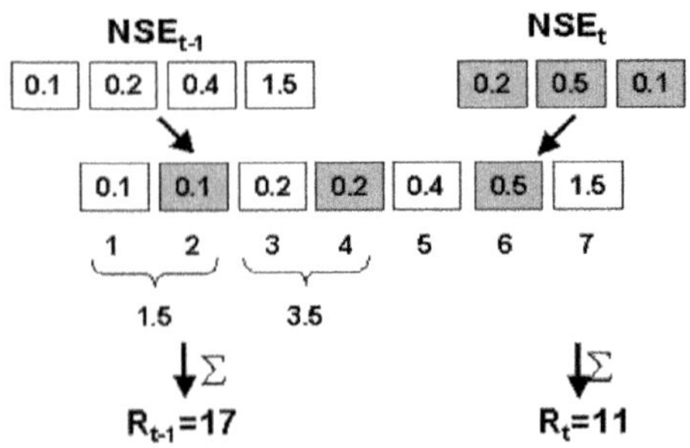

Fig. 10.8. Example of evaluating the rank with respect to the normalized square error in generation $t-1$ and generation t

Given the two ranks the quality measure $\rho^{(NERD)}$ can be determined by calculating the difference of the two normalized ranks. The ranks also have to be normalized by the number of λ^* individuals because λ^* might change if it is adapted during optimization:

$$\rho^{(NERD)} = \frac{R(t-1)}{\lambda^*_{t-1}} - \frac{R(t)}{\lambda^*_t}. \tag{10.5}$$

If R_t is smaller than R_{t-1}, the quality measure $\rho^{(NERD)}$ is less than 0 and the neural network should be used more often. On the other hand, if R_{t-1} is smaller than R_t, $\rho^{(NERD)}$ is bigger than 0 and the neural network should be used less often. λ^*_{t+1}, or rather λ'_{t+1} is adapted as follows:

$$\lambda^*_{t+1} = \lambda^*_t - \rho^{(NERD)} \cdot \Delta\lambda, \tag{10.6}$$

$$\lambda'_{t+1} = \lambda'_t + \rho^{(NERD)} \cdot \Delta\lambda. \tag{10.7}$$

The remaining difficulty is the choice of the correct free parameter $\Delta\lambda$, which might be problem-specific.

10.4.2 Selection Based Adaptation Mechanism (Sel)

From the evolutionary computation point of view, only the correct selection is of importance and not the approximation error of the model. The error of the neural network does not have direct influence on the evolutionary selection process. Therefore, as introduced in [15], a selection based quality measure can be considered to evaluate the quality of the model based selection process. For each correctly selected individual, based on the estimation of the model, it is given a rank of $(\lambda_t^* - i)$, if the individual has the i-th best fitness based on the true fitness. To calculate the rank, as shown in Fig. 10.9, we first pick out the μ^* best individuals based on the estimation of the neural network. Note that these individuals are chosen only to calculate the error measure.

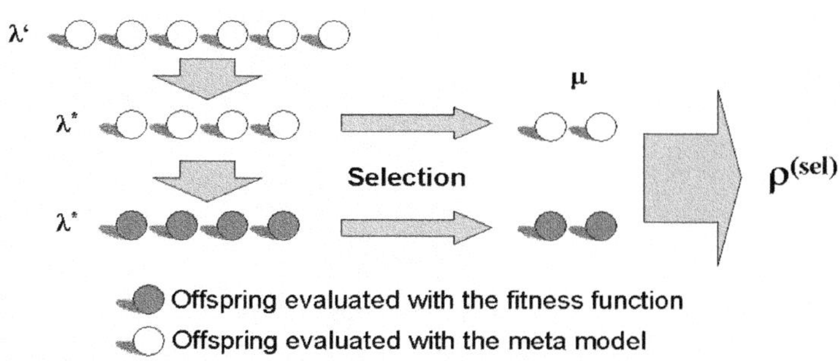

Fig. 10.9. Illustration of how to evaluate the selection-based quality measure

Afterwards, all λ^* individuals are evaluated using the original fitness function and μ individuals will be selected out of the λ_t^* (where t is the generation index) offspring and the sum of the ranks of all m correctly selected individuals measures the quality $\rho^{(sel)}$ in the current generation:

$$\rho^{(sel)} = \sum_{i=1}^{m} (\lambda_t^* - i). \tag{10.8}$$

If all individuals are selected correctly, the quality measure reaches its maximum:

$$\rho_{max}^{(sel)} = \sum_{i=1}^{\mu} (\lambda_t^* - i). \tag{10.9}$$

The idea is to compare the actual quality measure with the expected quality of a random selection process $\langle \rho^{(rand)} \rangle$:

$$\langle \rho^{(rand)} \rangle = \frac{\mu^2}{\lambda_t^*} \cdot \frac{2\lambda_t^* - \mu - 1}{2}. \tag{10.10}$$

If the quality in the current generation is better than the quality of a random selection process, then the neural network can be used to replace the original fitness function more often. Otherwise, the neural network should be less often used. The adaptation rule differs a little bit whether λ^* (Equation 10.12, 10.14) or λ' (Equation 10.11, 10.13) will be adapted.

$\rho_t^{(sel)} > \rho^{(rand)}$:

$$\lambda'_{t+1} = \lambda'_t + \frac{\rho_t^{(sel)} - \langle \rho^{(rand)} \rangle}{\rho^{(max)} - \langle \rho^{(rand)} \rangle} \cdot \Delta\lambda, \tag{10.11}$$

$$\lambda^*_{t+1} = \lambda^*_t - \frac{\rho_t^{(sel)} - \langle \rho^{(rand)} \rangle}{\rho^{(max)} - \langle \rho^{(rand)} \rangle} \cdot \Delta\lambda. \tag{10.12}$$

$\rho_t^{(sel)} < \rho^{(rand)}$:

$$\lambda'_{t+1} = \lambda'_t - \frac{\langle \rho^{(rand)} \rangle - \rho_t^{(sel)}}{\langle \rho^{(rand)} \rangle} \cdot \Delta\lambda, \tag{10.13}$$

$$\lambda^*_{t+1} = \lambda^*_t + \frac{\langle \rho^{(rand)} \rangle - \rho_t^{(sel)}}{\langle \rho^{(rand)} \rangle} \cdot \Delta\lambda. \tag{10.14}$$

It has to be considered that λ^*_{t+1} is bigger than the number of parents μ and as in all other adaptation methods λ'_{t+1} has to be equal or bigger than λ^*_{t+1}. As in the normalized squared error based adaptation framework, the free parameter $\Delta\lambda$ has also to be specified. One drawback of the selection based approach is probably the small number of μ individuals taken into account to measure the quality, which could result in strong oscillations in adaptation. Note that CMA-ES often uses a small population size.

10.4.3 Correlation Based Adaptation Mechanism (Rank, Corr)

Using the correlation based framework, all λ^*_t offspring are taken into account to evaluate the quality measure. Two different possibilities are suggested in [15] to evaluate the correlation between the λ^*_t individuals. The first correlation based quality measure is the *rank correlation (Rank)*. To evaluate the rank correlation measure, after estimation the λ^*_t individuals are sorted by their fitness and a given rank. The same is done after evaluating the fitness with the original fitness function. The rank correlation quality measure can now be calculated in the following way:

$$\rho^{(rank)} = 1 - \frac{6 \sum_{l=1}^{\lambda^*_t} (r_l - \widehat{r}_l)^2}{\lambda^*_t((\lambda^*_t)^2 - 1)}, \tag{10.15}$$

where $\widehat{r}_l$ is the rank of the l'th individual based on the estimated fitness and r_l is the rank based on the real fitness.

The second correlation based quality measure is the so called *continuous correlation (Corr)*. This quality measure calculates the correlation between the fitness values instead of the ranks. So the continuous correlation between the approximate model output and the original fitness function can be calculated by using Equation 10.16:

$$\rho^{(corr)} = \frac{\frac{1}{\lambda^*_t} \sum_{l=1}^{\lambda^*_t} \left(\phi_l^{(MLP)} - \bar{\phi}^{(MLP)} \right) \left(\phi_l^{(Orig)} - \bar{\phi}^{(Orig)} \right)}{\sigma^{(MLP)} \sigma^{(Orig)}}. \tag{10.16}$$

Here $\bar{\phi^{(m)}}$ and $\sigma^{(m)}$ are the mean value and the standard derivation of the fitness values evaluated using the neural network and $\bar{\phi^{(o)}}$ and $\sigma^{(o)}$ are the mean and the standard derivation of the real fitness values.

The rules to adapt λ^* and λ' are similar to the rules in the normalized squared error driven mechanism:

$$\lambda^*_{t+1} = \lambda^*_t - \rho \cdot \Delta\lambda, \tag{10.17}$$

$$\lambda'_{t+1} = \lambda'_t + \rho \cdot \Delta\lambda, \tag{10.18}$$

where ρ stands for $\rho^{(corr)}$ or $\rho^{(rank)}$.

10.4.4 Simulation Results

The empirical results are presented to investigate whether the adaptation mechanisms are able to improve the performance of the individual-based evolution control methods. To dynamically control the impact of the neural network on the individual-based evolutionary control strategy, two parameters can be adjusted during optimization.

Simulations have been conducted using the pre-selection strategy, where the number of pre-selected individuals is adapted using the introduced adaptation mechanisms. As mentioned, $\Delta\lambda$ has to be specified before starting the simulations. $\Delta\lambda$ was determined during some experiments and varies with the different adaptation mechanisms as listed in Table 10.3.

	Sel	NERD	Rank, Corr
$\Delta\lambda$	12	24	16

Table 10.3. Configuration of the free parameter $\Delta\lambda$ to adapt the number of pre-selected individuals

The initial value for λ' was fixed to 12 and the parameters for the evolution strategy are set to $(3, [\lambda'(t)]12[12])$. In all simulations, the structure of the neural network is optimized.

As one can see in Fig. 10.10, using the normalized error rank based (NERD) adaptation mechanism performs very poorly on all the test functions. In Fig.10.11, the change of λ' are shown on the left and the quality measure on the right. Using the error-based adaptation mechanism (NERD), it can be seen that λ' increases continuously on all test-functions. So the network error seems to decrease during optimization. This might indicate that the error of the neural network by itself is not directly correlated with the performance of the optimization process. Another reason might be that $\Delta\lambda$ has not been chosen correctly.

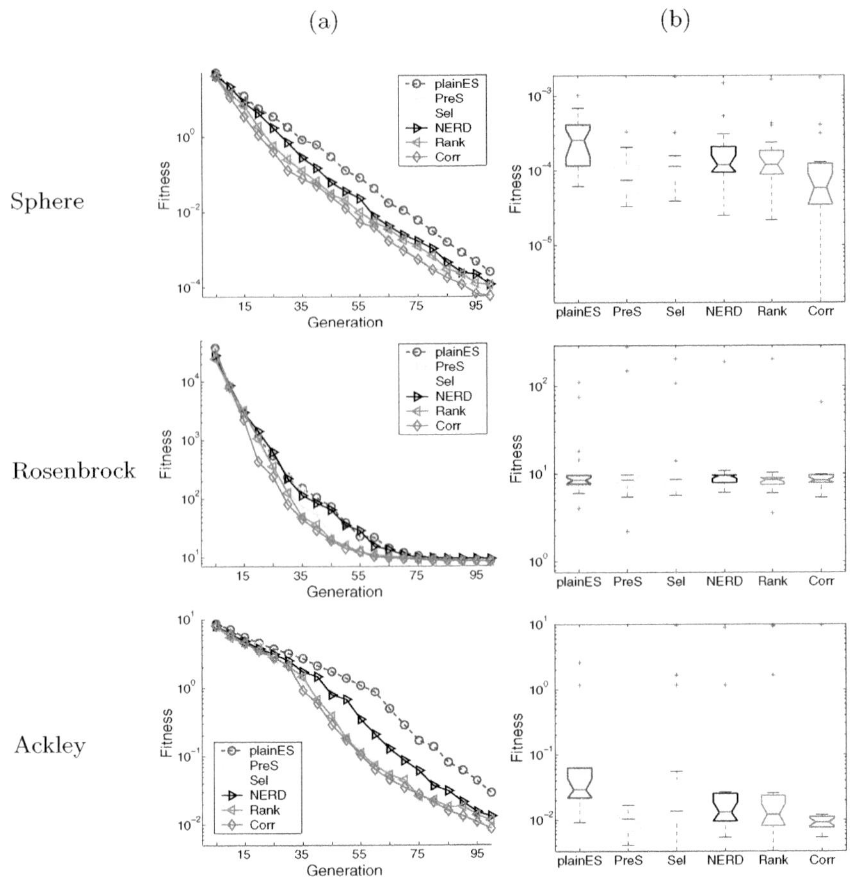

Fig. 10.10. Results using adaptation mechanisms to control the number of pre-selected individuals: (a) performance comparison and (b) boxplot over 20 runs after 100 generations

The selection based adaptation mechanism failed to perform well on all three test-functions too. The change of λ' and the quality measure oscillates dramatically, see Fig. 10.11. It is noticed that in [19], the selection-based quality measure has successfully been used to control the number of the pre-selected individuals. Note that the population size used in [19] is larger.

The best results in our simulations are obtained with the correlation-based adaptation mechanisms, especially the continuous correlation mechanism. But the correlation based adaptation mechanisms can not significant improve the pre-selection method without adaptation. Looking at the characteristic of λ' on Fig. 10.11, both adaptation mechanisms show the same trend. At the beginning the impact of the neural network to the optimization process steady increases. After some maximum is reached λ' decreases. The characteristics of

λ' appears to agree with the idea that the neural network should be less used until it is well trained. And if the optimization process comes closer to the optimum, the neural network should also be less used because the estimation is accurate enough to reach the real optimum.

From the above results, it seems that the adaptation mechanism is not successful in improving the performance of the pre-selection strategy and thus, the adaptation mechanisms are not tested on other model management frameworks.

10.5 3D Blade Design Optimization

In this section, we apply the pre-selection and the clustering with best strategy to 3D stator blade optimization.

10.5.1 Shape Representation

The 3D shape of the blade is approximated using three sections of 2D blades, as illustrated in Fig. 10.12(a). The hub section is directly connected to the hub at a radius of $R = 98.6mm$ from the engine axis. The tip section lies at a radius of $R = 130.0mm$. The 3D blade is built up by linear interpolation between these two sections. Another important section for the calculation of the design constraints is the casing section, which lies between the hub and the tip section at $R = 117.5mm$.

For each blade section, the blade length in axial direction is defined by the *axial chordlength*, refer to Fig. 10.12(b). The axial chordlength depends on the distance of the section to the engine axis.

Another parameter is the *axial solidity* s. For optimization, the solidity is measured at the casing section and the hub section. The final blade solidity is defined as the maximum value of these two measurements.

To describe the geometry of the blade, the *thickness* Θ can not be omitted. The thickness is also defined on a 2D section as the distance from the *medial axis* to a point p on the outline of the section. Depending on where the thickness is measured, the *trailing edge thickness* Θ_{TE}, near the trailing edge of the blade and the *leading edge thickness* Θ_{LE} near the leading edge are considered. Also important are the minimal Θ_{min} and maximal Θ_{max} values of the thickness.

The last parameter introduced here is the *throat area* (Fig. 10.12a). The throat area is defined as the area between two adjacent blades.

In this representation, 88 design parameters need to be optimized.

10.5.2 Performance Evaluation

Given the geometry of the 3D blade, the performance can be described by the *pressure loss*, which has to be minimized under certain constraints. This

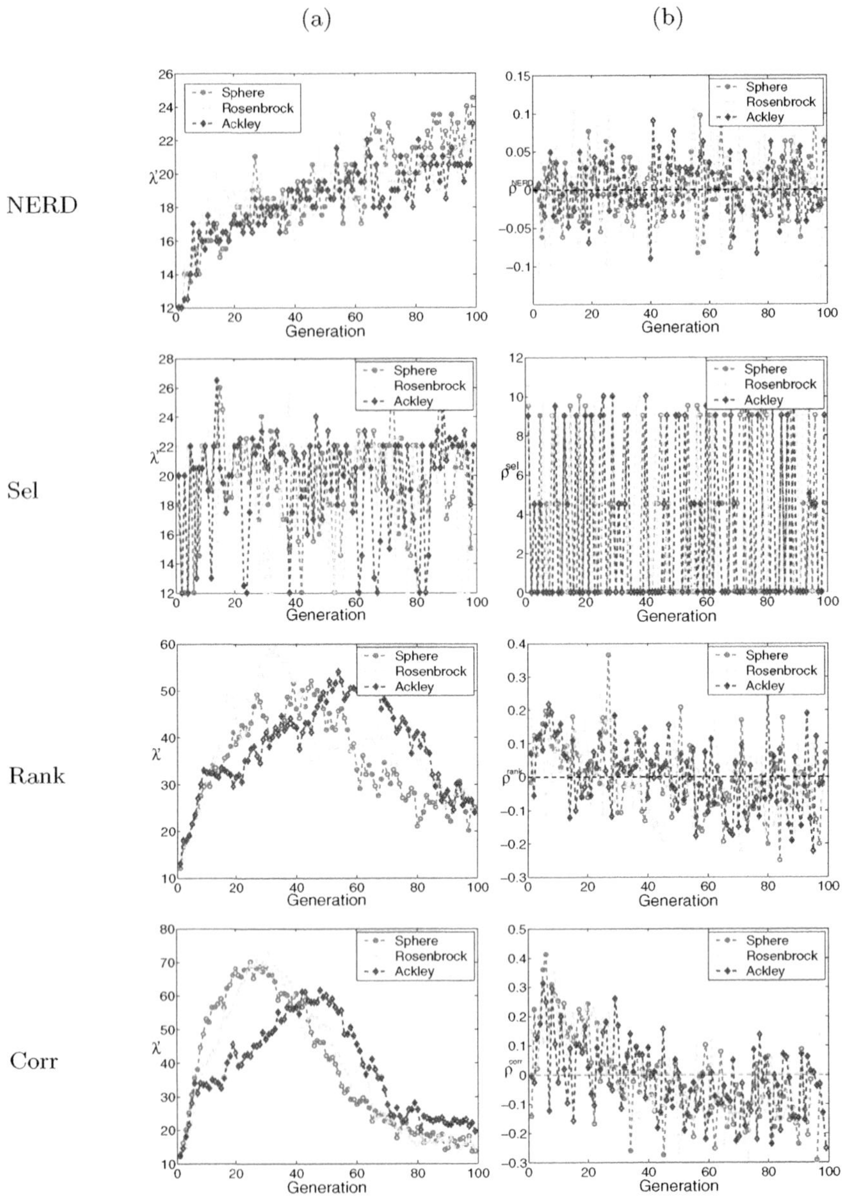

Fig. 10.11. Development of (a) λ' and (b) the quality measure during optimization

leads to the following objective or fitness function as a weighted sum of the pressure loss and the blade constraints. A penalty is applied in term of a big weight w_i so that the fitness becomes worse when a constraint is violated.

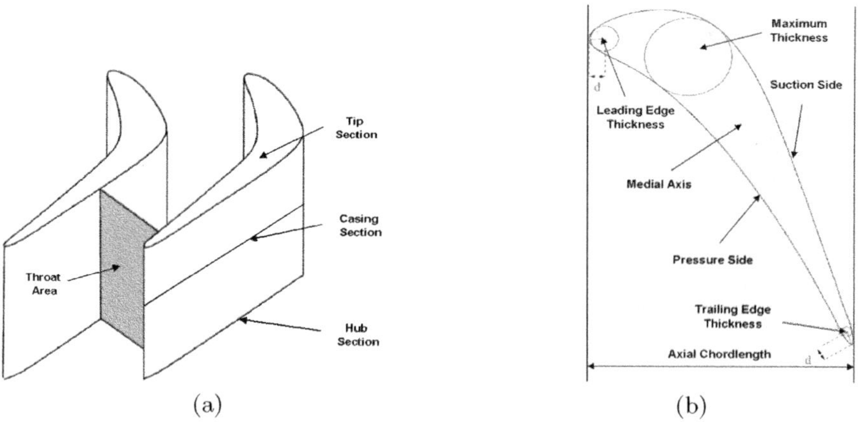

Fig. 10.12. Parameters and terminologies in 3D turbine blade design

$$f = w_0 t_0 + \sum_{i=1}^{4} w_i t_i^2 \longrightarrow min, \tag{10.19}$$

where t_0 is the pressure loss of the given blade, and t_i are the following constraints:

- $t_1 : max\big(0, |\beta_{2,design} - \beta_2| - \delta\beta_2\big)$,
- $t_2 : max\big(0, \Theta_{min,design} - \Theta_{min}\big)$,
- $t_3 : max\big(0, \Theta_{TE,min,design} - \Theta_{TE,min}\big)$,
- $t_4 : max\big(0, s_{max} - s_{max,design}\big)$,

where the following design values and tolerances are used:

- $\beta_{2,design} = 72.0deg$
- $\delta\beta_2 = 0.5deg$
- $\Theta_{min,design} = 0.72mm$
- $\Theta_{TE,min,design} = 0.9mm$
- $s_{max,design} = 0.706$.

The geometrical constraints like the minimal thickness Θ, trailing edge thickness Θ_{TE}, and the solidity s can all be determined directly from the geometry of the blade. However the *outlet angle* β_2 and the pressure loss can only be calculated from the results of the computational fluid dynamics (CFD) simulations. To simulate the fluid dynamics, the parallelized 3D Navier-Stokes flow solver HSTAR3D is used. The computational time of a CFD simulation varies between 2.5 and 6 hours on an AMD Opteron 2GHz dual processor. After flow analysis, each blade can be assigned a corresponding fitness value using Equation 10.19.

To investigate whether individual-based evolution control methods give a benefit to real world optimization problems, two methods are implemented

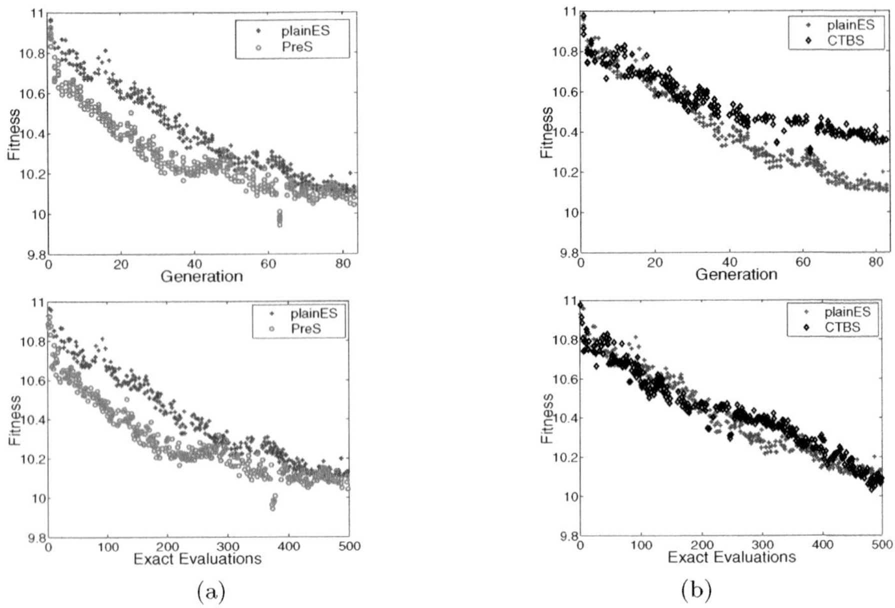

Fig. 10.13. Comparison of the performance between the plain evolution strategy and (a) using the pre-selection strategy or (b) using the clustering technique with best selection

in the 3D blade design optimization problem. The two methods are the $(1, [12]6[6])$ pre-selected strategy and the $(1, [6]6[4])$ clustering technique with best strategy. The performance of the methods are compared with the $(1, 6)$ plain evolution strategy. By use of the model assisted methods, the computationally expensive flow analysis was partially replaced by the neural network. The neural network consists of 88 input nodes, 4 hidden nodes and 2 output nodes. Using the clustering technique with best strategy, the performance of only 4 instead of 6 individuals each generation was determined by evaluating the CFD. Therefore the methods are compared by the fitness over the number of generations and also by the fitness over the number of exact fitness function evaluations, see Fig. 10.13.

As can be seen in 10.13(a), using the pre-selection strategy gives a benefit to the plain evolution strategy. It was possible to save up to about 20 generations to reach the same fitness value. The fitness of all individuals in each generation was evaluated in parallel. If the evaluation of the fitness takes about three hours, we saved about 60 hours of computational time. But the gap between the plain evolution strategy and the pre-selection strategy over the entire optimization process is nearly constant. There is no dramatic improvement in performance compared to the plain evolution strategy.

Using the clustering technique with best strategy, there is no improvement to the plain optimization process, refer to 10.13(b). Comparing the fitness over the number of generations, it might be clear that the (1, 6) plain evolution strategy can not be improved if only 4 individuals each generation are evaluated with the exact fitness function. Further it should be analyzed whether the clustering technique with best strategy might improve the (1, 4) evolution strategy. But comparing the fitness against the number of exact fitness function evaluations, there is also no improvement to the plain optimization process.

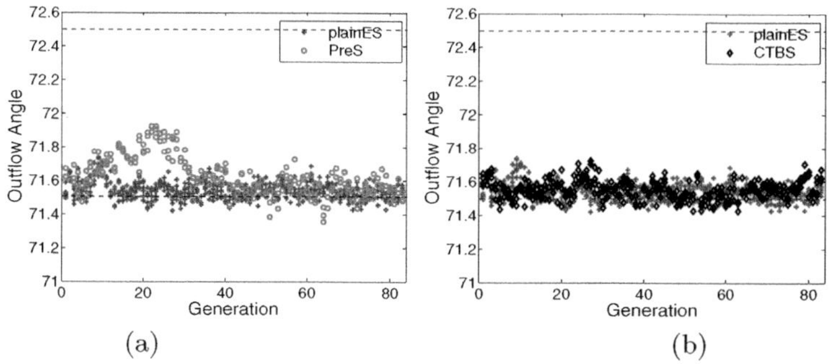

Fig. 10.14. Comparison of the outflow angle between the plain evolution strategy and (a) using the pre-selection strategy or (b) using the clustering technique with best selection

The fitness for each individual was determined by using equation 10.19. If the geometry of the blade does not violate any constraint, the pressure loss equals the fitness value because the weight for the pressure loss is set to one. Otherwise, if the geometry violates some constraints, a penalty of $10^2 2$ was given and the fitness becomes worse. Because the fitness values of invalid blade geometries are very large, these fitness values are not illustrated in Fig. 10.13.

A comparison of the outflow angle of the plain evolution strategy and the individual controlled strategies is shown in Fig. 10.14. It can be seen that the value for the outflow angle is near the lower bound. One exception occurs when the pre-selection strategy is used, where the outflow angle differs from the plain evolution strategy in the first generations.

To investigate how good the neural network substitute the time-expensive CFD-calculations, the estimation error of the two neural network outputs are illustrated in Fig. 10.15. The upper panel illustrates the estimation error of the pressure loss and the bottom panel shows the estimation error of the outflow angle. The squared error of the value evaluated with the CFD-calculation and the value estimated with the neural network has been plotted. It can be seen that the most values lie between 10^{-2} and 10^{-4}. Recall that the pressure loss is about 10 and the outflow angle is about 71.5, which indicates that the estimation of the neural network is quite good.

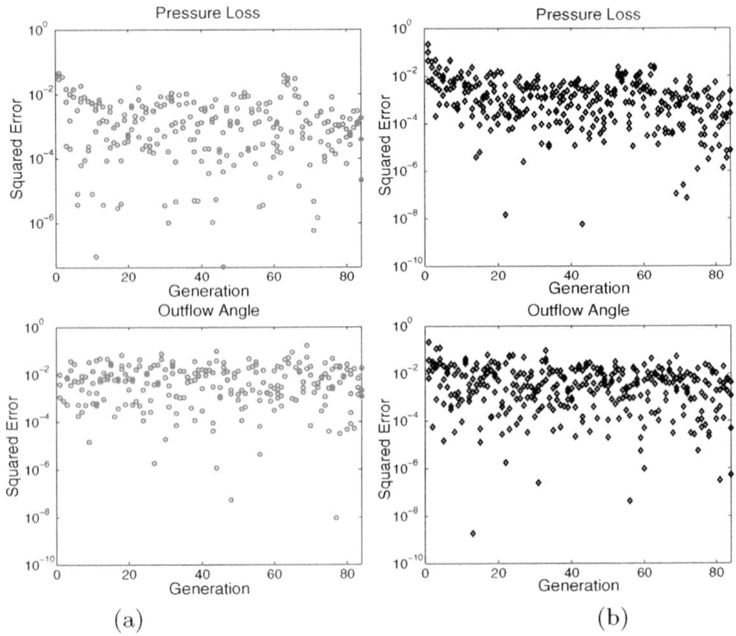

Fig. 10.15. Illustration of the neural network error for the pressure loss and the outflow angle (a) using the pre-selection strategy and (b) using the clustering technique with best selection

10.6 Conclusions

In this chapter, we have present four individual-based evolution control methods in order to reduce the number of expensive fitness evaluations. The performance of the methods is tested on three widely used benchmark functions. The pre-selection strategy shows the best results in both serial and in parallel computational environments, which improves the plain evolution strategy significant on mostly all test functions. The stability of the individual based evolution strategy might be improved if the parents for the next generation are selected out of the individuals evaluated with the original fitness function, as it is done in the pre-selection strategy.

To adaptively control the impact of the neural network on the evolutionary process, different adaptation mechanisms are investigated. The number of pre-selected individuals was controlled using the pre-selection strategy. Preliminary simulation results are not very promising. Further research should be done to see whether the free parameter $\Delta\lambda$ was chosen correctly. It should be analyzed why the results presented in [19] are much better than in our work.

To investigate whether individual-based evolution control methods are practical in real world optimization problems, the pre-selection strategy and the clustering technique with best strategy are implemented in the 3D blade design optimization. It turned out that the pre-selection strategy gives a benefit to the performance of the optimization process. The neural network sufficient substitutes the CFD-calculation.

There are a few possible reasons that lead to the relatively poor performance of the clustering methods. First, the population size used in this work is very small, which makes the clustering less sensible. Second, we only used a single neural network contrasting the work in [12], where a neural network ensemble has been used. In the future work, it should be investigated whether a neural network ensemble instead of a single neural network can improve the estimation accuracy of the model.

References

1. G. Cybenko. Approximation by superposition of a sigmoidal function. *Math. Control Signals Systems*, 2:303–314, 1989.
2. K. Foli, M. Olhofer T. Okabe, Y. Jin, and B. Sendhoff. Optimization of micro heat exchanger: CFD, analytical approach and multi-objective evolutionary algorithms. *International Journal of Heat and Mass Transfer*, 49:1090–1099, 2006.
3. L. Gräning, Y. Jin, and B. Sendhoff. Efficient evolutionary optimization using individual-based evolution control and neural networks: A comparative study. In *European Symposium on Artificial Neural Networks*, pages 273–278, 2005.
4. N. Hansen and S. Kern. Evaluating the cma evolution strategy on multimodal test functions. In *Eight International Conference on Parallel Poblem Solving from Nature PPSN VIII*, pages 282–291. Springer, 2004.
5. M. Hasenjäger, B. Sendhoff, T. Sonoda, and T. Arima. Three dimensional aerodynamic optimization for an ultra-low aspect ratio transonic turbine stator blade. In *ASME Turbo Expo*, 2005.
6. K. Hornik, M. Stinchcombe, and H. White. Multilayer feedforward networks are universal approximators. *Neural Networks*, 2(5):359–366, 1989.
7. M. Hüsken, Y. Jin, and B. Sendhoff. Structure optimization of neural networks for evolutionary design optimization. In *GECCO Workshop on Approximation and Learning in Evolutionary Computation*, pages 13–16, 2002.
8. Y. Jin. A comprehensive survey of fitness approximation in evolutionary computation. *Soft Computing*, 9(1):3–12, 2005.
9. Y. Jin, M. Huesken, and B. Sendhoff. Quality measures for approximate models in evolutionary computation. In *Proceedings of GECCO Workshops: Workshop on Adaptation, Learning and Approximation in Evolutionary Computation*, pages 170–174, Chicago, 2003.
10. Y. Jin, M. Olhofer, and B. Sendhoff. On evolutionary optimization with approximate fitness functions. In *Genetic and Evolutionary Computation Conference*, pages 786–792. Morgan Kaufmann, 2000.
11. Y. Jin, M. Olhofer, and B. Sendhoff. A framework for evolutionary optimization with approximate fitness functions. *IEEE Transactions on Evolutionary Computation*, 6(5):481–494, 2002.
12. Y. Jin and B. Sendhoff. Reducing fitness evaluations using clustering techniques and neural networks ensembles. In *Genetic and Evolutionary Computation Conference*, volume 3102 of *LNCS*, pages 688–699. Springer, 2004.

13. M. Olhofer, T. Arima, Y. Jin, T. Sonoda, and B. Senhoff. Optimisation of transonic gas turbine blades with evolution strategies. *Honda Technical Reviews*, pages 203–216, April 2002. documents/HTR02.pdf.
14. M. Olhofer, T. Arima, T. Sonoda, and B. Sendhoff. Optimisation of a stator blade used in a transonic compressor cascade with evolution strategies. In *Adaptive Computation in Design and Manufacture*, pages 45–54. Springer, 2000.
15. Y.S. Ong, P.B. Nair, and A.J. Keane. Evolutionary optimization of computationally expensive problems via surrogate modeling. *AIAA Journal*, 41(4):687–696, 2003.
16. Y.S. Ong, P.B. Nair, A.J. Keane, and K.W. Wong. Surrogate-assisted evolutionary optimization frameworks for high-fidelity engineering design problems. In Y. Jin, editor, *Knowledge Incorporation in Evolutionary Computation*, pages 307–331. Springer, 2005.
17. A. Oyama. Multidisciplinary optimization of transonic wing design based on evolutionary algorithms coupled with cfd solver. In *European Congress on Computational Methods in Applied Science and Engineering, ECCOMAS 2000*, 2000.
18. M. Riedmiller and H. Braun. A direct adaptive method for faster backpropagation learning: The rprop algorithm. *IEEE Int. Conference on Neural Networks*, pages 586–591, 1993.
19. H. Ulmer, F. Streichert, and A. Zell. Evolution strategies with controlled model assistance. In *Congress on Evolutionary Computation*, pages 1569–1576, 2004.
20. Cheng Xiang. Geometrical interpretation and architecture selection of MLP. *IEEE Transaction on Neural Networks*, 16(1):84–96, 2005.
21. X. Yao. Evolving artificial neural networks. *Proceedings of the IEEE*, 87(9):1423–1447, 1999.

11

Evolutionary Shape Optimization Using Gaussian Processes

Wenbin Song

School of Engineering Sciences, University of Southampton
University Road, Southampton SO17 1BJ, U.K.
w.song@soton.ac.uk

Summary. This chapter presents studies on structural shape optimization using evolutionary computation methods and Gaussian process based meta-modeling techniques. Methods of evolutionary computation have been used to solve design optimization problems in a wide range of areas in a process mimicking the evolution of biological life in natural world. Among various evolutionary computation methods, genetic algorithms have been attracting attentions due to its easiness to implement and its robustness in locating near global optimal solutions. However, the large number of iterations typically required by evolutionary search methods to converge has prompted research interests in the use of meta-models in the process. Gaussian process based meta-modeling technique is one of the popular choices since it not only can provide a prediction on the function value, but also can provide error estimate on the prediction. This chapter describes frameworks using genetic algorithms and Gaussian process based meta-modeling techniques for structural shape optimization problems. Application examples of such approaches are given in areas of firtree shape optimization using finite element method and engine nacelle optimization using computational fluid dynamics.

11.1 Introduction

The robustness in finding near-global optimal solutions using the evolutionary search methods, and in particular, genetic algorithms (GAs) [1], has attracted an increasing amount of research interests in the use of such methods in various optimization and design problems in fields such as engineering design, finance, and job scheduling [2–5]. GAs typically require large number of fitness evaluations for the results to converge to global optimal solutions. Therefore, improving the efficiency of GAs has become a key factor in their successful applications to real-world problems when fitness evaluations become computationaly expensive. In structural shape optimization problems, high fidelity analysis codes, such as finite element and computational fluid dynamics, are often used to compute fitnesses of individuals in the population. This presents

W. Song: *Evolutionary Shape Optimization Using Gaussian Processes*, Studies in Computational Intelligence (SCI) **51**, 251–267 (2007)
www.springerlink.com

particular challenge for the application of GAs to real-world structural shape optimization problems.

Two categories of techniques have been proposed to tackle the issue of efficiency of evolutionary search methods: the first is focused on devising more efficient variants of the canonical algorithms either by using a hybrid of local search and evolutionary methods or by using genetic operators customized to applications [6–9]. The former is also known as memetic algorithms which combine gradient based local search with GAs in the sense of Lamarckian or Baldwinian learning mechanisms [10]. The second type involves using meta-models, also called approximation models or surrogate models, in lieu of the exact and often expensive function evaluations [11]. One of the common features of these techniques is that both types of methods try to reduce the number of exact fitness evaluations to improve overall efficiency. As these two types of techniques can be easily combined to further speed up the process, research effort can be focused on them separately. The focus of the current chapter is on efficient frameworks for combining genetic algorithms with Gaussian process based approximation technique, which is also known as Kriging [12]. The emphasis is placed on devising effective frameworks which balance the needs for exploitation and exploration. The purpose of exploitation is to improve the accuracy of approximation models based on available data and the aim of exploration is to reliably locate optimal solutions on the approximation models.

Various strategies have been proposed and studied to tackle the use of meta-models in evolutionary frameworks in addition to researches into different meta-modeling techniques. For example, Ong [19] proposed the use of radial basis function (RBF) models in a trust region framework to reduce expensive function evaluations in local searches, which are combined with a genetic algorithm to assure global convergence. Song [14] illustrated the use of a 3σ update strategy in combining a global Kriging meta-model with GA on an industrial design problem. In addition to being used in expensive optimization studies, surrogate models also prove valuable in problems of multidisciplinary design where interactions between different disciplines can be studied more thoroughly using surrogates.

In the next section, different methods for building the meta-models are described to provide some background knowledge for commonly used approximation methods. Two frameworks incorporating these Gaussian process based meta-models in evolutionary search methods is described in section III. Two application examples are given in section IV. The first example involves the optimization of turbine blade firtree root using finite element method. The second example demonstrates the use of computational fluid dynamics on the shape optimization of civil aero-engine nacelle. A brief discussion is given at the end of the chapter.

11.2 General Methods for Building Metamodels

The first step in an optimization task is to build a parametric model which, given a set of values for the parameter, will produce a set of performance measures of the model. These performance measures can be used as either objectives or constraints in subsequent optimization studies. Such models could either be simple analytical ones built based on physical laws, or mathematical models built from experimental data, in graph or tabular form. However, as more and more increasingly sophisticated computer codes are used to produce such performance measures, the total number of simulations that can be done will be limited by available computing resources, or the time constraint. Typical examples of these scenarios include car design optimization using crash analysis code; or aerodynamic design of airplanes using computational fluid dynamics. Therefore, the use of meta-models becomes a preferred solution. The meta-models are mathematical models which are easy to evaluate but capable of capturing the characteristics of the original problems.

There exists a number of methods for building meta-models. The reader is referred to [9] for a comprehensive survey on various approximation methods available in the literature. Let $(\mathbf{x}_i, \mathbf{y}(\mathbf{x}_i), i = 1, ...m)$ denotes the data set collected by running the simulation codes on a number of data points selected randomly or using methods of design of experiment such as Latin hypercube samples, where x_i and $\mathbf{y}(x_i)$ are the input vector and output for the ith point, respectively. m is the number of points in the data set. Three most commonly used models are briefly described in the following sections, along with some comparisons on their strengths and weaknesses.

11.2.1 Polynomial Approximation

The polynomial models were originally developed to analyze data from physical experiments and to create empirical models of the observed input-output relations. The commonly used response surface model is low order polynomials which can be represented as

$$y = \beta_0 + \sum_{i=1}^{k} \beta_i x_i \tag{11.1}$$

for linear case, and

$$y = \beta_0 + \sum_{i=1}^{k} \beta_i x_i + \sum_{ii=1}^{k} \beta_{ii} {x_{ii}}^2 + \sum_{i<j} \sum \beta_{ij} x_i x_j \tag{11.2}$$

for quadratic approximations. Higher order polynomials are less commonly used due to its high fluctuation with the input variables. The coefficients in the equation can be solved using least square methods. Detailed discussions on these models can be found in many texts [9]. These models are less favorable when dealing with high dimensional optimization problems because of its inflexibility to capture more complex relationships.

11.2.2 Neural Network

Radial basis function (RBF) is a type of neural networks employing a hidden layer of radial units and an output layer of linear units and is characterized by its reasonably fast training process. Since our primary concern is to build approximation models on the data generated from deterministic computer simulations, an interpolation model using RBF is therefore used

$$y(\mathbf{x}) = \sum_{k=1}^{m} \alpha_k K(||\mathbf{x} - \mathbf{x_k}||) \tag{11.3}$$

where $K(||\mathbf{x}-\mathbf{x_k}||)$ is a radial basis kernel function and $(\alpha_k, k = 1, ..., m) \in R^m$ represents the weight vector for the design points. Given a suitable choice of the kernel function, the weight vector can be computed by solving the linear algebraic system of equations

$$\mathbf{K}\alpha = \mathbf{y} \tag{11.4}$$

where $\mathbf{y} = \{y_1, y_2, ..., y_m\}$ denotes the vector of outputs and $\mathbf{K}^{m\times m}$ is the Gram matrix computed using the training data, i.e., the (i, j)th element of $\mathbf{K}$ is computed as $K(||\mathbf{x}_i - \mathbf{x}_j||)$.

Typical choices for the kernel function $K(h)$, for $h \in R$, include linear splines, thin-plate splines, cubic splines, Gaussian and multiquadrics functions, as listed in Table 11.1. The Gram matrix is positive semi-definite. To guarantee the non-singularity of the Gram matrix, duplicate points should be excluded from the dataset (and it also makes sense when dealing with data from deterministic computer simulations) and positive definite kernels (such as Gaussian and inverse multiquadric) need to be used. However, it should be noted that the Gram matrix could become highly ill-conditioned for extreme values of θ in these kernels.

Table 11.1. Kernel Functions for Radial Basis Function Neural Network

Type of kernel functions	Definition
Linear	$(\lvert h\rvert)$
Gaussian	$(e^{-h^2/\theta})$
Multiquadratics	$(\sqrt{(1 + h^2/\theta)})$
Inverse multiquadratics	$(1/\sqrt{(1 + h^2/\theta)})$

The model parameter θ can be optimized to minimize the prediction error when more data becomes available or when leave-n-out schemes are used for cross-validation of the model. Regulating terms can also be used in some cases to avoid overfitting of the data.

11.2.3 Gaussian Process Models

Originally developed in geostatistical field for processing spatially correlated data [17], Gaussian process models, also known as Kriging, is a technique for building global approximation models which characterize local effects in a statistically meaningful sense. Typically, data points are interpolated in the Kriging model, but approximation models can also be developed to filter out noise in the data [18]. The interpolation Kriging model is here expressed as

$$y(\mathbf{x}) = \beta + Z(\mathbf{x}) \tag{11.5}$$

where β represents a constant term in the model, and $Z(\mathbf{x})$ is a Gaussian random process with zero mean and variance of σ^2. The covariance matrix of $Z(\mathbf{x})$ is given by

$$Cov(Z(\mathbf{x}^i), Z(\mathbf{x}^j)) = \sigma^2 R(\mathbf{x}^i, \mathbf{x}^j) \tag{11.6}$$

where σ^2 is the variance of the stochastic process and $R(.,.)$ is a correlation function between $\mathbf{x}^i$ and $\mathbf{x}^j$. Different types of correlation functions can be employed as noted in Jones et al. [12]. A commonly used type of correlation function can be expressed as

$$R(\mathbf{x}^i, \mathbf{x}^j) = \prod_{k=1}^{n} exp(-\theta_k |x_k^i - x_k^j|^p) \tag{11.7}$$

where $\theta_k > 0$ and $1 \leq p_k \leq 2$ are the hyperparameters, and n denotes the number of dimensions in the data set. Note that the above equation asserts that there is a complete correlation of a point with itself and this correlation decreases rapidly as the two points move away from each other in the parameter space. The choice of $p_k = 2$ would provide enough flexibility for modeling smooth but highly non-linear functions for most cases. The hyperparameters θ_k are estimated by maximizing the log-likelihood function given by

$$-\frac{1}{2}[n \ln \sigma^2 + \ln |\mathbf{R}| + \frac{1}{\sigma^2}(\mathbf{y} - \mathbf{1}\beta)^T \mathbf{R}^{-1}(\mathbf{y} - \mathbf{1}\beta)] \tag{11.8}$$

where σ^2 and β can be derived using the following equations once the θ_k are given

$$\hat{\beta} = (\mathbf{1}^T \mathbf{R}^{-1} \mathbf{1})^{-1} \mathbf{1}^T \mathbf{R}^{-1} \mathbf{y} \tag{11.9}$$

$$\hat{\sigma}^2 = \frac{1}{n}(\mathbf{y} - \mathbf{1}\beta)^T \mathbf{R}^{-1}(\mathbf{y} - \mathbf{1}\beta) \tag{11.10}$$

A numerical optimization procedure is required to obtain the Maximum Likelihood Estimates (MLE) of the hyperparameters. Once the hyperparameters are obtained from the training data, the function values at new points can be predicted by

$$\hat{y}(\mathbf{x}^*) = \hat{\beta} + \mathbf{r}^T \mathbf{R}^{-1}(\mathbf{y} - \mathbf{1}\hat{\beta}) \tag{11.11}$$

along with the posterior variance $s^2(\mathbf{x}^*)$ given by

$$s^2(\mathbf{x}^*) = \sigma^2[1 - \mathbf{r}\mathbf{R}^{-1}\mathbf{r} + \frac{(1 - \mathbf{1}^T\mathbf{R}^{-1}\mathbf{r})^2}{(\mathbf{1}^T\mathbf{R}^{-1}\mathbf{1})}] \tag{11.12}$$

where $\mathbf{r}(\mathbf{x}) = R(\mathbf{x}, \mathbf{x}^1), ..., R(\mathbf{x}, \mathbf{x}^n)$ is the correlation vector between the new point $\mathbf{x}$ and the training dataset. This quantity provides an indication on the accuracy of the prediction at new points and will be used in our framework to decide whether and where further exact analyses are required.

Fig. 11.1 compares the approximated response of a kriging model to that of a radial basis neural network, in which both methods have provided a good approximation because of the sufficient number of sample points on the curve and there is just one variable. But, in general, the Krig model is more time-consuming to build than the RBF model.

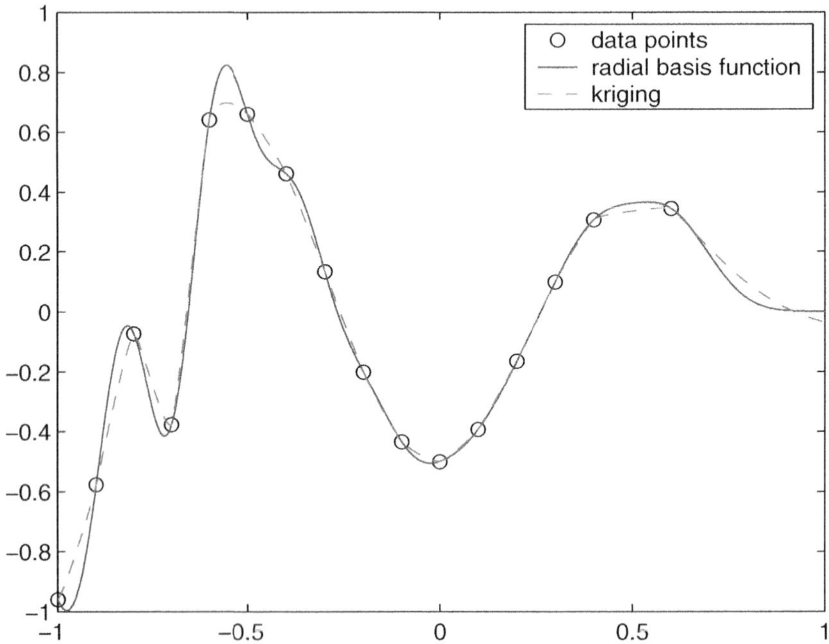

Fig. 11.1. Comparison of a radial basis neural network model and a kriging model for the same data

11.3 Evolutionary Frameworks Using Meta-models

In recent years, there has been a growing interest in methods using approximation models in optimization, in which, the complex simulation codes are treated as black-box functions [12, 19–21]. Data are first gathered via evaluations of expensive simulation codes at design points chosen by methods of design of experiments. Then mathematical surrogate models are built using the data. Various techniques for the construction of approximation models have been proposed, such as those discussed in section 2. Perhaps the most popular technique involves a polynomial approximation created by least squares curve fitting to a set of data. Another line of methods is based on interpolation techniques, which are proved to be more effective for data obtained by running deterministic computer simulation codes. Methods of this type cover radial-basis functions (RBF) developed in the field of neural networks [22], stochastic-based methods, developed in the field of geostatistics, commonly referred to as Gaussian process based method or Kriging as used by Sacks et al. [24]. An efficient procedure for global optimization using Kriging models was proposed by Jones et al. [12], in which a branch-and-bound algorithm was applied to find the point with potential maximum improvement for resampling. Other techniques using classifier systems and the concept of space mapping have also been proposed [23].

The primary aim of using approximations or meta-models is to reduce the cost of finding global optimal designs. This is equivalent to reducing the number of function calls to the exact simulation codes. Therefore the optimal points on a meta-model should converge to the optimal points on the exact model, which is the requirement of global convergence for meta-models. Global convergence will be naturally achieved if an exact meta-model can be built. However, due to the modeling errors inherent in the meta-models, balances need to be made in building an accurate meta-model and locating the global optimal points on it.

An explorative comparison was made by Guinta et al. [25] between polynomial and interpolation models using test problems. A comparative study between neural networks and response surface models was provided by Daberkow et al. [26] using a preliminary aircraft design problem. Among various methods, RBF and Kriging were identified by Jin et al. [27] to be able to produce better results compared to other methods under multiple modeling criteria. Moreover, the Kriging method is statistically more meaningful and allows the possibility of computing error estimates for untried data points. One of the drawbacks of the Kriging method is the relatively high computational cost in estimating the hyper-parameters in the model, especially when large numbers of sample points are involved.

Apart from differences in methods of constructing meta-models, a number of different frameworks for the management of approximation models have also been proposed. It has been argued that any effective framework for managing the surrogates in optimization is always associated with particular optimization algorithms. Therefore, in principle, there are at least as many possible frameworks as optimization algorithms. In general, these frameworks can be categorized according to the algorithms with which they are used. There are two broad classes of optimization algorithms, gradient-based methods and non-gradient-based methods in which evolutionary algorithms and direct search methods are among the most widely used. A rigorous framework was presented by Booker et al. [28] for the use of surrogate models with direct search methods. Frameworks based on trust regions and gradient-based search procedures have drawn most attention in the past few years [29–34]. One of features possessed by this type of rigorous framework is that they all guarantee convergence to a local optimum of the model. However, these work with non-linear programming techniques or direct search methods.

The number of studies on use of surrogate models with evolutionary algorithms, though relatively small, is growing in the past few years. In these cases, the high computational cost associated with the successive application of evolutionary algorithms to complex, high-dimensional engineering problems is a known problem, as evolutionary algorithms typically require far more number of function evaluations to converge to near optimal solutions compared to direct search. Therefore, the successive use of meta-models models is believed to be crucial to the practical application of evolutionary optimization methods to the design and optimization of complex engineering systems.

Several attempts have been made in the last several years to tackle the problem of using meta-models with evolutionary computation methods, particularly genetic algorithms (GAs). Robinson and Keane [35] employed variable-fidelity analysis models and approximation techniques to improve the efficiency of the Evolutionary Strategy (ES). A procedure of using a number of successive single-point approximation models with a genetic algorithm was proposed by Nair et al. [36], in which, some domain knowledge was employed to construct the meta-models and a simple generation-delay approach was used to control the use of exact models. Ratle [14] proposed a simple local convergence criterion to decide when the exact model should be resorted to in a procedure integrating a genetic algorithm with Kriging models. However, this does not prevent the search from converging to the false optima. A revised criterion was proposed by El Beltagy et al. [38] for the similar synthesis between GA and Kriging models, where a gradually reduced tolerance was used to control the switch between surrogate and exact models for each individual in the population. However, the specification of criteria values in the above methods depends largely on users and may not be appropriate for all problems. Nevertheless, the requirement for some sort of re-sampling was identified as necessary to overcome the inadequacy of the surrogate models.

A different type of effort was attempted by Liang et al. [39], where a hybrid search procedure was formulated with the evolutionary search working on the quadratic response surface constructed from many local optima obtained from local searches. This method essentially reduced the difficulties in building a global surrogate model without changing the overall landscape of the exact functions. However, the quality of the final global surrogate model depends very strongly on the accuracy of the local search results and the use of evolutionary search methods in finding global optima in a simplified smooth landscape with few local optima may not be necessary. Jin [40] also proposed a framework for coupling Evolutionary Strategy (ES) and neural network-based surrogate models. Two types of evolution control methods were presented to dictate the frequency at which the exact model was used. The common weakness in the above methods is that neither the historical search data nor the convergence properties of the evolutionary search method are fully utilized.

A number of methods can be used when it comes to updating the Kriging models with added data points evaluated using exact objective functions. The most straightforward technique is to use the best points found on the Kriging model. This approach depends on the accuracy of the Kriging model and quality of the search on it, and the process may miss the global optima as it lacks the ability to identify promising but unsearched areas due to prediction errors involved in the Kriging. A more promising method is to update the Kriging model with points where the expected improvement (EI) is large compared with available exact solutions, as shown in [12]. The update procedure in these two approaches is very much separated from the overall process of identifying the global optimum. Therefore, almost any optimization method can be applied in the search for optimum on the Kriging model or the expected improvements model, for example, the Branch-and Bound method was suggested by Jones et al. [12] in the search for EI.

In the use of the concept of expected improvement, such as in [12], the point with maximum expected improvement is found by using a branch-and-bound algorithm followed by re-sampling at that point and the re-construction of the surrogate. A simple 3σ principle is proposed by Song [14] and this value is used instead of the maximum expected improvement in determining whether or not an evaluation using the exact model is necessary. The method eliminates the need for optimization to find the point at which maximum expected improvement can be achieved. The principle is described as

$$\text{Evaluate } Y(\mathbf{x}^*) \text{ when}$$

$$\hat{y}(\mathbf{x}^*) - 3s(\mathbf{x}^*) < \frac{1}{q}\sum_{j=1}^{q} y_j^{best} (\text{minimization}) \tag{11.13}$$

where $\hat{y}(\mathbf{x}^*)$ and $s(\mathbf{x}^*)$ are computed using 11.11 and 11.12, y_j^{best}s represent the q best solutions in the dataset. Note that the right hand side reduces to y_{min} when $q = 1$ for minimization problems.

The 3σ principle is based on the fact that if the average fitness of top q designs lies outside the interval $[\hat{y}(\mathbf{x}^*) - 3s(\mathbf{x}^*), \hat{y}(\mathbf{x}^*) + 3s(\mathbf{x}^*)]$ computed at point $\mathbf{x}^*$, the probability of producing a better design at point $\mathbf{x}^*$ is very low. Two control parameters p and q are used to specify the number of design points to build the surrogate and number of design points used on the righthand side of equation 11.13. It is not difficult to understand the effect of these two parameters: increasing the value of parameter p will cause more points to be used in building the surrogate model and increasing the value of parameter q will lead to more new points falling into the p interval and therefore more exact evaluations.

It should also be mentioned that the basic Kriging model could further be extended to include derivative information as reported by [41] when derivatives are available either analytically or computed using automatic differentiation (AD) tools, adjoints, etc. The availability of efficient adjoint methods for sensitivity computations makes this expansion more attractive for complex simulation codes.

Some further development in [43] adopted GAs as the optimization algorithm for searching on the Krig model as well as in the basic framework, leading to a second framework involving a two-level GA optimization strategy with meta-models. The searches on the Krig are carried out twice: the first is to locate optimal objective function values, and the second is to locate points where maximum posterior errors occur. This is the reason why Krig is chosen as the approximated objective function and its posterior error can be obtained in one go. Suggested points from these two GA searches are evaluated in parallel before being aggregated for building an updated meta-model. Duplicate points are removed to assure nonsingularity of the Gram matrix. Adding these newly available exact points to the meta-model achieves the following two aims simultaneously: reduced approximation errors and improved search efficiency. The framework is shown in Fig. 11.2.

The process begins by generating an initial set of data points using methods of design of experiments. The population size is defaulted to ten times the number of dimensions of the design space. Objective functions for these data points are then evaluated in parallel. The initial Kriging model is constructed and searched twice using GAs to obtain update points, which are added to the data set for the subsequent update of the meta-model. This step is repeated until predefined computational budget in terms of number exact function evaluations is exhausted.

11.4 Application Examples

Two application examples has been used here to demonstrate the effectiveness of the strategies coupling genetic algorithm and Gaussian process based surrogate models. The focus in these examples has been on how to evolve more efficiently towards improved solutions by making use of information from the meta model to suggest update points.

11.4.1 Turbine Blade Firtree Shape Optimization

In this section, a turbine blade firtree root local notch profile optimization problem [42] using cubic NURBS (Non-Uniform Rational B-Spline) is chosen to illustrate the effectiveness of the first framework. This notch profile is part of the geometry at the base of a turbine blade that is used to attach the turbine blade to the rotating disc in turbine aeroengines. This is a critical region where high contact stresses occur. Here the firtree geometry has been modeled using

BEGIN

- Generate the initial population using design of experiments, the population size will be 10n;
- Evaluate the 10n points using high fidelity analysis codes, and store objective function values
- Build the initial response surface model using kriging, the hyperparemeters are tuned using GA/DHC (Dynamic Hill Climber) code;
- **While** (number of exact evaluations less than the number requested/ affordable)
    - Search the response surface model, return the best n points in terms of the best objective function on the response surface, and also search the response surface error model, return n points in terms of the maximum posterior prediction errors
    - Remove those points from combined set of design points based on correlation criteria or Euclidean distance criteria, this will produce the candidate points for exact evaluations
    - Evaluate these points using exact codes and place these points into database, otherwise if the aggregated data set is empty, terminate the search.
    - upate the response surface by updating the hyperparameters of the Kriging model
- **End While**

END

Fig. 11.2. Hybrid update scheme using genetic algorithm and Kriging

the rule-based geometry modeling capability provided by ICAD. The notch geometry is modeled using NURBS instead of simple circular arcs. The use of NURBS gives more flexibility in modeling local shape variations. However, it also introduces more design variables, therefore increases the cost of finding an optimum design. The definition of the local profile is illustrated in Fig. 11.3. Finite element analysis is carried out to evaluate the peak stress at the notch region and is used as the objective function in the optimization. Results are shown in Fig. 11.4, in which, two convergence curves were shown, one without using the surrogate model, and the other with. It can be seen that comparable results can be obtained with only one third of the number of exact function evaluations that would be required when exact analysis codes are used in optimization. Comparison between the original and optimized geometry is shown in Fig. 11.5. The worst principal stress at point B (where the maximum stress occurs) is reduced from 745.4Mpa to 685.6MPa - a reduction of over 8% without worsening the stresses at point A, using around one third of the computing effort which would be otherwise required using GA directly on finite element codes.

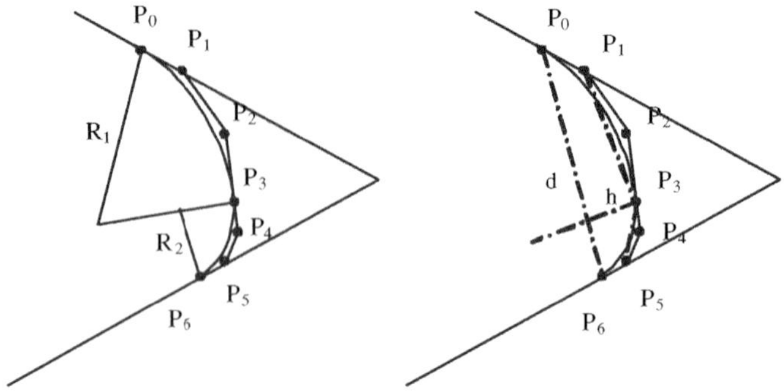

Fig. 11.3. NURBS representation of double-arc fillet using seven control points and its defining coordinates

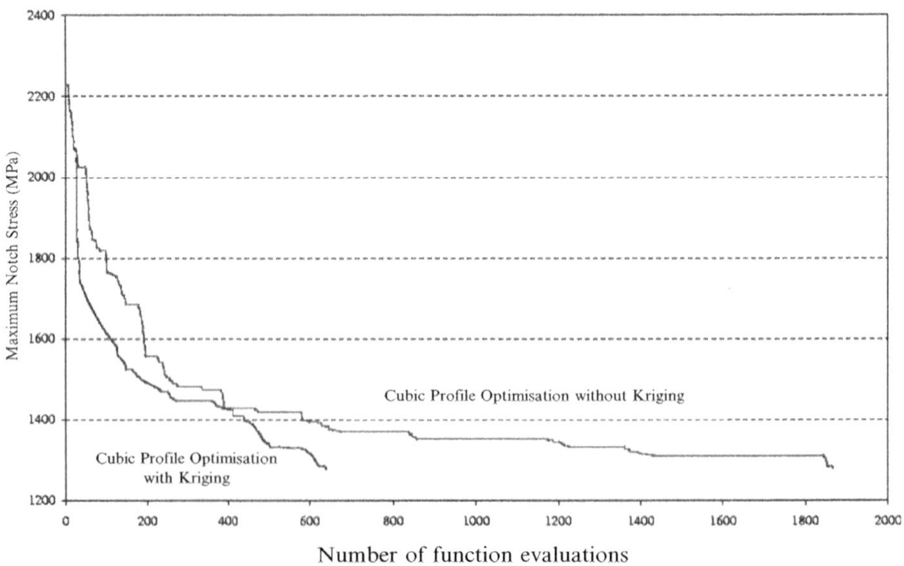

Fig. 11.4. Genetic algorithms with surrogate modeling on local profile optimization of firtree root

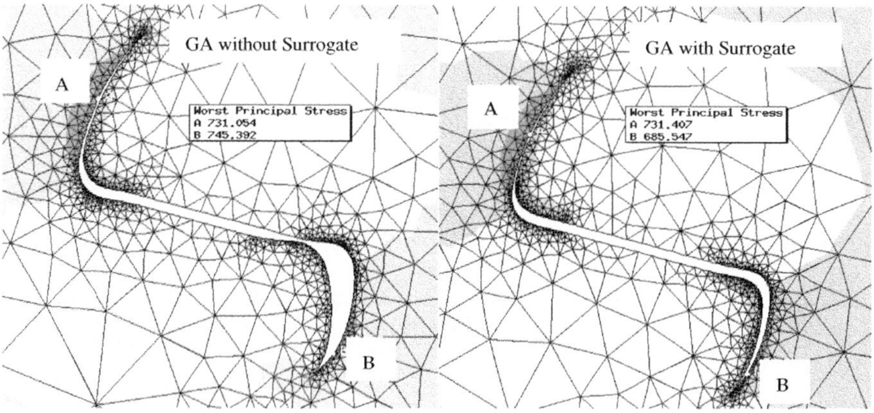

Fig. 11.5. Comparison between results of GA search without and with surrogates

11.4.2 Engine Nacelle Shape Optimization

An aero engine nacelle shape optimization problem is used to illustrate the effectiveness of the second framework [43]. The engine nacelle geometry is defined in ProEngineer with around 40 parameters, here six parameters that define the shape of top lip profile plus the scarf angle parameter are used to formulate a seven parameter problem. The aim is to study the aerodynamic effect of top lip profile under different scarf angles. The parameters are listed in Table 11.2, along with parameter ranges and reference values. The meshing package GAMBIT and flow solver FLUENT [44] are used in the study. First, an appropriate mesh density is determined based on the accuracy of the solution and computational time it requires. Here it was decided based on the number of available processors and licenses: the ideal situation using 30 fluent jobs on 8 processors means that each calculation can be finished within 10 hours, and around 200 jobs should be finished within a week if jobs are run immediately after placed into the job queue. In practice, it has taken a little more than three weeks to finish 200 calculations.

The geometries and pressure distributions for the base design and final design achieved using 200 calculations are presented in Fig. 11.6.

11.5 Conclusions

The high cost of evolutionary computation methods for design optimization using high-fidelity analysis codes has prompted the study into the use of efficient meta-models. This can provide efficient and effective solutions to problems of structural shape design using finite element or computational

Table 11.2. Design Variables for Engine Nacelle Geometry

Variables	x_l	x_0	x_u	Description
Scarf angle	-10	-5	25	Negative scarf angle (deg.)
Teaxis	5	12	20	Axial coordinate of top external profile (mm)
Telater	5	10	20	Radial coordinate of top external profile (mm)
Tiaxis	1.5	2	2.5	Ratio of top inner profile coordinate in axial direction against radial direction
Tilater	1	1.34	1.6	Coefficient used to determine top inner profile coordinate in lateral direction
Var_d225	22	25.4	28	Radial control length of top lip profile (mm)
Var_d226	22	25.4	28	Axial control length of top lip profile (mm)

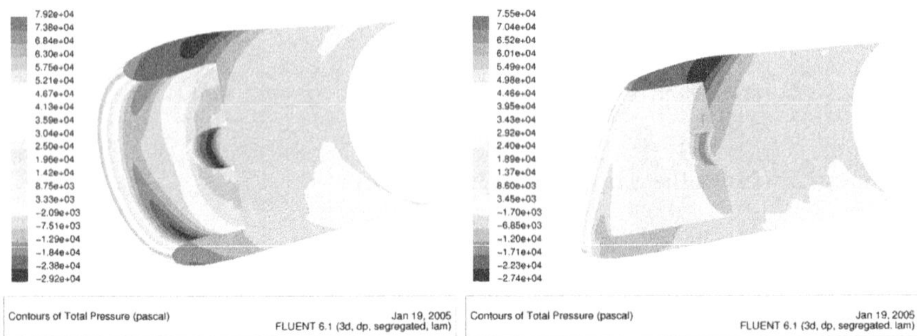

Fig. 11.6. Geometry and pressure distributions of base and final design

fluid dynamics codes. In this chapter, a brief description of several commonly used techniques for building meta-models is given, followed by an overview of different frameworks that couple optimization methods and meta-modeling methods. Two frameworks using genetic algorithms and Gaussian based meta-models are given particular attention in the chapter. In the first framework, a 3σ principle was applied to control when the surrogate model needs to be updated with additional data points. The second framework carried out a parallel search on the meta-model and error prediction model to decide where to place the update points for computationally costly analysis. Both approach have been applied to real-world shape optimization problems achieving a reduction in the number of exact evaluations of analysis codes, while delivering an improved design at the same time.

References

1. Goldberg D. (1989) Genetic Algorithms in Search, Optimization and Machine Learning, Addison-Wesley Longman Publishing Co., Inc. Boston, MA, USA.

2. Lin L., Cao L., Wang J., and Zhang C. (2004) The Applications of Genetic Algorithms in Stock Market Data Mining Optimisation, Proceedings of Fifth International Conference on Data Mining, Text Mining and their Business Applications, Malaga, Spain. September 15-17. 2004.
3. Obayashi S., Yamaguchi Y., and Nakamura T. (1997) Multiobjective genetic algorithm for multidisciplinary design of transonic wing platform. Journal of Aircraft, 34(5):690-693, 1997.
4. Ong Y. S., Nair P. B., Keane A. J., and Wong K. W. (2004) Surrogate-Assisted Evolutionary Optimization Frameworks for High-Fidelity Engineering Design Problems. In Y. Jin, editor, Knowledge Incorporation in Evolutionary Computation, Studies in Fuzziness and Soft Computing, pages 307–332. Springer, 2004.
5. Doğan A. and Özgüner F. (2004) Genetic Algorithm Based Scheduling of Meta-Tasks with Stochastic Execution Times in Heterogeneous Computing Systems, Cluster Computing 7, 177-190, Kluwer Academic Publishers. Manufactured in The Netherlands, 2004.
6. Hacker H. A., Eddy H. and Lewis K. E. (2002) Efficient Global Optimization Uisng Hybrid Genetic Algorithms, 9th AIAA/IMMSO Symposium on Multidisciplinary Analysis and Optimization, 4-6 September 2002, Altanta, Georgia, AIAA 2002-5429.
7. Xu Z.-B., Leung, K.-S., Liang Y., and Leung Y. (2003) Efficiency speed-up strategies for evolutionary computation: fundamentals and fast-GAs, Applied Mathematics and Computation Vol. 142, 341-388, 2003.
8. Salami M. and Hendtlass T. (2003) A fast evaluation strategy for evolutionary algorithms. Applied Soft Computing, 2:156–173, 2003.
9. Potter M. A. and De Jong K. A. (1994) A cooperative Coevolutionary Approach to Function Optimisation, The Third Parallel Problem Solving From Nature, Jerusalem, Israel, pp. 249-257, 1994.
10. Ong Y. S. and Keane A. J. (2004) Meta-Lamarckian Learning in Memetic Algorithm, IEEE Transactions On Evolutionary Computation, Vol. 8, No. 2, pp. 99-110, April 2004.
11. Branke J. and Schmidt C. (2003) Fast convergence by means of fitness estimation. Soft Computing Journal, 2003.
12. Jones, D. R., Schinlau, M., and Welch, W. J. (1998) Efficient Global Optimisation of Expensive Black-box Functions, *Journal of Global Optimization*, Vol. 13, pp. 455-492, 1998.
13. Ong Y. S., Nair P. B. and Keane A. J. (2003) Evolutionary Optimization of Computationally Expensive Problems via Surrogate Modeling, AIAA Journal, Vol. 41, No. 4, pp. 687-696, 2003.
14. Song W. and Keane A. J. (2005) An efficient evolutionary optimisation framework applied to turbine blade firtree root local profiles, Structural and Multidisciplinary Optimisation, Vol. 29 No. 5, 2005, pp. 382-390.
15. Jin Y. (2005) A comprehensive survey of fitness approximation in evolutionary computation. Soft Computing. Vol. 9, No. 1, pp. 3-12, Springer, 2005.
16. Myers R. (2002) Response Surface Methodology: Process and Product Optimization Using Designed Experiments, John Wiley & Sons Inc. 2002.
17. Cressie, N. A. C. (1993) Statistics for Spatial Data, rev., Wiley, New York, 1993.
18. Kleijnen J. P. C. and Van Beers W. (2002) Kriging for interpolation in random simulation. Journal of the Operational Research Society, **54**, No. 3, 255-262, 2003.

19. Ahn, J. A., Kim, H., Lee, D., Rho, O. (2001) Response Surface Method for Airfoil Design in Transonic Flow, *Journal of Aircraft*, Vol. **38**, No. 2, 2001.
20. Simpson T.W. (1998) Comparison of Response Surface and Kriging Models in the Multidisciplinary Design of an Aerospike Nozzle, NASA/CR-1998-206935, ICASE report No. 98-16, 1998.
21. Venter, G., Haftka, R. T., Starners, J. H. Jr. (1998) Construction of Response Surface Approximations for Design Optimisation, *AIAA Journal*, Vol. **36**, No. 12, 1998.
22. Bishop, C. (1995) Neural Networks for Pattern Recognition, Oxford University Press 1995.
23. Bandler, J. W., Cheng Q. S., Dakroury S. A., Mohamed A. S., Bakr M. H., Madsen, K. M. and Sondergaard J. (2004) Space Mapping: The State of the Art, IEEE Trasactions on Microwave Theory and Techniques. Vol. 52, No. 1, January 2004.
24. Sacks, J., Welch, W. J., Mitchell, J. J., Wynn, H. P. (1989) Design and Analysis of Computer Experiments, *Statistical Science*, Vol. **4**, No. 4, 1989, pp. 409-435.
25. Guinta, A. A., Watson, L. T. (2003) A Comparison of Approximation Modelling Techniques: Polynomial versus Interpolating Models, AIAA-98-4758, 1998.
26. Daberkow, D. D., Marris, D. N. (1998) New Approaches to Conceptual and Preliminary Aircraft Design: A Comparative Assessment of a Neural Network Formulation and A Response Surface Methodology, AIAA, 1998 World Aviation Conference, September 28-30, 1998, Anaheim, CA, 1998.
27. Jin, R., Chen, W., Simpson, T. W. (2000) Comparative Studies of Metamodelling Techniques under Multiple Modelling Criteria, AIAA-2000-4801, 2000.
28. Booker, A. J., Dennis, J. E., Frank, P. D., Serafini, D. B., Torczon, V., and Trosset, M. W. (1999) A Rigorous Framework for Optimization of Expensive Functions by Surrogates, *Structural Optimization*, Vol. **17**, No. 1, 1999, pp. 1-13.
29. Alexandrov, N. M., Dennis, J. E. Jr., Lewis, R. M. (1997) A Trust Region Framework for Managing the Use of Approximation Models in Approximation, NASA/CR-201745, 1997.
30. Alexandrov, N. M. and Lewis, R. M. (2003) First-Order Frameworks for Managing Models in Engineering Optimisation, 1st International Workshop on Surrogate Modelling and Space Mapping for Engineering Optimisation, 11/16-19/2000, TDU, 2003.
31. Guinta, A. A. and Eldred, M. S. (2000) Implementation of a Trust Region Model Management Strategy in the Dakota Optimisation Toolkit, AIAA-2000-4935, 2000.
32. Sellar, R. S., Batill, S. M., Renaud, J. E. (2003) Response Surface Based, Concurrent Subspace Optimisation for Multidisciplinary System Design, 2003.
33. Wujek, B. A. and Renaud, J. E. (1998) New Adaptive Move-limit Management Strategy for Approximate Optimization, Part 1, *AIAA Journal*, Vol. **36**, No. 10, 1998, pp. 1911-1921.
34. Alexandrov, N. M. (1998) On Managing the Use of Surrogates in General Nonlinear Optimization and MDO, AIAA-98-4798, 1998.
35. Robinson, G. M. and Keane, A. J. (1999) A Case for Multi-level Optimisation in Aeronautical Design, *Aeronautical Journal*, Vol. **103**, 1999, pp. 481-485.
36. Nair, P. B. and Keane, A. J. (1998) Combining Approximation Concepts with Genetic Algorithm-based Structure Optimisation Procedure, 1998.

37. Ratle, W. (1998) Accelerating the Convergence of Evolutionary Algorithms by Fitness Landscape Approximation, Parallel Problem Solving from Nature V, 1998, pp. 87-96.
38. El-Beltagy, M. A. and Keane, A. J. (1999) Evolutionary Optimisation for Computationally Expensive Problems Using Gaussian Processes, Proceedings of the Genetic and Evolutionary Computation Conference (GECCO99), Morgan Kaufman, 1999, pp. 196-203.
39. Liang, K. H., Yao, X., Newton, C. (2000) Evolutionary Search of Approximated N-dimensional Landscapes, International Journal of Knowledge-Based Intelligent Engineering Systems, Vol. 4, No. 3, 2000, pp. 172-183.
40. Jin, Y., Olhofer, M. and Sendhoff, B. (2000) A Framework for Evolutionary Optimisation with Approximate Fitness Functions, IEEE Transactions on Evolutionary Computation, 2000.
41. Morris, M.D., Mitchell, T.J. and Ylvisaker, D. (1993) Baysian Design and Analysis of Computer Experiments: Use of Derivatives in Surface Prediction, Technometrics, Vol. 35, 1993, pp. 243-255.
42. Song W., Keane A.J., Rees J., Bhaskar A. and Bagnall S. (2002) Local Shape Optimisation of a Firtree root using NURBS, 9th AIAA/ISSMO Symposium on Multidisciplinary Analysis and Optimization, Atlanta, Georgia 4-6 Sep 2002.
43. Song W. and Keane A.J. (2005) A New Hybrid Update Scheme for an Evolutionary Search Strategy Using Genetic Algorithm and Kriging, 46th AIAA/ASME/ASCE/AHS/ASC Structures, Structural Dynamics and Materials Conference, 13th AIAA/ASME/AHS Adaptive Structures Conference 7t, Austin, Texas, Apr. 18-21, 2005.
44. Fluent (2006) http://www.fluent.com, 2006.

12

A Study of Techniques to Improve the Efficiency of a Multi-Objective Particle Swarm Optimizer

Margarita Reyes-Sierra and Carlos A. Coello Coello

CINVESTAV-IPN (Evolutionary Computation Group)
Departamento de Ingeniería Eléctrica, Sección Computación
Av. IPN No. 2508, Col. San Pedro Zacatenco, México D.F. 07360, México
mreyes@computacion.cs.cinvestav.mx, ccoello@cs.cinvestav.mx

Summary. The use of particle swarm optimization (PSO) in multi-objective optimization has become widespread in the last few years. However, very few researchers have explored the use of techniques that allow to reduce the number of fitness evaluations of a PSO-based approach for multi-objective optimization. This chapter precisely explores the advantages and disadvantages of using fitness inheritance and approximation techniques to reduce the number of fitness evaluations into a PSO-based multi-objective algorithm previously proposed by the authors. Our study includes fifteen fitness inheritance techniques and four approximation techniques which are applied to a set of test functions taken from the specialized literature.

12.1 Introduction

Given the high computational cost of evaluating the fitness functions of many real-world applications, the total number of fitness function evaluations that an Evolutionary Algorithm (EA) may perform in such applications may become severely limited. In order to improve the performance of EAs, several enhancement techniques have been proposed in the past. In this chapter, we will focus on two of these enhancement techniques: fitness inheritance and approximation techniques. Fitness Inheritance is an enhancement technique [1] in which the fitness value of an offspring is obtained from the fitness values of its parents. On the other hand, approximation techniques [2] let us estimate the fitness of an individual using the previously calculated fitness of its neighbors (either including its parents or not). In fact, fitness inheritance is a particular case of fitness approximation. However, in general, the idea of using enhancement techniques, is that we do not need to evaluate every individual at each generation, such that the total computational cost can be reduced. In this chapter, we perform a study of different inheritance and approximation techniques applied to a real-coded PSO-based approach that has been previously

M. Reyes-Sierra and Carlos A. Coello Coello: *A Study of Techniques to Improve the Efficiency of a Multi-Objective Particle Swarm Optimizer*, Studies in Computational Intelligence (SCI) **51**, 269–296 (2007)
www.springerlink.com

proposed by the authors to solve multi-objective problems [3]. Since the main focus of our research is on fitness inheritance techniques, we are proposing a higher number of fitness inheritance techniques (fifteen) than approximation techniques (four). In our study, we use five well-known multi-objective test functions in an attempt to determine the best from the enhancement techniques analyzed.

In the final part of our study, once we have identified which were the best enhancement techniques, we compare these techniques with respect to other PSO-based multi-objective optimizers which are representative of the state-of-the-art, adopting different test functions.

The remainder of this chapter is organized as follows. An introduction to Fitness Inheritance and Fitness Approximation is given in Sections 12.2 and 12.3, respectively. Section 12.4 introduces the multi-objective PSO-based algorithm in which the proposed techniques are incorporated. The enhancement techniques proposed in this paper are presented in Section 12.5. In Sections 12.6 and 12.7 we present the obtained results and their discussion, respectively. A comparison against other PSO-based algorithms is presented in Section 12.8. Finally, our conclusions and some of the possible paths for future research are described in Section 12.9.

12.2 Fitness Inheritance

The use of fitness inheritance to improve the performance of Genetic Algorithms (GAs) was originally proposed by Smith et al. [1]. The authors proposed two possible ways of inheriting fitness: the first consists of taking the average fitnesses of the two parents and the other consists of taking a weighted (proportional) average of the fitnesses of the two parents. The second approach is related to how similar the offspring is with respect to its parents (this is done using a similarity measure). They applied inheritance to a very simple problem (the OneMax problem) [1] and found that the weighted fitness average resulted in a better performance and indicated that fitness inheritance was a viable alternative to reduce the computational cost of a genetic algorithm.

Zheng et al. [4] performed the first application of fitness inheritance on a GA for the problem of codebook design in data compression techniques. They concluded that the use of fitness inheritance didn't degrade the performance of the GA.

Sastry et al. [5] provided some theoretical foundations for fitness inheritance. They investigated convergence times, population sizing and the optimal proportion of inheritance for the OneMax problem. Chen et al. [6] investigated fitness inheritance as a way to speed up multi-objective GAs and EAs. They extended the analytical model proposed by Sastry et al. to multi-objective problems. Convergence and population-sizing models are derived and compared with respect to experimental results. The authors concluded that the number of function evaluations can be reduced with the use of fitness inheritance.

Salami et al. [7] proposed a "Fast Evolutionary Algorithm" in which a fitness and associated reliability value are assigned to each new individual that is only evaluated using the true fitness function if the reliability value is below a certain threshold. Also, they incorporated random evaluation and error compensation strategies. The authors obtained an average reduction of 40% in the number of evaluations while obtaining similar solutions. In the same work, they presented an application of fitness inheritance to image compression obtaining reductions between 35% and 42% of the number of evaluations.

Ducheyne et al. [8] tested the performance of average and weighted average fitness inheritance on a well-known test suite of multi-objective optimization problems [27], using a binary GA. They concluded that the fitness inheritance efficiency enhancement techniques can be used in order to reduce the number of fitness evaluations provided that the Pareto front is convex and continuous. They also concluded that if the Pareto surface is not convex or if it is discontinuous, the fitness inheritance strategies fail to reach the true Pareto front.

Pelikan et al. [10] used fitness inheritance to estimate fitness for a proportion of solutions in the Bayesian Optimization Algorithm (BOA). They concluded that fitness inheritance is a promising concept in BOA, because population-sizing requirements for building appropriate models of promising solutions lead to good fitness estimates even if only a small proportion of candidate solutions is evaluated using the true fitness function.

Bui et al. [11] performed a comparison of the performance of anti-noise methods, particularly probabilistic and re-sampling methods, using the NSGA-II [12]. They applied the concept of fitness inheritance to both types of methods in order to reduce computational time. The authors obtained a substantial amount of savings in the number of computations, reaching a peak of 30% of savings without deteriorating the performance.

In a previous work [13], we proposed the first attempt to incorporate the concept of fitness inheritance to a real-coded Multi-Objective PSO (MOPSO) previously proposed by us [3]. In [13], we tested the performance of weighted average fitness inheritance on a well-known test suite of multi-objective optimization problems [27]. Based on the obtained results, we concluded that fitness inheritance reduces the computational cost without decreasing the quality of the results in a significant way. Also, the fitness inheritance technique used was able to generate non-convex and discontinuous Pareto fronts. These conclusions were somewhat surprising given the conclusions (pointing in the exact opposite direction) previously obtained by Ducheyne et al. in [8].

12.3 Fitness Approximation

A promising possibility when an evaluation is very time-consuming or expensive is not to evaluate every individual, but just estimate the quality of some of the individuals based on an approximate model of the fitness landscape.

Approximation techniques estimate individual fitness on the basis of previously observed objective function values of neighboring individuals. There are many possible approximation models. In the simplest case, the fitness of a new individual is derived from its parents' fitnesses (fitness inheritance). However, some other more complicated methods have been used. A survey of approximation methods in evolutionary computation can be found in [2]. Here, we briefly mention a few examples of different approximation techniques commonly used.

Ratle [14] presented a new approach based on a classical real-encoded genetic algorithm for accelerating convergence of evolutionary optimization methods. With this aim, a reduction in the number of fitness function calls is obtained by means of an approximate model of the fitness landscape using kriging interpolation. The author builds a statistical model from a small number of data points obtained during one or more generations of the evolutionary method using the true fitness landscape. The model is updated every time a convergence criteria is reached.

Jin et al. [15] investigated the convergence property of an evolution strategy with neural network based fitness evaluations. In this work, the authors introduce the concept of *controlled evolution*, in which, the evolution proceeds using not only the approximate fitness function, but also the true fitness function. They also introduce two possibilities to combine the true with the approximate fitness function: the *controlled individuals* approach and the *controlled generation* approach. The authors define *controlled* as *true evaluated*. Both approaches are studied and some interesting conclusions/recommendations about the correct use of such techniques are provided.

Sano et al. [16] proposed a genetic algorithm for optimization of continuous noisy fitness functions. In this approach, the authors utilize the history of the search to reduce the number of fitness evaluations. The fitness of a novel individual is estimated using the fitness values of the other individuals as well as the sampled fitness values for it. So, as to increase the number of individuals adopted for evaluation, they use not only the current generation but also the whole history of the search. To utilize the history of the search, a stochastic model of the fitness function is introduced, and the maximum likelihood technique is used for estimation of the fitness function. The authors concluded that the proposed method outperforms a conventional GA in noisy environments.

Branke et al. [17] suggest the use of local regression for estimation, taking the fitness of neighboring individuals into account. Since in local regression is very important to determine which individuals belong to the neighborhood of a given individual, the authors studied two different approaches for setting the value of the size of the local neighborhood (*relative neighborhood* and *adaptive neighborhood*). The authors concluded that local regression provides better estimations than previously proposed approaches. In more recent work, Branke et al. [18] compared two estimation methods: interpolation and regression. They concluded that regression seems to be slightly preferable, particularly if only a very small fraction of the individuals in the population is evaluated. Their experiments also show that using fitness estimation, it is possible to either reach a better fitness level in a given time, or to reach a desired fitness level much faster (using roughly a half of the original number of fitness function evaluations).

Ong et al. [19] proposed a local surrogate modeling algorithm for parallel evolutionary optimization of computationally expensive problems. The proposed algorithm combines hybrid evolutionary optimization techniques, radial basis functions, and trust-region frameworks. The main idea of the proposed approach is using an EA combined with a feasible sequential quadratic programming solver. Each individual within an EA generation is used as an initial solution for local search, based on Lamarckian learning. The authors employ a trust-region framework to manage the interaction between the original objective and constraint functions and the computationally cheap surrogate models (which consist of radial basis networks constructed by using data points in the neighborhood of the initial solution), during local search. Extensive numerical studies are presented for some benchmark test functions and an aerodynamic wing design problem. The authors show that the proposed framework provides good designs on a limited computational budget. In more recent work, Ong et al. [19] present a study on the effects of uncertainty in the surrogate on Surrogate-Assisted Evolutionary Algorithms (SAEA). In particular, the authors focus on both the "curse of uncertainty" (impairments due to errors in the approximation) and "blessing of uncertainty" (benefits of approximation errors). The authors tested several algorithms: the Surrogated-Assisted Memetic Algorithm (SAMA) proposed in [19], a standard genetic algorithm, a memetic algorithm (considered as the standard hybridization of a genetic algorithm and the feasible sequential quadratic programming solver used in [19]) and the SAMA-Perfect algorithm (which is the SAMA algorithm but using the exact fitness function as surrogate model), on three multimodal benchmark problems (Ackley, Griewank and Rastrigin). The authors concluded that approximation errors lead to convergence at false global optima, but prove to be beneficial in some cases, accelerating the evolutionary search.

Gaspar-Cunha et al. [21] developed an algorithm using artificial neural networks to reduce the number of exact function evaluations for multi-objective optimization problems. The proposed method can save, in some cases, more than 50% of true function evaluations.

Regis and Shoemakes [22] developed an approach for the optimization of continuous costly functions that uses a space-filling experimental design and local function approximation to reduce the number of function evaluations in an evolutionary algorithm. The proposed approach estimates the objective function value of an offspring by means of a function approximation model over the k nearest previously evaluated points. The estimated values are used to identify the most promising offspring per function evaluation. A Symmetric Latin Hypercube Design (SLHD) is used to determine initial points for function evaluation, and for the construction of the function approximation models. The authors compared the performance of an Evolution Strategy (ES) with local quadratic approximation, an ES with local cubic radial basis function interpolation, an ES whose initial parent population is obtained from a SLHD, and a conventional ES (in all cases, the authors use a (μ, λ)-ES with uncorrelated mutations). The algorithms are tested on a groundwater bioremediation problem and on some benchmark test functions for global optimization (including Dixon-Szegö, Rastrigin and Ackley). The obtained results (which include analysis of variance to provide stronger and solid claims regarding the relative performance of the algorithms) suggest that the approach that uses SLHDs together with local function approximations has potential for success in enhancing EAs for computationally expensive real-world problems. Also, the cubic radial basis function approach appears to be more promising than the quadratic approximation approach on difficult higher-dimensional problems.

Lim et al. [23] presented a Trusted Evolutionary Algorithm (TEA) for solving optimization problems with computationally expensive fitness functions. The TEA is designed to maintain good trustworthiness of the surrogate models in predicting fitness improvements or controlling approximation errors throughout the evolutionary search. In this case, the authors are more interested in predicting search improvement as opposed to the quality of the approximation, which is regarded as a secondary objective. TEA begins its search using the canonical EA, with only exact function evaluations. During the canonical EA search, the exact fitness values obtained are archived in a central database together with the design vectors (to be used later for constructing surrogate models). After some initial search generations (specified by the user), the trust region approach takes place beginning from the best solution provided by the canonical EA. The trust region approach uses a surrogate model (radial basis neural networks) and contracts or expands the trust radius depending on the ability of the approximation model in predicting fitness improvements, until the termination conditions are reached. The authors performed an empirical study on two highly multi-modal benchmark functions commonly used in the global optimization literature (Ackley and Griewank). Numerical comparisons to the canonical EA and the original trust region line search framework are also reported. From the obtained results, the authors concluded that TEA converges to near-optimum solutions more efficiently than the canonical evolutionary algorithm.

Voutchkov and Keane [24] discussed the idea of using surrogate models for multi-objective optimization. That is, instead of using the expensive computational methods during the optimization, they used a much cheaper but accurate replica. The authors aim to overview the usage of several methods, using the NSGA-II algorithm [12] as their search engine. Several different surrogate models are used, including radial basis functions, regression models and kriging. The authors tested the surrogate models on four different test functions taken from the specialized multi-objective optimization literature. From the obtained results, the authors concluded that the performance of all methods depends on the features of the objective functions being optimized. Also, the authors discussed the weaknesses of these models on deceptive problems. In general, they concluded that the kriging method appears to perform well in most situations, however, it is much more computationally expensive than the rest.

Knowles [25] proposed an algorithm called ParEGO, which is a hybrid approach with on-line landscape approximation for expensive multi-objective optimization problems. ParEGO is an extension of the Efficient Global Optimization (EGO) algorithm, which is a frequently cited algorithm from the global optimization literature, designed for expensive functions of low dimension. The EGO algorithm makes use of kriging to model the search landscape from solutions visited during the search. Specifically, it exploits a version of the Design and Analysis of Computer Experiments (DACE) model, which is based on Gaussian processes. ParEGO extends the EGO algorithm for solving multi-objective problems by converting the k different cost values (objectives) of a solution into a single cost via a parameterized scalarizing weight vector (using the augmented Tchebycheff function). By choosing a different weight vector at each iteration of the search, an approximation to the whole Pareto front is built up gradually. The author tested the ParEGO algorithm on a test suite of nine difficult, but low-dimensional, multi-objective functions of limited ruggedness, over just 100 and 250 function evaluations. Also, the obtained results are compared against those obtained by the NSGA-II algorithm [12] (performing 100 and 260 function evaluations) and a random search approach (over 10000 function evaluations). From the obtained results, the author concluded that both ParEGO and NSGA-II outperform the random search, even over a very small number of function evaluations, and that ParEGO generally outperforms NSGA-II on the test functions adopted, at both 100 and 250 function evaluations (especially when the worst case performance is measured). Overall, the author concluded that ParEGO exhibits a promising performance for multi-objective optimization problems where evaluations are expensive or otherwise restricted in number.

In this chapter, we adopt very simple approximation techniques, based only on the objective values of the closest neighbors. Such techniques will be explained in Section 12.5.

12.4 Multi-Objective Particle Swarm Optimization

In this chapter, we incorporate several fitness inheritance and approximation techniques into a MOPSO that was previously proposed by us in [3] and updated in [13].

The PSO algorithm is a population-based search algorithm based on the simulation of the social behavior of birds within a flock [26]. In PSO, individuals, referred to as particles, are "flown" through a hyperdimensional search space. Changes to the position of the particles within the search space are based on the social-psychological tendency of individuals to emulate the success of other individuals.

A swarm consists of a set of particles, where each particle represents a potential solution. The position of each particle is changed according to its own experience and that of its neighbors. Let $\mathbf{x}_i(t)$ denote the position of particle i, at time step t. The position of particle i is then changed by adding a velocity $\mathbf{v}_i(t)$ to the current position, i.e.:

$$\mathbf{x}_i(t) = \mathbf{x}_i(t-1) + \mathbf{v}_i(t) \tag{12.1}$$

The velocity vector drives the optimization process and reflects the socially exchanged information. In the global best version (used here) of PSO, each particle uses its history of experiences in terms of its own best solution thus far (*pbest*) but, in addition, the particle uses the position of the best particle from the entire swarm (*gbest*). Thus, the velocity vector changes in the following way:

$$\mathbf{v}_i(t) = W\mathbf{v}_i(t-1) + C_1 r_1(\mathbf{x}_{pbest_i} - \mathbf{x}_i(t-1)) + C_2 r_2(\mathbf{x}_{gbest_i} - \mathbf{x}_i(t-1)) \tag{12.2}$$

where W is the inertia weight, C_1 and C_2 are the learning factors (usually defined as constants), and $r_1, r_2 \in [0,1]$ are random values. In this work, we use $W = random(0.1, 0.5)$ and $C_1, C_2 = random(1.5, 2.0)$.

The MOPSO proposed in [3, 13] is based on Pareto dominance, since it considers every nondominated solution as a new leader.[1] Additionally, the approach uses a crowding factor [12] as a second discrimination criterion which is also adopted to filter out the list of available leaders. For each particle, we select the leader in the following way: 97% of the time a leader is randomly selected, if and only if that leader dominates the current particle, and, the remaining 3% of the time, we select a leader by means of a binary tournament based on the crowding value of the available set of leaders. If the size of the set of leaders is greater than the maximum allowable size, only the best leaders are retained based on their crowding value. We also proposed the use of different mutation (or *turbulence*) operators which act on different subdivisions of the swarm. We proposed a scheme by which the swarm is subdivided in three parts (of equal size): the first sub-part has no mutation at all, the second sub-part

[1] A *leader* is used as the *gbest* solution in Equation 12.2.

uses uniform mutation and the third sub-part uses non-uniform mutation. The available set of leaders is the same for each of these sub-parts. Finally, the proposed approach also incorporates the ϵ-dominance concept [27] to fix the size of the set of final solutions produced by the algorithm. Figure 12.1 shows the pseudo-code of our proposed approach.

```
Begin
    Initialize swarm. Initialize leaders.
    Send leaders to ε-archive
    crowding(leaders), g = 0
    While g < gmax
        For each particle
            Select leader. Flight. Mutation.
⇒           If(p_i) Inherit Else Evaluation.
            Update pbest.
        EndFor
        Update leaders, Send leaders to ε-archive
        crowding(leaders), g++
    EndWhile
    Report results in ε-archive
End
```

Fig. 12.1. Pseudocode of our algorithm

In Figure 12.1, the symbol ($\Rightarrow$) indicates the line in which the concept of fitness inheritance (or approximation) is incorporated. The *inheritance* or *approximation proportion*, p_i, is the proportion of individuals in the population whose fitness is inherited or approximated. It is very important to note that a particle that has inherited its objective values **can not** enter into the final Pareto front, since a final solution must have true objective values.

12.5 Proposed Techniques

12.5.1 Fitness Inheritance

Since PSO has no recombination operator, we adopted as "parents" of a particle the previous position of the particle, its *pbest* and its leader.

Linear Combination Based on Distances (LCBD)

We propose to calculate the new position in the objective space of a particle by means of a linear combination of the positions of the particles that were considered to calculate the new position in the search space. We consider the position of the leader as the most important. Thus, the leader will be always considered.

Given a particle x_{old}, its personal best x_{pbest}, its assigned leader x_{ld} and the new particle x_{new}, we proceed to calculate the distance from x_{new} to its "parents" (as defined before): $d_1 = d(x_{new}, x_{old}), d_2 = d(x_{new}, x_{pbest}), d_3 = d(x_{new}, x_{ld})$, where d is an Euclidean distance. We propose variants of the same idea, based on the individuals that can be considered:

FI1 Previous position and leader: $r = \frac{d_1}{d_1+d_2}$,

$$f_i(x_{new}) = rf_i(x_{ld}) + (1-r)f_i(x_{old}), i = 1, ..., n.$$

FI2 *pbest* and leader. $r = \frac{d_2}{d_2+d_3}$,

$$f_i(x_{new}) = rf_i(x_{ld}) + (1-r)f_i(x_{pbest}), i = 1, ..., n.$$

FI3 Previous position, *pbest* and leader.
$r_1 = \frac{d_1}{d_1+d_2+d_3}$, $r_2 = \frac{d_2}{d_1+d_2+d_3}$, $r_3 = \frac{d_3}{d_1+d_2+d_3}$, $r_1 = 1/r_1, r_2 = 1/r_2, r_3 = 1/r_3$

$$f_i(x_{new}) = r_1 f_i(x_{old}) + r_2 f_i(x_{pbest}) + r_3 f_i(x_{ld}),$$

$i = 1, ..., n$. Where f_i is the value of the objective function i and n is the number of objective functions. See Figure 12.2 for an illustration of these techniques.

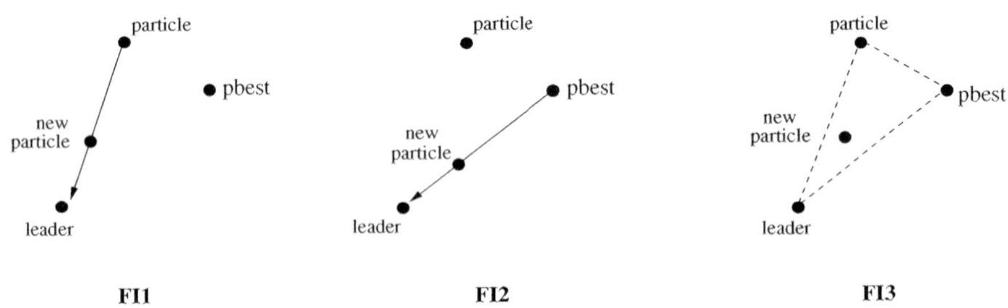

Fig. 12.2. Illustration of techniques FI1, FI2 and FI3

The technique FI1 is the one proposed in [13]. As in [13], in all the inheritance techniques, if the leader selected does not dominate the current particle, we will proceed to calculate the inherited position and to assign the objective values of the closest leader to that position. This procedure is used to avoid the generation of invalid particles in the case of non-convex Pareto fronts. See Figure 12.3.

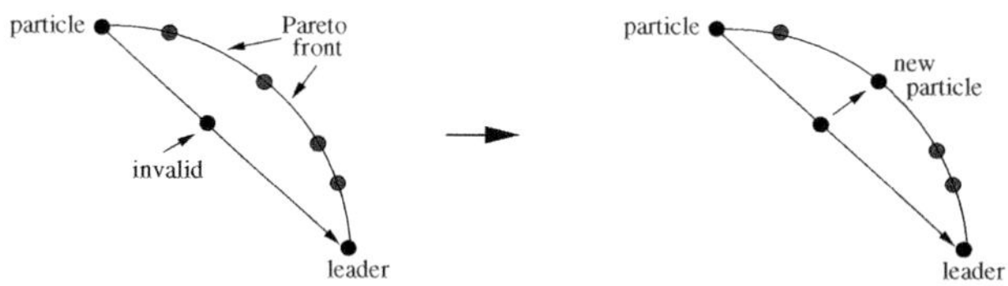

Fig. 12.3. Case in which "invalid" particles can be obtained and the method used to repair them

Flight Formula on Objective Space (FFOS)

As we mentioned before, in PSO, the position of each particle in the search space is updated using the formula:

$$\boldsymbol{x}_i(t) = \boldsymbol{x}_i(t-1) + \boldsymbol{v}_i(t)$$

$$\boldsymbol{v}_i(t) = W\boldsymbol{v}_i(t-1) + C_1 r_1(\boldsymbol{x}_{pbest_i} - \boldsymbol{x}_i(t-1)) + C_2 r_2(\boldsymbol{x}_{gbest_i} - \boldsymbol{x}_i(t-1))$$

In this case, we propose an analogous formula to update the position of each particle in objective function space:

$$\boldsymbol{f}_{\mathbf{i}}(t) = \boldsymbol{f}_{\mathbf{i}}(t-1) + \boldsymbol{vf}_{\mathbf{i}}(t)$$

$$\boldsymbol{vf}_{\mathbf{i}}(t) = W\boldsymbol{vf}_{\mathbf{i}}(t-1) + C_1 r_1(\boldsymbol{f}_{\mathbf{pbest_i}} - \boldsymbol{f}_{\mathbf{i}}(t-1)) + C_2 r_2(\boldsymbol{f}_{\mathbf{gbest_i}} - \boldsymbol{f}_{\mathbf{i}}(t-1))$$

where $\boldsymbol{f}_{\mathbf{i}}$, $\boldsymbol{f}_{\mathbf{pbest_i}}$ and $\boldsymbol{f}_{\mathbf{gbest_i}}$ are the values of the objective function i for the current particle, its *pbest* and *gbest*, respectively. We adopt the same values of W, C_1, r_1, C_2 and r_2 previously used for the flight in the decision variable space. We will consider the following variants based on the vectors considered:

FI4 Considering the whole formula:

$$\boldsymbol{vf}_{\mathbf{i}}(t) = W\boldsymbol{vf}_{\mathbf{i}}(t-1) + C_1 r_1(\boldsymbol{f}_{\mathbf{pbest_i}} - \boldsymbol{f}_{\mathbf{i}}(t-1)) + C_2 r_2(\boldsymbol{f}_{\mathbf{gbest_i}} - \boldsymbol{f}_{\mathbf{i}}(t-1))$$

FI5 Ignoring the previous direction:

$$\boldsymbol{vf}_{\mathbf{i}}(t) = C_1 r_1(\boldsymbol{f}_{\mathbf{pbest_i}} - \boldsymbol{f}_{\mathbf{i}}(t-1)) + C_2 r_2(\boldsymbol{f}_{\mathbf{gbest_i}} - \boldsymbol{f}_{\mathbf{i}}(t-1))$$

FI6 Ignoring the direction to the *pbest*:

$$\boldsymbol{vf}_{\mathbf{i}}(t) = W\boldsymbol{vf}_{\mathbf{i}}(t-1) + C_2 r_2(\boldsymbol{f}_{\mathbf{gbest_i}} - \boldsymbol{f}_{\mathbf{i}}(t-1))$$

Combination Using Flight Factors

Non-linear Combination (NLC)
In this case, we propose to calculate the new objective position of a particle using the elements of the flight formula:

$$\boldsymbol{f}_{\mathrm{i}}(t) = W\boldsymbol{f}_{\mathrm{i}}(t-1) + C_1 r_1 \boldsymbol{f}_{\mathbf{pbest_i}} + C_2 r_2 \boldsymbol{f}_{\mathbf{gbest_i}}$$

As in the previous cases, the variants considered are:

FI7 Considering the whole formula:

$$\boldsymbol{f}_{\mathrm{i}}(t) = W\boldsymbol{f}_{\mathrm{i}}(t-1) + C_1 r_1 \boldsymbol{f}_{\mathbf{pbest_i}} + C_2 r_2 \boldsymbol{f}_{\mathbf{gbest_i}}$$

FI8 Ignoring the previous position:

$$\boldsymbol{f}_{\mathrm{i}}(t) = C_1 r_1 \boldsymbol{f}_{\mathbf{pbest_i}} + C_2 r_2 \boldsymbol{f}_{\mathbf{gbest_i}}$$

FI9 Ignoring the position of the *pbest*:

$$\boldsymbol{f}_{\mathrm{i}}(t) = W\boldsymbol{f}_{\mathrm{i}}(t-1) + C_2 r_2 \boldsymbol{f}_{\mathbf{gbest_i}}$$

On the other hand, since $W \in (0.1, 0.5)$ and $C_1 r_1$, $C_2 r_2 \in (0.0, 2.0)$, we propose to modify the previous formula in the following way:

$$\boldsymbol{f}_{\mathrm{i}}(t) = \frac{W}{0.5}\boldsymbol{f}_{\mathrm{i}}(t-1) + \frac{C_1 r_1}{2.0}\boldsymbol{f}_{\mathbf{pbest_i}} + \frac{C_2 r_2}{2.0}\boldsymbol{f}_{\mathbf{gbest_i}}$$

As a result, we obtain the following variants:

FI10 Considering the whole formula:

$$\boldsymbol{f}_{\mathrm{i}}(t) = \frac{W}{0.5}\boldsymbol{f}_{\mathrm{i}}(t-1) + \frac{C_1 r_1}{2.0}\boldsymbol{f}_{\mathbf{pbest_i}} + \frac{C_2 r_2}{2.0}\boldsymbol{f}_{\mathbf{gbest_i}}$$

FI11 Ignoring the previous position:

$$\boldsymbol{f}_{\mathrm{i}}(t) = \frac{C_1 r_1}{2.0}\boldsymbol{f}_{\mathbf{pbest_i}} + \frac{C_2 r_2}{2.0}\boldsymbol{f}_{\mathbf{gbest_i}}$$

FI12 Ignoring the position of the *pbest*:

$$\boldsymbol{f}_{\mathrm{i}}(t) = \frac{W}{0.5}\boldsymbol{f}_{\mathrm{i}}(t-1) + \frac{C_2 r_2}{2.0}\boldsymbol{f}_{\mathbf{gbest_i}}$$

Linear Combination (LC)
We propose to use the previous formula but in such a way that the result is a linear combination of the elements considered:

$$\boldsymbol{f}_{\mathrm{i}}(t) = \frac{W}{r}\boldsymbol{f}_{\mathrm{i}}(t-1) + \frac{C_1 r_1}{r}\boldsymbol{f}_{\mathbf{pbest_i}} + \frac{C_2 r_2}{r}\boldsymbol{f}_{\mathbf{gbest_i}}$$

where $r = W + C_1 r_1 + C_2 r_2$. The corresponding variants are the following (note the changes in r):

FI13 Considering the whole formula, $r = W + C_1r_1 + C_2r_2$:

$$\boldsymbol{f}_\mathbf{i}(t) = \frac{W}{r}\boldsymbol{f}_\mathbf{i}(t-1) + \frac{C_1r_1}{r}\boldsymbol{f}_{\mathbf{pbest_i}} + \frac{C_2r_2}{r}\boldsymbol{f}_{\mathbf{gbest_i}}$$

FI14 Ignoring the previous position, $r = C_1r_1 + C_2r_2$:

$$\boldsymbol{f}_\mathbf{i}(t) = \frac{C_1r_1}{r}\boldsymbol{f}_{\mathbf{pbest_i}} + \frac{C_2r_2}{r}\boldsymbol{f}_{\mathbf{gbest_i}}$$

FI15 Ignoring the position of the *pbest*, $r = W + C_2r_2$:

$$\boldsymbol{f}_\mathbf{i}(t) = \frac{W}{r}\boldsymbol{f}_\mathbf{i}(t-1) + \frac{C_2r_2}{r}\boldsymbol{f}_{\mathbf{gbest_i}}$$

12.5.2 Fitness Approximation (FA)

As we could see in the previous section, fitness inheritance techniques assign the fitness value (objective values, in our case) of an individual using the fitness values of its parents. However, in the case of fitness approximation techniques, it is possible to use any set of neighboring particles to estimate the fitness of a particle, based on an approximate model of the fitness landscape. We propose four simple approximation techniques. In each case, the particle will take the objective values of the particle indicated:

FA1 The closest particle. Since we have the available set of leaders stored in an external list (archive), in this case we will search for the closest leader, either member of the swarm itself or member of the available set of leaders.
FA2 The closest leader. In this case, we will search for the closest particle member of the available set of leaders.
FA3 The closest particle member of the swarm. In this case, we will not consider the available set of leaders.
FA4 The average of the 10 closest particles. In this case, we will consider both the particles members of the swarm and the particles members of the available set of leaders.

We use the Euclidean distance in the decision variable space. In technique FA4, there are cases in which an invalid particle may be created. In this way, if among the 10 closest particles there are two or more leaders, or there is just one leader but this leader does not dominate the current particle, we will proceed as it was explained before. See Figure 12.3.

12.6 Comparison of Results

In order to compare the proposed techniques, we performed a study using five well-known test functions taken from the specialized literature on evolutionary multi-objective optimization:

- Test Function ZDT1 [27]:

$$\begin{aligned} \text{Minimize} \quad & (f_1(\boldsymbol{x}), f_2(\boldsymbol{x})) \\ f_1(\boldsymbol{x}) &= x_1 \\ f_2(\boldsymbol{x}) &= g(\boldsymbol{x})h(f_1, g) \\ g(\boldsymbol{x}) = 1 + 9\sum_{i=2}^{m} x_i/(m-1), & \; h(f1, g) = 1 - \sqrt{f_1/g} \end{aligned} \tag{12.3}$$

where $m = 30$, and $x_i \in$[0,1].

- Test Function ZDT2 [27]:

$$\begin{aligned} \text{Minimize} \quad & (f_1(\boldsymbol{x}), f_2(\boldsymbol{x})) \\ f_1(\boldsymbol{x}) &= x_1 \\ f_2(\boldsymbol{x}) &= g(\boldsymbol{x})h(f_1, g) \\ g(\boldsymbol{x}) = 1 + 9\sum_{i=2}^{m} x_i/(m-1), & \; h(f1, g) = 1 - (f_1/g)^2 \end{aligned} \tag{12.4}$$

where $m = 30$, and $x_i \in$[0,1].

- Test Function ZDT3 [27]:

$$\begin{aligned} \text{Minimize} \quad & (f_1(\boldsymbol{x}), f_2(\boldsymbol{x})) \\ f_1(\boldsymbol{x}) &= x_1 \\ f_2(\boldsymbol{x}) &= g(\boldsymbol{x})h(f_1, g) \\ g(\boldsymbol{x}) = 1 + 9\sum_{i=2}^{m} x_i/(m-1) \;\;, & \;\; h(f1, g) = 1 - \sqrt{f_1/g} - (f_1/g)sin(10\pi f_1) \end{aligned} \tag{12.5}$$

where $m = 30$, and $x_i \in$[0,1].

- Test Function ZDT4 [27]:

$$\begin{aligned} \text{Minimize} \quad & (f_1(\boldsymbol{x}), f_2(\boldsymbol{x})) \\ f_1(\boldsymbol{x}) &= x_1 \\ f_2(\boldsymbol{x}) &= g(\boldsymbol{x})h(f_1, g) \\ g(\boldsymbol{x}) = 1 + 10(m-1) + \sum_{i=2}^{m}(x_i^2 - 10cos(4\pi x_i)), & \; h(f1, g) = 1 - \sqrt{f_1/g} \end{aligned} \tag{12.6}$$

where $m = 10$, $x_1 \in$[0,1] and $x_i \in$[-5,5], $i = 2, ..., m$.

- Test Function DTLZ2 [28]:

$$\begin{aligned} \text{Minimize} \quad & (f_1(\boldsymbol{x}), f_2(\boldsymbol{x}), f_3(\boldsymbol{x})) \\ f_1(\boldsymbol{x}) &= (1 + g(\boldsymbol{x}))cos(x_1\pi/2)cos(x_2\pi/2) \\ f_2(\boldsymbol{x}) &= (1 + g(\boldsymbol{x}))cos(x_1\pi/2)sin(x_2\pi/2) \\ f_3(\boldsymbol{x}) &= (1 + g(\boldsymbol{x}))sin(x_1\pi/2) \\ g(\boldsymbol{x}) &= \sum_{i=3}^{m}(x_i - 0.5)^2 \end{aligned} \tag{12.7}$$

where $m = 12$ and $x_i \in [0,1]$.

Functions ZDT1 and ZDT4 have convex Pareto fronts, ZDT2 and DTLZ2 have non-convex Pareto fronts and ZDT3 has a non-convex and discontinuous Pareto front.

We performed experiments with different values of *inheritance (approximation) proportion* p_i. We experimented with: $p_i = 0.1, 0.2, 0.3, 0.4$. Note that this proportion of individuals indicates also the percentage by which the number of evaluations is reduced (e.g., $p_i = 0.1$ means that 10% less evaluations are performed). We performed 20 runs for each function and each technique. The parameters adopted for our MOPSO were: 100 particles, 200 generations and 100 particles in the external archive.

Next, we show the results corresponding to the following measure:

Success Counting (SCC): We define this measure based on the idea of the measure called *Error Ratio* proposed by Van Veldhuizen [29] which indicates the percentage of solutions (from the nondominated vectors found so far) that are not members of the true Pareto optimal set. In this case, we count the number of vectors (in the current set of nondominated vectors available) that are members of the Pareto optimal set:

$$SCC = \sum_{i=1}^{n} s_i,$$

where n is the number of vectors in the current set of nondominated vectors available; $s_i = 1$ if vector i is a member of the Pareto optimal set, and $s_i = 0$ otherwise. It should then be clear that $SCC = n$ indicates an ideal behavior, since it would mean that all the vectors generated by our algorithm belong to the true Pareto optimal set of the problem. Note that SCC avoids the bias introduced by the Error Ratio measure, which normalizes the number of solutions found (which belong to the true Pareto front) and, therefore, provides only a percentage of solutions that reached the true Pareto front. This percentage does not provide any idea regarding the actual number of nondominated solutions that each algorithm produced.

Tables 12.1, 12.2, 12.3, 12.4, 12.5 and 12.6 present a summary of the results obtained. In each case, we present the average of the SCC measure over the 20 runs, and the percentage of decrement or increment on the quality of the results. Also, we present the average of the percentages for each value of inheritance proportion, for each technique.

Table 12.1. Obtained results for different values of inheritance proportion, for techniques FI1, FI2 and FI3

FI1	Inheritance proportion p_i								
function	0.0	0.1 (-10%)		0.2 (-20%)		0.3 (-30%)		0.4 (-40%)	
ZDT1	71	77	(+8.5%)	64	(-9.9%)	62	(-12.7%)	61	(-14.1%)
ZDT2	89	83	(-6.7%)	86	(-3.4%)	79	(-11.2%)	77	(-13.5%)
ZDT3	68	73	(+7.4%)	65	(-4.4%)	64	(-5.9%)	59	(-13.2%)
ZDT4	80	81	(+1.3%)	81	(+1.3%)	60	(-25.0%)	68	(-15.0%)
DTLZ2	18	15	(-16.7%)	16	(-11.1%)	12	(-33.3%)	12	(-33.3%)
Average			**-1.2%**		**-5.5 %**		**-17.6 %**		**-17.8 %**
FI2	Inheritance proportion p_i								
function	0.0	0.1 (-10%)		0.2 (-20%)		0.3 (-30%)		0.4 (-40%)	
ZDT1	71	74	(+4.2%)	68	(-4.2%)	68	(-4.2%)	59	(-16.9%)
ZDT2	89	81	(-9.0%)	82	(-7.9%)	78	(-12.4%)	77	(-13.5%)
ZDT3	68	64	(-5.9%)	67	(-1.5%)	58	(-14.7%)	63	(-7.4%)
ZDT4	80	77	(-3.8%)	83	(+3.8%)	67	(-16.3%)	69	(-13.8%)
DTLZ2	18	15	(-16.7%)	18	(0.0%)	15	(-16.7%)	14	(-22.2%)
Average			**-6.2%**		**-2.0 %**		**-12.9 %**		**-14.8 %**
FI3	Inheritance proportion p_i								
function	0.0	0.1 (-10%)		0.2 (-20%)		0.3 (-30%)		0.4 (-40%)	
ZDT1	71	73	(+2.8%)	69	(-2.8%)	69	(-2.8%)	50	(-29.6%)
ZDT2	89	87	(-2.2%)	82	(-7.9%)	71	(-20.2%)	76	(-14.6%)
ZDT3	68	67	(-1.5%)	63	(-7.4%)	64	(-5.9%)	60	(-11.8%)
ZDT4	80	81	(+1.3%)	79	(-1.3%)	59	(-26.3%)	68	(-15.0%)
DTLZ2	18	17	(-5.6%)	18	(0.0%)	14	(-22.2%)	9	(-50.0%)
Average			**-1.0%**		**-3.9 %**		**-15.5 %**		**-24.2%**

12.7 Discussion of Results

Since comparing 19 different techniques is very difficult, we decided to represent each technique with a vector. The vector used is that containing the average of the change in the quality of results for each inheritance proportion value. For example, to represent technique FI1, we construct the following vector (see Table 12.1):

Inheritance proportion p_i	0.1	0.2	0.3	0.4
Average vector	-1.2	-5.5	-17.6	-17.8

Table 12.2. Obtained results for different values of inheritance proportion, for techniques FI4, FI5 and FI6

FI4	Inheritance proportion p_i								
function	0.0	0.1 (-10%)		0.2 (-20%)		0.3 (-30%)		0.4 (-40%)	
ZDT1	71	62	(-12.7%)	62	(-12.7%)	59	(-16.9%)	49	(-31.0%)
ZDT2	89	85	(-4.5%)	84	(-5.6%)	78	(-12.4%)	79	(-11.2%)
ZDT3	68	73	(+7.4%)	69	(+1.5%)	60	(-11.8%)	58	(-14.7%)
ZDT4	80	88	(+10.0%)	88	(+10.0%)	85	(+6.3%)	82	(+2.5%)
DTLZ2	18	18	(0.0%)	11	(-38.9%)	11	(-38.9%)	12	(-33.3%)
Average			**0.0%**		**-9.1 %**		**-14.7 %**		**-17.5%**
FI5	Inheritance proportion p_i								
function	0.0	0.1 (-10%)		0.2 (-20%)		0.3 (-30%)		0.4 (-40%)	
ZDT1	71	74	(+4.2%)	69	(-2.8%)	61	(-14.1%)	56	(-21.1%)
ZDT2	89	89	(0.0%)	79	(-11.2%)	84	(-5.6%)	77	(-13.5%)
ZDT3	68	72	(+5.9%)	70	(+2.9%)	55	(-19.1%)	58	(-14.7%)
ZDT4	80	87	(+8.8%)	85	(+6.3%)	85	(+6.3%)	82	(+2.5%)
DTLZ2	18	18	(0.0%)	12	(-33.3%)	13	(-27.8%)	12	(-33.3%)
Average			**+3.8%**		**-7.6 %**		**-12.1 %**		**-16.1%**
FI6	Inheritance proportion p_i								
function	0.0	0.1 (-10%)		0.2 (-20%)		0.3 (-30%)		0.4 (-40%)	
ZDT1	71	70	(-1.4%)	61	(-14.1%)	62	(-12.7%)	47	(-33.8%)
ZDT2	89	83	(-6.7%)	82	(-7.9%)	76	(-14.6%)	70	(-21.3%)
ZDT3	68	72	(+5.9%)	72	(+5.9%)	59	(-13.2%)	61	(-10.3%)
ZDT4	80	83	(+3.8%)	84	(+5.0%)	80	(0.0%)	79	(-1.3%)
DTLZ2	18	16	(-11.1%)	14	(-22.2%)	14	(-22.2%)	11	(-38.9%)
Average			**-1.9%**		**-6.7 %**		**-12.5 %**		**-21.1%**

In this way, in Table 12.7 we present the vectors of all techniques. Since every entry in each vector is a change in the quality of the obtained results given a value of inheritance proportion, the bigger the values of the vector, the better the corresponding technique is. Thus, we are interested on the vector or vectors that represent the solution to the problem of maximizing all the entries (i.e., each entry is considered as an objective).

The nondominated vectors among all the 19 techniques are the vectors corresponding to techniques FI2, FI3, FI5, FI9, FI11, FI14, FA1, FA3 and FA4. That is, these nine techniques are the best overall performers. For this reason, all these techniques are marked with a level of 1 in Table 12.7. As we can see, among the best techniques, 6 are inheritance techniques and 3 are approximation techniques. In fact, it is very interesting to note that four of the six best inheritance techniques ignore the previous position of the particle, in order to update the position in objective function space. On the other hand, the only approximation technique that doesn't figure as one of the best is the one that only considers the set of leaders to assign the objective function values of a particle.

Table 12.3. Obtained results for different values of inheritance proportion, for techniques FI7, FI8 and FI9

FI7	Inheritance proportion p_i								
function	0.0	0.1 (-10%)		0.2 (-20%)		0.3 (-30%)		0.4 (-40%)	
ZDT1	71	64	(-9.9%)	58	(-18.3%)	57	(-19.7%)	47	(-33.8%)
ZDT2	89	83	(-6.7%)	74	(-16.9%)	68	(-23.6%)	66	(-25.8%)
ZDT3	68	66	(-2.9%)	69	(+1.5%)	64	(-5.9%)	57	(-16.2%)
ZDT4	80	80	(0.0%)	74	(-7.5%)	57	(-28.8%)	44	(-45.0%)
DTLZ2	18	13	(-27.8%)	14	(-22.2%)	13	(-27.8%)	9	(-50.0%)
Average			**-9.5%**		**-12.7 %**		**-21.2 %**		**-34.2%**
FI8	Inheritance proportion p_i								
function	0.0	0.1 (-10%)		0.2 (-20%)		0.3 (-30%)		0.4 (-40%)	
ZDT1	71	69	(-2.8%)	62	(-12.7%)	53	(-25.4%)	47	(-33.8%)
ZDT2	89	85	(-4.5%)	84	(-5.6%)	66	(-25.8%)	65	(-27.0%)
ZDT3	68	71	(+4.4%)	67	(-1.5%)	61	(-10.3%)	52	(-23.5%)
ZDT4	80	80	(0.0%)	72	(-10.0%)	63	(-21.3%)	52	(-35.0%)
DTLZ2	18	16	(-11.1%)	14	(-22.2%)	13	(-27.8%)	12	(-33.3%)
Average			**-2.8%**		**-10.4 %**		**-22.1 %**		**-30.5%**
FI9	Inheritance proportion p_i								
function	0.0	0.1 (-10%)		0.2 (-20%)		0.3 (-30%)		0.4 (-40%)	
ZDT1	71	67	(-5.6%)	58	(-18.3%)	54	(-23.9%)	44	(-38.0%)
ZDT2	89	90	(+1.1%)	85	(-4.5%)	69	(-22.5%)	68	(-23.6%)
ZDT3	68	68	(0.0%)	67	(-1.5%)	61	(-10.3%)	51	(-25.0%)
ZDT4	80	83	(+3.8%)	76	(-5.0%)	72	(-10.0%)	58	(-27.5%)
DTLZ2	18	19	(+5.6%)	18	(0.0%)	19	(+5.6%)	18	(0.0%)
Average			**+1.0%**		**-5.9 %**		**-12.2 %**		**-22.8%**

12.8 Comparison with Other PSO Approaches

In the previous section, we found nine techniques to be the best from the set proposed. In this section, some of those nine techniques will be compared against two other PSO-based multi-objective approaches representative of the state-of-the-art: the Sigma-MOPSO [30] and the Cluster-MOPSO [31].

For our comparison, we chose only five from the nine best techniques to be compared. With this aim, we calculated the norm (distance to the origin) of the vector of each technique and we selected the five vectors with the lowest values. In this way, we selected the techniques from the *knee* of the Pareto front, that is, the compromise solutions from the central portion of the front. In Table 12.7, we show the norm values of the nine best techniques and their corresponding rank according to this value. As we can see, the five techniques with the lowest value of the norm are: FI2, FI5, FI11, FA1 and FA3.

For this comparative analysis, we will use the two following test functions:

- Test Function DTLZ4 [28]:

Table 12.4. Obtained results for different values of inheritance proportion, for techniques FI10, FI11 and FI12

FI10	Inheritance proportion p_i							
function	0.0	0.1 (-10%)		0.2 (-20%)		0.3 (-30%)		0.4 (-40%)
ZDT1	71	71	(0.0%)	58	(-18.3%)	59	(-16.9%)	48	(-32.4%)
ZDT2	89	78	(-12.4%)	78	(-12.4%)	69	(-22.5%)	58	(-34.8%)
ZDT3	68	70	(+2.9%)	63	(-7.4%)	61	(-10.3%)	47	(-30.9%)
ZDT4	80	78	(-2.5%)	81	(+1.3%)	58	(-27.5%)	52	(-35.0%)
DTLZ2	18	17	(-5.6%)	13	(-27.8%)	11	(-38.9%)	11	(-38.9%)
Average			**-3.5%**		**-12.9 %**		**-23.2 %**		**-34.4%**
FI11	Inheritance proportion p_i								
function	0.0	0.1 (-10%)		0.2 (-20%)		0.3 (-30%)		0.4 (-40%)	
ZDT1	71	62	(-12.7%)	63	(-11.3%)	55	(-22.5%)	37	(-48.0%)
ZDT2	89	84	(-5.6%)	87	(-2.2%)	81	(-9.0%)	76	(-14.6%)
ZDT3	68	69	(+1.5%)	60	(-11.8%)	57	(-16.2%)	44	(-35.3%)
ZDT4	80	82	(+2.5%)	81	(+1.3%)	73	(-8.8%)	73	(-8.8%)
DTLZ2	18	17	(-5.6%)	21	(+16.7%)	23	(+27.8%)	23	(+27.8%)
Average			**-4.0%**		**-1.5 %**		**-5.7 %**		**-15.8%**
FI12	Inheritance proportion p_i								
function	0.0	0.1 (-10%)		0.2 (-20%)		0.3 (-30%)		0.4 (-40%)	
ZDT1	71	66	(-7.0%)	56	(-21.1%)	55	(-22.5%)	48	(-32.4%)
ZDT2	89	87	(-2.2%)	85	(-4.5%)	74	(-16.9%)	80	(-10.1%)
ZDT3	68	66	(-2.9%)	64	(-5.9%)	55	(-19.1%)	53	(-22.1%)
ZDT4	80	80	(0.0%)	75	(-6.3%)	71	(-11.3%)	61	(-23.8%)
DTLZ2	18	18	(0.0%)	18	(0.0%)	16	(-11.1%)	16	(-11.1%)
Average			**-2.4%**		**-7.6 %**		**-16.2 %**		**-19.9%**

$$
\begin{aligned}
\text{Minimize} \quad & (f_1(\boldsymbol{x}), f_2(\boldsymbol{x}), f_3(\boldsymbol{x})) \\
f_1(\boldsymbol{x}) &= (1 + g(\boldsymbol{x}))cos(x_1^\alpha \pi/2)cos(x_2^\alpha \pi/2) \\
f_2(\boldsymbol{x}) &= (1 + g(\boldsymbol{x}))cos(x_1^\alpha \pi/2)sin(x_2^\alpha \pi/2) \\
f_3(\boldsymbol{x}) &= (1 + g(\boldsymbol{x}))sin(x_1^\alpha \pi/2) \\
g(\boldsymbol{x}) &= \sum_{i=3}^{m} (x_i - 0.5)^2
\end{aligned}
\tag{12.8}
$$

where α=100, $m = 12$ and $x_i \in$[0,1].

- Test Function DTLZ6 [28]:

Table 12.5. Obtained results for different values of inheritance proportion, for techniques FI13, FI14 and FI15

FI13	Inheritance proportion p_i								
function	0.0	0.1 (-10%)		0.2 (-20%)		0.3 (-30%)		0.4 (-40%)	
ZDT1	71	68	(-4.2%)	69	(-2.8%)	63	(-11.3%)	58	(-18.3%)
ZDT2	89	84	(-5.6%)	81	(-9.0%)	79	(-11.2%)	80	(-10.1%)
ZDT3	68	70	(+2.9%)	68	(0.0%)	63	(-7.4%)	54	(-20.6%)
ZDT4	80	79	(-1.3%)	81	(+1.3%)	64	(-20.0%)	59	(-26.3%)
DTLZ2	18	14	(-22.2%)	16	(-11.1%)	14	(-22.2%)	12	(-33.3%)
Average			**-6.1%**		**-4.3 %**		**-14.4 %**		**-21.7%**
FI14	Inheritance proportion p_i								
function	0.0	0.1 (-10%)		0.2 (-20%)		0.3 (-30%)		0.4 (-40%)	
ZDT1	71	75	(+5.6%)	66	(-7.0%)	58	(-18.3%)	59	(-16.9%)
ZDT2	89	88	(-1.1%)	79	(-11.2%)	83	(-6.7%)	72	(-19.1%)
ZDT3	68	74	(+8.8%)	69	(+1.5%)	63	(-7.4%)	60	(-11.8%)
ZDT4	80	81	(+1.3%)	79	(-1.3%)	73	(-8.8%)	67	(-16.3%)
DTLZ2	18	16	(-11.1%)	15	(-16.7%)	13	(-27.8%)	13	(-27.8%)
Average			**+0.7%**		**-6.9 %**		**-13.8 %**		**-18.4%**
FI15	Inheritance proportion p_i								
function	0.0	0.1 (-10%)		0.2 (-20%)		0.3 (-30%)		0.4 (-40%)	
ZDT1	71	69	(-2.8%)	63	(-11.3%)	69	(-2.8%)	56	(-21.1%)
ZDT2	89	86	(-3.4%)	81	(-9.0%)	72	(-19.1%)	73	(-18.0%)
ZDT3	68	72	(+5.9%)	70	(+2.9%)	64	(-5.9%)	58	(-14.7%)
ZDT4	80	81	(+1.3%)	78	(-2.5%)	63	(-21.3%)	70	(-12.5%)
DTLZ2	18	13	(-27.8%)	15	(-16.7%)	16	(-11.1%)	10	(-44.4%)
Average			**-5.4%**		**-7.3 %**		**-12.0 %**		**-22.1%**

$$\begin{aligned} \text{Minimize} \quad & (f_1(\boldsymbol{x}), f_2(\boldsymbol{x}), f_3(\boldsymbol{x})) \\ f_1(\boldsymbol{x}) &= x_1 \\ f_2(\boldsymbol{x}) &= x_2 \\ f_3(\boldsymbol{x}) &= (1 + g(\boldsymbol{x}))h(f_1, f_2, g) \end{aligned} \tag{12.9}$$

$$g(\boldsymbol{x}) = 1 + 9/(m-2)\sum_{i=3}^{m} x_i \quad , \quad h(f_1, f_2, g) = 3 - \sum_{i=1}^{2}[\frac{f_i}{1+g}(1 + sin(3\pi f_i))]$$

where $m = 22$ and $x_i \in [0,1]$.

As in previous experiments, we used different values of p_i. We performed 20 runs for each function and each approach. The approaches without fitness inheritance or approximation performed 20000 objective function evaluations. The parameters adopted for our MOPSO were the same as before. Cluster-MOPSO used 40 particles, 4 swarms, 5 iterations per swarm and a total number of iterations of 100. In the case of Sigma-MOPSO, 200 particles were used through 100 iterations (these values were suggested by the author of

Table 12.6. Obtained results for different values of approximation proportion, for techniques FA1, FA2, FA3 and FA4

FA1	Approximation proportion p_a								
function	0.0	0.1 (-10%)		0.2 (-20%)		0.3 (-30%)		0.4 (-40%)	
ZDT1	71	74	(+4.2%)	64	(-9.9%)	63	(-11.3%)	55	(-22.5%)
ZDT2	89	88	(-1.1%)	85	(-4.5%)	81	(-9.0%)	76	(-14.6%)
ZDT3	68	73	(+7.4%)	61	(-10.3%)	60	(-11.8%)	55	(-19.1%)
ZDT4	80	85	(+6.3%)	89	(+11.3%)	79	(-1.3%)	80	(0.0%)
DTLZ2	18	18	(0.0%)	13	(-27.8%)	14	(-22.2%)	12	(-33.3%)
Average			**+3.4%**		**-8.2 %**		**-11.1 %**		**-17.9%**
FA2	Approximation proportion p_a								
function	0.0	0.1 (-10%)		0.2 (-20%)		0.3 (-30%)		0.4 (-40%)	
ZDT1	71	75	(+5.6%)	57	(-19.7%)	54	(-23.9%)	46	(-35.2%)
ZDT2	89	83	(-6.7%)	72	(-19.1%)	63	(-29.2%)	76	(-14.6%)
ZDT3	68	63	(-7.4%)	58	(-14.7%)	58	(-14.7%)	56	(-17.6%)
ZDT4	80	86	(+7.5%)	87	(+8.8%)	81	(+1.3%)	83	(+3.8%)
DTLZ2	18	15	(-16.7%)	13	(-27.8%)	11	(-38.9%)	10	(-44.4%)
Average			**-3.5%**		**-14.5 %**		**-21.1 %**		**-21.6%**
FA3	Approximation proportion p_a								
function	0.0	0.1 (-10%)		0.2 (-20%)		0.3 (-30%)		0.4 (-40%)	
ZDT1	71	71	(0.0%)	67	(-5.6%)	63	(-11.3%)	50	(-29.6%)
ZDT2	89	88	(-1.1%)	87	(-2.2%)	85	(-4.5%)	76	(-14.6%)
ZDT3	68	65	(-4.4%)	65	(-4.4%)	55	(-19.1%)	57	(-16.2%)
ZDT4	80	89	(+11.3%)	91	(+13.8%)	86	(+7.5%)	87	(+8.8%)
DTLZ2	18	16	(-11.1%)	15	(-16.7%)	16	(-11.1%)	13	(-27.8%)
Average			**-1.1%**		**-3.0 %**		**-7.7 %**		**-15.9%**
FA4	Approximation proportion p_a								
function	0.0	0.1 (-10%)		0.2 (-20%)		0.3 (-30%)		0.4 (-40%)	
ZDT1	71	69	(-2.8%)	59	(-16.9%)	60	(-15.5%)	52	(-26.8%)
ZDT2	89	87	(-2.2%)	80	(-10.1%)	76	(-14.6%)	71	(-20.2%)
ZDT3	68	67	(-1.5%)	71	(+4.4%)	56	(-17.6%)	56	(-17.6%)
ZDT4	80	86	(+7.5%)	85	(+6.3%)	79	(-1.3%)	80	(0.0%)
DTLZ2	18	20	(+11.1%)	16	(-11.1%)	14	(-22.2%)	12	(-33.3%)
Average			**+2.4%**		**-5.5 %**		**-14.2 %**		**-19.6%**

the method). The PSO approaches will be identified with the following labels: sMOPSO refers to [30], cMOPSO refers to [31], and oMOPSO is our MOPSO. All the algorithms were set such that they provided Pareto fronts with 100 points. In this case, we also show the obtained results with respect to the following measure:

Inverted Generational Distance (IGD): The concept of generational distance was introduced by Van Veldhuizen & Lamont [32, 33] as a way of estimating how far are the elements in the Pareto front produced by our algorithm from those in the true Pareto front of the problem. This measure

Table 12.7. Vectors of change in quality for each technique, for each value of inheritance or approximation proportion

Group		0.1	0.2	0.3	0.4	level	norm	rank
	FI1	-1.2	-5.5	-17.6	-17.8			
LCBD	FI2	-6.2	-2.0	-12.9	-14.8	**1**	20.69	**3**
	FI3	-1.0	-3.9	-15.5	-24.2	**1**	29.62	9
	FI4	0.0	-9.1	-14.7	-17.5			
FFOS	FI5	3.8	-7.6	-12.1	-16.1	**1**	21.86	**4**
	FI6	-1.9	-6.7	-12.5	-21.1			
	FI7	-9.5	-12.7	-21.2	-34.2			
	FI8	-2.8	-10.4	-22.1	-30.5			
NLC	FI9	1.0	-5.9	-12.2	-22.8	**1**	26.54	8
	FI10	-3.5	-12.9	-23.2	-34.4			
	FI11	-4.0	-1.5	-5.7	-15.8	**1**	17.33	**1**
	FI12	-2.4	-7.6	-16.2	-19.9			
	FI13	-6.1	-4.3	-14.4	-21.7			
LC	FI14	0.7	-6.9	-13.8	-18.4	**1**	24.02	6
	FI15	-5.4	-7.3	-12.0	-22.1			
	FA1	3.4	-8.2	-11.1	-17.9	**1**	22.86	**5**
FA	FA2	-3.5	-14.5	-21.1	-21.6			
	FA3	-1.1	-3.0	-7.7	-15.9	**1**	17.95	**2**
	FA4	2.4	-5.5	-14.2	-19.6	**1**	24.94	7

is defined as:

$$GD = \frac{\sqrt{\sum_{i=1}^{n} d_i^2}}{n}$$

where n is the number of nondominated vectors found by the algorithm being analyzed and d_i is the Euclidean distance (measured in objective space) between each of these and the nearest member of the true Pareto front. It should be clear that a value of $GD = 0$ indicates that all the elements generated are in the true Pareto front of the problem. Therefore, any other value will indicate how "far" we are from the global Pareto front of our problem. In our case, we implemented an "inverted" generational distance measure (IGD) in which we use as a reference the true Pareto front, and we compare each of its elements with respect to the front produced by an algorithm. In this way, we are calculating how far are the elements of the true Pareto front, from those in the Pareto front produced by our algorithm. Computing this "inverted" generational distance value reduces the bias that can arise when an algorithm didn't fully cover the true Pareto front.

Tables 12.8 and 12.9 present a summary of the results obtained. In each case, we present the average and standard deviation of the Success Counting (SCC) and Inverted Generational Distance (IGD) measures over the 20 runs.

As we can see in Table 12.8, in function DTLZ4 our approach (oMOPSO) is outperformed by one of the other PSO-based approaches (sMOPSO).

Table 12.8. Obtained results for the test function DTLZ4, for sMOPSO, cMOPSO, oMOPSO, and oMOPSO with techniques FI2, FI5, FI11, FA1 and FA3 incorporated (p_i=0.1, 0.2, 0.3, 0.4)

FI2		sMOPSO	cMOPSO	oMOPSO	0.1	0.2	0.3	0.4
SCC	mean	39	1.4	11	10	9	10	12
	std. dev.	14.8	3.3	4.4	4.7	3.0	3.8	8.1
IGD	mean	0.0064	0.0223	0.0106	0.0121	0.0129	0.0111	0.0107
	std. dev.	0.0007	0.0074	0.0034	0.0052	0.0037	0.0043	0.0050
FI5		sMOPSO	cMOPSO	oMOPSO	0.1	0.2	0.3	0.4
SCC	mean	39	1.4	11	10	10	9	11
	std. dev.	14.8	3.3	4.4	6.8	4.9	4.7	6.9
IGD	mean	0.0064	0.0223	0.0106	0.0100	0.0119	0.0109	0.0108
	std. dev.	0.0007	0.0074	0.0034	0.0045	0.0050	0.0050	0.0046
FI11		sMOPSO	cMOPSO	oMOPSO	0.1	0.2	0.3	0.4
SCC	mean	39	1.4	11	11	13	14	12
	std. dev.	14.8	3.3	4.4	4.5	11.5	14.8	16.2
IGD	mean	0.0064	0.0223	0.0106	0.0107	0.0105	0.0111	0.0119
	std. dev.	0.0007	0.0074	0.0034	0.0047	0.0046	0.0059	0.0065
FA1		sMOPSO	cMOPSO	oMOPSO	0.1	0.2	0.3	0.4
SCC	mean	39	1.4	11	**13**	**10**	**12**	**8**
	std. dev.	14.8	3.3	4.4	**6.9**	**3.4**	**6.5**	**4.2**
IGD	mean	0.0064	0.0223	0.0106	**0.0093**	**0.0112**	**0.0102**	**0.0122**
	std. dev.	0.0007	0.0074	0.0034	**0.0045**	**0.0039**	**0.0042**	**0.0044**
FA3		sMOPSO	cMOPSO	oMOPSO	0.1	0.2	0.3	0.4
SCC	mean	39	1.4	11	11	8	9	7
	std. dev.	14.8	3.3	4.4	3.2	3.1	3.7	3.8
IGD	mean	0.0064	0.0223	0.0106	0.0120	0.0138	0.0113	0.0129
	std. dev.	0.0007	0.0074	0.0034	0.0033	0.0040	0.0040	0.0042

On the other hand, in Table 12.9 we can see that in function DTLZ6 our approach is clearly the best.

Table 12.8 shows that, in function DTLZ4, all the techniques have a very good performance with respect to the IGD measure, with technique FA1 being (marginally) the best (considering the standard deviation values). On the other hand, although with respect to the average values of the SCC measure, technique FI11 is the best, this technique has the highest standard deviations. Thus, we conclude that the best technique is FA1 also in this case. In fact, both approximation techniques have the lowest standard deviations, in both performance measures.

From Table 12.9 we can conclude that, as in function DTLZ4, in function DTLZ6 all the techniques have a very good performance with respect to the IGD measure. However, in this case technique FI11 is the best. Also, with respect to the SCC measure technique FI5 is the best, considering all cases.

Table 12.9. Obtained results for the test function DTLZ6, for sMOPSO, cMOPSO, oMOPSO, and oMOPSO with techniques FI2, FI5, FI11, FA1 and FA3 incorporated (p_i=0.1, 0.2, 0.3, 0.4)

FI2		sMOPSO	cMOPSO	oMOPSO	0.1	0.2	0.3	0.4
SCC	mean	1	0	62	60	53	50	47
	std. dev.	0.0	0.0	13.0	21.4	20.9	22.0	17.3
IGD	mean	0.0673	0.0373	0.0091	0.0082	0.0092	0.0101	0.0111
	std. dev.	0.0000	0.0172	0.0058	0.0060	0.0062	0.0065	0.0062
FI5		sMOPSO	cMOPSO	oMOPSO	0.1	0.2	0.3	0.4
SCC	mean	1	0	62	**61**	**60**	**61**	**43**
	std. dev.	0.0	0.0	13.0	**17.3**	**21.7**	**17.2**	**20.7**
IGD	mean	0.0673	0.0373	0.0091	0.0074	0.0089	0.0087	0.0101
	std. dev.	0.0000	0.0172	0.0058	0.0060	0.0060	0.0058	0.0058
FI11		sMOPSO	cMOPSO	oMOPSO	0.1	0.2	0.3	0.4
SCC	mean	1	0	62	54	62	54	50
	std. dev.	0.0	0.0	13.0	23.4	20.9	26.8	20.6
IGD	mean	0.0673	0.0373	0.0091	**0.0090**	**0.0064**	**0.0057**	**0.0058**
	std. dev.	0.0000	0.0172	0.0058	**0.0058**	**0.0052**	**0.0053**	**0.0052**
FA1		sMOPSO	cMOPSO	oMOPSO	0.1	0.2	0.3	0.4
SCC	mean	1	0	62	54	56	49	53
	std. dev.	0.0	0.0	13.0	10.7	20.0	21.6	21.5
IGD	mean	0.0673	0.0373	0.0091	0.0123	0.0089	0.0108	0.0082
	std. dev.	0.0000	0.0172	0.0058	0.0045	0.0062	0.0060	0.0068
FA3		sMOPSO	cMOPSO	oMOPSO	0.1	0.2	0.3	0.4
SCC	mean	1	0	62	62	51	54	53
	std. dev.	0.0	0.0	13.0	15.0	27.0	20.0	20
IGD	mean	0.0673	0.0373	0.0091	0.0109	0.0099	0.0100	0.0116
	std. dev.	0.0000	0.0172	0.0058	0.0056	0.0059	0.0061	0.0056

Table 12.10. Obtained results for test functions ZDT4 and DTLZ2, for oMOPSO with techniques FI2, FI5, FI11, FA1 and FA3 incorporated (p_i=0.1, 0.2, 0.3, 0.4)

ZDT4	Inheritance proportion p_i								
technique	0.0	0.1 (-10%)		0.2 (-20%)		0.3 (-30%)		0.4 (-40%)	
FI2	80	77	(-3.8%)	83	(+3.8%)	67	(-16.3%)	69	(-13.8%)
FI5	80	87	(+8.8%)	85	(+6.3%)	85	(+6.3%)	82	(+2.5%)
FI11	80	82	(+2.5%)	81	(+1.3%)	73	(-8.8%)	73	(-8.8%)
FA1	80	85	(+6.3%)	89	(+11.3%)	79	(-1.3%)	80	(0.0%)
FA3	80	89	**(+11.3%)**	91	**(+13.8%)**	86	**(+7.5%)**	87	**(+8.8%)**

DTLZ2	Inheritance proportion p_i								
technique	0.0	0.1 (-10%)		0.2 (-20%)		0.3 (-30%)		0.4 (-40%)	
FI2	18	15	(-16.7%)	18	(0.0%)	15	(-16.7%)	14	(-22.2%)
FI5	18	18	(0.0%)	12	(-33.3%)	13	(-27.8%)	12	(-33.3%)
FI11	18	17	**(-5.6%)**	21	**(+16.7%)**	23	**(+27.8%)**	23	**(+27.8%)**
FA1	18	18	**(0.0%)**	13	**(-27.8%)**	14	**(-22.2%)**	12	**(-33.3%)**
FA3	18	16	(-11.1%)	15	(-16.7%)	16	(-11.1%)	13	(-27.8%)

In general, fitness inheritance techniques seem to have a better performance in function DTLZ6 and fitness approximation techniques seem to have a better performance in function DTLZ4. These results agree with the results obtained before. As we can see in Table 12.10, the results obtained in the previous study show that technique FA3 was the best in function ZDT4. On the other hand, in function DTLZ2 technique FA1 is one of the two technique with the best results (the other is FI11). In fact, functions ZDT4 and DTLZ2 have the lowest number of variables. Function ZDT4 has 10 variables and function DTLZ2 has 12 variables, while functions ZDT1, ZDT2 and ZDT3 have 30 variables each, and DTLZ6 has 22 variables. In this way, we can conclude that, in general, fitness approximation techniques have better results when the test function has a low dimensional decision space and that fitness inheritance techniques have better results when the test function has a high dimensional decision space. This conclusion seems to agree with the results obtained in some of our previous work [13].

Finally, we should observe the results of technique FI11. In function DTLZ2, this technique obtained almost the best results. Also, in function DTLZ4, technique FI11 obtained the best average results. However, in the case of function DTLZ4, we could see that technique FI11 had the highest standard deviation values. In this way, technique FI11 is the only inheritance technique that had a very good performance in the test functions with a low number of variables, specifically from the test suite DTLZ. This denotes the importance of considering test functions from different test suites and with different characteristics. Technique FI11 almost always improved the results of the approach without inheritance even when a 40% of the number of evaluation was saved, in functions DTLZ2 and DTLZ4. Certainly, technique FI11 deserves further study, and this is already one of the priorities of our future work.

12.9 Conclusions

We proposed several fitness inheritance and approximation techniques and incorporated them into a Multi-Objective Particle Swarm Optimizer previously proposed by the authors. We studied the proposed techniques using several standard test functions taken from the multi-objective optimization literature.

From the nineteen techniques proposed, nine were found to be the best. Six of those nine techniques were inheritance techniques and the other three were approximation techniques. From the six best inheritance techniques, four techniques don't consider the previous position of a particle in order to compute the new objective position. On the other hand, the only approximation technique that didn't appear as one of the best was the one that only considers the set of leaders to assign the objective values of a particle. The results obtained indicate that the members of the swarm must be always considered in order to estimate the fitness value of a particle. Also, the importance of the *pbest* particle constitutes a topic that deserves further analysis, since it doesn't provide any useful information when the fitness value of a particle is being inherited.

Five of the nine best techniques found were selected to be tested on other functions and compared with respect to other PSO-based multi-objective algorithms. The obtained results show that the five enhancement techniques have a good performance and are very promising. In general, fitness inheritance techniques seem to be more appropriate for high-dimensional decision space problems and fitness approximation techniques seem more appropriate for low-dimensional decision space problems. Only one of the inheritance techniques had a very good performance when applied to functions with low number of variables, from a specified test suite. Such inheritance technique almost always improved the results of the approach without inheritance and it certainly deserves further study.

As part of our future work, we plan to improve the enhancement techniques that were found to be the best in this study, in order to minimize the decrement in quality of results while obtaining major savings in the number of evaluations performed.

Acknowledgments

The first author acknowledges CONACyT for granting her a scholarship to pursue graduate studies at the computer science section of CINVESTAV-IPN. The second author gratefully acknowledges support from CONACyT project number 45683.

References

1. Robert E. Smith, B. A. Dike, and S. A. Stegmann. Fitness Inheritance in Genetic Algorithms. In *SAC '95*, pages 345–350. ACM Press, 1995.
2. Yaochu Jin and Bernhard Sendhoff. Fitness Approximation in Evolutionary Computation: A Survey. In *GECCO 2002*, pages 1105–1112. Morgan Kaufmann, 2002.
3. Margarita Reyes Sierra and Carlos A. Coello Coello. Improving PSO-based Multi-objective Optimization using Crowding, Mutation and ϵ-Dominance. In *EMO 2005*, pages 505–519. Springer-Verlag, LNCS 3410, 2005.
4. Xiaowei Zheng, Bryant A. Julstrom, and Weidong Cheng. Design of Vector Quantization Codebooks Using a Genetic Algorithm. In *IEEE International Conference on Evolutionary Computation (ICEC'97)*, pages 525–529. IEEE Press, 1997.
5. Kumara Sastry, David E. Goldberg, and Martin Pelikan. Don't Evaluate, Inherit. In *GECCO 2001*, pages 551–558. Morgan Kaufmann, 7-11 2001.
6. Jian-Jung Chen, David E. Goldberg, Shinn-Ying Ho, and Kumara Sastry. Fitness Inheritance in Multi-Objective Optimization. In *GECCO'2002*, pages 319–326. Morgan Kaufmann Publishers, 2002.

7. Mehrdad Salami and Tim Hendtlass. A Fast Evaluation Estrategy for Evolutionary Algorithms. *Applied Soft Computing*, 2(3):156–173, 2003.
8. Els I. Ducheyne, Bernard De Baets, and Robert De Wulf. Is Fitness Inheritance Useful for Real-World Applications? In *EMO 2003*, pages 31–42. Springer. LNCS 2632, 2003.
9. Eckart Zitzler, Kalyanmoy Deb, and Lothar Thiele. Comparison of Multiobjective Evolutionary Algorithms: Empirical Results. *Evolutionary Computation*, 8(2):173–195, 2000.
10. Martin Pelikan and Kumara Sastry. Fitness Inheritance in the Bayesian Optimization Algorithm. Technical Report 2004009, Illinois Genetic Algorithms Laboratory, University of Illinois at Urbana-Champaign, Urbana, IL, 2004.
11. Lam T. Bui, Hussein A. Abbass, and Daryl Essam. Fitness Inheritance for Noisy Evolutionary Multi-Objective Optimization. In *Proceedings of the Genetic and Evolutionary Computation Conference (GECCO'2005)*, pages 25–29. ACM, 2005.
12. Kalyanmoy Deb, Amrit Pratap, Sameer Agarwal, and T. Meyarivan. A Fast and Elitist Multiobjective Genetic Algorithm: NSGA–II. *IEEE Transactions on Evolutionary Computation*, 6(2):182–197, April 2002.
13. Margarita Reyes Sierra and Carlos A. Coello Coello. Fitness Inheritance in Multi-Objective Particle Swarm Optimization. In *IEEE Swarm Intelligence Symposium*, pages 116–123, Pasadena, California, USA, 2005. IEEE Service Center.
14. Alain Ratle. Accelerating the Convergence of Evolutionary Algorithms by Fitness Landscape Approximation. In *Proceedings of the International Conference on Parallel Problem Solving from Nature (PPSN V), LNCS 3242*, pages 87–96. Springer-Verlag, 1998.
15. Yaochu Jin, Markus Olhofer, and Bernhard Sendhoff. On Evolutionary Optimization with Approximate Fitness Function. In *Proceedings of the Genetic and Evolutionary Computation Conference (GECCO'2000)*, pages 786–793. Morgan Kaufmann Publishers, 2000.
16. Yasuhito Sano and Hajime Kita. Optimization of Noisy Fitness Functions by Means of Genetic Algorithms Using History of Search. In *Proceedings of the International Conference on Parallel Problem Solving from Nature (PPSN VI)*, pages 571–580. Springer-Verlag, 2000.
17. Jürgen Branke anad Christian Schmidth and Hartmut Schmeck. Efficient Fitness Estimation in Noisy Environments. In *Proceedings of the Genetic and Evolutionary Computation Conference (GECCO'2001)*, pages 243–250. Morgan Kaufmann Publishers, 2001.
18. J. Branke and C. Schmidt. Faster convergence by means of fitness estimation. *Soft Computing*, 9(1):13–20, January 2005.
19. Yew S. Ong, Prasanth B. Nair, and Andrew J. Keane. Evolutionary optimization of computationally expensive problems via surrogate modeling. *AIAA Journal*, 41(4):687–696, April 2003.
20. Yew-Soon Ong, Zongzhao Zhu, and Dudy Lim. Curse and blessing of uncertainty in evolutionary algorithm using approximation. In *Proceedings of the 2006 Congress on Evolutionary Computation (CEC'2006)*, Vancouver, Canada, July 2006. IEEE Service Center.
21. A. Gaspar-Cunha, Armando S. Vieira, and Carlos M. Fonseca. Multi-Objective Optimization: Hybridization of an Evolutionary Algorithm with Artificial

Neural Networks for Fast Convergence. In *4th EU/ME Workshop on Design and Evaluation of Advanced Hybrid Meta-Heuristics*, page http://webhost.ua.ac.be/eume/welcome.htm?workshops/hybrid/index.php&1, Nottingham, UK, November 2004. European Association of OR Societies.
22. Rommel G. Regis and Christine A. Shoemaker. Local function approximation in evolutionary algorithms for the optimization of costly functions. *IEEE Transactions on Evolutionary Computation*, 8(5):490–505, October 2004.
23. Dudy Lim, Yew-Soon Ong, Yaochu Jin, and Bernhard Sendhoff. Trusted evolutionary algorithm. In *Proceedings of the 2006 Congress on Evolutionary Computation (CEC'2006)*, Vancouver, Canada, July 2006. IEEE Service Center.
24. Ivan Voutchkov and A. J. Keane. Multiobjective Optimization using Surrogates. In I. C. Parmee, editor, *Adaptive Computing in Design and Manufacture. Proceedings of the Seventh International Conference*, pages 167–175, Bristol, UK, April 2006. The Institute for People-centered Computation (IP-CC).
25. Joshua Knowles. ParEGO: A Hybrid Algorithm With On-Line Landscape Approximation for Expensive Multiobjective Optimization Problems. *IEEE Transactions on Evolutionary Computation*, 10(1):50–66, February 2006.
26. James Kennedy and Russell C. Eberhart. Particle Swarm Optimization. In *Proceedings of the 1995 IEEE International Conference on Neural Networks*, pages 1942–1948, Piscataway, New Jersey, 1995. IEEE Service Center.
27. Marco Laumanns, Lothar Thiele, Kalyanmoy Deb, and Eckart Zitzler. Combining Convergence and Diversity in Evolutionary Multi-objective Optimization. *Evolutionary Computation*, 10(3):263–282, 2002.
28. Kalyanmoy Deb, Lothar Thiele, Marco Laumanns, and Eckart Zitzler. Scalable Multi-Objective Optimization Test Problems. In *CEC 2002*, volume 1, pages 825–830, USA, 2002. IEEE Service Center.
29. David A. Van Veldhuizen. *Multiobjective Evolutionary Algorithms: Classifications, Analyses, and New Innovations*. PhD thesis, Department of Electrical and Computer Engineering. Graduate School of Engineering. Air Force Institute of Technology, Wright-Patterson AFB, Ohio, May 1999.
30. Sanaz Mostaghim and Jürgen Teich. Strategies for Finding Good Local Guides in Multi-objective Particle Swarm Optimization (MOPSO). In *IEEE Swarm Intelligence Symposium Proceedings*, pages 26–33, USA, 2003. IEEE Service Center.
31. Gregorio Toscano Pulido and Carlos A. Coello Coello. Using Clustering Techniques to Improve the Performance of a Particle Swarm Optimizer. In *GECCO 2004. Part I*, pages 225–237, USA, 2004. Springer-Verlag, LNCS 3102.
32. David A. Van Veldhuizen and Gary B. Lamont. Multiobjective Evolutionary Algorithm Research: A History and Analysis. Technical Report TR-98-03, Department of Electrical and Computer Engineering, Graduate School of Engineering, Air Force Institute of Technology, Wright-Patterson AFB, Ohio, 1998.
33. David A. Van Veldhuizen and Gary B. Lamont. On Measuring Multiobjective Evolutionary Algorithm Performance. In *2000 Congress on Evolutionary Computation*, volume 1, pages 204–211, Piscataway, New Jersey, July 2000. IEEE Service Center.

13

An Evolutionary Multi-objective Adaptive Meta-modeling Procedure Using Artificial Neural Networks

Kalyanmoy Deb and Pawan K.S. Nain

Kanpur Genetic Algorithms Laboratory (KanGAL), Department of Mechanical Engineering, Indian Institute of Technology Kanpur, Kanpur, PIN 208 016, India
{deb, pksnain}@iitk.ac.in

Summary. This paper explores the possibility of using approximate models in multi-objective optimization. A multi-objective genetic algorithm based optimizer, namely the elitist non-dominated sorting genetic algorithm or NSGA-II, is integrated with an artificial neural network (ANN) for this purpose. The proposed technique makes use of successive fitness landscape modeling for reducing the exact function evaluation calls while retaining the basic search capability of NSGA-II. To test the suggested procedure, several test problems and a couple of practical problems are considered. The simulation results show a considerable savings in exact function evaluations in achieving both tasks of converging close to the true Pareto-optimal frontier and maintaining a good diversity of solutions. Although a neural network is used in this study, other meta-modeling techniques, such as RSM or Kriging methods, can also be used with the proposed methodology.

13.1 Introduction

One of the main challenges ahead with applied optimization studies is the enormous computational time needed in evaluating real-world optimization problems. In order to reduce the overall computational time, researchers in the area of search and optimization look for efficient algorithms which demand only a few function evaluations to arrive at a near-optimal solution. Although successes in this direction have been achieved by using new and unorthodox techniques (such as evolutionary algorithms, tabu search, simulated annealing etc.) involving problem-specific operators, such techniques still demand a considerable amount of simulation time, particularly in solving computationally expensive problems. In such problems, the main difficulty arises in the large computational time required in evaluating a solution. This is because such problems either involve many decision variables or a computationally involved evaluation procedure, such as the use of finite element procedure or a network flow computation.

K. Deb and Pawan K.S. Nain: *An Evolutionary Multi-objective Adaptive Meta-modeling Procedure Using Artificial Neural Networks*, Studies in Computational Intelligence (SCI) **51**, 297–322 (2007)
www.springerlink.com

Although the use of a parallel computer is a remedy to these problems in reducing the overall computational time, in this paper, we suggest a fundamental algorithmic change to the usual optimization procedure which can be used either serially or parallely. Most search and optimization algorithms begin their search from one or more random guess solutions. Thus, the main task of a search algorithm in the initial few iterations is to provide a direction towards the optimal region in the search space. To achieve such a task, it may not be necessary to use an exact (or a very fine-grained) model of the optimization problem early on. An approximate model of the problem may be adequate to provide a reasonably good search direction. However, as the iterations progress, finer models can be used successively to converge closer to the true optimum of the actual problem. Although this idea of using an approximate model in the beginning of a search algorithm and refining the model with iterations is not new [10, 25, 31] and has been mainly tried in single-objective optimization problems, we suggest a generic procedure which can be tried to solve any arbitrary multi-objective optimization problem.

In the reminder of this paper, we describe the proposed coarse-to-fine grained meta-modeling procedure. Thereafter, we suggest an artificial neural network (ANN) based procedure, specifically to model an approximate version of the actual problem and show simulation results of the proposed technique applied to a number of test problems and to a couple of practical problems. Different variations of the ANN design and training procedures are compared. Simulation results show a large computational advantage of the proposed procedure, thereby suggesting the applicability of the proposed procedure in more complex real-world search and optimization problems.

13.2 Past Studies

Various groups have reported their studies on the use of approximation models in evolutionary algorithms. A complete survey on the use of fitness approximation in evolutionary algorithms is reported by Jin and Sendhoff [2]. These authors have broadly classified the approximation methods in three categories, namely, response surface methodology, Kriging models and the methodology of artificial neural networks. Other issues of concern are the efficient use of exact function evaluations (termed as model management) and the quality of model itself, which should improve with iterations. Jin et al. [19], while working with approximate models, have also demonstrated the controlled evolution in an evolution strategy. Here, the word control refers to the original fitness function evaluations. So if the entire generation is evaluated using the original fitness function, it is called as controlled generation. Similarly, if only few population members are evaluated using original fitness function, it is referred as controlled individuals. A framework which guarantee the correct convergence while reducing the computational cost is established. The idea of using less function evaluations in order to reach the optima using controlled individuals with the help of clustering and neural network is explored by Jin and Sendhoff [22]. The individuals near the center of the cluster is evaluated using expensive fitness function evaluation to create neural network ensemble which is used for fitness values of remaining individuals. The structure and parameters of the neural network ensemble are also optimized using a standard evolution strategy.

Branke and Schmidt [2] have used two estimation methods, namely, regression and interpolation, to achieve faster convergence to the optima. In their technique, at every generation, a fixed percentage of the population is evaluated with exact objective function. The fitness of the rest of the population is estimated. The individuals which are evaluated accurately are determined based on their estimated fitness and uncertainty. Savings in accurate function evaluations up to fifty percent are reported.

Farina [15] has also used radial basis neural network for objective function approximation. The algorithm has been successfully tested on test problem ZDT3 [4] which has a typical discrete Pareto-optimal front.

The attempt to reduce the number of function evaluations using fitness inheritance technique [1, 3] is also reported. Sastry et al. [1] have used inheritance combined with population sizing models and have reported a saving of 20% in function evaluations. In case of fixed population size, the study has reported a saving of 70% by employing a simple inheritance technique. Chen et al. [3] have extended the fitness inheritance concept for multi-objective optimization. The study has reported a 40% saving in terms of function evaluations for the case of fitness inheritance without fitness sharing, while in the case of fitness inheritance with fitness sharing, a saving of 25% is claimed.

Rasheed and Hirsh [26] have used informed-operators for speeding up the genetic algorithm procedure. They have modified the genetic operators (mutation and crossover) and have made them more knowledge-based (or informed) using reduced models. Instead of making a random choice of parents, they generate a number of candidates and rank them using inexpensive reduced models, and then take the best of the result. Naturally, it provides a speed-up in the genetic algorithm procedure. They have also successfully tested their method on a complex engineering design problem. In another study by Rasheed et al. [27], a comparison between two methods for using reduced models to speed up the search in genetic algorithm based engineering design optimization is presented. They have reported that the informed-operators approach is better than the genetic engineering approach. It is also found that least square approximation with any of the above two speed-up approach produces better results than neural network approximations.

Applications using surrogate models (or meta-models) which can be substituted for exact and costly evaluation tools is also discussed by Giannakoglu [16]. In this work, radial basis function neural networks are used for generating a surrogate model. The prediction capability of the surrogate model is dependent on the shape of the real response surface, availability of sufficient training data, number of hidden units and the activation function used.

Studies [16, 17] have also suggested a genetic algorithm which is based on inexact pre-evaluations (GA-IPE). This method starts with exact evaluations of all solutions in the first generation. Later, when a surrogate model is constructed, the entire population is evaluated approximately using the model. Only the best individuals identified based on the inexact evaluation are exactly evaluated. This process reduces the computational burden of the overall procedure. Hence, although the GA-IPE takes more generations, it has a low overall evaluation cost. It is successfully applied to solve a turbomachinery problem of airfoil design.

El-Beltagy et al. [11] have suggested a Kriging approach, in which local meta-models are employed instead of global meta-models. This is because of the computational cost associated with building a global meta-model. In another study, Nair et al. [24] have combined the function approximation concept with genetic algorithms for structural optimization applications. Investigators have tried to reduce the number of exact function evaluations while ensuring a convergence to the optima of the original problem. They have employed a dynamic optimization technique, wherein the fitness function changes over successive generations. They have controlled the generational delay before the approximation model is updated along with an adaptive selection operator. This technique is applied to solve a 10-bar truss design problem. They have also used a strategy by which the fitness function is changed during the run but the granularity of the optimization model is kept unchanged.

A reconstruction algorithm is proposed by Ratle [28] which uses a Kriging meta-model for fitness landscape approximation. The algorithm shows an overall reduction in the number of fitness calls. The algorithm is tested successfully for a two-dimensional problem and for a difficult 20-dimensional multi-modal optimization problem.

Recently, Emmerich et al. [12] have proposed a meta-model assisted evolution strategy. A local Kriging meta-model is built in which a fixed number of nearest neighbors are used. The method is tested on test problems and on an airfoil shape optimization problem. The results are quite encouraging for single objective optimization. In another study by Emmerich and Naujokes [13], the Kriging meta-models are used in multi-objective optimization. The local meta-models are used to decide the potential of a new population member, i.e. to decide whether it should be evaluated precisely or rejected. Three different rejection principles are also tested and compared with the original algorithm.

Although different approaches are suggested, not many studies are attempted to solve multi-objective optimization problems for finding a set of Pareto-optimal solutions. If the target is to find multiple solutions in the objective space, it is important to develop a meta-model which would capture the entire Pareto-optimal region. This is contrary to the focus to a narrow region needed in the case of a single-objective metal-modeling technique. This makes the multi-objective meta-modeling different, because certain portion of the Pareto-optimal front may be easier to approximate and certain portion of the trade-off frontier may be difficult to approximate. Before we discuss the issues related to meta-modeling in multi-objective optimization, we briefly describe different procedures which can be used to reduce the computational cost of evaluating a solution.

13.3 Procedures for Reducing Computational Cost

It is needless to say that the lion's share of the overall computational cost of applying a GA (or for that matter any optimization procedure) comes from the evaluation of solutions. In order to reduce the computational time required to execute one function evaluation, the following strategies can be used:

1. Use a partial evaluation of a solution,
2. Use a parallel computer, and
3. Use an approximation of the optimization problem

Certain search and optimization problems may be functionally decomposable into a number of subproblems. In such problems, the most important subproblems can only be evaluated in the initial GA generations. Although this procedure will introduce some error in evaluating a solution in early generations (since all subproblems are not evaluated), the computations can be performed quickly. In the early generations the task of an optimizer is to determine correct search directions towards the optimum, such errors may not cause a large deviation from the true search direction. However, as generations progress, less important subproblems can be included and more accurate function evaluations can be expected. In such a procedure, the computational advantage would come from the savings in the computational time in the early generations. However, it is obvious that the partial evaluation of a solution cannot be performed in problems where a functional decomposition is impossible.

Because of the availability of parallel computers, it may be plausible to take advantage of parallel computing of different tasks involved in a function evaluation. For example, to evaluate a solution involving FFT or finite element computations, the solution can be sent to multiple processors for a faster computation. Since GAs use a population of solutions in each generation, most parallel GA applications perform a distributed computing of allocating one complete solution to each available processor, thereby reducing the overall computational time to complete one generation. In such applications, although each solution can be evaluated with the help of multiple processors, usually this is not followed. Although faster solution procedures are developed using parallel migration and island models, most such studies have found a lower bound on the computational complexity achievable in terms of resorting to an optimum number of processors. Beyond the optimum number of processors, the computational advantage is overshadowed by the communication time involved among the processors.

The focus of this study is to use a successive approximation of the optimization problem, which we describe next.

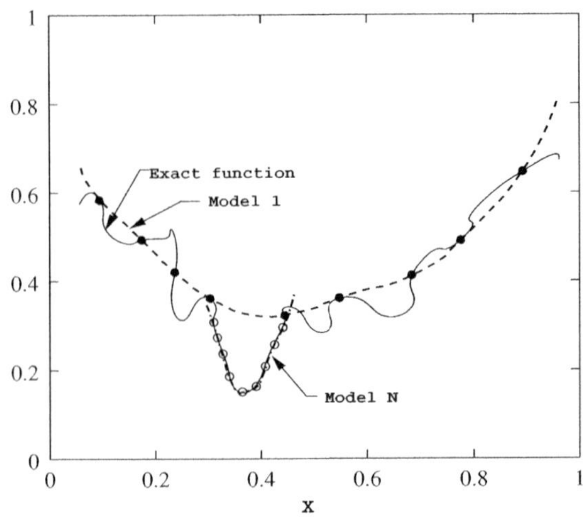

Fig. 13.1. Proposed successive meta-modeling technique

13.4 Proposed Meta-Modeling Procedure

Starting with a coarsely approximated model of the problem, GAs use successively fine-grained models with generations. Figure 13.1 depicts this procedure. The figure shows a hypothetical one-dimensional objective function for minimization in a solid line. Since this problem has a number of local minimum solutions (which is one of the difficulties often exist in a real-world optimization problem), it would be a difficult problem for any optimization technique. It is concluded elsewhere [18] that to find the global optimum in such a problem using a GA, a population of size $O(\gamma^2)$, where γ is the inverse of the signal-to-noise of the function, is needed. The signal being the difference between the global and the next-best local optimal values and the noise being equal to the variance of the function values. Thus, the objective function shown in the figure demands a large population size, if the GA has to start from an initial random population. If each function evaluation is expensive, the application of a GA in such problems becomes difficult. Although a remedy in such problems is to not use a random initial population, but to use problem information to create the initial population in the vicinity of the global optimum, in many problems such information is simply not available. However, there exist generic solutions to the problem which we discuss in the following paragraph.

Figure 13.1 also shows a coarsely approximated function in the entire range of the function with a dashed line. There could be a variety of ways such an approximated function can be obtained:

1. Linear or quadratic (or a response surface methodology) approximation of the true function
2. Approximation through a set of basis functions
3. Approximation through a chosen set of solutions

Classical methods often linearize non-linear optimization problems at suitable solutions and use a linear programming technique successively, such as the Frank-Wolfe method or successive linearization methods [29]. Besides, linearization techniques, non-linear problems can be approximated by quadratic or higher-order polynomial functions. Another way to approximate a function is to use a set of basis functions and find a weighted sum of such basis functions (finite or infinite numbers of them) as an approximation. Fourier approximation and Walsh function approximations are two such examples. Once such an approximation is known, the individual properties of the optimum of the basis functions may be analyzed to make a guess of the optimum of the approximating function. Although such techniques are used to decouple linkages among decision variables to convert the problem into a class of separable programming [29], such techniques are usually not used in optimization studies. However, such approximations can also be used nicely with the following approximation procedure.

The optimization problem can be evaluated exactly at a few finite number of pre-specified solutions in the entire range of the decision variables. Thereafter, an approximating function can be fitted through these function values using regression or some other sophisticated techniques such as an artificial neural network methodology. It is clear that if a large number of solutions are chosen, the approximating function will take a shape closer to the original function. On the other hand, if only a few solutions are chosen, the approximated function will ignore the local details and represent a coarse trend in variation of the function values. If this approximating function is optimized, it is likely that a GA will proceed in the right direction. However, as a GA tends to converge to the optimum of the approximating function, the approximating function needs to be modified to make a better approximated function from before. Since the population diversity will be reduced while approximating the first approximated function, the second approximating function need not be defined over the whole range of the decision variables, as shown in Figure 13.1. This process may continue till no further approximation results in an improvement in the function value. Although this successive approximation technique seems a reasonable plan, care must be taken to ensure that adequate diversity is retained while switching from one approximating function to a better one.

13.4.1 Specific ANN-Based Procedure

Figure 13.2 outlines a schematic of a plausible plan for the proposed procedure. The combined procedure begins with a set of randomly created N solutions, where N is the population size. Since an adequate size of solutions are required

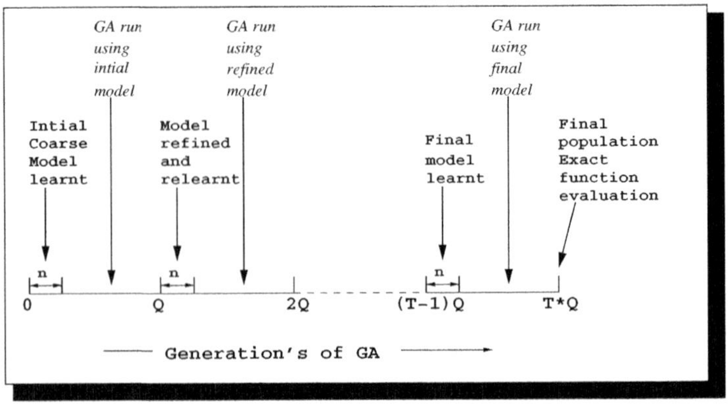

Fig. 13.2. A sketch of the proposed NSGA-II-ANN technique

to arrive at an approximated problem, we execute a GA with exact function evaluations for n generations, thereby collecting a total of $N' = nN$ solutions for approximation. At the end of n generations, the approximation technique is invoked with N' solutions and the first approximated problem is created. The GA is then performed for the next $(Q-n)$ generations with this approximated problem. Thereafter, the GA is performed with the exact function for the next n generations and a new approximated problem is created. This procedure is continued till the termination criterion is met. Thus, this procedure requires a fraction n/Q of total evaluations in evaluating the problem exactly. A better approach will be to decrease N' solutions towards the later generations of GA run as during this phase approximate model will not change significantly.

If a problem cannot be evaluated exactly, instead some approximations (such as involving FFT or finite element techniques) are needed to evaluate, the parameter n is set to zero and GAs are run for Q generations with the most coarse model (in the case of a finite element method only a few elements can be chosen to start with). Thereafter, the model is modified and GAs are run for another Q generations with the modified (and hopefully a better) model. This procedure will continue till a termination criterion is met. It is interesting to note that a set of basis functions with varying importance to local variations can be used as approximating functions here. In such cases, a GA may be started with an approximating function involving only a few basis functions, thereby allowing a quicker computation of the optimal solution. With generations, more and more basis functions can be added to make the model more similar to the actual optimization problem. In this study we concentrate on solving problems for which an exact evaluation method exists but is computationally expensive. However, similar methodology can also be used to solve problem for which no exact evaluation method exists.

We propose combining a GA with the artificial neural networks (ANN) as the basic approximating technique for fitness computation. The primary reason for using ANN as the basic approximating tool is its proven capabilities as function approximation tool from a given data set. The multilayer perceptron trained with the back-propagation algorithm may be viewed as a practical vehicle for performing a non-linear input output mapping of general nature [19]. It is important to note that first function evaluation should always be performed using the exact problem, as we do not have any database to train the neural network. Secondly, whenever we are performing exact function evaluations, we have to build or update the training database for training or retraining of the ANN. The frequency of training is exclusively controlled by the user. The parameters Q and n would be so chosen that even if a previous approximation function does not adequately approximate the true problem, subsequent approximations have a chance to correct the population and push it in the right direction.

Another advantage of the proposed technique is its adaptability. Initially the GA population will be randomly spread in the entire search space for the problem undertaken. Since the same information is available in the ANN training database the approximated model generated using this database, will also be very general in nature and hence may miss some finer details about the search space. However as the generations proceed, the GA population will start drifting and focusing on the important regions which it identifies based on the current model. So when the proposed technique updates its model using exact function evaluation for the current generation, it will have more information about the local regions of interest as more population members will now be available in those regions than earlier. The ANN will now retrain and update its weights making it learn and adapt to the new smaller regions in the search space. Hence it will give finer refined approximated model to direct the search of GA in the subsequent generations. Thus the proposed technique will adaptively refine the approximated model from coarse to fine approximated model of the optimization problem.

The reasons for choosing ANN for function approximation are (i) its ability to learn a multi-variable functional relationship on arbitrary functions, and (ii) its fast evaluation of a solution with the trained ANN. It is worth mentioning here that the computational time taken to train an ANN must be added to the overall computational time taken to solve the problem. However, using an incremental learning ANN, the ANN training phase and n generations of a GA run with exact evaluations can be continued parallely by using a dual processor machine. However, it is assumed in this study that the computational time for each function evaluation is so large that the time taken for ANN training is comparatively small. It is also worth mentioning that the proposed technique is equally applicable to single and multi-objective optimization and to constrained optimization, as an additional objective or constraint in the optimization problem is equivalent to addition of one more neuron in the output layer of ANN.

13.5 Proof-of-Principle Results

In all case studies performed here, we have used the NSGA-II procedure [5] as the basic multi-objective optimizer. The standard error back-propagation algorithm with sigmoidal activation function [19] is used for generating approximate model of the problem successively. For the ANN, input neurons represent problem variables and output neurons represent different objective functions and constraint functions. The combined procedure will be referred as NSGA-II-ANN simulation. The ANN is trained using two different models, namely, batch training and incremental training. If a simulation is performed with batch training in which after every Q generation the approximation model is refined and the training database is collected over n generations, It is called as the B-Q-n model. If the training method is incremental, then it is called as the I-Q-n model.

Here, we choose a set of available two and three-objective test problems, namely TNK, ZDT4, DTLZ2 [4, 6]. These problems are selected as they test different aspects of the optimizer, which will be highlighted in the subsequent sections. The learning rate of 0.3, a momentum factor of 0.1, permissible rms error of 0.001, SBX crossover probability of 0.9 and distribution index of 10 and the polynomial mutation (probability equal to the inverse of number of variables) with a distribution index of 50 are used for all test problems presented here.

13.5.1 Test Problem TNK

It is a two variable constrained test problem. The problem variables are real valued. The Pareto-optimal front is disconnected with three parts and falls on the constraint boundary. Since it is a difficult problem, 200 population members and a maximum of 2,500 generations are chosen for all simulations including the meta-modeled NSGA-II-ANN (Figures 13.3 and 13.4). The number of exact function evaluations taken by simulations of various models in order to reach the Pareto-optimal front are given in Table 13.1. Figure 13.3 shows that B-10-2 and B-10-3 NSGA-II-ANN simulations reach the Pareto-optimal front with a saving of about 50% of exact function evaluations. Figure 13.4 shows that I-10-2 NSGA-II-ANN simulations reach the Pareto-optimal front without any saving in exact function evaluations.

The spread metric calculation method is as suggested in Deb et al. [4]. In this metric, the Euclidean distance between two extreme ends of Pareto-optimal front as well as the uniformity of distribution for intermediate solutions is considered. A smaller value of the spread metric means a better spread. Batch model simulations show better spread results as well as savings in exact function evaluation. If we concentrate on the spread metric value, B-10-2 simulation is the best and is closely followed by the B-10-3 simulation. However, incremental mode simulations have difficulty in capturing the

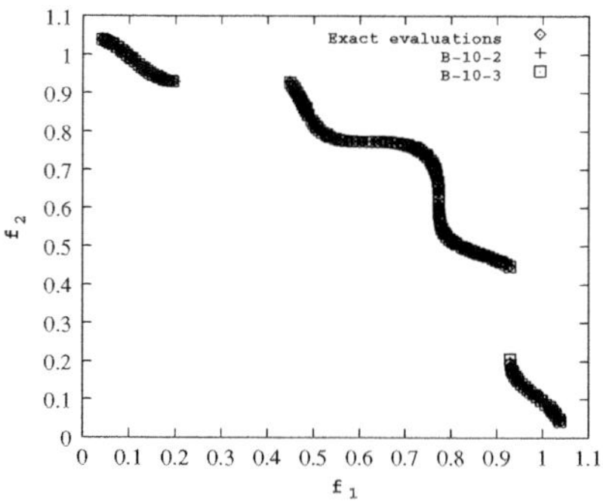

Fig. 13.3. Batch model simulation results for the TNK test problem

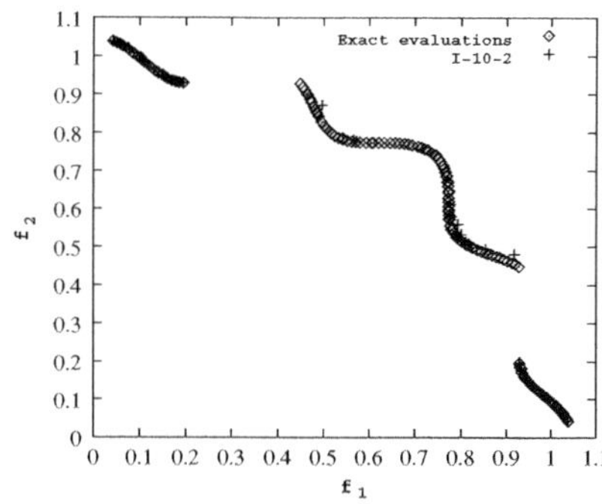

Fig. 13.4. Incremental model simulation results for the TNK test problem

Table 13.1. Spread metric results of NSGA-II-ANN simulations on the TNK test problem

Model Name	Hidden Neurons	Part-I	Part-II	Part-III	Exact Function Evaluations
NSGA-II	NA	0.644	0.872	0.704	5,00,400
B-10-2	9	0.597	0.810	0.613	2,50,183
B-10-3	11	0.588	0.833	0.724	2,50,256
I-10-2	9	0.540	1.047	0.507	4,99,933

middle portion of the Pareto-optimal front which is also evident from the worse spread metric value obtained for the *I*-10-2 simulation.

13.5.2 Test Problem ZDT4

It is a ten-variable, real valued problem having a convex global Pareto-optimal front. This is a difficult test problem as it exhibits 100 distinct local Pareto-optimal fronts, out of which only one is global. Figures 13.5 and 13.6 show that the NSGA-II-ANN in batch and incremental mode, respectively, are able to reach the global Pareto-optimal front in all simulations with a population size of 100 run for 300 generations. However, due to presence of local sub-optimal Pareto-optimal fronts, the saving in exact function evaluations is somewhat smaller. The number of exact function evaluations taken by simulations of various models in order to reach the Pareto-optimal front are given in Table 13.2. The saving in both batch models namely, *B*-10-2 and *B*-10-3 models are approximately 25%.

The best spread metric value is achieved by NSGA-II-ANN simulation in batch mode of training, namely, for *B*-10-3 model. The incremental models, namely, *I*-10-2 and *I*-10-3 also show better spread metric values. Only one

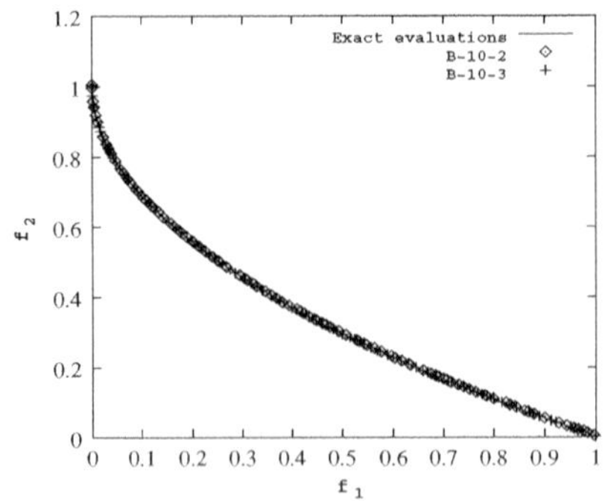

Fig. 13.5. Batch model simulation results for the ZDT4 test problem

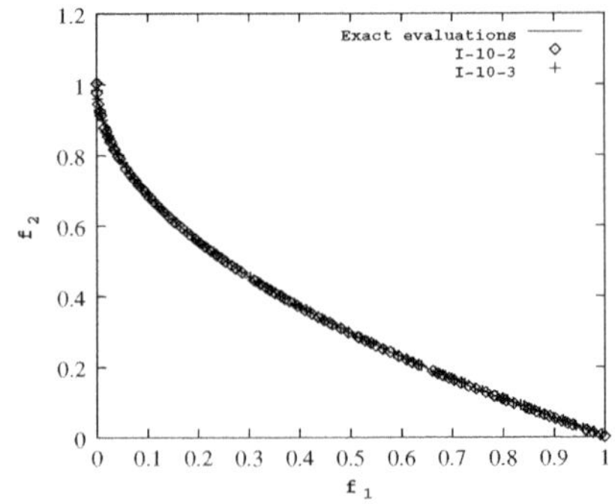

Fig. 13.6. Incremental model simulation results for the ZDT4 test problem

Table 13.2. Spread metric results of NSGA-II-ANN simulations on the ZDT4 test problem

Model Name	Hidden Neurons	Spread metric value	Exact Function Evaluations
NSGA-II	NA	0.386	30,200
B-10-2	17	0.422	22,675
B-10-3	15	0.332	22,730
I-10-2	13	0.344	22,675
I-10-3	17	0.388	21,781

simulation, namely *B*-10-2 gives a poor spread when compared to the NSGA-II simulation performed with exact function evaluations.

13.5.3 Test Problem DTLZ2

This 12-variable test problem has three objectives with a concave Pareto-optimal front. It is a problem where most of the multi-objective optimizers find difficulty in reaching the Pareto surface and maintaining a uniform distribution of the solutions on the Pareto-optimal surface. Figures 13.7 and 13.8 show the Pareto-optimal fronts for batch and incremental NSGA-II-ANN with 100 population members run for 300 generations. The number of exact function evaluations taken by both simulations are tabulated in Table 13.3. The batch model *B*-10-3 shows a saving of 62% in exact function evaluations. The incremental model *I*-10-3 provides a saving of 44% in exact function evaluations. The distribution of the solutions on the Pareto-optimal surface is evaluated by using a sparsity measure [7]. It is a method to quantify the distribution of solutions similar to an entropy measure or the grid diversity measure. A higher value of the sparsity measure means a better diversity among the solutions. Table 13.3 shows the sparsity measure for NSGA-II-ANN simulation.

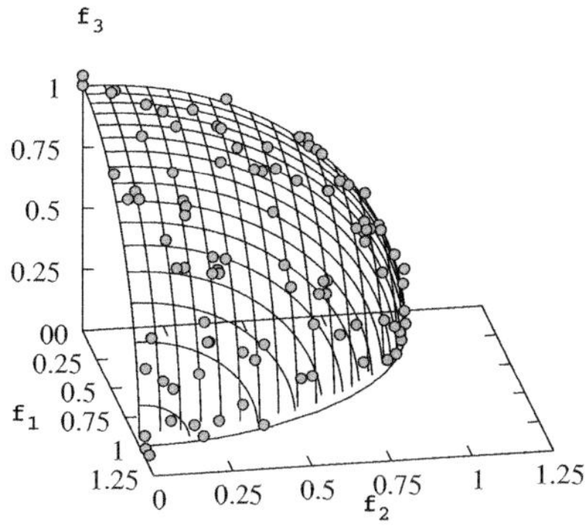

Fig. 13.7. Batch model simulation results for the DTLZ2 test problem

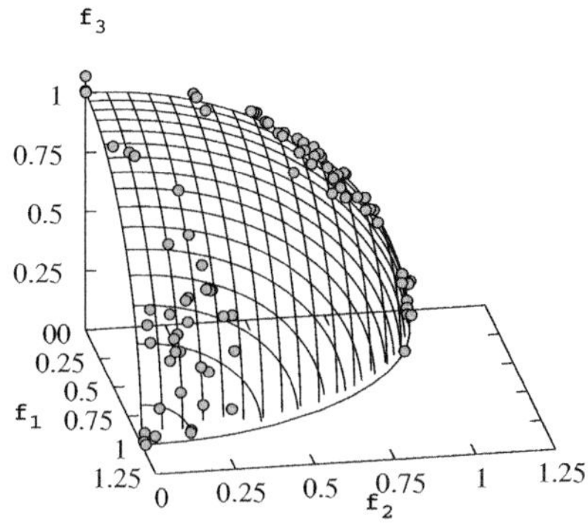

Fig. 13.8. Incremental model simulation results for the DTLZ2 test problem

Table 13.3. Sparsity measure results of NSGA-II-ANN simulations on the DTLZ2 test problem

Model Name	Hidden Neurons	Sparsity measure value	Exact Function Evaluations
NSGA-II	NA	0.951	30,200
B-10-3	17	0.964	11,496
I-10-3	7	0.908	17,120

The *B*-10-3 batch model shows the best diversity among Pareto-optimal solutions followed by NSGA-II run with only exact function evaluations. *I*-10-3 incremental model depicts the poorest distribution of solutions in this problem.

13.6 Case Study 1: A Curve Fitting Problem

The proposed technique is now applied to a B-spline curve fitting problem. A saw-tooth function with two teeth is taken as the basic curve to be fitted using B-splines (Figure 13.9). The tooth root height of 0.1, tooth peak height of 0.6, and tooth span of 0.5 are chosen here.

A B-spline with the parameter $k = 3$ produces a polynomial curve of degree two, with C^1 continuity for all curve segments and guarantees to pass through the starting and end control points and make tangents at the corresponding line segments.

The following two conflicting objectives are considered:

1. Minimize the error between the saw-tooth curve and the B-spline fitted curve, and

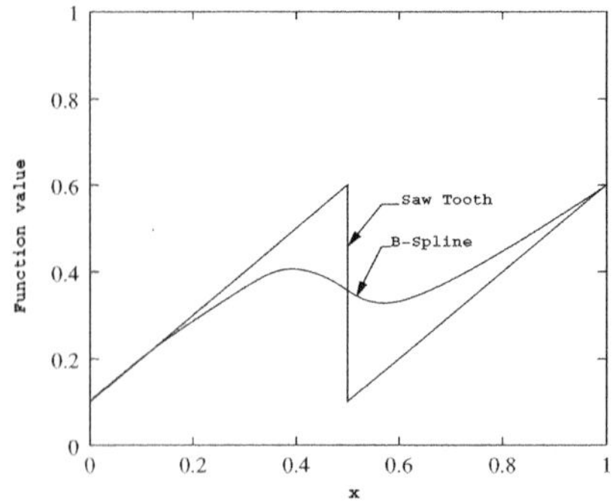

Fig. 13.9. The B-spline curve fitting problem

2. Minimize the maximum curvature of the B-spline fitted curve.

In the current problem, the number of control points are taken to be 41, thus dividing total x range in 40 equal divisions. However, in order to create meaningful solutions, the first and the last control points for the B-spline is fixed at tooth root height and tooth peak height, respectively, thereby leaving only 39 control points to be treated as decision variables. For any given B-spline curve $\mathcal{S}$, the exact evaluation of the first objective can be achieved in the following manner:

$$F_1(\mathcal{S}) = \int_{x=0}^{x=1} |f_{\text{saw-tooth}} - \mathcal{S}|dx. \tag{13.1}$$

Since such a computation is difficult to achieve exactly (mainly because of the non-differential modulus function used in the operand), we compute the above integral numerically by using the Trapezoidal rule. We have used 400 divisions in the entire range of x for the computation of two objectives. The second objective can be written as follows:

$$F_2(\mathcal{S}) = \max_{x=0}^{x=1} \frac{\frac{d^2\mathcal{S}}{dx^2}}{\left[1 + \left(\frac{d\mathcal{S}}{dx}\right)^2\right]^{3/2}}. \tag{13.2}$$

Since the B-spline curve $\mathcal{S}$ is defined piece-wise, the term for the curvature can be derived exactly for each segment. The term can then be optimized exactly using the first and second-order optimization criteria and the following location of the optimum is found in each B-spline segment:

$$u^* = 0, \text{ if } u_c \leq 0; \quad u^* = 1, \text{ if } u_c \geq 1; \quad u^* = u_c, \text{ otherwise}, \tag{13.3}$$

where the parameter u_c is calculated as follows:

$$u_c = \frac{x_{uu}(x_0 - x_1) + y_{uu}(y_0 - y_1)}{x_{uu}^2 + y_{uu}^2}, \quad x_{uu} = x_0 - 2x_1 + x_2, \quad y_{uu} = y_0 - 2y_1 + y_2.$$

Here, (x_0, y_0), (x_1, y_1) and (x_2, y_2) are three control points of each segment. Once the optimal u^* is calculated, the corresponding curvature can be calculated as follows:

$$R = \frac{x_u y_{uu} - x_{uu} y_u}{(x_u^2 + y_u^2)^{3/2}}, \tag{13.4}$$

where the first derivatives x_u and y_u are calculated as follows:

$$x_u = (u^* - 1)x_0 + (1 - 2u^*)x_1 + u^* x_2,$$
$$y_u = (u^* - 1)y_0 + (1 - 2u^*)y_1 + u^* y_2.$$

Such computations can be performed for all segments and the maximum curvature of the entire B-spline curve can be determined. For a large number of B-spline segments, many such computations are required, thereby involving a large computation time to evaluate the second objective. If such computations are extended to 3-D curve or surface fitting, the computations become even more expensive.

Here, we use a population of size 200, a crossover probability of 0.9 with a distribution index of 10, and the polynomial mutation with a distribution index of 50 [4]. The database size reduces linearly to 25% of its initial size in a span of 50% middle generations of EA. For initial and last 25% generations of EA, database size is kept constant. The permissible normalized rms error for ANN model is taken 0.005.

An extensive parametric study is performed by varying four parameters, (i) ANN learning rate, (ii) number of hidden neurons, (iii) model updating frequency parameter Q and (iv) the database size parameter n. In all cases, the number of overall exact function evaluations is kept the same to 750×200 or 1,50,000. All simulations are compared with an NSGA-II simulation performed with exact evaluations (200 population size and 750 generations). The trade-off frontiers corresponding to the best performing batch and incremental learning NSGA-II-ANNs are shown against the exact NSGA-II performance in Figure 13.10. In all simulations with batch and incremental learning models with different parameter settings, the obtained non-dominated front is found to be better than that obtained using exactly-evaluated NSGA-II. The best result for incremental training is found with the I-20-2 model, while in the case of batch training slightly better results are found with the B-10-3 model. This demonstrates that although approximate models are used, the combined NSGA-II-ANN procedure proposed here is able to find a better non-dominated front than the exact model. In order to investigate how many generations it would take by the NSGA-II with exact evaluations to obtain a front similar to that obtained using the proposed approaches, we have continued the NSGA-II run with exact evaluations for more than 750 generations. Figure 13.11 shows that the B-10-3 model reaches a similar front in about 1,100 generations, thereby making a saving of around 32% by the meta-modeled NSGA-II.

In order to visualize the B-spline curves obtained using the NSGA-II-ANN simulations, the two extreme solutions and a *knee* solution from the final non-dominated front of B-10-3 model are drawn in Figure 13.12. The original

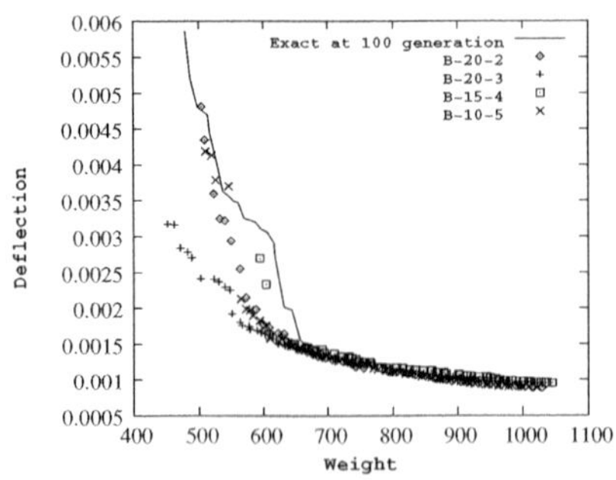

Fig. 13.10. Best of both incremental and batch models simulation results for curve fitting problem

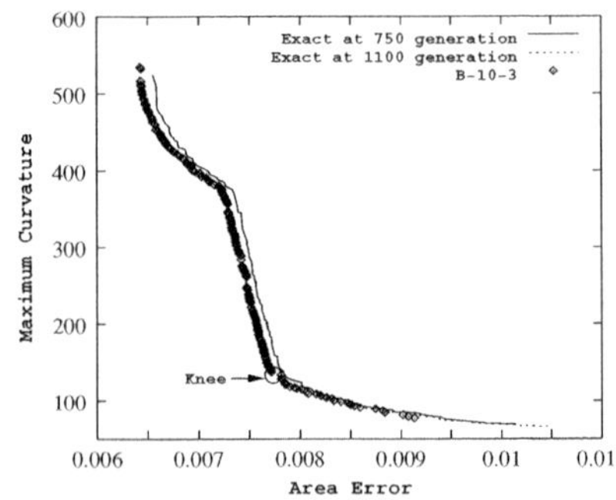

Fig. 13.11. Comparison of the overall best simulation result with exact solution at 1, 100 generation for curve fitting problem

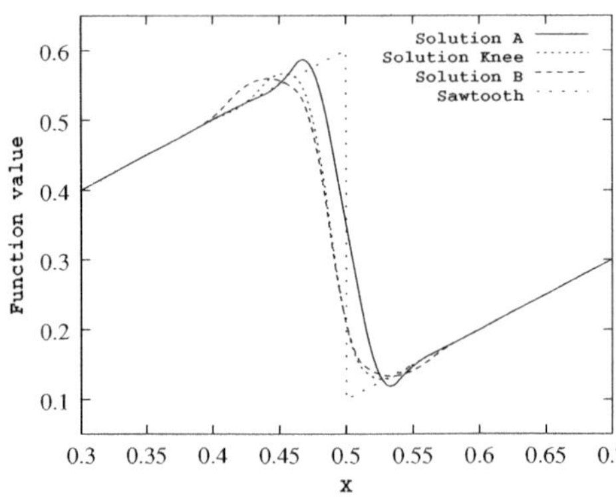

Fig. 13.12. Two extreme solutions and a knee solution for the curve fitting problem

saw-tooth function is also shown in dots. For clearly showing the region near $x = 0.5$, we plot them for $x \in [0.3, 0.7]$. The figure shows that one solution (marked as A) is a better fit with the saw-tooth function, whereas the other solution (marked as B) not a good fit in the vicinity of $x = 0.5$, but produces a smaller curvature. The knee solution is an intermediate solution from which if we move in either direction for a small improvement in one objective, a large sacrifice in other objective is to be made [1].

13.6.1 Effect of Permissible RMS Error

The issue of choosing a proper permissible normalized rms error for ANN also plays an important role in the proposed NSGA-II-ANN procedure. A large value of permissible normalized rms error may not adequately approximate the true problem as it will propagate a large error between the exact and the obtained model. Similarly, a very low value of permissible normalized rms error will lead to over-approximation of the true problem, thereby causing ANN to lose its generalization capability. This discussion indicates that there should be a critical value of permissible normalized rms error at which the proposed NSGA-II-ANN procedure should produce a better performance.

Various values of permissible normalized rms error are also tried with the B-10-3 model in the range $[0.001, 0.010]$. It is found that a permissible normalized rms error value less then 0.003 leads to over-approximation of the true problem. Hence NSGA-II-ANN procedure fails to converge close to the true Pareto-optimal front. Table 13.4 shows that permissible normalized rms error values above 0.003 and up to 0.004 are able to find a non-dominated front which is better converged than that obtained using the exact function evaluations (for 750 generations), indicating a saving of exact evaluations. At the permissible normalized rms error value of 0.005, the performance of proposed NSGA-II-ANN procedure is the best. However, a further increase in the permissible normalized rms error value above 0.006 shows poor convergence, with no savings in the exact evaluations. The obtained non-dominated front is inferior to that obtained with exact evaluations for 750 generations. The

Table 13.4. Convergence metric for different modeling error in curve fitting problem

Sl. No.	Model Name	Permissible Normalized RMS Error	Convergence Metric	Normalized Convergence Metric
1	B-10-3	0.003	0.012525	0.3779
2	B-10-3	0.004	0.012525	0.3779
3	B-10-3	0.005	0.000779	0.0235
4	B-10-3	0.006	0.033142	1.0000
5	B-10-3	0.007	0.033142	1.0000
6	B-10-3	0.008	0.033142	1.0000
7	B-10-3	0.009	0.033142	1.0000
8	B-10-3	0.010	0.033142	1.0000
9	I-20-2	0.005	0.003458	0.1043
10	$Exact$-750	N.A.	0.019351	0.5839
11	$Exact$-1100	N.A.	0.003795	0.1145

convergence metric computes the average distance of each obtained solution from a reference set of points. As the current problem is a practical problem, the true Pareto-optimal front is not known. As suggested in [8], we choose a reference set P^* containing 274 data points obtained from a combined pool of 11 simulations (shown in Table 13.4). For calculating the convergence metric value, first the non-dominated set F of the final generation of each simulation is identified. Then for each point in F, smallest normalized Euclidean distance to P^* is calculated. Next the convergence metric value is calculated by averaging the normalized distance of all points in the F. Lastly, in order to keep the convergence metric within $[0, 1]$, we divide the convergence metric value by the maximum value found among all simulations.

13.6.2 Comparing Both Convergence and Spread

Next, we make a detailed study comparing different batch and incremental learning NSGA-II-ANNs with exactly evaluated NSGA-II based on two performance measures: (i) convergence and (ii) diversity preservation. Since the true Pareto-optimal front is not known in this problem, we combined the solutions found in all simulation runs and treat the resulting non-dominated front as the target frontier.

Figure 13.13 shows the normalized convergence metric values of the different simulations. However it should be noted that maximum value of conver-

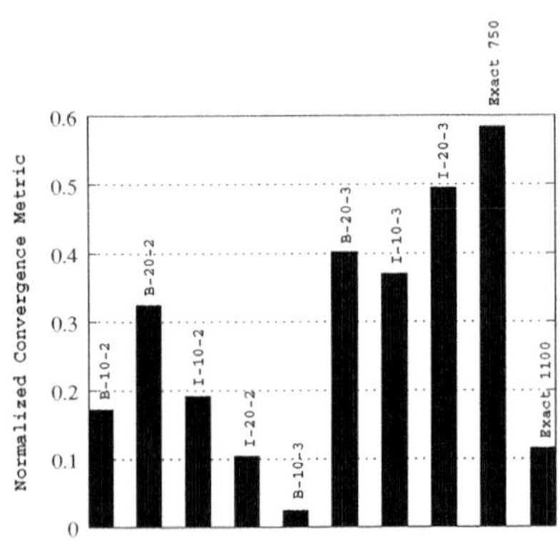

Fig. 13.13. Normalized convergence metric for different simulations for curve fitting problem

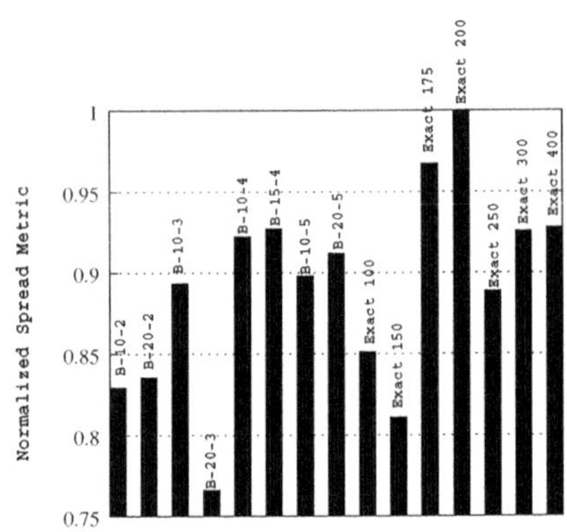

Fig. 13.14. Normalized spread metric for different simulations for curve fitting problem

gence metric obtained for the curve fitting problem is 0.033142 (as shown in Table 13.4). Hence the maximum ordinate in the Figure 13.13 is not unity. But still this figure clearly communicates the comprehensive picture of the convergence obtained in various simulations. It is observed from the figure that for the batch model running with two generational database size ($2N$), B-10-2 model performed better than B-20-2 model. In the incremental model with two generational database, I-20-2 model performed better than I-10-2 model. In case of three generation ($3N$) database size with batch model (results not shown here), B-10-3 model has also performed better than B-20-3 model. In case of incremental model with three generational database size, I-10-3 model has also performed better than I-20-3 model. The best incremental model is found to be I-20-2 and the best batch model is B-10-3. Both of these models performed better than NSGA-II simulation run with exact function evaluations for 1, 100 generations. The overall best performance is obtained in the B-10-3 batch model. So all simulations working with approximate function evaluations performed better than the NSGA-II simulation working with exact function evaluations for 750 generations and hence show an encouraging trend for the use of the technique presented here.

Next, the distributing ability of these simulations is examined. The spread metric is calculated by identifying two extremes of the reference set P^*. The simulation for which the spread metric is to be calculated, is taken and the best non-dominated front F is identified. Next the two extremes of this best non-dominated front F are identified. In the spread metric calculation two aspects are covered:

1. The closeness of two extremes of the reference set P^* to the corresponding extremes of best non-dominated front of simulation F, and
2. The uniformity of distribution of intermediate non-dominated points of the simulation F.

The smaller is the spread metric value, better is the distribution of the simulation under investigation. In order to visualize a comprehensive picture of the spread metric value for all the ten simulations, once again the spread metric values are normalized in the range of $[0, 1]$ and is plotted in Figure 13.14. The worst distribution is obtained for the benchmark simulation labeled Exact-750. All NSGA-II-ANN simulations performed better than this particular simulation. But the best simulation with respect to spread metric is Exact-1100 simulation with normalized spread metric value of 0.821. The worst batch model NSGA-II-ANN simulation is B-20-3 with 194 non-dominated points and normalized convergence metric value of 0.933. The best batch model NSGA-II-ANN simulation is B-10-3 with 187 non-dominated points and normalized convergence metric value of 0.864. Even though the number of non-dominated points in case of B-10-3 simulation are less than that with B-20-3 simulation, the spread is better for B-10-3 simulation. The worst incremental simulation is I-10-3 with 193 non-dominated points and normalized convergence metric value of 0.988. The best incremental simulation is I-20-3 with 193 non-dominated points and normalized convergence metric value of 0.837. So it can be concluded that in view of the spread measure, better performance of meta-medeled NSGA-II-ANNs is observed.

13.6.3 Effect of Initial Population and Attainment Curve

Next, in order to study the effect of initial population, we perform 21 different simulations of the best-performing batch model (B-10-3) with different initial populations. Independent non-dominated fronts are combined and 0%, 50% and 100% attainment curves [14] are plotted in Figure 13.15. The 0% and 100% attainment curves divide the objective space into three main regions. The first region located to the left of the 0% attainment curve (for minimizing

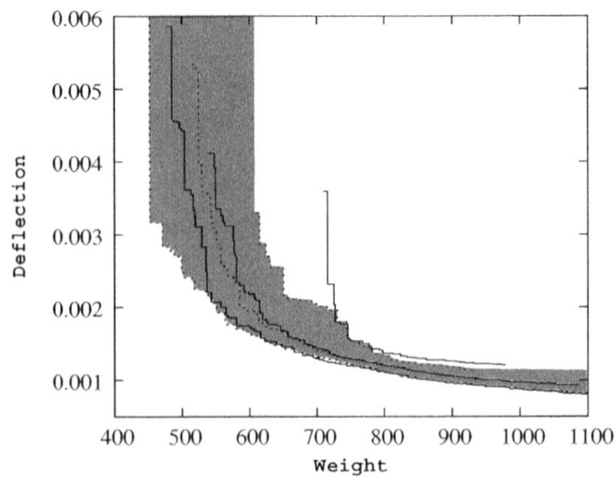

Fig. 13.15. Attainment curve plot for exact and approximate runs for the curve fitting problem

both objectives) is composed of the solution vectors which were never attained in any of the run. This region gives the idea about the best performance of the algorithm. The second region located to the right of the 100% attainment curve represents the solution vectors which were attained in all the simulation runs. This region gives the idea of the worst-case performance of the algorithm. The third region located within the 0% and 100% attainment curves is composed of solutions which are attained in some of the simulation runs but not in all simulations. This region can further be subdivided into sub-areas according to percentage of simulation runs in which the corresponding objective vectors are attained. A 50% attainment curve indicates the curve which is attained in half of the simulations.

It can be observed from the figure that the best performance (0% attainment curve) of NSGA-II simulations, working with exact evaluations for 750 generations, is better than the NSGA-II-ANN B-10-3 model. The 50% attainment curve for the NSGA-II working with exact function evaluation and is almost similar to that for the NSGA-II-ANN simulations working with B-10-3 model, suggesting that the median performance of the two approaches are more or less identical. The 100% attainment curves show that worst-case performance of NSGA-II-ANN approach is better than that of the exactly-evaluated NSGA-II. It is also observed from the figure that variation in the performance of NSGA-II with exact evaluations due to a change in initial population is slightly more than that in the batch learning NSGA-II-ANN.

13.7 Case Study 2: A Cantilever Plate Design for Optimal Shape

The design task is to find optimal shapes for two conflicting objectives: (i) minimize weight of the plate and (ii) minimize maximum deflection anywhere on the plate due to application of the load. A functional constraint restricts the maximum developed stress to be less than the specified yield strength of the plate material. The plate, loading, and initial meshes (of size 12×20) used in the study are shown in Figure 13.16. A load of 100 kN is applied at the free end of the cantilever. The plate material is taken to be steel. A binary variable represents presence or absence of the material at a grid element. Thus, a binary string of size 240 represents a shape. The continuity of materials from support to load is enforced and a finite element procedure is employed for exact function evaluations, thereby requiring a large computational overhead.

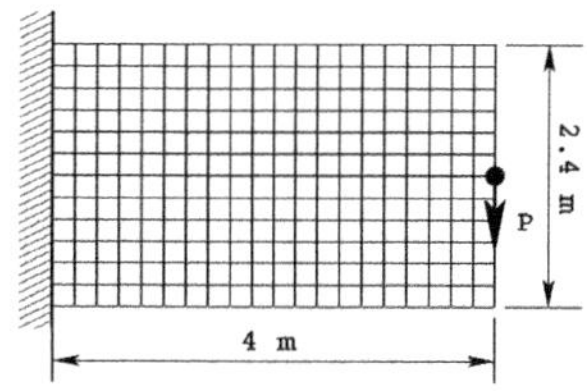

Fig. 13.16. A cantilever plate

Since in the previous examples, the batch learning NSGA-II-ANN performed the best, we apply this method only to this shape optimization problem. Several batch models are applied and the best performing results are shown in Figure 13.17. For a comparison, the trade-off frontier corresponding to the exactly-evaluated NSGA-II is also shown. In all simulations, 5,400

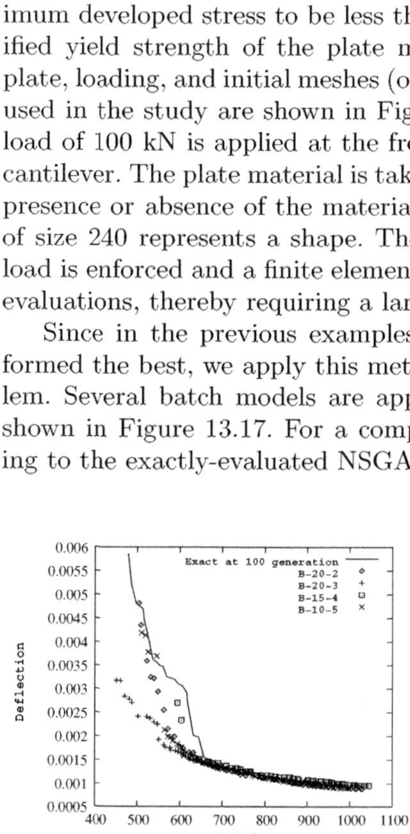

Fig. 13.17. Best of different batch models for the shape optimization problem

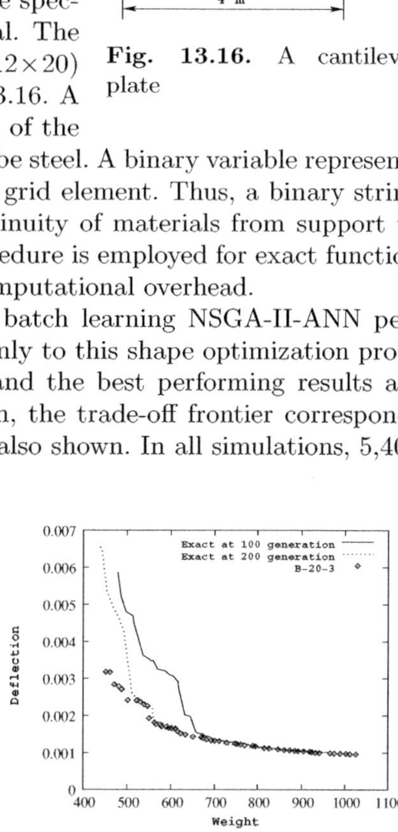

Fig. 13.18. Comparison of best batch simulation with exact NSGA-II runs for the shape optimization problem

exact function evaluations are allowed. It can be seen that the *B*-20-3 NSGA-II-ANN frontier fully dominates that obtained using exactly-evaluated NSGA-II simulation (Figure 13.18). Interestingly, even when the exactly-evaluated NSGA-II is continued to run for 200 generations (totaling 10,800 exact function evaluations), the obtained frontier is not as good as that obtained with the meta-modeled NSGA-II using 50% evaluations. Hence, the *B*-20-3 model saves at least 50% function evaluations.

13.7.1 Convergence and Spread of Solutions

To quantify the level of convergence achieved in each simulation, we compute a convergence and spread measures used earlier. Like before, we choose a reference set P^* containing 94 data points obtained from a combined pool of 15 simulations. Figure 13.19 shows the normalized convergence metric value calculated for various simulations by the NSGA-II-ANN procedure and the NSGA-II working with exact function evaluations (labeled as exact). This figure shows that convergence metric value for all the NSGA-II-ANN simulations is either near or better than NSGA-II run with exact function evaluations for 200 generations. The convergence metric value of the best simulation, namely B-20-3, is somewhere in between 250 generation and 300 generation. Hence, a saving larger than 50% is achieved. Next, we compute the spread measure

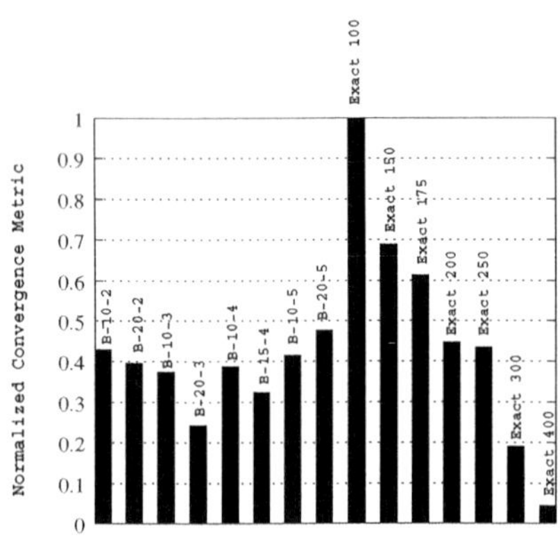

Fig. 13.19. Normalized convergence metric results for different simulations for shape optimization problem

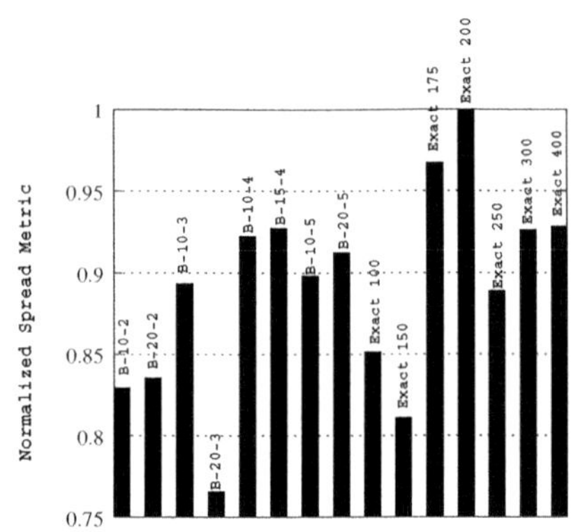

Fig. 13.20. Normalized spread metric results for different simulations for shape optimization problem

using the same reference set P^*. Figure 13.20 shows that the B-20-3 model finds the best spread.

13.7.2 Effect of Initial Population and Attainment Curves

We now study the effect of initial population on the best-performing B-20-3 meta-model. We use 5 hidden neurons and a learning rate of 0.3, which performed the best in the above discussion. Once again, 21 different simulations are run for this meta-model and for the exactly-evaluated NSGA-II and different attainment curves [14] are plotted in Figure 13.21. In both cases, the runs are terminated when a total of 5,400 exact evaluations are done. The NSGA-II attainment curves are shown by solid lines and NSGA-II-ANN simulation in

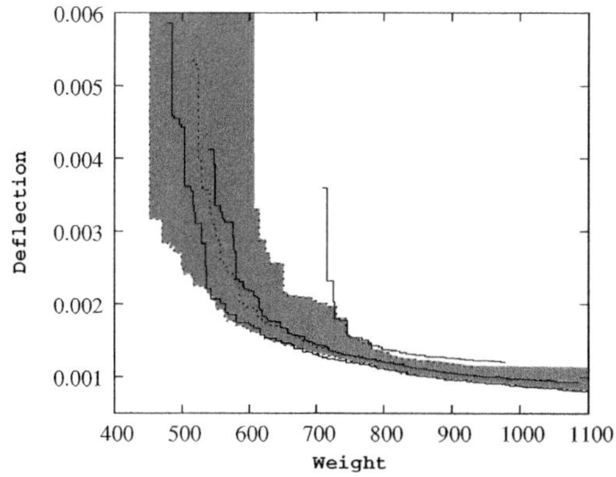

Fig. 13.21. Attainment curve plot for exact and approximate runs for the shape optimization problem

dashed lines. It can be observed that the B-20-3 NSGA-II-ANN model produces a better performance in the small-weight region (high-deflection region) compared to that obtained using the exactly-evaluated NSGA-II simulations. For larger-weight or smaller-deflection solutions, both NSGA-II and meta-modeled NSGA-II-ANN have achieved a similar level of performance. The median performance of the two algorithms are more or less same in the larger-weight or smaller-deflection region, while in smaller-weight or larger-deflection region, the performance of meta-modeled NSGA-II-ANN working with B-20-3 model is found to be better. Investigating the 100% attainment curve, it can be concluded that the worst-case performance of meta-modeled NSGA-II-ANN is better than that of the exactly-evaluated NSGA-II. To show the variation of performance due to a change in the initial population clearly, we shade the region between 0% and 100% attainment curves for meta-modeled NSGA-II-ANN simulations. It is observed from this figure that although variations in both approaches are substantial in the small-weight region of the trade-off curve, the performance of meta-modeled NSGA-II is somewhat better. Overall, the study also indicates that optimal solutions corresponding to smaller-weight (or larger-deflection) are difficult to achieve and more sensitive to GA parameter settings than the smaller-deflection (or larger-weight) solutions. Importantly, the reduced exact evaluations with the proposed scheme is found to be better or equivalent to the exactly-evaluated NSGA-II.

13.8 Conclusions

Most real-world optimization problems require a computationally expensive procedure of evaluating objective functions and constraints. This computational load is no less in multi-objective optimization. Hence there is a need for a generic multi-objective optimization procedure which can work reliably with approximate models. In this paper, we have suggested a meta-modeled multi-objective optimizer, namely NSGA-II-ANN in which the well-known elitist multi-objective optimizer NSGA-II has been combined with ANN. The NSGA-II-ANN procedure has been tested on a number of test problems and a couple of practical problems, involving a curve fitting problem and an engineering shape optimization problem. Test problems and case studies involve typical features of practical optimization problems: (i) a combination of real and discrete variables, (ii) unconstrained problem and constrained problems, (iii) discontinuous and continuous Pareto-optimal fronts, and (iv) non-convex and convex Pareto-optimal fronts. Simulation results have been analyzed for two aspects, namely, convergence to the trade-off frontier and the spread of obtained solutions. It can be concluded that NSGA-II-ANN simulations generally has shown a saving in exact function evaluations of about 25% to 62% to achieve a similar performance compared to the exactly-evaluated NSGA-II.

An important finding of this study is that a single strategy (the batch learning NSGA-II-ANN with the B-10-3 model) has been found to perform consistently better in all problems. The use of three generations of exactly-evaluated solutions for training and the use of the trained ANN for the next seven generations is a winning plan (providing adequate *exploitation-to-exploration* trade-off) found for all problems considered in this study. The problem difficulties are taken care of by the choice of an appropriate population size. The current study must now be extended and compared with other meta-modeling techniques, such as RSM and Kriging, and the superiority of one method over the other can be established.

References

1. Branke, J., Deb, K., Dierolf, H., and Osswald M.: Finding Knees in Multi-objective Optimization. In *Proceedings of Parallel Problem Solving from Nature (PPSN 2004)*. Springer, 2004, pp. 722–731.
2. Branke, J., and Schmidt, C.: Faster convergence by means of fitness estimation. In *Soft Computing Journal*. (in press).
3. Chen, J., Goldberg, D. E., Ho, S., and Sastry, K.: Fitness inheritance in multi-objective optimization. In *Proceedings, Genetic and Evolutionary computation Conference, 2002*. Morgan Kaufmann, 2002, pp. 319-326.
4. Deb, K.: *Multi-Objective Optimization Using Evolutionary Algorithms*. First Edition, Chichester, Uk: Wiley, 2001.
5. Deb, K., Agrawal, S., Pratap, A., and Meyarivan, T.: A fast and elitist multi-objective genetic algorithm: NSGA-II. IEEE Transactions on Evolutionary Computation, 6(2):182–197, 2002.
6. Deb, K., Thiele, L., Laumanns, M., and Zitzler, E.: Scalable test problems for evolutionary multi-objective optimization. In Abraham, A., Jain, L., and Goldberg, R., editors, *Evolutionary Multiobjective Optimization, 2005*, pages 105–145. London: Springer-Verlag.

7. Deb, K., Mohan, M., and Mishra, S.: Towards a quick computation of well-spread Pareto-optimal solutions. In *Proceedings, Evolutionary Multi-Criterion Optimization, 2003*. Springer, 2003, pp. 226-236.
8. Deb, K., and Jain, S.: Running performance Metrics for evolutionary multi-objective optimization. In *Proceedings, Fourth Asia-Pacific Conference on Simulated Evolution and Learning (SEAL'02)*. (Singapore), 2002, pp. 13-20.
9. Deb, K. and Gupta, H.: Searching for robust Pareto-optimal solutions in multi-objective optimization. In *Proceedings of Evolutionary Multi-Criterion Optimization (EMO 2005)*, Springer-Verlag, 2005, pp. 150–164.
10. Eby, D., Averill, R. C., Punch III, W. F., and Goodman, E. D.: Evaluation of injection island GA performance on flywheel design optimization. In *Proceedings, Third Conference on Adaptive Computing in Design and Manufacturing*. Springer, 1998.
11. El-Beltagy, M. A., Nair, P. B., and Keane, A. J.: Metamodeling techniques for evolutionary optimization of computationally expensive problems: promises and limitations. In *Proceedings of the Genetic and Evolutionary Computation Conference, 1999*. Morgan Kaufman, 1999, pp. 196-203.
12. Emmerich, M., Giotis, A., Ozdenir, M., Back, T., and Giannakoglou, K.: Metamodel-assisted evolution strategies. In *Proceedings, Parallel Problem Solving from Nature, 2002*. Springer, 2002, pp. 371-380.
13. Emmerich, M., and Naujoks, B.: Metamodel assisted multi-objective optimization strategies and their application in airfoil design. In *Proceedings, Adaptive Computing in Design and Manufacture VI, 2004*. Springer, 2004, pp. 249–260.
14. Fonseca, C. M. and Fleming, P. J.: On the performance assessment and comparison of stochastic multiobjective optimizers. In *Proceedings of Parallel Problem Solving from Nature IV (PPSN-IV)*, Springer, 1996, pp. 584–593.
15. Farina, M.: A neural network based generalized response surface multi-objective evolutionary algorithms. In *Congress on Evolutionary Computation, 2002*. IEEE Press, 2002, pp. 956-961.
16. Giannakoglou, K. C.: Design of optimal aerodynamic shapes using stochastic optimization methods and computational intelligence. In *Progress in Aerospace Science,* Vol. 38, pp. 43-76, 2002.
17. Giotis, A. P., and Giannakoglou, K. C.: Chapter 23: low cost GAs assisted by ANNs - applications in turbomachinery. edited by *Periaux, J. et al.*, Jhon Wiley & Sons, (to appear).
18. Goldberg, D. E., Deb, K., and Clark, J. H.: Genetic algorithms, noise, and the sizing of populations. *Complex System*, 6, pp. 333-362, 1992.
19. Haykin, S.: *Neural networks a comprehensive foundation.* second edition, Singapore: Addison Wesley, 2001. pp. 208.
20. Jin, Y., Olhofer, M., and Sendhoff, B.: A framework for evolutionary optimization with approximate fitness functions. In *IEEE Transactions on Evolutionary Computation,* 6(5), pp. 481-494, 2002.
21. Jin, Y., and Sendhoff, B.: Fitness approximation in evolutionary computation - A survey. In *Proceedings, Genetic and Evolutionary Computation Conference, 2002.* Morgan Kaufmann, 2002, pp. 1105-1112.
22. Jin, Y., and Sendhoff, B.: Reducing fitness evaluations using clustering techniques and neural network ensembles. In *Proceedings, Genetic and Evolutionary Computation Conference, 2004*. Springer, 2004, pp. 688-699.

23. Nain, P. K. S., and Deb, K.: Computationally effective search and optimization procedure using coarse to fine approximation. In *Proceedings, Congress on Evolutionary Computation, 2003.* IEEE Computer Society Press, 2003, pp. 2081-2088.
24. Nair, P. B., Keane, A. J., and Shimpi, R. P.: Combining approximation concepts with genetic algorithm based structural optimization procedures. In *Proceedings of the 39th AIAA/ASME/ASCE/AHS/ASC Structures, Structural Dynamics and Materials Conference*, 1998, pp. 1741-1751
25. Poloni, C., Giurgevich, A., Onesti, L., and Prdiroda, V.: Hybridization of a multi-objective genetic algorithm, a neural network and a classical optimizer for a complex design problem in fluid dynamics. In *Computer Methods in Applied Mechanics and Engineering*, volume 186, 2000, pp. 403-420.
26. Rasheed, K., and Hirsh, H.: Informed operators: speeding up genetic-algorithm-based design optimization using reduced models. In *Proceedings, Genetic and Evolutionary Computation Conference, 2000.* Morgan Kaufmann, 2000, pp. 628-635.
27. Rasheed, K., Vattam, S., and Ni, X.: Comparison of methods for using reduced models to speed up design optimization. In *Proceedings, Genetic and Evolutionary Computation Conference, 2002.* Morgan Kaufmann, 2002, pp. 1180-1187.
28. Ratle, A.: Accelerating the convergence of evolutionary algorithms by fitness landscape approximation. In *Proceedings, Parallel Problem Solving from Nature, 1998.* volume V, 1998, pp. 87–96.
29. Reklaitis, G. V., Ravindran, A. and Ragsdell, K. M. (1983). *Engineering Optimization Methods and Applications.* New York : Wiley.
30. Sastry, K., Goldberg, D. E., and Pelikan, M.: Don't evaluate, inherit. In *Proceedings, Genetic and Evolutionary computation Conference, 2001.* Morgan Kaufmann, 2001, pp. 551-558.
31. Sefrioui, M., and Périaux, J.: A hierarchical genetic algorithm using multiple models for optimization. In *Proceedings, 6th International Conference on Parallel Problem Solving from Nature - PPSN VI .* Lecture Notes in Computer Science 1917, Springer 2000.

14

Surrogate Model-Based Optimization Framework: A Case Study in Aerospace Design

Yolanda Mack[1], Tushar Goel[1], Wei Shyy[2], and Raphael Haftka[1]

[1] Mechanical and Aerospace Engineering Department, 231 MAE-A, P.O. Box 116250, University of Florida, Gainesville, FL 32611-6250, USA
{tiki,tusharg,haftka}@ufl.edu

[2] Department of Aerospace Engineering, François-Xavier Bagnoud Building, 1320 Beal Avenue, University of Michigan, Ann Arbor, MI 48109-2140, USA
weishyy@umich.edu

Summary. Surrogate-based optimization has proven very useful for novel or exploratory design tasks because it offers a global view of the characteristics of the design space, and it enables one to refine the design of experiments, conduct sensitivity analyses, characterize tradeoffs between multiple objectives, and, if necessary, help modify the design space. In this article, a framework is presented for design optimization on problems that involve two or more objectives which may be conflicting in nature. The applicability of the framework is demonstrated using a case study in space propulsion: a response surface-based multi-objective optimization of a radial turbine for an expander cycle-type liquid rocket engine. The surrogate model is combined with a genetic algorithm-based Pareto front construction and can be effective in supporting global sensitivity evaluations. In this case study, due to the lack of established experiences in adopting radial turbines for space propulsion, much of the original design space, generated based on intuitive ideas from the designer, violated established design constraints. Response surfaces were successfully used to define previously unknown feasible design space boundaries. Once a feasible design space was identified, the optimization framework was followed, which led to the construction of the Pareto front using genetic algorithms. The optimization framework was effectively utilized to achieve a substantial performance improvement and to reveal important physics in the design.

14.1 Introduction

With continuing progress in computational simulations, computational-based optimization has proven to be a useful tool in reducing the design process duration and expense. Numerous methods exist for conducting design optimizations. Popular methods include gradient-based methods [4, 14], adjoint methods [6, 10], and surrogate model-based optimization methods such as the response surface approximation (RSA) [9]. Gradient-based methods rely on a

Y. Mack et al.: *Surrogate Model-Based Optimization Framework: A Case Study in Aerospace Design*, Studies in Computational Intelligence (SCI) **51**, 323–342 (2007)
www.springerlink.com

step by step search for an optimum design using the method of steepest descent on the objective function according to a convergence criterion. Adjoint methods require formulations that must be integrated into the computational simulation of the physical laws. For a new design or a computationally expensive design, optimization based on an inexpensive surrogate, such as an RSA, is a good choice. Surrogate-based optimization allows for the determination of an optimum design, while at the same time providing insight into the workings of the design. A surrogate model not only provides the benefit of low-cost for function evaluations, but it can also help revise the problem definition of a design task, which is not unusual for new efforts. Furthermore, it can conveniently handle the existence of multiple desirable design points and offer quantitative assessments of trade-offs as well as facilitate global sensitivity evaluations of the design variables [7, 13].

In the present article, an effort in surrogate model-based multi-objective optimization and sensitivity analysis will be described. A case study in space propulsion will be used: a radial turbine for an expander cycle-type liquid rocket engine. The surrogate model is combined with a genetic algorithm-based Pareto front construction and facilitates global sensitivity evaluations. The framework steps include 1) modeling of the objectives using surrogate models, 2) refining the design space, 3) reducing the problem dimensionality, and 4) handling multiple objectives with the aids of a Pareto front and a global sensitivity evaluation method. Figure 14.1 illustrates the process used to develop optimal designs for the problems involving multiple and possibly conflicting objectives. The steps of the framework are detailed in the following sections.

14.2 Framework Details

The first steps in any optimization problem are to identify the performance criteria, the design variables and their allowable ranges, and the design constraints. These critical steps require expertise about the physical process. A multi-objective optimization problem is formulated as

$$Minimize\,\mathbf{F}(\mathbf{x}),\ where\quad \mathbf{F} = f_j : \forall j = 1, M;\quad \mathbf{x} = x_i : \forall i = 1, N$$

Subject to

$$\begin{aligned}\mathbf{C}(\mathbf{x}) &\leq 0, where\ \ \mathbf{C} = c_p : \forall p = 1, P\\ \mathbf{H}(\mathbf{x}) &= 0, where\ \ \mathbf{H} = h_k : \forall k = 1, K\end{aligned}$$

Once the problem is defined, the designs are evaluated through experiments or numerical simulations. For numerical simulations, the type of numerical model and design evaluations used varies with the goals of the study. For a simple preliminary design optimization, the use of an inexpensive 1-D solver may be sufficient. However, for the final detailed design, more complex solvers may be needed [11, 12]. The choice of a model has an important bearing

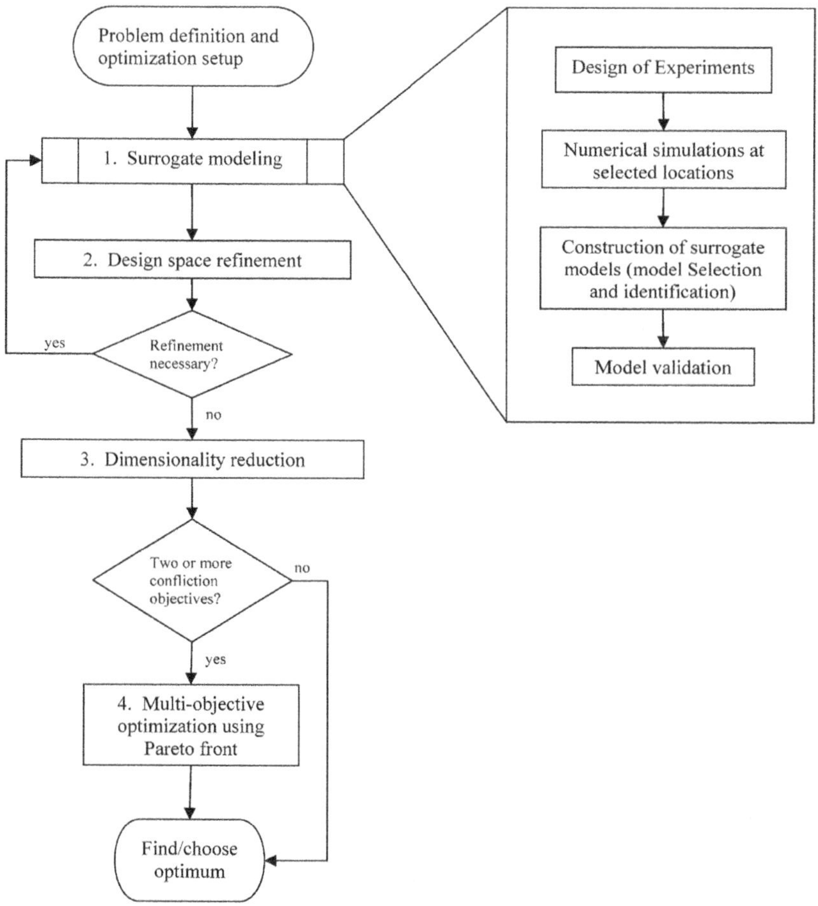

Fig. 14.1. Flowchart of optimization framework

on the computational expense of evaluating designs. When obtaining many design points is time-prohibitive, it is often more prudent to use an inexpensive surrogate model in place of the expensive numerical model.

14.2.1 Model objectives using surrogate models

This step in the framework involves developing alternate models based on a limited amount of data to analyze and optimize designs. The surrogate-based optimization (SBO) approach has been shown to be an effective approach for engineering design in aerospace systems, aerodynamics, structures, and propulsion, among other disciplines. The surrogates provide fast approximations of the system response making optimization and sensitivity studies possible. Response surface approximations, neural network techniques, spline, and kriging are examples of methods used to generate surrogates for simulations in the optimization of complex flows involving applications such as engine diffusers [8], rocket injectors [17], and supersonic turbines [11, 12].

The major benefit of surrogate models is the ability to quickly obtain any number of additional function evaluations without resorting to more expensive numerical models. In this aspect, surrogate models can be used for multiple purposes. Obviously, they are used to model the design objectives, but they can also be used to model the constraints and help identify the feasible region in design space. Key stages in the construction of surrogate models are shown in Figure 14.1.

For the construction of a surrogate model, it is necessary to first begin with a Design of Experiments (DOE). The DOE is an initial selection of points to be evaluated by the experiment or numerical model. A relatively small number of points are selected to effectively represent the entire design space. The results of the evaluation of these points are used to build the surrogate model.

Design of Experiments

The search space, or design space, is the set of all possible combinations of the design variables. If all design variables are real, the design space is given as$\mathbf{x} \in \mathbf{R}^N$, where N is the number of design variables. The feasible domain $\mathbf{S}$ is the region in design space where all constraints are satisfied.

To construct a surrogate model, the sampling of points in design space is facilitated by different design of experiment techniques. The key issues in the selection of an appropriate DOE include (*i*) the dimensionality of the problem, (*ii*) whether noise is important source of error, (*iii*) the number of simulations or experiments that can be afforded, (*iv*) the type of surrogate used to model the problem, and (*v*) the shape of the design space. If noise is the dominant source of error, DOEs that reduce the sensitivity to noise are commonly used. These include central composite designs, face-centered cubic designs, factorial designs, and Box-Behnken designs for box-shaped domains. D- or A-optimal designs are useful for irregular shaped domains when minimizing noise is important. When noise is not an issue, Latin-Hypercube Sampling (LHS), minimum bias designs, and Orthogonal Arrays (OAs) are preferred.

Surrogate Model Identification and Fitting

There are many types of surrogate models to choose from. There are parametric models that include polynomial response surfaces and kriging models, and there are non-parametric models such as projection-pursuit regression and radial basis functions. The parametric approaches assume the global functional form of the relationship between the response variable and the design variables is known, while the non-parametric ones use different types of simple local models in different regions of the design space to build up an overall model. In this study, the polynomial RSA will be highlighted.

The polynomial RSA assumes that the function of interest f, can be represented as a linear combination of N_c basis functions z_j and an error term ε. For a typical observation i, a response can be given in the form of a linear equation as

$$f_i(\mathbf{z}) = \sum_{j=1}^{N_c} \beta_j z_j^{(i)} + \varepsilon_i,\ E(\varepsilon_i) = 0,\ V(\varepsilon_i) = \sigma^2, \tag{14.1}$$

where the errors ε_i are considered independent with an expected value E equal to zero and a variance Vequal to σ^2. The coefficients β_j represent the quantitative relation among basis functions z_j. Monomials are the preferred basis functions.

The relationship between the coefficients β_j and the basis functions z_j is obtained using N_s sample values of the response f_i for a set of basis functions $z_j^{(i)}$ such that the error in the prediction is minimized in a least squares sense. For N_s sample points, the set of equations specified in Equation (14.2) can be expressed in matrix form as

$$\mathbf{f} = \mathbf{X}\beta + \varepsilon,\quad E(\varepsilon) = 0,\quad V(\varepsilon) = \sigma^2 I, \tag{14.2}$$

where $\mathbf{X}$ is a $N_s \times N_c$ matrix of basis functions, also known as a Gramian design matrix, with the design variable values as the sampled points. A Gramian design matrix for a quadratic polynomial in two variables ($N_s = 2$; $N_c = 6$) is shown in Equation (14.3).

$$\mathbf{X} = \begin{bmatrix} 1 & x_{11} & x_{21} & x_{11}^2 & x_{11}x_{21} & x_{21}^2 \\ 1 & x_{12} & x_{22} & x_{12}^2 & x_{12}x_{22} & x_{22}^2 \\ \vdots & \vdots & \vdots & \vdots & \vdots & \vdots \\ 1 & x_{1i} & x_{2i} & x_{1i}^2 & x_{1i}x_{2i} & x_{2i}^2 \\ \vdots & \vdots & \vdots & \vdots & \vdots & \vdots \\ 1 & x_{1N_s} & x_{2N_s} & x_{1N_s}^2 & x_{1N_s}x_{2N_s} & x_{2N_s}^2 \end{bmatrix} \tag{14.3}$$

The vector $\boldsymbol{b}$ of the estimated coefficients, which is an unbiased estimate of the coefficient vector β and has minimum variance, can then be found by

$$\mathbf{b} = \left(\mathbf{X}^{\mathbf{T}}\mathbf{X}\right)^{-1}\mathbf{X}^{\mathbf{T}}\mathbf{f} \tag{14.4}$$

At a new set of basis function vector $\boldsymbol{z}$ for design point P, the predicted response and the variance of the estimation are given by

$$\hat{f}_P(\mathbf{z}) = \sum_{j=1}^{N_c} b_j z_j^{(i)} \text{ and } V\left(\hat{f}_P(\mathbf{z})\right) = \sigma^2\left(\mathbf{z}^T\left(\mathbf{X}^{\mathbf{T}}\mathbf{X}\right)^{-1}\mathbf{z}\right) \tag{14.5}$$

Surrogate Model Validation

Once the surrogate models are available, it is imperative to establish the predictive capabilities of the surrogate model away from the available data. In the context of an RSA, several measures of predictive capability are given in the following subsections.

Adjusted root mean square error

The error ε_i at any design point i is given by

$$\varepsilon_i = f_i - \hat{f}_i \tag{14.6}$$

where f_i is the actual value and $\hat{f}_i$ is the predicted value. Hence, the adjusted root mean square (rms) error σ_a is given by

$$\sigma_a = \sqrt{\frac{\sum_{i=1}^{N_s} \varepsilon_i^2}{(N_s - N_c)}} \tag{14.7}$$

For a good fit, σ_a should be small compared to the data.

Root mean square error at test points

If N_t additional test data are used to test the quality of the approximation, the rms error σ is given by

$$\sigma = \sqrt{\frac{\sum_{i=1}^{N_t} \varepsilon_i^2}{N_t}} \tag{14.8}$$

A small rms error indicates a good fit.

Coefficient of multiple determination

The adjusted coefficient of multiple determination R^2_{adj} defines the prediction capability of the polynomial RSA as

$$R^2_{adj} = 1 - \left(\frac{\sigma_a^2 (N_s - 1)}{\sum_{i=1}^{N_s} (f_i - \bar{f})^2} \right); \quad where \ \ \bar{f} = \frac{\sum_{i=1}^{N_s} f_i}{N_s} \tag{14.9}$$

For a good fit, R^2_{adj} should be close to 1.

Prediction Error Sum of Squares

When there are an insufficient number of data points available to test the RSA, the prediction error sum of squares (PRESS) statistic is used to estimate the performance of the RSA. A residual is obtained by fitting an RSA over the design space after dropping one design point from the training set. The value predicted by the RSA at that point is then compared with the expected value. PRESS is given by

$$PRESS = \sqrt{\frac{\sum_{i=1}^{N_p} (y_i - \hat{y}_i^*)^2}{N_p}} \tag{14.10}$$

where $\hat{y}_i^*$ is the value predicted by the RSA for the ith point which is excluded while generating the RSA. If the PRESS value is close to σ_a, this indicates that the RSA performs well.

14.2.2 Design Space Refinement

The process of design space refinement is often a necessary step of an optimization process involving surrogate models. It is especially necessary when the exact relationship between the selected design variables, constraints, and objectives is unknown. If this is the case, bounds are set on the design variables, which are usually based on empirical data and experimental results. This can often lead to a larger than needed design space.

After populating this design space and obtaining computational results, several issues may arise. Upon closer inspection of the design space, it may be discovered that many points in the design space are grossly infeasible, meaning that they violate design constraints severely, or they represent unrealistic or poorly performing designs. The quality of the surrogate model may also be affected by the size of design space. If the size of the design space is too large, the surrogate model may provide a poor fit for the data. All of these issues can be alleviated by intelligently examining the design space and reducing it accordingly. Balabanov et al. applied the design space reduction technique to a high speed civil transport wing [1]. They discovered that 83% of the points in their original design space violated geometric constraints, while many of the remaining points were simply unreasonable. Reducing the size of the design space eliminated their unreasonable designs and improved the accuracy of the surrogate model. Roux et al. [15] found that the accuracy of polynomial RSA is sensitive to the size of the design space for structural optimization problems. They recommended the use of various measures to find a small "reasonable design space."

The design space can be refined by reducing the range of the design variables or by using functions to define irregular design space boundaries. When the new design space is set, the surrogate model step of the framework must be repeated. However, the original surrogate model may be used to screen the points before they are used in the experiment or numerical simulation. Using the surrogate model to predict if a point may land outside of the design space is particularly necessary when irregular design spaces exist. It can prevent the expensive and unnecessary simulation of undesirable points. The design space refinement should provide more accurate surrogate models and the reduction of poor performance points.

14.2.3 Dimensionality Reduction

At the beginning of a design optimization, several variables are chosen as design variables with the assumption that they are important to the optimization. However, having large numbers of design variables can greatly increase the cost of the optimization. It is of great benefit, therefore, to simplify the design problem by identifying variables that are unimportant and removing them from the analysis. The most efficient way of doing this is to perform a sensitivity analysis. A global sensitivity analysis can provide essential information on the sensitivity of a design objective to individual variables and variable interactions. By removing the variables that have negligible influence on the design objective, the dimensionality of the problem can be reduced.

Another way of reducing the dimensionality of a problem is through a correlation analysis of the objective functions. If two objectives are highly correlated, they may be dropped, and a representative objective can be used for all the correlated objectives, thus reducing the dimensionality of the problem in function space [3].

Global Sensitivity Analysis

Global sensitivity analyses enable the study of the behavior of different design variables. This information can be used to identify the variables which are the least important, thus, the number of variables can be reduced. The theoretical formulation of the global sensitivity analysis is described below.

A surrogate model $f(\mathbf{x})$ of a square integrable objective as a function of a vector of independent input variables $\mathbf{x}$ ($x_i \in [0,1]\ \forall i = 1, N$) is assumed and is modeled as uniformly distributed random variables. The surrogate model can be decomposed as the sum of functions of increasing dimensionality

$$f(\mathbf{x}) = f_0 + \sum_i f_i(x_i) + \sum_{i<j} f_{ij}(x_i, x_j) + \cdots + f_{12\ldots N}(x_1, x_2, \ldots, x_N) \tag{14.11}$$

where$f_0 = \int_{\mathbf{x}=\mathbf{0}}^{\mathbf{1}} f d\mathbf{x}$. If the following condition

$$\int_0^1 f_{i_1\ldots i_s} dx_k = 0 \tag{14.12}$$

is imposed for $k = i_1, \ldots, i_s$, then the decomposition described in Equation (14.11) is unique.

In the context of a global sensitivity analysis, the total variance denoted as $V(f)$ can be shown to be equal to

$$V(f) = \sum_{i=1}^{n} V_i + \sum_{1\le i\le j\le N} V_{ij} + \cdots + V_{1\ldots N} \tag{14.13}$$

where $V(f) = E\left((f - f_0)^2\right)$ and each of the terms in Equation (14.13) represent the partial variance of the independent variables (V_i) or set of variables to the total variance. This provides an indication of their relative importance. The partial variances can be calculated using the following expressions:

$$\begin{aligned} V_i &= V(E[f|x_i]) \\ V_{ij} &= V(E[f|x_i, x_j]) - V_i - V_j \\ V_{ijk} &= V(E[f|x_i, x_j, x_j]) - V_{ij} - V_{ik} - V_{jk} - V_i - V_j - V_k \end{aligned} \tag{14.14}$$

and so on, where V and E denote variance and the expected value respectively. Note that $E[f|x_i] = \int_0^1 f_i dx_i$ and $V(E[f|x_i]) = \int_0^1 f_i^2 dx_i$. Now the sensitivity indices can be computed corresponding to the independent variables and set of variables. For example, the first and second order sensitivity indices can be computed as

$$S_i = \frac{V_i}{V(f)}, \quad S_{ij} = \frac{V_{ij}}{V(f)} \tag{14.15}$$

Under the independent model inputs assumption, the sum of all the sensitivity indices is equal to one.

The first order sensitivity index for a given variable represents the main effect of the variable, but it does not take into account the effect of interaction of the variables. The total contribution of a variable on the total variance is given as the sum of all the interactions and the main effect of the variable. The total sensitivity index of a variable is then defined as

$$S_i^{total} = \frac{V_i + \sum\limits_{j, j \neq i} V_{ij} + \sum\limits_{j, j \neq i} \sum\limits_{k, k \neq i} V_{ijk} + \ldots}{V(f)} \tag{14.16}$$

The above referenced expressions can be easily evaluated using surrogate models of the objective functions. Sobol [16] proposed a variance-based non-parametric approach to estimate the global sensitivity for any combination of design variables using Monte Carlo methods. To calculate the total sensitivity of any design variable x_i, the design variable set is divided into two complementary subsets of x_i and Z $\quad (Z = x_j, \forall j = 1, N; j \neq i)$. The purpose of using these subsets is to isolate the influence of x_i from the influence of the remaining design variables included in Z. The total sensitivity index for x_i is then defined as

$$S_i^{total} = \frac{V_i^{total}}{V(f)} \tag{14.17}$$

$$V_i^{total} = V_i + V_{i,Z} \tag{14.18}$$

where V_i is the partial variance of the objective with respect to x_i and $V_{i,Z}$ is the measure of the objective variance that is dependent on interactions between x_i and Z. Similarly, the partial variance for Z can be defined as V_Z. Therefore, the total objective variability can be written as

$$V = V_i + V_Z + V_{i,Z} \tag{14.19}$$

While Sobol used Monte Carlo simulations to conduct the global sensitivity analysis, the expressions given above can be easily computed analytically once the RSA is available.

Correlation Analysis

To conduct a correlation analysis, a large number of designs in design variable space must be analyzed with the correlations computed among different objectives. All the pairs of objectives which have a very high correlation coefficient (close to 1) are highly correlated and are candidates for dimensionality reduction.

14.2.4 Multiple Objective Optimization

After developing a computationally inexpensive way of evaluating different designs, the final step is to perform the actual optimization. In the case of a single objective, this requires a simple search of the design space for the minimum value of the objective. For two or more objectives, additional treatment is needed. Highly correlated objectives can be combined into a single objective function. When the objectives are conflicting in nature, there may be an infinite number of possible solutions that will provide possible good combinations of objectives. These solutions are known as Pareto optimal solutions (POS). While there are numerous methods of solving multi-objective optimization problems, the use of evolutionary algorithms (EAs) is a natural choice to obtain many POS in a single simulation due to its population based approach and its ability to converge to global optimal solutions.

Pareto Optimal Front

A feasible design $\mathbf{x}^{(1)}$ dominates another feasible design $\mathbf{x}^{(2)}$ (denoted by $\mathbf{x}^{(1)} < \mathbf{x}^{(2)}$), if both of the following conditions are true:

1. The design $\mathbf{x}^{(1)}$ is no worse than $\mathbf{x}^{(2)}$ in all objectives, i.e., $f_j\left(\mathbf{x}^{(1)}\right) > f_j\left(\mathbf{x}^{(2)}\right)$ for all $j = 1,2,\ldots, M$ objectives.

$$\mathbf{x}^{(1)}\mathbf{x}^{(2)} \Rightarrow \forall j \in M\ f_j\left(\mathbf{x}^{(1)}\right) \not> f_j\left(\mathbf{x}^{(2)}\right) \quad or \quad \forall j \in M\ f_j\left(\mathbf{x}^{(1)}\right) \leq f_j\left(\mathbf{x}^{(2)}\right) \tag{14.20}$$

1. The design $\mathbf{x}^{(1)}$ is strictly better than $\mathbf{x}^{(2)}$ in at least one objective, or $f_j\left(\mathbf{x}^{(1)}\right) < f_j\left(\mathbf{x}^{(2)}\right)$for at least one $j \in \{1, 2, ..., M\}$.

$$\mathbf{x}^{(1)} < \mathbf{x}^{(2)} \Rightarrow \forall j \in M\ f_j(\mathbf{x}^{(1)}) < f_j(\mathbf{x}^{(2)}) \tag{14.21}$$

If two designs are compared, then the designs are said to be non-dominated with respect to each other if neither design dominates the other. A design

$\mathbf{x} \in \mathbf{S}$, where $\mathbf{S}$ is the set of all feasible designs, is said to be non-dominated with respect to a set $\mathbf{A} \subseteq \mathbf{S}$, if $\nexists \mathbf{a} \in \mathbf{A} : \mathbf{a} < \mathbf{x}$. Such designs in function space are called non-dominated solutions. All the designs $\mathbf{x}$ ($\mathbf{x} \in \mathbf{S}$) which are non-dominated with respect to any other design in set S, comprise the Pareto optimal set. The function space representation of the Pareto optimal set is the Pareto optimal front. When there are two objectives, the Pareto optimal front is a curve, and when there are three objectives, the Pareto optimal front is represented by a surface. If there are more than three objectives, it is represented by a hyper-surface.

Elitist Non-dominated Sorting Genetic Algorithm (NSGA-II) with Archiving

In this study, an elitist non-dominated sorting genetic algorithm NSGA-II [2] with a parallel archiving strategy to overcome the Pareto drift problem [3] is used as the multi-objective optimizer to generate Pareto optimal solutions. The description of the algorithm is given as follows:

1. Randomly initialize a population (designs in the design space) of size *npop*.
2. Compute objectives and constraints for each design.
3. Rank the population using non-domination criteria. Many individuals can have the same rank with the best individuals given the designation of *rank-1*. Initialize an archive with all the non-dominated solutions.
4. Compute the crowding distance. This distance finds the relative closeness of a solution to other solutions in the function space and is used to differentiate between the solutions on same rank.
5. Employ genetic operators (selection, crossover, and mutation) to create an intermediate population of size *npop*.
6. Evaluate objectives and constraints for this intermediate population.
7. Combine the two (parent and intermediate) populations, rank them, and compute the crowding distance.
8. Update the archive:
   a) Compare archive solutions with *rank-1* solutions in the combined population.
   b) Remove all dominated solutions from the archive.
   c) Add to the archive all *rank-1* solutions in the current population which are non-dominated with respect to the archive.
9. Select a new population *npop* from the best individuals based on the ranks and the crowding distances.
10. Return to step 3 and repeat until the termination criteria is reached, which in the current study is chosen to be the number of generations.

14.3 A Case Study: Response Surface-Based Multi-Objective Optimization of a Compact Liquid-Rocket Radial Turbine

The purpose of this case study is to increase the work output of a compact radial turbine in a liquid rocket expander cycle engine. If the turbine inlet temperature is held constant, an increase in turbine work is directly proportional to the increase in efficiency. The radial turbine design must provide maximum efficiency while keeping the overall weight of the turbine low. Thus, the goal of the design optimization is to maximize the turbine efficiency while minimizing the turbine weight. Using a response surface analysis, an accurate surrogate model is constructed to predict the radial turbine weight and the efficiency across the selected design space.

A total of six design variables were identified. The ranges of the design variables were set based on current design practices. Additionally, five constraints were identified. Two of the five constraints are structural constraints, two are geometric constraints, and one is an aerodynamic constraint. The aerodynamic constraint is based on general guidelines. The descriptions of all variables and the results of the baseline case simulation are given in Table 14.1 It was unknown in what way the constraints depended on the design variables. The possibility existed that certain combinations of design variables would cause a constraint violation. It was also unknown whether the selected ranges would result in feasible designs. Thus, RSAs were used to help clarify these unknown factors.

14.3.1 Phase 1: Problem Definition, Initial Design of Experiments, and Construction of Constraint Surrogates

The CFD solutions were obtained using a 1-D Meanline [5] code. A total of 77 points were selected using a standard face-centered cubic DOE. Of the 77 solutions, seven cases failed and 60 cases violated one or more of the five constraints resulting in only 10 feasible cases. The design variable ranges were reduced, and another 77 CFD cases were submitted. With the reduced ranges, no cases failed, but 90% of the cases still violated one or more of the output constraints. Before the optimization could be conducted, a reasonable design space first had to be identified. Because there was limited information on the dependency of the output constraints on the input variables, RSAs was used to determine this dependency. Response surfaces were used to properly scale the design variable ranges and identify irregular constraint boundaries.

Response surface approximations were fit to the output constraints. For each response surface, the effects of variables that contributed little to the RSA were removed. In this way, each RSA was simplified until variable dependencies could be accurately determined. The variable dependencies are shown in Equation (14.22).

Table 14.1. Variable names and descriptions

Objective Variable	Description	Baseline design		
Rotor Wt	Relative measure of "goodness" for overall weight	1.147		
Etats	Total-to-static efficiency	85%		
Design Variable		MIN	Baseline	MAX
RPM	Rotational Speed	80,000	122,000	150,000
React	Percentage of stage pressure drop across rotor	0.45	0.55	0.70
U/C isen	Isentropic velocity ratio	0.50	0.61	0.65
Tip Flw	Ratio of flow parameter to a choked flow parameter	0.30	0.25	0.48
Dhex %	Exit hub diameter as a % of inlet diameter	0.10	0.58	0.40
AnsqrFrac	Used to calculate annulus area (stress indicator)	0.50	0.83	1.0
Constraint Variable		**Desired Range**		
Tip Spd	Tip speed (ft/sec) (stress indicator)	$\leq$ 2500		
AN^2	Annulus area × speed^2 (stress indicator)	$\leq$ 850		
Beta1	Blade inlet flow angle	$0 \leq Beta1 \leq 40$		
Cx2/Utip	Recirculation flow coefficient (indication of pumping upstream)	$\geq$ 0.20		
Rsex/Rsin	Ratio of the shroud radius at the exit to the shroud radius at the inlet	$\leq$ 0.85		

$$\begin{aligned}
&AN^2 = AN^2\,(AnsqrFrac)\\
&Tip\ Spd = Tip\ Spd\,(U/C\ isen)\\
&Cx2/Utip = Cx2/Utip\,(RPM, U/C\ isen, AnsqrFrac)\\
&Beta1 = Beta1\,(React, U/C\ isen, Tip\ Flw)\\
&Rsex/Rsin = Rsex/Rsin\,(AnsqrFrac, U/C\ isen, Dhex\%)
\end{aligned} \tag{14.22}$$

The information obtained from the RSAs about the output variable dependences was further used to develop input variable constraints. An RSA was constructed for each input variable as a function of the output variable and the remaining input variables. The most accurate RSAs ($R^2_{adj} \geq 0.99$) were used to determine the input variable constraints. The output variables were then set to the constraint limits. For example, a constraint on *React* can be applied to coincide with the constraint $Beta1 \geq 0$:

$$\begin{aligned}
&Beta1 = Beta1\,(React, U/C\ isen, AnsqrFrac) \geq 0\\
&\Rightarrow\ \ React|_{Beta1=0} \leq React\,(Beta1 = 0, U/C\ isen, AnsqrFrac)
\end{aligned} \tag{14.23}$$

Constraint "surfaces" were developed in this manner for each constraint. It was discovered that two of the five constraints (AN^2 and *Tip Spd*) were simple one-dimensional constraints. For the one-dimensional constraints, the variable ranges could simply be reduced to match the constraint boundaries.

The remaining constraints were more complex. However, one of these constraints (*Rsex/Rsin*) was automatically satisfied by the reduction of the variable ranges for the one-dimensional constraints. The new valid region is shown in Figure 14.2.

The remaining constraints involved three variables each. It was discovered that many low values of *RPM* violated the *Cx2/Utip* constraint. The region of violation was a function of *RPM*, *U/C isen*, and *AnsqrFrac* as shown in Figure 14.3. The *Beta1* constraint was found to be the most demanding and resulted in a very irregular design space as shown in Figure 14.4. Much of the original design space violated this constraint. It was also discovered that the constraint surface for *Beta1* ≤ 40 lay outside of the original design variable range for *React*. In this case, the lower bound for *React* was sufficient to satisfy this constraint.

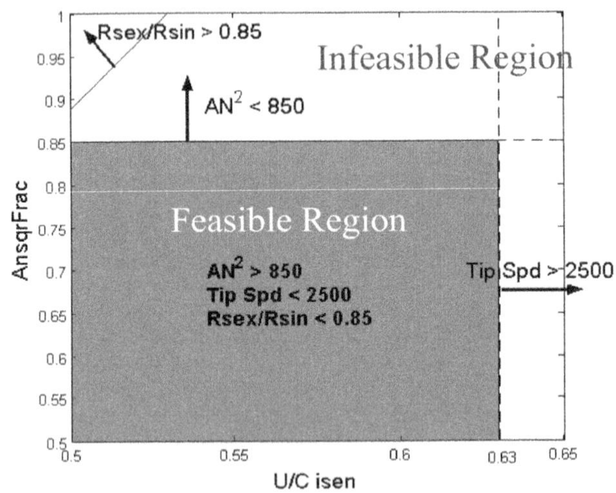

Fig. 14.2. Constraint regions for three constraints. Three of five constraints automatically satisfied by range reduction of two design variables.

The RSA-based constraints were tested on their predictive capability using the two available data sets. The design variable values were input into the RSAs for the constraints. The RSAs correctly identified all points that violated the output constraints. Now that the feasible design space was accurately identified, data points could be placed within it. The results and summary of the prediction of constraint violations are as follows:

1. Constructed RSAs to determine the relationship between output constraints and design variables
2. Adjusted the variable ranges based on information from constraint surfaces

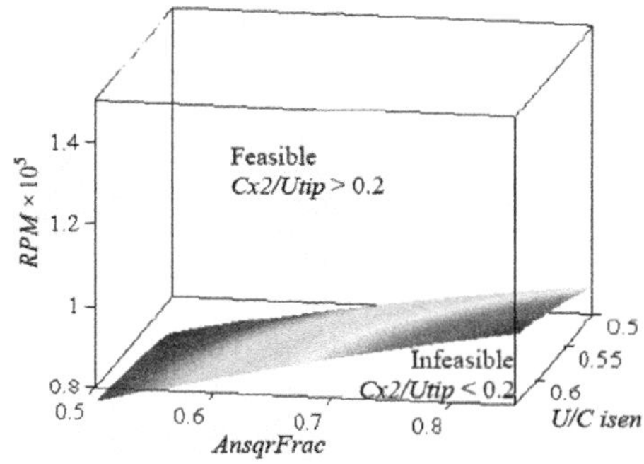

Fig. 14.3. Constraint surface for $Cx2/Utip = 0.2$. At higher values of *AnsqrFrac* and *U/C isen*, lower values of *RPM* are invalid.

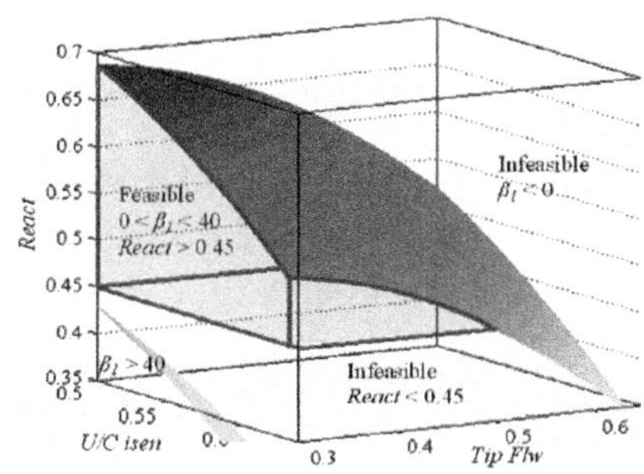

Fig. 14.4. Constraint surfaces for $Beta1 = 0$ and $Beta1 = 40$. Values of $Beta1 > 40$ lay outside of design variable ranges.

3. Applied a 3-level full factorial design (729 points) within the new variable ranges
4. Eliminated points that violated constraints (498/729 points) based on RSAs of the constraints

Using the constraint RSAs, 97% of the new data points obtained by the Meanline code lay in the feasible design space region. The points that did violate the output constraints often violated the $Beta1 \geq 0$ constraint, with the violation being only a slightly negative *Beta1* value, so that they were still useful for fitting.

14.3.2 Phase 2: Design Space Refinement, Intermediate Response Surface Approximations for Data Screening and New DOEs

Plotting the data points in function space reveals additional information about the nature of the response. A large area of function space contained data points with a lower efficiency than was desired. There also existed areas of high weight without improvement in efficiency. These undesirable areas could be eliminated thus reducing the design space to a reasonable design region. The density of points could then be increased within this reasonable region of interest. The region of interest is shown in Figure 14.5. The design variables regions would be reduced again to match the new design region. Response surface approximations were constructed for the turbine weight, *Rotor Wt*, and the turbine total-to-static efficiency, *Etats*. It was discovered that the fidelity of the *Etats* RSA was lower than desired ($R^2_{adj} = 0.91$) indicating that the RSA construction could also benefit from a reduced design space. The response surfaces were adequate enough to be used to screen for poorly performing points. Points predicted to lie outside of the new design space would be neglected.

For the reasonable design space, a third set of data was required, and the following steps were taken:

1. Only the portion of the design space with best performance was reserved to allow for a concentrated effort on the region of interest and to increase response surface fidelity
2. Latin Hypercube Sampling (LHS) was used over all six variables, and a 5-level factorial design was used over the three variables with the highest impact on the best performing points

The combination of the DOEs resulted in a total of 323 design points.

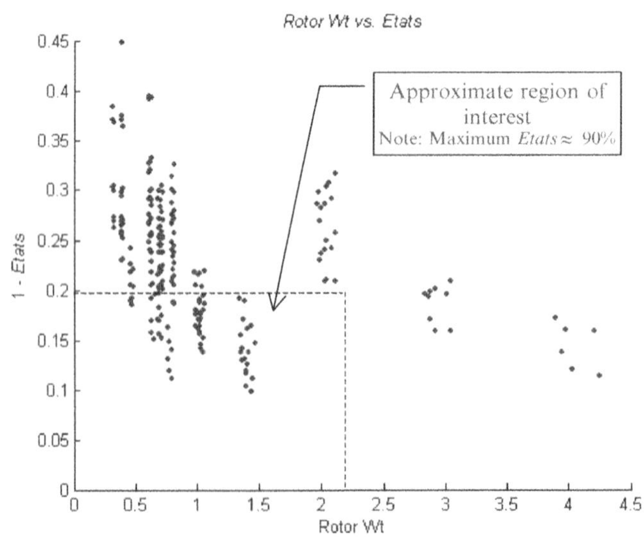

Fig. 14.5. Region of interest in function-function space (The quantity 1 – *Etats* is used for improved plot readability)

14.3.3 Phase 3: Global Sensitivity Analysis and Dimensionality Reduction Check

A global sensitivity analysis was conducted on the RSAs. The results are shown in Figure 14.6. It was discovered that the turbine rotational speed *RPM* had the largest impact on the variability of the resulting turbine weight *Rotor Wt*. The effects of the rotational speed *RPM* along with the isentropic velocity ratio *U/C isen* make up 97% of the variability in *Rotor Wt*. All other variables and variable interactions have minimal effect on *Rotor Wt*. For the total-to-static efficiency *Etats*, the effect of the design variables are more evenly distributed. Because of the more even distribution of the sensitivity indices for *Etats*, no variable can be completely eliminated from the computation.

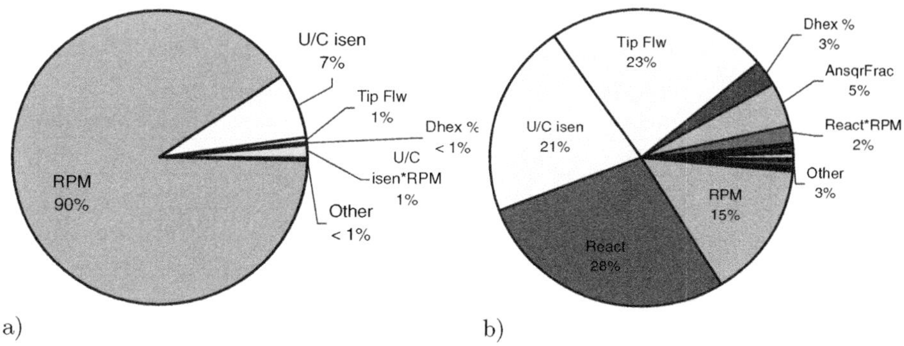

Fig. 14.6. Global sensitivity analysis. Effect of design variables on (a) *Rotor Wt* and (b) *Etats*.

14.3.4 Phase 4: Pareto Front Construction and Validation

Response surface approximations were constructed for each objective using the third set of data. Points found to be infeasible after the simulations were dropped. Function evaluations from the response surfaces were used with the genetic algorithm NSGA-II to construct the Pareto Front shown in Figure 14.7. In this study, the real-coded version of NSGA-II was used; that is, crossover and mutation operations were conducted in the real space rather than the binary space. For all simulations, a tournament selection operator with a tournament size of two was used. The parameters set for the Pareto set selection are shown in Table 14.2.

Within the Pareto front, a region was identified that would provide the best value in terms of maximizing efficiency and minimizing weight. This trade-off region was selected for validation of the Pareto front. The results of the subsequent simulation of the validation data indicated that the RSAs and corresponding Pareto front were very accurate, as shown in Table 14.3.

Table 14.2. Parameters for Pareto set selection in NSGA-II

Population size (npop)	100
Generations	250
Crossover probability (P_{cross})	1.00
Distribution parameter (for crossover)	20
Mutation probability (P_{mut})	0.20
Distribution parameter (for mutation)	200

A noticeable improvement was attained compared to the baseline radial turbine design. The selected design had the same weight as the baseline case with a 5% improvement in efficiency.

Table 14.3. RSA fit statistics before and after final design space reduction

	Rotor Wt	*Etats*	*Rotor Wt*	*Etats*
	Before design space reduction		After design space reduction	
R^2	0.987	0.917	0.996	0.995
R^2_{adj}	0.985	0.905	0.996	0.994
RMS Error	0.0940	0.0200	0.0235	0.00170
Mean of Response	1.04	0.771	1.04	0.844
Observations	224	224	310	310

14.4 Conclusions

An optimization framework can be used to facilitate the optimization of a wide variety design problems. The applicability of the framework was demonstrated using a liquid-rocket compact radial turbine.

1. **Surrogate Modeling**. The radial turbine optimization process began without a clear idea of the location of the feasible design region. Response surface approximations of output constraints were successfully used to identify the feasible design space.
2. **Design Space Refinement**. The feasible design space was still too large to accommodate the construction of an accurate RSA for the prediction of turbine efficiency. A reasonable design space was defined by eliminating poorly performing areas, thus improving RSA fidelity.
3. **Dimensionality Reduction**. A global sensitivity analysis provided a summary of the effects of design variables on objective variables, and it was determined that no variable could be eliminated from the analysis.
4. **Multi-objective Optimization using Pareto Front.** Using the Pareto front information constructed using genetic algorithms, a best trade-off

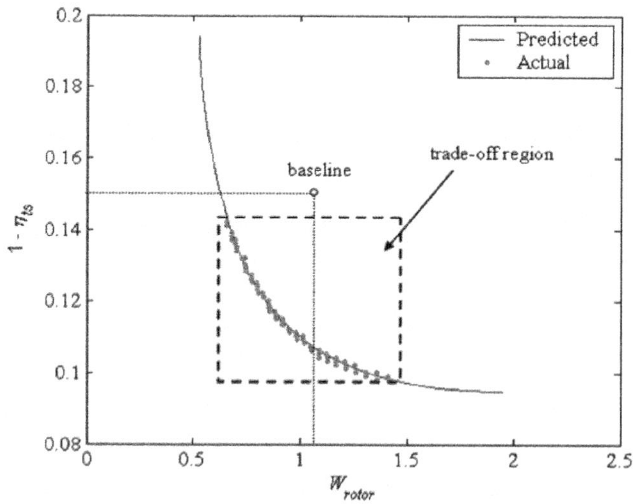

Fig. 14.7. Pareto Front with validation data. Deviations from the predictions are due to rounded values of the input variables (prediction uses more significant digits). The quantity 1 – *Etats* is used for improved plot readability.

region was identified within which the Pareto front and the RSAs used to create the Pareto front were validated. At the same weight, the response surface-based optimization resulted in a 5% improvement in efficiency over the baseline case.

Through this case study, a number of aspects from the framework were demonstrated, and the benefits of the various steps were made apparent. The framework provides an organized methodology for attacking several issues that arise in design optimization.

Acknowledgments

This research was performed in coordination with the Institute for Future Space Transport (IFST) under the Constellation University Institute Project (CUIP).

References

1. Balabanov VO, Giunta AA, Golovidov O, Grossman B, Mason W, Watson LT, Haftka RT (1999), Reasonable design space approach to response surface approximation. J Aircraft 36:1:308–315
2. Deb K, Agrawal S, Pratap A, Meyarivan T (2000) A fast and elitist multi-objective genetic algorithm for multi-objective optimization: NSGA-II. In: Nature VI Conference, Paris, pp 849–858

3. Goel T, Vaidyanathan R, Haftka RT, Queipo NV, Shyy W, Tucker PK (2004) Response surface approximation of Pareto optimal front in multi-objective optimization. In: 10th AIAA/ISSMO Multidisciplinary Analysis and Optimization Conference, Albany NY, Paper No. 2004-4501
4. He B, Ghattas O, Antaki JF (1997) Computational Strategies for Shape Optimization of Time-Dependent Navier-Stokes Flows. Technical Report CMU-CML-97-102, Computational Mechanics Lab, Department of Civil and Environmental Engineering, Carnegie Mellon University
5. Huber, F (2001) Turbine aerodynamic design tool development. In: Space Transportation Fluids Workshop, Marshall Space Flight Center, AL
6. Kim S, Leoviriyakit K, Jameson A (2003) Aerodynamic Shape and Planform Optimization of Wings Using a Viscous Reduced Adjoint Gradient Formula. Second M.I.T. Conference on Computational Fluid and Solid Mechanics at M.I.T, Cambridge, MA
7. Kontoravdi C, Asprey SP, Pistikopoulos EN. Mantalaris A (2005). Application of Global Sensitivity Analysis to Determine Goals for Design of Experiments – An Example Study on Antibody-producing Cell Cultures, Biotechnol Progr (in press).
8. Madsen JI, Shyy W, Haftka RT (2000) Response surface techniques for diffuser shape optimization. AIAA J 38:1512–1518
9. Myers RH and Montgomery DC (2002) Response Surface Methodology. John Wiley & Sons, Inc., New York
10. Nadarajah S, Jameson A, and Alonso JJ. (2002) Sonic boom reduction using an adjoint method for wing-body configurations in supersonic flow., 9th. AIAA/ISSMO Symposium on Multidisciplinary Analysis and Optimization Conference, Atlanta, GA, AIAA paper 2002-5547
11. Papila N, Shyy W, Griffin L, Huber F, Tran K. (2000) Preliminary design optimization for a supersonic turbine for rocket propulsion. In: AIAA/SAE/ASME/ASEE 35th Joint Propulsion Conference, Paper No. 2000–3242
12. Papila N, Shyy W, Griffin L, Dorney DJ (2002) Shape optimization of supersonic turbines using global approximation methods. J Propulsion and Power 18:509–518
13. Queipo N, Haftka RT, Shyy W, Goel T, Vaidyanathan R (2005) Surrogate-Based Analysis and Optimization. Accepted for publication in Prog in Aero Sci
14. Rodriguez DL (2002) A Multidisciplinary Optimization Method for Designing Boundary Layer Ingesting Inlets. 9th AIAA/ISSMO Symposium on Multidisciplinary Analysis and Optimization, Atlanta, Georgia, AIAA paper 2002-5665
15. Roux WJ, Stander N, Haftka RT (1998) Response surface approximations for structural optimization. Int J for Numer Methods in Eng 42: 517–534
16. Sobol IM (1993) Sensitivity analysis for nonlinear mathematical models. Mathematical Modeling & Computational Experiment 1:4:407-414
17. Vaidyanathan R, Papila N, Shyy W, Tucker KP, Griffin LW, Haftka RT, Fitz-Coy N (2000) Neural network and response surface methodology for rocket engine component optimization. In: 8th AIAA/USAF/NASA/ISSMO Symposium on Multidisciplinary Analysis and Optimization, Long Beach, CA, Paper No. 2000-4480

Part III

Handling Noisy Fitness Functions

15

Hierarchical Evolutionary Algorithms and Noise Compensation via Adaptation

Ferrante Neri[1,2] and Raino A. E. Mäkinen[1]

[1] Department of Mathematical Information Technology, P.O. Box 35 (Agora), FI-40014 University of Jyväskylä, Finland, neferran@cc.jyu.fi, rainom@it.jyu.fi

[2] Dipartimento di Elettrotecnica ed Elettronica, Politecnico di Bari, Via E. Orabona 4, 70125, Bari, Italy, neri@deemail.poliba.it

Summary. Hierarchical Evolutionary Algorithms (HEAs) are Nested Algorithms composed by two or more Evolutionary Algorithms having the same fitness but different populations. More specifically, the fitness of a Higher Level Evolutionary Algorithm (HLEA) is the optimal fitness value returned by a Lower Level Evolutionary Algorithm (LLEA). Due to their algorithmic formulation, the HEAs can be efficiently implemented in Min-Max problems. In this chapter the application of the HEAs is shown for two different Min-Max problems in the field of Structural Optimization. These two problems are the optimal design of an electrical grounding grid and an elastic structure. Since the fitness of a HLEA is given by another evolutionary algorithm (LLEA), it is noisy. This noise, namely Hierarchical Noise (HN) is distributed according to an asymmetrical and non-Gaussian probability function. A preliminary analysis of the HN is performed and a set of adaptive rules are then carried out in order to robustly handle this kind of noise. The Adaptive Higher Level Evolutionary Algorithm (AHLEA) is thus proposed. The AHLEA works on the sample size, the population size, and the survivor selection scheme in order to ensure the reliability of the optimization process in presence of the HN. The analysis of a benchmark problem proves the effectiveness of the adaptive rules carried out and the tests for grounding grids and elastic structures show the applicability of the AHEAs to real world problems.

15.1 Introduction

In applied science and engineering to perform a design means to determine that set of parameters defining the design (design parameters), such that a disadvantage is minimum (or an advantage is maximum). Since every design is supposed to work in various working conditions, there naturally arises the necessity of finding, for each set of design parameters, the worst conditions, that is the situation (e.g. the instant or the application point) such that the disadvantage is maximum. Then, among these possible designs it is necessary

F. Neri and Raino A.E. Mäkinen: *Hierarchical Evolutionary Algorithms and Noise Compensation via Adaptation*, Studies in Computational Intelligence (SCI) **51**, 345–369 (2007)
www.springerlink.com

to find that one which ensures that the most critical situation is minimum. In other words, the class of problems under study can be formalized as the minimization, in a certain domain, of the maximization, in another domain, of this disadvantage or, more technically named as Min-Max problems.

In order to solve the Min-Max problems the Authors propose the application of an evolutionary algorithm which consists of two hierarchically connected evolutionary algorithms [1], [2], [3], [4], [5]. The proposed Hierarchical Evolutionary Algorithm (HEA) is composed of a Lower Level Evolutionary Algorithm (LLEA) which performs the maximization of the disadvantage in order to find the worst case working conditions and a Higher Level Evolutionary Algorithm (HLEA) which performs the minimization of the worst case disadvantage working on a population made up of design parameters [4]. Since the HLEA has to minimize a fitness given by another evolutionary algorithm, the LLEA, it works with a noisy fitness. This kind of noise is asymmetrical and cannot be modelled by means of a normal distribution $N(0, \sigma^2)$. This chapter analyzes this noise and proposes some adaptive rules in order to defeat it. An Adaptive Higher Level Evolutionary Algorithm (AHLEA) and thus an Adaptive Hierarchical Evolutionary Algorithm (AHEA) are shown. Moreover, the authors apply the AHEA to two complex and delicate engineering problems, that is the design of electrical grounding grids and the design of an elastic structure.

15.2 Algorithmic Aspects

In this section the theoretical aspects concerning the Hierarchical Evolutionary Algorithms (HEAs) are analyzed. It is shown how the HEAs can be naturally applied to solve Min-Max and Max-Min problems [6]. Moreover, reasons of the presence of the algorithmic noise are explained and the behavior of this noise is shown in some cases. A characterization of the noise based on the statistics of the phenomenon is also shown.

15.2.1 Definitions and Notation

We consider the following abstract design optimization problem. Let $\mathcal{U}$ denote the set of admissible designs α. To each design $\alpha \in \mathcal{U}$ we associate a closed and bounded set $X_\alpha \subset \mathbb{R}^n$. Denoting $X = \cup_{\alpha \in \mathcal{U}} X_\alpha$, we define a function $\Phi : \mathcal{U} \times X \to \mathbb{R}$. For fixed α, the function $\Phi(\alpha, \cdot) : X_\alpha \to \mathbb{R}$ is the observation on the system. The cost or disadvantage caused by the design is measured by the maximum value of the observation. Our aim is to find a design that minimizes this cost. In specific cases α may represent a vector of physical dimensions of the system, or be just an integer referring to a fixed number of possible system configurations. The observation, on the other hand, may represent e.g. the temperature distribution in a structure.

Definition 1. *By a Min-Max problem we mean finding $\alpha^* \in \mathcal{U}$ that solves*

$$\min_{\alpha \in \mathcal{U}} \max_{x \in X_\alpha} \Phi(\alpha, x). \tag{15.1}$$

In other words, the problem consists in finding design α such that the worst case situation is minimum.

In what follows we assume that either $\mathrm{card}(X_\alpha) < \infty$, or $\Phi(\alpha, \cdot)$ is continuous on X_α. Thus the maximum in (15.1) is well-defined. Well-posedness of the minimization problem is clear in the combinatorial optimization case $\mathrm{card}(\mathcal{U}) < \infty$. Rigorous mathematical analysis of a more general case is beyond the scope of this chapter. In order to solve this kind of problem it is necessary to have an instrument which performs the maximization of the function $\Phi(\alpha, \cdot)$ within each set X_α and an instrument which finds the minimum value of $f(\alpha) := \max_x \Phi(\alpha, x)$ and the corresponding optimal design α^*. If the function Φ is not differentiable or the evaluation of its gradient is not practical, an approach which makes use of gradient-based information cannot be applied. Moreover, if this function can be multimodal a direct method can easily fail and an evolutionary approach can be more successful. If an evolutionary approach is applied, a lower level evolutionary algorithm working on a population of $\{x_j\}$ has to maximize the function $\Phi(\alpha, \cdot)$ and the returned maximum value $\Phi(\alpha, x^*)$ has to be used as the fitness of a higher level evolutionary algorithm which works on a population $\{\alpha_i\}$ of design parameters. The result of the whole optimization process is therefore a design which ensures that the most critical situation is minimized. According to the analysis performed the following definition is given.

Definition 2. *A Hierarchical Evolutionary Algorithm (HEA) is a nested evolutionary algorithm composed by two evolutionary algorithms: the Lower Level Evolutionary Algorithm (LLEA) and the Higher Level Evolutionary Algorithm (HLEA). The two algorithms work on different populations and the fitness of the HLEA is the optimal fitness returned by the LLEA. More specifically, each fitness evaluation performed by the HLEA requires an optimization process executed by the LLEA.*

The pseudocode of a HEA is shown in Fig. 15.1. It is obvious that a HEA can be used analogously to solve a max-min problem.

15.2.2 The Noise in Hierarchical Evolutionary Algorithms

As highlighted above, the HLEA is an evolutionary algorithm whose fitness $f(\alpha)$ at α is given by the optimization result of another evolutionary algorithm, i.e. $f(\alpha) = \max \Phi(\alpha, x)$, $x \in X_\alpha$. Due to its inner structure, an evolutionary algorithm returns an optimal value that is not deterministic but takes its own value according to a stochastic process. The robustness of an evolutionary algorithm obviously depends on both the problem (the fitness

```
begin-HLEA
  create initial HLEA population;
  while (conditions HLEA)
    recombination and mutation;
    for i=1 to number of fitness evaluations
      begin-LLEA
        create initial LLEA population;
        while (conditions LLEA)
          recombination and mutation;
          fitness evaluations of Φ(α_i, x_j);
          survivor selection;
        end-while
      end-LLEA and returns Φ(α_i, x*)
    end-for
    survivor selection;
  end-while
end-HLEA
```

Fig. 15.1. Pseudocode of the HEA

landscape) and the setting of parameters such as the mutation probability and the population size and so on. The HLEA, therefore, works in a noisy environment since the fitness returned by the LLEA is noisy.

The problem of the evolutionary optimization in noisy environment has been intensively discussed in recent years. Considering that the noise in fitness evaluations can come from measurement errors or from randomized simulations, the noisy fitness has been described in literature as follows (see [9] and [8]):

$$F(\alpha) = \int_{-\infty}^{\infty} [f(\alpha) + z] \, p(z) \, dz = f(\alpha), \quad z \sim N\left(0, \sigma^2\right), \tag{15.2}$$

where α is the variable to be optimized, $f(\alpha)$ is a time-invariant fitness function. The additive noise z, as shown in (15.2), is classically assumed to be normally distributed with zero mean (just in some cases it is assumed to have a Cauchy distribution [9]). Under these hypotheses, when an evolutionary algorithm is implemented, since $f(\alpha)$ is, as is obvious, not available the following approximated function is often used:

$$\tilde{F}(\alpha) = \frac{1}{n_s} \sum_{k=1}^{n_s} [f(\alpha) + z_k], \tag{15.3}$$

where n_s is the sample size. This sample size (of a candidate solution) is defined as the number of fitness evaluations performed for a given candidate solution. In literature, several approaches to approximate the fitness in presence of noise have been proposed. These approaches were making use of *Explicit* or *Implicit Averaging* techniques. The averaging techniques are defined as the following [9].

Definition 3. Explicit Averaging *techniques are those algorithmic strategies which, in order to assign a more reliable fitness value, calculate the average value among a certain set of numbers. Those techniques which average over a number of performed samples are called* Explicit Averaging Over Time. *Those techniques which calculate the fitness by averaging over the neighborhood of the point to be evaluated are called* Explicit Averaging Over Space.
Implicit Averaging *techniques are those algorithmic strategies which work on enlargement (static or dynamic) of the population size in order to defeat the noise.*

For the kind of noise analyzed in this chapter, the mathematical description in (15.2) and the approximation in (15.3) are not valid and a "classical" approach cannot be thus applied. In fact, in a Min-Max problem, the LLEA is supposed to find the global maximum of the function $\Phi(\alpha, \cdot)$. The LLEA could converge to the global maximum $f(\alpha)$ or could converge to a suboptimal solution. In the latter case, the LLEA would return a value $f(\alpha)' < f(\alpha)$. It is obvious that the LLEA cannot return a value higher than $f(\alpha)$ since it is the global maximum. Moreover, it is important to remark that, in the HEA noisy environment, the HLEA does not work with overestimated solutions but only with underestimated ones. It follows that noise due to the hierarchical algorithms is non-Gaussian, asymmetrical and an explicit averaging [1], [2], [12] or an implicit averaging [7], [14], [12], [16] technique cannot be applied since the fitness value nearest to the true value cannot be calculated by means of the average value among a certain number of samples n_s. Three examples of distribution of the noise in HEAs are given. Three different functions have been tested performing 500 samples each. Figure 15.2 shows the distribution of the samples in the case of the maximization of the Rastrigin's function [17] multiplied by -1, Fig. 15.3 shows the distribution in the case of the Ackley function [17] multiplied by -1, and Fig. 15.4 shows the distribution in the case of the current field problem [18]. As can be observed, in the latter case the LLEA has a much higher probability of finding a suboptimal solution than an optimal one. This behavior is due to the fact that there is a suboptimal solution having a "strong" genotype which tends to attract the optimization process. Taking into account the three shown examples, it is possible to conclude that in general the probability distribution related to this kind of noise can have several different shapes (depending on the problem and on the parameter setting of the LLEA), and it cannot be easily approximated by means of a classical distribution function.

Let us define this noise as "Hierarchical Noise" (HN) and let us define the set of distribution functions $\{H(f(\alpha), w)\}$ characterizing it. Here w is the amplitude of the interval where an infinite number of samples would fall within. For a finite number of samples n_s the noise is approximated in the

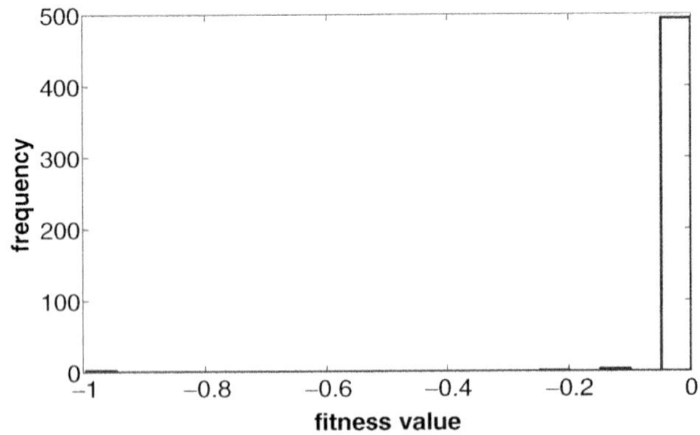

Fig. 15.2. Distribution of 500 samples performed by a LLEA for the Rastrigin's function

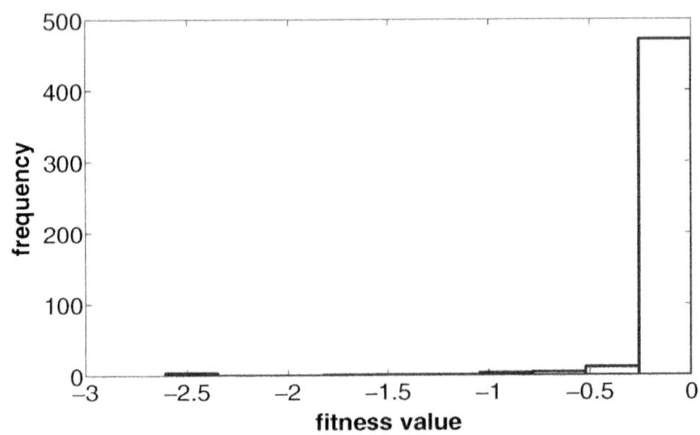

Fig. 15.3. Distribution of 500 samples performed by a LLEA for the Ackley function

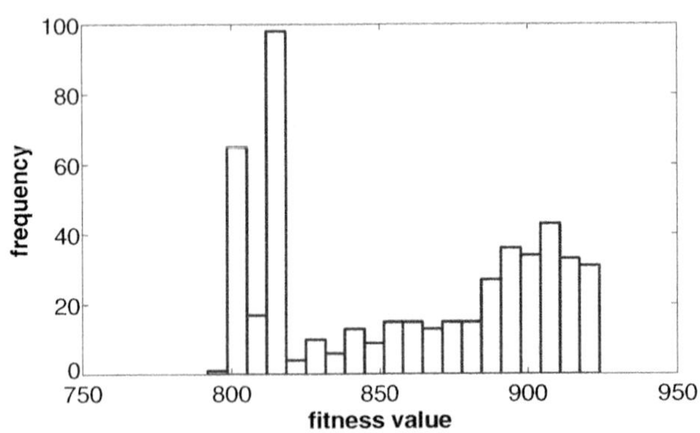

Fig. 15.4. Distribution of 500 samples performed by a LLEA in a current field problem

following way. If $S = \{F_1(\alpha), F_2(\alpha), ..., F_{n_s}(\alpha)\}$ is the set of samples obtained by means of the LLEA, the noise is approximated by

$$H(\gamma, \tilde{w}), \tag{15.4}$$

where $\gamma = \max(S)$ and $\tilde{w} = \max(S) - \min(S)$.
Under these hypotheses, Eq. (15.2) in the case of HN is thus replaced with the following

$$F(\alpha) = f(\alpha) + z, \; z \sim H(\gamma - f(\alpha), \tilde{w}), \; z \leq 0. \tag{15.5}$$

15.2.3 Noise Compensation via Adaption

In this section we consider that the LLEA is a "black box" fitness calculator which returns a noisy value (i.e. the algorithmic parameters of the LLEA cannot be tuned and are unknown) and our aim is to minimize this noisy fitness by means of the HLEA. In deterministic optimization problems, if the fitness landscape contains several minima, a standard evolutionary algorithm does not ensure convergence to the global optimum and therefore the algorithm could stagnate or prematurely converge. In the presence of uncertainties this problem is emphasized since noise in the optimization process introduces some "false" minima and maxima which clearly disturb functioning of the algorithm. Since the HN does not allow an individual to be overestimated but only underestimated, only "false" minima can be found in the fitness landscape. It is important to remark that an underestimation in the maximization logic of the LLEA corresponds to an overestimation in the minimization logic of the HLEA. Therefore, these "false" minima can easily lead to the convergence to a suboptimal solution or introduce wrong search directions and thus jeopardize the behavior of the algorithm [8].

All the evolutionary algorithms contain fitness-based pairwise comparisons which lead to a selection of the individuals which are going to generate offspring or survive to the subsequent generation [19]. In other words, every evolutionary algorithm executes explicitly or implicitly a fitness-based sorting of the individuals. When the fitness function is noisy the pairwise comparisons are disturbed and thus the algorithm could wrongly select the individuals. Since in the case of HN the HLEA could perform only overestimation, the optimization process could turn out unreliable.

It is therefore fundamental, in order to successfully optimize a noisy fitness, to synergically operate on two different aspects of the algorithm:

(a) assigning a proper fitness value at each candidate solution (by means of re-sampling and population size)

(b) sorting the solutions taking into account the possibility of wrong fitness estimations (by means of the selection scheme)

Thus, in order to smooth "false" minima in the fitness landscape and to defeat this noise without excessively increasing computational overhead, several algorithmic aspects are analyzed here and an adaptation is proposed.

Fitness

Since the true fitness is not available due to the presence of the noise, an approximated estimated fitness must be used. As highlighted above the formula (15.3) cannot be used for the HN. Considering that one candidate solution has been evaluated n_s times, the following approximated fitness is proposed here

$$\tilde{F}(\alpha) = \gamma + \frac{\beta}{n_s}, \tag{15.6}$$

where $\gamma = \max(S) = \max\left\{F_1(\alpha), F_2(\alpha), ..., F_{n_s}(\alpha)\right\}$ and β is a positive weight coefficient. The meaning of $\tilde{F}$ is that the estimated fitness value γ has to be corrected by a term which takes into account the reliability of the estimated fitness value. The corrective term $\frac{\beta}{n_s}$ is a penalty term which has a big influence with unreliable solutions (n_s small) and which, progressively, tends to have a negligible influence for reliable solutions (n_s large). Besides, in order to avoid extremely high computational overhead, a maximum number of samples n_s^{max} is established taking into account features of the noise under examination. As can be observed from (15.6), the fitness $\tilde{F}$ is dynamic since it is composed of two dynamic contributions. First the term $\frac{\beta}{n_s}$ decreases as the number of samples increase, secondly, the term γ is updated every time a higher fitness sample is found, since it is the maximum value among the performed samples.

Fitness Diversity of the Population

It is also important to observe that the influence of noise on the optimization process is not the same in all stages of the evolution. More specifically, a population containing a high fitness diversity is not significantly influenced by the noise; on the contrary a population made up of individuals having similar fitness values is heavily influenced by the noise. This statement can be visualized in the following way. If the distance between two individuals of the population is such that their two fitness values differ by a quantity bigger than the amplitude of the "false" minima, the pairwise comparison is efficiently executed. If this quantity is smaller than the amplitude of the "false" minima, it is necessary to sample the fitness values of both the individuals several times and then compare the two γ values. From this consideration it follows that in high diversity conditions there is no need to perform many samples *Explicitly* (i.e. re-sampling several times the same candidate solution) or *Implicitly* (i.e. using a large population size) but as the fitness diversity decreases an increasing sample and population size is required.

In order to estimate the fitness diversity of the population at a certain generation the following coefficient is used

$$\xi = \min\left\{1, \left|\frac{\tilde{F}_{best} - \tilde{F}_{avg}}{\tilde{F}_{best}}\right|\right\}, \tag{15.7}$$

where $\tilde{F}_{best}$ and $\tilde{F}_{avg}$ are respectively the best and average fitness among the fitness values of the individuals of the population in the generation under study. The coefficient ξ can be seen as a measurement of the current state of convergence of the algorithm [20]. More specifically, if $\xi \approx 1$, it means that there is a high diversity (in terms of fitness values) among individuals of the population and that the solutions are not exploited enough, on the contrary, if $\xi \approx 0$, it means that the diversity is low and thus convergence is going to happen and since this convergence can be premature, a higher search pressure is needed. Thus, the coefficient ξ can be used to dynamically set the algorithmic parameters [21] in order to prevent premature convergence and stagnation and therefore ensure a more robust behavior in presence of uncertainties.

Sample Size

Although an explicit averaging technique among n_s [1], [2], [7], [22], [23], [24], [25] is not applicable, it is clear from (15.6) that a large sample size allows the algorithm to have more chances of finding the real global maximum or a value near to it (i.e. $\gamma \approx f(\alpha)$). It is therefore useful to execute re-samplings in order to smooth the "false" minima. Obviously, the number of samples ensuring that $\gamma \approx f(\alpha)$ depends on the algorithmic parameters of the LLEA and on the fitness landscape with whom the LLEA is working. On the other hand it should be noted that each HLEA function evaluation comes from a LLEA evolutionary maximization and is then usually computationally expensive [25]. A strategy which simply enlarges the sample size for all individuals of the population is not acceptable in terms of computational overhead. It is therefore proposed not to re-sample all the candidate solutions but only those which "need" to be re-sampled. More specifically,it is proposed that the population undergoes a reevaluation cycle. For each individual the following value is calculated:

$$n_s^{add} = \text{round}\left(n_s^{max} \cdot \frac{(1-\xi)}{n_s}\right), \tag{15.8}$$

where n_s^{add} is the number of additional samples that have to be performed. Each individual is then associated with its own number of additional samples to be performed. As shown in (15.8), this number depends on the coefficient ξ and on the number n_s of samples already executed. If n_s is low and then the solution unreliable, some additional samples are performed; if, on the contrary, the solution has already been sampled several times, further samples will not lead to a significant improvement of the solution in terms of reliability. If $\xi \approx 0$ some additional samples are performed in order to avoid the risk of wrongly executing the sorting over individuals having similar fitness values; if $\xi \approx 1$ there is no need to recalculate the fitness since the noise does not heavily bias sorting of the individuals. The main idea behind this reevaluation cycle is the following. As explained above, a high number of samples gives more of a chance to the candidate solution to take a value $\gamma \approx f(\alpha)$ and therefore it can help in handling the noisy landscape but these additional samples are computationally expensive and it is thus important to avoid unnecessary samples. The proposed criterion is then oriented to execute additional samples just when they are necessary, analyzing the history of each individual and the fitness diversity of the whole population. When the additional samples have been established for each individual, these samples are executed and each fitness $\tilde{F}$ is consequently updated.

Population Size

In the case of the Gaussian noise, it has been shown that employment of a large population size is beneficial since it allows sampling of several solutions belonging to the same promising area thus executing an *Implicit Averaging* of the noise [9], [7], [12], [16]. Also for the HN a large population size can be beneficial because it samples many solutions in the promising areas of the decision space and thus gives a better chance that some of them are properly scored (i.e. $\gamma \approx f(\alpha)$). On the other hand, due to computational cost of the fitness function, it is important to avoid unnecessary fitness evaluations. A dynamic population size is therefore proposed which considers the population as the merge of two contributions adaptively determined at each generation (see (15.9) and (15.10)). The general idea is that the algorithm enlarges the population size when ξ decreases and shrinks it when ξ increases. In this way the algorithm executes a massive *Implicit Averaging* only when the fitness diversity is low i.e. $\xi \approx 0$, it is thus desirable that a proper fitness based sorting is performed.

Survivor Selection

The problem with choice of the reinserting strategy in presence of noise is definitely questionable [8] and a proper choice probably depends on the noise under consideration as well as the other algorithmic choices. In the case of the HN, the "prudent-daring" survivor selection mechanism is proposed [20] based on the following consideration. Due to the fact that only overestimations are allowed it is important that the survivor selection scheme takes into account the reliability of the solutions in order to avoid overestimated solutions offering wrong search directions. On the other hand, in order to try speeding up the optimization process, the unreliable performing individuals should not be excessively penalized. The proposed survivor selection scheme employs a competitive logic which makes use of two reinserting mechanisms. The main idea is that a "prudent" reinserting mechanism and a "daring" reinserting mechanism should compete and cooperate [26] following both the noisy fitness landscape and the state of the population evolution. Operatively the survivor selection mechanism consists of the following.

(a) The "prudent" mechanism selects, among parents and newly generated offspring a number, equal to S_{pru}, of the individuals with best performance according to the approximated fitness $\tilde{F}$. S_{pru} is given by

$$S_{pru} = S_{pru}^{f} + \text{round}\left(S_{pru}^{v}\left(1 - \xi\right)\right) \tag{15.9}$$

where S_{pru}^{f} is the minimum size of the population and S_{pru}^{v} is the maximum size of the variable population.

(b) The "daring" mechanism selects, among parents and newly generated offspring, a small number of individuals having the best performance according to γ. More specifically, at each generation, the "daring" mechanism calculates the value

$$S_{dar} = \text{round}\left(S_{dar}^{max} \cdot \xi\right) \tag{15.10}$$

where S_{dar} is the number of individuals to be selected according to γ and S_{dar}^{max} is the maximum number of allowed "daring" solutions.

Prudent and daring individuals thus constitute the population for the subsequent generation. A control filter avoids that the same individual being taken twice by both the reinserting mechanisms.

As equations (15.9) and (15.10) show, S_{pru} and S_{dar} depend on ξ. When ξ decreases S_{pru} is enlarged S_{dar} is shrunk, when ξ increases S_{pru} is shrunk S_{dar} is enlarged. According to the algorithmic philosophy the biggest contribution of the population is given by the prudent survivor selection but a limited number promising solutions are taken into consideration, although these solutions could have been overestimated. The quantity of prudent and daring individuals in the population is thus determined by means of the the coefficient ξ. In particular, when the fitness diversity is very low ($\xi \approx 0$) the population is made up of only prudent individuals since the introduction of unreliable solutions could mislead the search of the optimum.

The Adaptive Hierarchical Evolutionary Algorithm

On the basis of the considerations carried out, an Adaptive Hierarchical Evolutionary Algorithm (AHEA) is proposed in order to defeat the algorithmic noise HN. Considering that the LLEA returns a noisy fitness value, the Adaptive Higher Level Evolutionary Algorithm (AHLEA) has to handle the noise by adaptation. It is important to remark that a proper parent selection scheme, recombination and mutation strategy depend on the problem under study (e.g. the real value or discrete representation, the cardinality of the decision space etc.). The mutation probability here proposed is dynamic (see [27]) and depends on the coefficient ξ. The value 0.4 is a weight chosen based on a semi-empirical consideration that a mutation occurred on over 40% of the population could be too explorative an action spoiling the genotype of some good solutions (see [20] and [28]). This dynamic mutation probability has the role of increasing the explorative pressure in low diversity conditions and to decrease it in high diversity conditions. According to our study, we concluded this parameter and its adaptation does not have a big influence on the robustness of the algorithm in the presence of HN; nevertheless its employment is highly beneficial in terms of handling very high cardinality of the decision space and multimodality of the fitness landscapes. The pseudocode of a AHLEA is shown in Fig. 15.5.

Parameter Setting

The latest issue to be discussed is the parameter setting. As highlighted in [17], the algorithm performance of an adaptive evolutionary algorithm is not very sensitive to details of the meta-parameters unlike the case of the traditional evolutionary algorithms. For example when a classical Genetic Algorithm (GA) is designed, a proper choice of the population size is determinant for the success of the algorithm. In our case, although the AHLEA is proven to be very robust, a proper setting of the meta-parameters is required in order to efficiently handle the HN.

According to our empirical studies some rules in making a proper parameter setting are given. Concerning n_s^{max}, the choice must take into account that a HN noise with a wide range of variability $\tilde{w}$ requires a high number of samples and a HN with a narrow range of variability $\tilde{w}$ does not. This statement can be easily justified by means of an analogy with the Gaussian noise: a distribution with a big standard deviation value requires more samples to be characterized than one with a small standard deviation value. Thus, n_s^{max} has to be chosen according to the following formula:

$$n_s^{max} = K_1 \tilde{w}, \tag{15.11}$$

where $K_1 \in [2, 20]$. The boundaries of this interval have been empirically set on the basis of several tests.

Regarding β, we suggest that the penalty term $\frac{\beta}{n_s}$ does not penalize more than 10% the fitness value when only one sample is performed and does not penalize at all (less than 1%) when the maximum number of sampling has been performed. In other words from

$$\begin{cases} \beta = K_2 \tilde{F}(\alpha), \; K_2 \in [0.05, 0.3], \; \text{when } n_s = 1 \\ \frac{\beta}{n_s^{max}} = K_3 \tilde{F}(\alpha), \; K_3 \in]0, 0.01], \; \text{when } n_s = n_s^{max} \end{cases} \tag{15.12}$$

considering (15.6) and (15.11) it can be found

```
begin-AHLEA
  create initial AHLEA population;
  fitness evaluations of F by mean of the LLEA and
  calculation of F̃ = γ + β/n_s;
  calculation of ξ;
  while (conditions AHLEA)
    parents selection;
    recombination;
    calculation of the mutation probability p_m = 0.4(1 − ξ);
    mutation;
    fitness evaluations of F by the LLEA and calculation of F̃ = γ + β/n_s;
    survivor selection:
    1) prudent survivor selection:
      calculation of S_pru = S^f_pru + round(S^v_pru (1 − ξ));
      selection according to F̃ of the S_pru best performing candidate
      solutions among parents and newly generated offspring;
    2) daring survivor selection:
      calculation of S_dar = round(S^max_dar · ξ);
      selection according to γ of the S_dar best performing candidate
      solutions among parents and newly generated offspring;
    3) filtering in order to avoid that the same candidate
      solution is selected by both the mechanisms;
    4) merge of the solutions selected by the two mechanism and creation
      of the new population made up S_pru + S_dar individuals;
    re-sampling cycle:
      for (each individual of the population)
        calculation of n^add_s = round(n^max_s · (1−ξ)/n_s);
        execution of the additional evaluations;
      end-for
    calculation of ξ;
  end-while
end-AHLEA
```

Fig. 15.5. Pseudocode of the AHLEA

$$\begin{cases} \beta = \frac{\gamma}{9} \\ \beta = \frac{1}{99}\gamma\, n_s^{max} = \frac{5}{99}\gamma\, \tilde{w}. \end{cases} \tag{15.13}$$

Obviously it is not always possible to satisfy both equations in (15.13) and a compromise must be found.

Concerning the value of S_{dar}^{max} the following empirical formula is proposed:

$$S_{dar}^{max} = K_4 S_{pru}^{f}, \quad K_4 \in]0, 0.5]. \tag{15.14}$$

This formula can be justified in the following way. When $\xi \approx 1$ the prudent population size is S_{pru}^{f} and then the population size is minimum. In this condition the "daring" mechanism should work at the maximum but its contribution should never be dominant with respect to the "prudent" one.

15.2.4 Benchmark Test of the AHLEA

Let us consider the Ackley function:

$$Ack(\bar{x}) = -20\exp\left(-0.2\sqrt{\frac{1}{n}\sum_{i=1}^{n}x_i^2}\right) - \exp\left(\frac{1}{n}\sum_{i=1}^{n}\cos(2\pi x_i)\right) + 20 + e \quad (15.15)$$

with $n = 3$. Then, let us define $\tilde{Ack}(\bar{x}) = -Ack(\bar{x})$ and consider the following problem:

$$\min_{x_3} g(\bar{x}) = \min_{x_3}\max_{x_1,x_2} \tilde{Ack}(\bar{x}) \text{ in } [-5,5]^3 \quad (15.16)$$

Table 15.1 shows the parameter setting for the LLEA employed for the maximization part of the problem.

Table 15.1. LLEA Parameter Setting

representation	vectors of real numbers
population size	40 individuals
parent selection	proportionate
crossover	20 blend arithmetic crossover per generation
mutation	Gaussian mutation with mutation probability 0.1
survivor selection	fitness-based plus strategy
stop criterion	number of generations > 70 OR $\lvert f_{best} - f_{avg}\rvert < 0.0001$

The AHLEA has been tested for solving the minimization part of the problem (15.16) and it has been compared with the following algorithms:

(a) A standard Genetic Algorithm (GA)
(b) A standard Evolution Strategy (ES)
(c) A Re-sampling Evolution Strategy (RES) with a fixed sample size for each individual of the population [7]
(d) An Adaptive Re-sampling Evolutionary Algorithm (AREA) with the same ξ-based dynamical control of the population size and mutation probability implemented for AHLEA and a fixed sample size

Table 15.2 summarizes the algorithmic parameters chosen for the five methods under consideration.

Each algorithm has been run 50 times. Table 15.3 shows the results of the optimization x_3^* in the most successful run, the corresponding fitness values g_{min}, the fitness values in the most unsuccessful run g_{max}, the mean value g_{mean} over the 50 experiments and the related standard deviation σ.

Table 15.2. Methods for comparison

parameter	GA and ES	RES	AREA	AHLEA
initial population size	80	80	80	80
population size for subsequent generations	60	60	dynamic between 40 and 80	dynamic between 40 and 80
mutation probability	0.2	0.2	dynamic between 0 and 0.4	dynamic between 0 and 0.4
sample size	1	10	10	dynamic between 1 and 10
S_{dar}^{max}	-	-	-	10
β	-	-	-	0.2
fitness evaluations	6000	6000	6000	6000

Table 15.3. Comparison of the Min-Max results

Algorithm	x_3^*	g_{min}	g_{mean}	g_{max}	σ
GA	-4.7729	-10.2981	-9.2310	-8.8336	0.4919
ES	-4.6712	-10.9946	-9.3347	-9.0911	0.4201
RES	-4.5129	-10.7143	-10.5442	-10.2891	0.1191
AREA	-4.5302	-10.8312	-10.6626	-10.4813	0.1321
AHLEA	-4.5341	-10.9156	-10.7056	-10.5824	0.1254

Fig. 15.6 shows the algorithmic performance of the methods under study. The average best fitness is the average over the 50 runs of the best fitness values at the end of each generation. Since the algorithms work with different population sizes, the performance is expressed in terms of fitness evaluations.

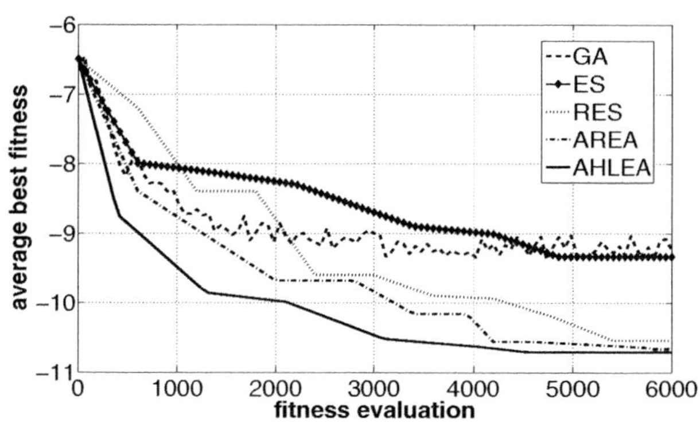

Fig. 15.6. Performance of the five algorithms under study

The results show that the non-re-sampling algorithms (GA and ES) cannot efficiently handle this kind of noisy fitness landscape. Both algorithms after having carried out some improvements, during the initial generations, converge to suboptimal solutions. According to our interpretation, this is due to the fact that the algorithms accept overestimated solutions which mislead the search. In addition, the values of σ in Table 15.3 prove that these algorithms are not robust for this class of problems. The three re-sampling algorithms converged to similar results and have a similar performance in terms of robustness, as the σ values in Table 15.3 show. On the other hand, the performance of these three algorithms differ a lot in terms of convergence velocity. In fact, the AREA outperformed the RES and the AHLEA outperformed both the RES and the AREA, in terms of convergence velocity. In light of the fact that the AREA can be seen as a RES with adaptive population size and the AHLEA can be seen as a RES with adaptive population and sample size, it follows that the comparison in Fig. 15.6 confirms the effectiveness of the applied adaptation.

Fig. 15.7 shows the behavior of the coefficient ξ in the most successful run of the AHLEA.

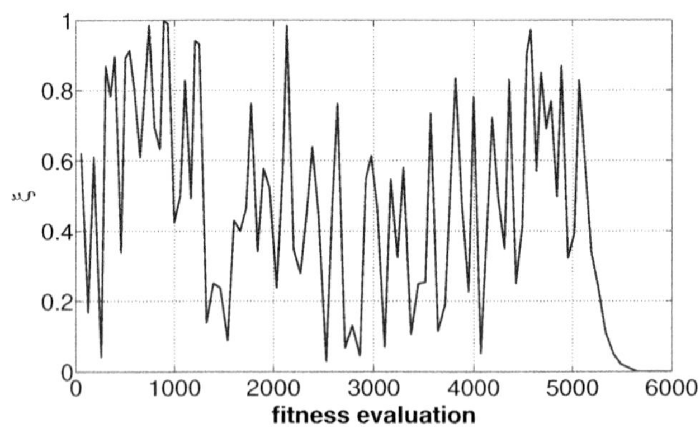

Fig. 15.7. Behavior of ξ during the optimization process

As expected this parameter has an oscillatory behavior at the beginning of the optimization process and after it settles down to the value 0 in convergence conditions (see [20]). Figure 15.8 shows the behavior of the prudent and daring population sizes. In presence of high fitness diversity, S_{pru} and S_{dar} have an oscillatory behavior and both cooperate in order to explore the decision space and to let promising solutions generate offspring without using a very large population size. In presence of very low diversity (see the last generations in Figure 15.8), just the prudent mechanism works in order to avoid overestimated solutions spoiling the search directions.

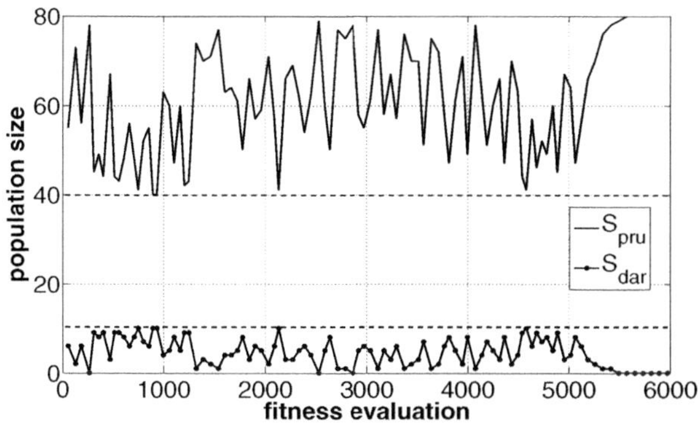

Fig. 15.8. Behavior of S_{pru} and S_{dar} during the optimization process

15.3 Applications

In this section, two applications of the AHLEA are shown for two structural optimization problems: the optimal design of a grounding grid and of an elastic structure.

15.3.1 Grounding Grid Optimal Design

Grounding grids are important countermeasures to assure the safety and reliability of the power systems and apparatus. When a fault occurs, the grounding grid leaks a fault current I_F which, flowing in the grounding grid, generates a current field and then touch voltages U_T on the points P of the soil surface. In order to guarantee the safety level conditions, these touch voltages must be lower than the values fixed by Standards [29]. This requirement causes a significant increase in the cost of both conductor material and ditching [30]. It is therefore fundamental, when a grounding grid has to be designed, to choose a criterion which ensures respect of the safety conditions without an extremely high installation cost. If the number of horizontal and vertical conductors and thus the installation cost is fixed by the designer, the problem can be described as the following

$$\min_{G} \max_{P} U_T \left(G, P\right), \tag{15.17}$$

where P is the point of the soil surface (i.e. two variables determining the position) and G is the the set of parameters characterizing the design of the grounding grid. In other words, we are aiming to find that grounding grid design which ensures that the most critical condition on U_T is minimized. In order to perform this design an AHEA is employed. With reference to (15.17), a LLEA is implemented to find the most dangerous point of the soil surface and then the maximum touch voltage, and an AHLEA is implemented to find the structure which ensures that the most dangerous condition (i.e. the maximum value of U_T) is minimized. The parameters characterizing the problem are summarized in Table 15.4. Table 15.5 shows the LLEA parameter setting

Table 15.4. Grounding Grid Problem

parameter	symbol	value
horizontal length	L_x	80 m
vertical length	L_y	60 m
burying depth	d	0.5 m
fault current	I_F	5 kA
resistivity of the soil	ρ	100 Ω/m
horizontal conductors	N_{cy}	5
vertical conductors	N_{cx}	7
conductors section	S_c	69 mm^2

for the maximization part of (15.17). The noise introduced by this LLEA

Table 15.5. LLEA Parameter Setting

representation	vectors of two real numbers
population size	100 individuals
parent selection	tournament
crossover	50 blend arithmetic crossover per generation
mutation	Gaussian mutation with mutation probability 0.2
survivor selection	fitness-based plus strategy
stop criterion	number of generations > 100 OR $\lvert U_{Tbest} - U_{Tavg}\rvert < 0.1$

is distributed as shown in Fig. 15.4. Moreover, it is important to observe that in order to determine each fitness value for the LLEA (the touch voltage U_T in a point P) it is necessary to solve a system of partial differential equations describing the current field. Since this system cannot be in general analytically solved, a numerical method is required [49], [32]. It follows that the fitness function of the LLEA is computationally rather expensive and then the fitness function of the AHLEA requires a high computational effort. It is therefore fundamental to avoid unnecessary fitness evaluations. The AHLEA with the parameter setting shown in Table 15.2 has been implemented. An individual representation made up of integer nonnegative numbers has been chosen since, as shown in [4], it turned out more convenient than classical binary representations in terms of cardinality of the decision space [33]. Briefly, according to this representation, an individual is described by two chromosomes in whose genes is encoded the distance of each conductor from the reference axis (see Fig. 15.9). The two point crossover technique and the random mutation explained in [4] have been set. The optimal grounding grid design is shown in Fig. 15.9. The maximum touch voltage generated by the

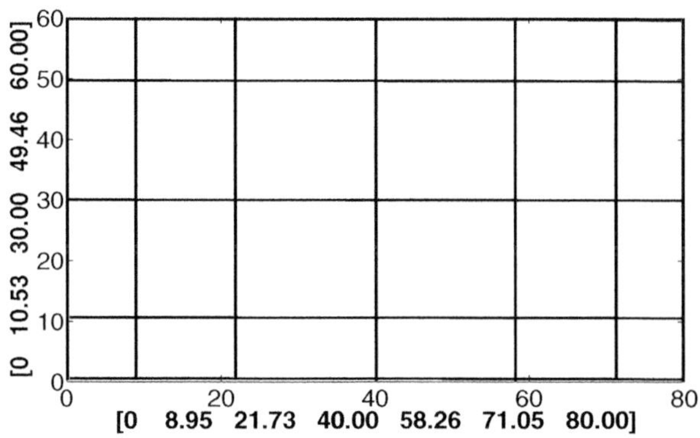

Fig. 15.9. Optimal grounding grid

structure in Fig. 15.9 is equal to 761.9968 V and is in the point having coordinates (30.6718, 40.5562). Obviously, due to the central symmetry of the structure, there are 3 more points taking the same value of touch voltage (within the other 3 central meshes). As expected, the result obtained is an "unequally spaced" grounding grid which as shown in [30], [34], [35] and [36] offers better performance than the traditional equally spaced one.

15.3.2 Topology Optimization of an Elastic Structure

We consider optimization of the topology of a linear elastic structure under static loading. Topology optimization differs from the classical shape optimization of structures [37] in such a way that the shape of the optimal structure is not obtained by a continuous transformation of an initial shape. The most popular deterministic approximate methods to solve this kind of optimization problems are the homogenization method and the so-called power-law approaches [38], [39]. Without a doubt, evolutionary optimization methods for structural topology optimization problems [40] are very slow compared to deterministic methods. However, they still deserve attention as they can overcome some of the weaknesses (e.g. local optimums, "intermediate" materials) of the homogenization type methods.

Assume that the structure in question is made from linear elastic material with Poisson's ratio equal to 0.3 and lies inside the rectangle $[0, N_1] \times [0, N_2]$. (In what follows nondimensional units are used.) On the lower left corner of the rectangle both displacements are restricted and in the lower right corner the vertical displacement is restricted. Let the rectangle be made discrete by $N_1 \times N_2$ square finite elements. Each element has either zero or unit thickness, i.e. void or material. The design parameter is thus the $N_1 \times N_2$ binary array $\{d_{k\ell}\}$. We require that $\sum_{k,\ell} d_{k\ell} = \varphi N_1 N_2$, where φ is the prescribed volume fraction. We consider a unit point load applied at $(P, 0)$ into direction $(\cos(\theta), \sin(\theta))$, $\theta \in [1.25\,\pi, 1.75\,\pi]$. If $\boldsymbol{K}_d$ denotes the stiffness matrix of the structure and $\boldsymbol{f}_x$ the force vector, then the nodal displacements are obtained as the solution of the linear algebraic system of equations $\boldsymbol{K}_d \boldsymbol{u} = \boldsymbol{f}_x$. For a structure defined by the binary array $\{d_{k\ell}\}$ we look for the "worst" load case, i.e. the value of the parameter vector $x = (P, \theta)$ that maximizes the compliance $C(d, x) := \boldsymbol{u}^{\mathrm{T}} \boldsymbol{K}_d \boldsymbol{u}$. Finally, the design optimization problem consists of finding structural layout that minimizes the worst loading compliance, i.e., finding d that solves

$$\min_d \max_x C(d, x). \tag{15.18}$$

In order to find the vector x of the "worst" load case, a LLEA has been implemented. This LLEA is a steady-state evolutionary algorithm which works with a mixed discrete-continuous representation. More specifically the vector x is composed by two components: the position P of the load and the angle of application θ of such load. The position P of the load is naturally made discrete by means of the grid of the Finite Element Method, on the contrary, the angle θ is continuous. It follows that the gene θ is recombined by means of the blend crossover alpha (BLX-α) with $\alpha = 0.5$ [6] and a Gaussian mutation along with gene P is recombined according to a blend crossover with rounding of both its terms and the mutation is operated adding a small random value at the individual. Table 15.6 shows the parameter setting for the LLEA in the case of optimization of the elastic structure. Concerning the AHLEA a given

Table 15.6. LLEA Parameter Setting

representation	vectors of one real number and one integer number
population size	40 individuals
parent selection	tournament
crossover	20 blend arithmetic crossover per generation
mutation	Gaussian mutation with mutation probability 0.2
survivor selection	fitness-based steady state
stop criterion	number of generations > 100 OR $\lvert C_{best} - C_{avg} \rvert < 0.01$

structure is represented by a 5×15 matrix of binary numbers where the zero defines the absence of material in a certain element and the one represents the presence of material. The *two point standard* crossover [42] and a simple 2-bit flip mutation have been applied. For the problem under study it has been chosen that the structure should contain 60% of material. The AHLEA with a dynamic population varying between 40 and 120 individuals has been applied. The optimal structure found by the algorithms and its deformation under the "worst" load is shown in Fig. 15.10. The "worst" load case found by the LLEA is for $P = 9$ and $\theta = 1.6564\,\pi$. Under this load $C(d, x) = 21.9940$. Finally, the

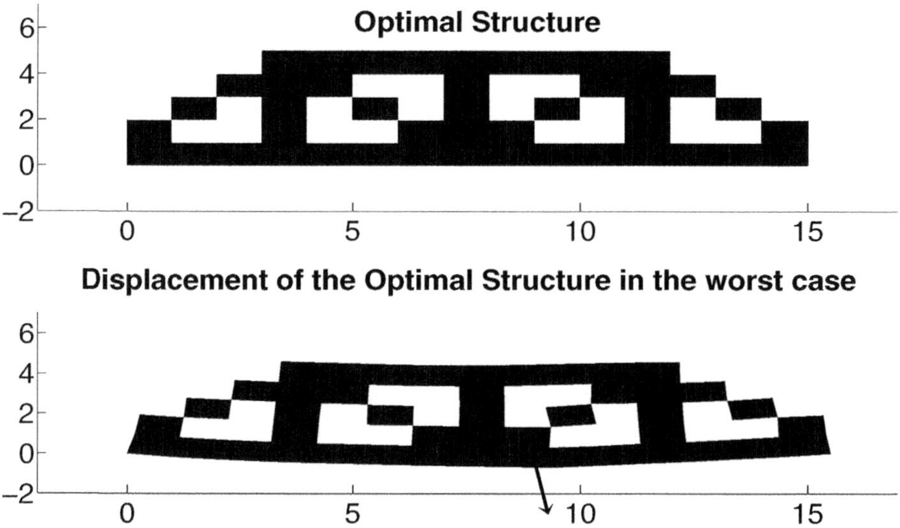

Fig. 15.10. Optimal structure and its deformation under the "worst" load case

distribution of the 500 samples by the LLEA for the found structure is shown in Fig. 15.11.

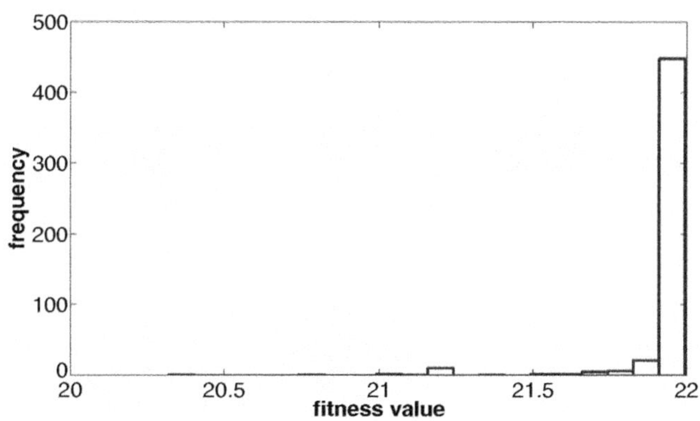

Fig. 15.11. Distribution of 500 samples for the elastic structure problem

15.4 Conclusions

In this chapter the problem of noisy fitness in hierarchial evolutionary algorithms for Min-Max problems is analyzed. The noise under examination, namely Hierarchical Noise (HN), is non-Gaussian and non-zero-mean and it has a distribution which significantly varies with the problem and the parameter setting of the Lower Level Evolutionary Algorithms (LLEA). In addition, for this kind of noise an *Implicit* or *Explicit Averaging* technique is not applicable since the value nearest to the true value is the largest among a certain number of samples. In order to propose some algorithmic components that are able to defeat the HN several aspects of the problem are analyzed. Thus, an Adaptive Higher Level Evolutionary Algorithm (AHLEA) is designed.

The AHLEA is characterized by an adaptive re-sampling mechanism and an adaptive population sizing mechanism. The combination of these two algorithmic components aims to execute fitness evaluations only when it is necessary. These two mechanisms decide both sample and population sizing taking into account fitness diversity of the population of the algorithm and the fitness evaluations already performed for each individual. In addition, two cooperating-competing survivor selection schemes ("prudent-daring") synergically work in order to allow a proper balance of the explorative efforts and the necessity of not making the population any worse by inserting an overestimated individual into the population.

The numerical results on a benchmark problem (Ackley function) show the efficiency of the adaptive rules. The AHLEA outperformed, for the benchmark problem under study, other standard methods in terms of convergence velocity and reliability of the algorithm. In this kind of noisy environment, the AHLEA can quickly converge to the optimal solution and maintain high performance in terms of reliability and robustness.

Two Min-Max structural optimization problems are also shown as an example: the optimal design of an electrical grounding grid generating minimum values of maximum touch voltages and and the optimal design of an elastic structure for the "worst" load case. The numerical results in these two cases show that the AHLEA can be successfully applied to different real-world problems.

References

1. Gulsen M, Smith AE (1999) A Hierarchical Genetic Algorithm for System Identification and Curve Fitting with a Supercomputer Implementation. In: Davis L.D. et al. (eds) Evolutionary Algorithms, Springer, Berlin Heidelberg New York
2. De Jong ED, Thierens D, Watson RA (2004) Hierarchical Genetic Algorithms. In: Yao X et al. (eds) Proceedings of the 8th International Conference on Parallel Problem Solving from Nature PPSN-VIII, Lecture Notes in Computer Science, Vol. 3242, 232–241, Springer-Verlag, Berlin Heidelberg New York
3. Neri F (2004) A New Evolutionary Method for Designing Grounding Grids by Touch Voltage Control. Proceedings of IEEE International Symposium on Industrial Electronics, ISIE 2004, Vol. 2, 1501–1505, Ajaccio France
4. Neri F, Kononova AV, Delvecchio G, Sylos Labini M, Uglanov A (2005) A Hierarchical Evolutionary Algorithm with Noisy Fitness in Structural Optimization

Problems. In: Rothlauf F et al. (eds) Applications of Evolutionary Computation, Lecture Notes in Computer Science, Vol. 3449, 610–616, Springer, Berlin Heidelberg New York
5. Zhou ZZ, Ong YS, Nair PB (2004) Hierarchical Surrogate-Assisted Evolutionary Optimization Framework. Proceedings of the IEEE Congress on Evolutionary Computation 2004, 20–23
6. Ong YS, Nair PB, Lum KY (2005) Max-Min Surrogate-Assisted Evolutionary Algorithm for Robust Aerodynamic Design. IEEE Transactions on Evolutionary Computation, Vol. 10, No. 4, August 2006.
7. Jin Y, Branke J (2005) Evolutionary Optimization in Uncertain Enviroments-A Survey. IEEE Transactions on Evolutionary Computation, Vol. 9, No. 3, 303–317
8. Branke J (2001) Evolutionary Optimization in Dynamic Environments. Kluwer Academic Publisher, 125–172
9. Arnold DV, Beyer HG (2003) On Effect of Outliers on Evolutionary Optimization. In: Intelligent Data Engineering and Automated Learning, Lecture Notes in Computer Science, Vol. 2690, 151–160, Springer-Verlag, Berlin Heidelberg New York
10. Aizawa AN, Wah BW (1993) Dynamic Control of Genetic Algorithms in Noisy Environment. In: Proc. Conf. Genetic Algorithms, 48–55
11. Aizawa AN, Wah BW (1994) Scheduling of Genetic Algorithms in a Noisy Environment. Evolutionary Computation, Vol. 2, No. 2, 97–122
12. Branke J, Schmidt C, Schmeck H (2001) Efficient Fitness Estimation in Noisy Environment. In: L. Spector et al. (eds) Genetic and Evolutionary Computation, 243–250, Morgan Kauffman, San Mateo
13. Fitzpatrick JM, Grefenstette JI (1988) Genetic Algorithms in Noisy Environments. Machine Learning, Vol. 3, 101–120
14. Goldberg DE, Deb K, Clark J (1992) Genetic Algorithms, Noise, and the Sizing of the Population. Complex Systems, Vol. 6, 333–362
15. Miller BL, Goldberg DE (1996) Genetic Algorithms, Selection Schemes and the Varying Effect of the Noise. Evolutionary Computation, Vol. 4, No. 2, 113–131
16. Rattray LM, Shapiro J (1997) Noisy Fitness Evaluations in Genetic Algorithms and the Dynamics of Learning. In: R.K. Belew and M.D. Vose (eds) Foundations of Genetic Algorithms, 117–139, Morgan Kauffman, San Mateo
17. Eiben AE, Smith JE (2003) Introduction to Evolutionary Computing. Springer-Verlag, Berlin Heidelberg New York
18. Sylos Labini M, Delvecchio G, Neri F (2003) A Genetic Algorithm Method for Determining the Maximum Touch Voltage Generated by a Grounding System. In: Rudnicki M, Wiak S (eds) Optimization and Inverse Problems in Electromagnetism, 85–92, Kluwer Academic Publisher
19. Schmidt C, Branke J, Chick SE (2006) Integrating Techniques from Statistical Ranking into Evolutionary Algorithms. In: Rothlauf F. et al. (eds.) Applications of Evolutionary Computing, Lectures Notes in Computer Science, Vol. 3907, 752–763, Springer
20. Neri F, Cascella GL, Salvatore N, Kononova AV, Acciani G (2006) Prudent-Daring vs Tolerant Survivor Selection Schemes in Control Design of Electric Drives. In: Rothlauf F. et al. (eds.) Applications of Evolutionary Computing, Lectures Notes in Computer Science, Vol. 3907, 805–809, Springer

21. Eiben AE, Hinterding R Michaelwicz Z (2000) Parameter Control. In: Bäck T, Fogel DB, Z. Michaelwicz Z (eds) Evolutionary Computation 2, Advanced Algorithms and Operators, 170–187, Institute of Physics Publishing
22. Branke J, Schmidt C (2004) Sequential Sampling in Noisy Environments. In: Parallel Problem Solving in Nature VIII PPSN, Lecture Notes in Computer Science, Vol. 3242, 202–211, Springer, Berlin Heidelberg New York
23. Cantu-Paz E (2004) Adaptive sampling for noisy problems. In: Genetic and Evolutionary Computation Conference GECCO2004, 947–958, Springer, Berlin Heidelberg New York
24. Stagge P (1998) Averaging Efficiently in Presence of Noise. In: Eiben AE et al.(eds) V Parallel Problem Solving from Nature, Lectures Notes in Computer Science, Vol. 1498, 188–197, Springer-Verlag, Berlin Heidelberg New York
25. Di Pietro A, While L, Barone L (2004) Applying Evolutionary Algorithms to Problems with Noisy, Time-Consuming Fitness Functions. Proceeding of the Conference on Genetic Algorithms,1254–1261
26. Ong YS, Keane AJ (2004) Meta-Lamarkian Learning in Memetic Algorithms. IEEE Transactions on Evolutionary Computation, Vol. 8, No. 2, 99–110
27. Yang S (2003) Adaptive Mutation using Statistics Mechanism for Genetic Algorithms. In: Coenen F, Preece A, Macintosh A, (eds.) Research and Development in Intelligent Systems XX, Springer-Verlag, 19–32
28. Caponio A, Cascella G L, Neri F, Salvatore N, Sumner M (2006) A Fast Adaptive Memetic Algorithm for Off-line and On-line Control Design of PMSM Drives, to appear IEEE Transactions on Systems, Man and Cybernetics Part B, Special Issue on Memetic Algorithms
29. IEEE Standard 80 - 2000 (2000) IEEE Guide for Safety in AC Substation Grounding
30. Huang L, Chen L, Yan H (1995) Study of Unequally Spaced Grounding Grids. IEEE Transactions on Power Delivery, Vol. 10, No. 2, 716–722
31. Yuan J, Yang H, Zhang L, Cui X, Ma X (2000) Simulation of Substation Grounding Grids with Unequal potential. IEEE Transactions of Magnetics, Vol. 36, No. 4, 1468–1471
32. Delvecchio G, Di Sciascio E, Grassi S, Neri F, Sylos Labini M (2005) Some Geometrical and Evolutionary Procedures for Optimizing the Calculation Times of 3-D Current Fields by the Finite Element Method. COMPEL: International Journal for Computation and Mathematics in Electrical and Electronic Engineering, MCB University Press, Vol. 24, No. 3, 984–996
33. Otero AF, Cidras J, Garrido C (1998) Genetic Algorithm Based Method for Grounding Grids Design. Proceedings of the IEEE International Conference on Evolutionary Computation, World Congress of Computational Intelligence, 120–123
34. Phithakwong B, Kraisnachinda N, Bayjomgjit S, Chompo-Inwai C, Kando M (2000) New Techniques the Computer-Aided Design for Substation Grounding. IEEE Power Engineering Society Winter Meeting, Vol. 3, 2011–2015
35. El-Dessouky SS, El Aziz MA, Khamis A (1998) An Accurate Design of Substation Grounding System Aid Expert System Methodology. Conference Record of the IEEE International Symposium on Electrical Insulation, Vol. 2, 411–414
36. Sun W, He J, Gao Y, Zeng R, Wu W, Su Q (2000) Optimal Design Analysis of Grounding Grids for Substations built in non-uniform soil. Proceedings of Powercon. International Conference on Power System Technology, Vol. 3, 1455–1460

37. Haslinger J, Mäkinen RAE (2003) Introduction to Shape Optimization: Theory, Approximation, and Computation. SIAM, Philadelphia
38. Bendsøe MP (1995) Optimization of Structural Topology, Shape and Material. Springer, Berlin Heidelberg New York
39. Bendsøe MP, Sigmund O (1999) Material Interpolations in Topology Optimization. Arch. Appl. Mech., Vol. 69, 635–654
40. Kane C, Schoenauer M (1996) Topological Optimum Design Using Genetic Algorithms. Control and Cybernetics, Vol. 25, 1059–1088
41. Eshelman LJ, Shaffer JD (1993) Real-coded Genetic Algorithms and Interval-Schemata. In: Fondations of Genetic Algorithms 2, 187–202
42. Schoneauer M (1995) Shape Representation for Evolutionary Optimization and Identification in Structural Mechanics. Proceedings of EUROGEN 1995, 5–30, John Wiley and Sons Ltd

16

Evolving Multi Rover Systems in Dynamic and Noisy Environments

Kagan Tumer[1] and Adrian Agogino[2]

[1] NASA Ames Research Center, Mailstop 269-4, Moffett Field, CA 94035, USA
ktumer@mail.arc.nasa.gov
[2] UC Santa Cruz, NASA Ames Research Center, Mailstop 269-3, Moffett Field, CA 94035, USA
adrian@email.arc.nasa.gov

Summary. In this chapter, we address how to evolve control strategies for a collective: a set of entities that collectively strives to maximize a global evaluation function that rates the performance of the full system. Addressing such problems by directly applying a global evolutionary algorithm to a population of collectives is unworkable because the search space is prohibitively large. Instead, we focus on evolving control policies for each member of the collective, where each member is trying to maximize the fitness of its own population. The main difficulty with this approach is creating fitness evaluation functions for the members of the collective that induce the collective to achieve high performance with respect to the system level goal. To overcome this difficulty, we derive member evaluation functions that are both aligned with the global evaluation function (ensuring that members trying to achieve high fitness results in a collective with high fitness) and sensitive to the fitness of each member (a member's fitness depends more on its own actions than on actions of other members).

In a difficult rover coordination problem in dynamic and noisy environments, we show how to construct evaluation functions that lead to good collective behavior. The control policy evolved using aligned and member-sensitive evaluations outperforms global evaluation methods by up to a factor of four. In addition we show that the collective continues to perform well in the presence of high noise levels and when the environment is highly dynamic. More notably, in the presence of a larger number of rovers or rovers with noisy sensors, the improvements due to the proposed method become significantly more pronounced.

16.1 Introduction

In this chapter we show how to extend evolutionary control methods to dynamic domains that contain many devices that need to be controlled in the presence of noise. These methods are applicable to many distributed domains such as coordinating multiple robots, controlling constellations of satellites,

K. Tumer and A. Agogino: *Evolving Multi Rover Systems in Dynamic and Noisy Environments*, Studies in Computational Intelligence (SCI) **51**, 371–387 (2007)
www.springerlink.com

and routing over a data network promises significant application opportunities [3, 11, 14]. In this chapter we specifically look at the problem of coordinating multiple rovers in such a way that they maximize their collective observational capability.

The challenge in this problem is efficiently evolving control policies for the rovers, such that they collectively maximize a single objective function, which is in general a non-linear function over the actions of all the rovers. The straight-forward approach to this problem is to directly evolve the entire collective by using a population of collectives and having the evolutionary operators work directly on the collective to produce a solution with high global fitness. Unfortunately this method is impractical at best and impossible at worst, since the search space for such an approach is simply too large for all but the simplest problems. Instead we will approach this problem by having each rover in the collective have its own population of control policies, using its own fitness evaluation function to evaluation these control policies. The key issue in such an approach is to ensure that the rover fitness evaluation function possesses the following two properties: (i) it is aligned with the global evaluation function, ensuring that the rovers that maximize their own fitness do not hinder one another and hurt the fitness of the collective; and (ii) it is sensitive to the fitness of the rover, ensuring that it provides the right selective pressure on the rover (i.e., it limits the impact of other rovers in the fitness evaluation function).

In this chapter we show how to create fitness evaluation functions that have these properties. We then show how to use them in a multi-rover domain that is challenging in the following two ways:

1. The environment is dynamic, meaning that the conditions under which the rovers evolve changes with time. The rovers need to evolve general control policies, rather than specific policies tuned to their current environment.
2. The rovers' sensors and actuators are noisy, meaning that the signals they receive from the environment are not reliable and the control signals that the robot sends out are not reliably carried out. The rovers need to demonstrate that the control policies are not sensitive to such fluctuations in sensor readings and control outputs.

This domain is modeled reflect the important properties evolutionary control systems need to have to be deployed. Significantly this domain does not have any "episodes" or "trials." The environment changes continuously and the rovers move continuously where neither the environment or rovers' positions are ever reset. The rovers must evolve in-situ and be able to use the control policies in an environment different from what they were evolved in.

In Section 16.2 we discuss the properties needed in a collective, how to evolve rovers using evaluation functions possessing such properties along with a discussion of related work. In section 16.3 we present the "Rover Problem" where a planetary rovers in a collective use neural networks to determine their movements based on a continuous-valued array of sensor inputs. Section 16.4

presents the performance of the rover collective evolved using rover evaluation functions in dynamic, noisy and communication limited domains. The results show the effectiveness of the rovers in gathering information is 400% higher with properly derived rover fitness functions than in rovers using a global evaluation function. Finally Section 16.5 we discuss the implication of these results and their applicability to different domains.

16.2 Evolving a Collective

In general, one has three possible approaches based on evolutionary computation to design control policies for collectives.

1. One can operate directly on the collective, treating it as an instance of a solution and operate on populations of collectives. In this case, the standard evolutionary algorithms are used to select for the collective that best satisfies a predetermined global evaluation function.
2. One can operate on members in the collective, treating each rover as an instance of a solution and operate of populations of rovers. In this case, the evolutionary algorithms are used to select the rovers constituting the collective based on how a given rover satisfies *the predetermined global evaluation function.*
3. One can operate on members in the collective, treating each rover as an instance of a solution and operate of populations of rovers. In this case, the evolutionary algorithms are used to select the rovers constituting the collective based on how a given rover satisfies *a specialized rover evaluation function tuned to the fitness of that rover.*

The first method presents a computationally daunting task in all but the simplest problems. Finding good control strategies is difficult enough for single controllers, but the search space become prohibitively large when they are concatenated into an "individual" representing the full collectives. Even if good rovers are present in the collective, there is no mechanism for isolating and selecting them when the collective to which they belong performs poorly. As a consequence, this approach is practically unworkable in large continuous domains.

The second method addresses part of the issue: Because the rovers in the collective are evolved independently, it avoids the explosion of the state space. However, this method introduces a new problem: How is a rover's evolution guided when the evaluation function depends on the fitness of all the other rovers? In small collectives, this method provides good solutions, but as the collectives size increases, this problem becomes more and more acute. As a consequence, this approach, though preferable to the first approach in some ways, is unlikely to provide good solutions in large collectives.

The third method provides a specialized rover evaluation function for each rover. This approach, cleans up the fitness signal a rover receives, but introduces a new twist to the problem: How does one ensure that the specialized rover evaluation functions are aligned with the global evaluation function? In other words, the fundamental question is how to guarantee that the collective evolved using rover evaluation functions will have a high fitness with respect to the global evaluation function. In this chapter we discuss the second and third approaches, focusing on how to select rover evaluation function in a formal manner as discussed below.

16.2.1 Rover Evaluation Function Properties

Given a global evaluation function to be maximized, this section presents the desirable properties that rover-specific evaluation functions must have. Let the **global evaluation function** be given by $G(z)$, where z is the state of the full system (e.g., the position of all the rovers in the system, along with their relevant internal parameters and the state of the environment). Let the **rover evaluation function** for rover i be given by $g_i(z)$. The first desirable property the rover evaluation functions of each agent needs to have is high *factoredness* with respect to G, intuitively meaning that an action taken by an agent that improves its rover evaluation function also improves the global evaluation function (i.e. G and g_η are aligned). Formally, the degree of factoredness between g_i and G is given by:

$$\mathcal{F}_{g_i} = \frac{\int_z \int_{z'} u[(g_i(z) - g_i(z'))\,(G(z) - G(z'))]dz'dz}{\int_z \int_{z'} dz'dz} \tag{16.1}$$

where z' is a state which only differs from z in the state of rover i, and $u[x]$ is the unit step function, equal to 1 when $x > 0$. Intuitively, a high degree of factoredness between g_i and G means that a rover evolved to maximize g_i will also maximize G.

The second property the rover evaluation function needs to have is high sensitivity to changes in its own fitness and low sensitivity to changes in the fitness of other rovers in the collective. Formally we quantify the *rover-sensitivity* of evaluation function g_i, at z as:

$$\lambda_{i,g_i}(z) = E_{z'}\left[\frac{\|g_i(z) - g_i(z - z_i + z'_i)\|}{\|g_i(z) - g_i(z' - z'_i + z_i)\|}\right] \tag{16.2}$$

where $E_{z'}[\cdot]$ provides the expected value over possible values of z', and $(z - z_i + z'_i)$ notation specifies the state vector where the components of rover i have been removed from state z and replaced by the components of rover i from state z'. So at a given state z, the higher the rover-sensitivity, the more $g_i(z)$ depends on changes to the state of rover i, i.e., the better the associated signal-to-noise ratio for i. Intuitively then, higher rover-sensitivity means there is "cleaner" (e.g., less noisy) selective pressure on rover i.

As an example, consider the case where the rover evaluation function of each rover is set to the global evaluation function, meaning that each rover is evaluated based on the fitness of the full collective (e.g., approach 2 discussed in Section 16.2). Such a system will be fully factored by the definition of Equation 16.1. However, the rover fitness functions will have low rover-sensitivity (the fitness of each rover depends on the fitness of all other rovers).

16.2.2 Difference Evaluation Functions

Let us now focus on improving the rover-sensitivity of the evaluation functions. To that end, consider **difference** evaluation functions [17], which are of the form:

$$D_i \equiv G(z) - G(z_{-i} + c_i) \tag{16.3}$$

where z_{-i} contains all the states on which rover i has no effect, and c_i is a fixed vector. In other words, all the components of z that are affected by rover i are replaced with the fixed vector c_i. Such difference evaluation functions are fully factored no matter what the choice of c_i, because the second term does not depend on i's states [17] (e.g., D and G will have the same derivative with respect to z_i). Furthermore, they usually have far better rover-sensitivity than does a global evaluation function, because the second term of D removes some of the effect of other rovers (i.e., noise) from i's evaluation function. In many situations it is possible to use a c_i that is equivalent to taking rover i out of the system. Intuitively this causes the second term of the difference evaluation function to evaluate the fitness of the system without i and therefore D evaluates the rover's contribution to the global evaluation.

Though for linear evaluation functions D_i simply cancels out the effect of other rovers in computing rover i's evaluation function, its applicability is not restricted to such functions. In fact, it can be applied to any linear or non-linear global utility function. However, its effectiveness is dependent on the domain and the interaction among the rover evaluation functions. At best, it fully cancels the effect of all other rovers. At worst, it reduces to the global evaluation function, unable to remove any terms (e.g., when z_{-i} is empty, meaning that rover i effects all states). In most real world applications, it falls somewhere in between, and has been successfully used in many domains including rover coordination, satellite control, data routing, job scheduling and congestion games [3, 15, 17]. Also note that the computation of D_i is a "virtual" operation in that rover i computes the impact of its not being in the system. There is no need to re-evolve the system for each rover to compute its D_i, and computationally it is often easier to compute than the global evaluation function [15]. Indeed in the problem presented in this chapter, for rover i, D_i is easier to compute than G is (see details in Section 16.4).

16.2.3 Related Work

Evolutionary computation has a long history of success in single agent and multi-agent control problems [1, 2, 7, 10, 16]. Advances in evolutionary computation methods in single agent domains tend to result from improvements in search methods. In [10] this is accomplished through fuzzy rules in a helicopter control problem, while in [16] cellular encoding is used to improve performance on pole-balancing control. Similarly [7] addresses planetary rover control by having genetic algorithms search through a space of plans generated from a planning algorithm.

Many advances in evolutionary computation for multi-agent control have been accomplished through the use of domain specific fitness functions. Ant colony algorithms [6] solve the coordination problem by utilizing "ant trails" that provide implicit fitness functions resulting in good performance in path-finding domains. In [2], the algorithm takes advantage of a large number of agents to speed up the evolution process in certain domains, but uses greedy fitness functions that are not generally factored. Also outside of evolutionary computation, coordination between a set of mobile robots has been accomplished through the use of hand-tailored rewards designed to prevent greedy behavior [12]. While highly successful in many domains all of these methods differ from the methods used in this chapter in that they lack a general framework for efficient evolution in multi-agent systems.

16.3 Continuous Rover Problem

In this section, we show how evolutionary computation with the difference evaluation function can be used effectively in the Rover Problem[3]. In this problem, there is a collective of rovers on a two dimensional plane, which is trying to observe points of interests (POIs). Each POI has a value associated with it and each observation of a POI yields an observation value inversely related to the distance the rover is from the POI. In this chapter the distance metric will be the squared Euclidean norm, bounded by a minimum observation distance, δ_{min}:[4]

$$\delta(x, y) = min\{\|x - y\|^2, \delta_{min}^2\} \ . \tag{16.4}$$

The global evaluation function is given by:

$$G = \sum_t \sum_j \frac{V_j}{\min_i \delta(L_j, L_{i,t})} \ , \tag{16.5}$$

where V_j is the value of POI j, L_j is the location of POI j and $L_{i,t}$ is the location of rover i at time t. Intuitively, while any rover can observe any

[3] This problem was first presented in [3].

[4] The square Euclidean norm is appropriate for many natural phenomenon, such as light and signal attenuation. However any other type of distance metric could also be used as required by the problem domain. The minimum distance is included to prevent singularities when a rover is very close to a POI.

POI, as far as the global evaluation function is concerned, only the closest observation matters[5].

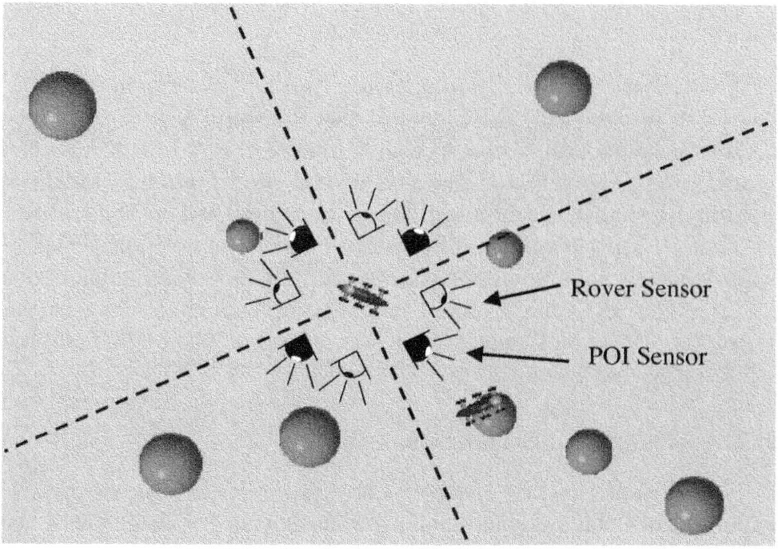

Fig. 16.1. Diagram of a Rover's Sensor Inputs. The world is broken up into four quadrants relative to rover's position. In each quadrant one sensor senses points of interests, while the other sensor senses other rovers.

16.3.1 Rover Capabilities

At every time step, the rovers sense the world through eight continuous sensors. From a rover's point of view, the world is divided up into four quadrants relative to the rover's orientation, with two sensors per quadrant (see Figure 16.1). For each quadrant, the first sensor returns a function of the POIs in the quadrant at time t. Specifically the first sensor for quadrant q returns the sum of the values of the POIs in its quadrant divided by their squared distance to the rover and scaled by the angle between the POI and the center of the quadrant:

$$s_{1,q,j,t} = \sum_{j \in J_q} \frac{V_j}{\delta(L_j, L_{i,t})} \left(1 - \frac{|\theta_{j,q}|}{90}\right) \tag{16.6}$$

where J_q is the set of observable POIs in quadrant q and $|\theta_{j,q}|$ is the magnitude of the angle between POI j and the center of the quadrant. The second sensor

[5] Similar evaluation functions could also be made where there are many different levels of information gain depending on the position of the rover. For example 3-D imaging may utilize different images of the same object, taken by two different rovers.

returns the sum of square distances from a rover to all the other rovers in the quadrant at time t scaled by the angle:

$$s_{2,q,i,t} = \sum_{i' \in N_q} \frac{1}{\delta(L_{i'}, L_{i,t})} \left(1 - \frac{|\theta_{i',q}|}{90}\right) \quad (16.7)$$

where N_q is the set of rovers in quadrant q and $|\theta_{i',q}|$ is the magnitude of the angle between rover i' and the center of the quadrant.

The sensor space is broken down into four regions to facilitate the input-output mapping. There is a trade-off between the granularity of the regions and the dimensionality of the input space. In some domains the tradeoffs may be such that it is preferable to have more or fewer than four sensor regions. Also, even though this chapter assumes that there are actually two sensors present in each region at all times, in real problems there may be only two sensors on the rover, and they do a sensor sweep at 90 degree increments at the beginning of every time step.

16.3.2 Rover Control Strategies

With four quadrants and two sensors per quadrant, there are a total of eight continuous inputs. This eight dimensional sensor vector constitutes the state space for a rover. At each time step the rover uses its state to compute a two dimensional output. This output represents the x, y movement relative to the rover's location and orientation. Figure 16.2 displays the orientation of a rover's output space.

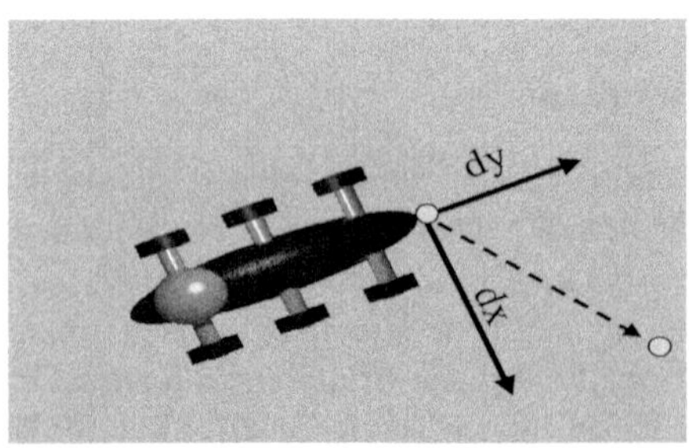

Fig. 16.2. Diagram of a Rover's Movement. At each time step the rover has two continuous outputs (dx, dy) giving the magnitude of the motion in a two directional plane relative to the rover's orientation.

The mapping from rover state to rover output is done through a Multi Layer Perceptron (MLP), with eight input units, ten hidden units and two output units[6]. The MLP uses a sigmoid activation function, therefore the

[6] Note that other forms of continuous reinforcement learners could also be used instead of evolutionary neural networks. However neural networks are ideal for this domain given the continuous inputs and bounded continuous outputs.

outputs are limited to the range $(0, 1)$. The actual rover motions dx and dy, are determined by normalizing and scaling the MLP output by the maximum distance the rover can move in one time step. More precisely, we have:

$$dx = 2d_{max}(o_1 - 0.5)$$
$$dy = 2d_{max}(o_2 - 0.5)$$

where d_{max} is the maximum distance the rover can move in one time step, o_1 is the value of the first output unit, and o_2 is the value of the second output unit.

16.3.3 Rover Selection

The MLP for a rover is selected using an evolutionary algorithm as highlighted in approaches two and three in Section 16.2. In this case, each rover has a population of MLPs. At each N time steps (N set to 15 in these experiments), the rover uses epsilon-greedy selection ($\epsilon = 0.1$) to determine which MLP it will use (e.g., it it selects the best MLP from its population with 90% probability and a random MLP from its population with 10% probability). The selected MLP is then mutated by adding a value sampled from the Cauchy Distribution (with scale parameter equal to 0.3) to each weight, and is used for those N steps. At the end of those N steps, the MLP's performance is evaluated by the rover's evaluation function and re-inserted into its population of MLPs, at which time, the poorest performing member of the population is deleted. Both the global evaluation for system performance and rover evaluation for MLP selection is computed using an N-step window, meaning that the rovers only receive an evaluation after N steps.

While this is not a sophisticated evolutionary algorithm, it is ideal in this work since our purpose is to demonstrate the impact of principled evaluation functions selection on the performance of a collective. Even so, this algorithm has shown to be effective if the evaluation function used by the rovers is factored with G and has high rover-sensitivity. We expect more advanced evolutionary computation algorithms used in conjunction with these same evaluation functions to improve the performance of the collective further.

16.3.4 Evolving Control Strategies in a Collective

The key to success in this approach is to determine the correct rover evaluation functions. In this work we test three different evaluation function for rover selection. The first evaluation function is the global evaluation function (G), which when implemented results in approach two discussed in Section 16.2:

$$G = \sum_t \sum_j \frac{V_j}{\min_i \delta(L_j, L_{i,t})} \tag{16.8}$$

The second evaluation function is the "perfectly rover-sensitive" evaluation function (P):

$$P_i = \sum_t \sum_j \frac{V_j}{\delta(L_j, L_{i,t})} \tag{16.9}$$

The P evaluation function is equivalent to the global evaluation function in the single rover problem. In a collective of rover setting, it has infinite rover-sensitivity (in the way rover sensitivity is defined in Section 16.2). This is because the P evaluation function for a rover is not affected by the states of the other rovers, and thus the denominator of Equation 16.2 is zero. However the P evaluation function is not factored. Intuitively P and G offer opposite benefits, since G is by definition factored, but has poor rover-sensitivity. The final evaluation function is the difference evaluation function. It does not have as high rover-sensitivity as P, but is still factored like G. For the rover problem, the difference evaluation function, D, becomes:

$$D_i = \sum_t \left[\sum_j \frac{V_j}{\min_{i'} \delta(L_j, L_{i',t})} - \sum_j \frac{V_j}{\min_{i' \neq i} \delta(L_j, L_{i,t})} \right] .$$

The second term of the D is equal to the value of all the information collected if rover i were not in the system. Note that for all time steps where i is not the closest rover to any POI, the subtraction leaves zero. As mentioned in Section 16.2.2, the difference evaluation in this case is easier to compute as long as rover i knows the position and distance of the closest rover to each POI it can see. In that regard it requires knowledge about the position of fewer rovers than if it were to use the global evaluation function.

16.4 Results

We performed extensive simulation to test the effectiveness of the different rover evaluation function under a wide variety of environmental conditions and rover capabilities. In these experiments, each rover had a population of MLPs of size 10. The world was 75 units long and 75 units wide. All of the rovers started the experiment at the center of the world. Unless otherwise state as in the scaling experiments, there were 30 rovers in the simulations. The maximum distance the rovers could move in one direction during a time step, d_{max}, was set to 3. The rovers could not move beyond the bounds of the world. The minimum observation distance, δ_{min}, was equal to 5.

In these experiments the environment was dynamic, meaning that the POI locations and values changed with time. There were as many POIs as rovers, and the value of each POI was set to between three and five using a uniformly random distribution. In these experiments, each POI disappeared with probability 2.5%, and another one appeared with the same probability at 15 time step intervals. Because the experiments were run for 3000 time steps, the initial

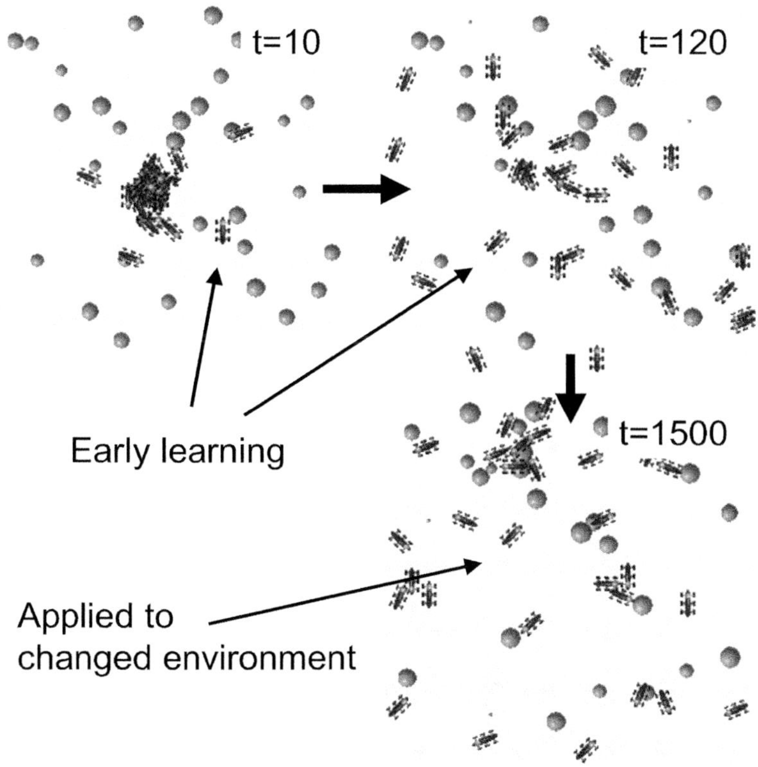

Fig. 16.3. Sample POI Placement. Left-Top: Environment at time = 10. Right-Top: Environment at time = 120. Bottom: Environment at time = 1500. Environment at time step 10 is similar to environment at time step 120, but significantly different than environment at time step 1500. Rovers must to able to use their control policies evolved from earlier time step, in future changed environments.

and final environments had little similarities. All results were averaged over at least one hundred independent trials (except for the seventy agent runs where there were thirty trials). For each experiment and trial the weights of the neural network were set to random using the Cauchy distribution (parameter of 0.5).

Results for episodic environments where the agents were restored to their initial state at the end of each trial were reported in [3]. In such a case the rovers evolve specific control policies tuned to the particular environment in which they are trained. Though useful in domains where the simulated environment closely matches the environment in which the rovers will operate, this approach has limited applicability in general. A more desirable approach is for the rovers to evolve efficient policies that are solely based on their sensor inputs and not on the specific configuration of the POIs. The dynamic environment experiments reported here explore this premise and provide rover control policies that can be generalized from one set of POIs to another, regardless of how significantly the environment changes. Figures 16.3 shows an instance of change in the environment throughout a simulation. The final POI set is not particularly close to the initial POI set and the rovers are forced to focus on the sensor input-output mappings rather than focus on regions in the (x, y) plane.

16.4.1 Evolution in Noise Free Environment

The first set of experiments tested the performance of the three evaluation functions in a dynamic noise-free environment for 30 rovers. Figure 16.4 shows the performance of each evaluation function. In all cases, performance is measured by the same global evaluation function, regardless of the evaluation function used to evolve the system. All three evaluation functions performed adequately in this instance, though D_i outperformed both P and G.

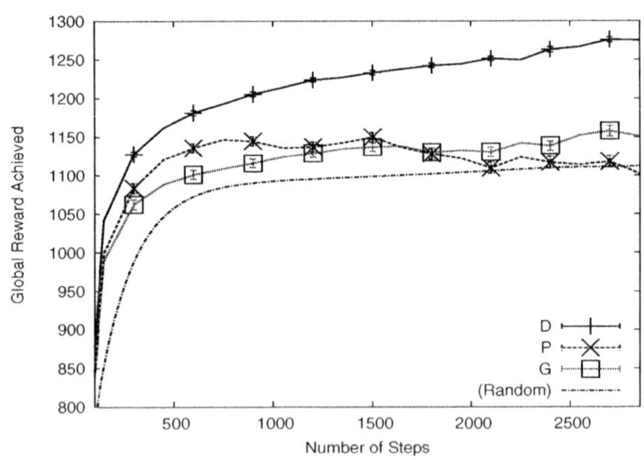

Fig. 16.4. Performance of a 30-rover collective for all three evaluation functions in noise-free environment. Difference evaluation function provides the best collective performance because it is both factored and rover-sensitive.

The evolution of this system also demonstrate the different properties of the rover evaluation functions. After initial improvements, the system with the G evaluation function improves slowly. This is because the G evaluation function has low rover-sensitivity. Because the fitness of each rover depends on the state of all other rovers, the noise in the system overwhelms the evaluation function. P on the other hand has a different problem: After an initial improvement, the performance of systems with this evaluation function decline. This is because though P has high rover-selectivity, it is not fully factored with the global evaluation function. This means that rovers selected to improve P do not necessarily improve G. D on the other hand is both factored and has high rover-sensitivity. As a consequence, it continues to improve well into the simulation as the fitness signal the rovers receive are not swamped by the states of other rovers in the system. This simulation highlights the need for having evaluation function that are both factored with the global evaluation function and have high rover-sensitivity. Having one or the other is not sufficient.

16.4.2 Scaling in Noise Free Environments

The second set of experiments focuses on the scaling properties of the three evaluation functions in a dynamic noise-free environment. Figure 16.5 shows the performance of each evaluation function at t=3000 for a collective of 10 to 70 rovers (where the number of POIs is equal to the number of rovers). For each case, the results are qualitatively similar to those reported above, except where there are only 5 rovers, in which case P performs as well as G. This is not surprising since with so few rovers, there are almost no interactions among the rovers, and in as large a space as the one used here, the 5 rovers act almost independently.

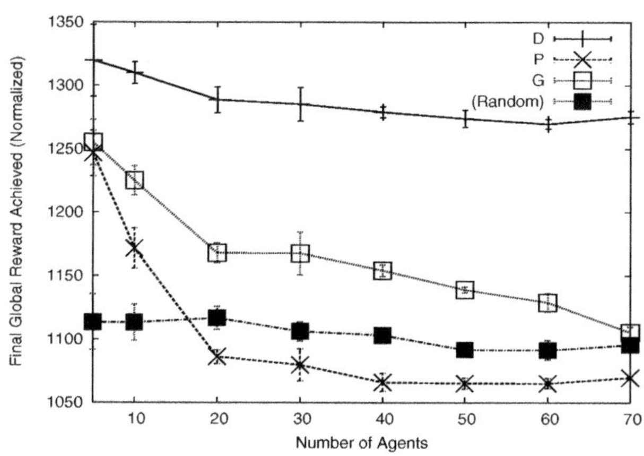

Fig. 16.5. Scaling properties of the three evaluation functions. The D evaluation function not only outperforms the alternatives, but the margin by which it outperforms them increases as the size of the collective goes up.

As the size of the collective increases though, an interesting pattern emerges: The performance of both P and G drop at a faster rate than that of D. Again, this is because G has low rover-sensitivity and thus the problem becomes more pronounced as the number of rovers increases. Similarly, as the number of rovers increases, P becomes less and less factored. In fact the performance of rovers using P is even worse than random when there are many rovers because the rovers' greedy actions make them focus on only a few POIs, while the random rovers at least distribute themselves among the POIs. D on the other hand handles the increasing number of rovers quite effectively. Because the second term in Equation 16.3 removes the impact of other rovers from rover i, increasing the number of rovers does very little to limit the effectiveness of this rover evaluation function. This is a powerful result suggesting that D is well suited to evolve large collectives in this and similar domains where the interaction among the rovers prevents both G and P from performing well. This result also supports the intuition expressed in Section 16.2 that approach 2 (i.e., evolving rovers based on the fitness of the full collective) is ill-suited to evolving collectives in all but the smallest examples.

16.4.3 Evolution in Noisy Environment

The third set of experiments tested the performance of the three evaluation functions in a dynamic environment for 30 rovers with noisy sensors. Figure 16.6 shows the performance of each evaluation function when both the input sensors and the output values of the rovers have 5% noise added. All three evaluation functions handle the noise well. This result is encouraging in that it shows that not only simple evaluation functions such as P can handle moderate amounts of noise in their sensors and outputs, but so can D. In other words, considering the impact of other rovers to yield a factored evaluation function does not cause to compound moderate noise in the system and overwhelm the rover evaluation.

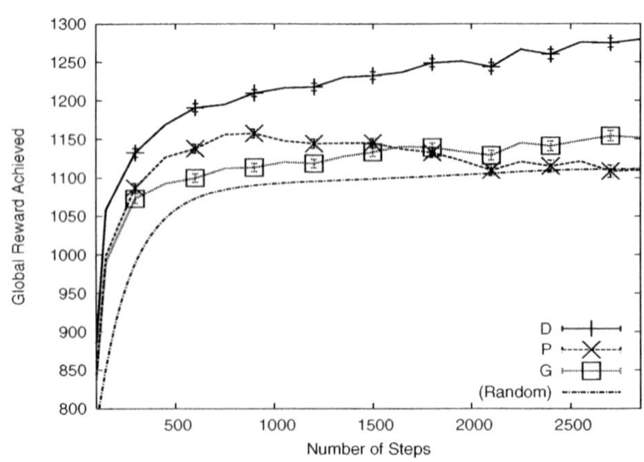

Fig. 16.6. Performance of a 30-rover collective for all three evaluation functions when the rover sensors and outputs have 5% noise

Figure 16.7 shows the noise sensitivity of the three different evaluation functions. The performance is reported as a function of additive noise to sensors as the percentage shown on the x-axis (e.g., 0.5 means the magnitude of the added noise is half that of the sensor value.) The results are shown as the D is the most sensitive to high levels of noise, though even at 80% noise it still far outperforms both G and P. This is an encouraging result in the power of the D evaluation function is that it "cleans up" the evaluation function for a rover (e.g., it has high rover-sensitivity). Adding noise, starts to cancel this property of D, but even when half the signal being noise does not prevent D from far outperforming D and P. Interestingly, rovers using P actually perform marginally better as noise increases, demonstrating the importance of factoredness. Adding noise to the system actually hindered these rovers from learning some counter-productive actions.

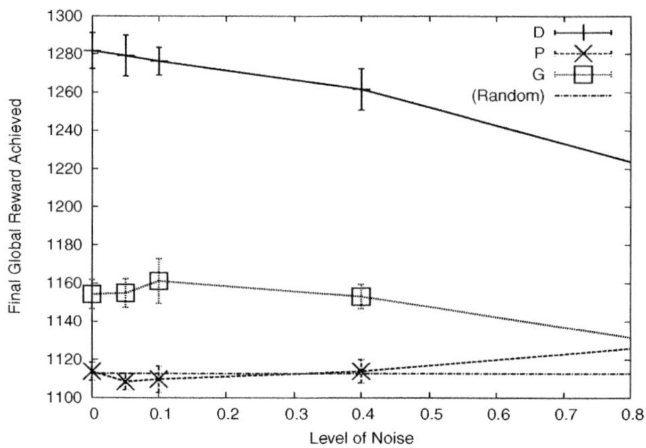

Fig. 16.7. Sensitivity of the three evaluation functions to the degree of noise in their sensors

16.5 Discussion and Conclusions

Extending the success of evolutionary algorithms in continuous single-controller domains to large, distributed multi-controller domains has been a challenging endeavor. Unfortunately the direct approach of having a population of collectives and applying the evolutionary algorithm to that population results in a prohibitively large search space in most cases. As an alternative, this chapter presents a method for providing rover specific evaluation functions to directly evolve individual rovers in collective. The fundamental issue that needs to be addressed in this approach is determining the rover specific evaluation functions that are both aligned with the global evaluation function and are as sensitive as possible to changes in the fitness of each member.

In dynamic, noise-free environments rovers using the difference evaluation function D, derived from the theory of collectives, were able to achieve high levels of performance because the evaluation function was both factored and highly rover-sensitive. These rovers performed better than rovers using the non-factored perfectly rover-sensitive evaluation and more than 400% better (over random rovers) than rovers using the hard to learn global evaluations.

We then extended these results both to rovers with noisy sensors, and to larger collectives. In each instance the collectives evolved using D performed better than alternative and in most cases (e.g., larger collectives) the gains due to D increase as the conditions worsened. These results show the power of using factored and rover-sensitive fitness evaluation functions, which allow evolutionary computation methods to be successfully applied to large distributed systems in real world applications where the rover sensors and actuators cannot be noise-free. Currently, we are investigating the impact of communication restriction and partial observability on rover collectives.

References

1. A. Agah and G. A. Bekey. A genetic algorithm-based controller for decentralized multi-agent robotic systems. In *In Proc. of the IEEE International Conference of Evolutionary Computing*, Nagoya, Japan, 1996.
2. A. Agogino, K. Stanley, and R. Miikkulainen. Online interactive neuro-evolution. *Neural Processing Letters*, 11: 29–38, 2000.
3. A. Agogino and K. Tumer. Efficient evaluation functions for multi-rover systems. In *The Genetic and Evolutionary Computation Conference*, pages 1–12, Seatle, WA, June 2004.
4. T. Balch. Behavioral diversity as multiagent cooperation. In *Proc. of SPIE'99 Workshop on Multiagent Systems*, Boston, MA, 1999.
5. G. Baldassarre, S. Nolfi, and D. Parisi. Evolving mobile robots able to display collective behavior. *Artificial Life*, 9: 255–267, 2003.
6. M. Dorigo and L. M. Gambardella. Ant colony systems: A cooperative learning approach to the travelling salesman problem. *IEEE Transactions on Evolutionary Computation*, 1(1):53–66, 1997.
7. S. Farritor and S. Dubowsky. Planning methodology for planetary robotic exploration. In *ASME Journal of Dynamic Systems, Measurement and Control*, volume 124, pages 4: 698–701, 2002.
8. D. Floreano and F. Mondada. Automatic creation of an autonomous agent: Genetic evolution of a neural-network driven robot. In *Proc. of Conf. on Simulation of Adaptive Behavior*, 1994.
9. F. Gomez and R. Miikkulainen. Active guidance for a finless rocket through neuroevolution. In *Proceedings of the Genetic and Evolutionary Computation Conference*, Chicago, Illinois, 2003.
10. F. Hoffmann, T.-J. Koo, and O. Shakernia. Evolutionary design of a helicopter autopilot. In *Advances in Soft Computing - Engineering Design and Manufacturing, Part 3: Intelligent Control*, pages 201–214, 1999.
11. A. Martinoli, A. J. Ijspeert, and F. Mondala. Understanding collective aggregation mechanisms: From probabilistic modelling to experiments with real robots. *Robotics and Autonomous Systems*, 29:51–63, 1999.

12. M. J. Mataric. Coordination and learning in multi-robot systems. In *IEEE Intelligent Systems*, pages 6–8, March 1998.
13. K. Stanley and R. Miikkulainen. Efficient reinforcement learning through evolving neural network topologies. In *Proceedings of the Genetic and Evolutionary Computation Conference (GECCO-2002)*, San Francisco, CA, 2002.
14. K. Tumer and A. Agogino. Overcoming communication restrictions in collectives. In *Proceedings of the International Joint Conference on Neural Networks*, Budapest, Hungary, July 2004.
15. K. Tumer and D. H. Wolpert. Collective intelligence and Braess' paradox. In *Proceedings of the Seventeenth National Conference on Artificial Intelligence*, pages 104–109, Austin, TX, 2000.
16. D. Whitley, F. Gruau, and L. Pyeatt. Cellular encoding applied to neurocontrol. In *International Conference on Genetic Algorithms*, 1995.
17. D. H. Wolpert and K. Tumer. Optimal payoff functions for members of collectives. *Advances in Complex Systems*, 4(2/3): 265–279, 2001.

17

A Memetic Algorithm Using a Trust-Region Derivative-Free Optimization with Quadratic Modelling for Optimization of Expensive and Noisy Black-box Functions

Yoel Tenne[1] and Steven William Armfield[2]

[1] School of Aerospace, Mechanical and Mechatronic Engineering, University of Sydney, Sydney NSW 2006, Australia
joel.tenne@aeromech.usyd.edu.au
[2] School of Aerospace, Mechanical and Mechatronic Engineering, University of Sydney, Sydney NSW 2006, Australia
armfield@aeromech.usyd.edu.au

Summary. A novel algorithm integrates evolutionary optimization, clustering, and the trust-region derivative-free optimization framework for global minimization of black-box functions whose evaluation is computationally resource intensive and where uncertainty exist in the objective function value, i.e. the latter contains noise. On the global scale the EA efficiently explores the search space; no global model of the objective function is generated. On the local scale the objective function is modeled by a series of quadratic models which are checked for agreement with the objective function and are updated if necessary. The algorithm incorporates numerous new techniques to enhance both its global and its local search stages. The performance of the algorithm was evaluated by using functions of dimension 2–20, with and without noise. The algorithm performed well; its performance in the presence of noise in the objective function is attributed both to the mild effect of noise on the evolutionary algorithm and to mechanics of the trust-region algorithm. The latter uses quadratic models and an interpolation technique which generates spatially spaced points; both of these diminish the effect of noise in derivatives-based trust-region minimization. Accordingly, the memetic algorithm presented here efficiently minimized black-box functions with up to 20 variables which also contain noise in the objective function value.

17.1 Introduction

17.1.1 Problem Definition

Applications in engineering and in science often require obtaining the global minimizer of a function

Y. Tenne and S.W. Armfield: *A Memetic Algorithm Using a Trust-Region Derivative-Free Optimization with Quadratic Modelling for Optimization of Expensive and Noisy Black-box Functions*, Studies in Computational Intelligence (SCI) **51**, 389–415 (2007)
www.springerlink.com

$$\boldsymbol{x}_{\mathrm{G}} : \min\{f(\boldsymbol{x}) , \quad \boldsymbol{x} \in \mathcal{D} \subseteq \mathcal{R}^n\} \tag{17.1}$$

where $\mathcal{D}$ is the prescribed feasible domain (search space). In many applications no expression is available for f or its derivatives and each evaluation of $f(\boldsymbol{x})$ is computationally resource intensive. In these settings the objective function is termed an *expensive black-box function*. It follows that during a minimization of such a function only the function values are available; for example, the value of $f(\boldsymbol{x})$ may be obtained from an experiment or from running a computer code for a candidate solution $\boldsymbol{x}$, i.e. a *point* in $\mathcal{D}$ [31, 41]. Furthermore, often an uncertainty exists regarding the value of the objective function, e.g. due to discretization errors in the computer code or due to measurement errors in an experiment. In both cases the obtained function value can be represented as

$$f_{\mathrm{noise}}(\boldsymbol{x}) = f(\boldsymbol{x}) + \nu \, , \tag{17.2}$$

where f_{noise} is the obtained function value, f is the true objective function value and ν is the added noise (which is assumed to be stochastic hence otherwise it could be eliminated). Minimization of expensive black-box functions in the presence of noise contains four major difficulties:

- No information is available on f or its derivatives, except for its values at a set of points where it has been previously evaluated. In particular, f may have a complicated and/or multimodal landscape.
- Since function evaluations are expensive only a small number of such evaluations can be used.
- Since the value of $f(\boldsymbol{x})$ is obtained from a complicated procedure, e.g. by running a computer code, there may exist *infeasible* points for which it is impossible to obtain a function value, i.e. f is undefined for some points in the search space.
- Noise distorts the landscape of the objective function and it also adversely affects approximations of derivatives.

Accordingly, this study is concerned with optimization of expensive black-box functions which contain noise and which are possibly multimodal and/or undefined at some points.

17.1.2 Modeling

The high evaluation cost and the restricted number of function evaluations has motivated the use of *surrogates* or *models* in optimization, i.e. functions which approximate the objective function and which can be evaluated by using much smaller computational resources [2, 14, 30, 35, 50]. The model is generated based on points where the objective function has been evaluated; the model is then used to approximate the value of the objective function at a new (i.e. previously unsampled) point.

Several techniques for generating a model have been studied, e.g. the response surface methodology, Kriging, artificial neural networks and radial basis functions [9, 14, 15, 19, 31, 35, 41]. The 'goodness' of the model, i.e. how small is the difference between its predicted value and the true value of the objective function, depends on many factors. In particular, the model can be a poor approximation of the objective function when the latter has a complicated landscape or is highly multimodal, the problem is high-dimensional or only a small number of points are available [3, 12, 19]. Models which are a poor approximation of the objective function can lead the optimization algorithm to converge to a false minimizer, i.e. a minimizer of the model but which is not a minimizer of the objective function [21]. These difficulties suggests that:

- It may be computationally too expensive to construct a good global model (i.e. which models the objective function well over the entire search space).
- Instead, a small number of function evaluations can be allocated for efficient domain exploration so as to identify regions where minimizers are likely to exist.
- After such regions have been identified *local* models can be generated so as to approximate the objective function in a small neighbourhood. Such models are likely to be more accurate in this neighbourhood than a global model would be.
- A mechanism is required to measure the 'goodness' of a model and to improve it, if necessary.
- The 'goodness' of the model should be only marginally affected by noise.

Discussions on these issues and pertinent algorithms are available in [1, 4, 11, 12, 14, 19, 21, 24, 30, 31, 50].

17.1.3 The Proposed Algorithm

Based on the above considerations we present a memetic algorithm for efficient global optimization of expensive black-box functions which contain noise and which are possibly multimodal and/or undefined at some points. The algorithm implements the clustering approach in optimization [37], [48, p. 95–116]. It integrates an evolutionary algorithm (EA), a cluster analysis algorithm and a trust-region derivative-free optimization algorithm. The EA explores the search space and concentrates points in clusters around minima. Next, the cluster analysis algorithm identifies the formed clusters such that points of the same cluster are considered to be in the same region of attraction. Lastly, from these clusters a derivative-free algorithm, which uses the trust-region approach with local quadratic models of the objective function, is used to efficiently converge to a minimizer; the models are local, i.e. no attempt is made to model the objective function accurately over the entire search space. In evolutionary optimization such an algorithm, which integrates an EA with one or more additional search techniques, is termed a *hybrid* or a *memetic* algorithm [17, 23, 29, 30, 36, 40] . Our choice of combining the EA with the trust-region algorithm in the clustering framework is based on the following considerations:

- The EA efficiently explores the search space, i.e. it requires only a small number of function evaluations, and since it does not rely on derivatives it remains efficient in the presence of noise.
- The quadratic models approximate the objective function locally and hence are likely to approximate it better than a global model would, they are generated using a relatively small number of points (relevant for expensive functions), they are efficiently minimized by gradient-based algorithms and they retain their goodness in the presence of noise thanks to the interpolation technique used and since, being quadratic, they smooth out noise fluctuations [3, 10, 19].
- The trust-region algorithm provides a mechanism to measure the 'goodness' of the models and to improve them if necessary [1, 8, 11, 30, 42].
- The trust-region algorithm used in this study is globally convergent, i.e. it is guaranteed to converge to a stationary point of f (under mild conditions on the boundness of the model gradient and Hessian) [7, 8].

The main novelties in the proposed memetic algorithm are:

- The combination of an EA, a cluster analysis algorithm and a trust-region derivative-free algorithm: this combination of algorithms, besides being new, also motivates the use of the trust-region algorithm to complete the convergence to a minimizer since the former efficiently exploits local information.
- The use of the density cluster analysis with EAs for global optimization: other uses of clustering techniques with EAs have been studied such as diversity safeguarding [16] and fitness approximation [22]. Also, a scaling technique has been introduced so as to improve the resultant clusters.
- The trust-region algorithm presented in this study uses new techniques for generating the quadratic models and for management of the models.
- The trust-region information generated during the local searches are used to bias the exploration of the search space by the EA. Thus, Lamarckian learning and past information are used.

A pseudo-code of the memetic algorithm presented in this study is given in Fig. 17.1.

The chapter is structured as follows: Sect.17.2 describes the EA, Sect.17.3 describes the cluster analysis algorithm, Sect.17.4 describes the trust-region derivative-free optimization algorithm, Sect.17.5 describes the overall operation of the memetic algorithm and Sect.17.6 provides results and analysis for the performance of the memetic algorithm without and with noise.

17.2 The Evolutionary Algorithm

The purpose of this section is to describe the EA implemented in the memetic algorithm.

---

begin
  generate an initial set of points covering $\mathcal{D}$
  repeat
    Global stage: employ an EA for a prescribed number of generations
    Cluster analysis: employ a cluster analysis algorithm on the resultant population to identify clusters
    Local stage: generate local quadratic models and use a trust-region algorithm to converge to minimizers
  until the termination condition is met
end

Fig. 17.1. A pseudo-code for the memetic algorithm

The EA is used to efficiently explore the search space and to concentrate points(candidate solutions) around minima. It is based on well-known variants [18, p. 49–65], [25, p. 33–44] but it has been modified to handle infeasible points and to improve the exploration of the search space. The search space $\mathcal{D} \in \mathcal{R}^n$ is assumed to be box shaped (hyper-box for $n > 3$), i.e. it is defined by vectors of lower and of upper bounds $\boldsymbol{l}$ and $\boldsymbol{u}$, respectively. Also, it is assumed that an initial population of `POPSIZE` feasible points is available, i.e.

$$\boldsymbol{C} = (\boldsymbol{x}_1 \ , \ \boldsymbol{x}_2 \ , \ \ldots \ , \ \boldsymbol{x}_{\texttt{POPSIZE}}) \ , \quad \boldsymbol{x}_i \in \mathcal{D} \ , \ i = 1 \ldots \texttt{POPSIZE} \ . \tag{17.3}$$

The EA is comprised of four steps:

- Selection: this operator chooses the best `PARSIZE` points (i.e. whose function value is the lowest) and they are designated as *parents*. Values of the objective function are used without ranking. Also, Elitism selection is used where the best `ELITESIZE` points are copied to the offspring population.
- First recombination: this operator generates *offspring* using the chosen parents. For each offspring $\texttt{NP} > 1$ parents are chosen using roulette wheel selection and an offspring is generated as a linear combination of the parents. To enhance the domain exploration of the operator a weight matrix

$$\boldsymbol{W} \ : \ W_{i\ ,\ j} \in [-1 \ , \ 1] \ , \ \boldsymbol{W} \in \mathcal{R}^{\texttt{NP} \times \texttt{NP}} \ , \tag{17.4}$$

is generated whose element $W_{i\ ,\ j}$ is a random variable obtained by using a uniform distribution. Denoting by $\boldsymbol{P}^{n \times \texttt{NP}}$ the matrix whose row vectors are the chosen parents then the new offspring $\boldsymbol{x}_{\text{off}}$ is the vector comprised of the diagonal elements of $\boldsymbol{PW}$, i.e.

$$\boldsymbol{x}_{\text{off}} = \text{diag}(\boldsymbol{PW}) \ . \tag{17.5}$$

This allows for an offspring to be generated outside the bounds of the parents and for each parent to have a different weight contribution for each allele. The procedure is repeated so as to generate `POPSIZE` offspring.

To avoid expensive function evaluations the offspring generated by the recombination operator are not evaluated since in the next step (mutation) some will be changed.

- Mutation: this operator produces μ random changes to components (alleles) of points in the offspring population. The number of mutations is proportional to the function dimension, the population size and the prescribed mutation rate, i.e.

$$\mu = n \times \texttt{POPSIZE} \times \texttt{MUTERATE}\,, \qquad \texttt{MUTERATE} \in [0.05\,,\, 0.2]\,. \tag{17.6}$$

For each mutation an offspring is chosen at random and a component of this candidate solution is chosen at random. The ith component of a candidate solution $\boldsymbol{x}$ is mutated by setting it to

$$x_i = l_i + U[0\,,\,1](u_i - l_i)\,, \tag{17.7}$$

where $U[0\,,\,1]$ is a random variable between [0, 1] which is chosen using a uniform probability distribution. To minimize the number of function evaluations all μ mutations are applied to the population before the points(candidate solutions) are evaluated, so a mutated point is evaluated only once. If mutated points are infeasible they are mutated again, until μ mutations result in feasible points. Since this operator increases the population diversity it is not used in the generation before the cluster analysis.

- Second recombination: the offspring which have not been mutated are now evaluated. If an infeasible point is found then another recombination is made until a feasible one is found to replace it.

After these four steps the offspring population consists of `POPSIZE` feasible candidate solutions. These four steps are applied for a prescribed number of generations `EAGEN` or until a function evaluations threshold is reached. To clarify the operation of the EA a pseudo-code is given in Fig 17.2.

17.3 Cluster Analysis

The purpose of this section is to describe the cluster analysis algorithm implemented in the memetic algorithm.

The EA concentrates points(candidate solutions) in clusters around minima in the search space and these clusters are identified by the cluster analysis algorithm (for the reasons mentioned in Sect.17.1). Due to the randomness in the EA the initial distribution of points is changed and the final cluster configuration is unknown a-priori. Accordingly, in the memetic algorithm the cluster analysis algorithm of Törn [46, 47] is used since this algorithm is insensitive to the distribution of points and since it does not require any input from the user.

Require: an initial population of candidate solutions: C
begin
  initialize generation counter: $gc \leftarrow 1$
  repeat
    initialize offspring population $C_{\text{off}} \leftarrow \{\emptyset\}$
    selection:
      copy the best `ELITESIZE` candidate solutions to C_{off}
      select the best `PARSIZE` candidate solutions of C as parents
    first recombination:
      for $i =$ `ELITESIZE` $+ 1 \ldots$ `POPSIZE` **do**
        generate $\boldsymbol{x}_{\text{off}} = \text{diag}(PW)$ and add it to C_{off}
      end for
    mutation:
      set number of mutations to perform: $\mu = \mu_2 \leftarrow n \times$ `POPSIZE` $\times$ `MUTERATE`
      $C_f \leftarrow \{\emptyset\}$
      repeat
        apply μ_2 mutations to candidate solutions in C which are not in C_f
        evaluate mutated candidate solutions
        retain acceptable candidate solutions; re-mutate infeasible
        $\mu_2 \leftarrow \mu_2 -$ number of mutations in retained candidate solutions
      until μ mutations resulted in feasible candidate solution
    second recombination:
      repeat
        evaluate unmutated offspring
        replace infeasible candidate solutions by new ones obtained by using another recombination
      until offspring population consists of `POPSIZE` feasible candidate solutions
    $C \leftarrow C_{\text{off}}$
    $gc \leftarrow gc + 1$
  until termination condition is met
  return final population of feasible candidate solutions: C
end

Fig. 17.2. A pseudo-code of the EA in the memetic algorithm

The density cluster analysis algorithm operates as follows: clusters are identified as regions where the points density (points per volume) is higher than a domain-averaged value κ_{av}. Each cluster is initiated from a *seed point* which is the point with the smallest function value not yet assigned to a cluster. A sphere (hyper-sphere for $n > 3$) is grown around the seed point until the point density inside the sphere (the number of points inside the sphere divided by the sphere volume) is equal or below κ_{av}; then all points inside the sphere are considered a cluster. The process resumes with the remaining unclustered points until all of them have been clustered. An example is shown in Fig. 17.3.

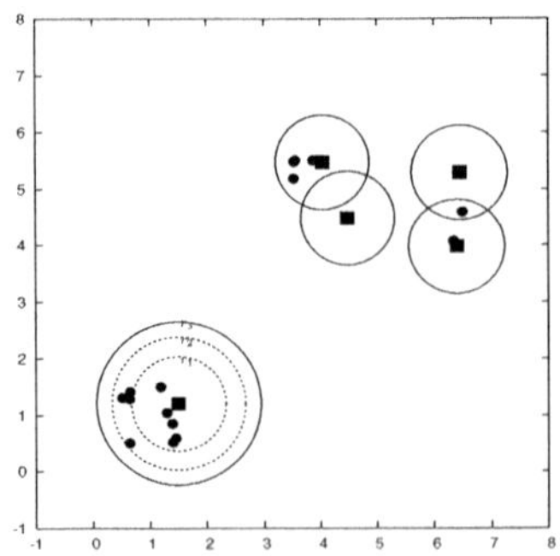

Fig. 17.3. An example of density cluster analysis. ■ are the seed points and • are the remaining points. Initially the points in the lower-left region are identified as a cluster. It is expanded, as shown from the radii r_1, r_2 and r_3. Its expansion stops after r_3 and the remaining points at the upper-right region are then clustered.

The point density measure κ_{av} is critical since a value too low results in very large and sparse clusters while a value too high results in many small clusters. For this reason a reference volume is used $V_{ref} = 4\prod_{i=1}^{d'} \lambda_i$, where λ_i are the dominant eigenvalues of the points' covariance matrix . This assists to identify whether the points reside in a subspace and so the volume enclosing them can be estimated accurately [46].

If components of the points differ by orders of magnitude then the resultant clusters may be unsatisfactory, since the large magnitude components dominate the distance calculations and variations in the remaining components are ignored. Thus, in the proposed memetic algorithm a scaling technique was introduced so as to improve the resultant clusters. Clustering is performed on a scaled pseudo-population $\boldsymbol{C}_{\text{scaled}} = \boldsymbol{D}_{scale}\boldsymbol{C}$, where the scaling matrix $\boldsymbol{D}_{scale}$ is such that in $\boldsymbol{C}_{\text{scaled}}$ the order of magnitude of the mean magnitude of each component in the population is the same. This has been found to yield satisfactory clusters. An example is shown in Fig. 17.4.

17.4 The Trust-region Derivative-free Optimizer

The purpose of this section is to describe the trust-region derivative-free optimization algorithm implemented in the memetic algorithm.

After identifying the clusters the trust-region algorithm is used to converge to the minimizers of the objective function. Convergence is iterative, where at each iteration the algorithm generates a quadratic model which locally approximates the objective function and it computes a restricted step towards the minimizer of the model. The merits of this algorithm in the framework of minimizing expensive black-box functions with noise are given in Sect.17.1.

The derivative-free trust-region algorithm is an extension of the classical (non-derivative-free) trust-region algorithm; for completeness the classical algorithm is described first followed by the derivative-free extension.

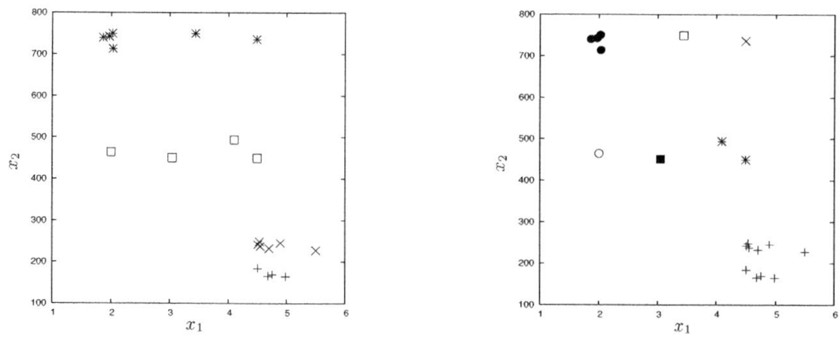

(a) Without scaling

(b) With scaling.

Fig. 17.4. An example of clusters generated by using the scaling technique. Figure **(a)** shows the resultant clusters when no scaling is performed hence changes in x_1 are ignored. Figure **(b)** shows the scaling technique yields better clusters.

17.4.1 The Classical Trust-region Algorithm

Trust-region algorithms iteratively converge along a path composed of restricted Newton steps. A general (non-restricted) Newton step is computed by assuming f can be locally approximated by a quadratic function around a point $\boldsymbol{x}_\mathrm{c}$, i.e.

$$f(\boldsymbol{x}_\mathrm{c} + \delta\boldsymbol{x}) \simeq \hat{f}(\boldsymbol{x}_\mathrm{c} + \delta\boldsymbol{x}) = f(\boldsymbol{x}_\mathrm{c}) + \langle \delta\boldsymbol{x} \,,\, \nabla f(\boldsymbol{x}_\mathrm{c}) \rangle + \frac{1}{2}\langle \nabla^2 f(\boldsymbol{x}_\mathrm{c})\delta\boldsymbol{x} \,,\, \delta\boldsymbol{x} \rangle \,, \tag{17.8}$$

where $\nabla f(\boldsymbol{x}_\mathrm{c})$ and $\nabla^2 f(\boldsymbol{x}_\mathrm{c})$ are the gradient and the Hessian of f at $\boldsymbol{x}_\mathrm{c}$, respectively. Accordingly, a non-restricted Newton step towards a minimizer (a root of $\nabla f(\boldsymbol{x}) = 0$) is given by

$$\delta\boldsymbol{x} = -\nabla^2 f(\boldsymbol{x}_\mathrm{c})^{-1} \nabla f(\boldsymbol{x}_\mathrm{c}) \,. \tag{17.9}$$

Since the approximation (17.8) may be inadequate far from $\boldsymbol{x}_\mathrm{c}$, by using non-restricted Newton steps the iterates may diverge and hence the technique is not globally convergent. To ensure global convergence a trust-region algorithm restricts the Newton step to a region where the quadratic approximation of f is trusted to be valid, i.e. it computes a restricted step [26, 28, 49]. This region is a sphere centered at $\boldsymbol{x}_\mathrm{c}$ and of radius Δ, i.e.

$$\mathcal{T} = \{\boldsymbol{x} : \|\boldsymbol{x} - \boldsymbol{x}_\mathrm{c}\|_2 \leqslant \Delta\} \,, \tag{17.10}$$

and is termed a 'trust-region'. The Newton step restricted to $\mathcal{T}$ is denoted by $\delta\boldsymbol{x}_\mathrm{m}$, and the predicted minimizer of f is

$$\boldsymbol{x}_\mathrm{m} = \boldsymbol{x}_\mathrm{c} + \delta\boldsymbol{x}_\mathrm{m} \,, \quad \boldsymbol{x}_\mathrm{m} \in \mathcal{T} \,. \tag{17.11}$$

Based on the value of the objective function at $\boldsymbol{x}_{\mathrm{m}}$ the validity of the approximation (17.8) in the current trust-region is checked. This is done by comparing the predicted reduction in the value of f to the true reduction, i.e.

$$\rho = \frac{f(\boldsymbol{x}_{\mathrm{c}}) - f(\boldsymbol{x}_{\mathrm{m}})}{\hat{f}(\boldsymbol{x}_{\mathrm{c}}) - \hat{f}(\boldsymbol{x}_{\mathrm{m}})} \,, \tag{17.12}$$

where $\rho = 1$ represents an exact agreement (the predicted reduction matches the actual reduction in the objective function). If the agreement is good (ρ is sufficiently large) then the trust-region is enlarged (since the approximation may be valid in a larger region) and $\boldsymbol{x}_{\mathrm{m}}$ becomes the new trust-region centre. Otherwise, the trust-region is reduced and another attempt is made. Accordingly, given the prescribed parameters η_+ , $\delta_+ > 1$, $0 < \delta_- < 1$, $\Delta_{\max} > 0$, a typical update of the trust-region is

$$\boldsymbol{x}_{\mathrm{c}}^{(t+1)} = \begin{cases} \boldsymbol{x}_{\mathrm{m}}^{(t)} & \text{if } \rho \geqslant \eta_+ \\ \boldsymbol{x}_{\mathrm{c}}^{(t)} & \text{if } \rho < \eta_+ \end{cases} , \qquad \Delta^{(t+1)} = \begin{cases} \min\left\{\delta_+ \Delta^{(t)} , \Delta_{\max}\right\} & \text{if } \rho \geqslant \eta_+ \\ \delta_- \Delta^{(t)} & \text{if } \rho < \eta_+ \end{cases} . \tag{17.13}$$

Convergence is declared when $\Delta < \Delta_{\min}$, where $\Delta_{\min}$ is prescribed. Typical values are $\eta_+ \in [0.25 , 0.75]$ and $\Delta_{\min} \in [10^{-3} , 10^{-2}]$. Further details on trust-region algorithms are available in [6, 26, 28, 49].

17.4.2 Derivative-free Extension

To use the trust-region algorithm when f is a black-box, i.e. $\nabla f(\boldsymbol{x}_{\mathrm{c}})$ and $\nabla^2 f(\boldsymbol{x}_{\mathrm{c}})$ are unavailable, the following modification is used: a multivariate polynomial interpolant is generated based on points(candidate solutions) where f has been evaluated and the interpolant's gradient and Hessian are taken as approximations to $\nabla f(\boldsymbol{x}_{\mathrm{c}})$ and to $\nabla^2 f(\boldsymbol{x}_{\mathrm{c}})$, respectively. This approach is termed a trust-region derivative-free optimization (TR-DFO) algorithm. The quadratic model functions are generated by interpolation [8, 10, 33, 42, 43]. A linear polynomial was found to be inferior to a quadratic one [32, 34], while polynomials of degree higher than two [10] were rejected since they may require more interpolation points which is undesirable when function evaluations are expensive, and since they may contain oscillations [3]. While the gradient and Hessian obtained from the interpolant are likely to be inaccurate (which differs from the classical trust-region algorithm) the TR-DFO algorithm *is* globally convergent (as with the classical trust-region algorithm) under mild conditions on the boundness of the model gradient and Hessian [7, 8].

The three steps of the TR-DFO algorithm, constructing a quadratic model of the objective function, obtaining the minimizer of the model and updating the model and the trust-region, are now explained.

Constructing the Quadratic Model

The quadratic model function $\hat{f}$ is an interpolant generated such that it satisfies the Lagrange interpolation condition

$$\hat{f}(\boldsymbol{x}_i) = f(\boldsymbol{x}_i)\ , \ i = 1 \dots s\ , \ s \leqslant k_{\max}\ , \quad k_{\max} = \tfrac{(n+1)(n+2)}{2}\ , \tag{17.14}$$

where $k_{\max}$ is the maximal number of interpolation points required to uniquely define a quadratic polynomial; $\mathcal{X}$ will represent the set of interpolation points. The quadratic $\hat{f}$ is generated as a weighted sum of polynomial basis functions in Π_2^n–the space of all n-variate polynomials which are at most quadratic, i.e.

$$\hat{f}(\boldsymbol{x}) = \sum_{j=1}^{s} \gamma_j \boldsymbol{p}_j(\boldsymbol{x})\ , \quad \boldsymbol{p}_i(\boldsymbol{x}) \in \Pi_2^n\ , \quad \gamma_i \in \mathcal{R}\ , \quad i = 1 \dots s\ . \tag{17.15}$$

A difficulty in generating $\hat{f}$ is that not every set of points uniquely defines the interpolant. The interpolation points must not all lie on a quadratic curve in $\mathcal{R}^n$ since otherwise any multiple of the curve can be added to the interpolant while (17.14) would remain valid, hence the interpolant would not be unique. This implies suitable points must satisfy certain geometric conditions [13, 38]. A set of points which uniquely defines the quadratic interpolant is termed *poised* with respect to Π_2^n [8].

Given a set of points and their corresponding function values

$$\mathcal{X} = \{\boldsymbol{x}_1\ , \dots\ , \boldsymbol{x}_s\}\ , \quad \mathcal{Y} = \{f(\boldsymbol{x}_1)\ , \dots\ , f(\boldsymbol{x}_s)\}\ , \tag{17.16}$$

respectively, then the algorithm of Sauer–Xu is used to extract the set of poised points (each poised point is from $\mathcal{X}$)

$$\mathcal{X}_{\mathrm{p}} = \{\boldsymbol{x}_{\mathrm{p},1}\ , \dots\ , \boldsymbol{x}_{\mathrm{p},\mathrm{k}}\}\ , \quad \mathcal{X}_{\mathrm{p}} \subseteq \mathcal{X}\ , \quad k \leqslant s\ , \tag{17.17}$$

and to generate the quadratic interpolant [8, 39]. The polynomial basis function generated by the Sauer-Xu algorithm are termed the Newton Fundamental Polynomials.

The algorithm starts with a set of s monomials, e.g. for $n = 2\ , \ s = 5$ a possible initial set is $\{1\ , x_1\ , x_2\ , {x_1}^2\ , x_1x_2\}$. The algorithm uses a Gram–Schmidt process over Π_2^n to identify the set of k points which are poised and it transforms k monomials to the Newton Fundamental Polynomials [39]. The magnitude of one of the latter at a point is called its *pivot* (as in the Gram–Schmidt process). The Sauer–Xu algorithm identifies a poised point as such where the pivot does not vanish. If such a point is found than the non-vanishing polynomial is normalized with respect to this point and all other polynomials are orthogonalized with respect to this point. Thus, each Newton Fundamental Polynomial is associated with a unique poised point. The process resumes until all points have been examined. The Newton Fundamental Polynomials are used since they can be efficiently computed and since the magnitude of their pivot is a measure of how close the interpolant is to becoming non-unique(a larger pivot magnitude is better).

In practice, and particularly in the presence of noise in the function, it is preferable to identify a *well-poised* set of points such that even with small changes to the points of this set the interpolant would still remain unique. The points of such a set are well-spaced and accordingly the resultant interpolant is likely to model f better than an interpolant based on points which are nearly non-poised. Furthermore, since well-poised points are well-spaced this diminishes the effect of noise on the resultant model. Mathematically, a well-poised set is such that the pivots of its corresponding points are bounded away from zero. This motivated the *pivot threshold strategy* where a point is accepted into the poised set only if it corresponds to a pivot which is larger than a positive lower-bound θ [8, 39].

After obtaining the poised points and the Newton Fundamental Polynomials the coefficients γ_i are obtained from the latter's Vandermonde matrix based on the poised points. The Sauer–Xu algorithm for generating the Newton Fundamental Polynomials is given in Fig. 17.5 and further details are available in [39].

In the proposed memetic algorithm the following issues are addressed:

- The orthogonalization in the Sauer–Xu algorithm implies that if $\mathcal{X}_{\mathrm{p}}$ changes then a point $\boldsymbol{x}_{\mathrm{p},\mathrm{i}}$ with an index $i > 1$ may be not well-poised with respect to the new set and hence it would be discarded. Accordingly, $\boldsymbol{x}_{\mathrm{p},1} = \boldsymbol{x}_{\mathrm{c}}$ is used so as to retain $\boldsymbol{x}_{\mathrm{c}}$.
- If $n + 1 < k < \frac{(n+1)(n+2)}{2}$, i.e. the cardinality of $\mathcal{X}$ is submaximal but higher than linear so some quadratic monomials are processed, then initially the non-mixed monomials are chosen (e.g. ${x_1}^2$, ${x_2}^2$) so as to obtain a model with a diagonal Hessian which approximates the curvature of $\nabla^2 f$ along the main axis.

In the following sections the interpolation set is poised and for clarity the subscript p is dropped; a point $\boldsymbol{x}_i$ corresponds to a Newton Fundamental Polynomial $\boldsymbol{p}_i$, but otherwise the partition to blocks is ignored.

Obtaining the Minimizer

In the next step the minimizer of the model $\hat{f}$ in the current trust-region is obtained. This minimizer is denoted $\boldsymbol{x}_{\mathrm{m}}$. Since $\hat{f}$ is quadratic efficient gradient-based algorithms exist for this constrained optimization problem. The algorithm implemented in the proposed memetic algorithm is that of Moré–Sorensen [28] since it is both efficient and since it handles all the singular cases arising in this optimization problem. The parameters of this algorithm (defined in [28]) are set so as to ensure at least a 99% of the maximal reduction possible which is sufficiently accurate and avoids excessive iterations.

Require: A set of s feasible points $\mathcal{X}$, $\mathcal{Y}$ and a set of s monomials
begin
Arrange monomials in blocks such that each block $h = 0, 1, 2$ contains polynomials of maximal degree h such that $\boldsymbol{p}_{\tilde{g}}^{[\tilde{h}]}$ is the gth monomial in the hth block
$\tilde{h} \leftarrow 0$
$\tilde{g} \leftarrow 1$
$\mathcal{X}_\mathrm{p} \leftarrow \{\emptyset\}$
$\mathcal{P} \leftarrow \{\emptyset\}$
repeat
if there is a point $\boldsymbol{x}_i \in \mathcal{X}$ for which $\boldsymbol{p}_{\tilde{g}}^{[\tilde{h}]}(\boldsymbol{x}_i) \neq 0$ **then**
Normalize the current polynomial: $\hat{\boldsymbol{p}}_{\tilde{g}}^{[\tilde{h}]}(\boldsymbol{x}) = \boldsymbol{p}_{\tilde{g}}^{[\tilde{h}]}(\boldsymbol{x}) / \boldsymbol{p}_{\tilde{g}}^{[\tilde{h}]}(\boldsymbol{x}_i)$
Normalize polynomials in current block and in higher blocks:

$$\text{for v} \neq \tilde{g} \, : \, \boldsymbol{p}_v^{[\tilde{h}]}(\boldsymbol{x}) = \boldsymbol{p}_v^{[\tilde{h}]}(\boldsymbol{x}) - \boldsymbol{p}_v^{[\tilde{h}]}(\boldsymbol{x}_i)\hat{\boldsymbol{p}}_{\tilde{g}}^{[\tilde{h}]}(\boldsymbol{x}) \tag{17.18}$$

$$\text{for v} \geq 1 \, , \, w > \tilde{h} \, : \, \boldsymbol{p}_v^{[w]}(\boldsymbol{x}) = \boldsymbol{p}_v^{[w]}(\boldsymbol{x}) - \boldsymbol{p}_v^{[w]}(\boldsymbol{x}_i)\hat{\boldsymbol{p}}_{\tilde{g}}^{[\tilde{h}]}(\boldsymbol{x}) \tag{17.19}$$

Increment $\tilde{g}$; if all polynomials in current blocks have been processed then increment $\tilde{h}$ and set $\tilde{g} = 1$
Remove $\boldsymbol{x}_i$ from $\mathcal{X}$, i.e. $\mathcal{X} \leftarrow \mathcal{X} \setminus \{\boldsymbol{x}_i\}$
Add $\boldsymbol{x}_i$ to $\mathcal{X}_\mathrm{p}$, i.e. $\mathcal{X}_\mathrm{p} \leftarrow \mathcal{X}_\mathrm{p} \cup \{\boldsymbol{x}_i\}$
Add $\boldsymbol{p}_{\tilde{g}}^{[\tilde{h}]}$ to $\mathcal{P}$, i.e. $\mathcal{P} \leftarrow \mathcal{P} \cup \{\boldsymbol{p}_{\tilde{g}}^{[\tilde{h}]}\}$
end if
until all polynomials have been processed or no suitable point is found
return The poised set of points $\mathcal{X}_\mathrm{p} = \{\boldsymbol{x}_{\mathrm{p},1}, \ldots, \boldsymbol{x}_{\mathrm{p},\mathrm{k}}\}$ and their corresponding Newton Fundamental Polynomials $\mathcal{P}$
end

Fig. 17.5. A pseudo-code for the Sauer–Xu algorithm for generating the Newton Fundamental Polynomials

Updating the Trust-Region, the Interpolation Set and the Model

After $\boldsymbol{x}_\mathrm{m}$ has been found the trust-region, the interpolation set and the model are updated. The steps are similar to those in the classical trust-region algorithm but with modifications so as to update $\mathcal{X}$ and to handle infeasible points.

Initially $f(\boldsymbol{x}_\mathrm{m})$ is evaluated. If $\boldsymbol{x}_\mathrm{m}$ is infeasible then the trust-region is temporarily reduced and another attempt is made. This is done by the heuristic that since $\boldsymbol{x}_\mathrm{c}$ is feasible then it may be possible to find another feasible point in a smaller neighbourhood around it [9]. The procedure is repeated until a feasible $\boldsymbol{x}_\mathrm{m}$ is found or until the temporarily reduced Δ is smaller than $\Delta_{\min}$.

Based on $\boldsymbol{x}_\mathrm{m}$ the model is updated as follows:

- If $\boldsymbol{x}_\mathrm{m}$ is feasible and $\rho \geqslant \eta_+$: then $\boldsymbol{x}_\mathrm{m}$ is a good minimizer and is added to $\mathcal{X}$. If the cardinality of $\mathcal{X}$ is already maximal ($k = k_{\max}$) then $\boldsymbol{x}_\mathrm{m}$ replaces an existing point of $\mathcal{X}$.

- If $\boldsymbol{x}_{\mathrm{m}}$ is feasible and $\rho < \eta_+$: then $\boldsymbol{x}_{\mathrm{m}}$ is not a good minimizer, but it is added to $\mathcal{X}$ if it is poised with respect to the points in $\mathcal{X}$ (i.e. it satisfies the pivot threshold criterion). This is done since function evaluations are expensive and so are utilized as much as possible. If the cardinality of $\mathcal{X}$ is already maximal ($k = k_{\max}$) then $\boldsymbol{x}_{\mathrm{m}}$ replaces an existing point of $\mathcal{X}$.
- If $\boldsymbol{x}_{\mathrm{m}}$ is infeasible *or* if $\rho < \eta_+$ and the number of points in $\mathcal{T}$ is smaller than `CRMODAD`: then a new point is generated and it replaces an existing point in $\mathcal{X}$. The new point, $\boldsymbol{x}_{\mathrm{N}}$, is taken to be the maximizer of a Newton Fundamental Polynomial of an existing point from $\mathcal{X}$. This is done since a point which increases the pivot of its corresponding Newton Fundamental Polynomial improves the poisedness of the interpolation set $\mathcal{X}$ [8].

Next, the trust-region radius Δ is updated as follows:

- If $\rho \geqslant \eta_+$: then Δ is increased as

$$\Delta^{(t+1)} = \min\left\{\delta_+ \times \Delta^{(t)} \, , \, \Delta_{\max}\right\} . \tag{17.20}$$

- If $\rho < \eta_+$ *and* there are at least `CRMODAD` ('cardinality of adequate model') points in $\mathcal{T}$: then Δ is decreased as

$$\Delta^{(t+1)} = \delta_- \times \Delta^{(t)} . \tag{17.21}$$

  The condition on the cardinality in $\mathcal{T}$ ensures that Δ is decreased only when $\hat{f}$ is based on enough points, so as to avoid reducing Δ based on an inaccurate model.

The last step checks whether a better minimizer than $\boldsymbol{x}_{\mathrm{c}}$ exists in $\mathcal{X}$, i.e. for $i = 2 \ldots k$ the condition

$$\tilde{\rho} = \frac{f(\boldsymbol{x}_i) - f(\boldsymbol{x}_1)}{\hat{f}(\boldsymbol{x}_i) - \hat{f}(\boldsymbol{x}_1)} \geqslant \eta_- \, , \quad \eta_- \leqslant \eta_+ \, , \tag{17.22}$$

is checked, and the point for which the condition is met and $\tilde{\rho}$ is maximal becomes the new trust-region centre, i.e. it is interchanged with $\boldsymbol{x}_1$. A merit of using quadratic models and the Newton Fundamental Polynomials in the trust-region framework is that this allows not only to generate the models but also to evaluate their goodness and to identify new points which would improve the latter. Several issues are noted:

- The technique described for temporarily reducing Δ is applied also when finding $\boldsymbol{x}_{\mathrm{N}}$.
- When the cardinality of $\mathcal{X}$ is maximal and a candidate point ($\boldsymbol{x}_{\mathrm{m}}$ and/or $\boldsymbol{x}_{\mathrm{N}}$) is to replace an existing point $\boldsymbol{x}_i \in \mathcal{X}$ then the latter is chosen such that its corresponding Newton Fundamental Polynomial satisfies the pivot threshold criterion at the candidate point, e.g. $\|\boldsymbol{p}_i(\boldsymbol{x}_{\mathrm{m}})\| \geqslant \theta$. When seeking a point to replace first $\boldsymbol{x}_k$ is checked then $\boldsymbol{x}_{k-1}$ etc. This is done so as to prefer replacing a point with a high index as possible since this reduces the chances of causing existing points in $\mathcal{X}$ becoming unpoised and hence being discarded.

After updating $\mathcal{X}$ and $\mathcal{T}$ the model $\hat{f}$ is regenerated and another iteration is performed. Figure 17.6 shows an example of a TR-DFO algorithm minimization.

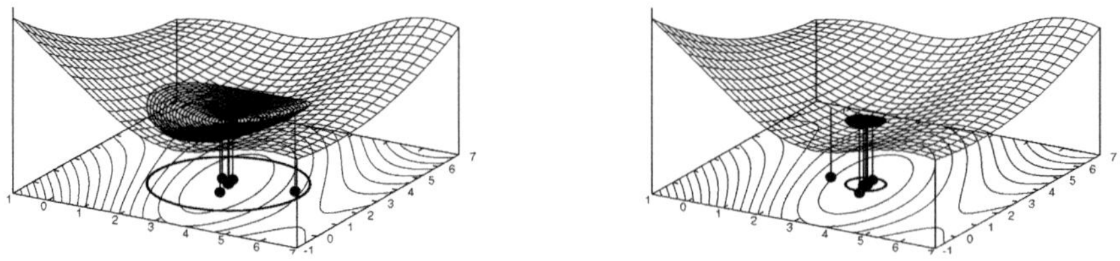

(a) (b)

Fig. 17.6. An example of the TR-DFO algorithm minimization of the Branin function. The objective function surface, the quadratic model function and the trust-region are shown. Figure **(a)** shows the model at an early stage of the minimization when the trust-region is large; Figure **(b)** shows near convergence when the trust-region has been reduced and encloses $\boldsymbol{x}_\mathrm{G}$.

At the beginning of each iteration if the cardinality of $\mathcal{X}$ is smaller than the prescribed limit `NPMIN` then new points are sought to add to the set. New points are introduced by maximizing the $(k+1)$th Newton Fundamental Polynomial (as is done when obtaining $\boldsymbol{x}_\mathrm{N}$). If an acceptable point cannot be found then the algorithm still proceeds by the heuristic that it may still be possible to find a minimizer better than $\boldsymbol{x}_\mathrm{c}$.

The TR-DFO algorithm terminates when one of the following criteria is met:

- $\Delta < \Delta_{\min}$.
- $f(\boldsymbol{x}_1) < \mathtt{MINF}\,, \quad \mathtt{MINF} \geqslant -\infty$.
- $\|\nabla \hat{f}\|_2 \leq \mathtt{GMIN}$ *and* there are at least $n+1$ points in $\mathcal{T}$, i.e. a stationary point is considered valid only when the model gradient is based on a sufficient number of points.

For clarification a pseudo-code of the TR-DFO algorithm is given in Fig. 17.7.

17.5 The Overall Algorithm

The purpose of this section is to describe the overall operation of the proposed memetic algorithm.

The algorithm begins by generating a set of random points in $\mathcal{D}$ by using a uniform probability distribution. Initially (the global stage) the EA is used

Require: A non-empty set of feasible initial points $\mathcal{X}^{(0)}$,
$\eta_+ > 0, \delta_+ > 1, \texttt{GMIN} > 0, 0 < \Delta_{\min} < \Delta_{\max}$
$\eta_- \leqslant \eta_+, 0 < \delta_- < 1, 0 < \texttt{CRMODAD} \leqslant k_{\max}, 1 \leqslant \texttt{NPMIN} \leqslant k_{\max}$
begin
  repeat
    if $k <$ NPMIN attempt to add points until $k =$ NPMIN or no acceptable point is found
    Generate the quadratic model $\hat{f}$
    Obtain $\boldsymbol{x}_{\mathrm{m}}$;
      if $\boldsymbol{x}_{\mathrm{m}}$ is feasible calculate ρ
      if $\boldsymbol{x}_{\mathrm{m}}$ is feasible and $\rho \geqslant \eta_+$ add $\boldsymbol{x}_{\mathrm{m}}$ to $\mathcal{X}$
      if $\boldsymbol{x}_{\mathrm{m}}$ is feasible and $\rho < \eta_+$ but $\boldsymbol{x}_{\mathrm{m}}$ satisfies the pivot threshold criterion add $\boldsymbol{x}_{\mathrm{m}}$ to $\mathcal{X}$
      if $\boldsymbol{x}_{\mathrm{m}}$ is infeasible or insufficient points in $\mathcal{T}$ replace a point from $\mathcal{X}$ with a new point in $\mathcal{T}$
      for each point $\boldsymbol{x}_i \in \mathcal{X}$, $i = 2 \ldots k$ calculate $\tilde{\rho}$ and check if to replace $\boldsymbol{x}_{\mathrm{c}}$
  until a termination criterion is met
end

Fig. 17.7. A pseudo-code for the TR-DFO implemented in the memetic algorithm

for EAGEN generation. Next, the cluster analysis algorithm is used so as to identify clusters. Lastly (the local stage) the TR-DFO algorithm is initiated from clusters and obtains the minimizers of f . Such a sequence is termed a *cycle*. Several issues are addressed in the algorithm:

- The mutation operator is not applied in the generation prior to the cluster analysis so as to assist cluster formation.
- For very expensive functions the TR-DFO is initiated using points only from the first cluster since the latter contains the current best point.
- After the TR-DFO is stopped the points of clusters from which the TR-DFO was initiated are removed from the population. To preserve the population size new points are added to the latter.
- Trust-regions where a good minimizer was found are used to guide how the EA explores the search space. The heuristic is that in these trust-regions f was modeled well and hence the global search should be biased away from them. Accordingly, when the EA generates points(candidate solutions) by mutation or by introducing new points to the population then these points are accepted only if they are outside such trust-regions. Thus, information from previous local searches is used to bias the EA search; by the taxonomy in [23] the proposed algorithm incorporates *historic information.* Figure 17.8 shows an example of this technique.

The proposed memetic algorithm terminates when one of the following criteria is satisfied:

- The number of function evaluations exceeds a prescribed threshold.

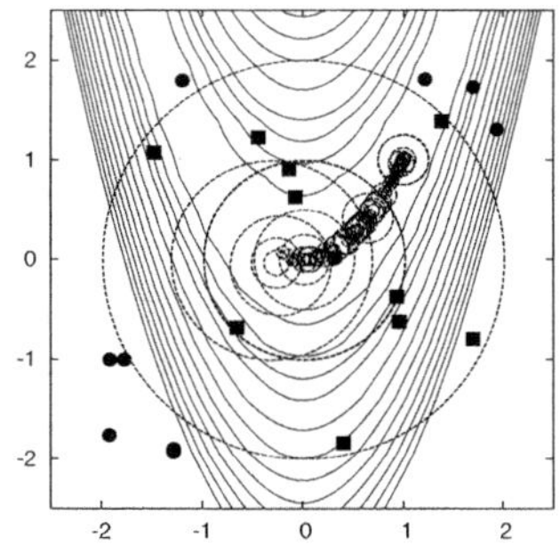

Fig. 17.8. An example of using the trust-regions to bias the search of the EA, taken from minimization of Rosenbrock's function. The ■ are the initial random points. After the first cycle the trust-regions for which a good minimizer was found (*dashed line*). In the second cycle the new points (•) are generated outside these trust-regions so the global search is biased to regions in the search space where f has not been modelled well or not at all.

- The number of EA generation exceeds a prescribed threshold.
- A minimizer $\boldsymbol{x}_{\mathrm{L}}$ has been found for which the objective function value is sufficiently small, i.e. $f(\boldsymbol{x}_{\mathrm{L}}) \leqslant \mathtt{MINF}\,, \quad \mathtt{MINF} \geqslant -\infty\,.$

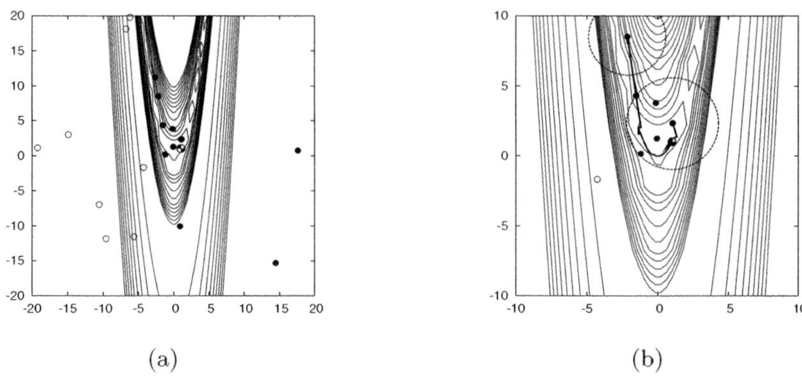

Fig. 17.9. An example of the operation of the memetic algorithm taken from minimization of Rosenbrock's function. Figure **(a)** shows the global stage where initial random points were generated (○) and those found by the EA (•). Figure **(b)** shows the local stage where clusters were identified followed by convergence of the TR-DFO algorithm to the minimizer $\boldsymbol{x}_{\mathrm{G}} = (1\,,\,1)$.

17.6 Results and Analysis

The purpose of this section is to provide results and analysis for the performance of the proposed memetic algorithm without and with noise.

For its performance evaluation the memetic algorithm was used to minimize a range of well-established benchmark functions of dimension $n = 2 \ldots 20$

[5, 27, 34, 45]. They range from quadratic to multimodal and hence are adequate test cases (Appendix A provides a detailed description of these functions). For all functions $f(\boldsymbol{x}_G) = 0$, except for the BrownDennis function where $f(\boldsymbol{x}_G) \simeq 8.582 \times 10^{+4}$, and the Bdqrtic10D, Bdqrtic20D functions where $f(\boldsymbol{x}_G) \simeq 1.187 \times 10^{+1}$, $3.542 \times 10^{+1}$, respectively.

Since for expensive functions the number of evaluations is likely to be limited then in all tests the number of function evaluations was limited to $\texttt{MAXFE} = 50n$ (n is the function dimension). The performance of the proposed memetic algorithm was evaluated by the following metrics:

- f_{best} : the best function value found.
- $\|\Delta\boldsymbol{x}_{\text{best}}\|_2 = \|\boldsymbol{x}_{\text{best}} - \boldsymbol{x}_G\|_2$: the Euclidean norm of the difference between the best minimizer found after MAXFE function evaluations ($\boldsymbol{x}_{\text{best}}$) and the global minimizer ($\boldsymbol{x}_G$). This is a measure of the algorithm's ability to locate $\boldsymbol{x}_G$.

In all tests the following parameter settings were used:

- For the EA:

$$\texttt{POPSIZE} = \min(10\,,\ \max(n\,,\ 5)), \texttt{PARSIZE} = \texttt{POPSIZE}/2, \texttt{NP} = 2$$
$$\texttt{ELITESIZE} = \min(5\,,\ \max(n/2)), \texttt{EAGEN} = \max(5\,,\ n/2), \texttt{MUTERATE} = 0.1$$

- For the TR-DFO algorithm:

$$\eta_+ = 0.75, \delta_+ = 2, \texttt{CRMODAD} = 3, \eta_- = 0.1, \delta_- = 0.5$$
$$\Delta_{\min} = 10^{-3}, \texttt{NPMIN} = n + 1, \texttt{GMIN} = 10^{-3}, \Delta_{\max} = 10$$

- Stopping criteria:

$$\texttt{MINF} = -\infty, \texttt{MAXFE} = 50n$$

For the Rosenbrock function and for the Chained Rosenbrock functions the settings $\eta_+ = 0.25$, $\eta_- = 0$ were used based on results from [44]. Each test function was minimized for ten times and the mean of each metric is given (indicated by the prefix M in Tables 17.1–17.3). The EA was also used by itself for MAXFE evaluations so as to see whether adding the cluster analysis and the TR-DFO algorithm improves the performance. The EA used the same initial population and the same parameters as the memetic algorithm.

Initially the performance of the proposed memetic algorithm was evaluated without noise. Test results are given in Table 17.1, from which the following points arise:

- The memetic algorithm obtained a good approximation to $\boldsymbol{x}_G$ within the function evaluations threshold in all tests.
- The use of the EA allowed the memetic algorithm to remain efficient as n increased since although the search space was enlarged the EA efficiently

explored it and located points which were closer to a minimizer than the initial random population.

- The memetic algorithm obtained a final result which was significantly better than that of the EA. This is attributed to the TR-DFO which uses accurate local quadratic models and gradient-based minimization.

Table 17.1. Benchmark Results for Test Functions without Noise

		Memetic Algorithm		EA	
Function	n	$\mathrm{M}\,(f_{\text{best}})$	$\mathrm{M}\,(\Vert\Delta\boldsymbol{x}_{\text{best}}\Vert_2)$	$\mathrm{M}\,(f_{\text{best}})$	$\mathrm{M}\,(\Vert\Delta\boldsymbol{x}_{\text{best}}\Vert_2)$
Beale	2	$\mathbf{7.661 \times 10^{-7}}$	$\mathbf{9.923 \times 10^{-4}}$	3.134×10^{-1}	5.551×10^{-1}
Rosenbrock	2	$\mathbf{6.049 \times 10^{-2}}$	$\mathbf{3.076 \times 10^{-1}}$	6.074×10^{-1}	$1.159 \times 10^{+0}$
Box3D	3	8.502×10^{-3}	$1.312 \times 10^{+1}$	$\mathbf{3.174 \times 10^{-3}}$	$\mathbf{1.251 \times 10^{+1}}$
BrownDennis	4	$\mathbf{8.582 \times 10^{+4}}$	$\mathbf{1.700 \times 10^{-3}}$	$1.964 \times 10^{+5}$	$8.625 \times 10^{+0}$
KowalikOsborne	4	$\mathbf{7.283 \times 10^{-3}}$	$\mathbf{8.170 \times 10^{+0}}$	2.838×10^{-2}	$6.152 \times 10^{+0}$
ArrowHead10D	10	$\mathbf{9.928 \times 10^{-5}}$	$\mathbf{2.754 \times 10^{-3}}$	$1.947 \times 10^{+1}$	$2.490 \times 10^{+0}$
Bdqrtic10D	10	$\mathbf{1.188 \times 10^{+1}}$	$\mathbf{7.467 \times 10^{-2}}$	$1.564 \times 10^{+1}$	8.301×10^{-1}
CRosenbrock10D	10	$\mathbf{1.037 \times 10^{+0}}$	$\mathbf{1.164 \times 10^{+0}}$	$8.398 \times 10^{+0}$	$2.992 \times 10^{+0}$
ArrowHead20D	20	$\mathbf{7.597 \times 10^{-4}}$	$\mathbf{9.755 \times 10^{-3}}$	$4.563 \times 10^{+1}$	$3.788 \times 10^{+0}$
Bdqrtic20D	20	$\mathbf{3.551 \times 10^{+1}}$	$\mathbf{2.556 \times 10^{-1}}$	$4.594 \times 10^{+1}$	$1.140 \times 10^{+0}$
CRosenbrock20D	20	$\mathbf{7.130 \times 10^{-1}}$	$\mathbf{7.407 \times 10^{-1}}$	$2.167 \times 10^{+1}$	$4.296 \times 10^{+0}$

To test the performance of the memetic algorithm in the presence of noise a procedure similar to the one just described was used, but the value of the objective function was modified to

$$f_{\text{noise}} = f\,(1 + U[0\,,\,1]A_{\text{noise}})\ , \tag{17.23}$$

where f_{noise} is the objective function value with noise, f is the objective function value without noise, $U[0\,,\,1]$ is a random number in the range $[0, 1]$ obtained by using a uniform probability distribution and A_{noise} is the relative noise amplitude set to either 1% or 10%. Test results for the two noise levels are given in Tables 17.2, 17.3, from which the following conclusions arise:

- The proposed memetic algorithm was only marginally affected by the 1% noise level and it obtained an accurate minimizer even in 10% noise level.
- The TR-DFO (used in the memetic algorithm) typically obtained a good minimizer despite of using a gradient-based approach. This is attributed to two factors: the interpolation technique (Sect.17.4.2) which resulted in points which were spatially spaced and so the effect of noise was diminished, and the use of quadratic models which smoothed out some of the multimodality caused by the noise.
- For the 1% noise level the proposed memetic algorithm outperformed the EA in all tests. In the 10% noise level the relative performance of the EA

improved which is attributed to the function becoming more multimodal. However, when the minimizer found by the memetic algorithm was better then it was significantly more accurate than that found by the EA alone.

- The proposed memetic algorithm often converged in fewer function evaluations when compared to the noiseless tests. This is since in latter stages of the optimization the points became concentrated and the noise effectively created a 'local minimum' and so the TR-DFO converged to an approximate minimizer.

Table 17.2. Benchmark Results for Test Functions with 1% Noise

		Memetic Algorithm		EA	
Function	n	M (f_{best})	M ($\|\Delta \boldsymbol{x}_{\text{best}}\|_2$)	M (f_{best})	M ($\|\Delta \boldsymbol{x}_{\text{best}}\|_2$)
Beale	2	$\mathbf{4.822 \times 10^{-9}}$	$\mathbf{1.041 \times 10^{-4}}$	4.255×10^{-1}	8.868×10^{-1}
Rosenbrock	2	$\mathbf{4.042 \times 10^{-1}}$	$\mathbf{3.593 \times 10^{-1}}$	6.774×10^{-1}	$1.246 \times 10^{+0}$
Box3D	3	$\mathbf{1.457 \times 10^{-3}}$	$\mathbf{7.986 \times 10^{+0}}$	1.589×10^{-4}	$2.148 \times 10^{+1}$
BrownDennis	4	$\mathbf{8.735 \times 10^{+4}}$	$\mathbf{6.794 \times 10^{-1}}$	$1.791 \times 10^{+5}$	$9.269 \times 10^{+0}$
KowalikOsborne	4	$\mathbf{1.030 \times 10^{-2}}$	$\mathbf{1.020 \times 10^{+1}}$	1.280×10^{-2}	$1.003 \times 10^{+1}$
ArrowHead10D	10	$\mathbf{1.247 \times 10^{-5}}$	$\mathbf{1.274 \times 10^{-3}}$	$1.912 \times 10^{+1}$	$2.443 \times 10^{+0}$
Bdqrtic10D	10	$\mathbf{1.187 \times 10^{+1}}$	$\mathbf{6.461 \times 10^{-2}}$	$1.504 \times 10^{+1}$	7.530×10^{-1}
CRosenbrock10D	10	$\mathbf{3.630 \times 10^{-2}}$	$\mathbf{2.561 \times 10^{-1}}$	$8.377 \times 10^{+0}$	$2.988 \times 10^{+0}$
ArrowHead20D	20	$\mathbf{5.060 \times 10^{-1}}$	$\mathbf{1.382 \times 10^{-1}}$	$4.921 \times 10^{+1}$	$4.009 \times 10^{+0}$
Bdqrtic20D	20	$\mathbf{4.483 \times 10^{+1}}$	$\mathbf{1.334 \times 10^{+0}}$	$4.626 \times 10^{+1}$	$1.113 \times 10^{+0}$
CRosenbrock20D	20	$\mathbf{1.109 \times 10^{+1}}$	$\mathbf{2.789 \times 10^{+0}}$	$1.854 \times 10^{+1}$	$4.328 \times 10^{+0}$

Table 17.3. Benchmark Results for Test Functions with 10% Noise

		Memetic Algorithm		EA	
Function	n	M (f_{best})	M ($\|\Delta \boldsymbol{x}_{\text{best}}\|_2$)	M (f_{best})	M ($\|\Delta \boldsymbol{x}_{\text{best}}\|_2$)
Beale	2	$\mathbf{6.694 \times 10^{-3}}$	$\mathbf{1.317 \times 10^{-1}}$	4.910×10^{-1}	7.529×10^{-1}
Rosenbrock	2	$\mathbf{2.855 \times 10^{-1}}$	$\mathbf{3.085 \times 10^{-1}}$	8.376×10^{-1}	$1.293 \times 10^{+0}$
Box3D	3	2.984×10^{-2}	$1.100 \times 10^{+1}$	$\mathbf{9.163 \times 10^{-4}}$	$\mathbf{1.692 \times 10^{+1}}$
BrownDennis	4	$5.709 \times 10^{+5}$	$1.681 \times 10^{+1}$	$\mathbf{2.109 \times 10^{+5}}$	$\mathbf{9.316 \times 10^{+0}}$
KowalikOsborne	4	$\mathbf{6.521 \times 10^{-3}}$	$\mathbf{8.051 \times 10^{+0}}$	1.510×10^{-2}	$4.321 \times 10^{+0}$
ArrowHead10D	10	$\mathbf{4.730 \times 10^{+0}}$	$\mathbf{7.373 \times 10^{-1}}$	$2.235 \times 10^{+1}$	$2.472 \times 10^{+0}$
Bdqrtic10D	10	$1.975 \times 10^{+1}$	$1.031 \times 10^{+0}$	$\mathbf{1.650 \times 10^{+1}}$	$\mathbf{6.242 \times 10^{-1}}$
CRosenbrock10D	10	$\mathbf{2.025 \times 10^{+0}}$	$\mathbf{9.607 \times 10^{-1}}$	$9.521 \times 10^{+0}$	$3.062 \times 10^{+0}$
ArrowHead20D	20	$\mathbf{4.011 \times 10^{+1}}$	$\mathbf{3.253 \times 10^{+0}}$	$5.494 \times 10^{+1}$	$3.993 \times 10^{+0}$
Bdqrtic20D	20	$8.944 \times 10^{+1}$	$1.966 \times 10^{+0}$	$\mathbf{5.089 \times 10^{+1}}$	$\mathbf{1.135 \times 10^{+0}}$
CRosenbrock20D	20	$2.350 \times 10^{+1}$	$3.855 \times 10^{+0}$	$\mathbf{2.076 \times 10^{+1}}$	$\mathbf{4.411 \times 10^{+0}}$

17.7 Conclusions

A memetic algorithm has been proposed for optimization of black-box functions whose evaluation is computationally resource intensive and where uncertainty exists in the objective function, i.e. the latter contains noise. The algorithm relies on the global-local search framework with clustering for global optimization; on the global scale an EA is used for efficient exploration of the search space. On the local scale a trust-region derivative-free algorithm is used to efficiently converge to a stationary point by using local quadratic surfaces. The algorithm has several merits such as global convergence to a stationary point and a mechanism to safeguard the accuracy of the local quadratic surfaces.

Extensive test cases with varied test functions of dimension 2–20 show the memetic algorithm obtained a good approximation to a minimizer of the objective function within a small number of function evaluations, with and without noise. In the presence of 1% noise the algorithm was only marginally affected when compared to minimization without noise. In the presence of 10% noise the algorithm often converged in fewer function evaluations since the objective function landscape becomes multimodal and this affected the gradinet-based trust-region algorithm; however, a good approximation to a minimizer of the objective function was still found. Overall, the good performance in the presence of noise is attributed both to the use of the EA and the TR-DFO algorithm: the former does not rely on derivatives and remains efficient in the presence of noise while the latter uses quadratic models which smooth out some of the multimodality induced by the noise and it uses an interpolation technique which generates spatially separated points; both of these diminish the effect of noise. In summary, the approach implemented in the proposed memetic algorithm, i.e. combining an EA for a global search stage with the trust-region derivative-free algorithm described which uses quadratic models for the local search stage efficiently minimized black-box functions with up to 20 variables which also contain noise in the function value.

A Details of the Test Functions

For each test function used in this study this appendix gives its dimension, the reference where it was defined, its global minimizer (**1** denotes a vector whose components are all 1) and the function expression. Numbers are rounded to three significant digits.

1. Beale ($n = 2$) [27]:
   A sixth order polynomial whose global minimizer is inside a curved landscape and hence is difficult to approach.
   $\boldsymbol{x}_{\mathrm{G}} = (3, 0.5)^{\top}$, $f(\boldsymbol{x}_{\mathrm{G}}) = 0$.
   $f(\boldsymbol{x}) = \sum_{i=1}^{3}\{y_i - x_1(1 - {x_2}^i)\}^2$,
   $y_1 = 1.5$, $y_2 = 2.25$, $y_3 = 2.625$.

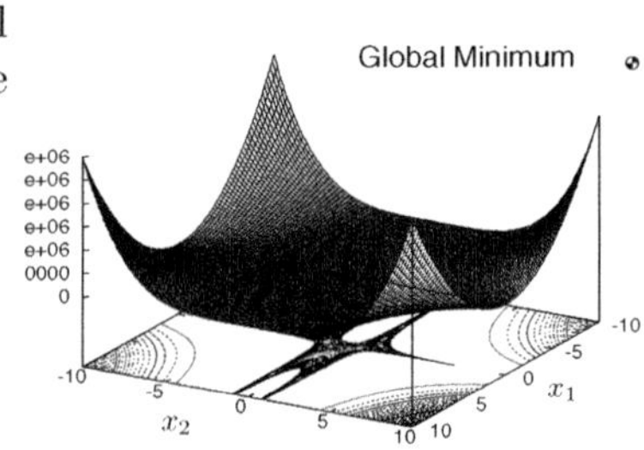

2. Rosenbrock ($n = 2$) [27]
   A fourth order polynomial function whose minimizer is located inside a tight and curved valley making it difficult to approach. Function values increases rapidly away from the minimizer.
   $\boldsymbol{x}_{\mathrm{G}} = \mathbf{1}$, $f(\boldsymbol{x}_{\mathrm{G}}) = 0$.
   $f(\boldsymbol{x}) = \{10(x_2 - {\boldsymbol{x}_1}^2)\}^2 + (1 - x)^2$.

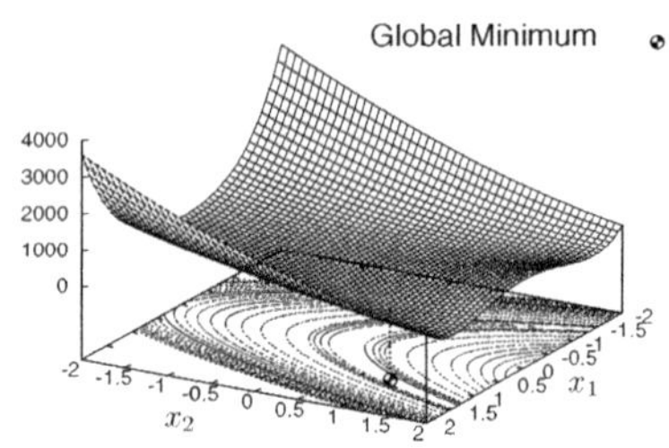

3. Box 3D ($n = 3$) [27]:
   A sum of squared decreasing exponents and a mixed monomial-exponent term. The function is multimodal, having the two minima $\boldsymbol{x}_{\mathrm{G}} = (1, 10, 1)$, $(10, 1, -1)$ and an infinite number of minima along the valley $x_1 = x_2$, $x_3 = 0$. $f(\boldsymbol{x}_{\mathrm{G}}) = 0$.
   $$f(\boldsymbol{x}) = \sum_{i=1}^{m}\{\exp(-t_i x_1) - \exp(-t_i x_2) - x_3\big(\exp(-t_i) - \exp(-10t_i)\big)\}^2,$$
   $t_i = (0.1)^i$, $\quad m > n\,(m = 10$ was used in this study).
4. Brown-Dennis ($n = 4$) [27]
   A sum of squared exponents, monomials and sinusoidal functions. The function resembles a convex quadratic and is unimodal.
   $\boldsymbol{x}_{\mathrm{G}} = (-11.594, 13.203, -0.403, 0.236)^{\top}$, $f(\boldsymbol{x}_{\mathrm{G}}) = 85822.2$.
   $$f(\boldsymbol{x}) = \sum_{i=1}^{m}\Big\{\big(x_1 + t_i x_2 - \exp(t_i)\big)^2 + \big(x_3 + x_4\sin(t_i) - \cos(t_i)\big)^2\Big\}^2,$$
   $t_i = i/5$.

5. Kowalik-Osborne ($n = 4$) [27]:
A sum of squared polynomial ratios. The function is multimodal and has an asymptotic global minimizer at $(+\infty, -14.7, -\infty, -\infty)$ and several local minimizers.
$f(\boldsymbol{x}_\mathrm{G}) = 1.027 \times 10^{-3}$.
The plot shows a projection of the function landscape for $x_3 = x_4 = 0$.

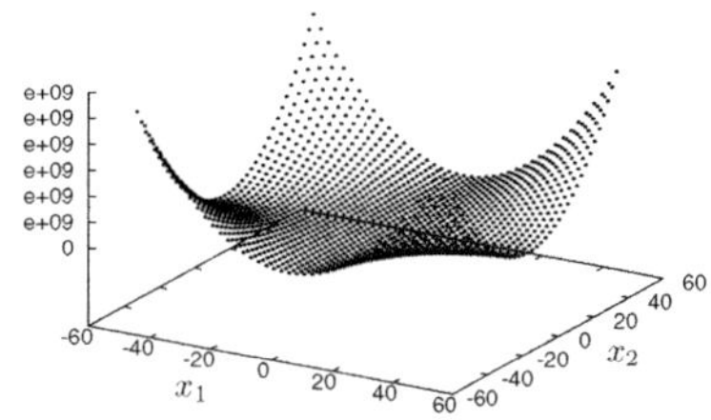

$$f(\boldsymbol{x}) = \sum_{i=1}^{11} \left\{ y_i - \frac{x_1({u_i}^2 + u_i x_2)}{{u_i}^2 + u_i x_3 + x_4} \right\}^2 ,$$

i	y_i	u_i	i	y_i	u_i
1	0.1957	4.0000	7	0.0456	0.1250
2	0.1947	2.0000	8	0.0342	0.1000
3	0.1735	1.0000	9	0.0323	0.0833
4	0.1600	0.5000	10	0.0235	0.0714
5	0.0844	0.2500	11	0.0246	0.0625
6	0.0627	0.1670			

6. Arrowhead (n is variable) [5]:
A variable-dimension fourth order multivariate polynomial. The function is convex quadratic and unimodal.
$\boldsymbol{x}_\mathrm{G} = (1, 1, \ldots 1, 0)^\top$, $f(\boldsymbol{x}_\mathrm{G}) = 0$.
$f(\boldsymbol{x}) = \sum_{i=1}^{n-1} \{(x_i^2 + x_n^2)^2 - 4x_i + 3\}$.

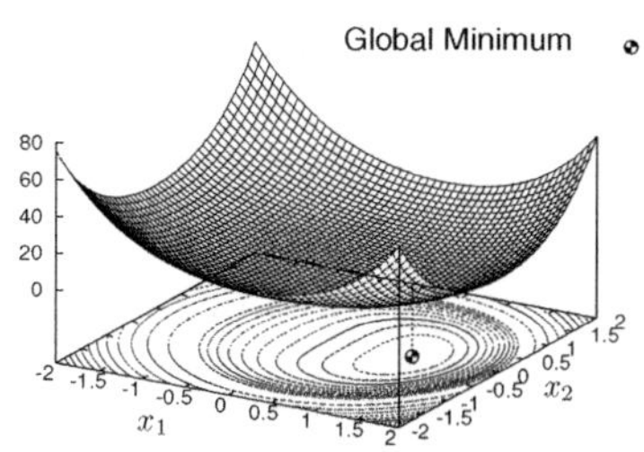

7. Bdqrtic ($n > 4$ and is variable) [5]:
A variable-dimension multivariate quadratic function whose minimizer varies with n. For $n = 10$ $f(\boldsymbol{x}_\mathrm{G}) = 11.865$; for $n = 10$ $f(\boldsymbol{x}_\mathrm{G}) = 35.420$.
$f(\boldsymbol{x}) = \sum_{i=1}^{n-4} \{({x_i}^2 + 2{x_{i+1}}^2 + 3{x_{i+2}}^2 + 4{x_{i+3}}^2 + {x_{i+4}}^2) - 4x_i + 3\}$.

8. Chained Rosenbrock (CRosenbrock) (n is variable) [45]:
The function is a multivariate extension of the bivariate Rosenbrock function so the global minimizer is located in an n-dimensional tight and curve valley making it difficult to locate.
$\boldsymbol{x}_\mathrm{G} = \boldsymbol{1}$, $f(\boldsymbol{x}_\mathrm{G}) = 0$.
$f(\boldsymbol{x}) = \sum_{i=2}^{n} \{4(x_{i-1} - x_i^2)^2 + (1 - x_i)^2\}$.

References

1. N. Alexandrov, J. Dennis, Jr, R. M. Lewis, and V. Torczon. A trust-region framework for managing the use of approximation models in optimization. *Structural Optimization*, 15(1):16–23, 1998.
2. J.-F. Barthelemy and R. Haftka. Approximation concepts for optimum structural design—a review. *Structural Optimization*, 5:129–144, 1993.
3. R. R. Barton. Metamodels for simulation input-output. In J. Swain, D. Goldsman, R. Crain, and J. Wilson, editors, *Proceedings of the 1992 Winter Simulation Conference*, pages 289–299, New York, NY, USA, 1992. ACM Press.
4. A. J. Booker, J. E. Dennis, Jr, P. D. Frank, D. B. Serafini, V. Torczon, and M. W. Trosset. A rigorous framework for optimization of expensive functions by surrogates. *Structural Optimization*, 17(1):1–13, 1998.
5. A. R. Conn, N. I. Gould, M. Lescrenier, and P. L. Toint. Performance of a multifrontal scheme for partially separable optimization. In S. Gomez and J. Hennart, editors, *Advances in Optimization and Numerical Analysis*, pages 79–96. Kluwer Academic Publishers, 1994.
6. A. R. Conn, N. I. Gould, and P. L. Toint. *Trust-Region Methods*. SIAM, Philadelphia, PA, 2000.
7. A. R. Conn, K. Scheinberg, and P. L. Toint. On the convergence of derivative-free methods for unconstrained optimization. In A. Iserles and M. Buhmann, editors, *Approximation Theory and Optimization: Tributes to M.J.D. Powell*, pages 83–108. Cambridge University Press, Cambridge; New York, 1997.
8. A. R. Conn, K. Scheinberg, and P. L. Toint. Recent progress in unconstrained nonlinear optimization without derivatives. *Mathematical Programming*, 79:397–414, 1997.
9. A. R. Conn, K. Scheinberg, and P. L. Toint. A derivative free optimization algorithm in practice. In *Proceedings of the Seventh AIAA/USAF/NASA/ISSMO Symposium on Multidisciplinary Analysis and Optimization, St.Louis, MO.* American Institute of Aeronautics and Astronautics, American Institute of Aeronautics and Astronautics, 1998.
10. A. R. Conn and P. L. Toint. An algorithm using quadratic interpolation for unconstrained derivative free optimization. In G. Di Pillo and F. Giannessi, editors, *Nonlinear Optimization and Applications*, pages 27–47. Plenum Press, New York, N.Y., 1996.
11. J. Dennis, Jr and V. Torczon. Managing approximation models in optimization. In N. M. Alexandrov and M. Y. Hussaini, editors, *Multidisciplinary Design Optimization: State of the Art*, pages 330–347. SIAM, Philadelphia, 1997.
12. P. D. Frank. Global modeling for optimization. *SIAG/OPT Views-and-News*, (7):9–12, 1995.
13. M. Gasca and T. Sauer. Polynomial interpolation in several variables. *Advances in Computational Mathematics*, 12:377–410, 2000.
14. K. Giannakoglou. Design of optimal aerodynamic shapes using stochastic optimization methods and computational intelligence. *Progress in Aerospace Science*, 38(1):43–76, 2002.
15. A. A. Giunta and L. T. Watson. A comparison of approximation modeling techniques: polynomial versus interpolating models. In *AIAA/USAF/NASA/ISSMO Seventh Symposium on Multidisciplinary Analysis and Optimization*, volume 1, St. Louis, MO, Sept. 2–4 1998. AIAA, AIAA. AIAA-1998-4758.

16. V. Hanagandi and M. Nikolaou. A hybrid approach to global optimization using a clustering algorithm in a genetic search framework. *Computers and Chemical Engineering*, 22(12):1913–1925, 1998.
17. W. E. Hart and R. K. Belew. Optimization with genetic algorithm hybrids that use local search. In R. K. Belew and M. Mitchell, editors, *Adaptive Individuals in Evolving Populations: Models and Algorithms, Santa Fe Institute Studies in the Sciences of Complexity*, volume 26, chapter 27, pages 483–496. Addison-Wesley Publishing Company, Reading, MA, 1996.
18. R. L. Haupt and S. E. Haupt. *Practical Genetic Algorithms.* John Wiley and Sons, New York, 1998.
19. R. Jin, W. Chen, and T. Simpson. Comparative studies of metamodelling techniques under multiple modelling criteria. *Structural and Multidisciplinary Optimization*, 23:1–13, 2001.
20. Y. Jin. A comprehensive survey of fitness approximation in evolutionary computation. *Soft Computing*, 9(1):3–12, 2005.
21. Y. Jin, M. Olhofer, and B. Sendhoff. A framework for evolutionary optimization with approximate fitness functions. *IEEE Transactions on evolutionary computation*, 6(5):481–494, 2002.
22. H.-S. Kim and S.-B. Cho. An efficient genetic algorithm with less fitness evaluation by clustering. In *Proceedings of 2001 IEEE Congress on Evolutionary Computation*, volume 2, pages 887–894. IEEE, IEEE, 2001.
23. N. Krasnogor and J. Smith. A tutorial for competent memetic algorithms: Model, taxonomy and design issues. *IEEE Transactions on Evolutionary Computation*, 9(5):474–488, 2005.
24. K.-H. Liang, X. Yao, and C. Newton. Evolutionary search of approximated n-dimensional landscapes. *Journal of Knowledge-Based Intelligent Engineering Systems*, 4(3):172–183, 2000.
25. Z. Michalewicz. *Genetic Algorithms + Data Structures = Evolution Programs.* Springer-Verlag, Berlin; New York, third edition, 1996.
26. J. J. Moré. Recent developments in algorithms and software for trust region methods. In A. Bachem, M. Grötschel, and B. Korte, editors, *Mathematical Programming Bonn 1982 - The State of the Art*, pages 258–287. Springer-Verlag, Berlin, 1983.
27. J. J. Moré, B. S. Garbow, and K. E. Hillstrom. Testing unconstrained optimization software. *ACM Transactions on Mathematical Software*, 7(1):17–41, 1981.
28. J. J. Moré and D. Sorensen. Computing a trust region step. *SIAM Journal on Scientific and Statistical Computing*, 4(3):553–572, 1983.
29. Y. S. Ong and A. J. Keane. Meta-lamarckian learning in memetic algorithms. *IEEE Transactions on Evolutionary Computation*, 8(2):99–110, 2004.
30. Y. S. Ong, P. B. Nair, and A. J. Keane. Evolutionary optimization of computationally expensive problems via surrogate modeling. *American Institute of Aeronautics and Astronautics Journal*, 41(4):687–696, 2003.
31. Y.-S. Ong, P. B. Nair, and K. Y. Lum. Max-min surrogate assisted evolutionary algorithms for robust aerodynamic design. *IEEE Transactions on Evolutionary Computation*, 10(4):392–404, 2006.
32. M. J. Powell. Direct search algorithms for optimization calculations. *Acta Numerica*, pages 287–336, 1998.
33. M. J. Powell. UOBYQA: unconstrained optimization by quadratic approximation. *Mathematical Programming, Series B*, 92(3):555–582, 2002.

34. M. J. Powell. On trust region methods for unconstrained minimization without derivatives. *Mathematical Programming, Series B*, 97(3):605–623, 2003.
35. A. Ratle. Accelerating the convergence of evolutionary algorithms by fitness landscape approximations. In A. Eiben, B. Thomas, M. Schoenauer, and H.-P. Schwefel, editors, *Proceedings of the 5th International Conference on Parallel Problem Solving from Nature - PPSN V*, volume 1498 of *Lecture Notes in Computer Science*, pages 87–96, Berlin Heidelberg, 1998. Springer-Verlag.
36. J.-M. Renders and S. P. Flasse. Hybrid methods using genetic algorithms for global optimization. *IEEE Transactions on Systems, Man and Cybernetics–Part B:Cybernetics*, 26(2):243–258, 1996.
37. A. Rinnooy Kan and G. Timmer. Stochastic global optimization methods part I: Clustering methods. *Mathematical Programming*, 39:27–56, 1987.
38. T. Sauer. Computational aspects of multivariate polynomial interpolation. *Advances in Computational Mathematics*, 3(3):219–237, 1995.
39. T. Sauer and Y. Xu. On multivariate Lagrange interpolation. *Mathematics of Computation*, 64(211):1147–1170, 1995.
40. G. Seront and H. Bersini. A new GA-local search hybrid for continuous optimization based on multi level single linkage clustering. In H. Beyer, E. Cantu-Paz, D. Goldberg, I. Parmee, L. Spector, and D. Whitley, editors, *Proceedings of GECCO-Genetic and Evolutionary Computation Conference 2000*, pages 90–95. Morgan Kaufmann, 2000.
41. T. W. Simpson, J. J. Korte, T. M. Mauery, and F. Mistree. Comparison of response surface and Kriging models for multidisciplinary design optimization. In *Proceedings of the 7th AIAA/USAF/NASA/ISSMO Symposium on Multidisciplinary Analysis and Optimization, St.Louis, MO*, volume 1, pages 381–391. American Institute of Aeronautics and Astronautics, American Institute of Aeronautics and Astronautics, 1998. AIAA-1998-4755.
42. Y. Tenne and S. Armfield. Efficiently minimizing expensive black-box functions by a combination of an evolutionary algorithm, density cluster analysis and a trust-region derivative-free optimizer. *Journal of Global Optimization*, 2005. Accepted, To Appear.
43. Y. Tenne and S. Armfield. A novel evolutionary algorithm for efficient minimization of expensive black-box functions with assisted-modelling. In *Proceedings of the IEEE World Congress on Computational Intelligence–WCCI 2006*, 2006. Accepted, To Appear.
44. Y. Tenne and S. Armfield. Computational aspects and performance of a trust-region derivative-free algorithm for minimization of black-box functions. In Preparation.
45. P. L. Toint. Some numerical results using a sparse matrix updating formula in unconstrained optimization. *Mathematics of Computation*, 32(143):839–851, 1978.
46. A. A. Törn. Cluster analysis using seed points and density-determined hyperspheres as an aid to global optimization. *IEEE Transactions on Systems, Man and Cybernetics*, 7(8):610–616, 1977.
47. A. A. Törn. A search-clustering approach to global optimization. In L. Dixon and G. Szegö, editors, *Towards Global Optimization 2*, pages 49–62. North-Holland Publishing Company, Amsterdam; New York; Oxford, 1978.
48. A. A. Törn and A. Žilinskas. *Global Optimization.* Number 350 in Lecture Notes In Computer Science. Springer-Verlag, Berlin; Heidelberg; New York; London, 1989.

49. Y.-x. Yuan. A review of trust region algorithms for optimization. In J. Ball and J. C. Hunt, editors, *ICIAM 99 : proceedings of the Fourth International Congress on Industrial and Applied Mathematics, Edinburgh*, pages 271–282, New York; Oxford, 2000. Oxford University Press.
50. Z. Zhou, Y. S. Ong, P. B. Nair, A. J. Keane, and K. Y. Lum. Combining global and local surrogate models to accelerate evolutionary optimization. *IEEE Transactions on Systems, Man and Cybernetics–Part C(Applications and Reviews)*, To Appear.

18

Genetic Algorithm to Optimize Fitness Function with Sampling Error and its Application to Financial Optimization Problem

Masaru Tezuka[1], Masaharu Munetomo[2], and Kiyoshi Akama[2]

[1] Research and Development Section, Hitachi East Japan Solutions, Ltd.
2-16-10, Honcho, Aoba, Senadi, 980-0014, Japan
tezuka@hitachi-to.co.jp

[2] Information Initiative Center, Hokkaido University
Kita 11 Nishi 5, Sapporo, 060-0811, Japan
{munetomo, akama}@iic.hokudai.ac.jp

Summary. Fitness function of financial optimization problem is often evaluated by Monte-Carlo method which is based on stochastic sampling. In this chapter we deal with the trade-off between the accuracy of the fitness estimation and its computational overheads, and introduce a method to decide the number of samples that maximizes the efficiency of genetic algorithms for financial problems. We define an index called selection efficiency which shows how close the population approaches to takeover in a fixed amount of time. When selection efficiency is maximized, the population converges to good solution rapidly, and optimization progresses most efficiently. Selection efficiency is generally difficult to calculate analytically. Thus bootstrapping approach is employed to calculate selection efficiency. The method estimates selection efficiency from the empirical distribution. The method is applied to the optimization of the procurement plan problem, and it optimizes Value at Risk efficiently.

18.1 Introduction

In many real-world optimization problems, fitness is estimated from a number of samples and has sampling error. Increasing the number of samples does contribute reducing percentage of sampling errors, but computational time is subject for increase. Since the amount of computation time available is limited, there is a problem of the balance between accuracy of fitness evaluation and computation time. We would like to challenge this problem by introducing a new approach of deciding the number of samples that maximizes the efficiency of genetic algorithms (GAs) for financial problems.

In financial field, optimization criteria such as Value at Risk are unable to be calculated by analytical methods. Thus, Monte-Carlo method is widely used to evaluate the criteria.

M. Tezuka et al.: *Genetic Algorithm to Optimize Fitness Function with Sampling Error and its Application to Financial Optimization Problem*, Studies in Computational Intelligence (SCI) **51**, 417–434 (2007)
www.springerlink.com

In order to estimate the optimization criteria, Monte-Carlo method perform stochastic sampling in repetitive manner. The given objective of the optimization is estimated from the samples. With regard to obtaining precise evaluation of the criteria, abundant volume of samples and considerably long computational time are required. Therefore, we have a trade-off between the accuracy of the estimation and its computational overheads.

Computation time has to be allocated into two conflicting functions such as calculation of criteria and searching for optimal solution. If longer time is allocated for the calculation of criteria, then estimation has tendency to become more precise. However, searching time would have to be reduced accordingly which means less points are searched, and then it would become impossible to obtain sufficient and acceptable solution. On the other hand, if less time is allocated for the calculation, accuracy of the estimates becomes lower. It may increase the possibility of the search going to wrong direction.

The method we are introducing in this chapter is based on the idea that high accuracy estimation of the optimization criteria may not always be necessary. Of course, final output of the optimization must be precise enough. However, during the search progressing, lower accuracy may just be enough to drive optimization.

We must balance of the allocation of time between the calculation of criteria and the search of optimal solution. We discuss the case that fitness is estimated with samples, so the decision of the time allocation is equivalent to the decision of the number of samples.

In order to decide the best number of samples, we investigate the efficiency of selection operation affected by the sampling error from the perspective of takeover time, and propose *selection efficiency*. Selection efficiency depends on the number of samples. The method we introduce in this chapter is to decide the number of samples achieving maximum selection efficiency. When selection efficiency is maximized, optimization progresses most rapidly.

Selection efficiency is difficult to calculate analytically except for the special cases such as the case the sampling error follows normal distribution. To realize the optimization of financial problems, we have to deal with general cases. Thus we propose a method utilizing bootstrapping approach to calculate selection efficiency.

In Sect. 18.2, we briefly review the works related to the optimization of noisy fitness function. We define selection efficiency in Sect. 18.3. In Sect. 18.4, the number of samples which yields the highest selection efficiency is investigated for expected value and variance optimization on the assumption that the samples follow normal distribution. In general, it is difficult to calculate the best number of samples. Therefore we show you the method to estimate the number based on bootstrap method in Sect. 18.5 and show the results of numerical experiments. Section 18.6 is the conclusion of this chapter.

18.2 Brief Review of Optimization of the Fitness Function with Sampling Error

On the optimization problems we deal with, evaluation of the optimization criteria have sampling error. In other words, these are nothing but the optimization of noisy fitness function. We should point out that the error may follow not only normal distribution but also various kinds of distributions when financial criteria such as volatility and value at risk are coped First of all, we would like to review the works which are related to noisy fitness function briefly.

GA is one of the multipoint search methods and is considered to be robust to noisy fitness function. Nissen and Propach [14] compared population-based optimization approaches with point-based approaches on several problems with noise. GA and Evolutionary Strategy (ES) were used as population-based approaches and Pattern Search and Threshold Accepting were used as point-based approaches. In this experiment, population-based approaches were superior to point-based approaches on the optimization of the fitness function with noise. Beker and Hadany reported that in some cases the existence of noise is beneficial to GA [4]. Noise can be considered as a variation generating force for maintaining population's diversity.

When GA is used for optimization and fitness is estimated with samples, total computation time is,

$$T_e = nt\,(\alpha + s\beta) \tag{18.1}$$

where s is the number of samples used to evaluate one individual, t is the number of generation alternation, n is population size, α is the computation time to create one individual, i.e. the time to execute crossover and mutation, β is the computation time to draw one sample from the individual. T_e is usually limited by some practical constraints.

Fitzpatrick and Grefensette investigated the effect of the noise to the optimization performance in a fixed amount of time T_e varying s and n [7]. They added Gaussian noise to the fitness function. Thus, estimation error of the fitness function followed normal distribution. It was shown that the performance was the highest with one or two samples when α/β was zero. When α/β was about 3, the highest performance was achieved with 5 to 20 samples. With larger α/β, more samples were needed to achieve good performance.

With utilization of ANOVA(analysis of variance) [11], Aizawa and Wah proposed the method to dynamically decide s for the selection to be done correctly [1, 2]. They took variation of individuals in population as between-group variance of ANOVA and sampling error as within-group variance. If within-group variance is larger than between-group variance, it becomes impossible to distinguish each individual. Thus the number of samples is decided such that the ratio of between-group variance to within-group is large enough to distinguish each individual. However, their assumption was that sampling error followed normal distribution and the fitness variation of individuals in a population also followed normal distribution.

Miller et al. analyzed the selection intensity and the convergence of the population on noisy OneMax problems [12]. However, they did not pay attention to more general problems.

As a different approach to suppress computational cost for evaluation, fitness inheritance was proposed [15]. In the approach, some offspring are not evaluated but their fitness are estimated as the average of their parents' fitness. The approach enables to reduce computational cost. However, it only worked when the average or the weighted average of the parents' fitness is good estimation of their offspring.

As we pointed out, on the financial optimization problems, the estimation error may follow various kinds of distribution. For example, when the objective is to minimize volatility, the estimation error follows chi-squared distribution. When the objective is the to maximize value at risk, the error is considered to follow extreme value distribution. Thus, a new method is required to achieve efficient optimization of financial problems.

18.3 Selection Efficiency and Decision of Sample Number

Here, we consider the case that the computation time is limited. In such case, T_e of (18.1) is fixed.

α depends on the operators such as crossover, mutation, and selection you chose. β depends of the problem to be optimized. We assume that the population size n is decided in one way or another. Thus, you can control only t and s. If you assign large value to s, t becomes small, and vice versa. Controlling t and s, we want to bring the search close to a good solution in the fixed amount of time T_e.

Figure 18.1 illustrates the progress of search with small and large number of samples. If s, the number of samples, is small, sampling error becomes large and the error leads the search to a wrong direction as shown in Fig. 18.1 (a). However, t can be large and many points can be searched. On the other hand, If s is large as shown in the Fig. 18.1 (b), search goes toward right direction. However, the computation time is consumed to draw the samples and only a few points can be searched. Thus, the search may not progress enough.

As shown in Fig. 18.2, we consider that there is the optimal number of samples which can obtain the highest search efficiency.

In order to decide the optimal number of samples, we describe *selection efficiency* which represents the search progress rate in a limited time [18]. Selection efficiency is based on takeover time which is a performance measure of the convergence of GA population.

Takeover time is defined as the generation when the population reaches to contain $n-1$ best individuals where n is the size of population. The takeover time of probabilistic binary tournament selection was calculated by Goldberg

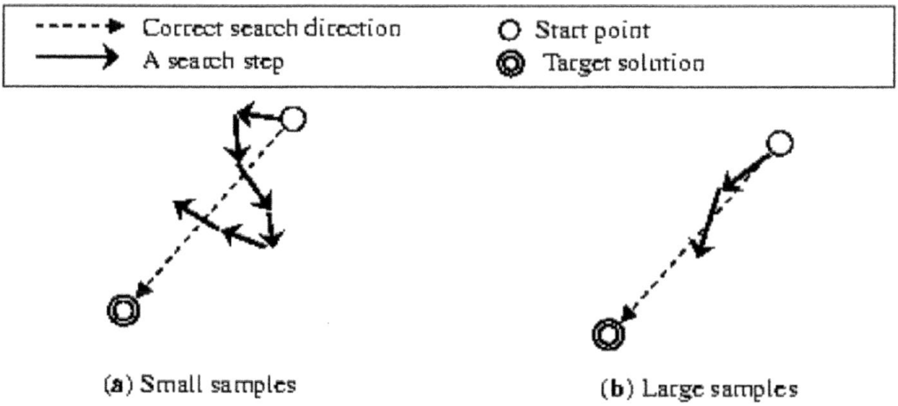

Fig. 18.1. Progress of search

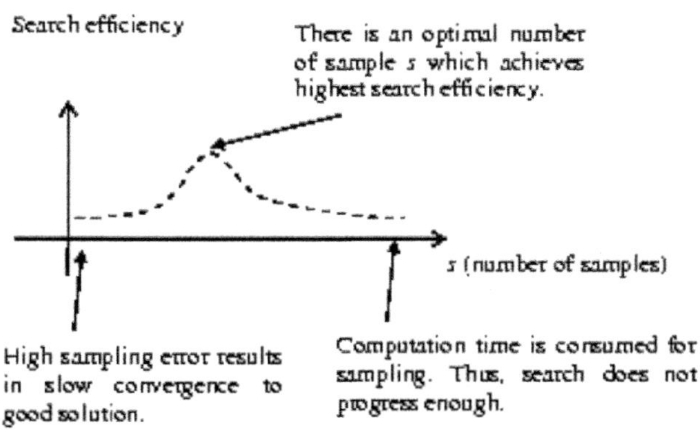

Fig. 18.2. The number of samples and search progress rate

and Deb [9]. In the selection, two individuals are chosen randomly from the population and the better one is selected with probability p. The takeover time t^* is,

$$t^* = \frac{2}{2p-1} \log_2 (n-1) . \tag{18.2}$$

In the case of probabilistic tournament selection, p is a parameter you set to control selection pressure. In optimization problems we deal with, the fitness of the individual is estimated by s samples and the probability p depends on sampling error. Thus we substitute p for $p(s)$.

We call $p(s)$ selection precision. The definition of selection precision $p(s)$ is the probability that the superiority between two individuals is correctly estimated with s samples. When s is set to 0, $p(s)$is 0.5, that is, one of two individuals is randomly selected. Usually, as s increases, accuracy of estimation increases and $p(s)$ approaches to 1.0.

From (18.1), t, the number of generations which can be altered in a fixed amount of time T_e is,

$$t = \frac{T_e}{n\left(\alpha + s\beta\right)}. \tag{18.3}$$

t/t^* shows how close the population approaches to takeover in the limited time T_e

$$\begin{aligned}\frac{t}{t^*} &= \frac{\left(2p\left(s\right)-1\right)T_e}{2\left(\alpha + s\beta\right)n\log_2\left(n-1\right)} \\ &= \frac{2p\left(s\right)-1}{\alpha + s\beta} \times \frac{T_e}{2n\log_2\left(n-1\right)}\end{aligned} \tag{18.4}$$

Taking the part related to sample number s from (18.4), we define *selection efficiency* η as

$$\eta = \frac{2p\left(s\right)-1}{\alpha + s\beta}, \tag{18.5}$$

and we consider the search progress rate is higher when η is larger.

To avoid confusion, let us make it clear the difference between selection precision $p(s)$ and selection efficiency η. Selection precision $p(s)$ is the probability that the superiority between two individuals is correctly estimated with s samples. Selection efficiency $\eta \sim t/t^*$ is the index that shows how close the population approaches to takeover in a limited time.

Takeover time is an index of the convergence time to good solution. Thus, we consider that the closer the population approaches to takeover, the better solution we have. When selection efficiency is maximized, optimization progresses most rapidly.

Since $p(s)$ runs from 0.5 through 1, the numerator of (18.5) takes 0 when $s = 0$ and approaches to 1 as s goes close to infinity. At the same time, the denominator of the equation does not have upper limit and approaches to infinity as s increases. Thus, η may be a monotonic decrease function or a function with the shape as shown in Fig. 18.2. That means too many samples results in low selection efficiency, and we consider search progress rate in a limited computation time is low in such cases.

18.4 Decision on the Number of Samples Following Normal Distribution

In this section, we deal with the case that uncertainty is modeled by normal distribution. Normal distribution, sometimes also called Gaussian distribution, is defined by two parameters, mean and standard deviation. Its probability density function (PDF) is,

$$f_{\mu,\sigma}(x) = \frac{1}{\sqrt{2\pi}\sigma} \mathrm{e}^{-\frac{(x-\mu)^2}{2\sigma^2}} \tag{18.6}$$

where μ is mean and σ is standard deviation. When $\mu = 0$ and $\sigma = 1$, it is called standard normal distribution.

It is considered that many natural phenomena follow normal distribution due to central limit theorem. Thus, normal distribution is widely used to model the dynamics and uncertainty in optimization problems such as the noise of control system, stock price, demand for home appliances, and so on.

18.4.1 Maximization and Minimization of the Expected Value of Normal Distribution

In this section, assuming samples retrieved from individual i follow normal distribution $N(\mu_i, \sigma_i)$ where μ_i is the mean and σ_i is the standard deviation, we consider maximization and minimization problem of the expected value.

Here, we also assume $\alpha \ll \beta$ and regard α as zero. This assumption is true in many real-world optimization problems. Then, substituting 0 for α of (18.5) we have

$$\eta = \frac{2p(s) - 1}{s} \times \frac{1}{\beta} = \frac{\acute{\eta}}{\beta} \tag{18.7}$$

where

$$\acute{\eta} = \frac{2p(s) - 1}{s}. \tag{18.8}$$

Since β, the computation time to obtain one sample, depends on the problem to be solved, it can be considered as a given constant. So we consider $\acute{\eta}$ as selection efficiency in this section.

Unbiased estimation of the expected value of individual i with s samples is

$$\hat{\mu}_i = \frac{1}{s} \sum_{k=1}^{s} h_{ik} \tag{18.9}$$

where h_{ik} is k-th sample taken from individual i.

The estimated expected value $\hat{\mu}_i$ follows normal distribution $N(\mu_i, \sigma/\sqrt{s})$ since the samples follow normal distribution. In tournament selection, individual i and j are randomly chosen from the population and the better one is selected. When true mean of individual i is larger than individual j, i.e. $\mu_i > \mu_j$, the better individual i is correctly selected if unbiased estimate of the expected value of individual i is also larger than individual j, i.e. $\hat{\mu}_i > \hat{\mu}_j$. Therefore $p(s, i, j)$, selection precision between individual i and j, is

$$\begin{aligned} p(s,i,j) = {} & P(\hat{\mu}_i > \hat{\mu}_j | \mu_i > \mu_j) \times P(\mu_i > \mu_j) \\ & + P(\hat{\mu}_j > \hat{\mu}_i | \mu_j > \mu_i) \times P(\mu_j > \mu_i). \end{aligned} \tag{18.10}$$

Since individual i and j are randomly chosen,

$$P(\mu_i > \mu_j) = P(\mu_j > \mu_i) = \frac{1}{2}. \tag{18.11}$$

Thus,

$$\begin{aligned} p(s,i,j) &= \frac{1}{2}\left[P(\hat{\mu}_i > \hat{\mu}_j | \mu_i > \mu_j) + P(\hat{\mu}_j > \hat{\mu}_i | \mu_j > \mu_i)\right] \\ &= \frac{1}{2}\left[\Phi\left(\frac{\mu_i - \mu_j}{\sqrt{\frac{\sigma_i^2+\sigma_j^2}{s}}}\right)\Bigg|_{\mu_i>\mu_j} + \Phi\left(\frac{\mu_j - \mu_i}{\sqrt{\frac{\sigma_j^2+\sigma_i^2}{s}}}\right)\Bigg|_{\mu_j>\mu_i}\right] \\ &= \frac{1}{2}\left[\Phi\left(\lambda_{ij}\sqrt{s}\right)\big|_{\lambda_{ij}>0} + \Phi\left(\lambda_{ji}\sqrt{s}\right)\big|_{\lambda_{ji}>0}\right] \end{aligned} \tag{18.12}$$

where $\Phi(z)$ is cumulative distribution function (CDF) of standard normal distribution $N(0,1)$,

$$\Phi(z) = \frac{1}{\sqrt{2\pi}}\int_{-\infty}^{z} \mathrm{e}^{-x^2/2}dx, \tag{18.13}$$

and $\lambda_{ij} = (\mu_i - \mu_j)/\sqrt{\sigma_i^2 + \sigma_j^2}$.

Again, i and j are interchangeable since individual i and j are randomly chosen. Thus, we name the individual whose true fitness is better as individual i, and rewrite λ_{ij} as λ for simplicity, then, $p(s,i,j) = \Phi(\lambda\sqrt{s})$, $\lambda \geq 0$, and (18.8) is written as

$$\acute{\eta} = \frac{2\Phi(\lambda\sqrt{s}) - 1}{s}. \tag{18.14}$$

$s = 1$ maximizes $\acute{\eta}$ under the condition that $s \geq 1$ and $\lambda \geq 0$.

Proof. From (18.13) and (18.14), we get

$$\frac{\partial}{\partial s}\acute{\eta} = \frac{\lambda \mathrm{e}^{-\lambda^2 s/2}}{\sqrt{2\pi s^3}} - \frac{2\Phi(\lambda\sqrt{s}) - 1}{s^2} \tag{18.15}$$

and

$$\frac{\partial}{\partial \lambda}\frac{\partial}{\partial s}\acute{\eta} = -\frac{(1 + s\lambda^2)}{\sqrt{2\pi s^3}}\mathrm{e}^{-\frac{s\lambda^2}{2}}. \tag{18.16}$$

From (18.16) and $s \geq 1$ it follows that

$$\frac{\partial}{\partial \lambda}\frac{\partial}{\partial s}\acute{\eta} < 0. \tag{18.17}$$

giving

$$\frac{\partial}{\partial s}\acute{\eta}\bigg|_{\lambda>0} < \frac{\partial}{\partial s}\acute{\eta}\bigg|_{\lambda=0} = 0. \tag{18.18}$$

Thus, under the condition,

$$\partial\acute{\eta}/\partial s \leq 0, \tag{18.19}$$

Since $\acute{\eta}$ always decreases as s increases and $s \geq 1$, $\acute{\eta}$ takes the highest value when s is 1. □

s is a positive integer value, i.e. $s = 1, 2, 3, ...$, since s is the number of samples. As described above we name the individual whose true fitness is better as individual i and the other as j. $\mu_i \geq \mu_j$ results in $\lambda \geq 0$. Thus the condition always satisfies.

On expected value maximization and minimization problems on which the samples following normal distribution and $\alpha \ll \beta$, selection efficiency is the highest when s, the number of samples, is set to 1.

Fitzpatrick and Grefensette [7] showed by numerical experiments that the performance of GA was the best with $s = 1$ or 2 when $\alpha/\beta = 0$. Our argument agrees with the result of them.

18.4.2 Minimization of the Variance of Normal Distribution

In this section, assuming samples retrieved from individual i follow normal distribution $N(\mu_i, \sigma_i)$, we describe minimization of the variance of samples. We also assume $\alpha \ll \beta$.

The variance reflects the volatility of samples. When stability is the most important, the variance is minimized.

Unbiased estimate of the variance is obtained by following equation.

$$\hat{\sigma}_i^2 = \frac{1}{s-1}\sum_{k=1}^{2}(h_{ik} - \hat{\mu}_i) \tag{18.20}$$

It is known that stochastic variable F shown in (18.21) follows F distribution with $[s-1, s-1]$ degrees of freedom.

$$F = \frac{\hat{\sigma}_i^2}{\hat{\sigma}_j^2}\frac{\sigma_j^2}{\sigma_i^2} \tag{18.21}$$

PDF of F distribution with $[n, m]$ degrees of freedom is,

$$f_{[n,m]}(x) = \frac{\Gamma\left(\frac{n+m}{2}\right) n^{n/2} m^{m/2}}{\Gamma\left(\frac{n}{2}\right)\Gamma\left(\frac{m}{2}\right)} \frac{x^{n/2-1}}{(nx+m)^{(n+m)/2}} \tag{18.22}$$

where $\Gamma(z)$ is the gamma function,

$$\Gamma(z) = \int_0^\infty t^{z-1}\mathrm{e}^{-t}dt. \tag{18.23}$$

When true variance of individual i is smaller than individual j, i.e. $\sigma_i^2/\sigma_j^2 < 1$, the better individual i is correctly selected if unbiased estimate of the variance of individual i is also smaller than individual j, i.e. $\hat{\sigma}_i^2/\hat{\sigma}_j^2 < 1$. Therefore, the probability of correct selection is,

$$p(s,i,j) = P\left(\frac{\hat{\sigma}_i^2}{\hat{\sigma}_j^2} < 1 \middle| \frac{\sigma_i^2}{\sigma_j^2} < 1\right) \times P\left(\frac{\sigma_i^2}{\sigma_j^2} < 1\right) + P\left(\frac{\hat{\sigma}_j^2}{\hat{\sigma}_i^2} < 1 \middle| \frac{\sigma_j^2}{\sigma_i^2} < 1\right) \times P\left(\frac{\sigma_j^2}{\sigma_i^2} < 1\right). \tag{18.24}$$

Since individual i and j are randomly chosen, i and j are interchangeable. Here we name the individual with smaller true variance as individual i, then,

$$p(s,i,j) = P\left(F_{[s-1,s-1]} < \frac{\sigma_j^2}{\sigma_i^2} \middle| \frac{\sigma_i^2}{\sigma_j^2} < 1\right)$$

and is the cumulative probability of F distribution with $[s-1, s-1]$ degrees of freedom from 0 to σ_j^2/σ_i^2. In the case that the objective is the minimization of variance and the samples follow normal distribution, selection precision does not depend of the expected values of the distribution but only on the variances.

As shown in (18.22), PDF of F distribution is very complex. It is difficult to obtain analytically the value of s which maximize selection efficiency on variance minimization problems. Thus we calculated the selection efficiency $\acute{\eta}$ with various s and the variance ratio of two individuals σ_j^2/σ_i^2 by numerical computation. Table 18.1 shows the selection efficiency. At least two samples are required for estimating variance. Thus, s takes an integer larger than 1. When variance ratio is smaller than 3, $s = 3$ yields the highest selection efficiency, otherwise $s = 2$ yields the highest.

Table 18.1. $\acute{\eta} = (2p(s) - 1)/s$ on minimization of variance

			σ_j^2/σ_i^2			
s	1.001	1.01	1.1	2	3	5
2	1.59	1.58	1.52	1.08	**1.67**	**2.32**
3	**1.67**	**1.66**	**1.59**	**1.11**	**1.67**	2.22
4	1.59	1.58	1.52	1.04	1.52	1.95
6	1.41	1.41	1.35	0.89	1.24	1.50
12	1.08	1.07	1.02	0.61	0.77	0.82
	$(\times 10^{-4})$	$(\times 10^{-3})$	$(\times 10^{-2})$	$(\times 10^{-1})$	$(\times 10^{-1})$	$(\times 10^{-1})$

18.4.3 Numerical Experiments on Samples Following Normal Distribution

In this experiments, we employ a real-coded GA [6, 8, 21]. The chromosome of individual is real-valued vector $\boldsymbol{x} \in \Re^d$ where d is the dimension of a problem.

We employ binary tournament selection. Two individuals are chosen randomly from the population and the individual with better unbiased estimate is selected. BLX-α [6] is used as the crossover operator. Parameter α of BLX is set to 0.366 which is the theoretical optimal value [10]. Crossover rate is 0.6. Mutation rate is 0.1. Each element of chromosomes is mutated by adding Gaussian noise $N(0, \sigma_{\text{mut}})$ and σ_{mut} is set to 1.0×10^{-2}. Population size is 40.

On the experiments, samples drawn from individual i are independent and identically distributed. On the minimization of expected value, k-th sample follows

$$h_{ik} \overset{\text{iid}}{\sim} N\left(f_{\text{sp}}(\boldsymbol{x}_i), 10\right) = f_{\text{sp}}(\boldsymbol{x}_i) + 10\ N(0,1) \tag{18.25}$$

where $\boldsymbol{x}_i$ is the chromosome of individual i and f_{sp} is Sphere function,

$$f_{\text{sp}}(x_1, ..., x_d) = \sum_{c=1}^{d} x_c^2. \tag{18.26}$$

The optimal solution is 0 when $\boldsymbol{x} = \boldsymbol{0}$.

On the experiment of the minimization of variance, the sample follows

$$h_{ik} \overset{\text{iid}}{\sim} N\left(100, \sqrt{f_{\text{sp}}(\boldsymbol{x}_i) + 100}\right) = 100 + \sqrt{f_{\text{sp}}(\boldsymbol{x}_i) + 100}\ N(0,1). \tag{18.27}$$

σ_i^2, true variance of individual i, is $f_{\text{sp}}(\boldsymbol{x}_i) + 100$. $\boldsymbol{x} = \boldsymbol{0}$ achieves the optimal solution where variance is 100 (or standard deviation is 10).

50 trial runs are performed for each s on each experiment. We choose the individual with the best true mean or true variance at each generation as the best individual while optimization is done according to unbiased estimate. The average of the best individual of 50 trials are shown in Fig. 18.3 and Fig. 18.4. x-axis shows total sampling number $n \times s \times t$. y-axis shows expected value in Fig. 18.3, and standard deviation (square root of variance) in Fig. 18.4.

On the experiment of minimization of expected value, as mentioned in previous section, optimization progresses fastest when the number of samples drawn from one individual is 1.

On the experiment of minimization of variance, in early stage, $s = 2$ is the fastest. After total sampling number reaches to about 1,400 times, $s = 3$ is the fastest. We consider that in early stage $s = 2$ is the fastest because population is randomly initialized and variance ratio of randomly chosen individuals is higher. As optimization progresses, population converges and the ratio gets smaller, then $s = 3$ is the fastest.

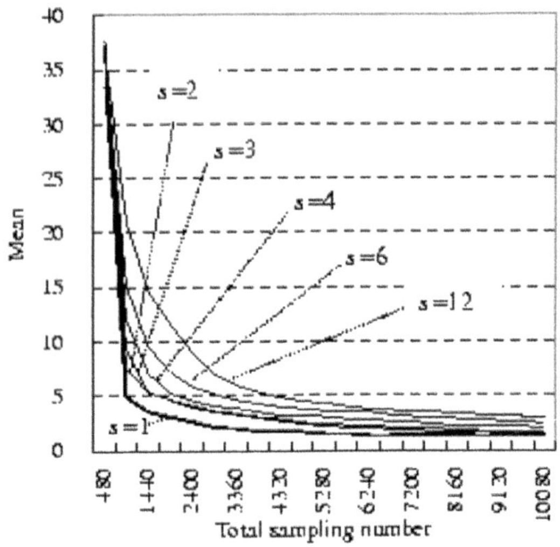

Fig. 18.3. Minimaization of expected value of normal samples with various s

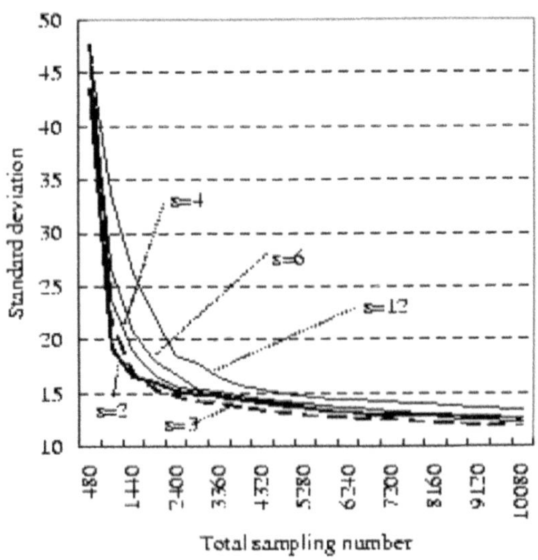

Fig. 18.4. Minimaization of variance of normal samples with various s

18.5 Application of Selection Efficiency on the Optimization of Financial Criteria

In this section we apply selection efficiency to the optimization of financial criteria. We take up Value-at-Risk (VaR) as a financial criterion which is getting popular in business. VaR of financial problems does not follow normal distribution and it is hard to calculate selection efficiency in such cases. In this section we employ bootstrap method to calculate selection efficiency. Then we show an application example on a financial problem.

18.5.1 Value at Risk

Value-at-Risk (VaR) is a well-known measure of the financial risk. $100\alpha\%$ VaR is the lower $100(1-\alpha)$ percentile of the profit distribution as shown in Fig. 18.5. For example, the probability of the profit falling below the VaR at 99% is 1%. Therefore VaR is able to be considered as the profit in the worst case.

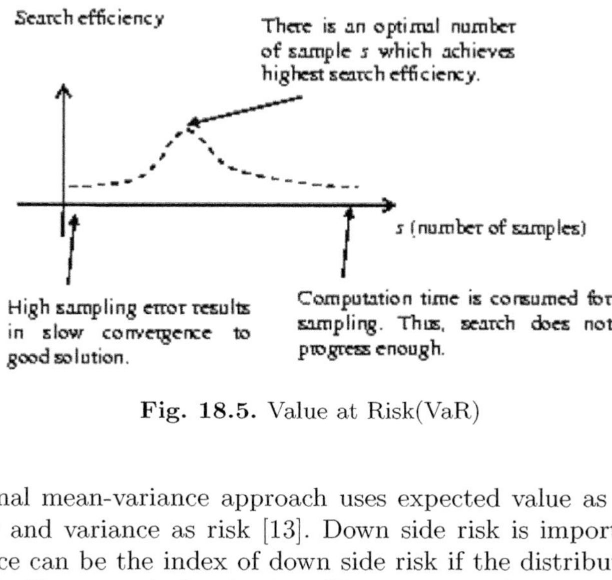

Fig. 18.5. Value at Risk(VaR)

Traditional mean-variance approach uses expected value as the index of profitability and variance as risk [13]. Down side risk is important in practice. Variance can be the index of down side risk if the distribution of profit is symmetric like normal distribution. However, it is known empirically by business practitioners that the distribution of the profit is skew and asymmetric. So VaR is the practically important index of down side risk. Bank for International Settlements introduced VaR as standard risk measure [3].

Another point of VaR is that it represents both profitability and risk as one index. Maximizing profit at VaR, profitability is also maximized and the risk of shortfall is minimized. It is intelligible to and useful for practitioners.

Many financial applications are introducing VaR as risk measure [16, 17] these days.

VaR is difficult to be calculated by analytical methods. Thus, Monte-Carlo method is widely used to estimate VaR in real-world problems. The method generates a lot of random numbers to simulate uncertainty in a profit model. s samples of the profit is drawn by the method, and VaR is estimated from the samples.

v_k denotes k-th sample of the profit drawn by the method. Sort the samples in ascending order and name them $\{v_{(1)}, v_{(2)}, ..., v_{(s)}\}$. 100α % VaR of the profit is estimated as

$$v_\kappa = \left(v_{\lfloor\kappa\rfloor+1} - v_{\lfloor\kappa\rfloor}\right)\left(\kappa - \lfloor\kappa\rfloor\right) + v_{\lfloor\kappa\rfloor} \tag{18.28}$$

where $\kappa = 1 + (s-1)(1-\alpha)$.

18.5.2 Bootstrapping Approach to Estimate Selection Precision

Selection precision $p(s)$ is difficult to obtain analytically except when the samples follow specific distribution like Normal distribution. It is also difficult to know which distribution the samples follow in real-world optimization problems. Thus, bootstrap method [5] is utilized to estimate selection precision $p(s)$ in this section.

$Y_1, ..., Y_s$ denotes the samples randomly selected from parent population following distribution function $F(y)$.

$$F_s(y) = \frac{1}{s}\sum_{k=1}^{s}\delta\left(Y_k \leq y\right) \tag{18.29}$$

where

$$\delta\left(Y_k \leq y\right) = \begin{cases} 1, & Y_k \leq y \\ 0, & \text{otherwise} \end{cases} \tag{18.30}$$

is called empirical distribution function. Bootstrap method resampling from the empirical distribution function estimates the parameters of the distribution of parent population. We employ bootstrap method to estimate selection precision $p(s)$ [20].

In the method described here, GA is employed for optimization. All the individuals in the first generation population are evaluated with s_L samples. s_L is set to large enough value to estimate fitness precisely. h_{ik} denotes k-th sample of the profit drawn from individual i. Then the empirical distribution function of the profit of individual i is,

$$F_{i,s_L}(y) = \frac{1}{s_L}\sum_{k=1}^{s_L}\delta\left(h_{ik} \leq y\right). \tag{18.31}$$

Here, $\theta_i(s)$ denotes the fitness of individual i estimated with s samples following empirical distribution function $F_{i,s_L}(y)$. For example, $\theta_i(s)$ is VaR and so on.

As the estimate of selection precision,

$$
\begin{aligned}
&\hat{p}(s\ ,i\ ,j) \\
&\quad = P\left(\theta_i(s) > \theta_j(s) \middle|\, \theta_i(s_L) > \theta_j(s_L)\right) \times P\left(\theta_i(s_L) > \theta_j(s_L)\right) \\
&\quad\quad + P\left(\theta_i(s) < \theta_j(s) \middle|\, \theta_i(s_L) < \theta_j(s_L)\right) \times P\left(\theta_i(s_L) < \theta_j(s_L)\right), \quad (18.32)
\end{aligned}
$$

is used. That is the probability that superiority between $\theta_i(s)$ and $\theta_j(s)$ is the same as the superiority between $\theta_i(s_L)$ and $\theta_j(s_L)$. Since s_L is large enough to estimate fitness precisely, we used the estimate with s_L samples as true fitness.

In practice, we use the best and the second best individual in the first generation population to estimate selection precision $p(s)$ because it is essential to distinguish the best individual from the others to drive optimization. Thus, the estimate of selection precision is,

$$\hat{p}(s) = P\left(\theta_{\mathrm{best}}(s) > \theta_{2^{\mathrm{nd}}\mathrm{best}}(s)\right) \quad (18.33)$$

on the maximization. On the minimization case, the inequality sign is reversed. Then,

$$s^* = \arg\max_{s} \frac{2\hat{p}\,(s) - 1}{\alpha + s\beta} \quad (18.34)$$

is the number of samples which maximizes the selection efficiency, that is the most efficient sample number. α and β are fixed by actual measurement.

From the second generation to the last, individuals are evaluated with s^* samples. Finally in the last generation, individuals are evaluated with s_L samples in order to output precise results.

18.5.3 Numerical Experiments

We show a result from numerical experiments to solve the procurement plan optimization problem on an electronic equipment manufacturer [19] in this section. The manufacturer purchases many kinds of materials, converts them to the products to sell. Procurement planning is to decide the quantity of each material to be procured, the date and time the order for each material is placed, and which material is to be procured. The plan is created according to the demand forecast for the products. Since the problem has uncertainty in the demand for the products, the profit of the manufacturer is stochastic. The objective of the optimization is the maximization of 97.5% VaR of the profit, and is estimated by Monte-Carlo method.

In this numerical experiments, s_L, the number of samples in the first generation is set to 1,000. VaR as the fitness of each individual is estimated with s_L samples. Then, s^*, the most efficient sample number is calculated by the bootstrap approach.

s^* samples is used to estimate the fitness of each individual from the second generation to the generation before the last. In this experiment, the computation time is limited to two hours. When two hours elapsed since the optimization started, the optimization is aborted and each individual in the last generation population is evaluated with s_L samples.

Figure 18.6 shows the estimated selection efficiency in a trial run. In this case, s^* is 2.

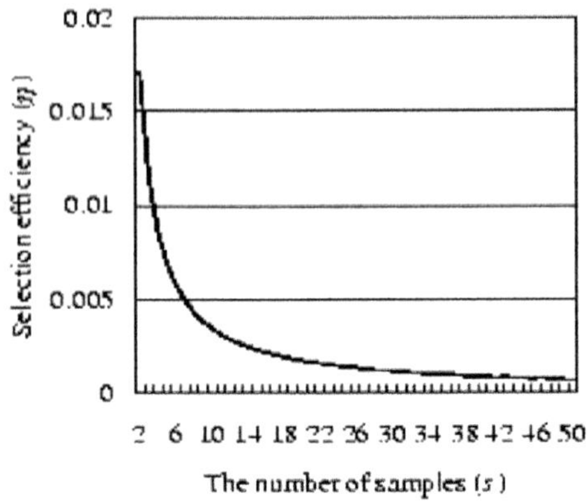

Fig. 18.6. Selection efficiency estimated by bootstrap approach

Table 18.2 shows the results of the optimization of VaR. 10 trial runs are performed with the method and the fixed sample number method. The mean and variance of the best individual of ten trials are shown in the table. The method which decides the number of samples according to selection efficiency outperforms the fixed sample number method.

Table 18.2. The result of Value at Risk optimization

	Sample number maximizing selection efficieny	Sample number fixed to s_L
Mean	498.2×10^6	488.8×10^6
Variance	2.838×10^{13}	1.863×10^{12}

The mean and variance of VaR of 10 trials.

t-test shows that there is a significant difference between the methods. The significance probability $P(|T| \leq t)$ is 0.0226.

18.6 Conclusions

Many financial optimization problems are found in the area of evaluating objective function where huge computation time is consumed. This is due to the nature of Monte-Carlo method which is based on stochastic sampling. We have to balance the trade-off between the accuracy of the estimation and its computational overheads. In this chapter, we have introduced a method of deciding the most efficient number of samples for evaluating objective function.

At first, we have defined an index called selection efficiency. It is to show how close the population approaches to takeover in a fixed amount of time. Selection efficiency depends on the number of samples for estimating optimization criteria. When selection efficiency is maximized, the population converges to good solution rapidly, and optimization progresses most efficiently.

With the assumption of the samples follow normal distribution and the computational cost to reproduce one individual of GA is negligible, it has been shown that the most efficient sampling number for expected value optimization is one, and two or three for minimizing variance. We have performed numerical experiments and have found that the optimization with the number of samples obtained by this method has achieved the best result.

Selection efficiency is difficult to calculate analytically except for the special cases such as the case the sampling error follows normal distribution. For example, when the objective is to minimize volatility, the estimation error follows chi-squared distribution. When the objective is to maximize value at risk, the error is considered to follow extreme value distribution. Thus we have utilized bootstrapping approach to calculate selection efficiency. The method estimates selection efficiency from the empirical distribution.

The method has been applied to the optimization of the procurement plan problem, and it has optimized Value at Risk efficiently. Except at the last generation, only two samples have been used for evaluating one individual. At the last generation, enough number of samples has been used to calculate precise solution. The number of samples used in the experiment is surprisingly low. However, it also has been shown that optimization with such small samples outperforms the conventional fixed sample number method which traditionally required huge number of samples.

References

1. A. N. Aizawa and B. W. Wah. Dynamic control of genetic algorithms in a noisy environment. In *Proceedings of the Fifth Intl. Conference on Genetic Algorithms*, pages 48–55, 1993.
2. A. N. Aizawa and B. W. Wah. Scheduling of genetic algorithms in a noisy environment. *Evolutionary Computation*, 2(2):97–122, 1994.
3. Basle Committee on Banking Supervision. *Amendment to the Capital Accord to Incorporate Market Risk*. Bank for International Settlements, 1996.

4. T. Beker and L. Hadany. Noise and elitism in evolutionary computation. In *Soft Computing Systems – Design, Management and Applications, HIS2002*, volume 87, pages 193–201, 2002.
5. A. C. Davison and D. V. Hinkley. *Bootstrap Methods and Their Application.* Cambridge University Press, 1997.
6. L. J. Eshelman and J. D. Schaffer. Real-coded genetic algorithms and interval-schemata. In L. D. Whitley, editor, *Foundations of Genetic Algorithms 2.* Morgan Kaufman, 1993.
7. J. M. Fitzpatrick and J. J. Grefensette. Genetic algorithms in noisy environments. *Machine Learning*, 3:101–120, 1988.
8. D. B. Fogel. Real-valued vectors. In T. Bäck, D. B. Fogel, and Z. Michalewicz, editors, *Handbook of Evolutionary Computation*, pages C1.3:1–1. Institute of Physics Publishing and Oxford University Press, 1997.
9. D. E. Goldberg and K. Deb. A comparative analysis of selection schemes used in genetic algorithms. In G. J. E. Rawlins, editor, *Foundations of Genetic Algorithms.* Morgan Kaufman, 1991.
10. T. Higuchi, S. Tsutsui, and M. Yamamura. Simplex crossover for real-coded genetic algolithms. *Transactions of the japanese Society for Artificial Intelligence*, 16(1):147–155, 2001.
11. B. F. J. Manly. *Multivariate Statistical Methods.* Chapman and Hall Ltd., 1986.
12. B. L. Miller and D. E. Goldberg. Genetic algorithms, selection schemes and the varying effects of noise. *Evolutionary Computation*, 4(2):113–131, 1996.
13. J. M. Mulvey. Introduction to financial optimization: Mathematical programming special issue. *Mathematical Programming*, 89(B):205–216, 2001.
14. V. Nissen and J. Propach. Optimization with noisy function evaluations. In *Parallel Problem Solving from Nature V*, pages 159–168, 1998.
15. K. Sastry, D. E. Goldberg, and M. Pelikan. Don't evaluate, inherit. In *Proceedings of the Genetic and Evolutionary Computation Conference 2001*, pages 551–558, 2001.
16. M. Tezuka, M. Hiji, Y. Ito, and Y. Kuwajima. Decision support for financing portfolio using genetic algorithm with simulation-based evaluation. In *Proceedings of the 4th Asia-Pacific conference on Simulated Evolution and Learning (SEAL'02)*, volume 2, pages 750–754, 2002.
17. M. Tezuka, M. Hiji, M. Munetomo, and K. Akama. Risk visualization and decision support for supply planning under uncertain demand. *IPSJ Journal*, 47(3):701–710, 2006.
18. M. Tezuka, M. Munetomo, and K. Akama. Selection efficiency and sampling error on genetic algorithms optimization under uncertainty. In *Proceedings of the 5th international conference on Simulated Evolution and Learning (SEAL 04)*, 2004.
19. M. Tezuka, M. Munetomo, K. Akama, and M. Hiji. Risk analysis and decision making on the combination strategy of planned and spot procurement. In *Proceeding of the Annual Conference of Japan Society for Management Information 2005 Autumn*, pages 180–183, 2005.
20. M. Tezuka, M. Munetomo, K. Akama, and M. HIJI. Genetic algorithm to optimize fitness function with sampling error and its application to financial optimization problem. In *2006 IEEE Congress on Evolutionary Computation*, pages 388–394, 2006.
21. A. H. Wright. Genetic algorithms for real parameter optimization. In G. J. E. Rawlins, editor, *Foundations of Genetic Algorithms.* Morgan Kaufman, 1991.

Part IV

Search for Robust Solutions

19

Single/Multi-objective Inverse Robust Evolutionary Design Methodology in the Presence of Uncertainty

Dudy Lim[1], Yew-Soon Ong[1], Meng-Hiot Lim[2], and Yaochu Jin[3]

[1] School of Computer Engineering, Nanyang Technological University, Nanyang Avenue, Singapore 639798
{dlim,asysong}@ntu.edu.sg
[2] School of Electrical and Electronics Engineering, Nanyang Technological University, Nanyang Avenue, Singapore 639798
emhlim@ntu.edu.sg
[3] Honda Research Institute Europe, Carl-Legien Strasse 30, 63073 Offenbach am Main, Germany
yaochu.jin@honda-ri.de

Summary. Many existing works for handling uncertainty in problem-solving rely on some form of a priori knowledge of the uncertainty structure. However, in reality, one may not always possess the necessary expertise or sufficient knowledge to identify suitable bounds of the uncertainty involved. Rather, it is more likely that specifications of the realistic performance desired are derived, which may be based on the maximum degradation tolerable or worst-case performance permissible in the final solution. In this chapter we present a Single/Multi-objective Inverse Robust Evolutionary (SMIRE) optimization methodology. In contrast to conventional forward robust optimization, an inverse approach based on non-probabilistic methods is introduced to avoid making possible erroneous assumptions about the uncertainty when insufficient field data exists for accurately estimating its structure. Further, since uncertainty is practically impossible to avoid, we consider the possible benefits as the uncertainty prevails by introducing an opportunity criterion in the inverse search scheme. Four inverse schemes are presented to include the different objectives possibly considered in robust evolutionary optimization. The inverse schemes are applied on synthetic test functions to illustrate their utility.

19.1 Introduction

Evolutionary Algorithm (EA) [1] is a modern stochastic optimization technique that has emerged as a prominent contender for global optimization in complex engineering problem-solving. Its popularity lies in the ease of implementation and the ability to arrive close to the global optimum solution with reasonable computational budget. Most early studies in the literature on the application of EAs to complex

D. Lim et al.: *Single/Multi-objective Inverse Robust Evolutionary Design Methodology in the Presence of Uncertainty*, Studies in Computational Intelligence (SCI) **51**, 437–456 (2007)
www.springerlink.com

engineering design have mainly emphasized on locating the global optimal design using deterministic computational models. However, in many real-world problems, uncertainties are often present and practically impossible to avoid. In the case where a solution is very sensitive to small variations either in the system's variables or the operating conditions, it may not be desirable to put it into practice. Hence optimization without taking uncertainty into consideration generally leads to solutions that should not be labeled as optimal as they are likely to perform differently when put into practice.

Various classifications of uncertainty have been suggested over the recent years [2–8]. In [2], four types of uncertainty were described. They are *1)* noise at fitness function, *2)* uncertainty at design variables or environmental parameters, *3)* approximation errors, and *4)* time varying fitness function. Similar categorization can also be found in [3]. Others [4, 5] classify uncertainty as either aleatory or epistemic. Aleatory uncertainty refers to naturally irreducible variability, e.g. quantities that are inherently variable over time and space. In contrast, epistemic uncertainty is caused by incomplete knowledge about the designs to be optimized and should be reducible if greater knowledge can be acquired. In [6–8], uncertainty is defined as the gap between the known and unknown facts. In this chapter, we follow the categorization of uncertainty in [2] and [3]. In particular, we focus on uncertainty in the system's variables and/or environmental parameters. To date, many approaches exist for coping with uncertainty in complex engineering design optimization. These include the One-at-a-Time Experiments, Taguchi Orthogonal Arrays, bounds-based, fuzzy and probabilistic methods [9]. In the context of EA, a number of prominent new studies on handling the presence of uncertainty in engineering designs have emerged recently. In [10], a Genetic Algorithm with Robust Searching Scheme (GA/RS3) was introduced. In this work, a probabilistic noise vector is added to the genotype before fitness evaluation. In biological terms, this means that part of the phenotypic features of an individual is determined by the decoding process of the genotypic code of genes in the chromosomes. The study of an (1+1)-Evolutionary Strategy (ES) with isotropic normal mutations using the noisy phenotype scheme has also been reported in [11]. An evolutionary algorithm based on max-min optimization strategy using a Baldwinian trust-region framework that employs surrogate models was also recently proposed in [3] for robust design. Recent applications of these robust EA strategies to engineering design problems include 2D aerodynamic airfoil [3, 12], lightweight space structures [13] and multilayer optical coating design [14].

In this chapter, we present a Single/Multi-objective Inverse Robust Evolutionary (SMIRE) design search methodology. In contrast to conventional forward robust optimization, the inverse approach avoids making assumptions about the uncertainty structure in the formulation of the optimization search process. Making assumptions about the uncertainty that are not backed up by strong evidence in evolutionary design optimization can possibly lead to erroneous designs that could have catastrophic consequences. Further, most existing schemes for handling uncertainty in evolutionary design optimization have focused on probabilistic methods [10–14]. Since probability theory may be inappropriate when insufficient field data exists for accurately estimating the structure of the uncertainty, we consider non-probabilistic methods, particularly, convex modeling, in the SMIRE. We begin with a single objective IRE approach in search for robust designs that are resilient to maximum uncertainty, given the worst-case performance permissible by the designers. Further, since uncertainty is practically impossible to avoid, we consider the possible benefits as the uncertainty prevails by introducing an opportunity criterion in the design search. To provide a trade-off between nominal, robustness, and/or opportunity in the final design solution, various multi-objective IRE schemes are introduced.

A motivating application for the proposed methodology perhaps is in the area of finance where portfolio design relies on the estimation of the expected returns on securities invested. The estimation is usually based on historical valuations of the securities and the deviation from the expected return on investment is crudely quantified as the risk level. One underlying assumption is that for longer investment horizon, the estimated return based on historical valuations is a good approximation of future returns. Such an investment planning scenario originated from Markovitz's pioneering work on portfolio optimization [15] and is usually hard to put into practice. One reason is that the uncertainty in estimation of the expected returns as mentioned. Although it appears that taking a very long term perspective on a portfolio may circumvent this source of uncertainty, it is seldom adhered to for practical reasons. Furthermore, with the current advance information technology and the dynamically changing macro-economics landscape, a 'sit-and-wait' attitude towards a portfolio is no longer viable. To the best of our knowledge, there has not been any work the attempts to apply evolutionary algorithms based on an inverse optimization approach to portfolio structuring based on an inverse optimization approach.

The rest of this chapter is organized as follows. In Section 26.2, we provide a brief overview of robust evolutionary design optimization. Four inverse schemes for evolutionary design search in the presence of uncertainty are presented in Section 26.3. To illustrate their applications, Section 26.4 provides an empirical study on a series of test functions with different complexities. Finally, Section 26.5 concludes this chapter.

19.2 Evolutionary Optimization in the Presence of Uncertainty

This section presents a brief overview on the fundamental aspects of evolutionary design in the presence of uncertainties. *Forward optimization* refers to those schemes where an optimal solution is sought based on some prior knowledge about the structure of the uncertainty. *Inverse optimization*, on the other hand, locates the target solution that satisfies some criteria specified by the designers. Here, we consider the general bound constrained nonlinear programming problem of the forms:

Forward Optimization:

$$\begin{array}{rl}\textbf{Optimize} & : \; f(x) \\ \textbf{Subject to} & : \; x_l \leq x \leq x_u\end{array} \tag{19.1}$$

or

Inverse Optimization:

$$\begin{array}{rl}\textbf{Optimize} & : \; f(x) - T \\ \textbf{Subject to} & : \; x_l \leq x \leq x_u\end{array} \tag{19.2}$$

where $f(x)$ is a scalar-valued objective function, T is the targeted output performance, $x \in \Re^d$ is the vector of design variables or environmental parameters, while x_l and x_u are vectors of lower and upper bounds for x.

Here, our focus is on EAs for robust engineering design optimization under uncertainties that arise in:

1. design vector, x
$$F(x) = f(x + \delta) \tag{19.3}$$
2. operating/environmental conditions, c
$$F(x) = f(x, c + \xi) \tag{19.4}$$

where $c = (c_1, c_2, \ldots, c_n)$ is the nominal value of the environmental parameters and ξ is a random vector used to model variability in the operating conditions. Since both forms of uncertainty may be treated equivalently, we do not differentiate uncertainty in design variables and the operating conditions. In the rest of this chapter we refer both the uncertain design variables and environmental parameters as uncertain parameters for the sake of brevity.

19.2.1 Probabilistic and Non-Probabilistic Schemes

Evolutionary techniques for handling uncertainty based on probabilistic schemes usually assume prior knowledge about the structure of the uncertainty. For example, the uncertainties, δ and/or ξ , are often assumed to be Gaussian (normal), Cauchy, or uniformly distributed. Very often, a Gaussian distribution with zero mean and variance σ^2, $N\left(0, \sigma^2\right)$ is considered for the uncertainty, by virtue of the central limit theorem. Consequently, the effective fitness function $F(x)$ for forward and inverse optimization can then be described as:

Forward probabilistic optimization:
$$F(x) = \int_{-\infty}^{\infty} f(x + \delta)\, \Phi(\delta)\, d\delta \tag{19.5}$$

or

Inverse probabilisic optimization:
$$F(x) = \int_{-\infty}^{\infty} \left(f(x + \delta) - T\right) \Phi(\delta)\, d\delta \tag{19.6}$$

where $\Phi(\delta)$ is the probability distribution of δ.

On the other hand, it is often the case in many real world engineering design problems that very little knowledge about the structure of the uncertainty involved is available. Making assumptions about the uncertainty that are not backed up by strong evidence in evolutionary design optimization can possibly lead to erroneous designs that could have catastrophic consequences. Instead of focusing on making any probably unjustifiable mathematical model out of the uncertainty, non-probabilistic methods may be used. Non-probabilistic approaches have attracted increasing attention in the engineering design community in recent years. They include evidence theory, possibility theory, interval analysis, and convex modeling. For example, interval analysis and convex modeling studies the uncertain parameters x for some intervals $[x_l, x_u]$, where x_l and x_u are the lower and upper bound and how this range affects the design solutions. Nevertheless, while non-probabilistic approaches generally require minimum assumption about the uncertainty involved, they can incur a high computational cost [3, 16]. For the details of non-probabilistic approaches in design optimization, the reader is referred to [3, 6, 16].

19.2.2 Benefits of Uncertainty

In most design optimization schemes, uncertainty has always been viewed upon as harmful to the final design solution. More specifically, the performance of the final design is believed to deteriorate in practice as the result of uncertainty. Since uncertainties are practically impossible to avoid, it is worth asking whether possible benefits can be derived from the presence of uncertainty. In [6–8], such an observation is termed as possible opportunity or windfall brought about by uncertainty. In Figure 19.1, it is shown that at $x = 5.0$, it is possible to obtain a better performance when x deviates to 4.0 on account of the uncertainty. The same can be explained for $x = 7.3$, which can even reach the global optimum at $x = 7.0$. In this chapter we will consider the benefits of uncertainty in our SMIRE algorithm in Section 19.3.2.

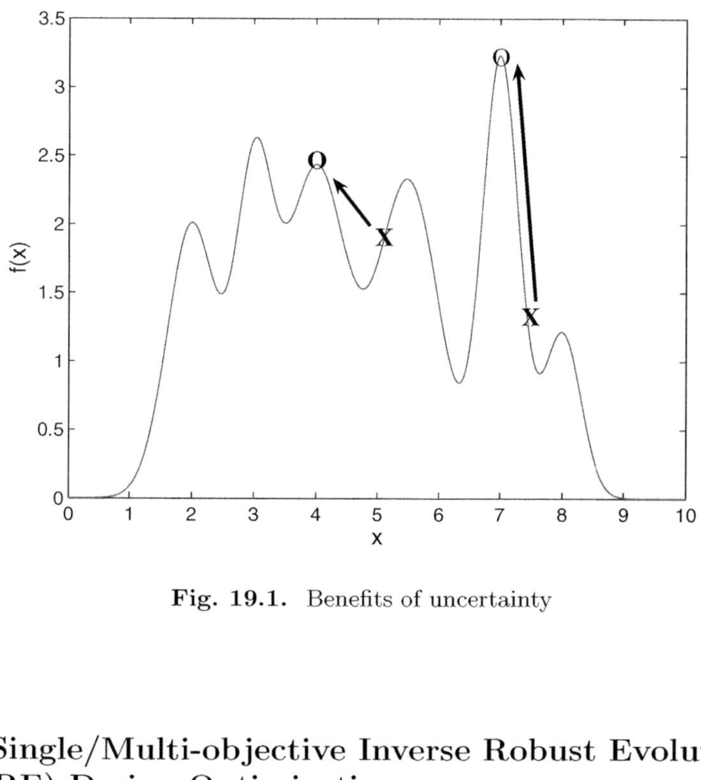

Fig. 19.1. Benefits of uncertainty

19.3 Single/Multi-objective Inverse Robust Evolutionary (SMIRE) Design Optimization

In this section, we present the Single/Multi-objective Inverse Robust Evolutionary algorithm for design optimization. In particular, four inverse optimization schemes are introduced.

19.3.1 Single and Bi-objective SMIRE Design Optimization

Here, we present a scheme for single and bi-objective inverse robust evolutionary design in the presence of uncertainty. The basic steps of the proposed algorithm are outlined in Figure 19.2. In the first step, the worst-case performance permissible for the final design f_t, and step size $\triangle$ used to conduct nested searches are defined and initialised by the designers. The robustness fitness $R_f(x)$ is then defined as the maximum uncertainty a design variable x can handle before violation of f_t. Hence, a design with a larger $Rf(x)$ represents one that is more robust to uncertainty.

Subsequently, a population of designs is created randomly or using Design of Experiments (DOE) techniques such as Latin hypercube sampling or minimum discrepancy sequences [21]. Each individual in the population is evaluated to determine its nominal fitness $f(x)$ and undergoes a sequence of nested searches across a family of nested search regions parameterized by the uncertainty. The aim of the nested searches is to determine the maximum amount of uncertainty that a design solution guarantees to handle before violating the worst-case fitness permissible as defined by f_t.

BEGIN SMIRE

Initialization Phase:
- Initialize worst-case permissible performance, f_t
- Initialize step size Δ for the inner search
- Generate a population of design vectors

Search Phase:

While (termination condition is not satisfied)

For (each individual i in the population)
- Objective-1: Obj-1 $= f(x_i)$, applicable for bi-objective SMIRE
- Objective-2:
  - ∘ Assign $k = 0$
  - ∘ **Repeat**
    - ◇ $k = k + 1$
    - ◇ Minimize: $f^k(x)$ subject to : $x_l^k \leq x \leq x_u^k$, where $x_l^k = x_i - k\Delta$, $x_u^k = x_i + k\Delta$
    - ◇ Obtain x_{opt}^k and $f\left(x_{opt}^k\right)$
    - ◇ Store $f\left(x_{opt}^k\right)$ and associate it with $k\Delta$ in database Db

    until $f\left(x_{opt}^k\right) \leq f_t$
  - ∘ Estimate maximum uncertainty δ_{max}^i by interpolating from $f\left(x_{opt}^k\right)$ and $k\Delta$ in Db
  - ∘ Obj-2 $= R_f(x_i) = \delta_{max}^i$

End for
- Apply MOEA (if multi-objective), or EA (if single objective) evolutionary operators to create a new population.

End while

END SMIRE

Fig. 19.2. Single and bi-objective SMIRE design algorithm

More specifically, for each chromosome, we solve a sequence of bound constrained optimization sub-problems of the form:

$$\begin{array}{rl}\textbf{Minimize} : & f^k(x) \\ \textbf{Subject to} : & x_l^k \leq x \leq x_u^k\end{array} \tag{19.7}$$

where x_l^k and x_u^k are the appropriate bounds on the uncertain parameters. Each kth optimization sub-problem locates the worst-case fitness in the direct neighborhood of individual x within the bounds, x_l^k and x_u^k which are updated with step size:

$$\begin{aligned} x_l^k &= x_i - k\Delta \\ x_u^k &= x_i + k\Delta \end{aligned} \tag{19.8}$$

In single objective inverse robust evolutionary design optimization, only the robustness function $R_f(x_i)$ is considered. The bi-objective IRE on the hand considers both $R_f(x_i)$ and $f(x)$ is in the design search to locate the pareto-optimum set.

Note that by conducting a sequence of nested searches across a family of ascending nested bounds parameterized by the uncertainty vector, we arrive at a monotonic increasing function of worst-case fitness versus uncertainty such that:

$$x_l^{k+1} \leq x_l^k, x_u^k \leq x_u^{k+1} \rightarrow f\left(x_{opt}^{k+1}\right) \leq f\left(x_{opt}^k\right) \tag{19.9}$$

where x_{opt}^k denotes the optimum at the kth iteration and $f\left(x_{opt}^k\right)$ is the corresponding worst-case fitness obtained for $x_l^k \leq x \leq x_u^k$. In addition, the $f\left(x_{opt}^k\right)$ found and associated $k\Delta$ for each search iteration is then stored to create a database of uncertainties and corresponding worst-case fitness. The sequences of iterative searches are terminated when the optimal solution of the kth sub-problem violates the inequality constraint in eq. (19.10), i.e.

$$f\left(x_{opt}^k\right) > f_t \tag{19.10}$$

At the end of the sequences of searches for a chromosome, the maximum uncertainty δ_{max} that a design may handle given a defined f_t can be determined by interpolating from the database of uncertainties and worst-case fitness values previously archived, i.e., $k\Delta$ and $f\left(x_{opt}^k\right)$. The search then proceeds with the standard evolutionary operators to create a new population and stops when the termination conditions are reached. Here, we further illustrate the procedure to locate the robustness fitness using an example in Figure 19.3. Consider the design point at $x = 4$ in the figure. Here, f_t and Δ are configured as 1.0 and 1.0, respectively. A sequence of bound-constrained optimization sub-problems are then conducted for $x = 4$, which is terminated upon violations of the constraint in eq. (19.10). The labels A, B and C correspond to $f\left(x_{opt}^k\right)$ for $k = 1$, 2 and 3. At $k = 3$, the worst-case fitness, indicated by C, has satisfied the termination condition. Then, the estimated maximum amount of uncertainty, δ_{max} the design point $x = 4$ can handle is determined by interpolating from A, B, and C.

The computational complexity for establishing the robustness fitness, $R_f(x)$ in SMIRE is of $O(gnkl)$. Here, g is the number of generations for the EA search, n is the number of chromosomes evolved, k is the average number of iterations to reach for each chromosome and l represents the average number of function evaluations incurred in each bound constrained optimization sub-problems. It is worth noting that $k \propto \frac{1}{\Delta}$.

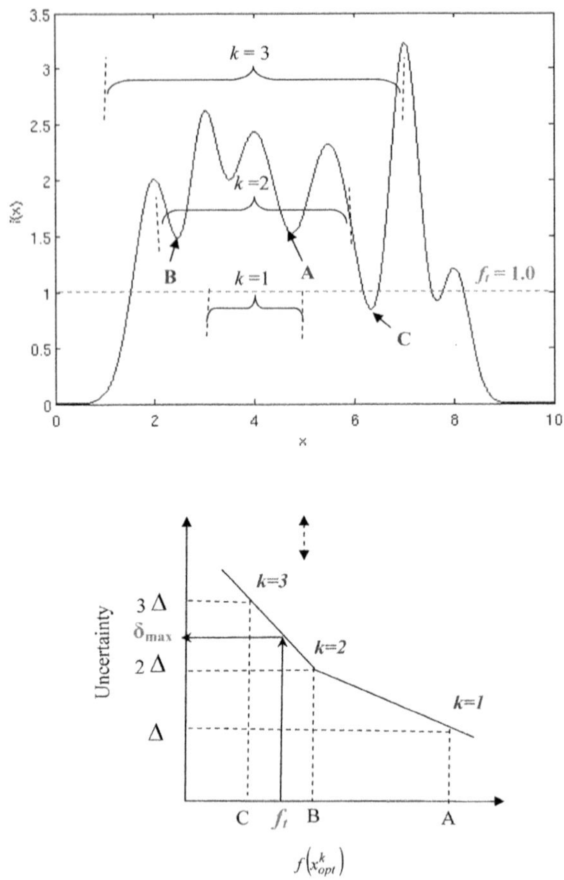

Fig. 19.3. Robustness fitness in SMIRE for $x_i = 4.0$ and $\Delta = 1.0$

19.3.2 Tri-objective SMIRE Design Optimization

In this subsection, we proceed from single and bi-objective SMIRE design optimization to consider higher number of objectives. In particular, we search for pareto-optimal solutions when considering nominal, robustness, and opportunity fitness simultaneously. Here we present two tri-objective schemes for exploiting the benefits of uncertainty, i.e. opportunity in the design search.

Tri-objective SMIRE Scheme I.

A straightforward approach as in [18–20] is to formulate the search problem as a tri-objective scheme which treats all three objectives equally. The procedures to obtain the first two objectives, i.e., nominal $f(x)$ and robustness fitness $R_f(x)$ for each design vector remain the same as described in Section 19.3.1 for the bi-objective IRE. The third objective, i.e. the opportunity fitness, is determined using an approach similar to establishing $R_f(x)$ involving a series of nested optimization search as illustrated in Figure 19.4.

BEGIN SMIRE

Initialization Phase:

- Initialize worst-case permissible performance, f_t and minimum performance improvement, f_v.
- Initialize step size Δ for the inner search.
- Generate a population of design vectors.

Search Phase:

While (termination condition is not satisfied)

For (each individual i in the population)

- Objective-1: Obj-1 $= f(x_i)$.
- Objective-2: Obj-2 $= R_f(x_i) = \delta^i_{max}$, as outlined in Objective-2 of Fig. 19.3.
- Objective-3:
  - ○ Assign $k = 0$
  - ○ **Repeat**
    - ◇ $k = k + 1$
    - ◇ Maximize: $d(x) = f^k(x) - f(x_i)$ subject to : $x^k_l \leq x \leq x^k_u$, where $x^k_l = x_i - k\Delta$, $x^k_u = x_i + k\Delta$
    - ◇ Obtain x^k_{opt} and $d\left(x^k_{opt}\right)$
    - ◇ Store $d\left(x^k_{opt}\right)$ and associate it with $k\Delta$ in database Db

    until $\{d^k(x) = f\left(x^k_{opt}\right) - f(x_i)\} \geq f_v$
  - ○ Estimate minimum uncertainty β^i_{max} by interpolating from $d\left(x^k_{opt}\right)$ and $k\Delta$ in Db
  - ○ Obj-3 $= O_f(x_i) = \beta^i_{min}$

End for

- Apply MOEA evolutionary operators to create a new population.

End while

END SMIRE

Fig. 19.4. Tri-objective SMIRE design algorithm, scheme I

The fitness of opportunity function $O_f(x)$ is defined as the minimum uncertainty the design variable x would require in order to acquire a performance improvement defined by f_v. In particular, for each chromosome, we solve a sequence of bound constrained optimization sub-problems in the form:

$$\begin{array}{rc} \textbf{Maximize} : d(x) = f^k(x) - f(x_i) \\ \textbf{Subject to} : \quad x_l^k \leq x \leq x_u^k \end{array} \tag{19.11}$$

where x_l^k and x_u^k are the appropriate bounds on the uncertain parameters.

Each kth optimization sub-problem locates the best-case fitness in the direct neighborhood of individual x within the bounds, x_l^k and x_u^k which are updated using eq. (19.8). The search for minimum uncertainty possible at x terminates when the best performance improvement possibly achieved has reached f_v, i.e. $\{d^k(x) = f\left(x_{opt}^k\right) - f(x_i)\} \geq f_v$. The minimum uncertainty β_{min}^i is then interpolated from the database of uncertainties and best fitness improvement recorded previously, i.e., $k\Delta$ and $d\left(x_{opt}^k\right)$.

The expected computational costs for obtaining $R_f(x)$ and $O_f(x)$ are equivalent. In effect, for the same number of search generations made, the tri-objective IRE scheme is approximately two times more computational expensive than the bi-objective IRE scheme described in Section 19.3.1.

Tri-objective SMIRE Scheme II.

Alternatively, one may consider an opportunity fitness which is bound-constrained by the maximum uncertainty obtained from objective-2, i.e. the robustness. As a result, the opportunity fitness is defined as the maximum performance improvement permissive in the presence of uncertainty and given by:

$$\begin{array}{rc} \textbf{Maximize} : & O_f(x_i) = f(x) - f(x_i) \\ \textbf{Subject to} : & x_i - \delta_{max}^i \leq x \leq x_i + \delta_{max}^i \end{array} \tag{19.12}$$

where δ_{max}^i is the maximum uncertainty defined by $R_f(x)$. The pseudo-code for scheme II is also outlined in Figure 19.5.

The computational complexity of Tri-objective SMIRE scheme II may be determined as $O(gnl(k+1))$. In comparison to other SMIRE schemes introduced earlier, scheme II incurs a higher computational cost than both the single and bi-objective IREs, i.e., $O(gnkl)$. But it is less computational expensive than the tri-objective IRE scheme I counterpart which has a complexity of $O(2gnkl)$.

19.4 Empirical Study

In this section, we present an empirical study on the proposed SMIRE schemes using five synthetic test functions plotted in Figure 19.6. These include a three 1D functions (f_1, f_2 and f_3) and two 2D functions (f_4 and f_5) having the characteristics described in Table 19.1. It is worth noting that the choice of low dimensional functions is only for academic purposes as the methodology introduced here applies also to high dimensional problems.

BEGIN SMIRE

Initialization Phase:

- Initialize worst-case permissible performance, f_t
- Initialize step size Δ for the inner search.
- Generate a population of design vectors.

Search Phase:

While (termination condition is not satisfied)

For (each individual i in the population)

- Objective-1: Obj-1 $= f(x_i)$.
- Objective-2: Obj-2 $= R_f(x_i) = \delta^i_{max}$, as outlined in Objective-2 of Fig. 19.3.
- Objective-3:
  - ∘ **Maximize :** $O_f(x_i) = f(x) - f(x_i)$
    Subject to : $x_i - \delta^i_{max} \leq x \leq x_i + \delta^i_{max}$
  - ∘ Obj-3 $= O_f(x_i)$

End for

- Apply MOEA evolutionary operators to create a new population.

End while

END SMIRE

Fig. 19.5. Tri-objective SMIRE design algorithm, scheme II

In the numerical studies, we employ a 16-bit binary coded standard GA for single objective SMIRE and NSGA [17] for multi-objective SMIRE. The population size is kept at 100, and maximum generation count is set to 100 generations. Uniform crossover and mutation are applied at probabilities of 0.9 and 0.01, respectively. The step size is chosen empirically to be 3% of the search bounds to minimize large interpolation errors. In the nested optimizations, a local search based on Sequential Quadratic Programming (SQP) is considered.

19.4.1 Discussions on Different SMIRE Schemes

Figures 19.7 and 19.8 show the obtained pareto-optimum solution sets in the parameter space, when using the four different SMIRE schemes to search on test functions $f1$ and $f2$, respectively. Using the single objective IRE scheme, the search is expected to converge to the most robust peak having nominal fitness greater than f_t. For instance, the most robust point of $f1$ and $f2$ are around $x = 4.0$ and $x = 8.0$, respectively. On the other hand, the solution set obtained based on bi-objective IRE are a compromise or trade-off between the two objectives considered in the scheme. On $f1$, the solution set has only two members with extremely good nominal or robustness and are located at different peaks of the search space. However, due to the higher modality of $f2$ fitness space, greater trade-off solutions between nominal and robustness would exist, leading to a larger pareto-optimum solution set than that of $f1$. Employing the tri-objective schemes, the solution set is even more varied as more trade-offs between nominal, robustness, and opportunity fitness are expected.

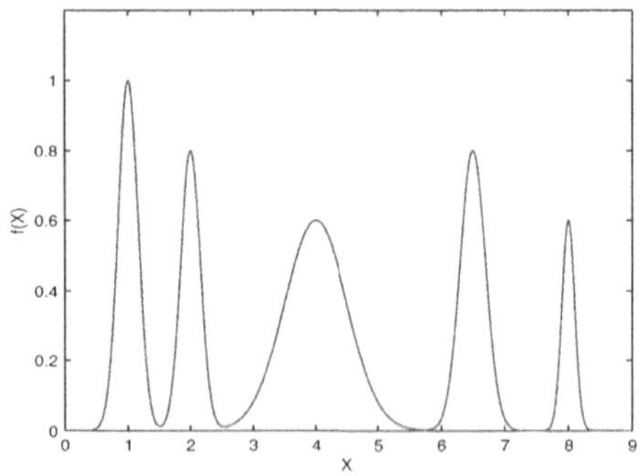

(a) $f1$

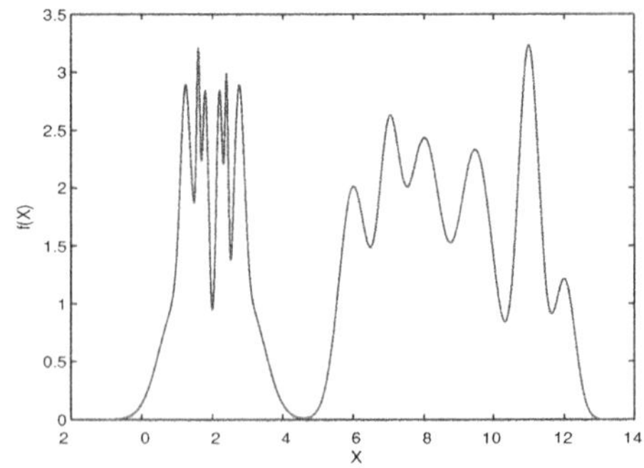

(b) $f2$

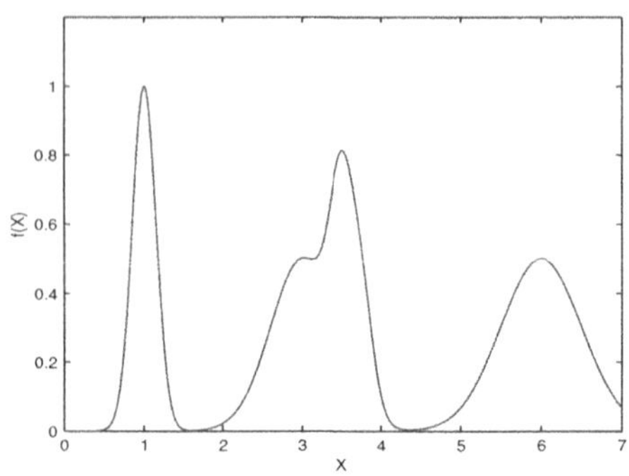

(c) $f3$

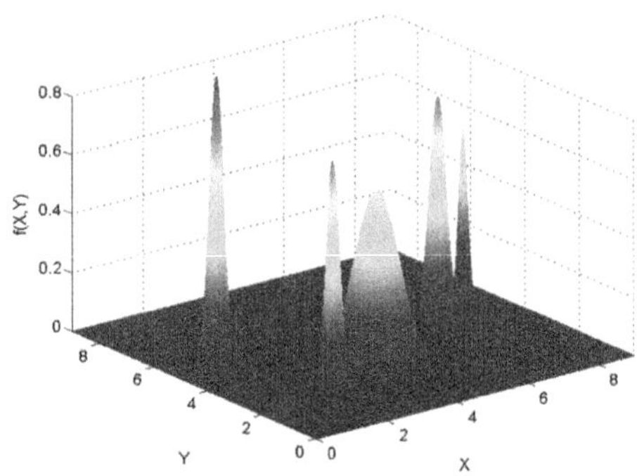

(d) $f4$

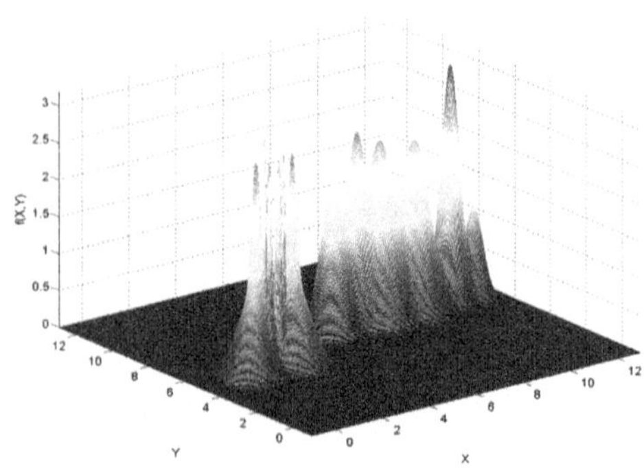

(e) $f5$

Fig. 19.6. 2D and 3D plots of the test functions considered

Test Function	Characteristics of the functions
$f1$	$f(x)=e^{-(x-1)^2/0.05}+0.8e^{-(x-2)^2/0.05}+0.6e^{-(x-4)^2/0.5}+0.8e^{-(x-6.5)^2/0.08}+0.6e^{-(x-8)^2/0.02}$, $0\le x\le 9,\ f_t=0.4\,and\,f_v=0.2$
$f2$	$f(x)=e^{-(x-1)^2/0.5}+2e^{-(x-1.25)^2/0.045}+0.5e^{-(x-1.5)^2/0.0128}+2e^{-(x-1.6)^2/0.005}+2.5e^{-(x-1.8)^2/0.02}+2.5e^{-(x-2.2)^2/0.02}+2e^{-(x-2.4)^2/0.005}+2e^{-(x-2.75)^2/0.045}+e^{-(x-3)^2/0.5}+2e^{-(x-6)^2/0.32}+2.2e^{-(x-7)^2/0.18}+2.4e^{-(x-8)^2/0.5}+2.3e^{-(x-9.5)^2/0.5}+3.2e^{-(x-11)^2/0.18}+1.2e^{-(x-12)^2/0.18}$ $-1\le x\le 13,\ f_t=1.5\,and\,f_v=1$
$f3$	$f(x)=e^{-(x-1)^2/0.045}+0.5e^{-(x-3)^2/0.32}+0.5e^{-(x-3.5)^2/0.045}+0.33e^{-(x-3.75)^2/0.045}+0.5e^{-(x-6)^2/0.5}$, $0\le x\le 7,\ f_t=0.3\,and\,f_v=0.2$
$f4$	$f(x,y)=e^{-\left[(x-1)^2+(y-5)^2\right]/0.05}+0.8e^{-\left[(x-2)^2+(y-2)^2\right]/0.05}+0.6e^{-\left[(x-4)^2+(y-3)^2\right]/0.5}+0.8e^{-\left[(x-6.5)^2+(y-4)^2\right]/0.08}+0.6e^{-\left[(x-8)^2+(y-5)^2\right]/0.02}$, $0\le x,y\le 9,\ f_t=0.4\,and\,f_v=0.2$
$f5$	$f(x,y)=e^{-\left[(x-1)^2+(y-5)^2\right]/0.5}+2e^{-\left[(x-1.25)^2+(y-5)^2\right]/0.045}+0.5e^{-\left[(x-1.5)^2+(y-5)^2\right]/0.0128}+2e^{-\left[(x-1.6)^2+(y-5)^2\right]/0.005}+2.5e^{-\left[(x-1.8)^2+(y-5)^2\right]/0.02}+2.5e^{-\left[(x-2.2)^2+(y-5)^2\right]/0.02}+2e^{-\left[(x-2.4)^2+(y-5)^2\right]/0.005}+2e^{-\left[(x-2.75)^2+(y-5)^2\right]/0.045}+e^{-\left[(x-3)^2+(y-5)^2\right]/0.5}+2e^{-\left[(x-6)^2+(y-7)^2\right]/0.32}+2.2e^{-\left[(x-7)^2+(y-7)^2\right]/0.18}+2.4e^{-\left[(x-8)^2+(y-7)^2\right]/0.5}+2.3e^{-\left[(x-9.5)^2+(y-7)^2\right]/0.5}+3.2e^{-\left[(x-11)^2+(y-7)^2\right]/0.18}+1.2e^{-\left[(x-12)^2+(y-7)^2\right]/0.18}$ $-1\le x,y\le 13,\ f_t=1.5\,and\,f_v=1$

Table 19.1. Five synthetic test functions considered in the empirical study

The difference between the solutions from tri-objective scheme I and II will be discussed further in next subsection.

19.4.2 Comparison on Tri-objective IRE Scheme I and II

To better illustrate the differences of tri-objective IRE scheme I to II, we study the pareto-optimum solutions obtained in $f3$, $f4$, and $f5$ at their parameter space. For $f3$, the pareto-optimum solutions obtained for both schemes are presented in Figures 19.9 and 19.10.

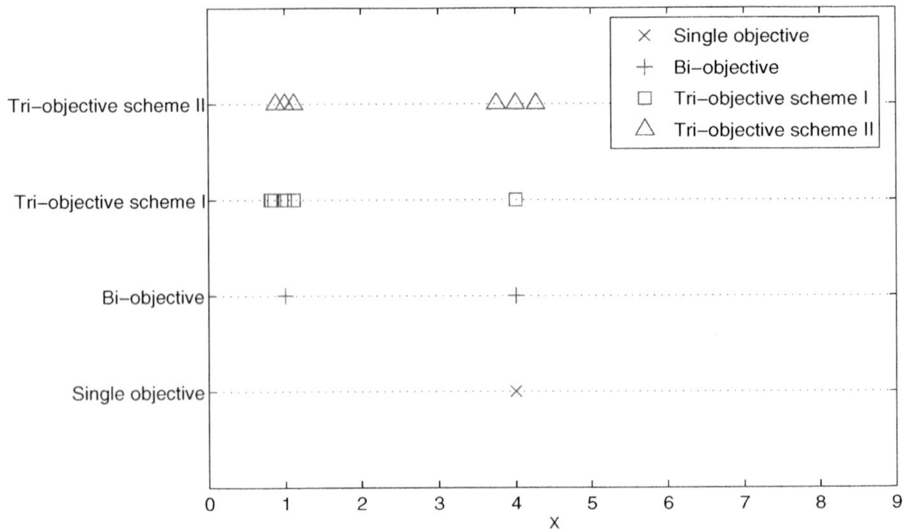

Fig. 19.7. Pareto-optimum solutions of SMIRE on $f1$

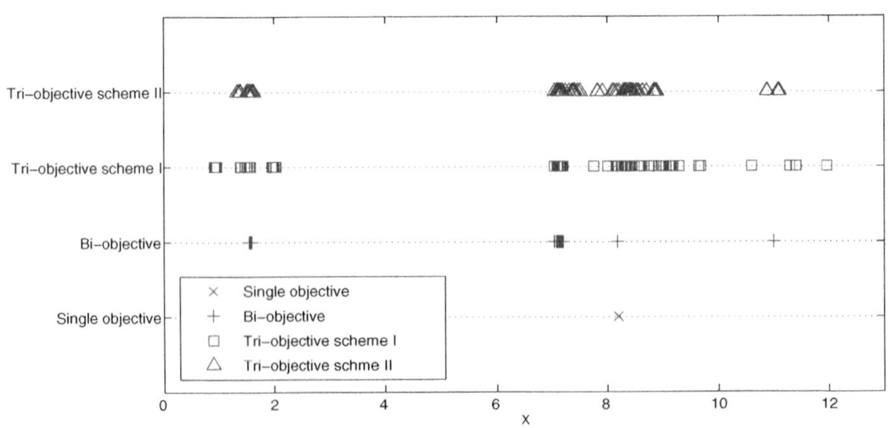

Fig. 19.8. Pareto-optimum solutions of SMIRE on $f2$

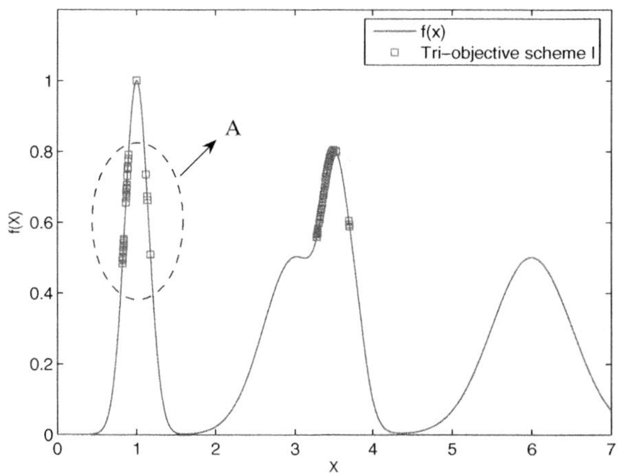

Fig. 19.9. Pareto-optimum solutions of tri-objective IRE scheme I on $f3$

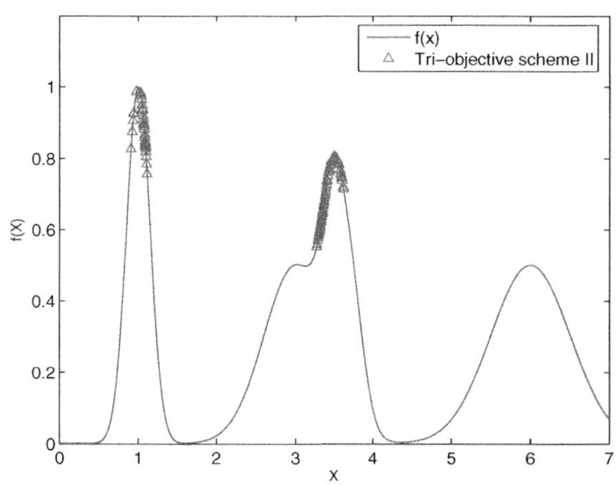

Fig. 19.10. Pareto-optimum solutions of tri-objective IRE scheme II on $f3$

In tri-objective scheme I, the opportunity fitness is treated as equals to both robustness and nominal, hence some solutions may display good opportunity properties but fares significantly poor on the other objectives, see Figure 19.9. Such solutions usually exist on steep peaks which is labelled as A in the figure. Figures 19.11 and 19.13 show that in tri-objective scheme I, many of the pareto-optimum solutions are those near the steep peaks, since the robustness property of the solutions are ignored.

In contrast, scheme II (see Figure 19.10) defines an opportunity fitness which is bound-constrained by the maximum uncertainty obtained from objective-2, i.e. the robustness. By considering robustness as a constraint in the opportunity objective, there is a greater chance for more robust solutions to have higher opportunity fitness. Hence, the tri-objective search in this scheme mostly converges to a set of solution that still lies on the robust peaks, as depicted in Figures 19.12 and 19.14.

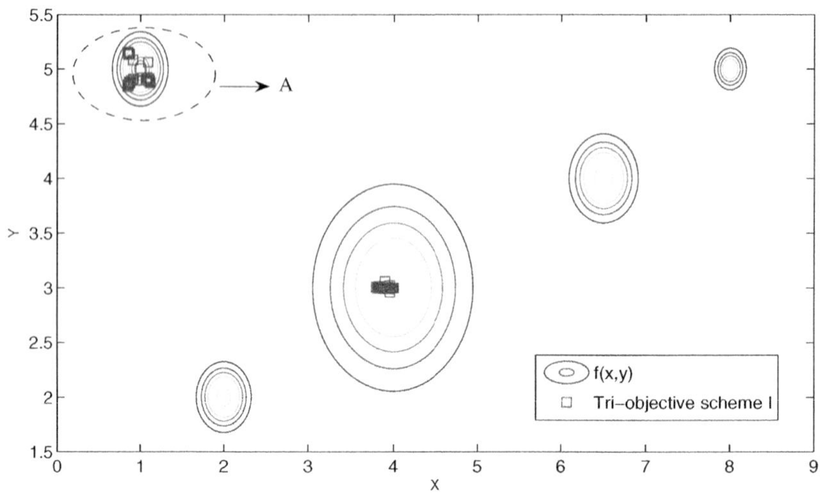

Fig. 19.11. Pareto-optimum solutions of tri-objective IRE scheme I on $f4$

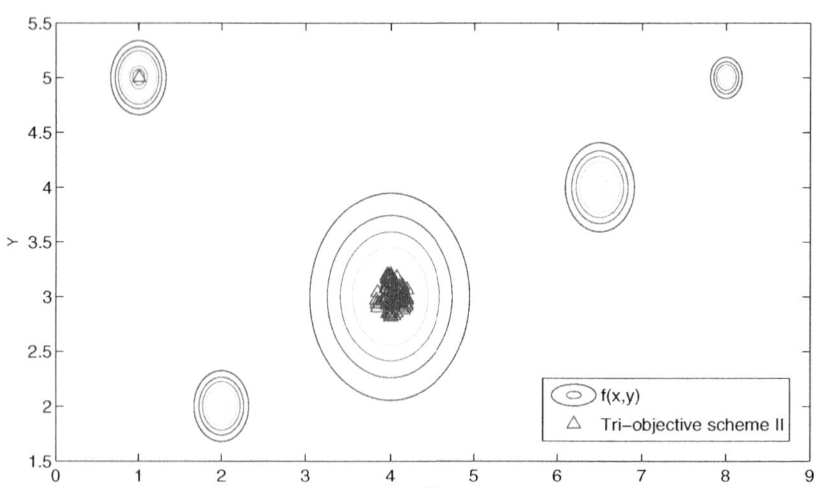

Fig. 19.12. Pareto-optimum solutions of tri-objective IRE scheme II on $f4$

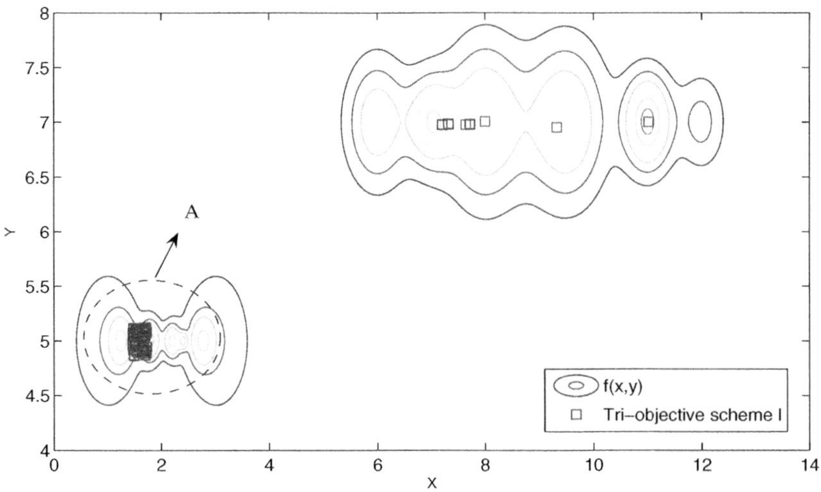

Fig. 19.13. Pareto-optimum solutions of tri-objective IRE scheme I on $f5$

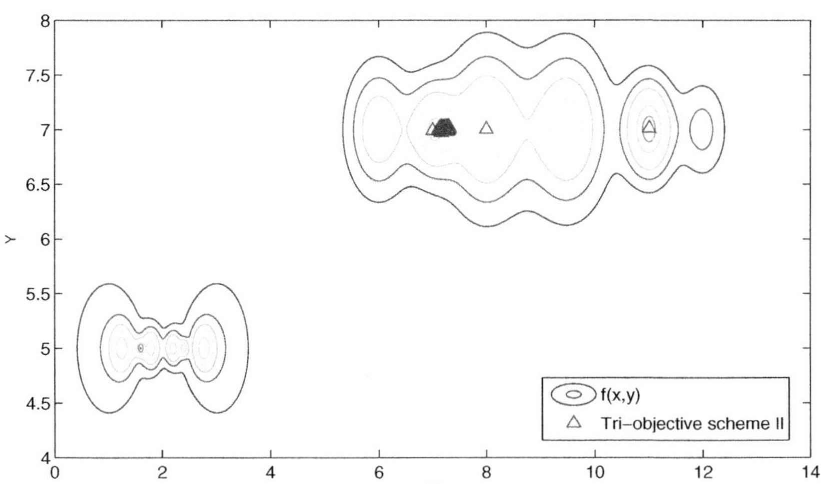

Fig. 19.14. Pareto-optimum solutions of tri-objective IRE scheme II on $f5$

Next we consider the computational complexity of the two tri-objective IRE schemes. Table 19.2 tabulates the computational costs incurred by the schemes, in terms of the total objective function calls, when used to search on each of the five test functions considered. From the results, it is shown that scheme I requires a higher computational effort than II.

Test Function	**Ratio of Computational Cost between Tri-objective Scheme I and II**
$f1$	1:0.78
$f2$	1:0.66
$f3$	1:0.75
$f4$	1:0.81
$f5$	1:0.62

Table 19.2. Ratio of computational cost incurred in IRE scheme I and II

19.5 Conclusions

In this chapter, we have presented the single and multi-objective inverse robust evolutionary design optimization in the presence of uncertainty. It proposes the use of inverse optimization technique in the field of robust optimization. Another important feature of the proposed methodology is that the solutions obtained were discovered without any requirement to make possible untrue assumptions about the structure of the uncertainties involved as what usually available in probabilistic methods. In addition, we have also presented the incorporation of the opportunity fitness to possibly explore the benefit of having uncertainty in design. Hence, the final solutions obtained provide the decision-makers or designers with more design options considering the trade-offs up to three objectives: robustness, nominal fitness, and opportunity fitness.

In evolutionary algorithms, many thousands of calls to the objective function are often required to locate a near optimal solution. While the algorithm proposed offers an effective approach to modeling of uncertainty in engineering design, a compelling limitation of the theory is the massive computational efforts incurred in the nested evolutionary design search. A simple solution to this issue would be to replace the nested global optimization with local search which is much cheaper as has been done in [16]. The computational efforts incurred would be even more devastating if the objective function is computationally expensive which is very common in complex engineering design problems. Nevertheless, it is worth noting here that a promising and intuitive way to reduce the search time incurred in solving the sequences of bound constrained sub-problems is to replace as much as possible the computationally expensive high-fidelity analysis solvers with lower-fidelity models that are computationally less expensive. The readers are referred to [22–24] for greater details on the available to achieve this cost savings.

References

1. Goldberg D.E. (1989) Genetic Algorithms in Search, Optimization and Machine Learning. Addison-Wesley, Reading, Massachusetts
2. Jin Y., Branke J. (2005) Evolutionary Optimization in Uncertain Environment- A Survey. In: IEEE Transactions on Evolutionary Computation, Vol. 9, No. 3

3. Ong Y.S., Nair P.B., Lum K.Y. (2006) Min-Max Surrogate Assisted Evolutionary Algorithm for Robust Aerodynamic Design. In: Special Issue on Evolutionary Optimization in the Presence of Uncertainties, IEEE Transactions on Evolutionary Computation, Vol. 10, No. 4
4. Oberkampf, W.L., et al. (2000) Estimation of Total Uncertainty in Modeling and Simulation. In: Sandia Report SAND2000-0824
5. Oberkampf, W., Helton, J., Sentz, K. (2001) Mathematical Representation of Uncertainty. In: AIAA Proceedings of Non-Deterministic Approaches Forum, Reston, VA
6. Ben-Haim Y. (2001) Information Gap Decision Theory. Academic Press, California
7. Ben-Haim Y. (2004) Uncertainty, Probability, and Information-Gaps. In: Reliability Engineering and System Safety 85, pp. 249-266
8. Ben-Haim Y. (1996) Robust Reliability in Mechanical Sciences. Springer-Verlag, Berlin
9. Huyse L. (2001) Solving Problems of Optimization Under Uncertainty as Statistical Decision Problems. In: AIAA-2001-1519
10. Tsutsui S. and Ghosh A. (1997) Genetic Algorithms with a Robust Solution Searching Scheme. In: IEEE Transaction on Evolutionary Computation, Vol. 1, No. 3, pp. 201-208
11. Arnold D.V. and Beyer H.G. (2002) Local Performance of the (1+1)-ES in a Noisy Environ-ment. In: IEEE Trans. Evolutionary Computation, Vol. 6, No. 1., pp 30-41
12. Huyse L. and Lewis R.M. (2001) Aerodynamic Shape Optimization of Two-dimensional Air-foils Under Uncertain Operating Conditions. ICASE NASA Langley Research Centre, Hampton, Virginia
13. Anthony D.K. and Keane A.J. (2003) Robust Optimal Design of a Lightweight Space Structure Using a Genetic Algorithm. In: AIAA Journal 41(8), pp. 1601-1604
14. Wiesmann D., Hammel U. and Back T. (1998) Robust design of multilayer optical coatings by means of evolutionary algorithms. In: IEEE Trans Evolutionary Computation, Vol 2, No 4, pp 162-167
15. Markovitz, H.M. (1952) Portfolio Selection. In: Journal of Finance, Vol. 7, pp. 77-91
16. Lim D., Ong Y.S., Lee B.S. (2005) Inverse Multi-Objective Robust Evolutionary Design Optimization in The Presence of Uncertainty. In: Workshop on Evolutionary Algorithm for Dynamic Optimization Problems, Genetic and Evolutionary Computation Conference (GECCO), Washington
17. Srinivas N. and Deb K. (1994) Multiobjective Optimization Using Nondominated Sorting in Genetic Algorithm. In: Evolutionary Computation, 2(3):221-248
18. Ray T. (2002) Constrained Robust Optimal Design Using a Multiobjective Evolutionary Algorithm. In: Proceedings of Congress on Evolutionary Computation
19. Jin Y. and Sendhoff B. (2003) Trade-Off between Performance and Robustness: An Evolution-ary Multiobjective Approach. In: Proceedings of Second International Conference on Evolu-tionary Multi-criteria Optimization. LNCS 2632, Springer, pp.237-251, Faro
20. Li M., Azarm S., Aute V. (2005) A Multi-Objective Genetic Algorithm for Robust Design Optimization. In: Genetic and Evolutionary Computation Conference (GECCO), Washington

21. Fang K.T., Ma C.X., and Winker P. (2000) Centered L2-Discrepancy of Random Sampling and Latin Hypercube Design, and Construction of Uniform Designs. In: Mathematics of Computation, Vol. 71, No. 237, pp. 275-296
22. Jin Y. and Sendhoff B. (2002) Fitness approximation in evolutionary computation: A survey. In: Proceedings of the 2002 Genetic and Evolutionary Computation Conference, pp. 1105-1112
23. Ong Y.S., Nair P.B. and Keane A.J. (2003) Evolutionary Optimization of Computationally Expensive Problems via Surrogate Modeling. In: AIAA Journal, Vol. 41, No. 4, pp 687-696
24. Zhou Z.Z., Ong Y.S., Nair P.B., Keane A.J. and Lum K.Y. (2006) Combining Global and Local Surrogate Models to Accelerate Evolutionary Optimization. In: IEEE Transactions On Systems, Man and Cybernetics - Part C

20

Evolving the Tradeoffs between Pareto-Optimality and Robustness in Multi-Objective Evolutionary Algorithms

Chi Keong Goh and Kay Chen Tan

Department of Electrical and Computer Engineering
National University of Singapore
4 Engineering Drive 3, Singapore 117576
{ckgoh, eletankc}@nus.edu.sg

Summary. Many real-world applications involve the simultaneous optimization of several competing objectives and are susceptible to decision or environmental parameter variation which results in large or unacceptable performance variation. While several studies on robust optimization have been presented in the domain of single-objective (SO) problems, the evolution of robust solutions is rarely studied in the context of evolutionary multi-objective optimization (EMOO). This chapter presents a robust multi-objective evolutionary algorithm for constrained multi-objective optimization. The proposed algorithm, incorporating the features of micro-GA which performs a local search for the worst case scenario of each candidate solution, the memory-based feature of tabu restriction to guide the evolutionary process and periodic re-evaluation of archived solutions to reduce uncertainty of evolved solutions, is capable of evolving the tradeoffs between Pareto optimality and robustness. The effectiveness of the algorithm is validated upon two benchmark with different properties and the I-beam design problem.

20.1 Introduction

Multi-objective evolutionary algorithms (MOEAs) are a class of stochastic optimization techniques that simulates biological evolution to solve problems with multiple objectives. Multi-objective (MO) optimization is a challenging research topic because it involves the simultaneous optimization of several complex objectives in the Pareto optimal sense and requires researchers to address many issues that are unique to MO problems. The advances made in MOEAs design is the result of two decades of intense research examining topics such as fitness assignment [9, 18], diversity preservation [4, 16], balance between exploration and exploitation [1], and elitism [17] in the context of MO optimization. Although the application of evolutionary multi-objective optimization (EMOO) to real-world problems [8, 24] has been gaining significant attention from researchers in different fields, there is a distinct lack in studies investigating the issues of uncertainties in the literature [11].

C.K. Goh and K.C. Tan: *Evolving the Tradeoffs between Pareto-Optimality and Robustness in Multi-Objective Evolutionary Algorithms*, Studies in Computational Intelligence (SCI) **51**, 457–478 (2007)
www.springerlink.com

Uncertainties in real-world optimization problems are due to factors such as data incompleteness and uncertainties, mathematical model inaccuracies, environmental conditions uncertainties, and solutions that cannot be implemented exactly. According to [14, 19], the various forms of uncertainty can be classified as 1) noisy fitness functions [11], 2) uncertainty of design variables or environmental parameters [5, 22], 3) approximation errors, and 4) time varying fitness functions. This chapter considers the optimization of a set of Pareto optimal solutions that remain satisfactory in the face of parametric variations, i.e. type 2 uncertainty. The optimization of robust solutions is of particular importance in real-world problems, such as scheduling, vehicle routing and engineering design optimization, where certain characteristics of the environment may not be known with absolute certainty and the decision spaces can be sensitive to parametric variations. Furthermore, variation in variable or environment may affect the quality and performance adversely and robust optimization considers the effects explicitly and seeks to minimize the consequences without eliminating efficiency.

A number of studies concerning evolutionary single-objective (SO) optimization of robust solutions have been reported in the literature. Tsutsui and Ghosh [12] proposed a genetic algorithm with robust solution searching scheme GAs/RS^3 where individual genotype is perturbed before evaluation and presented a mathematical model expressing the relationship between noise and fitness reduction. This scheme is later extended to consider the mean fitness of a number of perturbed individuals in [26]. On the other hand, Branke [2] explored different techniques to improve algorithmic performance and computational efficiency. More recently, an evolutionary algorithm based on max-min optimization strategy using a Baldwinian trust-region framework that employs surrogate models is proposed for robust aerodynamic design [19]. Robust SO (rSO) has also been successfully applied to engineering design problems, scheduling and multi-layer optical coating design. Nonetheless, the issue of robust optimization for MO problems is rarely considered in the literature until recently [5, 13].

In addition, the fact that there is an inherent tradeoff between robustness and optimality has some interesting implications for robust evolutionary optimization. This chapter is concerned with the development of an evolutionary technique for constrained robust MO (rMO) optimization and presents a robust multi-objective evolutionary algorithm (RMOEA) which is capable of evolving the tradeoffs between Pareto optimality and robustness. In particular, it incorporates the proposed worst performance measure and a local search process that performs the search for the worst case scenario for each candidate solution. In addition, it implements the memory-based feature of tabu search (TS) to improve the computational efficiency and constraints under uncertainty as well as periodic re-evaluation of archived solutions to reduce uncertainty of evolved solutions.

The organization of the chapter is as follows: Section 20.2 provides a brief description of MO optimization, rMO optimization and the basic MOEA framework. Based on the MOEA framework, Section 20.3 presents the proposed RMOEA for rMO optimization. After which, the performance of the proposed RMOEA is validated through empirical studies upon three different problems in Section 20.4. Finally, conclusions are drawn in Section 20.5.

20.2 Background Information

This section introduces relevant rMO optimization concepts and definitions. MO optimization and Pareto definitions are first described followed by the formulation of the rMO optimization problem. Finally, a short overview of the MOEA paradigm is presented.

20.2.1 Multi-objective Optimization and Pareto Optimality

Many real-world applications involve complex optimization tasks with various competing specifications and constraints. These optimization tasks are typically represented by its mathematical model and the specification of MO criteria captures more information about the modeled problem as several problem characteristics are taken into consideration. Without any loss of generality, this paper considers a minimization problem and the MO problem can be formally defined as

$$\min_{\boldsymbol{x} \in \boldsymbol{X}^{n_x}} \boldsymbol{f}(\boldsymbol{x}) = \{f_1(\boldsymbol{x}), f_2(\boldsymbol{x}), ..., f_M(\boldsymbol{x})\} \tag{20.1}$$

$$\text{s.t. } \boldsymbol{g}(\boldsymbol{x}) \geq 0, \boldsymbol{h}(\boldsymbol{x}) = 0$$

where $\boldsymbol{X}$ represents the n_x-dimensional continuous or discrete feasible solution space defined by $\boldsymbol{g}$ and $\boldsymbol{h}$, $\boldsymbol{x}$ is the decision vector and $\boldsymbol{f} \in \boldsymbol{F}^M$.

In contrast to SO optimization where an unique solution exist, a set of solutions representing the tradeoffs between the different objectives is sought in MO optimization. This set of solutions is also known as the Pareto optimal set. The concepts of Pareto dominance and Pareto optimality are fundamental in MO optimization, with Pareto dominance forming the basis of solution quality. In the total absence of information regarding the preferences of objectives, it is commonly regarded that a ranking scheme based upon the Pareto optimality is the most appropriate approach to represent the fitness of each individual in an evolutionary algorithm for MO optimization [10, 21].

Definition 1. *Pareto Dominance:* $\boldsymbol{f_1} \in \boldsymbol{F}^M$ *dominates* $\boldsymbol{f_2} \in \boldsymbol{F}^M$, *denoted by* $\boldsymbol{f_1} \prec \boldsymbol{f_1}$ $iff x_{1,i} \leq x_{2,i}\ \forall i \in \{1, 2, ..., M\}$ *and* $x_{1,j} < x_{2,j}\ \exists j \in \{1, 2, ..., M\}$

Definition 2. *Pareto Optimal Front: The Pareto optimal front, denoted as* $\boldsymbol{PF}^*$, *is the set of nondominated solutions with respect to the objective space such that* $\boldsymbol{PF}^* = \{\boldsymbol{f}_i^* | \boldsymbol{f}_i^* \prec \boldsymbol{f}_j,\ \forall\ \boldsymbol{f}_j \in \boldsymbol{F}^M\}$

Definition 3. *Pareto Optimal Set: The Pareto optimal set, denoted as* $\boldsymbol{PS}^*$, *is the set of nondominated solutions with respect to the decision space such that* $\boldsymbol{PS}^* = \{\boldsymbol{x}_i^* | \boldsymbol{F}(\boldsymbol{x}_i^*) \prec \boldsymbol{F}(\boldsymbol{x}_j),\ \forall\ \boldsymbol{F}(\boldsymbol{x}_j) \in \boldsymbol{F}^M\}$

A solution is said to be a nondominated solution if no other solution dominates it. Each objective component of any non-dominated solution in the Pareto optimal set can only be improved by degrading at least one of its other objective components [21].

20.2.2 Robust Multi-objective Optimization

In order to avoid any confusion in the subsequent discussions, it will be instructive to make a distinction between the solution sets based on rMO and MO optimization. The set of tradeoffs and solutions discovered under a deterministic setting will be referred as the efficient or deterministic Pareto front ($\boldsymbol{PF_{det}}$) and Pareto solution set ($\boldsymbol{PS_{det}}$) respectively. Likewise, $\boldsymbol{PF^*}$ and $\boldsymbol{PS^*}$ will be denoted as $\boldsymbol{PF^*_{det}}$ and $\boldsymbol{PS^*_{det}}$. On the other hand, the robust Pareto front ($\boldsymbol{PF_{rob}}$) and solution set ($\boldsymbol{PS_{rob}}$) will refer to the set of tradeoffs and solutions discovered with explicit robust considerations respectively. Given a particular set of desirable robust properties, the optimal robust Pareto front and solution set will be denoted as $\boldsymbol{PF^*_{rob}}$ and $\boldsymbol{PS^*_{rob}}$ respectively.

The MO problem formulated in the previous section reflects the conventional methodology adopted in the vast majority of the optimization literature which assumes that the MO problem is deterministic and the core optimization concern is the maximization of solution set quality. However, Pareto optimality does not necessarily mean that any of the solutions along the tradeoff is desirable or even implementable in practice. This is primarily because such a deterministic approach neglects the fact that real-world problems are characterized by uncertainty. Furthermore, with the increasing complexity of modern applications, it is not possible to model any problem with absolute certainty even with the state-of-the-arts modeling techniques.

In order to reduce the consequences of uncertainty on optimality and practicality of the solution set, factors such as decision variable variation, environmental variation and modeling uncertainty have to be considered explicitly. Therefore, the minimization MO problem is redefined as the following.

$$\min_{\boldsymbol{x}\in\boldsymbol{X}^{n_x}} \boldsymbol{f}(\boldsymbol{x},\boldsymbol{\sigma_x},\boldsymbol{\sigma_e}) = \{f_1(\boldsymbol{x},\boldsymbol{\sigma_x},\boldsymbol{\sigma_e}), f_2(\boldsymbol{x},\boldsymbol{\sigma_x},\boldsymbol{\sigma_e}), ..., f_M(\boldsymbol{x},\boldsymbol{\sigma_x},\boldsymbol{\sigma_e})\} \tag{20.2}$$

$$\text{s.t. } \boldsymbol{g}(\boldsymbol{x},\boldsymbol{\sigma_x},\boldsymbol{\sigma_e}) \geq 0, \boldsymbol{h}(\boldsymbol{x},\boldsymbol{\sigma_x},\boldsymbol{\sigma_e}) = 0$$

where σ_x and σ_e represent the uncertainty associated with $\boldsymbol{x}$ and environmental conditions. Both forms of uncertainty may be treated equivalently and this chapter will be focused on uncertainties in the decision variables. Different noise models such as normal, Cauchy, and uniform distribution have been considered in the literature. Nonetheless, only uniformly distributed noise in the form of $[lb, up]$ will be considered in this chapter.

20.2.3 MOEA framework

In contrast to SO optimization where convergence is the main concern, there are three optimization goals for MOEA: 1) proximity to $\boldsymbol{PF_{det}}$ ($\boldsymbol{PF_{rob}}$), 2) a diverse Pareto front, and 3) uniform distribution of solutions along the discovered Pareto front. The motivation of the second goal is to provide the decision maker with many choices while the third goal is to provide the decision maker with information about the tradeoffs between the different solutions.

In extending the conventional MOEA framework for rMO optimization, it is essential to understand the key components of MOEA and consider the role that each component plays in this context. Many different MOEAs have been proposed over the years with the aim of fulfilling the three optimization goals. Based on the different cost assignments and selection strategies, EAs for MO optimization can be broadly classified into two classes: non-Pareto approaches and Pareto-based approaches. Although Pareto-based MOEA is the dominant approach, recent studies have revealed the deficiencies of such techniques in handling high-dimensional problems. Nevertheless, most of the evolutionary techniques for MO optimization exhibit common characteristics and can be represented in the pseudocode shown in Fig. 20.1.

Based on the state-of-the-arts, MOEAs are characterized by some form of diversity preservation mechanisms and elitism. Diversity preservation operators are conducted to ensure diversity and a uniform distribution of solutions. Instances of diversity mechanisms include niche sharing, crowding, grid mapping, clustering, and etc. In the case of elitism, it involves 1) the storage of the best solutions found along the evolutionary process and/or 2) the reinsertion of elitist solutions into the evolving population to improve convergence. A variety of evolutionary operators such as mutation and crossover have been used in the literature. Note that variation can also be performed by some form of local search operators or heuristics.

```
Population Initialization
Fitness Evaluation
Update Archive
While (Stopping criteria not satisfied)
        Selection based on fitness and diversity
        Elitist Selection
        Evolutionary variation
        Fitness Evaluation
        Update Archive
End While
Output Archive
```

Fig. 20.1. Framework of MOEA

20.3 A Robust Multi-Objective Evolutionary Algorithm

From the discussion of the MOEA framework in the previous section, it is apparent that it is necessary to deal with the issues of fitness assignment, selection and elitism for robust optimization. This section describes the the proposed robust multi-objective evolutionary algorithm (RMOEA) which addresses these issues and is capable of evolving the tradeoffs between Pareto optimality and robustness.

20.3.1 Algorithmic Flow of RMOEA

In order to explore the tradeoffs between robustness and Pareto optimality for constrained MO problems, the RMOEA considers robustness as an independent optimization criteria and implements the features of micro-genetic algorithm (μGA), Tabu restriction, and archival re-evaluation. The flowchart of RMOEA is shown in Fig. 20.2. The evolutionary process starts with the initialization of the initial population of candidate solutions. The initialization process can either be random or created by means of Design of Experiment (DOE) techniques. After which, all individuals are evaluated based on the respective objective functions, the robust measure as well as the constraint violation. In particular, this chapter utilizes a measure that calculated based on the worst case situation and will be described in the next section.

The evaluation process involves the selection of neighboring points for the assessment of each candidate solutions. In order to minimize the effects of stochasticity associated with the random and DOE creation of neighboring points, RMOEA applies a μGA to search for the worst solution within a specified bound defined by the noise model considered. In addition, Tabu restriction is applied to guide the search and reduce computational load. After the evaluation process, archiving is performed to store the nondominated solutions found along the evolutionary process. Since robust optimization is inherently uncertain, archived solutions are re-evaluated at periodic intervals. The archiving process will be described in greater detail in a later section.

After which, the selection of individuals into the mating pool is conducted. The selection process is elitist in nature and is conducted in a two stage process: 1) the archive and evolving population are combined and 2) binary tournament selection of this combined population is conducted to fill up the mating pool. In order to promote the exploration and exploitation of robustness and optimality, the selection criterion adopted in each tournament is randomly based on either a constrained Pareto ranking scheme or the robust measure. In the event of a tie, i.e. same rank or same robust measure, the individual with the lower niche count [12] will be selected. Note that the mechanism of niche sharing is used in the tournament selection and diversity maintenance in the archive. After the selection process, the individuals will undergo the process of uniform crossover and adaptive mutation operator (AMO) [23].

20.3.2 MO Robust Measure

The RMOEA implements a MO approach, which allows the algorithm to explore the tradeoffs between robustness and Pareto-optimality. The issues considered in the design of the MO approach includes 1) the selection of an appropriate robust measure and 2) an appropriate means of incorporating the selected robust measure.

Many different robust measures such as the expected fitness, the fitness variance and the worst case fitness have been applied in the literature and each of them reflect a different aspect of robustness. In this chapter, the worst case scenario is considered and the robust measure for the $i-th$ objective can be written as,

$$f_i'(\boldsymbol{x}) = \frac{\max_{\boldsymbol{x}\in\boldsymbol{X}^{n_x}}\left(f_i(\boldsymbol{x}')\right) - f_i(\boldsymbol{x})}{f_i(\boldsymbol{x})} \tag{20.3}$$

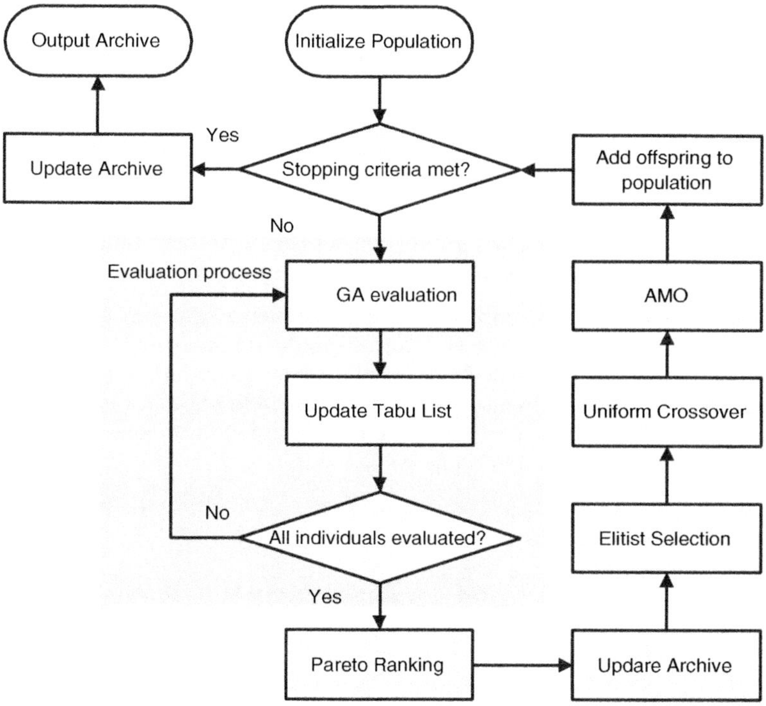

Fig. 20.2. Algorithmic flow of RMOEA

where $\boldsymbol{x}' \in [\boldsymbol{x}' - lb, \boldsymbol{x}' + up]$. From (20.3), it can be noted that the measure reflects the degree of variation resulting from the worst objective value.

With respect to the latter issue of incorporating the robust measure, the presence of multiple competing objectives in MO optimization leads to the issue of how the robust measure can be implemented without a substantial increase in problem dimensionality. In particular, the objective space will be doubled if the worst case situation for objective variation is considered independently, i.e. a 20-objective problem will become a 40-objective problem! Therefore, it will be more appropriate to consider the minimization of the worst possible case al all objectives and the MO problem to be solved can be now written as

$$\min_{\boldsymbol{x} \in \boldsymbol{X}^{n_x}} \boldsymbol{f}(\boldsymbol{x}, \boldsymbol{\sigma_x}, \boldsymbol{\sigma_e}) = \{f_1(\boldsymbol{x}), f_2(\boldsymbol{x}), ..., f_M(\boldsymbol{x}), f_{M+1}(\boldsymbol{x})\} \tag{20.4}$$
$$\text{where } f_{M+1}(\boldsymbol{x}) = \max_{i=1,2,..,M} \left(f_i'(\boldsymbol{x})\right)$$

20.3.3 Constrained Pareto Ranking

In order to evolve solutions that remains valid under perturbations, the feasibility of each individual for each constraint will be represented by the worst case violation in its neighborhood. The constrained Pareto ranking presented in this chapter is based on a constrain-dominance scheme that is similar to the scheme proposed in [3]. In particular, an individual $\boldsymbol{f_1}$ constrain-dominates the individual $\boldsymbol{f_2}$ if,

1. $\boldsymbol{f_1}$ is feasible and $\boldsymbol{f_2}$ is infeasible.
2. Both solutions are infeasible and $\boldsymbol{f_1}$ dominates $\boldsymbol{f_2}$ in terms of the different constraint violations.
3. Both solutions are infeasible, incomparable in terms of the different constraint violations and $\boldsymbol{f_1}$ dominates $\boldsymbol{f_2}$ in terms of the objective values.
4. Both solutions are feasible and $\boldsymbol{f_1}$ dominates $\boldsymbol{f_2}$ in terms of the objective values.

Actual ranking is based on the scheme presented by Fonseca and Fleming [10] which assigns the same smallest cost for all non-dominated individuals, while the dominated individuals are ranked according to how many individuals in the population dominating them. Therefore, the rank of an individual in the population will be given by $1 + D_C$ where D_C is the number of individuals constrain-dominating the particular individual in the objective domain.

20.3.4 Tabu Restriction

Tabu search is a single-point local search procedure which exploits memory of past movements in the form of tabu restriction to explore new regions of the search space. While Tabu search may not be applied effectively for the simultaneous discovery of Pareto-optimal solutions, the concept of Tabu restriction can be easily extended to guide the evolutionary optimization process to reduce computational effort. This feature is particularly useful in the context of robust optimization where the assessment of any individual requires the evaluation of neighboring points as well.

In this chapter, Tabu restriction is used in conjunction with the evaluation process to reduce the number of evaluations required for solutions that are similar to those that have been assessed to be infeasible. In addition, it avoids the classification of a previously known infeasible solution as feasible solution due to the uncertainties involved in the evaluation process. During the evaluation process of each individual, each of the neighboring point will be examined against the Tabu list. Similar points will not be evaluated and will inherit the attributes of the matching Tabu solution in the Tabu list. At the same time, the Tabu list will be updated with new solutions that violate the constraints. Similar to the archive, the most crowded member will be removed when the maximum size is reached.

20.3.5 μGA Evaluation of Worst Case Performance

The evaluation of each individual for robust optimization problems requires the sampling of multiple neighboring points for the computation of the various robust measures. The convention approach involves the creation of such points, either randomly or by means of DOE methods. However, the stochastic nature of such approaches results in an inefficient evaluation process and may provide a bad estimation of the robust measure. Therefore, this chapter incorporates a μGA to perform a directed search for each individual's worst case performance and constrain violation within the neighborhood of the particular individual.

The pseudocode of the μGA is shown in Fig. 20.3. The μGA begins with the creation of POP_SIZE_{GA} neighboring points about the individual under evaluation. The Tabu restriction assisted assessment presented in the previous section is implemented for the evaluation of these individuals. Since the μGA sought to discover the worst case performance, the μGA is actually optimizing a MO problem that maximizes 1) the worst case objective in (20.3) and 2) the worst constrain violation. At every generation, the worst performance found will be updated and the best individual will be stored. Elitism is applied by reinserting the best individual into the mating pool. Binary tournament selection is then conducted on the evolving population to fill up the mating pool. After which, the mating pool will undergo uniform crossover and AMO.

INITIALIZE POP_{GA} : Create POP_SIZE_{GA} neighbors

EVALUATE POP_{GA}

STORE best individual

STORE worst case performance

REPEAT UNTIL gen_req reached **DO**

    SELECTION: Insert best individual into mating pool. Binary tournament selection of POP_SIZE_{GA} - 1 individuals from POP_{GA}

    CROSSOVER: Perform uniform crossover

    MUTATION: Perform AMO

    EVALUATE: Assess each individual assisted by Tabu restriction

    STORE best individual

    STORE worst case performance

END

RETURN worst case performance

Fig. 20.3. Pseudocode of μGA performing local search for worst case performance

20.3.6 Archival Re-evaluation

This section describes the technique of archival re-evaluation to reduce the degree of uncertainty present in the archive. The idea of re-evaluating archived solution is to avoid the long term presence of solutions that have been wrongly perceived to be robust. An implication of such solutions on elitist MOEAs is that it will mislead the evolutionary optimization process as well as keeping the true robust solutions out of the archive.

A simple re-evaluation scheme is implemented in this section whereby all individuals that have been residing in the archive beyond a certain threshold, T_H will be re-evaluated. In this paper, T_H is set as two generations. Thereafter, all dominated solutions will be removed from the archive. During normal optimization process, the archive is updated at each cycle, i.e., a candidate solution will be added to the archive if it is not dominated by any members in the archive. Likewise, any archive members dominated by this solution will be removed from the archive. In order to preserve a diverse solution set, a recurrent truncation process based on the niche count is used to remove the most crowded archive member when the predetermined archive size is reached.

20.4 Empirical Studies

This section starts with Section 20.4.1 that describes the three test problems, including two new benchmarks and one practical engineering problem, used for the empirical studies. Section 20.4.2 presents the performance of RMOEA in finding the tradeoffs between Pareto optimality and robustness, the effects of the various features as well as the algorithm's ability to handle the constrained engineering problem.

20.4.1 Benchmark Problem

In order to investigate the tradeoffs between Pareto optimality and robustness, this section presents two new test problems with the following functional form.

$$\min f_1(\boldsymbol{x}_I) = x_1 + 15 \tag{20.5}$$

$$\min f_2(\boldsymbol{x}_I, \boldsymbol{x}_{II}) = g(\boldsymbol{x}_I) + r(\boldsymbol{x}_{II}) + 2 \tag{20.6}$$

where $\boldsymbol{x}_I$ and $\boldsymbol{x}_{II}$ represents the subset of decision variables associated with solution optimality and solution robustness respectively. The functions $g()$ and $r()$ controls the difficulty and landscape sensitivities of the problem respectively. To minimize any complications arising from problem characteristics, a simple $g()$ is used

$$g(\boldsymbol{x}_I) = 1 - \frac{1}{|\boldsymbol{x}_I| - 1} \cdot \sqrt{\frac{f_1 - 15}{\sum_{i=2}^{|\boldsymbol{x}_I|} x_i + 1}} \tag{20.7}$$

Test problem 1

Test problem 1 is an instantiation of a Type III problem described by Deb and Gupta [5]. The $r()$ function is defined as

$$r(\boldsymbol{x}_{II}) = 1 - \max_{i=1,\ldots,D} \left(d_i \cdot \exp\left(- \sum_{j}^{|\boldsymbol{x}_{II}|} \left(\frac{x_j - \nu_{i,j}}{w_i}\right)^2\right)\right) \tag{20.8}$$

where $D = 2$. The various parameter settings are shown in Table 20.1. Therefore (20.8) for test problem 1 defines two minima, $\boldsymbol{PF}^*_{det}$ at $x_i = 0.75\ \forall x_i \in \boldsymbol{x}_{II}$ and the local Pareto front at $x_i = 0.25\ \forall x_i \in \boldsymbol{x}_{II}$ as shown in Fig. 20.4. Being less sensitive to parametric variations, the local Pareto set corresponds to the global effective Pareto set and the different Pareto fronts are illustrated in Fig. 20.5.

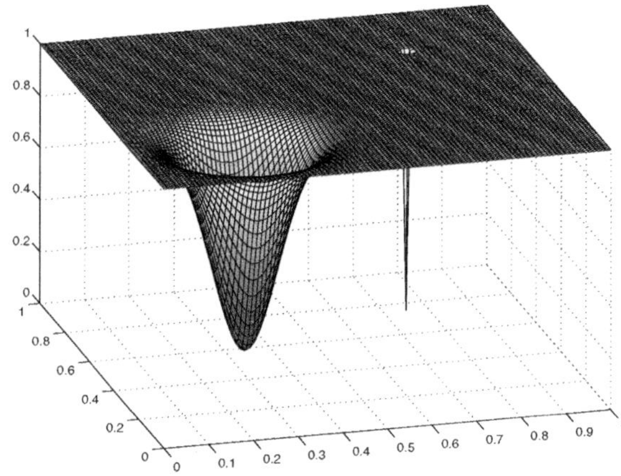

Fig. 20.4. Landscape sensitivities of test problem 1

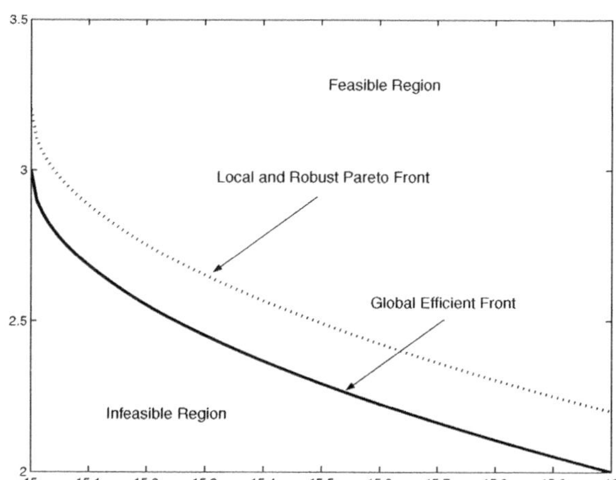

Fig. 20.5. Efficient and Robust Pareto front of test problem 1

Test problem 2

Test problem 2 is similar to test problem 1 except that five troughs are defined in this problem. The parameter settings are given in Table 20.1. The $\boldsymbol{PF}^*_{det}$ is located at $x_1 \in \boldsymbol{x}_{II} = 0.3$ and $x_2 \in \boldsymbol{x}_{II} = 0.5$. While the most robust front is located at $x_1 \in \boldsymbol{x}_{II} = 0.6$ and $x_2 \in \boldsymbol{x}_{II} = 0.8$, the most desirable front will depend on the noise level. The different troughs and Pareto fronts are illustrated in Fig. 20.6 and Fig. 20.7 respectively. The only unfavorable peak, regardless of noise intensity, is located at $x_1 \in \boldsymbol{x}_{II} = 0.8$ and $x_2 \in \boldsymbol{x}_{II} = 0.2$. The main difficulty is the discovery of the different robustness tradeoffs. In this chapter, $|\boldsymbol{x}_{II}| = 2$ for both problems.

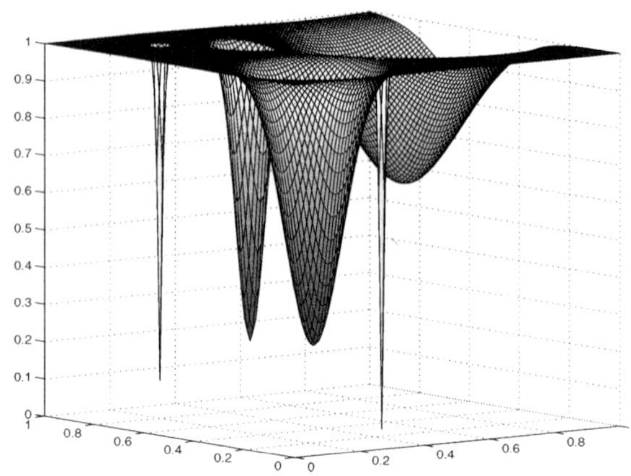

Fig. 20.6. Landscape sensitivities of test problem 2

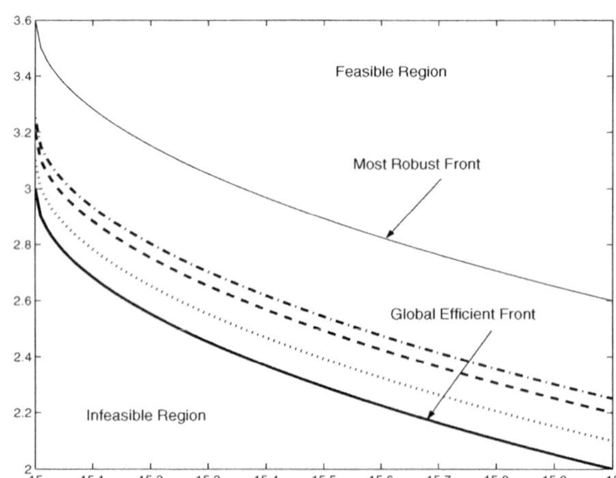

Fig. 20.7. Efficient and Robust Pareto front of test problem 2

Table 20.1. Parameter settings for Test Problems 1 and 2

Problem	No. of Troughs (D)	Trough Depth (d_i)	Trough Location ($\nu_{i,j}$)	Trough Width (w_i)
Test Problem 1	2	$d_1 = 0.8$	$\nu_{1,1} = \nu_{1,2} = 0.25$	$w_1 = 0.1$
		$d_2 = 1.0$	$\nu_{2,1} = \nu_{2,2} = 0.75$	$w_2 = 0.006$
Test Problem 2	5	$d_1 = 0.75$	$\nu_{1,1} = \nu_{1,2} = 0.25$	$w_1 = 0.1$
		$d_2 = 0.8$	$\nu_{2,1} = 0.7, \nu_{2,2} = 0.4$	$w_2 = 0.05$
		$d_3 = 0.4$	$\nu_{3,1} = 0.6, \nu_{3,2} = 0.8$	$w_3 = 0.2$
		$d_4 = 1.0$	$\nu_{4,1} = 0.3, \nu_{4,2} = 0.5$	$w_4 = 0.01$
		$d_5 = 0.9$	$\nu_{5,1} = 0.8, \nu_{5,2} = 0.2$	$w_5 = 0.01$

I-Beam Design

The design of the I-beam involves four decision variables, two objectives as well as geometric and strength constraints. The two minimization objectives are given by,

1. f_1, the cross section area of the beam,

$$f_1(\boldsymbol{x}) = 2x_1x_4 + x_3 \cdot (x_1 - 2x_4) \tag{20.9}$$

2. f_2, the static deflection of the beam,

$$f_2(\boldsymbol{x}) = \frac{60,000}{x_3 \cdot (x_1 - 2x_4)^3 + 2x_2x4 \cdot \left(4x_4^2 + 3x_1 \cdot (x_1 - 2x_4)\right)} \tag{20.10}$$

where x_1, x_2, x_3 and x_4 are the decision variables. The geometric constraints are given as,

$$10 \le x_1 \le 80 \tag{20.11}$$
$$10 \le x_2 \le 50 \tag{20.12}$$
$$0.9 \le x_3 \le 5 \tag{20.13}$$
$$0.9 \le x_1 \le 5 \tag{20.14}$$

while the strength constraint is,

$$16 - \frac{180,000x_1}{x_3 \cdot (x_1 - 2x_4)^3 + 2x_2x4 \cdot \left(4x_4^2 + 3x_1 \cdot (x_1 - 2x_4)\right)} - \frac{15,000x_2}{x_3^3 \cdot (x_1 - 2x_4) + 2x_2^3x4} \ge 0 \tag{20.15}$$

For more information about this problem, interested readers are referred to [7].

20.4.2 Simulation Results

The simulations are implemented in C++ on an Intel Pentium 4 2.8 GHz computer and thirty independent runs are performed for each of the test functions in order to obtain the statistical information, such as consistency and robustness. For comparison, an instance of RMOEA without the incorporation of μGA that optimizes the effective objective values is included in the study. This particular algorithm will be denoted as mRMOEA. Both algorithms adopt a binary chromosome representation and the various parameter settings are tabulated in Table 20.2. The noise model is selected to be [-0.025, 0.025].

Table 20.2. Parameter settings for RMOEA and mRMOEA

Parameter	Settings
Population Size	Evolving population and archive size: 100, μGA size: 4
Tabu list	100
Chromosome length	15 bits for each variable
Selection	Binary tournament selection
Crossover rate	0.8
AMO	$a = 0.8,\ b = 0.01,\ \alpha = 0.5 \cdot gen\_req$
Niche radius	Dynamic
Generations	RMOEA gen_req=500, μGA gen_req=5

Test Problem 1 and 2

The evolved tradeoffs and the associated decision variables of an arbitrary run for mRMOEA and RMOEA on test problem 1 are plotted in Fig. 20.8 and Fig. 20.9. Likewise, the results for test problem 2 are plotted in Fig. 20.10 and Fig. 20.11.

As expected, it can be observed from Fig. 20.8 and Fig. 20.10 that the final tradeoff evolved by mRMOEA is located at the most robust deep for both test problems 1 and 2. Nonetheless, it should be noted that the "most robust"front is actually dependent on the intensity of noise implemented. For instance, the most robust front for test problem 2 will lie at $x_1 \in \boldsymbol{x}_{II} = 0.7$ and $x_2 \in \boldsymbol{x}_{II} = 0.4$ if the noise model is selected to be [-0.01, 0.01]. Therefore, as mentioned before, it is necessary to discover the different tradeoffs by means of RMOEA.

For test problem 1, we expect RMOEA to discover two Pareto fronts, corresponding to the global efficient front and the robust front when robustness is considered. Interestingly, from Fig. 20.9(b), a third Pareto front that is considerably further away is discovered and maintained by the RMOEA. This Pareto front actually corresponds to the set of solutions that are located at the flat regions of the fitness landscape as observed in Fig. 20.9(a).

The number of tradeoffs for test problem 2 is considerably more as compared to the previous problem. The RMOEA is capable of finding the four different Pareto front located at the various troughs. Notice that the front associated at $x_1 \in \boldsymbol{x}_{II} = 0.8$ and $x_2 \in \boldsymbol{x}_{II} = 0.2$ is not favorable in terms of robustness or optimality and hence not considered by RMOEA. Similar to the situation for test problem 1, RMOEA discovered a fifth Pareto front that correspond to the solutions located at the flat regions of the fitness landscape.

Effects of Re-evaluation, Elitist Selection and μGA evaluation

It can be observed from the previous section that RMOEA is capable of evolving the different tradeoffs for the two benchmark problems. In this section, the effects of archive re-evaluation, elitist selection and μGA evaluation are examined. Test problem 2 is used in this section since it pose a greater challenge to the algorithm. The indices of RMOEA without the various features used in this section is shown in Table 20.3. In the case where the feature of μGA is removed, a sample size of $POP\_SIZE_{GA} \cdot gen\_req$ neighbors are generated by means of Latin hypercube sampling while selection based on pure Pareto-ranking is applied in the fifth case.

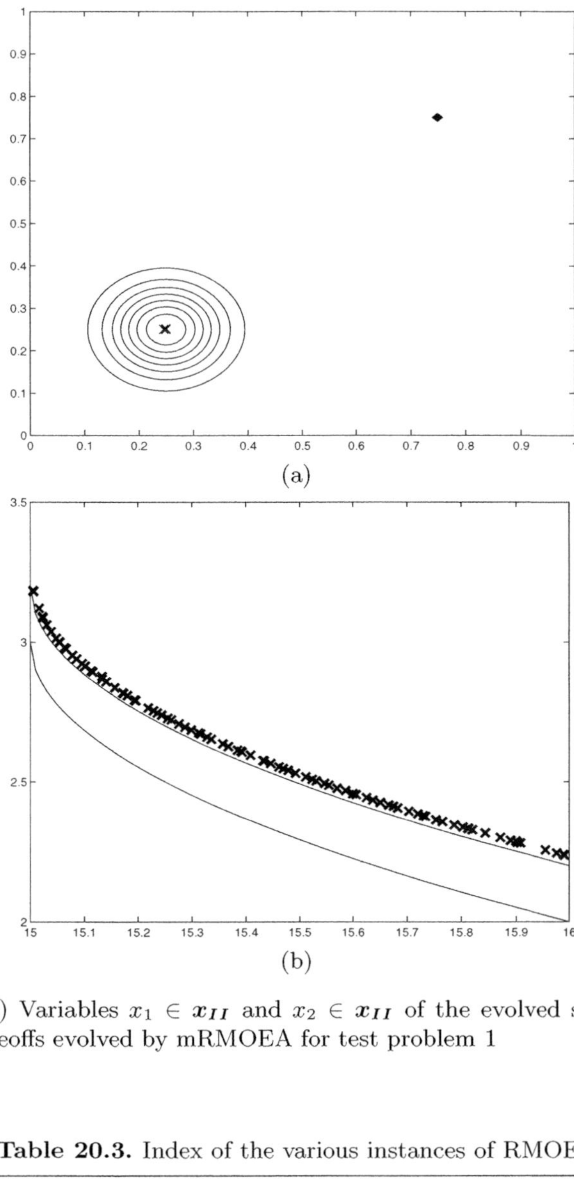

Fig. 20.8. (a) Variables $x_1 \in \boldsymbol{x}_{II}$ and $x_2 \in \boldsymbol{x}_{II}$ of the evolved solution set and (b) Final tradeoffs evolved by mRMOEA for test problem 1

Table 20.3. Index of the various instances of RMOEA

Index	1	2	3	4	5
Algorithm	RMOEA	w/o AMO	w/o μGA	w/o re-eval.	w/o elit. sel

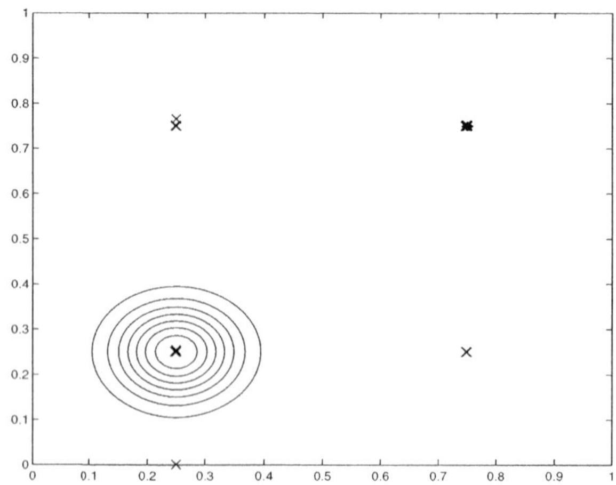

(a)

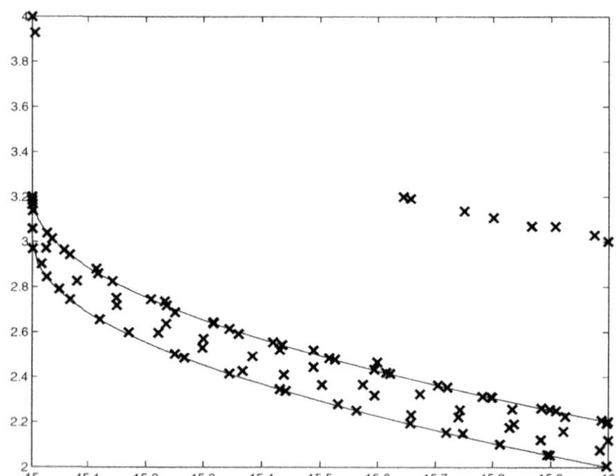

(b)

Fig. 20.9. (a) Variables $x_1 \in \boldsymbol{x}_{II}$ and $x_2 \in \boldsymbol{x}_{II}$ of the evolved solution set and (b) Final tradeoffs evolved by RMOEA for test problem 1

The performance of the various settings, in terms of the number of Pareto fronts failed to be evolved by the various settings and the distance to the ideal solution set in the decision space, are shown in Fig. 20.12(a) and Fig. 20.12(b) respectively. From the results, it is clear about the role that AMO plays in allowing RMOEA locate the various Fronts. Furthermore, the absence of archival re-evaluation or μGA also prevent the algorithm from evolving the various tradeoffs. Note that the one peak that RMOEA failed to detect is mostly associated with the unfavorable front mentioned before. The importance of re-evaluation is also clear from Fig. 20.12(b) where uncertainty prevents the algorithm from finding a near optimal front.

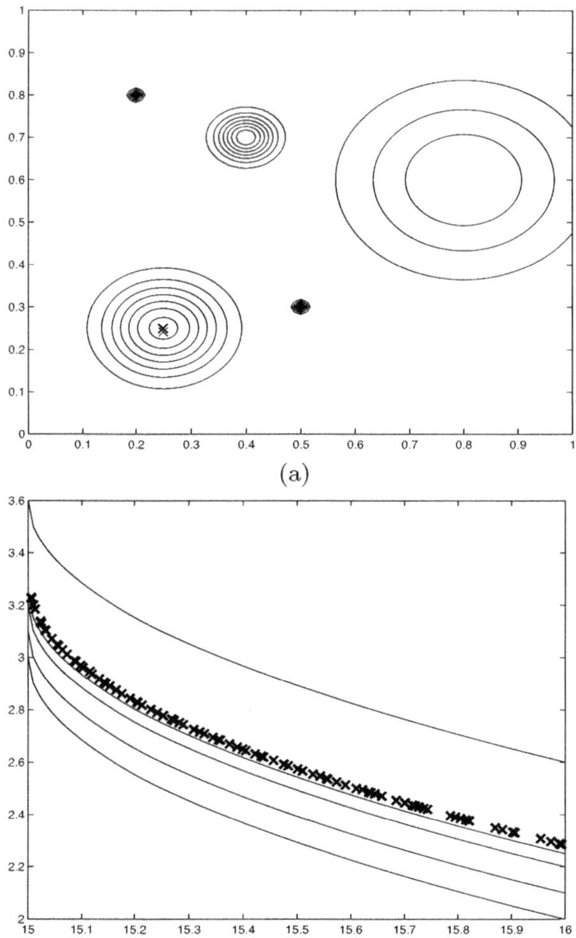

(b)

Fig. 20.10. (a) Variables $x_1 \in \boldsymbol{x}_{II}$ and $x_2 \in \boldsymbol{x}_{II}$ of the evolved solution set and (b) Final tradeoffs evolved by mRMOEA for test problem 2

I-Beam Design

This section examines the performance of RMOEA on the practical problem of I-Beam design. One of the key issues is to maintain the feasibility of the solution under uncertainties. For comparison, an instance of RMOEA that optimizes the efficient objective values is applied and denoted as MOEA. The evolved tradeoffs of MOEA and RMOEA for the I-Beam problem is plotted in Fig. 20.13 and Fig. 20.14 respectively. The crosses represent the feasible solutions while the open circles represent the infeasible solutions.

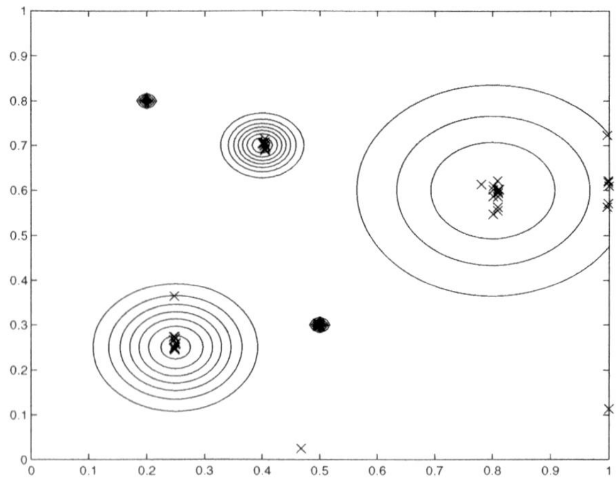

(a)

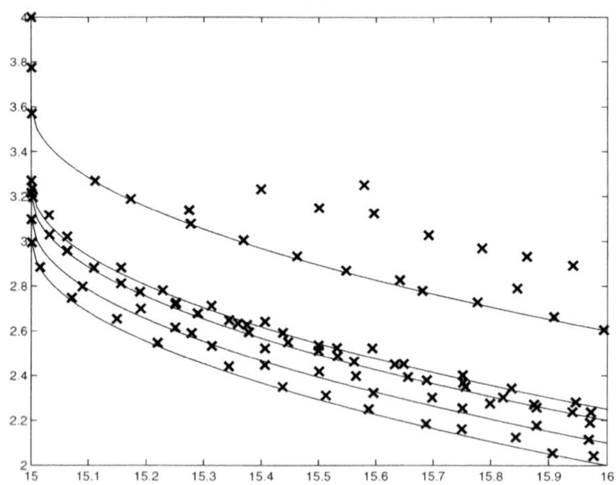

(b)

Fig. 20.11. (a) Variables $x_1 \in \boldsymbol{x}_{II}$ and $x_2 \in \boldsymbol{x}_{II}$ of the evolved solution set and (b) Final tradeoffs evolved by RMOEA for test problem 2

From the two figures, it can be noted that the I-Beam problem represents an instance of a Type 1 problem [5], with $\boldsymbol{PF}_{rob}$ coinciding with $\boldsymbol{PF}_{det}$. Nonetheless, part of the $\boldsymbol{PF}_{det}$ is now infeasible. Apart from demonstrating that RMOEA is capable of discovering solutions that remain feasible under the noise level applied, the importance of considering robustness explicitly during the optimization process is clear by comparing the two figures.

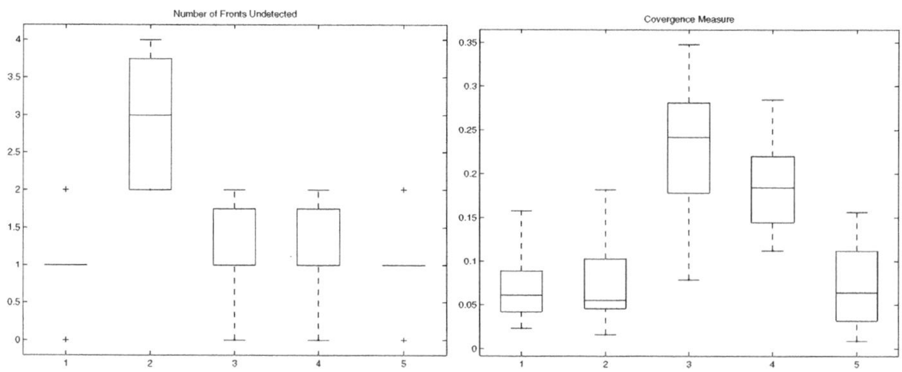

Fig. 20.12. (a) Number of fronts undiscovered and (b) Euclidean distance from the ideal solution for test problem 2

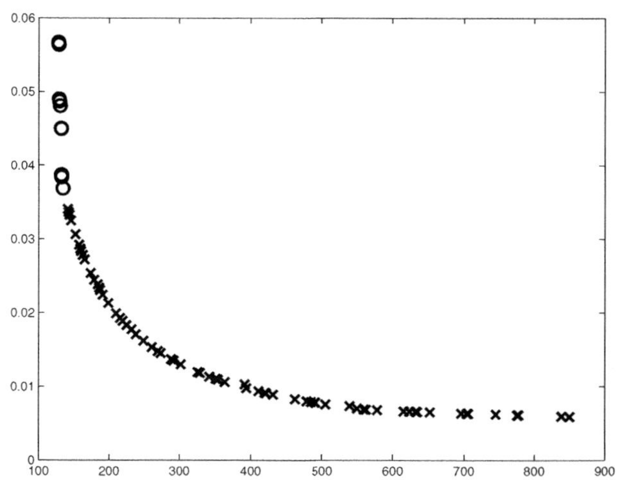

Fig. 20.13. Evolved Tradeoff of MOEA for I-BEAM problem

20.5 Conclusions

This chapter presents a robust multi-objective evolutionary algorithm for constrained multi-objective optimization. Since the effective Pareto front is sensitive to the noise intensity applied, the proposed RMOEA considers explicitly the tradeoffs between Pareto optimality and worst case performance. In particular, RMOEA incorporates the features of μGA, Tabu restriction and periodic re-evaluation. It has been demonstrated that RMOEA, augmented with the additional objective of the proposed worst case measure, is capable of evolving the tradeoffs between optimality and robustness. In addition, the contributions of the various features to the

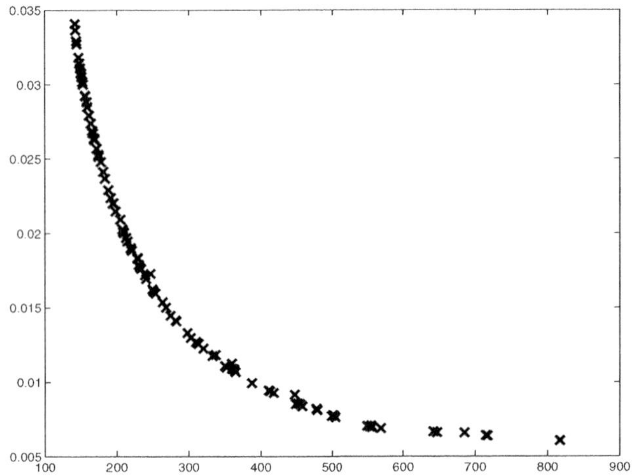

Fig. 20.14. Evolved Tradeoff of RMOEA for I-BEAM problem

robust optimization process are validated upon a benchmark problem. Furthermore, RMOEA's ability to evolve feasible solutions under uncertainty is validated on the practical problem of I-Beam design.

Acknowledgments

The authors would like to thank Yew-Soon Ong for discussions on robustness in evolutionary computation.

References

1. P. Bosman and D. Thierens, " The balance between proximity and diversity in multiobjective evolutionary algorithms," IEEE Transactions on Evolutionary Computation, vol. 7, no. 2, pp. 174-188, 2003.
2. J. Branke, "Creating robust solutions by means of evolutionary algorithms," in Proceedings of the 5th International Conference on Parallel Problem Solving from Nature, pp. 119-128, 1998.
3. K. Deb, Multi-objective Optimization Using Evolutionary Algorithms, John Wiley & Sons, New York, 2001.
4. K. Deb and D. E. Goldberg, "An investigation on niche and species formation in genetic function optimization," in Proceedings of Third International Conference on Genetic Algorithms, pp. 42-50, 1989.
5. K. Deb and H. Gupta, "Introducing robustness in multiobjective optimization," Kanpur Genetic Algorithms Lab. (KanGAL), Indian Institue of Technology, Kanpur, India, Technical Report 2004016, 2004.

6. D. Buche, P. Stoll, R. Dornberger and P. Koumoutsakos, "Multiobjective Evolutionary Algorithm for the Optimization of Noisy Combustion Processes," IEEE Transactions on Systems, Man, and CyberneticsPart C: Applications and Reviews, vol. 32, no. 4, pp. 460-473, 2002.
7. C. A. Coello Coello, "An empirical study of evolutionary techniques for multi-objective optimization in engineering design," Ph.D. dissertation, Department of Computer Science, Tulane University, New Orleans, LA, 1996.
8. C. A. Coello Coello and A. H. Aguirre, "Design of combinational logic circuits through an evolutionary multiobjective optimization approach," Artificial Intelligence for Engineering, Design, Analysis and Manufacture, Cambridge University Press, vol. 16, no. 1, pp. 39-53, 2002.
9. M. Farina and P. Amato, "A fuzzy definition of "optimality" for many-criteria optimization problems," IEEE Transactions on Systems, Man, and Cybernetics-Part A: Systems and Humans, vol. 34, no. 3, pp. 315-326, 2004.
10. C. M. Fonseca and P. J. Fleming, "Genetic algorithm for multiobjective optimization, formulation, discussion and generalization," in Proceedings of the Fifth International Conference on Genetic Algorithms, pp. 416-423, 1993
11. C. K. Goh and K. C. Tan, "An investigation on noisy environments in evolutionary multiobjective optimization," IEEE Transactions on Evolutionary Computation, in press.
12. D. E. Goldberg and J. Richardson, "Genetic algorithms with sharing for multimodal function optimization," in Proceedings of the Second International Conference on Genetic Algorithms, pp. 41-49, 1987.
13. H. Gupta and K. Deb, "Handling constraints in robust multi-objective optimization" in Proceedings of the 2005 IEEE Congress on Evolutionary Computation, pp. 25-32, 2005.
14. Y. Jin and J. Branke, "Evolutionary Optimization in Uncertain Environments A Survey," IEEE Transactions on Evolutionary Computation, vol. 9, no. 3, pp. 303-317, 2005.
15. Y. Jin and B. Sendhoff, "Tradeoff between performance and robustness: An evolutionary multiobjective approach," in Proceedings of the Second Conference on Evolutionary Multi-Criterion Optimization, pp. 237251, 2003.
16. E. F. Khor, K. C. Tan, T. H. Lee and C. K. Goh, "A study on distribution preservation mechanism in evolutionary multi-objective optimization," Artificial Intelligence Review, vol. 23, no. 1, pp. 31-56, 2005.
17. M. Laumanns, E. Zitzler and L. Thiele, "A unified model for multi-objective evolutionary algorithms with elitism," in Proceedings of the 2000 Congress on Evolutionary Computation, vol. 1, pp. 46-53, 2000.
18. H. Lu and G. G. Yen, "Rank-based multiobjective genetic algorithm and benchmark test function study," IEEE Transactions on Evolutionary Computation, vol. 7, no. 4, pp. 325-343, 2003.
19. Y. S. Ong, P. B. Nair and K. Y. Lum, "Min-Max Surrogate Assisted Evolutionary Algorithm for Robust Aerodynamic Design," IEEE Transactions on Evolutionary Computation, vol. 10, no. 4, pp.392-404, 2006.
20. T. Ray, "Constrained robust optimal design using a multiobjective evolutionary algorithm," in Proceedings of the 2002 Congress on Evolutionary Computation, pp. 419424, 2002.
21. N. Srinivas and K. Deb, "Multiobjective optimization using non-dominated sorting in genetic algorithms," Evolutionary Computation, vol. 2, no. 3, pp. 221-248, 1994.

22. K. C. Tan, C. Y. Cheong and C. K. Goh, "Solving multiobjective vehicle routing problem with stochastic demand via evolutionary computation," European Journal of Operational Research, in press.
23. K. C. Tan, C. K. Goh, Y. J. Yang and T. H. Lee, "Evolving better population distribution and exploration in evolutionary multi-objective optimization," European Journal of Operational Research, vol. 171, no. 2, pp. 463-495, 2006.
24. K. C. Tan, T. H. Lee, E. F. Khor and D. C. Ang, "Design and real-time implementation of a multivariable gyro-mirror line-of-sight stabilization platform," Fuzzy Sets and Systems, vol. 128, no. 1, pp. 81-93, 2002.
25. S. Tsutsui and A. Ghosh, "Genetic algorithms with a robust solution searching scheme," IEEE Transactions on Evolutionary Computation vol. 1, no. 3, pp. 201-208, 1997.
26. S. Tsutsui and A. Ghosh, "A comparative study on the effects of adding perturbations to phenotypic parameters in genetic algorithms with a robust solution searching scheme," in Proceedings of the 1999 IEEE International Conference on Systems, Man, and Cybernetics, pp. 585-591, 1999.
27. E. Zitzler, K. Deb, and L. Thiele, "Comparison of multiobjective evolutionary algorithms: empirical results," Evolutionary Computation, vol. 8, no. 2, pp. 173-195, 2000.
28. E. Zitzler, L. Thiele, M. Laumanns, C. M. Fonseca and V. G. Fonseca, "Performance assessment of multiobjective optimizers: An analysis and review," IEEE Transactions on Evolutionary Computation, vol. 7, no. 2, pp. 117-132, 2003.

21

Evolutionary Robust Design of Analog Filters Using Genetic Programming

Jianjun Hu[1], Shaobo Li[2], and Erik Goodman[3]

[1] MCB 403D, University of Southern California, Los Angeles, CA 90089, USA
jianjunh@usc.edu
[2] CAD/CIMS Institute, Guizhou University, Guiyang, Guizhou 550003, China
lishaobo@gzu.edu.cn
[3] 2120 Engineering Building, Michigan State University, East Lansing, MI 48824, USA
goodman@egr.msu.edu

Summary. This chapter proposes a robust design approach that exploits the open-ended topological synthesis capability of genetic programming (GP) to evolve robust lowpass and highpass analog filters. Compared with a traditional robust design approach based on genetic algorithms (GAs), the open-ended topology search based on genetic programming and bond graph modeling (GPBG) is shown to be able to evolve more robust filters with respect to parameter perturbations than what was achieved through parameter tuning alone, for the test problems.

21.1 Introduction

Topologically open-ended computational synthesis by genetic programming (GP) has been used as an attracting approach for engineering design innovation in a variety of domains, including design of analog circuits, digital circuits, chemical molecules, mechatronic systems, etc. [15] [18]. These works employ GP as a topologically open-ended search technique for functional design innovation – achieving given behaviors without pre-specifying the design topology – and has achieved considerable success. However, in practical engineering system design, there is another criterion in addition to functional specifications that should be considered during the design process. Robustness, as the ability of a system to maintain its function even with changes in internal structure (including variations of parameters from nominal values) or external environment [5], [10], is also critical to engineering design decisions. Engineering design systems, in reality, do not normally take into account all the types of uncertainties or variations to which the engineered artifacts are subject, such as manufacturing variation, degradation or non-uniformity of material properties, environmental changes, and changing operating conditions. There are two types of robustness of dynamic systems in robust engineering design. One is the robustness of systems with respect to perturbation of their parameters. This is the most commonly investigated type of robustness in the traditional robust design community and also

J. Hu et al.: *Evolutionary Robust Design of Analog Filters Using Genetic Programming*, Studies in Computational Intelligence (SCI) **51**, 479–496 (2007)
www.springerlink.com

in evolutionary robust design. Another type of system robustness is with respect to topological perturbation – for example, accidental removal or failure of components. Reliable systems, having the least sensitivity of performance to variations in the system components or environmental conditions, are very desirable. However, there are relatively few studies that explore how GP-based open-ended topology search may contribute to design of robust systems such that they can withstand internal or external perturbations. In traditional robust design, optimizing robustness is usually regarded as a step in the detailed design stage, in which the parameters of a system with a given functional structure are tuned to achieve better robustness.

In this chapter, genetic programming is applied to automatically synthesize robust analog filter which shows that topologically open-ended innovation capability of GP can allow us to design more robust dynamic systems with respect to parameter variations or uncertainty of the design variables. A set of experiments is conducted to show: that dynamic systems with high performance evolved by GP without considering a robustness criterion during the evolutionary process may have unacceptably low robustness with respect to parameter perturbation, 2) that the robustness of a system may be constrained by its topological/functional structure, and the amount of robustness improvement available through parameter tuning is limited as well, and 3) that topologically open-ended synthesis by GP may allow evolution of more robust solutions than the traditional robust design approaches with parameter tuning.

To examine the role of topology search in designing robust systems, two analog filter design problems, including low-pass and high-pass filters, are to be synthesized using genetic programming. For each synthesis problem, three types of experiments are conducted: a) evolutionary synthesis using GP without considering a robustness criterion, b) improving robustness of these evolved filters by tuning their parameters using a genetic algorithm (GA), and c) evolving robust filters (topological structure and parameters) using GP with a robustness criterion in the fitness function. These filter design problems are selected as they are perhaps the most popular problems in evolutionary synthesis research by either GA or GP [14] [16].

The rest of the chapter is organized as follows. Section 2 presents a survey of applications of evolutionary algorithms in robust design. Section 3 introduces the GPBG methodology, which exploits Genetic Programming and Bond Graphs for automated synthesis of dynamic systems. This section introduces some new features that improve standard developmental GP for bond graph synthesis. Section 4 discusses two approaches to evolving robust dynamic systems – the parameter search approach (by genetic algorithm) and the simultaneous topology and parameter search approach by genetic programming. Section 5 compares experimental results of these approaches. Finally, the conclusions and future research are highlighted in Section 6.

21.2 Related Work

Robust design as originally proposed by Taguchi [19] has been intensively investigated in the engineering design community since the 1980s and remains an important topic. In traditional robust design, a designer seeks to determine the control parameter settings that produce desirable values of the mean (nominal) performance, while at the same time minimizing the variance of the performance [19]. However, most of these robust design studies assume that there already exists a design solution for a system and the task of robust design is to determine its robust operating parameters with respect to various kinds of variations. The relationship between the topological or functional structure of a system and its robustness is often not treated. Especially, how robustness criteria should be incorporated into conceptual functional design stage is not addressed. This is partially because of the prevailing approach for engineering system designs is a top-down procedure starting from function design to detailed design and the robustness criterion is hard to evaluate without detailed design parameters which are only available after the detailed design stage.

Application of evolutionary algorithms to traditional parametric robust design has been attracting increasing attention in the past decade [20] Wiesmann:1998 [8] [11]. Tsutsui et al. [20] proposed a GA-based Robust Solution Searching Scheme (RS^3) to evolve robust solutions. This approach works by adding perturbation noise to the design variables before fitness evaluation. In Wiesmann et al.'s approach [21], however, each individual is simulated t times to estimate its expected loss function (fitness). Their experiments showed that the evolved designs were substantially more robust to parameter variations than the reference design, but usually at the cost of reduced performance in undisturbed situations. This observation motivated the later work of using an evolutionary multi-objective approach to figure out the trade-off map between robustness and optimal functional performance [8] [17] [11]. Forouraghi [8] introduced an interval computation method to avoid artificial insertion of Gaussian noise to parameter variables in order to build tolerance against internal or external perturbations. Ray [17] expressed the robust design problems as a three-criterion multi-objective problem, simultaneously optimizing an individual's performance without perturbation, the mean performance of its neighbors resulting from perturbations, and the standard deviation of its neighbors' performances. Jin et al. [11] proposed two methods for estimating the robustness measures of an individual – by exploiting its neighbor individuals in the current population or by using all evaluated individuals. Jin's robustness estimation approach can greatly reduce the number of function evaluations, when it is applicable. However, it is difficult to apply this method for evolving robust designs with variable structures as in topologically open-ended automated synthesis using GP because of the difficulty to define a neighborhood for a given individual in the structural space.

We chose analog filter synthesis problems as our benchmarks since they are the most widely used test problems in electric circuit optimization using GA or GP [14] [16] [7]. The pioneering work of Koza in automated analog circuit synthesis, including low-pass, high-pass, and asymmetric band-pass filters, is described in [14] [13]. Lohn and Colombano [16] proposed a linear representation approach to evolve analog circuits in which several low-pass filters were used as test problems. However, they did not specifically work on evolving robust circuits. In our previous work [7], we applied GP to the lowpass analog filter design problem using bond graphs as the modeling and simulation tool.

A lot of work has been done in both evolutionary robust design and analog circuit synthesis. However, there are few studies that specifically address how GP-based topologically open-ended synthesis may provide a new way for open-ended robust design. This may enable us to move robust design forward to the conceptual/functional design stage and thus achieve design for robustness at the very beginning, which will augment the current practice of design for robustness in parametric design.

21.3 Analog Filter Synthesis using Bond Graphs and GP

In this section, we present an improved methodology for open-ended computational synthesis of multi-domain dynamic systems based on bond graphs [12] and genetic programming–the GPBG approach = Genetic Programming+Bond Graphs.

21.3.1 Bond Graphs

The bond graph is a multi-domain modeling tool for analysis and design of dynamic systems, especially hybrid multi-domain systems, including mechanical, electrical, pneumatic, hydraulic, etc., components. Details of notation and methods of system analysis related to the bond graph representation can be found in [12]. Fig. 21.1 illustrates a small bond graph that represents the accompanying electrical system. Fig. 21.2 shows the complex bond graph model of a low-pass filter. A typical simple bond graph model is composed of inductors (I), resistors (R), capacitors (C), transformers (TF), gyrators (GY), 0-Junctions (J0), 1-junctions (J1), sources of effort (SE), and sources of flow (SF). In this chapter, we are only concerned with linear dynamic systems and did not include transformers and gyrators as components.

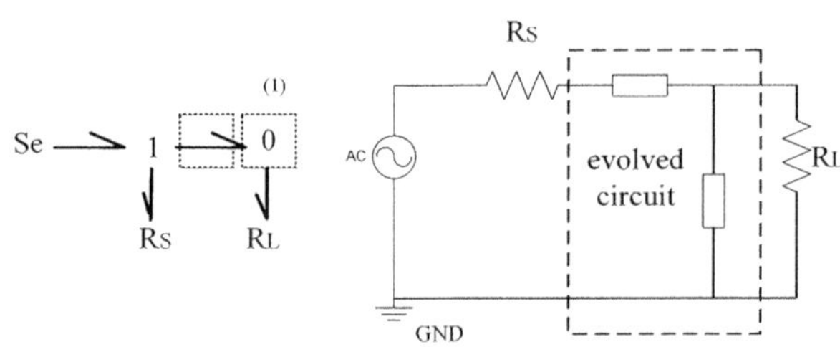

Fig. 21.1. A bond graph and its equivalent circuit. The dotted boxes in the left bond graph indicate modifiable sites at which further topological manipulations can be applied (to be explained in next section).

21.3.2 Evolving Analog Filters using Bond Graphs and GP: the GPBG framework

Automated synthesis of bond graphs involves two basic searches: the search for a good topology and the search for good parameters for each topology, in order to be able to evaluate its performance. Based on the pioneering work of Koza [13] on automated synthesis of electronic circuits, we created a developmental GP system for synthesizing mechatronic systems represented as bond graph [18]. This GPBG framework enables us to do simultaneous topology and parameter search.

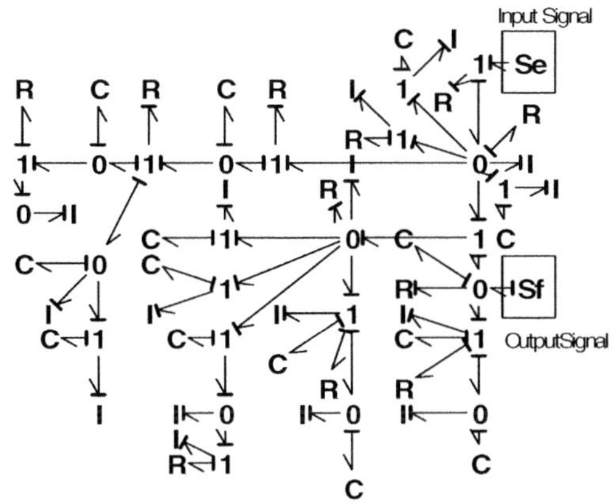

Fig. 21.2. Bond graph structure of low-pass filter evolved in 500,000 function evaluations. Filter has 39 components beyond embryo (Component sizes omitted for simplicity.).

The GPBG framework includes the following major components: 1) an embryo bond graph with modifiable sites at which further topological operations can be applied to grow the embryo into a functional system, 2) a GP function set, composed of a set of topology manipulation and other primitive instructions which will be assembled into a GP tree by the evolutionary process (execution of this GP program leads to topological and parametric manipulation of the developing embryo bond graph), and 3) a fitness function to evaluate the performance of candidate solutions.

In this chapter, we have improved the basic function set in [7] and developed a hybrid function set to reduce redundancy while retaining flexible topological exploration:

F={`Insert_J0E, Insert_J1E, Add_C/I/R, EndNode, EndBond, ERC`}

where the Insert_J0E, Insert_J1E (Fig. 21.3) functions insert a new 0/1-junction into a bond while attaching at least one and at most three elements (from among C/I/R). EndNode and EndBond terminate the development (further topology manipulation) at junction modifiable sites and bond modifiable sites, respectively; ERC represents a real number that can be changed by Gaussian mutation. In addition, the number and type of elements attached to such junctions are controlled by three bits. A flag mutation operator is used to evolve these flag bits, each representing the presence or absence of corresponding C/I/R components. This hybrid approach does not create the many bare (and unnecessary) junctions generated by the basic approach. At the same time, Add_C/I/R still provide the flexibility needed for broad topology search. For any of the three C/I/R components attached to each junction, there is a corresponding parameter to represent the component's value, which is evolved by a Gaussian mutation operator in the modified GP system used here. Fig. 21.4 shows a GP tree that develops an embryo bond graph into a complete bond graph solution. The comparison experiments of [9] showed that this function set was more effective on both an eigenvalue and an analog filter test problem, so the new set was used in this chapter.

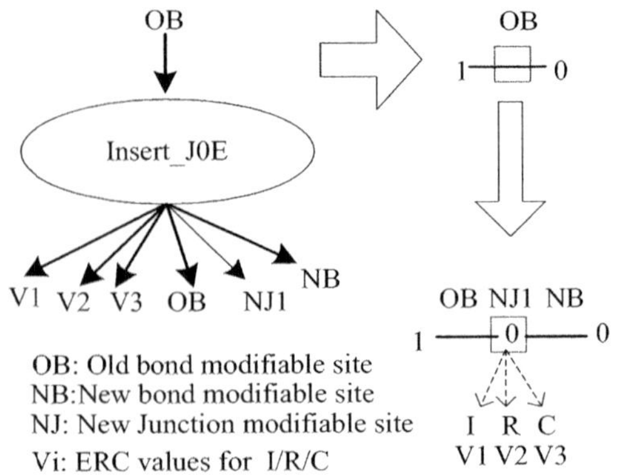

Fig. 21.3. The Insert_J0E function inserts a new junction into a bond along with a certain number of attached components

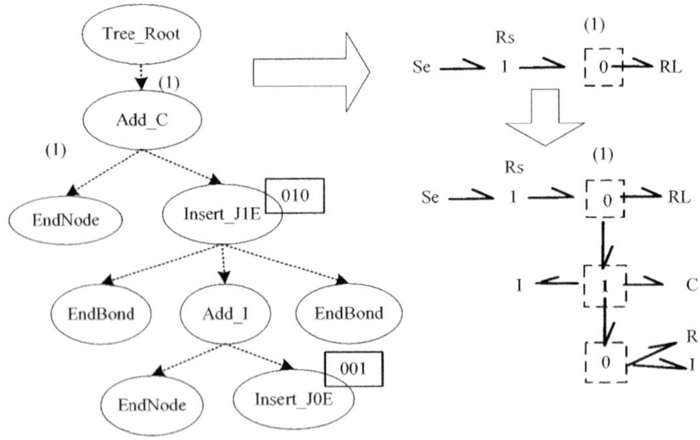

Fig. 21.4. Sample GP tree evolved by applying topology operators to embryo, generating a bond graph after depth-first execution (numeric ERC nodes omitted). Flag bits 010 and 001 show presence or absence of attached C/I/R components.

As a case study, we are interested in evolving two types of analog filters including low-pass and high-pass filters (Fig. 21.5). In these GPBG based filter design problems [7], a bond-graph-represented analog filter composed of capacitors, resistors, and inductors is to be evolved such that the magnitude of its frequency response approximates a specified filter frequency response specifications. An embryo bond graph and its equivalent circuit are illustrated in Fig. 21.1. This embryo bond graph is used in all three filter design problems. Note that the 0-junction is the modifiable site, where further topological developments can proceed as instructed by a GP program tree. The voltage of this 0-junction is the output signal.

Rather than using a sophisticated SPICE simulator as is often done in analog filter synthesis [14] [1], calculation of frequency response from a bond graph was done by automatically formulating the state equations (yielding A, B, C, and D matrices), then using MATLAB 3.0-derived C++ code to simulate the circuit behavior.

Detailed specifications of the filter synthesis problems are:

- The low-pass filter synthesis problem is extracted from [14], in which the frequency response performance of a candidate filter is defined as the weighted sum of deviations from ideal frequency response magnitude over 101 points:

$$F_{koza}(t) = \sum_{i=0}^{100} [W(d(f_i), f_i) * d(f_i)] \tag{21.1}$$

  Where f_i is the sampling frequency. $d(x)$ is the absolute deviation of candidate frequency response from target response at frequency x. W(x,y) is the weight function. The sampling points range from 1Hz to 100K Hz, logarithmically distributed. If the deviation from ideal magnitude is less than 0.03V, the weight is 1. If the deviation is more than 0.03V, the weight is 10. The pass band is [1,1K] Hz; the stop band is [2K,10K] Hz. A "don't" care band between 1K Hz and 2K Hz neglects any deviation from the target response.
- The high-pass problem is similar, except for the complementary definitions of the pass and stop bands. The passband is now defined as [2K,10K]Hz, while the stopband is [1,1K]Hz. A "don't care" band from 1KHz to 2KHz neglects any deviation from the target response.

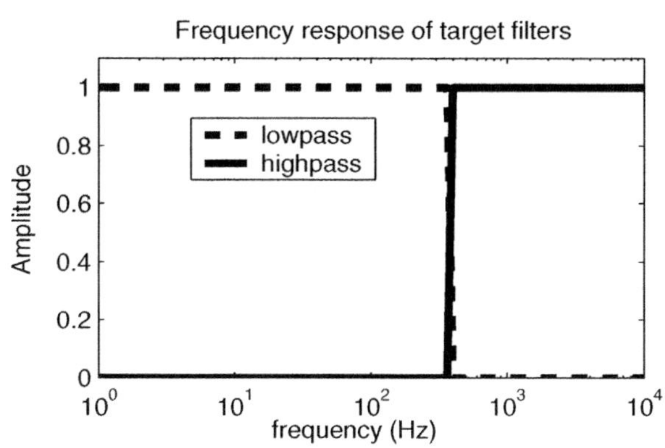

Fig. 21.5. Specification of analog low-pass and high-pass analog filter synthesis problem

To evolve an analog filter without considering robustness, the final fitness of a candidate solution is defined as follows:

First, calculate the raw fitness of a candidate solution defined as the average absolute deviation between the frequency response magnitude of the candidate solution and the target frequency response over all 101 sampling frequencies:

$$f_{raw} = \frac{1}{100} \cdot F_{koza}(t) = \frac{1}{100} \cdot \sum_{i=0}^{100} [W(d(f_i), f_i) * d(f_i)] \tag{21.2}$$

Note that this f_{raw} definition differs from Koza's raw fitness definition in Equ.21.1 by a multiplier equal to the number of sampling frequencies. We use the average deviation rather than sum of deviations to remove the influence of the number of sampling points.

Then calculate the final fitness as:

$$f_{norm} = \frac{NORM}{NORM + f_{raw}} \tag{21.3}$$

where NORM adjusts f_{norm} into the range [0,1]. This transforms the problem of minimizing deviation from target frequency response into a maximization problem appropriate for our GP system. Since tournament selection is used, NORM can be an arbitrary positive number (here set to 10, yielding fitness ranges around [0, 1]).

21.4 Evolving Robust Analog Filters Using Bond Graphs and Evolutionary Algorithms

We used three methods to evolve robust or non-robust lowpass and highpass filters. The first is a standard GP approach without considering robustness requirements. The second is a hybrid GP/GA robust design method (GP-GARMS). This approach first uses a standard GP to evolve a high-performance filter without incorporating any robustness criterion in the fitness function. And then the state-of-art G3PCX-GA is used to improve the robustness of the GP solution using the multi-simulation method to evaluate the robustness performance. The third is a standard GP with multi-simulation (GPRMS), which uses multiple simulations to estimate the robustness fitness of a candidate solution.

21.4.1 Evolving Robust Filters Against Parameter Variation: the Unified Approach

The typical approach for evolving robust designs [3] is to use multiple Monte Carlo samplings with different environmental or system configurations (e.g., perturbation of parameter values of the system) to calculate a worst-case or an average fitness for a given candidate solution as shown in Equ. (21.3). This robust-by-multiple-simulation (RMS) method is used in [21]. Another method is to simply add a perturbation to the design variable before evaluation. This perturbation, however, is not incorporated into the genome, making it different from normal parameter mutation operator or Larmarckian style evolution algorithms. This robust-by-perturbed-evaluation (RPE) method is used in [20] and is suggested to be more efficient by Jin et al. [11]. Both methods are tested in this work. For RPE method, no special fitness definition is needed. For the RMS multi-simulation method, our raw fitness for a design solution with robustness criterion is defined as follows:

$$f_{robustraw} = \sum_{k=1}^{SPI} f_{raw}^{k} \tag{21.4}$$

where SPI is the number of Monte Carlo sampling evaluations for each individual, f_{raw}^{k} is the raw fitness of the kth sampled evaluation with a different Monte Carlo perturbation of the parameters as defined in 21.2. With this raw robustness fitness, we then calculate the final fitness according to Equ. 21.3.

The perturbation of the component values during evolution in the experiments reported below is implemented by adding to each component parameter Gaussian noise $N(\mu, \sigma)$ with mean μ of 0 and standard deviation σ set at 10% of the parameter value. This perturbation model is widely used by previous researchers [20] [21] and may not be appropriate for all manufacturing processes. However, it is good enough for our purpose as an approximation to the real component value degradation model in some situations. If the parameter value is ever 0, σ is set to 1.

In the evolution stage of RMS method, the number of Monte Carlo samplings for fitness evaluation of each individual with respect to parameter perturbation is set as SPI =10. After the robust solutions are evolved, their robustness with respect to parameter perturbation is evaluated against a series of perturbation magnitudes: Gaussian noise $N(\mu, \sigma)$ with mean μ set at 0 and standard deviation σ set at 10% to 50% of parameter values in steps of 10%, each tested with 10000 samplings with different configurations of the component parameter perturbations.

21.4.2 Evolving Robust Analog Filters Using GA: the Traditional Robust Design

Evolutionary algorithms have been increasingly applied to evolve robust designs [2], [21] or for optimization in noisy environments [3], [4]. Most such research follows the practice of traditional robust design: given a system with a specific functional structure, tune its parameters using evolutionary algorithms to improve robustness.

We shall contrast the traditional approach described in this section with the new approach. We shall first evolve a high-performance analog filter with the improved GPBG approach as described in Section 21.3.2. No requirement for robustness is enforced during this evolution. Then we shall apply to the result a state-of-art real parameter genetic algorithm –the G3PCX-GA proposed by Deb [6]–to tune the parameters of this filter to improve its robustness with respect to parameter perturbation while keeping its functional structure unchanged.

In the minimization G3PCX-GA, we used the robust raw fitness defined in Equ. 21.4 as the final fitness of an individual. We believe this fitness measure is better than the average of normalized final fitness of each sampling evaluation for its lower distortion of the optimization objective values.

21.4.3 Evolving Robust Analog Filters Using GP: A New Robust Design Paradigm

This new approach aims at exploiting the topology search capability of GP to evolve more robust designs. The configuration of this approach is the same with standard GP-based synthesis except that the robustness criteria is incorporated in the fitness function. The final fitness of an individual, calculated from the sampling fitnesses, is the same as defined in (21.2), where f_i^k is defined as $Fitness_{norm}$ in (21.3).

21.5 Experiments and Results

For all experiments below, a fixed number of function evaluations is allocated to ensure fairness of comparison. For the lowpass and highpass filter design problem, the computation budget is 1,000,000 function evaluations. Note that for methods that use multiple simulations to estimate the robustness fitness, each simulation is counted as one function evaluation. In addition, for the hybrid GP-GARMS method, we allocate 500,000 for GP evolution and the remaining 500,000 for GA evolution for robustness.

All experiments described below used the same embryo bond graph shown in Fig. 21.1. The component values of source resistor R_s and load resistor R_{load} are both 1 Ω for lowpass and highpass filter synthesis. Our GPBG based system is implemented with C++. The GP code is based on modified Open Beagle. The bond graph simulator was developed in our lab. All experiments were run on a single Linux machine with a 3.0GHz CPU and 1GB memory. On average, for an experiment run with 1,000,000 fitness evaluations, it took about 10-20 hours depending on the complexity of the GP trees. To make our results to be practical, we intentionally used a single set of parameters as much as possible to run all experiments with little tuning effort.

To assess the statistical significance of the performance differences, for each target filter type and each synthesis method, 10 runs were conducted. This size of experiments is determined by the computing resources available. However, since the results are quite stable across runs, it is sufficient for the purposes of this chapter.

21.5.1 Evolving Analog Filters using GP Without Considering Robustness

In this experiment, ten analog lowpass and highpass filters were evolved using standard GP without incorporating a robustness criterion in the fitness function (21.3). The following common running parameters were used throughout all GP experiments in this chapter, as shown in Table 21.1.

Note that the maximum tree size is 8 rather than the commonly used value of 17. This parameter is selected by considering the reduced tree sizes due to our simplified parameter representation method and available computational resources. The maximum tree depth of 8 allows synthesis of analog filters with up to at least 100 components, which is sufficient for our purpose.

We select the evolved filter with the highest performance to test its noise tolerance over the degradation or variation of the component parameters with different

Table 21.1. Experimental parameters for analog filter synthesis without robustness criterion

Total population size: 2000(400*5)	Number of subpopulations: 5
migration interval: 5 generation	Migration size: 30 individuals
Max tree depth: 8	Crossover probability: 0.7
InitTreeDepth: 3-5	Standard mutation probability: 0.1
flag bit mutation rate: 0.1	swapping-tree mutation rate: 0.1
Tournament size: 7	Parametric mutation probability: 0.5
Max evaluations: 1,000,000	Flag mutation probability: 0.3
Pool size of elite individuals: 20	Elite pool update frequency: 5 generations

perturbation magnitudes. As described above, the evaluation of robustness with respect to parameter perturbation is conducted by running 10000 simulations of the configurations of the Gaussian parameter perturbations.

Fig. 21.2 and Fig. 21.6 show the topologies of the evolved lowpass and highpass filters with highest performance out of ten runs. The evolved best lowpass and highpass filters have 39 and 27 components, respectively. The lowpass and highpass filters approximate the ideal frequency response closely, with the sum of deviation over 101 points being only 6.43 and 0.32, respectively.

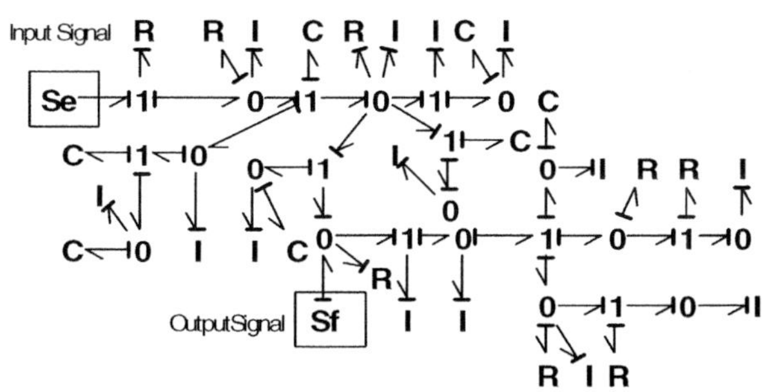

Fig. 21.6. Topology of best highpass analog filters evolved with standard GP with 500,000 function evaluations without considering robustness requirement.This filter has 27 C/I/R components beyond original embryo. The best evolved lowpass filter is shown in Fig. 21.2. This topology is generated by a simplification procedure which removes redundancy in the original evolved bond graphs while maintaining their functional behaviors.

21.5.2 Evolving Robust Analog Filters Using GA: the Classical Robust Design

In this experiment, the state-of-the-art genetic algorithm G3PCX is used to improve the robustness of the best analog filter evolved in the previous section through parameter tuning while keeping functional structure unchanged. As we can see from Fig. 21.2 and Fig. 21.6, these two filter models are very complex, with 39 and 27 parameters to search. As the objective function of this optimization is highly multi-modal, this is a hard optimization even for G3PCX-GA, as the experiment demonstrates. The running parameters for this experiment are summarized in Table 21.2.

Table 21.2. Experimental parameters for robust design by G3PCX-GA

Total population size: 200	Max evaluations: 500,000
Number of parents in crossover: 3	family size: 2
$sigma_zeta$: 0.1	$sigma_eta$: 0.1
SPI: 10	Perturbation noise percentage: 20%

Fig. 21.11 and Fig. 21.12 show the twenty frequency response curves of these robust lowpass and highpass filters with parameter perturbations of 30% of nominal values. Compared with the result in Fig. 21.7 and Fig. 21.8 without considering robustness, the G3PCX GA indeed improves the robustness. However, one needs to be cautious when interpreting these frequency response figures. As specified in our synthesis problems, we have a shallow "don't care" region for both lowpass and highpass target frequency responses. The robust filters evolved by G3PCX have better performance in the two-end regions, while they have large variation in the "don't care" region.

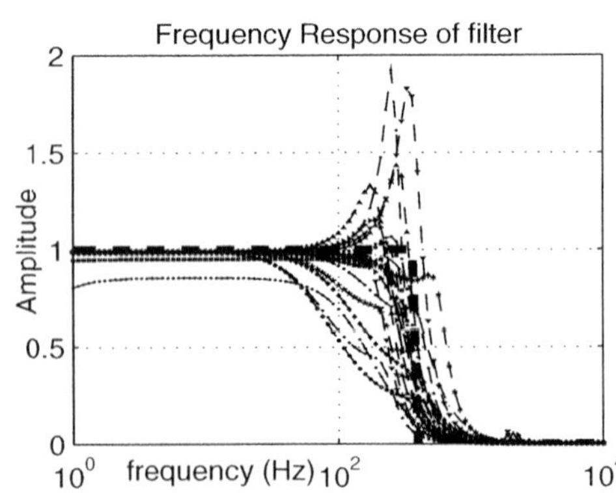

Fig. 21.7. Frequency responses of the lowpass filter in face of 30% Gaussian perturbation of their nominal parameters, evolved using normal GP without robustness requirements

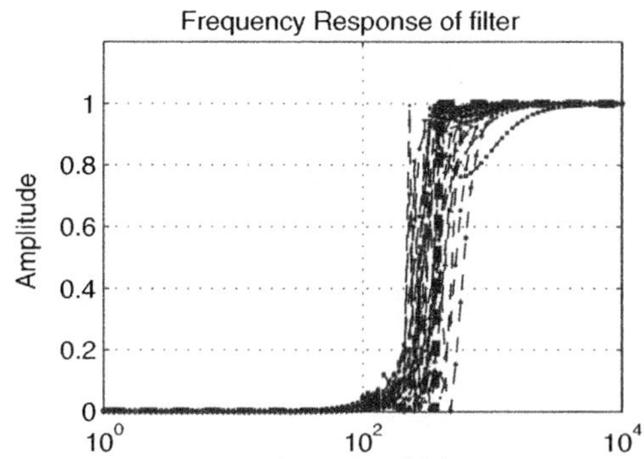

Fig. 21.8. Frequency responses of the highpass filter in face of 30% Gaussian perturbation of their nominal parameters, evolved using normal GP without robustness requirements

21.5.3 Evolving Robust Filters Using GP: Open-Ended Topology Search for Robust Design

In this section, we try to evolve robust analog filters that have higher tolerance of the variation of component values and have graceful performance degradation. The configuration of this experiment is the same as those used in Section 21.5.1.

The topology of the evolved robust lowpass and highpass filters are shown in Fig. 21.9 and Fig. 21.10. It is very interesting to compare the complexity of these two filters to those evolved with standard GP without considering robustness (Fig. 21.2 and Fig. 21.6). The robustness requirement drives the GP to evolve much simpler structures since large structures expose more components to perturbation noise. Of course, this depends on the perturbation model. In our model, we applied the perturbation to ALL components, so large filters with more components tend to suffer from more perturbation.

We can also compare the frequency responses of the robust filters with that of filters evolved by standard GP and GA with robustness. It appears that GP with robustness outperforms the other two approaches by allowing more variation in the "don't care" region while keeping tight control in the two boundary regions where stringent functional requirements are imposed.

21.5.4 Statistical Comparison of Three Methods

For the highpass filter problem, a t-test is conducted to compare the robustness of the evolved solutions by GPGARMS and standard GP in terms of fitness variation at 0.2 perturbation level. A significance level of $P =< 0.001$ is achieved which strongly indicates that GPGARMS improved the robustness of the evolved filters by standard GP. However, we found that this improvement is at the cost of degraded functional performance at the normal condition (without perturbation). A t-test was also applied to compare GPGARMS and GPRMS. The 95 percent confidence

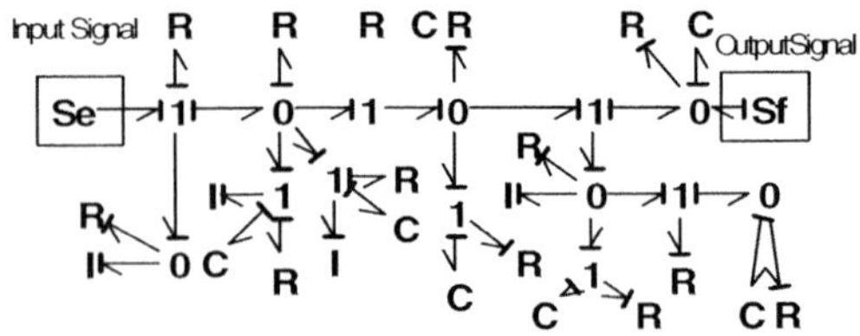

Fig. 21.9. Robust lowpass filter evolved using GP with robustness

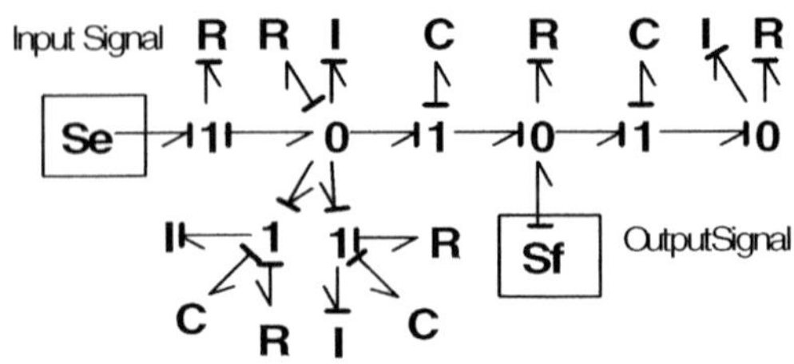

Fig. 21.10. Robust highpass filter evolved using GP with robustness

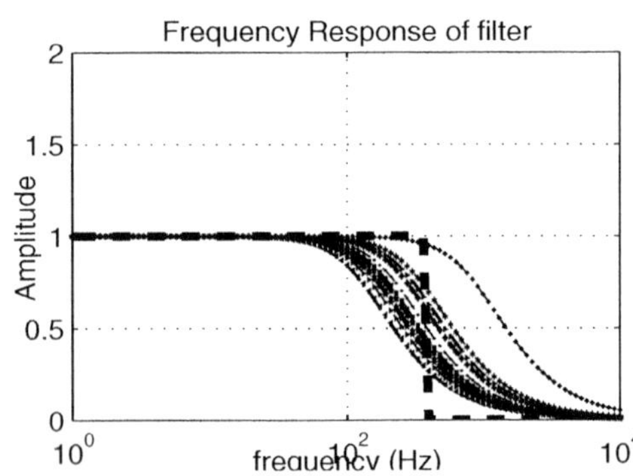

Fig. 21.11. Frequency responses of the lowpass filter in face of 30% Gaussian perturbation of their nominal parameters, evolved first using normal GP without robustness requirements and then fine-tuned using GA with robustness requirements

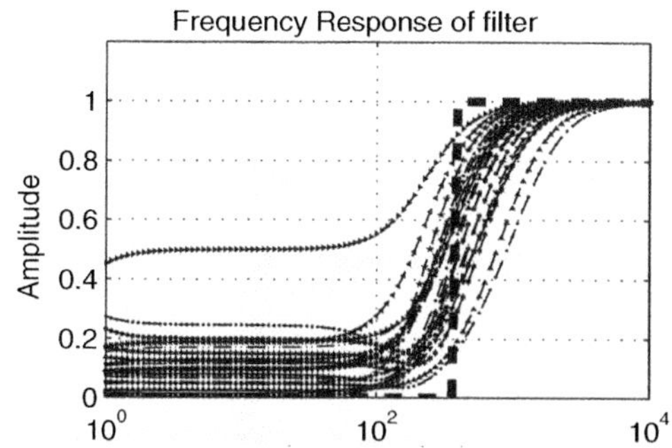

Fig. 21.12. Frequency responses of the highpass filter in face of 30% Gaussian perturbation of their nominal parameters, evolved first using normal GP without robustness requirements and then fine-tuned using GA with robustness requirements

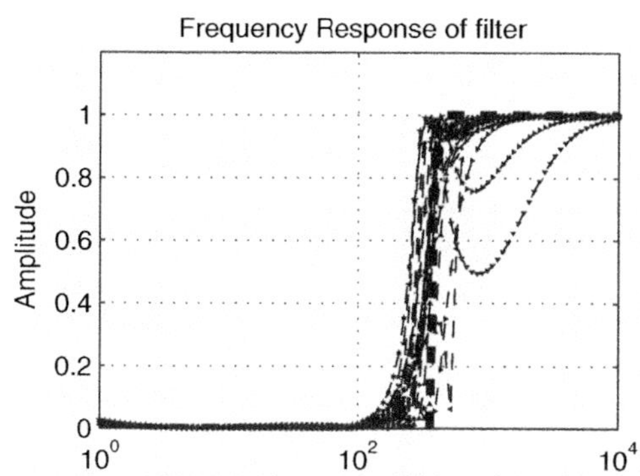

Fig. 21.13. Frequency responses of the highpass filter in face of 30% Gaussian perturbation of their nominal parameters, evolved using normal GP with robustness requirements

interval for difference of means of robustness fitness is -51.617 to -39.841, showing that GPGARMS degraded robustness. The difference in the mean values of the two groups is greater than would be expected by chance ($P =< 0.001$).

21.6 Conclusions and Future Work

This chapter applies genetic programming and bond-graph system modeling – the GPBG approach – to topologically open-ended synthesis of robust analog filters. It is shown that the traditional approach of robust design, in which the functional/conceptual design is conducted without considering a robustness requirement,

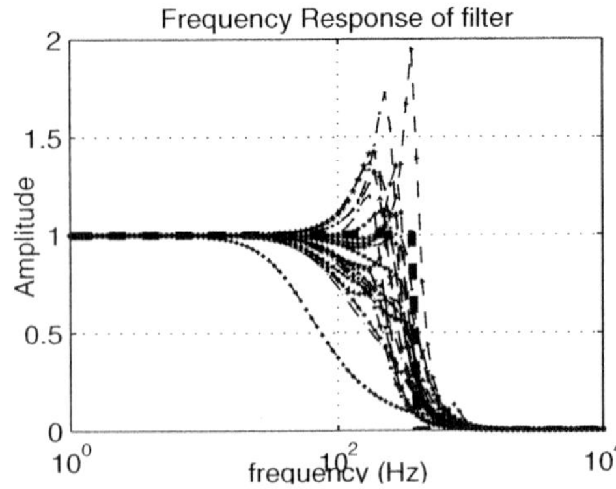

Fig. 21.14. Frequency responses of the lowpass filter in face of 30% Gaussian perturbation of their nominal parameters, evolved using normal GP with robustness requirements

may put severe limits on the possible robustness achievable through parameter-tuning-based robust design during the detailed design stage. It thus proposes that robust design in engineering should start from the conceptual stage, and that the open-ended topology search capability of GP can be exploited for this purpose. We find that our GP system enables us to find more robust analog filters with respect to the variations in their parameters compared to existing parameter-tuning-type evolutionary algorithms for robust design of fixed functional structures.

Evolving robustness is a rich research theme and there are several interesting topics to be further investigated such as the robustness with respect to topology perturbation or component failures, which may be important in many environments, especially on space missions. Our ongoing work shows that selection pressures for robustness with respect to parameter perturbation versus with respect to component faults lead to different topological patterns. It would be interesting to investigate how simultaneous requirements for both types of robustness would affect topological structures. In this chapter, only simple robustness estimation method based on multiple samplings is used. However, in the simultaneous topology and parameter search process, more effective approach can be devised to reduce the computational effort for estimating robustness of individuals.

Acknowledgments

This research is partially supported by the National Natural Science Foundation of China under Grant 50575047.

References

1. S. Ando and H. Iba. Linear genome methodology for analog circuit design. Technical report, Information and Communication Department, School of Engineering, University of Tokyo, 2000.
2. J. Branke. Creating robust solutions by means of an evolutionary algorithms. In A. Eiben, T.Back, M. Schoenauer, and H.-P. Schwefel, editors, *Proc. Parallel Problem Solving from Nature. number 1498 in LNCS. Springer*, 1998.
3. J. Branke. *Evolutionary Optimization in Dynamic Environments.* Kluwer, 2001.
4. J. Branke and H. Schmeck. Designing evolutionary algorithms for dynamic optimization problems. In S.Tsutsui and A. Ghosh, editors, *Theory and Application of Evolutionary Computation: Recent Trends.* Springer, 2002.
5. J. M. Carlson and J. Doyle. Complexity and robustness. *Proceedings of National Academy of Science (PNAS)*, 99(1):2538–2545, 2002.
6. K. Deb, A. Anand, and D. Joshi. A computationally efficient evolutionary algorithm for real-parameter optimization. *Evolutionary Computation*, 10(4):345–369, 2002.
7. Z. Fan, J. Hu, K. Seo, E. D. Goodman, R. C. Rosenberg, and B. Zhang. Bond graph representation and GP for automated analog filter design. In E. D. Goodman, editor, *2001 Genetic and Evolutionary Computation Conference Late Breaking Papers*, pages 81–86, San Francisco, California, USA, 9-11 July 2001.
8. B. Forouraghi. A genetic algorithm for multiobjective robust design. *Applied Intelligence*, 12:151–161, 2000.
9. J. Hu, E. Goodman, K. Seo, Z. Fan, and R. Rosenberg. The hierarchical fair competition (hfc) framework for sustainable evolutionary algorithms. *Evolutionary Computation*, 13(2), 2005.
10. E. Jen. Definitions of robustness. santa fe institute robustness site, rs-2001-009. 2001.
11. Y. Jin and B. Sendhoff. Trade-off between optimality and robustness: An evolutionary multi-objective approach. In C. F. et al., editor, *Proceeding of the Second Int. Conf. on Evolutionary Multi-criterion Optimization*, pages 237–251. Springer, 2003.
12. D. Karnopp, D. L. Margolis, and R. C. Rosenberg. *System Dynamics: Modeling and Simulation of Mechatronic Systems. Third Edition.* John Wiley & Sons, Inc., New York, 2000.
13. J. R. Koza, D. Andre, F. H. Bennett III, and M. Keane. *Genetic Programming 3: Darwinian Invention and Problem Solving.* Morgan Kaufman, Apr. 1999.
14. J. R. Koza, F. H. Bennett III, D. Andre, M. A. Keane, and F. Dunlap. Automated synthesis of analog electrical circuits by means of genetic programming. *IEEE Transactions on Evolutionary Computation*, 1(2):109–128, July 1997.
15. J. R. Koza, M. A. Keane, M. J. Streeter, W. Mydlowec, J. Yu, and G. Lanza. *Genetic Programming IV: Routine Human-Competitive Machine Intelligence.* Kluwer Academic Publishers, 2003.
16. J. Lohn and S. Colombano. A circuit representation technique for automated circuit design. *IEEE Transactions on Evolutionary Computation*, 3(3):205–219, 1999.
17. T. Ray. Constrained robust optimal design using a multi-objective evolutionary algorithm. In *Proceeding of Congress on Evolutionary Computation*, pages 419–424. IEEE press, 2002.

18. K. Seo, Z. Fan, J. Hu, E. D. Goodman, and R. C. Rosenberg. Toward an automated design method for multi-domain dynamic systems using bond graph and genetic programming. *Mechatronics*, 13(8-9):851–885, 2003.
19. G. Taguchi. *Taguchi on Robust Technology Development: Bringing*. ASME, ASME, 1993.
20. S. Tsutsui and A. Ghosh. Genetic algorithms with a robust solution searching scheme. *IEEE Trans. Evolutionary Computation*, 1(3):201–208, 1997.
21. D. Wiesmann, U. Hammel, and T. Back. Robust design of multilayer optical coatings by means of evolutionary algorithms. *IEEE Transactions on Evolutionary Computation*, 2(4):162–167, 1998.

22

Robust Salting Route Optimization Using Evolutionary Algorithms

Hisashi Handa[1], Lee Chapman[2], and Xin Yao[3]

[1] Graduate School of Natural Science and Technology, Okayama University, Tsushima-Naka 3-1-1, Okayama, 700-8530, JAPAN
handa@sdc.it.okayama-u.ac.jp
[2] School of Geography, Earth, and Environmental Science
The University of Birmingham, Edgbaston, Birmingham B15 2TT, U.K.
l.chapman@bham.ac.uk
[3] CERCIA, School of Computer Science, The University of Birmingham, Edgbaston, Birmingham B15 2TT, U.K.
x.yao@cs.bham.ac.uk, www.cs.bham.ac.uk/~xin

Summary. In winter, roads need to be salted and gritted when temperature drops to around the freezing point, in order to ensure the safety of road users (especially motor vehicles). In the UK, there are approximately 3000 salting routes covering about 120,000km (approximately 30% of the road network). Given limited resources and severe time constraints, it is imperative that salting routes are planned in advance for efficient and effective treatment. Unfortunately, there is no automatic route optimization system for salting trucks that can deal with different road conditions and constraints. Almost all published systems make unrealistic assumptions that do not hold in practice. This chapter describes a novel route optimization system based on newly proposed memetic algorithms. The system is designed with dynamic problems in mind. That is, given different road temperatures and different temperature distributions in a road network, the system can produce optimised routes for a fleet of salting trucks. The system has been evaluated using real world data from the South Gloucestershire council in England and obtained 10% improvement over their existing solution in terms of distances travelled by the salting trucks.

22.1 Introduction

In countries with a marginal winter climate, highway authorities are responsible for the precautionary salting of the road network. In the UK, there are approximately 3000 salting routes covering about 120,000km (approximately 30% of the road network). With limited resources and time constraints, it is imperative that salting routes are planned in advance for efficient and effective treatment. To aid this process, a Salting Route Optimization system (SRO) [1–3] which combines evolutionary algorithms with the neXt generation Road Weather Information System (XRWIS) has been developed [1]. The SRO system can cope with large-scale instances in the

H. Handa et al.: *Robust Salting Route Optimization Using Evolutionary Algorithms*, Studies in Computational Intelligence (SCI) **51**, 497–517 (2007)
www.springerlink.com

real world within reasonable computation times, to the extent that daily dynamic salting route optimization can be realised. However, a dynamic system [3] which responds to daily changes in temperature forecasts, will result in salting routes that also change each day. There is a danger that such a complicated system may confuse maintenance engineers and ultimately result in errors in the treatment regime. Therefore, before the dynamic approach can be used in the real world, there is a need to phase in the technology using an operationally simpler system.

In this chapter, a new evolutionary SRO which generates robust solutions to the problem is presented. Instead of treating each night on its own merits, the emphasis is instead placed on 'thermally ranking' static optimised routes. The aim is to amalgamate roads with similar thermal characteristics onto the same route so that the 'warmer' routes could be left untreated on marginal nights. Although the financial savings using a robust solution are potentially less than a dynamic solution, this is still a considerable advance on existing techniques.

22.2 Salting Route Optimization with XRWIS

The SRO system outlined in this chapter represents a synergy of evolutionary algorithms with XRWIS. XRWIS essentially provides an archive of past and forecast temperature distributions to the evolutionary algorithms module for evolution and operation. The system is fronted by an intuitive GUI which displays the resultant robust routes (Fig. 22.1). Although most of the road network data is inputted into the SRO as simple vector routing data, additional 'local' information can also be entered using the GUI, such as mandatory turns, one way streets, new roads and driver preferences. This section introduces XRWIS and explains how the temperature data is combined with the vector routing data to translate the salting route optimization problem into a Capacitated Arc Routing Problem (CARP).

22.2.1 XRWIS

Throughout the winter season, the decision of whether to salt the road network is taken by consulting a Road Weather Information System (RWIS) which produces a road condition forecast by combining weather forecast data with thermally mapped road temperature data. The first generation of RWIS, which is still largely in use, relies on methods and tools developed in the 1980s. However, as technology has moved on, it is now being superseded by the neXt generation RWIS (XRWIS) [4, 5]. XRWIS is a new route-based forecast system which accurately predicts road temperatures to a high spatial and temporal resolution. Instead of modelling road condition at a single site and interpolating temperatures by thermal maps, XRWIS models road surface temperatures at thousands of sites around the road network by considering the influence of the local geography on the road surface [4, 5]. Data is collected along each salting route by conducting a survey of the sky-view factor. This is a measure of the degree of sky obstruction by buildings and trees and is the dominant control on road surface temperatures. [6, 7]. The sky-view factor is then combined with other geographical parameters (latitude, longitude, altitude, slope, aspect, road construction, landuse and traffic volume) to produce a high resolution geographical parameter database.

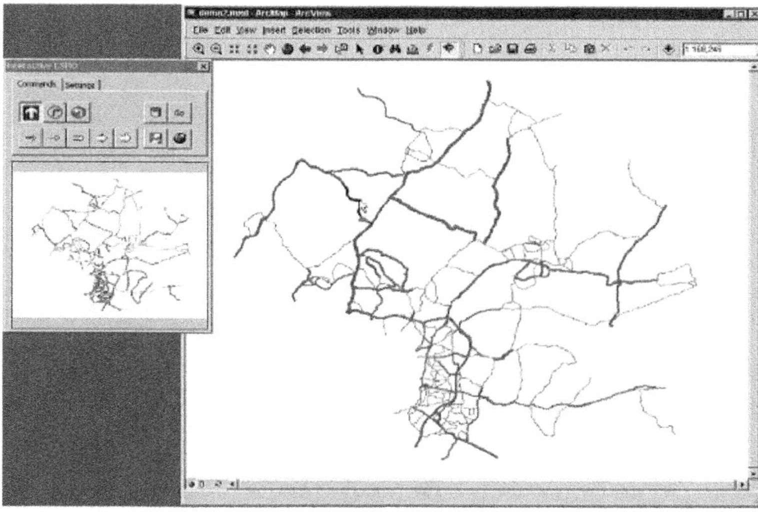

Fig. 22.1. The graphical user interface of the developed Salting Route Optimization System

The geographical data is combined with mesoscale meteorological data in an energy balance model to predict road conditions at typical spatial and temporal resolutions of 20 metres and 20 minutes respectively. The output is displayed as a colour-coded map of road temperature and condition that is disseminated over the Internet to the highway engineer (Fig. 22.2). Fig. 22.3 shows an example of the changes of temperature, predicated by the XRWIS, at a single site (point) during a day. From this forecast, actions are suggested as to whether or not an individual salting route needs treating.

22.2.2 Capacitated Arc Routing Problems (CARP)

SRO can be regarded as an instance of the Capacitated Arc Routing Problem (CARP) [8, 10]. Suppose that a graph $G = (V, E)$ is given, where V and E are sets of vertices and edges, respectively. Each edge e in E has a cost C_e. A set R ($\subset E$) of required edges is defined in the CARP and a demand D_e is defined to each edge e in R. There are several vehicles to fill the demands, where each vehicle has a predefined service capacity. A depot is also defined in V from which all vehicles must depart from and return to at the end of their service tour. The problem is to find a set of tours which have a minimum total cost for all vehicles, ensuring the demands of all required edges are filled by at least one vehicle, whilst ensuring the total services capabilities of each vehicle are not exceeded.

The road network is divided into vertices and edges. Vertices are set on intersections or branch points of roads, whereas edges are defined as roads between vertices. For example, using Fig. 22.2 as an example, there are 419 vertices and 597 edges. The cost of an edge is defined as the length of the feature where as the demand is

Fig. 22.2. Temperature distributions for two nights in South Gloucestershire: on a cold night (UPPER) and on a marginal night (LOWER)

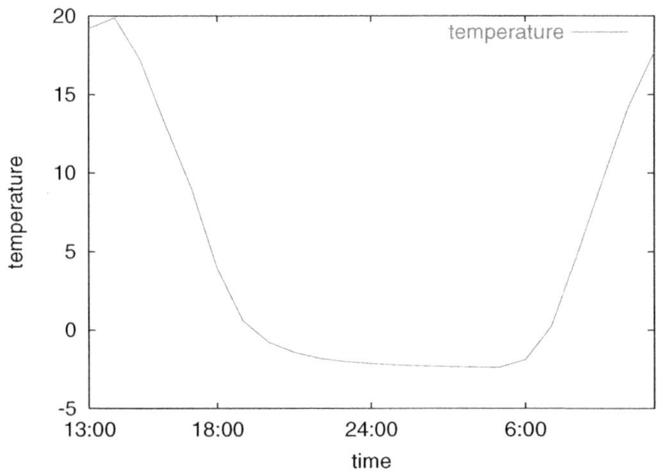

Fig. 22.3. The changes of predicted temperature at a single site over a 24 hour period

the amount of salt required to treat it. The set of required edges and their demands, is then defined by referring to the predicted temperature distribution provided by XRWIS. If a road section (edge) is predicted to go below freezing, then a minimum 10g/m^2 salt is required to be spread on that section before ice forms (although more salt will be required on wider, multi-laned roads). Thus, the amount of salt $S(e)$ on an edge e is defined as follows:

$$S(e) = \sum_{o \in e} d(o, \text{succ}(o, e)) \times w(e) \times f(t(o) - \theta),$$

where $\text{succ}(o, e)$ and $w(e)$ denote the succeeding prediction point of the prediction point o on the edge e, and width of the edge e, respectively. $f(x)$ is the threshold function such that $f(x)$ returns 1 if $x < 0$, otherwise 0. $t(o)$ and θ are the predicted temperature at o and threshold value fixed in advance. If $S(e)$ is greater than 0, the edge e is regarded as a member of the set of required edges. Fig. 22.5 shows the acquired CARP instances from temperature distributions in Fig. 22.2.

22.2.3 Representation and Evaluation of Tours

Each salting operation will see a number of trucks N in operational use. Each edge is assigned a unique ID and the tours T_i for trucks i $(i = 1, \ldots N)$ can be denoted as a sequence of the edge IDs. For instance, assume $N = 3$ and a set of ten edge IDs $\{1, 2, 3, 4, \ldots, 10\}$. A set of tours could be represented as follows:

T_1	2 3 5 6
T_2	1 8 4
T_3	10 9 7

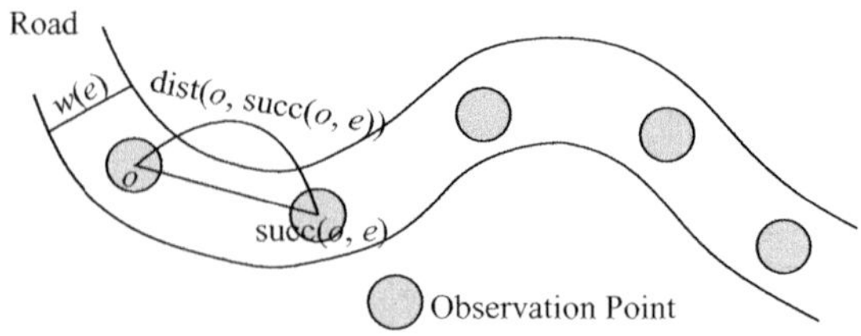

Fig. 22.4. Translation of SRO to a CARP instance

Now, by introducing a temperature distribution a, a set of edges requiring treatment $R(a)$ can be defined. Hence, the set of tours is then rewritten by neglecting non-members of the set of required edges $R(a)$. Let $R(A) = \{2, 3, 4, 6, 8\}$. The set of tours is rewritten for the temperature distribution a as follows:

$T_1(a)$ 2 3 6
$T_2(a)$ 8 4
$T_3(a)$

In this example, $T_3(a)$ is not required and hence, only two trucks are required to complete the operation.

The evaluation of a set of tours X is represented as follows:

$$E(X, a) = \sum_{T_i(a) \in X} C(T_i(a)) + C_p \times P(T_i(a), a), \tag{22.1}$$

where $\dot{C()}$ denotes the cost (distance) function of a tour as described in the appendix and C_p is a predefined coefficient for the penalty term. $\dot{P()}$ indicates the quantity of constraint violation in each truck and is defined as follows:

$$P(T, a) = \begin{cases} D(T, a) - L(T) & \text{if } D(T, a) - L(T) > 0 \\ 0 & \text{Otherwise,} \end{cases}$$

where $D(T, a)$ denotes the total services in tour T at a temperature distribution a, and $L(T)$ is a limitation subject to a truck for tour T.

22.3 Robust Solution of Salting Route Optimization

22.3.1 Robust Solution

Searching for robust solutions is one of the most significant topics of evolutionary optimization in uncertain environments [13]. Robust solutions are often used for problems where the decision variables or environmental parameters[4] are subject to

[4] The environmental parameters indicate parameters which characterise the fitness function

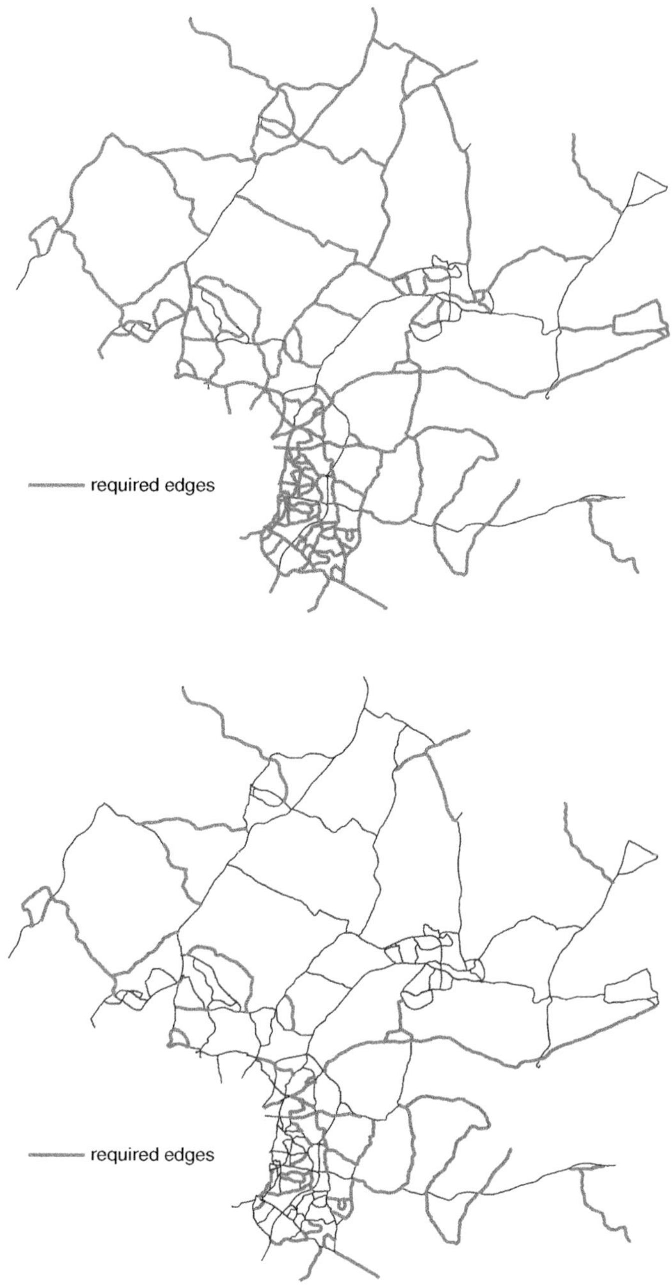

Fig. 22.5. Required edges for Fig. 22.2: on a cold night (UPPER) and on a marginal night (LOWER)

perturbation. The notion of effective fitness function is often used in this research area [11, 12].

$$F(X) = \int_{-\infty}^{\infty} f(X + \delta)p(\delta)d\delta, \tag{22.2}$$

where $p(\delta)$ indicates the probability distribution of perturbation δ. In the case of where a perturbation is added to the environmental parameters a,

$$F(X) = \int_{-\infty}^{\infty} f(X, a + \delta)p(\delta)d\delta. \tag{22.3}$$

In practise, an approximation of the effective fitness function is used. The approximation of the effective fitness function in the former is written by

$$\hat{F}(X) = \sum_{i=0}^{N} \frac{1}{N} f(X + \delta_i), \tag{22.4}$$

where N denotes the number of samplings to estimate $f(X)$. That is, evolutionary algorithms tackle to solve for $\hat{F}(X)$ instead of $f(X)$. Similar approximation is applied for the latter case.

22.3.2 Robust Solution of Salting Route Optimization

In the case of salting route optimization, a robust solution can be represented by an optimal design value X for the following function:

$$F(X) = \int E(X, a)p(a)da, \tag{22.5}$$

where a indicates a possible temperature distribution. Essentially, the optimization of $F(X)$ does not make sense if a, such that $p(a) \neq 0$, is uncorrelated with X in $E(X, a)$. In the case of salting route optimization, 'warmer' and 'colder' roads exist due to microclimatological effects caused by the local geography. However, although the distribution in temperature will vary daily across the road network, warmer sections are nearly always relatively warm and colder sections are nearly always relatively cold. As a result, even on cold nights, some warmer sections of road will still not require salt where as the coldest sections of road may need treating on even the least marginal of nights.

It is difficult to compute equation (22.5) exactly since the number of possible values in a is large and the probability distribution $p(a)$ is yet unknown. Hence, as in inductive learning, the number of typical temperature distributions is prepared for evolution. Let A_e is a set of temperature distributions for evolution. The following function is useful to evaluate the robustness of salting routes:

$$\hat{F}(X) = \sum_{a_i \in A_e} \frac{1}{|A_e|} E(X, a_i). \tag{22.6}$$

```
Procedure Robust Solutions of SRO by using EAs
begin
  translate SRO for temperature distributions a_i to CARP instance I_{a_i}
  initialise population
  evaluate population
  until Stopping criterion is reached
    select a CARP instance I_{a_i}
    pick up two parents
    generate 30 offspring from the parents by EAX for I_{a_i}
    select the best offspring
    apply local search to the copy of the best offspring for I_{a_i}
    evaluate the best offspring and the improved offspring by F̂(X)
    replacement of new individuals
  end
end
```

Fig. 22.6. Pseudo-code of the proposed method

22.4 Evolutionary Algorithms for Robust Salting Route Optimization

22.4.1 Overview

A new memetic algorithm for generating robust solutions is proposed (Fig. 22.6). The basic procedure of the algorithm is ordinal: selecting parents, reproducing offspring, applying local search methods to the offspring and finally replacing the resultant offspring if the new offspring is better than the worst individual in the population. A particular feature of the proposed method is that crossover operations and local search methods are applied to only one CARP instance at every generation, whilst the fitness function is composed of an ensemble of the evaluations of several CARP instances. Hence, at the beginning of each generation, a single CARP instance is selected for use by referring to the weights. The weights are then updated for predefined intervals of generations as described in 22.4.3 and are used for fitness evaluation.

22.4.2 Coding Method and Fitness Evaluation

A naive permutation encoding method for solving SRO is employed. The chromosome of an individual is composed of several special symbols and edge IDs. Special symbols $\mathtt{s}_1$ are used to indicate the beginning of tours for each truck. Using the following chromosome as an example, tours are yielded for two trucks (A permutation representation is employed in this coding method).: $T_1 = \{\mathtt{5\ 4\ 7\ 1}\}$, and $T_2 = \{\mathtt{8\ 3\ 2\ 6}\}$.

$\mathtt{2\ 6\ s_1\ 5\ 4\ 7\ 1\ s_2\ 8\ 3}$

As evolutionary algorithms have a tendency to find optimal solutions $E(X, a_i)$ in the case of equation (22.6), the normalised function $E_N(X, a_i)$ is employed as a fitness function of our evolutionary algorithms:

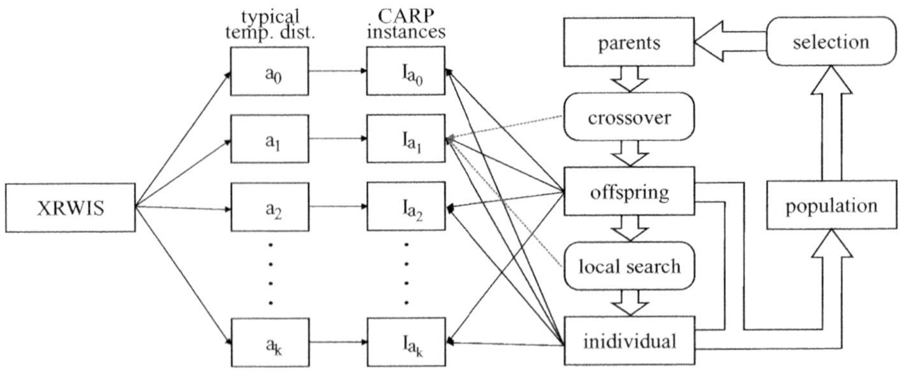

Fig. 22.7. The graphical user interface of the developed Salting Route Optimization System

$$\hat{F}(X) = \sum_{a_i \in A_e} w_i E_N(X, a_i), \tag{22.7}$$

where w_i $(0 < w_i < 1, \sum_{a_i \in A_e} w_i = 1)$ denotes a weight for each temperature distribution a_i. The normalised function $E_N(X, a_i)$ is defined as follows:

$$E_N(X, a_i) = \frac{E(X, a_i) - E^*(a_i)}{E^*(a_i)}, \tag{22.8}$$

where $E^*(a_i)$ is a real number indicating the difficulty of solving a CARP instance I_{a_i}, (e.g. lower bounds which is the distance searched by other algorithms). In this algorithm, $E^*(a_i)$ is defined as the best distance for a CARP instance I_{a_i} searched by using the Memetic Algorithm [1].

22.4.3 Weights and their Update

In this section, two types of weights w_i are examined: fixed weights and weights updated for every predefined interval. The values of the fixed weights and the initial value of changed weights are set to $1/|A_e|$.

Weight updates correspond to changes in the directions of evolution. For every predefined interval of generations L, the weight w_j is updated by using the best tour evaluation $E_N^b(X, a_i)$ which is searched by evolutionary algorithms:

$$w_j = \frac{\exp E_N^b(X, a_j)}{\sum_{k=1}^{m} \exp E_N^b(X, a_k)} \tag{22.9}$$

Genetic operators including crossover and local search methods are applied to only one CARP instance in every generation. A CARP instance is selected by referring to the weights in the beginning of every generation and applying genetic operators. For example, the following proportion s_i is used to randomly decide which CARP instance is applied to genetic operators, as in roulette wheel selection:

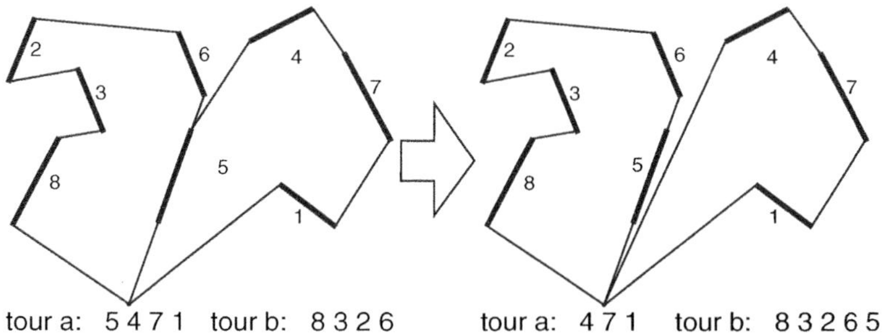

Fig. 22.8. A depiction of the repair procedure

$$s_i = \frac{w_i}{\sum w_j}. \tag{22.10}$$

This application method is employed because the genetic operators adopted use information in problem instances, for example, distance and amount of services on edges, to improve the fitness of individuals. Conflicts between the improvement for problem instances by genetic operators might occur if genetic operators are applied to several problem instances simultaneously and therefore a conflict resolution mechanism may be required. Furthermore, evolutionary algorithms tend to converge the easier instances at first which can cause Evolutionary Algorithms to become trapped into the local minimum of the total problem as represented by equation (22.6) This is because it is difficult to find out an improved solution for the more difficult problem instances whilst the solution quality of easier problems is kept constant.

22.4.4 Genetic Operators

In order to cope with large scale problems, the edge assembly crossover (EAX) operator proposed by Nagata *et al.* is used due to its search ability [15, 16]. However, since this operator is specifically designed for Travelling Salesman Problems, it can often yield an infeasible solution in our case. Hence, a repair operator for offspring individuals is incorporated in our memetic algorithm:

1. A counter variable *count* is set to 0.
2. Find a tour a which has maximum violation with respect to the constraint of the service capacity.
3. Randomly choose a required edge r in the tour a.
4. Find a tour b, which has an opening for the required edge t, such that the required edge must be traversed as a deadheading path (Figure 22.8). If no tour is found, increment *count* and go to 7. Otherwise go to the next step.
5. Move the required edge r from the tour a to the tour b.
6. Increment *count* and recalculate the total amount of services for the tours a and b.
7. Loop back to 2. until there is no violation in all tours or *count* exceeds 30.

22.4.5 Local Search Method

Local search methods are carried out with a probability. Because the EAX operator has similar characteristics to the k-opt operator, three naive local search methods are used in the Memetic Algorithms:

Move an edge
```
Before: 4 s1 1 3 5 8 s2 2 6 7
 After: 4 s1 1 3 5   s2 2 6 8 7
```
Move sequential two edges
```
Before: 4 s1 1 3 5 8 s2 2 6 7
 After: 4 s1 1 3     s2 2 6 5 8 7
```
Swap two edges
```
Before: 4 s1 1 3 5 8 s2 2 6 7
 After: 4 s1 1 3 5 6 s2 2 8 7
```

Upper and lower lines in each local search operation indicate an individual before and after applying the operation respectively. For all possible pairs of variables, above local search operations are applied. A pair of variables with the best improvement is then selected.

22.4.6 Initialisation of Population

Almost all the individuals in the population are generated as a random permutation. However, two types of additional sophisticated individuals are also inserted: Firstly, individuals generated by path scanning heuristics [9, 10] are applied to problem instances with large numbers of required edges. Secondly, individuals indicating the best solution $E^*(a_i)$ for the corresponding CARP instance I_{a_i} is solved solely by memetic algorithms in [1]. These individuals are represented by the IDs of required edges for the corresponding CARP instance I_{a_i}. Therefore, individuals in the proposed method require the IDs of all edges. In order to insert missed IDs to the best individuals, simple heuristics based on distance is employed: (1) insert the missed edge into the nearest tour providing the service capacity of the truck is not exceeded. (2) if no edge is found in (1), another tour is built by using the path scanning heuristics.

22.5 Experiments

22.5.1 Evolutionary Process

A series of experiments using the real data in South Gloucestershire, England, were conducted. Ten different temperature distributions were used for evolution (shown in Table 22.1) along with an 'ideal' temperature distribution a_{ideal}, where all temperature points are below 0 degree. In each case, the number of trucks refers to the minimum number of trucks which can cover all the demands of required edges in corresponding CARP instance I_{t_i}. The best distance is searched by the Memetic algorithm in [1] and is used for the normalised evaluation $E_N(X, a_i)$ shown in equation (22.8) The number of edges in the area and trucks needed to serve in the CARP

Table 22.1. Ten different temperature distributions were used in our experiments

	a_0	a_1	a_2	a_3	a_4
No. required Edges	97	257	297	323	354
No. trucks	2	4	4	5	6
Best distance $E^*(a_i)$	268029	366410	436006	463263	509717
	a_5	a_6	a_7	a_8	a_9
No. required Edges	437	428	519	568	578
No. trucks	7	8	9	11	11
Best distance $E^*(a_i)$	576513	640502	665371	727262	747710

Table 22.2. The average and best distances over 20 runs using our proposed method

Initialisation	unincorporated				incorporated			
Change ivl.	n/a	100	500	1000	n/a	100	500	1000
Ave.	0.3225	0.3233	0.3183	0.3342	0.3002	**0.2860**	**0.2769**	0.2916
Best	0.2894	0.2696	0.2883	0.2985	0.2730	0.2652	0.2525	0.2668

instance $I_{a_{ideal}}$ for the 'ideal' temperature distribution are 595 and 11 respectively. As a result, the string length is set to 696. The probability of applying local search method, the population size, and the number of generations in each run are 0.1, 300, and 100,000, respectively.

Table 22.2 shows the average and the best distances over 20 runs by using the proposed method. Three kinds of changing intervals for weights are examined: 100, 500, and 1000. Bold font denote results that show a statistical significance against uncharged weight method. Initialisation by incorporation of the best individuals for each CARP instance (which are searched by former memetic algorithms), gives good results. For all weight change intervals and unchanged weights, the averages and the bests are improved.

Fig. 22.9 depicts the changes of weights during evolutionary process for the changing interval of 500 generations. The horizontal axis denotes the temperature distribution a_i where as the depth axis and the vertical axis indicate the number of generations and corresponding weight values. Fig. 22.9 shows that the proposed method appears to find better solutions in CARP instances with less required edges.

22.5.2 Test of the Resultant Robust Solutions

A further 11 temperature distributions t_i were used to investigate the generalisation property of the acquired robust solutions. The best solutions for the algorithms in Table 22.2 are evaluated on the CARP instances t_i in Table 22.3. Table 22.4 summarises the results for the test. Results shown in bold indicate the best performance among variable changing interval of weights (including uncharged weights). 'Average' refers to the average value of $E_N(X, t_i)$ over 11 CARP instances I_{t_i}. For the average values, the proposed method with changing interval $L = 500$ and best individual incorporation outperform the other algorithms. In particular, for CARP instances with less total demands, e.g., $I_{t_0}, I_{t_1}, I_{t_2}, I_{t_3}$, the algorithm shows increased performance as just 10 trucks are now required.

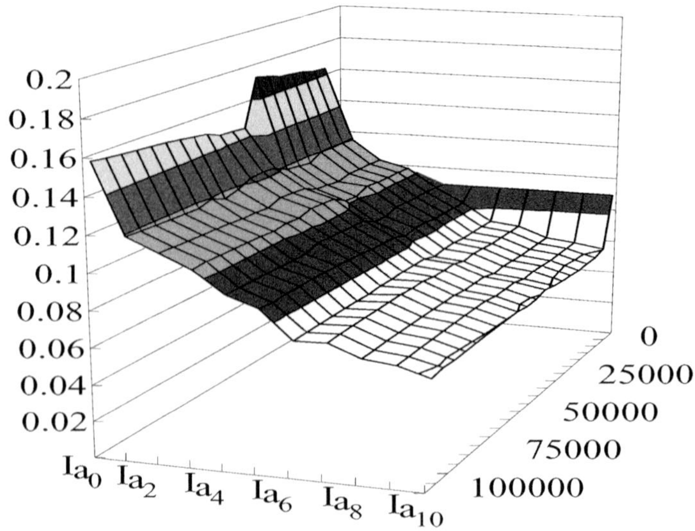

Fig. 22.9. Change of weights during evolution process

Table 22.3. Eleven CARP instances used to evaluate our algorithm

	t_0	t_1	t_2	t_3	t_4	t_5
No. required Edges	129	246	262	259	291	339
No. trucks	3	4	4	4	4	5
Best distance $E^*(t_i)$	299189	349094	376647	377277	422974	481216
	t_6	t_7	t_8	t_9	t_{10}	
No. required Edges	344	385	448	468	537	
No. trucks	5	6	8	8	10	
Best distance $E^*(t_i)$	482925	532018	578740	668918	671725	

22.5.3 Comparison with Conventional Routes

The acquired robust solutions were compared with conventional routes which are presently used by the South Gloucestershire council (England). Fig. 22.10 and Fig. 22.11 are plotted routes of the CARP instances I_{t_1} and I_{t_8}, respectively.

The conventional routes were generated manually by roughly allocating a region to each truck. Hence, the robust solution looks untidy in comparison but does offer an improvement of total distance travelled by more than 10%.

22.6 Conclusions

In this chapter, a robust solution of salting route optimization system, which combines evolutionary algorithms with the neXt generation Road Weather Information

Table 22.4. Results of our algorithm for the eleven CARP instances given in Table 22.3. Results shown in bold indicate the best performance among variable changing interval of weights (including uncharged weights). 'Average' refers to the average value of $E_N(X, t_i)$ over 11 CARP instances I_{t_i}.

	Initialisation	unincorporated				incorporated			
	Change ivl.	n/a	100	500	1000	n/a	100	500	1000
	Average	0.3225	0.3233	0.3183	0.3342	0.3002	0.2860	0.2769	0.2916
t_0	$E_N(X, t_0)$	0.8702	**0.8383**	0.8607	0.9086	0.8936	0.8191	**0.7997**	0.8302
	Distance	559550	549999	556708	571026	566551	544241	538455	547583
	No. Trucks	11	10	11	11	11	10	10	10
t_1	$E_N(X, t_1)$	0.5290	0.5610	**0.5281**	0.6207	0.6395	0.5677	**0.5416**	0.5822
	Distance	533775	544953	533453	565787	572348	547276	538166	552332
	No. Trucks	10	10	10	10	11	10	10	11
t_2	$E_N(X, t_2)$	0.6039	**0.5393**	0.6030	0.6887	0.6063	0.5660	**0.5041**	0.5802
	Distance	604089	579757	603767	636040	605015	589845	566504	595163
	No. Trucks	11	10	11	11	11	10	10	11
t_3	$E_N(X, t_3)$	0.6012	**0.5367**	0.6003	0.6859	0.6033	0.5632	**0.5013**	0.5762
	Distance	604089	579757	603767	636040	604891	589749	566408	594680
	No. Trucks	11	10	11	11	11	10	10	11
t_4	$E_N(X, t_4)$	0.5089	**0.4543**	0.5138	0.5917	0.5079	0.5268	**0.4949**	0.5199
	Distance	638232	615144	640314	673239	637787	645780	632284	642882
	No. Trucks	11	10	11	11	11	11	11	11
t_5	$E_N(X, t_5)$	0.3826	0.3916	**0.3802**	0.4652	0.3737	0.3714	0.3692	**0.3648**
	Distance	665324	669677	664168	705057	661054	659956	658886	656745
	No. Trucks	11	10	11	11	11	11	11	11
t_6	$E_N(X, t_6)$	0.3801	0.3906	**0.3753**	0.4624	0.3700	0.3677	0.3655	**0.3611**
	Distance	666486	671561	664168	706219	661615	660517	659447	657306
	No. Trucks	11	10	11	11	11	11	11	11
t_7	$E_N(X, t_7)$	0.3206	0.3418	**0.3055**	0.4186	0.2936	0.3086	0.2904	**0.2874**
	Distance	702595	713887	694553	754733	688219	696179	686511	684918
	No. Trucks	11	11	11	11	11	11	11	11
t_8	$E_N(X, t_8)$	0.2498	0.2592	**0.2439**	0.3451	0.2261	0.2249	**0.2141**	0.2174
	Distance	723293	728747	719915	778443	709600	708890	702662	704530
	No. Trucks	11	11	11	11	11	11	11	11
t_9	$E_N(X, t_9)$	0.1043	0.1260	**0.0971**	0.1868	0.0890	0.0882	**0.0850**	0.0880
	Distance	738676	753212	733860	793891	728464	727936	725755	727766
	No. Trucks	11	11	11	11	11	11	11	11
t_{10}	$E_N(X, t_{10})$	0.1081	0.1355	**0.1065**	0.1975	0.0882	0.0903	**0.0858**	0.1069
	Distance	744370	762735	743280	804358	730994	732408	729375	743514
	No. Trucks	11	11	11	11	11	11	11	11

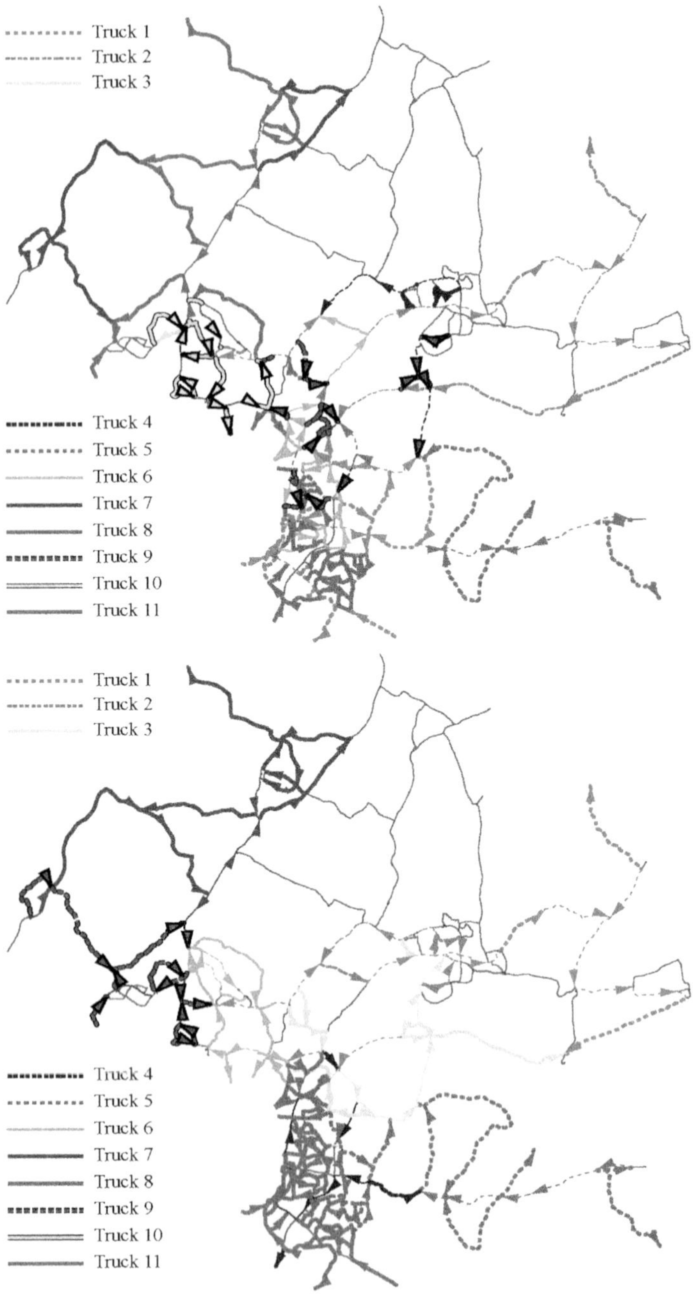

Fig. 22.10. Acquired routes (UPPER) and conventional routes (LOWER) for a CARP instance I_{t_1}. Gray line, coloured thick line and coloured thin line denote edges with no trucks, edges with service by corresponding truck, and deadheading edges, respectively.

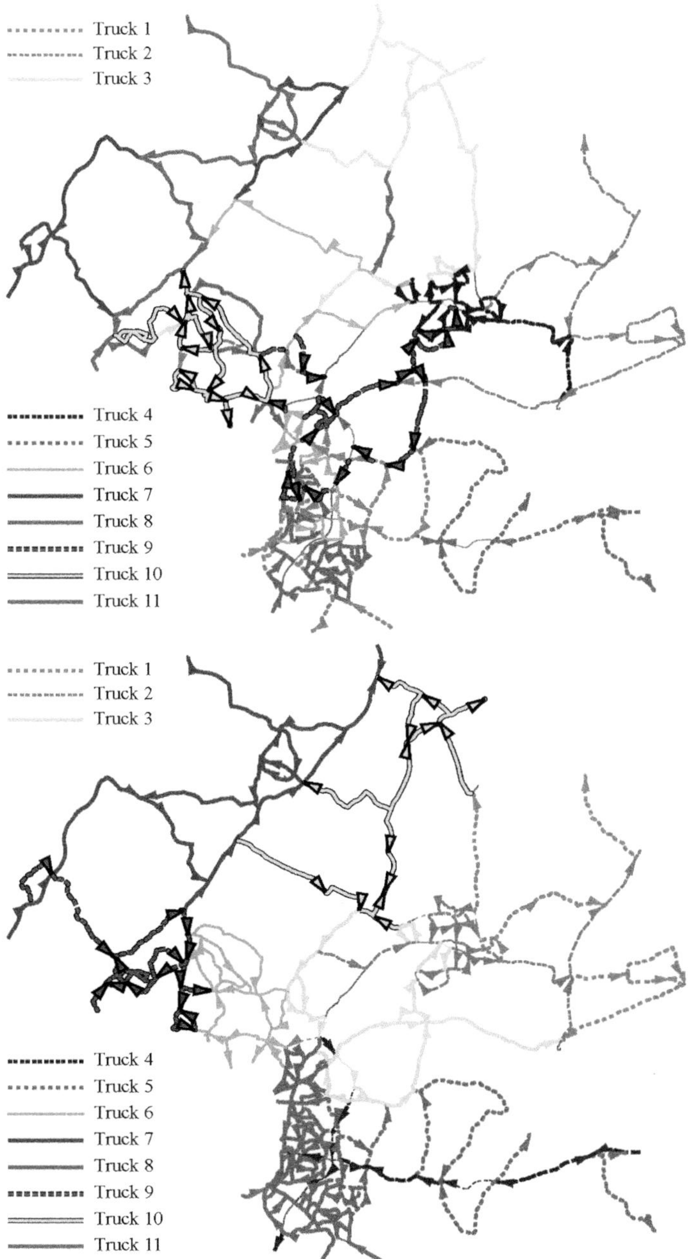

Fig. 22.11. Acquired routes (UPPER) and conventional routes (LOWER) for a CARP instance I_{t_8}. Gray line, coloured thick line and coloured thin line denote edges with no trucks, edges with service by corresponding truck, and deadheading edges, respectively.

System, was described. A new framework of robust solutions, not for perturbation, but for various values of environmental parameters was introduced. The weighting approach was examined in our proposed system in order to prevent the algorithm from converging to easier CARP instances. In addition, the insertion of the best individual for each temperature distribution was also incorporated into the proposed systems. Our proposed system has been tested on a real world case using the South Gloucestershire council data, more than 10% improvement was achieved in terms of the total distances travelled by trucks, in comparison with the routing that is currently being used.

The work presented here represents only the first step towards a practical solution to the salting route optimization problem. There are a number of difficult constraints that we need to incorporate into the system, such as different capacities of different trucks, multiple trips for a single truck, multiple depots, one-way streets, invalid turns, etc. Such new constraints have made CARP even more difficult to tackle. No existing algorithms can deal with all these constraints satisfactorily. New memetic algorithms are currently being developed by us. We hope to report our new results soon.

References

1. Handa, H, Chapman, L, Yao, X (2005) Dynamic Salting Route Optimisation using Evolutionary Computation, Proceedings of the 2005 Congress on Evolutionary Computation 1:158–165
2. Handa, H, Chapman, L, Yao, X (2006) Robust route optimisation for gritting/salting trucks: A CERCIA experience, IEEE Computational Intelligence Magazine, 1(1):6-9.
3. Handa, H, Lin, D, Chapman, L, Yao, X (2006) Robust Solution of Salting Route Optimisation Using Evolutionary Algorithms, Proc. of the 2006 IEEE Congress on Evolutionary Computation (CEC'06), Vancouver, Canada, 16-21 July 2006. pp.10455-10462, IEEE Press, USA.
4. Chapman, L, Thornes, JE, and Bradley, AV (2001) Modelling of road surface temperature from a geographical parameter database. Part 1: Statistical, Meteorological Applications 8:409–419
5. Chapman, L, Thornes, JE, and Bradley, AV (2001) Modelling of road surface temperature from a geographical parameter database. Part 2: Numerical, Meteorological Applications 8:421–436
6. Chapman, L, Thornes, JE, Bradley, AV (2002) Sky-view factor approximation using GPS receivers, International Journal Climatology 22(5):615–621
7. Chapman, L, Thornes, JE (2004) Real-Time Sky-View Factor Calculation and Approximation Journal of Atmospheric and Oceanic Technology 21:730–741
8. Golden, B, and Wong, RT (1981) Capacitated arc routing problem, Networks 11:305–315
9. Golden, B, and DeArmon, JS, and Baker, EK (1983) Computational Experiments with Algorithms for a Class of Routing Problems, Computers and Operational Research 10(1):47–59
10. Lacomme, P, Prins, C, and Ramdane-cherif, W (2004) Competitive Memetic Algorithms for Arc Routing Problems, Annals of Operations Research 131:159–185

11. Tsutsui S, Ghosh A (1996) A Robust Solution Searching Scheme in Genetic Search, Parallel Problem Solving from Nature IV, 201–208
12. Tsutsui S, Ghosh A (1997) Genetic Algorithms with a Robust Solution Searching Scheme, IEEE Trans on Evolutionary Computation 1(3):201–208
13. Jin Y, Branke J (2005) Evolutionary Optimization in Uncertain Environments, IEEE Trans on Evolutionary Computation 9(3):303–317
14. Moscato, P (1989) On Evolution, Search, Optimization, Genetic Algorithms and Martial Arts: Towards Memetic Algorithms, Caltech Concurrent Computation Program, C3P Report 826
15. Nagata, Y, Kobayashi, S (1997) Edge Assembly Crossover: A High-power Genetic Algorithm for the Traveling Salesman Problem, Proceedings of the 7th International Conference on Genetic Algorithms 450–457
16. Nagata, Y (2004) The EAX Algorithms Considering Diversity Loss Proceedings of the 8th International Conference on Parallel Problem Solving from Nature – PPSN VII 332–341

Appendix: Calculation of the Distance of Tours

It is difficult to predefine the deadheading costs between required edges. A tour for a vehicle is defined as a sequence of required edges, whereas the actual path of the tour is defined as a sequence of all edges which the vehicles must traverse. For example, assuming that there are four required edges A, B, C, and D (Fig. 22.12a). The minimum actual paths of tour "BC", "ABC", and "ABCD", which are the minimum length of all possible actual paths, are those as shown in Fig. 22.12 (b), (c), and (d), respectively. Note that deadheading edges between required edges B and C in these figure are depicted in dashed lines. The deadheading costs between adjacent required edges varies in accordance with a sequence of required edges in a tour. Therefore, all the required edges in a tour need to be taken into account in order to calculate the total cost.

A distance matrix is employed between vertices which is unchanged during evolutionary search. The distance matrix is calculated by using Dijkstra's Algorithm before evolutionary computation is run. The cost of tours is calculated as follows: by considering the graph in Fig. 22.13, which consists of a depot node and a sequence of required edges whose order is the same as a sequence in tours (Note that although there are two depot nodes in this figure, these nodes are the same). The minimum cost $d_{min}(t^0_{e_1})$, $d_{min}(t^1_{e_1})$ to arriving at the terminal nodes $t^0_{e_1}$, $t^1_{e_1}$ of the first required edge e_1 in the tour is set to $D(depot, t^0_{e_1}))$ and $D(depot, t^1_{e_1}))$, respectively, where $D(\cdot,\cdot)$ denotes the element of the distance matrix, and *depot* stands for the depot node. The minimum cost $d_{min}(t^0_{e_i})$, $d_{min}(t^1_{e_i})$ to arriving at the terminal nodes $t^0_{e_i}$, $t^1_{e_i}$ of required edge e_i is recursively calculated by using the minimum cost $d_{min}(t^0_{e_{i-1}})$, $d_{min}(t^1_{e_{i-1}})$ of previous required edge e_{i-1}. That is,

$$\begin{aligned} d_{min}(t^0_{e_i}) &= \min(d_{min}(t^0_{e_{i-1}}) + C_{e_{i-1}} + D(t^1_{e_{i-1}}, t^0_{e_i}), \\ &\qquad d_{min}(t^1_{e_{i-1}}) + C_{e_{i-1}} + D(t^0_{e_{i-1}}, t^0_{e_i})) \\ d_{min}(t^1_{e_i}) &= \min(d_{min}(t^0_{e_{i-1}}) + C_{e_{i-1}} + D(t^1_{e_{i-1}}, t^1_{e_i}), \\ &\qquad d_{min}(t^1_{e_{i-1}}) + C_{e_{i-1}} + D(t^0_{e_{i-1}}, t^1_{e_i})), \end{aligned}$$

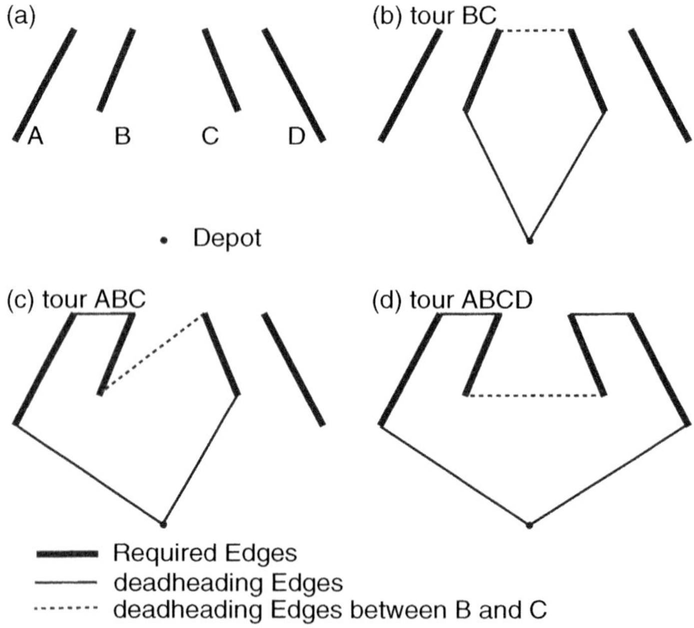

Fig. 22.12. (a) An alignment of required edges A, B, C, D, actual paths of tour (b) "BC", (C) "ABC", and (d) "ABCD"

where C_{e_i} is the cost of ith required edge. Finally, the cost C_{T_i} of tours T_i is defined as

$$C_{T_i} = \min(d_{min}(t^0_{e_{last}}) + C_{e_{i-1}} + D(t^1_{e_{last}}, depot) \\ d_{min}(t^1_{e_{last}}) + C_{e_{i-1}} + D(t^0_{e_{last}}, depot)),$$

where e_{last} indicates the last required edge in the tour T_i.

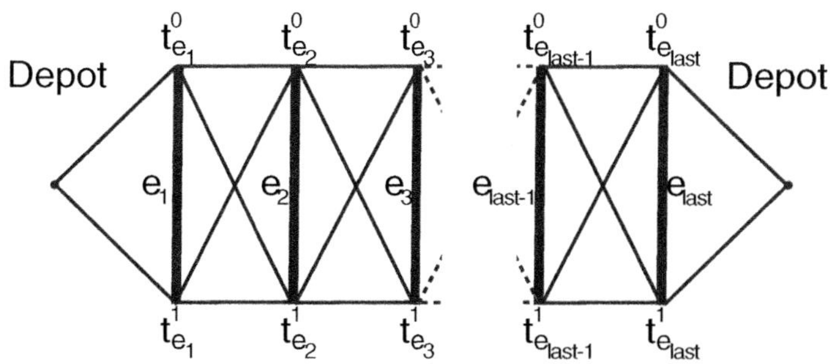

Fig. 22.13. A graph for calculating the cost of tours. Note the identical depot nodes.

23

An Evolutionary Approach For Robust Layout Synthesis of MEMS

Zhun Fan[1], Jiachuan Wang[2], Min Wen[3], Erik Goodman[4], and Ronald Rosenberg[5]

[1] Technical University of Denmark, Department of Mechanical Engineering
Lynby, 2800, Denmark
zf@mek.dtu.dk

[2] Systems Department, United Technologies Research Center
East Hartford, 06128, USA
WangJ2@utrc.utc.com

[3] Technical University of Denmark, Department of Informatics and Mathematical Modelling, Lynby, 2800, Denmark
mw@imm.dtu.dk

[4] Michigan State University, Department of Electrical and Computer Engineering
East Lansing, 48823, USA
goodman@egr.msu.edu

[5] Michigan State University, Department of Mechanical Engineering
East Lansing, 48823, USA
rosenben@egr.msu.edu

Summary. The chapter introduces a robust design method for layout synthesis of micro-electro-mechanical (MEM) resonators subject to inherent geometric uncertainties such as the fabrication error on the sidewall of the structure. The robust design problem is formulated as a multi-objective constrained optimisation problem with certain assumptions and treated by a special constrained genetic algorithm. Case studies based on layout synthesis of a crableg resonator and a comb-driven micro-resonator show that the approach proposed in this chapter can lead to design results that meet the target performance and are less sensitive to geometric uncertainties than typical designs.

23.1 Introduction

Micro-electro-mechanical systems (MEMS) are tiny mechanical devices that are built onto semiconductor chips and are measured in micrometers. They usually integrate across different physical domains, including fluidics, optics, mechanics and electronics, and are used to make devices such as pressure sensors, gyroscopes, engines, and accelerometers etc. The sucess of MEMS designs are usually highly dependent on the designers' knowledge and experience. One reason for this is the complexity involved in the modeling, design and fabrication of MEMS - there are many constraints in designing and fabricating MEM devices due to the limitations of current

Z. Fan et al.: *An Evolutionary Approach For Robust Layout Synthesis of MEMS*, Studies in Computational Intelligence (SCI) **51**, 519–542 (2007)
www.springerlink.com

fabrication techniques. As a result, many design issues are still not modeled and cannot be detected by the simulation software. However, as process technologies become more stable, research emphasis has been shifted from developing specific process technologies towards the design of systems with a large number of reusable components, such as resonators, accelerometers, gyroscopes, and micro-mirrors. It is obvious that performance of individual components will influence the quality of the whole system [14]. For example, frequency stability of a MEM resonator can directly affect the quality of the MEM RF system in which it serves as a component of a filter or an oscillator. It greatly benefits the MEMS designers if the routine design of frequently used components can be optimized automatically by computer programs, while the designers can take more time in contemplating the more creative conceptual designs. The research of layout synthesis of microresonators has been carried out by many researchers. Some notable research include both deterministic numeric approaches [6, 17, 23] and meta-heuristic approaches such as evolutionary computation [13, 15] and simulated annealing [18]. However, little has been done to account for the uncertainties and most of previous work has not considered another major dimension of MEMS design - robustness [12, 20]. Actually, with current micromachining techniques, the fabrication process variation in MEMS is inevitable and will continue when devices are miniaturized to the point of process limitations [14]. For example, it is reported in [9] that the width of a typical suspension beam has a fabrication tolerance of about 10%. Applications of robust design research include robust optical coating design [8, 22]. Hwang [10] used axiomatic design to do robust design of a vibratory micro gyroscope. In this chapter, we present a robust design approach for MEMS subject to process-induced geometrical uncertainties. In this approach, we first formulate the robust design problem as a multi-objective constrained optimization problem [19], and then solve it using a special constrained genetic algorithm. Case studies based on layout synthesis of a crableg resonator and a comb-driven MEM resonator show that the robust designs nominally meet the target performance and are less sensitive to geometric uncertainties.

23.2 Formulation of the Robust Design Problem

Let $\overrightarrow{x} = \{x_1, x_2, ..., x_n\}$ be an array of design variables of a given design problem. We assume that the uncertainty, $\overrightarrow{\delta} = \{\delta_1, \delta_2, ..., \delta_n\}$, can be characterized as a random vector with the following statistics

$$E(\overrightarrow{\delta}) = 0_{n\times 1} \tag{23.1}$$

$$E(\overrightarrow{\delta}\,\overrightarrow{\delta}^T) = \Omega \in \Re^{n\times n} \tag{23.2}$$

where Ω is the covariance matrix and is positive semi-definite. If the uncertainties are uncorrelated then Ω is diagonal, otherwise the off-diagonal entries are non-zero when correlation exists.

Given a function $f(x)$ describing the performance of a design merit, the robust design problem that we aim to solve is to minimize the expected value of the squared error between the actual and target performance. We can write this as:

$$\min_x E(\overrightarrow{f}(x,\delta) - \overrightarrow{f})^2 \text{ subject to } g_i(\overrightarrow{x}) \leq 0 \tag{23.3}$$

where is the target performance, and the expectation is taken over the random vector $\overrightarrow{\delta}$.

The problem posed in (24.3) is a difficult robust optimization problem to solve in general. To simplify the problem, we choose to approximate $f(\overrightarrow{x}, \overrightarrow{\delta})$ with a first order Taylor series expansion in $\overrightarrow{\delta}$ as

$$f(\overrightarrow{x}, \overrightarrow{\delta}) \cong f(\overrightarrow{x}, 0) + \nabla_{\delta} f(\overrightarrow{x}, 0)\overrightarrow{\delta} \tag{23.4}$$

where $\nabla_{\delta} f(\overrightarrow{x}, 0)$ is the gradient of $f(\overrightarrow{x}, \overrightarrow{\delta})$ with respect to $\overrightarrow{\delta}$. Using this approximation, we can expand the expression of $f(\overrightarrow{x}, \overrightarrow{\delta}) - \overrightarrow{f})^2$ into

$$\begin{aligned}(f(\overrightarrow{x}, \overrightarrow{\delta}) - \overrightarrow{f})^2 \cong f(\overrightarrow{x}, 0) - \overrightarrow{f})^2 + 2(f(\overrightarrow{x}, 0) - \overrightarrow{f})^2 \nabla_{\delta} f(\overrightarrow{x}, 0)\overrightarrow{\delta} \\ + \nabla_{\delta} f(\overrightarrow{x}, 0)\overrightarrow{\delta}\,\overrightarrow{\delta}^T \nabla_{\delta}^T f(\overrightarrow{x}, 0)\end{aligned} \tag{23.5}$$

Taking the expectation of the above equation, we can get

$$\begin{aligned}E(f(\overrightarrow{x}, \overrightarrow{\delta}) - \overrightarrow{f})^2 \cong f(\overrightarrow{x}, 0) - \overrightarrow{f})^2 + 2(f(\overrightarrow{x}, 0) - \overrightarrow{f})^2 \nabla_{\delta} f(\overrightarrow{x}, 0) E(\overrightarrow{\delta}) \\ + \nabla_{\delta} f(\overrightarrow{x}, 0) E(\overrightarrow{\delta}\,\overrightarrow{\delta}^T) \nabla_{\delta}^T f(\overrightarrow{x}, 0)\end{aligned} \tag{23.6}$$

By reducing equation (24.6), based on our assumptions about the mean and covariance of $\overrightarrow{\delta}$ according to (24.1) and (24.2), we obtain

$$\begin{aligned}E(f(\overrightarrow{x}, \overrightarrow{\delta}) - \overrightarrow{f})^2 \cong f(\overrightarrow{x}, 0) - \overrightarrow{f})^2 \\ + \nabla_{\delta} f(\overrightarrow{x}, 0) \Omega \nabla_{\delta}^T f(\overrightarrow{x}, 0)\end{aligned} \tag{23.7}$$

Substituting the approximation in (24.7) back into the original design problem posed in (24.3) yields

$$\begin{aligned}&\min_{x} f(\overrightarrow{x}, 0) - \overrightarrow{f})^2 + \nabla_{\delta} f(\overrightarrow{x}, 0) \Omega \nabla_{\delta}^T f(\overrightarrow{x}, 0) \\ &\text{subject to } g_i(\overrightarrow{x}) \leq 0\end{aligned} \tag{23.8}$$

To non-dimensionalize the cost function, we decide to divide through by $\overline{f}^2$. We then refer to the following expression as our robust design problem

$$\begin{aligned}&\min_{x}\left(\left(\frac{f(\overrightarrow{x}, 0) - \overrightarrow{f}}{\overrightarrow{f}}\right)^2 + \frac{1}{\overrightarrow{f}^2}\left(\nabla_{\delta} f(\overrightarrow{x}, 0) \Omega \nabla_{\delta}^T f(\overrightarrow{x}, 0)\right)\right) \\ &\text{subject to } g_i(\overrightarrow{x}) \leq 0\end{aligned} \tag{23.9}$$

It is now easy to see that the expression we want to minimize has two distinct terms. For notational convenience, we will label the two terms as

$$N(\overrightarrow{x}) \equiv \left(\frac{f(\overrightarrow{x}, 0) - \overrightarrow{f}}{\overrightarrow{f}}\right)^2 \tag{23.10}$$

$$D(\vec{x}, \Omega) \equiv \frac{1}{\vec{f}^{2}}(\nabla_{\delta} f(\vec{x}, 0)\Omega \nabla_{\delta}^{T} f(\vec{x}, 0)) \tag{23.11}$$

With the above definitions the robust design problem posed in (23.9) becomes

$$\min_{x}(N(\vec{x}) + D(\vec{x}, \Omega))\text{subject to} g_i(\vec{x}) \leq 0 \tag{23.12}$$

The first term, $N(\vec{x})$, penalizes deviation of the nominal solution, $f(\vec{x}, 0)$, from the target, $\bar{f}$, while the second term, $D(\vec{x}, \Omega)$, penalizes the sensitivity of the design with respect to $\vec{\delta}$. The first term is a performance index, while the second term is a robustness index. Since there are two objectives in the formation of the cost function to be minimized, a trade-off is usually needed to be made by the designer to either focus on minimizing the squared error of the nominal design or on reducing the sensitivity.

23.3 Modeling Uncertainty in MEMS Layout Design

In this research, we assume that the uncertainty in the fabrication process is introduced by etch-induced variations in line-width, and the structure is etched uniformly.

Figure 23.1 illustrate the two uniform etch scenarios on a structure - overetch and underetch. Take the under-etch situation for example, after process variation is introduced, some design variables may increase (such as L_1 and L_2), other design variables (such as L_3) may decrease, while some others may stay unchanged (such as L_4).

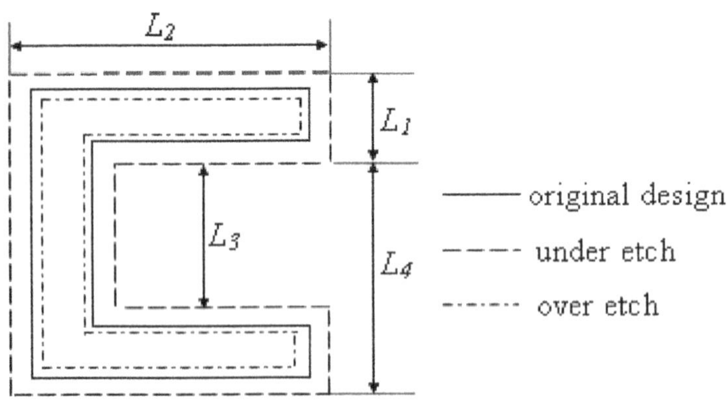

Fig. 23.1. Under- and over-etch of a MEM structure

We can model the geometric process variations using a simple additive uncertain model

$$\tilde{x} = \vec{x} + \vec{\delta} \tag{23.13}$$

where $\tilde{x}$ is the uncertain (actual) design vector and for the above simple example $\vec{\delta} = [\delta_{L_1}, \delta_{L_2}, \delta_{L_3}, \delta_{L_4}]$.

Since the structure is etched uniformly, we can assume that the standard deviation of each term of $\overrightarrow{\delta}$ is equal. If we define ρ to be a normal random variable with standard deviation of σ, then we can write

$$\overrightarrow{\delta} = \rho\zeta \tag{23.14}$$

where $\zeta = [1, 1, -1, 0]^T$, and is called a variation vector. Note that in the condition of underetch, L_1 and L_2 increase, L_3 decreases, and L_4 is not changed. Also note that in this case, ρ is positive, the above facts can easily be verified by (23.14).

Because ρ is a normal random variable, it can also be used to model the overetch situation, in which ρ will take a negative value. According to (24.2), we can obtain

$$\Omega = E(\overrightarrow{\delta}\overrightarrow{\delta}^T) = \sigma^2\zeta\zeta^T \tag{23.15}$$

23.4 Robust Design Optimization Using Evolutionary Approaches

Genetic algorithms with a robust solution searching scheme was first presented by Tsutsui [21]. Jin and Branke [11] made a thorough survey of applying evolutionary computation in uncertain environments. One advantage of using genetic algorithms is its convenience to solve the optimisation problem with both discrete and continuous design variables. While it is very difficult for many numerical optimization approaches (for example, gradient-based approaches) to include considerations of feature size constraints [6], it is quite convenient for genetic algorithms to do so. We need to modify the objective function only slightly, mapping real values of design variables to integer multiples of the feature size λ before using them in formulations of constraints and objectives. No modifications to the genetic algorithm are needed. In this research, we always set the feature size, λ, as $0.1\mu m$.

In this research, we use a special constrained GA that exploits pair-wise comparisons in a tournament selection operator to devise a penalty function approach that does not require any penalty parameter [1]. Careful comparisons among feasible and infeasible solutions are made so as to provide a search direction towards the feasible region. Once sufficient feasible solutions are found, a niching method (along with a controlled mutation operator) is used to maintain diversity among feasible solutions. This allows a real-parameter GA's crossover operator to continuously find better feasible solutions, gradually leading the search nearer to the true optimum solution.

As shown in equation (23.10) and (23.11), there are two design objectives to minimize in the robust design problem. The first objective relates to the design performance, while the second objective reflects robustness of the design. To verify that taking into account of the robustness objective in the optimization process can help to reduce the sensitivity of the resulting designs to variations of the design variables, we carry out a comparative study. In the first set of runs of genetic algorithm, we only consider the first design objective, i.e. the performance objective. In the second set of runs of genetic algorithm, we consider both design objectives, including performance objective and robustness objective.

In the second set of runs, we also need to use a special constraint handling strategy for the robust design [2]. Besides the original constraints in the design problem, the strategy considers additional constraints to make sure that the resulting designs are feasible even after variations of the design variables. The additional constraints can be handled using the constrained-domination principle [3], in which two other paramters, extent of neighbourhood (δ), and the number of neighboring points (H), are used to compute the mean effective objectives and the feasibility of a solution. In our research, we set the extend of neighbourhood, δ, to $0.1\mu m$, and the number of neighboring points, H, to 5. Because we assume a uniform etch process in the research, variations of the design variables are not independent. Instead, the variations of the design variables are interrelated with each other according to equation (23.14), which means we only need to know the variation of one design variable to predict the variations of all other design variables. As a result, we can have a much simpler implementation of the constrained-domination principle, i.e., for each individual design solution (represented by $\overrightarrow{x}_i$) we only need to have a one dimensional partition of the δ-neighbourhood of ρ into H equal segments. Then for each segment, we sample a point to represent ρ. For each of the samplings of ρ, we get a deviation of the design solution according to equation (23.13). We can then calculate the mean effective objective values of the H neighbourhood of the individual and use it to represent the individual's fitness.

For design results obtained from both sets of runs, Monto-Carlo simulations are carried out with the design variables deviated by predefined variations. It is expected that significant differences in design performances be observed between the two sets of results. The set of results with considerations of robustness should be less sensitive to uncertainties and have narrower distributions of performances due to parameter uncertainties.

23.5 Case Studies

Two case studies in the area of MEMS design were carried out to verify the effectiveness of the above robust design method using the constrained genetic algorithm. The first design problem is a six design variables crab-leg resonator. The second design problem is a comb-drive microresonator, with fifteen mixed-type design variables.

23.5.1 Crab-Leg Resonator Design

The Crab-Leg problem was originally taken from [19]. The goal of the design is to robustly match the resonant frequency to the predefined target frequency in the presence of geometric process variations. We are going to show that results obtained with robustness consideration are much more insensitive to geometric process variations through a comparative study.

Crab-Leg Resonator Design Variables and Constraints

The crab-leg resonator is shown in Figure 23.2. The structure exhibits four-fold symmetry and therefore there are six design variables of interest. We assume that the resonator is fabricated with surface micro-machining technique and the entire structure has a fixed thickness of t. The variables X and Y represent the maximum dimensions of the overall structure.

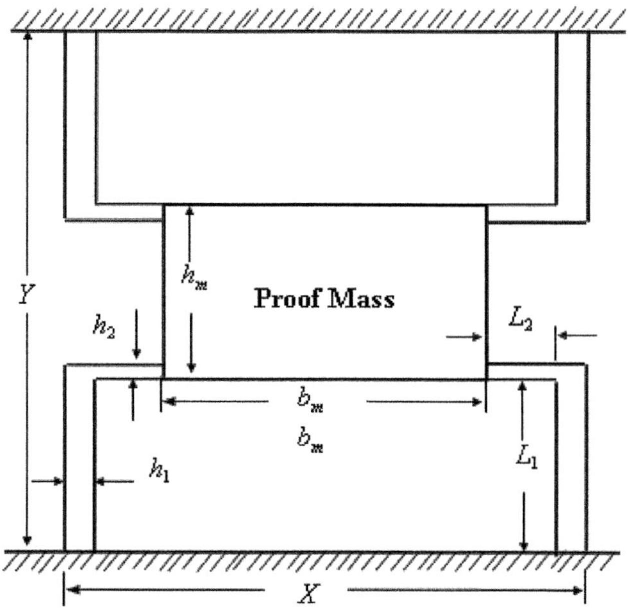

Fig. 23.2. Diagram of a crab-leg resonator

If we neglect the mass of the four legs, the expression for M is

$$M = \rho h_m b_m t \tag{23.16}$$

The expressions for k_x and k_y are given by

$$k_x = 16Et(\frac{h_1}{L_1})^3(\frac{h_1^3 L2 + h_2^3 L1}{4h_1^3 L2 + h_2^3 L1}) \tag{23.17}$$

$$k_y = 16Et(\frac{h_2}{L_2})^3(\frac{h_1^3 L_2 + h_2^3 L_1}{h_1^3 L_2 + 4h_2^3 L_1}) \tag{23.18}$$

We assume that

$$\Omega_n = \sqrt{k_n/M} \tag{23.19}$$

then

$$\Omega_n^2 = \frac{16Eh_1^3(h_1^3 L_2 + h_2^3 L_1)}{\rho h_m b_m L_1^3(4h_1^3 L_2 + h_2^3 L_1)} \tag{23.20}$$

We define the design vector to be $\overrightarrow{x} = [L_1, L_2, h_1, h_2, h_m, b_m]$. Now we need to define the constraints for the problem. They can be categorized into five groups:

1. $h_1 \geq h_{min}$, $h_2 \geq h_{min}$, $b_m \geq h_{min}$, $h_m \geq 2h_2 + h_{min}$
2. $L_1 \geq h_1$, $L_2 \geq 10h_2$
3. $X \geq 2L_2 + 2h_1 + b_m$, $Y \geq 2L_1 + h_m$
4. $k_y \geq \alpha k_x$
5. $\beta_{max} \geq \beta$

The first group of constrains was imposed to comply with the design rules of minimum line and space requirements used by MEMS processes such as MUMPs. The second group ensures that we can model the legs as thin-beams. The third group specifies the requirements on the overall size of the structure, which is usually imposed by the designer. The fourth group is to seperate the two resonant frequencies, thus reducing motion in the y-direction. Therefore we expect that the parasital y-direction resonant frequency to be much greater than that in the x-direction. The last constraint is to put a maximum limitation on the stress of the joint, β so that the leg will not break at a point where the two beams join to form a leg. The stress can be calculated using the following equation

$$\beta = \frac{12Eh_1^3h_2^3L_1D}{h_2^2L_1^2(4h_1^3L_2 + h_2^3L_1)} \tag{23.21}$$

where D is the maximum allowable deflection of the structure in the x-direction.

All constraints can be written as polynomials in the form $g_i(\overrightarrow{x}) \leq 0$. Table 23.1 lists design parameters for this case study. All of them are chosen to be realistic.

Table 23.1. Design parameter for crab-leg resonator

Design parameter	Description	Value
t	thickness of proof mass and beams	$2.0\mu m$
ρ	density of silicon	$2.3 \times 10^{-12} gm/\mu m^3$
E	Elastic Modulus of silicon (1600Gpa)	$1.6 \times 10^8 gm/\mu ms^2$
h_{min}	minimum dimension	$2.0\mu m$
D	maximum deflection in the x-direction	$2.0\mu m$
β_{max}	maximum allowable stress	$1.6 \times 10^6 gm/\mu ms^2$
X	maximum size of structure in the x-direction	$600\mu m$
Y	maximum size of structure in the y-direction	$600\mu m$
α	minimum stiffness ratio k_y/k_x	16

Crab-Leg Resonator Design Objectives

The first design objective is to match the natural frequency of the crab-leg resonator with a predefined natural frequency, which we call a performance objective. Given the design variables and constraints discussed in the preceding section, it is important to find an allowable range of natural frequencies. If the predefined natural frequency is outside of the allowable range, then there may exist no feasible solution to the design objective.

It is conveninet to use the constrained genetic algorithm to find the lower and upper bound of the achievable resonant frequencies given the above design constraints, if we define the objective to be minimized as natural frequency ω_n and its reciprocal $1/\omega_n$ respectively. In this case study, the allowable range we found is $\Theta_{crab} = [\omega_{min}, \omega_{max}] = [13K, 21.6M]Hz$. We therefore selected a target frequency:

$$\omega_{target} = 200KHz \in \Theta_{crab} \tag{23.22}$$

Our design objective is not only matching the natural frequency of the crab-leg resonator with the target frequency. In addition, we hope that the natural frequencies of our design will not deviate too much from the target frequency even under the condition of variations of the design variables. This will be our second design objective, and is called a robustness objective.

A Comparative Study of Robust and Non-Robust Design

Robust design takes into consideration of both the performance and the robustness objectives. Non-robust design only considers the performance objective.

According to equation (23.9), we know that the first term, $N(\overrightarrow{x})$, penalizes deviation of the nominal solution, $f(\overrightarrow{x},0)$, from the target, $\bar{f}$; while the second term, $D(\overrightarrow{x},\Omega)$, penalizes the sensitivity of the design with respect to $\overrightarrow{\delta}$. The first term relates to the performance objective, and the second term relates to the robustness objective. A comparative study was carried out to verify that involving the second term in the design objectives increases the robustness of the design solutions.

Results of Non-Robust Design

In the first set of runs, we only considered the performance objective, and define $f_{obj} = N(\overrightarrow{x})^{0.5}$. In the case study of crab-leg resonator, $N(x) \equiv (\frac{\omega_n(x,0)-\omega_{target}}{\omega_{target}})^2$. The performance objective can be easily derived from Equation (23.20) and Equation (23.22). The parameters used for genetic algorithm are listed in Table 23.2.

Table 23.2. Parameters for the constrained genetic algorithm for non-robust design

Parameter	Value
Population size	200
Total no. of generations	200
Cross over probability	0.9000
Mutation probability (real)	0.1500
Niching to be done	1(Yes)
Niching parameter value	0.9000
Exponent (n for SBX)	2.00
Exponent (n for Mutation)	50.00

Ten runs were repeated and the experimental data was obtained in Table 23.3. It can be seen from the table that all 10 resulting design candidates have obtained very good performances. The resonant frequecies of them are 199.9KHz. The performance objectives were minized to the level of 3E-7 for all candidates.

However, it can be easily noticed that the design results are non-robust. For instance, design variable h_1 is found to be $2.0\mu m$ for all design candidates, which lies exactly in the lower limit of the constraint for h_1: $h_1 \geq h_{min} = 2\mu m$. Obviously, if there is any decreasing variation of the design variable, which corresponds to an

Table 23.3. Parameters for the constrained genetic algorithm for non-robust design

Run No.	1	2	3	4	5	6	7	8	9	10
$L_1(\mu m)$	188.3	165.2	157.0	177.3	178.7	166.1	148.4	178.4	195.7	168.6
$L_2(\mu m)$	47.5	37.2	32.1	43.9	42.6	34.0	47.4	27.5	51.3	39.9
$h_1(\mu m)$	2.0	2.0	2.0	2.0	2.0	2.0	2.0	2.0	2.0	2.0
$h_2(\mu m)$	2.1	2.0	2.0	2.0	2.0	2.0	2.0	2.4	2.3	2.0
$h_m(\mu m)$	20.3	30.0	40.1	24.7	21.4	31.5	29.0	28.1	19.5	21.5
$b_m(\mu m)$	27.1	26.9	24.1	25.7	29.3	25.9	34.5	28.4	26.8	34.8
$N(\overrightarrow{x})$	3E-7	3E-7	3E-7	3E-7	3E-7	3E-7	3E-7	3E-7	3E-7	3E-7
f_{obj}	5E-4	5E-4	5E-4	5E-4	5E-4	5E-4	5E-4	5E-4	5E-4	5E-4

overetch situation in the fabrication process, the design will no longer be feasible. The robust design solution reported in [19] has the same feasibility problem (Table 23.4), in which design variable h_2 is optimized to be $2.0\mu m$ and lies exactly in the lower limit of the constraint for h_2: $h_2 \geq h_{min} = 2\mu m$.

A reason for this problem is that although robust design method in [19] treated the robustness of designs in terms of maintaining good performance, it did not, however, attempt to consider the robustness of the designs in terms of satisfying constraints. In the next section, we are going to see that using a special constrained genetic algorithm can help the design results to avoid infringing design constraints, and that the design results obtained with the robust design method using the constrained genetic algorithm is less sensitive to parameter variations.

Table 23.4. The robust design for crab-leg resonator in [17]

Parameter	$L_1(\mu m)$	$L_2(\mu m)$	$h_1(\mu m)$	$h_2(\mu m)$	$h_m(\mu m)$	$b_m(\mu m)$
value	265.64	22.84	14.09	2.00	68.71	392.60

Results of Robust Design

In the second set of runs, we considered both the performance objective $N(\overrightarrow{x})$ and the robustness objective $D(\overrightarrow{x}, 0)$. Because both objectives are normalized, we can simply sum them together to form a new objective

$$f_{obj} = N(\overrightarrow{x})^{0.5} + D(\overrightarrow{x}, 0)^{0.5} \tag{23.23}$$

where

$$N(x) \equiv (\frac{\omega_n(x, 0) - \omega_{target}}{\omega_{target}})^2 \tag{23.24}$$

and

$$D(\overrightarrow{x}, \Omega) \equiv \frac{1}{\omega_{target}^2} \nabla_\delta \omega_n(\overrightarrow{x}, 0) \Omega \nabla_\delta^T \omega_n(\overrightarrow{x}, 0) \tag{23.25}$$

As discussed above, $N(\overrightarrow{x})$ can be easily obtained. But to calculate $D(\overrightarrow{x}, \Omega)$ is not an easy task. First, we need to decide the varation vector

$$\zeta = [-1, -1, 1, 1, 1, 1]^T \tag{23.26}$$

Then according to Equation (23.15), we get

$$\Omega = E(\overrightarrow{\delta}\overrightarrow{\delta}^T) = \sigma^2 \zeta \zeta^T = \sigma \begin{pmatrix} 1 & 1 & -1 & -1 & -1 & -1 \\ 1 & 1 & -1 & -1 & -1 & -1 \\ -1 & -1 & 1 & 1 & 1 & 1 \\ -1 & -1 & 1 & 1 & 1 & 1 \\ -1 & -1 & 1 & 1 & 1 & 1 \\ -1 & -1 & 1 & 1 & 1 & 1 \end{pmatrix} \tag{23.27}$$

Here we need to make an assumption about σ. In this chapter, we assume $\sigma = 0.1\mu m$. We can also use a further normalized term

$$D(\overrightarrow{x}, \Omega)_{norm} = D(\overrightarrow{x}, \Omega)/\sigma^2 \tag{23.28}$$

so that $D(\overrightarrow{x}, \Omega)_{norm}$ will not be influenced by our assumptions. The parameters of the constrained genetic algorithm are listed in Table 23.5.

Table 23.5. Parameters for the constrained genetic algorithm for robust design

Parameter	Value
Population size	200
Total no. of generations	200
Cross over probability	0.9000
Mutation probability (real)	0.1500
Niching to be done	1(Yes)
Niching parameter value	0.9000
Exponent (n for SBX)	2.00
Exponent (n for Mutation)	50.00
extent of neighborhood (Δ)	$0.1\mu m$
number of neighborhood points (H)	5

Ten simulation runs were repeated and the experimental data was recorded in Table 23.6. It can be seen from the table that all resulting design candidates have satisfactory performances. The average resonant frequecy of them is 198.2KHz. The performance objectives were minized to the range of 8E-7 to 1E-4 for all candidates.

Comparison of Robust Design and Non-Robust Design

In the robust design process, we tried very hard to minimize the robustness objective, $D(\overrightarrow{x}, \Omega)$ which ranges from 2E-4 to 9E-4 in this set of runs. But will doing this help the resulting designs to increase their insensitivity to geometric uncertainties? To answer this question, we designed a comparative study as the following: we put two designs in one group for comparison, by selecting one design from the robust design group, and the other from the non-robust design group. We then ran Monte Carlo simulations to model uncertain MEMS fabrication processes.

Table 23.6. Experimental results for robust design of crab-leg resonator

Run No.	1	2	3	4	5	6	7	8	9	10
$L_1(\mu m)$	277.0	255.5	274.0	274.0	268.8	272.7	268.0	270.6	254.2	269.0
$L_2(\mu m)$	38.1	21.1	62.7	25.5	44.2	58.0	50.7	64.4	35.4	47.1
$h_1(\mu m)$	11.6	15.3	11.1	13.9	12.1	11.5	12.3	11.2	12.5	13.0
$h_2(\mu m)$	2.9	2.1	6.2	2.2	3.7	5.7	4.7	6.4	3.3	4.7
$h_m(\mu m)$	35.3	74.8	44.8	49.1	41.1	42.8	53.4	47.2	51.1	56.8
$b_m(\mu m)$	397.7	519.2	397.3	495.7	457.9	432.0	395.5	403.5	459.1	414.4
$N(\overrightarrow{x})$	8E-7	2E-6	2E-4	2E-4	4E-4	7E-5	2E-4	1E-4	1E-4	8E-6
$D(\overrightarrow{x}, \Omega)$	7E-4	2E-4	8E-4	3E-4	6E-4	9E-4	7E-4	9E-4	6E-4	7E-4
f_{obj}	0.029	0.017	0.042	0.029	0.044	0.038	0.039	0.040	0.035	0.029

We introduced the same variations to the design variables of both designs to emulate uniform overetch and/or underetch situations. To represent the variations in the process we generated 10,000 gaussian random vectors with a standard deviation, σ, of $0.1\mu m$. The natural frequencies of both the robust design and the non-robust design were calculated, and histograms of them plotted as shown in Figure 23.3 and 23.4.

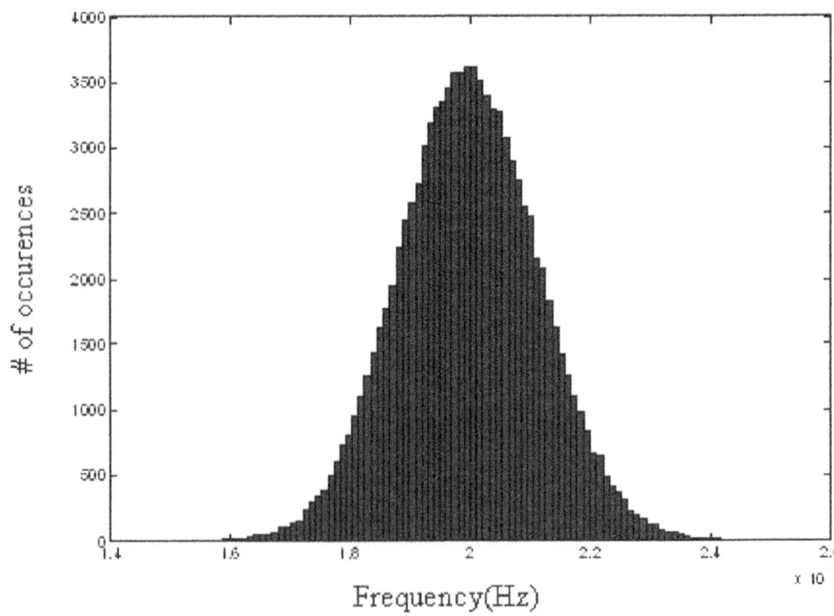

Fig. 23.3. Histogram of natural frequencies of the non-robust design of crab-leg resonator subject to uncertainties

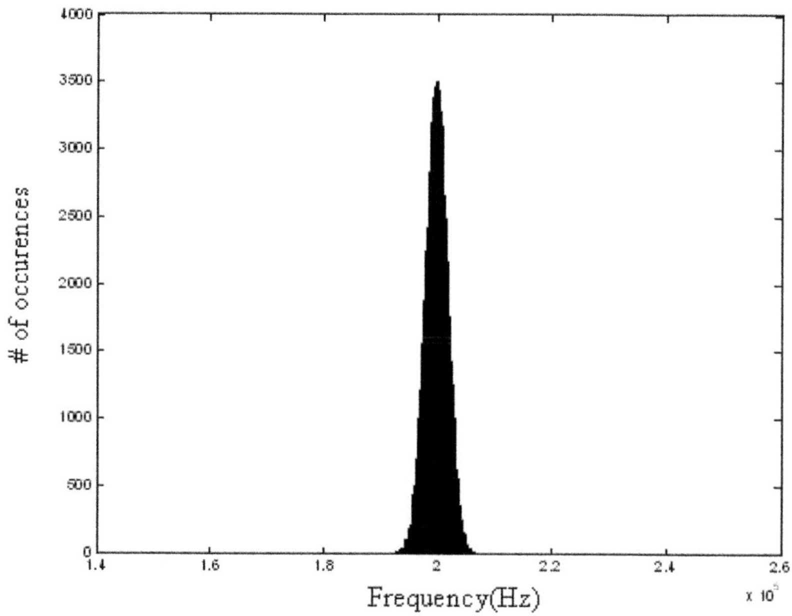

Fig. 23.4. Histogram of natural frequencies of the robust design of crab-leg resonator subject to uncertainties

According to Figure 23.3 and 23.4, we can see that robust design has a much tighter distribution of natural frequencies, and therefor is much less sensitive to geometric variations. If we assume that the natural frequencies have a normal distribution with a normal probability density functin (pdf) as the following:

$$y = f(x \mid \mu, \sigma) = \frac{1}{\sigma\sqrt{2\pi}} e^{\frac{-(x-\mu)^2}{2\sigma^2}} \tag{23.29}$$

In this equation, μ stands for the mean value and σ is the standard deviation. Through parameter estimation using a MATLAB funtion 'normfit', we got the following results: For the robust design, $\mu_{robust} = 199.7KHz$, $\sigma_{robust} = 1.95KHz$. For the non-robust design, $\mu_{non\_robust} = 199.9KHz$, $\sigma_{non\_robust} = 11.3KHz$.

From these results we can see that the mean value of the robust design and the non-robust design are both very close to the target frequency. However, the standard deviation of robust design is much lower than that of the non-robust design, which means that the robust design is much less sensitive to uncertainties. Statistically, under the condition of variations, its performances degrade much less compared with those of the non-robust design. The probability density function curves of both robust design and non-robust design are also shown in Figure 23.5 and 23.6. Tests of other design candidates from both robust design group and non-robust design group revealed similar results.

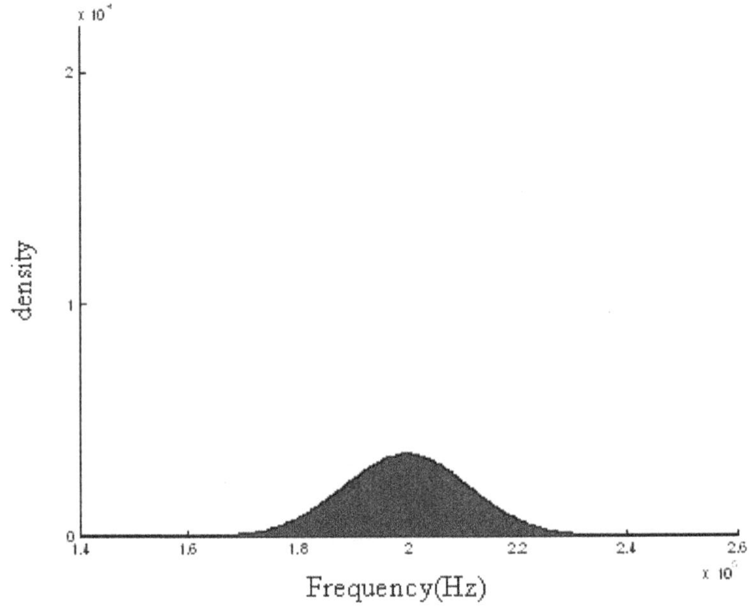

Fig. 23.5. Probability density function of natural frequencies of the non-robust design of crab-leg resonator subject to uncertainties

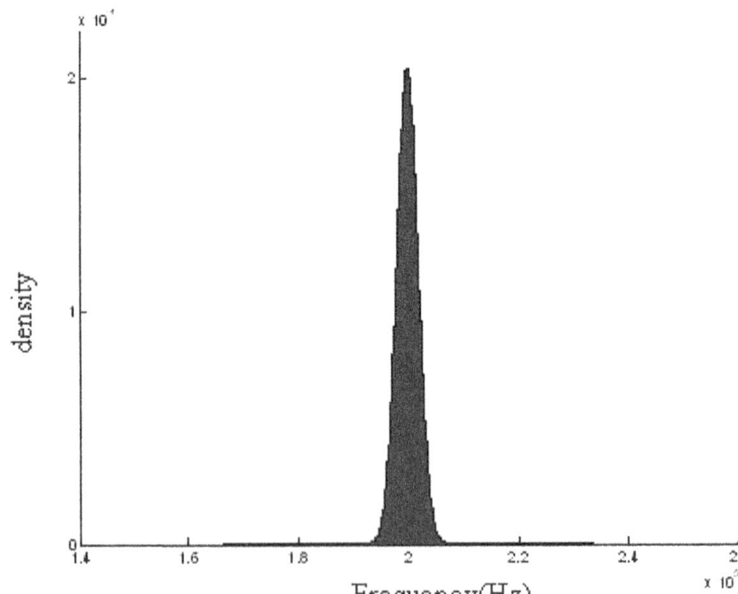

Fig. 23.6. Probability density function of natural frequencies of the robust design of crab-leg resonator subject to uncertainties

23.5.2 Comb-Driven Micro-Resonator Design

The comb-driven micro-resonator problem was originally taken from [6]. The goal of the design is also to robustly match the resonant frequency to the predefined target frequency in the presence of geometric process variations. The case study is more complex and difficult than the first case study in the following senses: 1) it has fifteen design variables, while the crableg problem only has six design variables. 2) its design variables are of mixed type, including thirteen discrete variables, one continuous variable, and one integer variable.

Comb-Driven Micro-Resonator Design Variables and Constraints

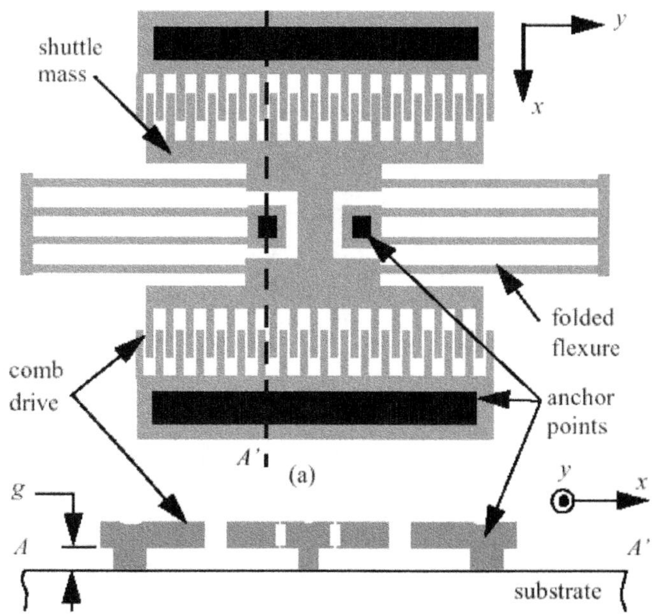

Fig. 23.7. A folded-flexure comb-drive microresonator fabricated in a polysilicon surface microstructural process a) Layout b) Cross-section A-A'

We decided to use 15 design variables for the folded-flexure comb-drive microresonator fabricated in a polysilicon surface microstructural process (Figure 23.7) [7]: in this research. The vector of design variables is defined as follows (Figure 23.8)

$\overrightarrow{x} = [L_b, w_b, L_t, w_t, L_{sy}, w_{sy}, w_{sa}, w_{cy}, L_{cy}, L_c, w_c, L_{sa}, x_o, V, N]$

L_b and w_b are length and width of flexture beam respectively. L_t and w_t are length and width of truss beam respectively. L_{sy} and w_{sy} are length and width of shuttle yoke respectively. L_{cy} and w_{cy} are length and width of comb yoke respectively. L_c and w_c are length and width of comb fingers respectively. w_{sa} represents the width of shuttle axle. g is the gap between comb fingers. s_o represents the comb finger overlap. N is number of rotor comb fingers. V is voltage amplitude.

It is noted that the first 13 design variables have units of μm. They are discrete variables because they can only be integer multiples of the feature size λ, which is set to be $0.1\mu m$ in this research. The fourteenth design variable has units of volts, and is a continuous variable. The fifteenth variable has no unit and is an integer variable.

The constraints for the design variables are also listed:

$2 \leq L_b \leq 400$, $2 \leq w_b \leq 20$, $2 \leq L_t \leq 400$, $2 \leq w_t \leq 20$, $2 \leq L_{sy} \leq 400$,
$10 \leq w_{sy} \leq 400$, $10 \leq w_{sa} \leq 400$, $10 \leq w_{cy} \leq 400$, $2 \leq L_{cy} \leq 700$, $8 \leq L_c \leq 400$, $2 \leq w_c \leq 20$,
$2 \leq L_{sa} \leq 400$, $4 \leq x_o \leq 400$, $0 \leq V \leq 50$, $3 \leq N \leq 50$.

In addition, we assume $t = w_c = g = d$ in our design, for the special case of equal comb finger width, gap, thickness and spacing above the substrate. Some design variables are predefined: they are $w_{ba} = 11$, $w_{ca} = 14$, $\gamma = 4$, in which w_{ba} is the width of beam anchors, and w_{ca} is the width of stator comb anchors, γ is the penetration depth of airflow above the structure.

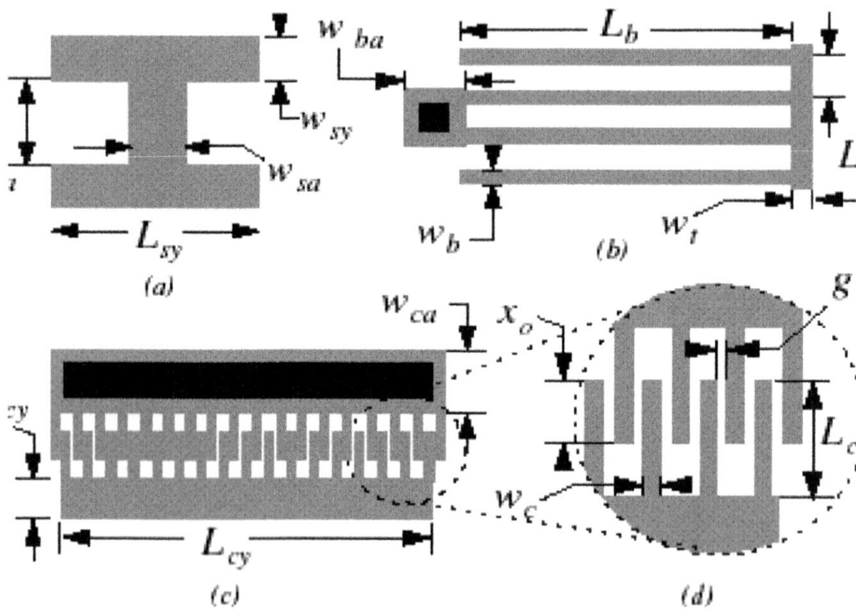

Fig. 23.8. Major design variables for microresonators

There are a number of design constraints for the microresonator cell component, including both geometric constraints and functional constraints. In this chapter, without loss of generality, we consider the following constraints:

$$g_1(x) : -(L_{cy} + 2g + 2w_c) \leq 0$$

$$g_2(x) : L_{cy} + 2g + 2w_c - 700 \leq 0$$

$g_3(x) : -(L_{sy} + 2L_b + 2w_t) \leq 0$

$g_4(x) : L_{sy} + 2L_b + 2w_t) - 700 \leq 0$

$g_5(x) : -(3L_t + w_{sy} + 4L_c - 2x_0 + 2w_{cy} + 2w_{ca}) \leq 0$

$g_6(x) : 3L_t + w_{sy} + 4L_c - 2x_0 + 2w_{cy} + 2w_{ca} - 700 \leq 0$

$g_7(x) : L_c - (x_0 + x_{disp}) - 200 \leq 0$

$g_8(x) : 4 - L_c + (x_0 + x_{disp}) \leq 0$

Among them, the first six are linear constraints, but the last two are nonlinear constraints because the term x_{disp} is highly nonlinear:

$$x_{disp} = QF_{e,x}/K_x \tag{23.30}$$

where is quality factor and can be represented as

$$Q = \sqrt{m_x K_x / B_x^2} \tag{23.31}$$

$F_{e.x}$ is the force generated by the comb drive. The force is proportional to the square of the voltage, V, applied across the comb fingers

$$F_{e,x} = 1.12\varepsilon_0 N \frac{t}{g} V^2 \tag{23.32}$$

where ε_0 is the permittivity of air. We also have

$$K_x = \frac{2EtW_b^3}{L_b^3} \frac{L_t^2 + 14\alpha L_t L_b + 36\alpha^2 L_b^2}{4L_t^2 + 41\alpha L_t L_b + 36\alpha^2 L_b^2} \tag{23.33}$$

where

$$\alpha = (W_t/W_b)^3 \tag{23.34}$$

and

$$B_x = \mu(A_s + 0.5A_t + 0.5A_b)(\frac{1}{d} + \frac{1}{r}) + \frac{A_c}{g} \tag{23.35}$$

where μ is the viscosity of air, and A_s, A_t, A_b, and A_c are bloated layout area of the shuttle, truss beams, flexure beams, and comb finger sidewalls, respectively.

Also we know

$$m_x = m_s + \frac{1}{4}m_t + \frac{12}{35}m_b \tag{23.36}$$

where $m_s = \rho A_s$, $m_t = \rho A_t$, $m_b = \rho A_b$

$$A_s = w_{sa}L_{sa} + 2w_{sy}L_{sy} \tag{23.37}$$

$$A_t = 2w_{ca}L_{cy} \tag{23.38}$$

$$A_b = 8L_b w_b + 2w_t(2L_t + w_a + 2w_b) \tag{23.39}$$

$$A_c = 2Nw_cL_c \tag{23.40}$$

The natural frequency ω_n is defined as

$$\omega_n = \sqrt{\frac{K_x}{m_x}} \tag{23.41}$$

Comb-Driven Micro-Resonator Design Objective

The design objective of comb-driven micro-resonator is to robustly match the natural frequency of the comb-driven micro-resonator with a predefined natural frequency.

Through a similar procedure as in the first case study of crab-leg design, we found the allowable range of natural frequency of the comb-driven micro-resonator to be $\Theta = [\omega_{min}, \omega_{max}] = [2.5K, 6.8G]Hz$. We therefore selected a target frequency :

$$\omega_{target} = 200KHz \in \Theta_{comb} \tag{23.42}$$

A Comparative Study of Robust and Non-Robust Design

A comparative study was carried out to verify that involving the robustness objective in the design objectives can increase the robustness of the design results in this more complex case study.

Results of Non-Robust Design

In the first set of runs, we only considered the performance objective $f_{obj} = N(\overrightarrow{x})^{0.5}$. Again, The parameters for the constrained genetic algorithm are the same as those used in crableg resonator design problem, and listed in Table 23.2. Ten runs were repeated and the experimental data was obtained in Table 23.7.

The table shows that all resulting design candidates have very good performances. The resonant frequecies of them are 199.2KHz. The performance objectives were minized to the level of 2E-5 for all candidates. However, the design results of run no.1 and no.10 can be easily observed to be non-robust. The design variable w_c is found to be $2.0\mu m$ for both design candidates, which lies exactly in the lower limit of the constraint for $w_c : 2 \leq w_c \leq 20$. Obviously, if there is any decreasing variation of the design variable, which corresponds to an overetch situation in the fabrication process, the design will become infeasible.

Results of Robust Design

We used the same robust design procedure as used in crableg resonator design. For the example of comb-driven micro-resonator, the variation vector is set to $\xi =$ [0 1 0 1 1 1 1 1 1 0 1 -1 1 0 0]. Ten runs were repeated and the experimental data was obtained in Table 23.8.

Comparison of Robust Design and Non-Robust Design

Table 23.7. Experimental results for non-robust design of comb-driven micro-resonator

Run No.	1	2	3	4	5	6	7	8	9	10
$L_b(\mu m)$	63.3	108.8	73.1	95.7	103.7	47.9	57.1	117.4	83.4	104.1
$w_b(\mu m)$	3.0	2.7	4.2	5.5	3.9	2.1	3.1	5.2	3.7	4.7
$L_t(\mu m)$	35.9	70.3	41.2	33.7	79.4	13.4	86.3	33.2	51.9	33.5
$w_t(\mu m)$	4.1	8.1	7.1	3.4	7.6	17.7	10.3	4.5	6.0	2.3
$L_{sy}(\mu m)$	224.1	11.1	146.8	96.5	5.3	39.9	146.9	68.4	142.1	47.0
$w_{sy}(\mu m)$	14.9	59.3	128.8	94.1	26.7	109.7	209.6	41.5	81.0	49.8
$w_{sa}(\mu m)$	110.5	23.0	269.5	67.9	51.6	42.0	185.2	201.4	113.4	135.1
$w_{cy}(\mu m)$	78.4	337.6	66.1	23.4	91.8	210.1	293.6	138.2	52.9	136.6
$L_{cy}(\mu m)$	333.7	58.0	372.7	124.4	401.4	440.9	355.7	254.1	155.7	95.8
$L_c(\mu m)$	74.7	191.5	116.9	57.5	95.4	175.2	194.2	230.5	130.3	39.1
$w_c(\mu m)$	2.0	3.7	3.7	2.2	2.9	2.5	6.3	3.8	4.1	2.0
$L_{sa}(\mu m)$	59.8	63.4	60.6	3.1	141.0	147.9	118.0	67.0	25.5	5.1
$x_o(\mu m)$	52.2	13.2	73.1	20.9	45.0	45.1	81.9	76.6	29.3	30.7
$V(volt)$	15.1	12.8	15.0	6.5	18.1	28.2	13.3	23.8	12.1	4.0
N	10	19	7	5	6	10	6	9	6	6
$N(\overrightarrow{x})$	2E-5	2E-5	2E-5	2E-5	2E-5	2E-5	2E-5	2E-5	2E-5	2E-5
f_{obj}	5E-3	5E-3	5E-3	5E-3	5E-3	5E-3	5E-3	5E-3	5E-3	5E-3

Table 23.8. Experimental results for non-robust design of comb-driven micro-resonator

Run No.	1	2	3	4	5	6	7	8	9	10
$L_b(\mu m)$	170.4	241.2	183.7	206.9	267.6	206.0	222.4	214.4	209.5	211.8
$w_b(\mu m)$	14.1	15.3	15.7	17.6	18.7	19.2	19.8	19.7	17.2	16.7
$L_t(\mu m)$	127.2	213.3	85.4	118.3	153.0	110.7	110.9	124.6	194.1	121.1
$w_t(\mu m)$	16.6	18.6	15.1	19.6	18.9	17.3	17.8	18.5	19.9	18.9
$L_{sy}(\mu m)$	311.4	141.4	217.9	172.8	102.3	234.3	211.4	123.8	218.1	194.3
$w_{sy}(\mu m)$	142.4	118.2	286.5	265.3	123.0	228.5	174.8	141.7	221.0	325.8
$w_{sa}(\mu m)$	370.4	321.2	294.6	387.8	51.2	345.2	218.9	264.4	376.5	314.9
$w_{cy}(\mu m)$	308.8	331.7	273.7	381.6	311.6	379.0	348.0	362.3	391.9	351.8
$L_{cy}(\mu m)$	595.8	335.7	577.9	630.0	536.3	655.1	592.1	595.1	629.1	635.3
$L_c(\mu m)$	244.7	192.4	185.8	236.8	247.3	226.7	215.3	242.5	221.6	189.6
$w_c(\mu m)$	5.6	8.4	5.0	6.2	2.3	4.7	3.4	2.4	7.5	5.9
$L_{sa}(\mu m)$	313.0	284.6	210.6	399.0	79.0	305.1	248.1	227.5	396.1	188.4
$x_o(\mu m)$	161.0	180.9	61.8	143.9	184.4	159.5	122.7	70.1	167.3	84.1
$V(volt)$	32.8	17.7	25.0	39.2	27.6	39.1	21.4	45.5	42.8	40.1
N	14	23	14	23	9	13	12	29	12	11
$N(\overrightarrow{x})$	2E-5	2E-5	2E-5	2E-5	2.1E-5	2E-5	2E-5	2E-5	2E-5	2.1E-5
$D(\overrightarrow{x}, \Omega)$	4E-3	3E-3	3E-3	2E-3	1.6E-3	2E-3	2E-3	2E-3	2E-3	2.6E-3
f_{obj}	0.065	0.060	0.063	0.054	0.045	0.052	0.049	0.049	0.054	0.056

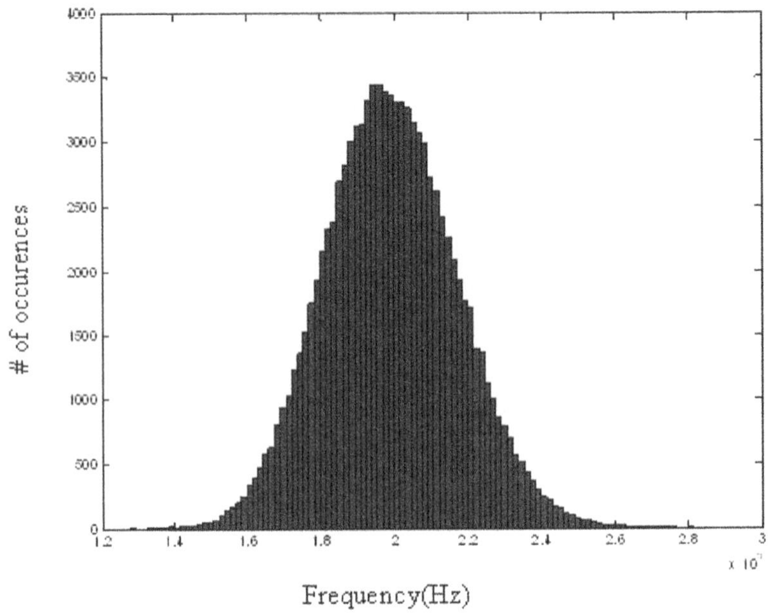

Fig. 23.9. Histogram of natural frequencies of the non-robust design of comb-driven resonator subject to uncertainties

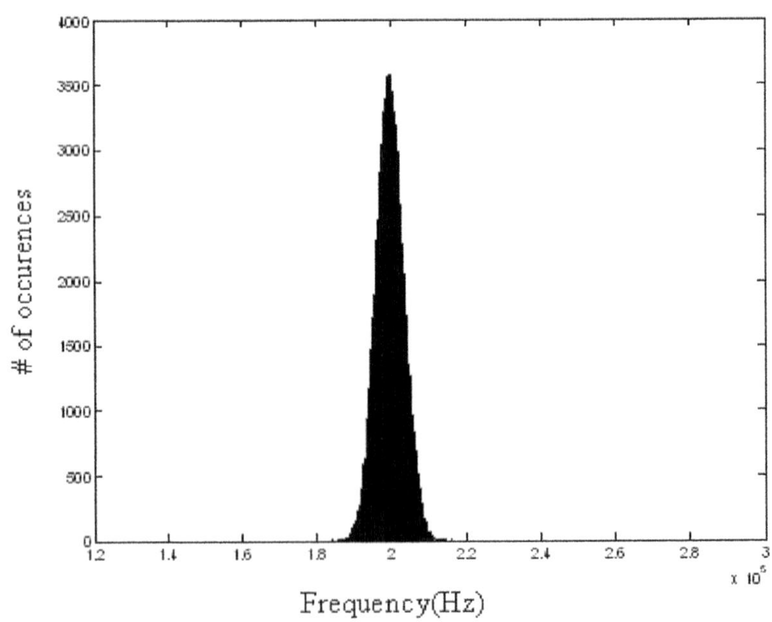

Fig. 23.10. Histogram of natural frequencies of the robust design of comb-driven resonator subject to uncertainties

Monte carlo simulations were run to make a comparison and results shown in Figure 23.9 and 23.10.

The natural frequencies of both the robust design and the non-robust design were calculated, and histograms of them plotted in Figure 23.9 and 23.10. Figure 23.9 and shows histograms of the first design candidate of the ten non-robust designs, and Figure 23.10 the first design candidate of the robust design group. According to the figures, we can see that robust design has a much tighter distribution of natural frequencies, and therefor is much less sensitive to geometric variations. If we assume that the natural frequencies have a normal distribution, we can get the following parameters for the probability density function:

For the robust design, $\mu_{robust} = 199.1KHz$, $\theta_{robust} = 2.96KHz$. For the non-robust design, $\mu_{non\_robust} = 199.2KHz$, $\theta_{non\_robust} = 11.4KHz$.

From this we can see that the mean value of both the robust design and the non-robust design are very close to the target frequency. However, the standard deviation robust design is much lower than that of the non-robust design, which again verifies that the robust design is much less sensitive to uncertainties. Statistically speaking, under the condition of variations, the performances of robust design degrade much less compared with those of the non-robust design. The probability density function curves of both robust design and non-robust design are also shown in Figure 23.11 and 23.12. Tests of other design candidates from both robust design group and non-robust design group revealed similar results.

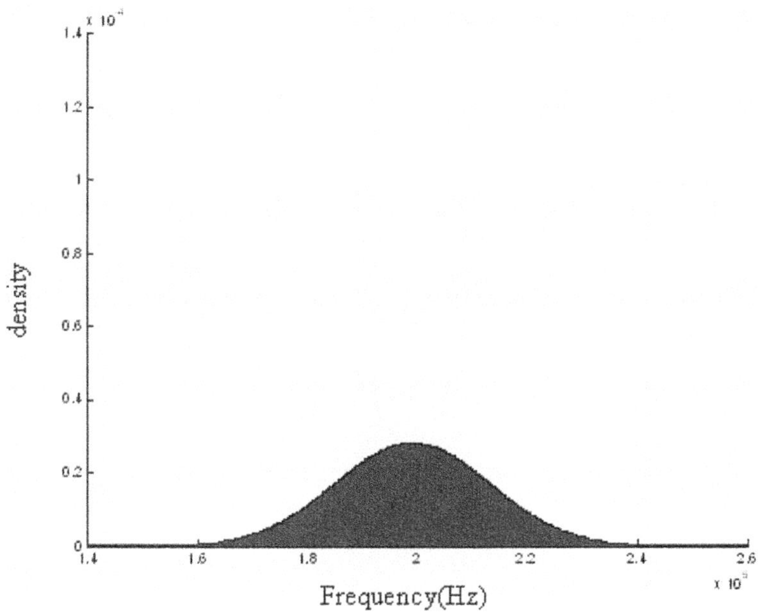

Fig. 23.11. Probability density function of natural frequencies of the non-robust design of comb-driven resonator subject to uncertainties

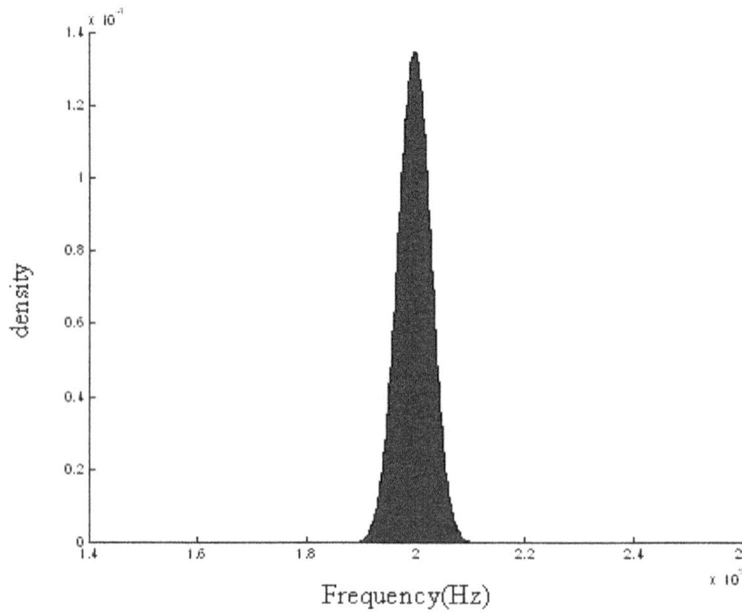

Fig. 23.12. Probability density function of natural frequencies of the robust design of comb-driven resonator subject to uncertainties

23.6 Conclusions

Layout synthesis is an important stage for structured design of MEMS [5, 16], after the stage of the system-level design [4]. The chapter reports a method of robust layout synthesis of MEMS that transforms the robust design problem into a multi-objective constrained optimisation problem, and then solve it using a constrained genetic algorithm. Simulation results based on case studies of layout synthesis of a crableg resonator and a comb-driven micro-resonator show that the robust design solutions obtained using the robust design method presented in this chapter are much less sensitive to process induced uncertainties. Using the constrained genetic algorithm to solve the robust design problem, we can address the issue of design robustness in terms of not only maintaining good design performance, but also satisfying design constraints and therefore keeping designs feasible. Although we have had effective ways of modelling robustness related to design performances, it is still a challenge to find an effecient way to model robustness related to design constraints.

While it is important to study more efficient robust design method and more effective approaches to model uncertainties, it is also an interesting research topic to investigate the relationship between the robustness subject to parametric variations (such as the layout synthesis) and the topology-related robustness in the system level design.

References

1. K. Deb, "An efficient constraint handling method for genetic algorithms," Comput. Methods Appl. Mech. Engrg., Vol. 186, pp. 311-338, 2000.
2. K. Deb, and H. Gupta, "A constraint handling strategy for robust multi-criterion optimization," KanGAL Report No. 2005001, 2005.
3. K. Deb, and H. Gupta, "Introducing Robustness in Multi-Objective Optimization," KanGAL Report No. 2004016, 2004.
4. Z. Fan, K. Seo, R. Rosenberg, J. Hu, E. Goodman, "System-Level Synthesis of MEMS via Genetic Programming and Bond Graphs," in Proc. 2003 Genetic and Evolutionary Computing Conference, Chicago, Springer, Lecture Notes in Computer Science, July, 2003, pp. 2058-2071.
5. G. Fedder, "A Vision of Structured CAD for MEMS," in Proceedings of the Fifth ACM/SIGDA Physical Design Workshop, April, 1996, pp. 76-80.
6. G. Fedder, S. V. Iyer and T. Mukherjee, "Automated Optimal Synthesis Of Microresonators," in Proceedings of 9th International Conference on Solid State Sensors and Actuators (TRANSDUCERS '97), June 16-19, 1997, Chicago, IL, USA pp, 1109-1112.
7. G. Fedder and T. Mukherjee, "Physical Design for Surface-Micromachined MEMS," in Proceedings of the Fifth ACM/SIGDA Physical Design Workshop, April 1996, pp. 53-60.
8. H. Greiner, "Robust Optical Coating Design With Evolutionary Strategies," Applied Optics, Vol. 35, No. 28, pp. 5477 - 5483, 1996.
9. Y. S. Hong, J. H. Lee, and S. H. Kim, "A laterally driven symmetric microresonator for gyroscopic applications," Journal of Micromechanics and Microengineering, vol. 10, pp 452-458, 2000.
10. K. Hwang, K. Lee, G. Park, B. Lee, Y. Cho, and S. Lee, "Robust Design of a Vibratory Gyroscope With an Unbalanced Inner Torsion Gimbal Using Axiomatic Design," Journal of Micromechanics and Microengineering, No. 13, pp 8-17, 2003.
11. Y. Jin and J. Branke, "Evolutionary optimization in uncertain environments - A survey. IEEE Transactions on Evolutionary Computation," Vol. 9, No. 3, pp. 303-317, 2005
12. Y. Jin and B. Sendhoff, "Trade-off between Optimality and Robustness: An Evolutionary Multiobjective Approach," Proceedings of the Second International Conference on Evolutionary Multi-criterion Optimization, C. M. Fonseca et al (Eds.) LNCS 2632, Springer, 2003, pp 237-251.
13. R. Kamalian, H. Takagi, and A. M. Agogino, "Optimized Design of MEMS by Evolutionary Multi-objective Optimization with Interactive Evolutionary Computation," in K. Deb et al. (Eds.): GECCO 2004, LNCS 3103, Springer-Verlag Berlin Heidelberg, 2004, pp. 1030-1041.
14. R. Liu, B. Paden, and K. Turner, "MEMS Resonators That are Robust to Process-Induced Feature Width Variations," Journal of Microelectromechanical Systems, vol. 11, No. 5, pp. 505-511,October 2002.
15. L. Ma and E. K. Antonsson, "Automated Mask-Layout and Process Synthesis for MEMS," in Technical Proceedings of the 2000 International Conference on Modeling and Simulation of Microsystems , 2000, pp. 20-23.
16. T. Mukherjee and G. Fedder, "Structured Design Of Microelectromechanical Systems," in Proceedings of 34th Design Automation Conference (DAC '97), Anaheim, CA, USA, June 9-13, 1997, pp. 680-685.

17. T. Mukherjee, S. Iyer and G. K. Feeder, "Optimization-based synthesis of micro-resonators," Sensors Actuators, A 70, pp. 118-127, 1998.
18. A. Ongkodjojo, F. E. H. Tay, "Global Optimization and Design for Micro-electromechanical Systems devices based on Simulated Annealing," Journal of Micromechanics and Microengineering, vol. 12, pp. 878-897, 2002.
19. P. J. Sedivec, "Robust Optimization: Design in MEMS," Master Thesis, University of California, Berkeley, 2002
20. G. Taguchi, "Taguchi on Robust Technology Development: Bringing Quality Engineering Upstream," ASME Press, New York, 1993.
21. S. Tsutsui, and A. Ghosh, "Genetic Algorithms with a Robust Solution Searching Scheme," IEEE Transactions on Evolutionary Computation, Vol. 1, No. 3, pp 201-208, September 1997.
22. D. Wiesmann, U. Hammel, and T. Back, "Robust Design of Multilayer Optical Coatings by Means of Evolutionary Algorithms," IEEE Transactions on Evolutionary Computation, Vol. 2, No. 4, pp. 162-16, November 1998.
23. W. Yeh, S. Mukherjee, and N. C. MacDonald, "Optimal Shape Design of An Electrostatic Comb Drive in MEMS," J. Microelectromech. Syst. vol. 7, pp. 16-26, 1998.

24

A Hybrid Approach Based on Evolutionary Strategies and Interval Arithmetic to Perform Robust Designs

Claudio M. Rocco S.[1] and Daniel E. Salazar A.[2]

[1] Facultad de Ingeniería, Universidad Central de Venezuela
Apartado Postal 47937, Los Chaguaramos, Caracas 1041A, Venezuela
crocco@reacciun.ve

[2] División de Computación Evolutiva (CEANI), Instituto de Sistemas Inteligentes y Aplicaciones Numéricas en Ingeniería (IUSIANI), Universidad de Las Palmas de Gran Canaria, Edif. Central del Parque Científico y Tecnológico, 2 planta, Campus de Tafira Baja, Las Palmas 35017, Spain
danielsalazaraponte@gmail.com

Summary. This chapter proposes an approach based on the use of Evolutionary Algorithms (EA) and Interval Arithmetic (IA) as an alternative technique to obtain a robust system design. EA are heuristics to optimise a given function, while IA is used as a checking technique to guarantee the feasibility of the design. IA is able to consider simultaneously the effects of uncertainty of all of the parameters on a performance function providing strict bounds (minimum and maximum values) with only one evaluation. EA and IA are used to obtain, by an iterative process, a robust design, that is, the maximum size of each variable deviation that allows complying with a set of specifications. The proposed approach is an indirect method based on optimization instead of a direct method based on mapping from the output into the input space. Numerical examples from the reliability field illustrate the application of the approach using single and multiple objective formulations.

24.1 Introduction

Sensitivity analysis, uncertainty propagation and uncertainty analysis are techniques that have been used for examining the effects of uncertain inputs within a model [19]. These techniques are usually carried out by determining which parameter or parameters have significant effects on the results of a study. An attempt is then made to increase the precision of these parameters in order to reduce the danger of serious error. In system design, mathematical models are used to describe the properties of the system to be designed. As an example, consider the Life-Support System in a Space Capsule [51] shown in figure 1. The following equation (Eq. 24.1) is the symbolic reliability expression of the system, where R_i is the reliability of component i.

Claudio M. Rocco S. and Daniel E. Salazar A.: *A Hybrid Approach Based on Evolutionary Strategies and Interval Arithmetic to Perform Robust Designs*, Studies in Computational Intelligence (SCI) **51**, 543–564 (2007)
www.springerlink.com

$$R_S = 1 - R_3(\bar{R}_1\bar{R}_4)^2 - \bar{R}_3[1 - \bar{R}_2(1 - \bar{R}_1\bar{R}_4)]^2 \qquad (24.1)$$

We can evaluate, for example, what are the effects on R_S if the value of one (local analysis) or more (global analysis) reliability components is changed. For example, we can conclude that a variation of ±10% in R_1 causes a variation of ±z% on R_S. These effects can provide information on how to modify component values. This type of analysis, from input to output, helps Decision-Makers (DM) figure out what to do in the presence of uncertainty in input values. Indeed, if the DM identifies the group of variables whose uncertainties have more influence on the system, they can formulate strategies to overcome the undesired behaviours. When such uncertainty analyses, as the abovementioned ones are incorporated during the design stage, we talk about "robust solutions" or simply "robust design". The way robust design is formulated depends on the information that can be retrieved from the DM and from the problem itself. Therefore, there are many concepts of robustness as well as ways of measuring it. Some references in robustness are the works of Kouvelis and Yu [34], Roy [56] [57], Vincke [72] [73], Davenport [12], Averbakh et al [2] [3] and Rosenhead [55] among others. In evolutionary computation a common assumption is that the uncertainty is already elicitated or represented somehow by means of an uncertainty function, thus the main concerns are the identification of robust solutions and the efficiency of the evolutionary search. An overview of the efficiency issue can be seen in [26]. The first issue is usually tackled by means of a Monte Carlo evaluation. For that matter the value of the design function F(X) is replaced by an average over n disturbed values of the design vector X, namely $X + \delta_i$, where δ_i is the ith sample of the uncertainty function, yielding $F_{eff} = \frac{1}{n}\sum_{i=1}^{n} F(X + \delta_i)$.

The above expression is called Effective Function and assesses the expected value of the design function in the presence of uncertainty. This approach was introduced first by Tsutsui et al [69] [70] and was followed by other researchers like Sevaux and Sörensen [66] [65] in single objective optimization and recently by Deb in multiple objective problems [13]. Notice that aim of this approach is to optimise the expected or robust value of F(X), i.e. F_{eff}(X), but in practice this goal should not be achieved disregarding the variance of F_{eff}(X) since it can mislead the search. In consequence some researchers treat the variance as an additional objective, minimising it, whereas others set some variance threshold for accepting or rejecting the solutions. However, the search for robust alternatives can be formulated in different way if the input's uncertainty is unknown or cannot be represented appropriately to perform the aforementioned approach. In this case the robustness of a design is defined as the maximum size of the deviation from this design that can be tolerated whereby the product still meets all requirements [23]. For example, what are the maximum possible deviations for each component in the Fig. 24.1 consistent with $0.990 \leq R_S \leq 0.999$?

Notice that the problem we address is different from variability or sensitivity analysis [9] [61], where, knowing the component parameters deviation, the method evaluates the variation on the properties due to variation on the inputs, that is moving from the input to the output. In this chapter we reverse the process, assessing the input parameters uncertainty for one (local) or all (global) components, which maintains the output performance function within specified bounds. This chapter presents an approach to obtain a robust design based on the use of Evolutionary Algorithms (EA) and Interval Arithmetic (IA). The remainder of the chapter is organized as follows: Section 2 gives an over-view of Interval Arithmetic. The

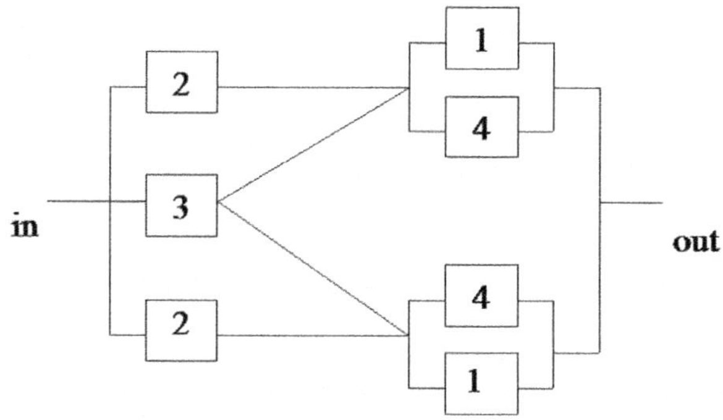

Fig. 24.1. Example of Complex System [51]

Evolutionary approach for solving the single-objective formulations is presented in Section 3. Section 4 describes the general approach used to solve the robust design problem. A single-objective example is considered in Section 5, whereas Section 6 introduces the multiple-objective formulation for robust design as well as a short description of the heuristic tool employed. An analysis of results for the different types of robust design formulations is also addressed in this section. Finally, Section 7 presents the conclusions.

24.2 Interval Arithmetic

Interval arithmetic originates from the recognition that frequently there is uncertainty associated with the parameters used in a computation [42] [46]. This form of mathematics uses interval "numbers", which are actually an ordered pair of real numbers representing the lower and upper bound of the parameter range. For example, if we know that the reliability value R_1 is between 0.8 and 0.9, the corresponding interval number would be written as follows: $R_1 = [0.8, 0.9]$. Interval arithmetic is built upon a basic set of axioms. If we have two interval numbers $T = [a, b]$ and $W = [c, d]$ with $a \leq b$ and $c \leq d$ then [22] [42] [46]:

$$T + W = [a, b] + [c, d] = [a + c, b + d] \tag{24.2}$$

$$T + W = [a, b] + (-[c, d]) = [a - d, b - c] \tag{24.3}$$

$$T * W = [min\{ac, ad, bc, bd\}, max\{ac, ad, bc, bd\}] \tag{24.4}$$

$$1/T = [1/b, 1/a], 0 \ni [a, b] \tag{24.5}$$

$$T/W = [a, b]/[c, d] = [a, b] * [1/d, 1/c], 0 \ni [c, d] \tag{24.6}$$

$$kT = k[a, b] = [ka, kb], k = ctte \geq 0 \tag{24.7}$$

Only some of the algebraic laws valid for real numbers remain valid for intervals. It is easy to show that interval addition and multiplication are associative as well as commutative. However, the distributive law does not always hold for interval arithmetic [42]. As an example: $[0, 1] \cdot (1 - 1) = 0$, while $[0, 1] - [0, 1] = [-1, 1]$.

An important property referred to as sub-distributivity does hold. It is given mathematically by the set inclusion relationship: $T(W+Z) \subseteq TW+TZ$. The failure of the distributive law often causes overestimation. It is interesting to notice that $T-T \neq 0$ and $T/T \neq 1$. In general, when a given variable occurs more than once in an interval computation, it is treated as a different variable in each occurrence; thus, T-T is the same as T-W with W equal to but independent of T, causing the widening of computed interval. This effect is called the "dependency problem". However, there are expressions where dependence does not lead to overestimation, e.g., T+T [46].

An interval function is an interval-valued function of one or more interval arguments. Consider a real valued function f of real variables $t-1, t_2, \ldots, t_n$ and an interval function F of interval variables $T_1, T_2, \ldots, T_n$. The interval function F is said to be an interval extension of f, if: $F(t_1, t_2, \ldots, t_n) = f(t_1, t_2, \ldots, t_n)$.

The range of a function f of real variables over an interval can be calculated from the interval extension F, changing t_i by T_i. Moore [42] states that (Inclusion Property): $f(t_1, t_2, \ldots, t_n) \subseteq F(T_1, T_2, \ldots, T_n), \forall t_i \in T_i (i = 1, \ldots, n)$.

This means that the range of the function $f(t_1, t_2, \ldots, t_n)$ can be calculated using interval arithmetic. In many practical problems, it could be difficult to obtain an expression in which each variable occurs not more than once. In these cases, a single computation of the interval extension of $f(t_1, t_2, \ldots, t_n)$ only yields an interval $F(T_1, T_2, \ldots, T_n)$, that could be wider than the exact range for $f(t_1, t_2, \ldots, t_n)$. Several techniques that have been developed to avoid the dependence problems: Quadratic approximation and Gauss Transformation [21], Deconvolution [17], Generalized Interval Arithmetic [22], Modal Interval [1] [71] and Central Forms [50].

Consider the system shown in Fig. 24.1 Initial reliability intervals for the components are $R_i = [0.80, 0.90]$. Evaluating R_S at the midpoint interval, we obtain that the reliability of the system is $R_S = 0.995279$.

Table 1 shows the interval associated to output R_S calculated with interval arithmetic, considering the interval variation for a single parameters (nominal range sensitivity) while fixing the other parameters at its midpoint. For example, if $R_1 = [0.80, 0.90]$; $R_2 = R_3 = R_4 = 0.85$, then $R_S = [0.9946150, 0.9958356]$. Last row of Table 24.1 shows the R_S interval when all variables are permitted to vary simultaneously in $R_i = [0.80, 0.90]$. Note that this interval is obtained, in interval arithmetic, with only one calculation.

Table 24.1. Sensitivity Analysis using Interval Arithmetic

VARIABLE	INTERVAL	R_S
R_1	[0.80, 0.90]	[0.9946150, 0.9958356]
R_2	[0.80, 0.90]	[0.9924411, 0.9974007]
R_3	[0.80, 0.90]	[0.9938743, 0.9966840]
R_4	[0.80, 0.90]	[0.9946150, 0.9958356]
All	[0.80, 0.90]	[0.9879552, 0.9987219]

24.3 Evolutionary Approach

Evolutionary Algorithms (EA) have been applied to a wide range of problems especially in those cases where traditional optimization techniques have shown poor performances or simply have failed [32].

A population of individuals each of which representing one point of the solution space collectively evolves towards better solutions by means of a parent's selection process, usually by means of a randomised recombination and/or mutation operators and a substitution strategy. Parent selection determines which individuals participate in the reproduction strategy. Recombination allows the exchange of already existing genes and mutation introduces new genetic material. The substitution strategy defines the individuals for the next generation population.

Evolution Strategies (ES) were developed as a powerful tool for parameter optimization tasks [32] [33]. Two basic types of ES have been developed.

In a two-member or (1+1) evolution strategy (ES(1+1)), one 'parent' produces one offspring per generation by applying a normally distributed perturbation, until a 'child' performs better than its ancestor and take its place.

Schwefel and Bäck [63] generalised these strategies to the multimember evolution strategy now denoted by ES($\mu+\lambda$) and ES(μ,λ). In a ($\mu+\lambda$) strategy, the μ best of all ($\mu+\lambda$) individuals survive to become parents of the next generation. Using the (μ,λ) strategy, selection takes place only among offspring. The second scheme is more realistic and therefore more successful, because no individual may survive forever, which could at least theoretically occur using the plus variant [39]. Furthermore the ($\mu+\lambda$) tends to emphasise local rather than global search properties. For this reason, modern ES use the (μ,λ) substitution strategy. A ratio of $\mu/\lambda \approx 7$ is recommended as a good setting for the relation between parent and offspring population size [63].

Cellular Evolution Strategies (CES) [38] are an approach that combines the ES(μ,λ) techniques with concepts from Cellular Automata [74] for the parent's selection step, using the concepts of neighbourhood. In the CES approach each individual is located randomly in a cell of a two-dimension array. To update a specific individual, the parents' selection is performed looking only at determined cells (its neighbourhood) in contrast with the general ES, which search parents in the whole population. As an example, Figure 24.2 shows the array for 20 n-components individuals ($a_1 \ldots a_{20}$) and the Von Neumann neighbourhood for individual a_{13} (radius=1). The Von Neumann neighbourhood is formed by the individuals a_8, a_{12}, a_{14} and a_{18}. To update a_{13}, the parents' selection is determined at random, with the same mating probabilities and, only taking into account a_8, a_{12}, a_{13}, a_{14} and a_{18}. Seven new individuals are then generated by means of the mutation scheme previously described and using an arithmetic crossover. The best of them replaces a_{13} (The number seven follows a suggestion made in [63]: $\mu/\lambda \approx 7$). The rest of the individuals are update in the same way. In CES the neighbourhood type and the radius are parameters to be selected. In [38] it was found that for optimization problems, the best results were obtained with a Von Neumann neighbourhood with radius equal to one. Using those parameters, the diversity among individuals is maintained during several evolutionary iterations. A bigger radius tends to simulate a conventional ES while a different neighbourhood (for example the Moore neighbourhood) tends to homogenise the population. While the ES were originally designed with the parameter optimization problem in mind, the CES were designed to find the global optimum or "near" optimum for complex multi-modal functions.

Its use is then suggested in the case of high dimensional problems [38]. Due to the stochastic nature of Cellular Evolutionary Strategies, it is important that problems solved using this techniques show their consistency, i.e. how many times a run is close to the global solution. Results from several runs are required to assess CES consistency.

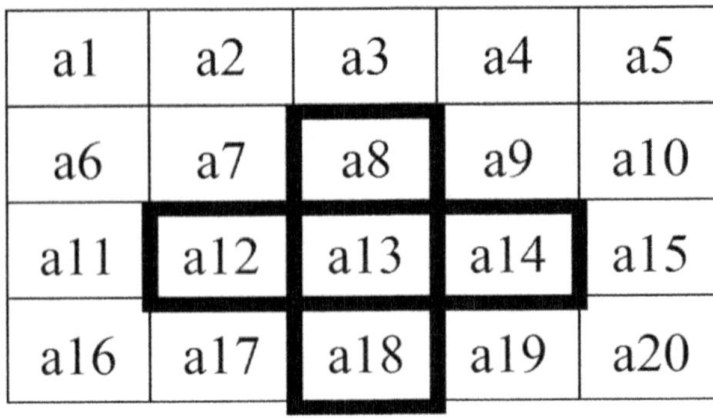

Fig. 24.2. Two-dimension cell array

24.4 Robust Design Methodology

24.4.1 Problem Description [23]

The robustness of a system design is defined as the maximum size of component deviations from their design values that can be tolerated such that the system still meets all defined specifications. The designer formulates target values on the quality of the product by setting lower and upper bounds on the property $y_i(x)$. The problem is to find a product x in the experimental region X which fulfils the requirements on the properties.

We define the following slack function $g_i(x)$:
$g_i(x) = UB_i - y_i(x)$ when there is an upper bound (UB) requirement or
$g_i(x) = y_i(x) - LB_i$ when there is an lower bound (LB) requirement. This results in a product design problem mathematically formulated as: find an element of $F \bigcap X$, with: $F := \{x \in \Re^n | g_i(x) \geq 0, i = 1, \ldots, m\}$.

24.4.2 Problem Definition

Suppose the area shown in Fig. 24.3a is the feasible zone for a generic design with variables R and R_a. Within the feasible zone any pair (R, R_a) satisfies the specifications. An exact description of the Feasible Solution Set (FSS) (Fig. 24.3a) is in general not simple, since it may be a very complex set. Moreover, the FSS could be limited by non-linear functions. For this reason, approximate descriptions are often looked for, using simply shaped sets like boxes or ellipsoids containing -*outer bounding (Fig. 24.3b)*- or contained in -*inner bounding (Fig. 24.3c and 24.3d)*- the set of interest [40]. In particular Minimum Volume Outer box (MOB) (Fig. 24.3b) and Maximum Volume Inner Box (MIB) (Fig. 24.3c and 24.3d) are of interest. In this study only the MIB determination is considered.

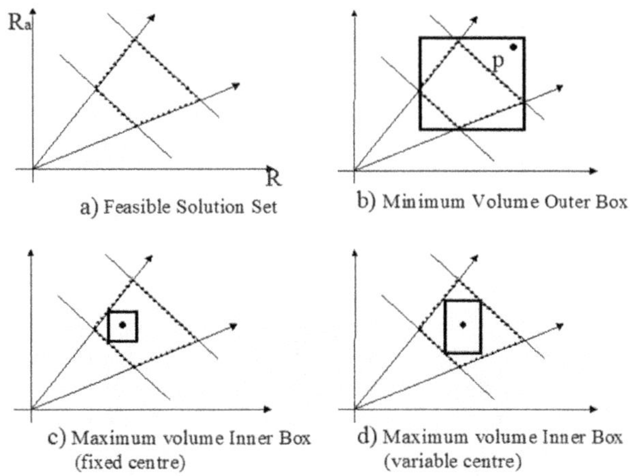

Fig. 24.3. Feasible Solution Set and approximate descriptions [54]

The maximum ranges of possible variations of the feasible values are the sizes (along co-ordinate axis) of the axis-aligned box of minimum volume containing FSS. To obtain the MIB it is required that all the points inside the generated box satisfy the constraints. Then, the mathematical formulation is [54]:
Let B the box defined by:

$$B := \{x, C \in \Re^n | x_i \in [x_{i,lower}, x_{i,upper}], C_i = (x_{i,upper} + x_{i,lower})/2\}$$

1. Centre Specified:

$$\max_{x} \prod_{i=1}^{n} |x_i - C_i| \quad s.t.\ x \in F \bigcap B$$

2. Centre Unspecified:

$$\max_{x,C} \prod_{i=1}^{n} |x_i - C_i| \quad s.t.\ x, C \in F \bigcap B$$

The objective functions represent a quantity that is proportional to the MIB hyper-volume. Note that x represents a vertex of the optimal MIB. From here, the range of each variable is easily determined, as described in the next section.

24.4.3 General Approach [52]

Given r constraint functions $g_i(x_1, x_2, \ldots, x_n)$, lower and upper bounds (LB_i, UB_i), generate an initial random point $x = x_{1_0}, x_{2_0}, \ldots, x_{n_0}$ (initial vertex). Afterwards, check if this point is a feasible point, evaluating all the constraint. If the point is infeasible, then generate a new point and check it again for feasibility. Otherwise x is feasible and there are two possible actions depending on the desired goal:

1) Inner box, centre specified

Given a feasible vertex, produce a symmetrical "box" (hyper-rectangle) using the point C as symmetry centre and check for the feasibility of the generated box. If the box is feasible, calculate the associated volume. If the box is not feasible, discard the generated box, and repeat the process with a new feasible initial vertex.

2) Inner box, centre unspecified

In this case, the centre of co-ordinates is considered as additional variables. So the initial random centre of co-ordinate $C = C_{1_0}, C_{2_0}, \ldots, C_{n_0}$ must be generated along with the initial random vertex,. As in the previous case, produce a symmetrical "box" using C as symmetry centre, and then check the feasibility of the generated box; next if the box is feasible, calculate the associated volume. In both cases, the goal is to maximise the inner volume using CES as the optimization technique. While checking the box feasibility, two problems may occur: 1) The function should be evaluated in the 2^n vertices of the box, and 2) The extreme values are not necessarily at the vertices of the generated box. To overcome these two drawbacks, constraint functions are evaluated as interval functions. This means that only one "interval" evaluation is required for each constraint and the exact range of the constraint functions inside the generated box is obtained. Note that if the FSS is non-convex, feasibility check using IA will consider this.

24.5 Example: Life-Support System in a Space Capsule [51]

The proposed approach is applied to define a robust design for the system [51] shown in Fig. 24.1. The problem is to obtain the ranges for each R_i such as $0.99 \leq R_S \leq 1$, subject to: $0.80 \leq R_i \leq 1$, with centre of each interval as: $C_i = 0.90$, i=1,2,3,4. Because of the stochastic nature of CES, 20 trials were performed and the best solution from among the 20 trials was used as the final solution. All CES runs were performed using 30 generations, 49 individuals in a 7×7 grid, Von Neumann neighbourhood with radius=1 and asynchronous substitution.

Table 24.2 shows the result obtained using the hybrid approach CES and IA, with a MIB volume of $7.21851 \cdot 10^{-5}$. These ranges produce an output R_S belonging to: [0.9900, 0.9999].

The MIB volume obtained using a non-linear optimization program, was $9.31811 \cdot 10^{-5}$, i.e. only 0.163% greater than the obtained with the CES-IA approach. The average relative error obtained in 20 runs was only 0.272%. Normally the design problem seeks to constraint a cost function, such as: $C_S = 2\sum K_i R_i^{\alpha_i}$. For example in [51]: K1=100, K2=100, K3=100, K4=150; $\alpha_i = 0.6\ \forall i$. Using the above values and the ranges shown in Table 24.2 for R_i, the range for C_S is [791.369485, 896.123195].

As previously mentioned, our approach can consider several constraints simultaneously. For example, we can solve the problem to obtain the ranges for each R_i

Table 24.2. Robust design using CES and IA approach: A single constraint for R_s

VARIABLE	Starting Interval	Final Interval
R_1	[0.80, 1.00]	[0.8001, 0.9998]
R_2	[0.80, 1.00]	[0.8220, 0.9779]
R_3	[0.80, 1.00]	[0.8032, 0.9967]
R_4	[0.80, 1.00]	[0.8652, 0.9347]

such as $0.99 \leq R_S \leq 1$ and $C_{min} \leq C_S \leq C_{max}$. In this case, if we define: $C_{min} = 800$ and $C_{max} = 870$, then the new solution is shown in Table 24.3.

Table 24.3. Robust design using CES and IA approach: Constraints on R_S and C_S

VARIABLE	Starting Interval	Final Interval
R_1	[0.80, 1.00]	[0.854327, 0.945673]
R_2	[0.80, 1.00]	[0.853171, 0.946829]
R_3	[0.80, 1.00]	[0.846696, 0.953304]
R_4	[0.80, 1.00]	[0.870334, 0.929666]

Using the new ranges for R_i, intervals for R_S and C_S are: R_S = [0.995628, 0.999835] and C_S = [820.8062, 868.4570].

24.6 Multiple Objective Formulation

In the example previously presented, DM could be interested on how to balance reliability and cost. For example, a design with higher reliability and higher cost or a design with lower cost sacrificing reliability could be chosen [18]. Using the a single-objective (SO) formulation, DM must solve several problems by varying a group of constraints to obtain a group of alternatives from which to choose the final solution; nevertheless, with a multiple-objective (MO) formulation, it is possible to determine the Pareto frontier which provides more information to DM.

For example, in the reliability field, many authors have analysed SO problems [7, 15, 16, 24, 27, 29–31, 35, 41, 43–45, 49, 51, 53, 68] while others have recognised the advantages of a MO formulation [8, 15, 16, 36, 43, 47, 49, 58, 59, 64, 75, 76].

In general there are two approaches to solve MO problems. The first approach formulates the problem as a multiple-objective optimization problem (MOP) based on the specified criteria (e.g. maximise reliability and minimise cost) to be solved directly. The second approach transforms the original MOP into several single-objective optimization problems (SOP) to be solved sequentially [37, 49].

In this section we consider the robust design using a MOP formulation, solved by Evolutionary Algorithms (EA). This approach belongs to the Multiple Objective Evolutionary Algorithms (MOEAs) family and is able to effectively handle constraints. The approach does not guarantee the determination of the exact Pareto frontier nor does any heuristic approach for that matter. However an important number of comparisons [14, 77, 78] performed in Evolutionary Multicriteria Optimization (EMO) on benchmark problems have shown that results obtained using different instances of MOEAs are very close to the exact solution. The MO formulation is analysed in the context of the ability of MOEAs to find a set of solutions and the quality of the information provided to the DM. Some important issues like MOEA comparisons and parameter tuning will not be examined in this chapter.

24.6.1 Multiple Objective Problem Description

A MO optimization problem consists of optimising a vector of functions:

$$Opt\,(F(x) = (f_1(x), f_2(x), \ldots, f_k(x)))$$

s.t.: $g_j(x) \leq 0, j = 1, 2, \ldots, q$; $h_j(x) = 0, j = 1, 2, \ldots, r$; $(q + r = m)$ where $x = (x_1, x_2, \ldots, x_n)^t \in X$ is the solution vector, or vector of decision variables, and X is the feasible domain.

The concept of optimality in single objective cannot be directly extrapolated to multiple-objective problems. For this reason a classification of the solutions is introduced in terms of Pareto optimality, according to the following definitions [79]. In terms of minimisation:

Definition 1: Pareto Optimal: A solution vector $x^* \in X$ is Pareto Optimal solution iff $\neg\exists x \in X : f_i(x) \leq f_j(x^*)$; $i = \{1, 2, \ldots, k\}$. These solutions are also called true Pareto solutions.

Definition 2: Pareto Dominance: x_1 dominates x_2, denoted $x_1 \succ x_2$, iff $f_i(x^1) \leq f_j(x^2) \wedge \exists j : f_j(x^1) < f_j(x^2)$; $i, j \in \{1, 2, \ldots, k\}$. If there are no solutions which dominates x^1, then x^1 is non-dominated.

Definition 3: Pareto Set: A set of non-dominated solutions $\{x^* | \neg\exists x : x \succ x^*\}$ is said to be a Pareto set.

Definition 4: Pareto Front: the set of vectors in the objective space that are image of a Pareto Set, i.e. $\{F(x^*) | \neg\exists x : x \succ x^*\}$.

The robust design problem previously presented can be formulated as MOP, transforming one or more constraints into objectives. For example:

1. Centre Specified:

$$\max_{x} \prod_{i=1}^{n} |x_i - C_i| \wedge \min_{x} C_S$$
$$s.t.\ x \in F \bigcap B$$

2. Centre Unspecified:

$$\max_{x,C} \prod_{i=1}^{n} |x_i - C_i| \wedge \min_{x} C_S$$
$$s.t.\ x, C \in F \bigcap B$$

Notice that even if the two objectives considered in these MOP types are MIB and cost, the MOP approach is general and can be used for any type and number of objectives (for example MIB and weight). The selection of such objectives clearly depends on the problem under study and the DM criteria.

Both robust design formulations can be solved using the MOEA approach discussed in the next section. The characteristics of each formulation as well as the methodology to solve them will be studied in further sections.

24.6.2 Multiple Objective Evolutionary Algorithms

Multiple-Objective Evolutionary Algorithms (MOEAs) is the term employed in the Evolutionary Multicriteria Optimization (EMO) field to refer to a group of evolutionary algorithms especially formulated to deal with MOP. This group of algorithms conjugates the basic concepts of dominance described in the later section with the general characteristics of evolutionary algorithms. MOEAs are able to deal with non-continuous, non-convex and/or non-linear spaces, as well as problems whose objective functions are not explicitly known (e.g. the output of Monte Carlo simulation runs).

Since the first recognized MOEA (Schaffer's VEGA (1984) [62]), the development of MOEAs has successfully evolved, producing better and more efficient algorithms. The existing MOEAs may be classified into two groups [6], according to its characteristics and efficiency. On the one hand there is a first group known as "first-generation" which includes all the early MOEAs (Weighted Sum [4], VEGA [62], MOGA [4], NPGA [25], NSGA [67]). On the other hand there is a second group named "second-generation MOEAs", which comprises very efficient optimizers like SPEA [77] / SPEA2 [78], M-PAES [28], PESA [10] / PESA-II [11] and NSGA-II [14], among others.

Basically, the main features that distinguish second-generation MOEAs from the first-generation group are:

Mechanism of adaptation assignment in terms of dominance: between one non-dominated solution and another dominated, the algorithm will favour the non-dominated one. Moreover, when both solutions are equivalent in dominance, the one located in a less crowded area will be favoured. Finally, the extreme points, (i.e. the solutions that have the best value in one particular objective) of the non-dominated population are preserved and their adaptation is better than any other non-dominated point, to promote maximum front expansion.

Incorporation of elitism: the elitism is commonly implemented using a secondary population of non-dominated solutions previously stored. When performing recombination (selection-crossover-mutation), parents are taken from this archive in order to produce the offspring.

In general terms, current MOEAs based on genetic algorithms follow a sequence similar to the flow chart depicted in Fig. 24.4. Notice that there are two populations: P^t, which represents the current population during generation t, and P_A^t, which consists of non-dominated solutions. Many state-of-the-art MOEAs keep a constant size M for the population and N for the file. Initially, M individuals are generated randomly and the archive is set to empty. In each generation all non-dominated individuals from both the archive and the population are selected and assigned to the archive P_A^{t+1}. Afterwards a reproductive selection of individuals is accomplished (typically by binary tournament) and a mating pool is filled up. At this stage a new population is generated following some $(\mu + \lambda)$ recombination strategy (μ parents are combined to produce λ offspring). Finally, the new population replaces the old one and the process continues until the maximum number of generations is reached. At the end, the non-dominated solutions from the archive are reported as output.

It is important to mention that in most of the cases, the number of non-dominated solutions found during each generation could be less than or greater than the archive size (N). Thus some criteria or methods for classifying and selecting are needed in order to set the archive size to N solutions. This can be achieved

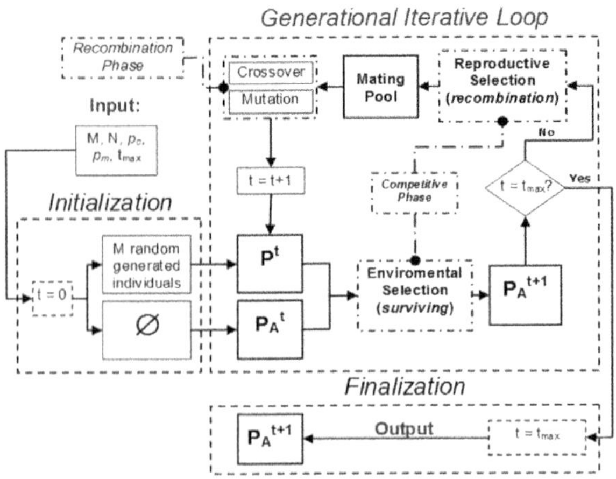

Fig. 24.4. Flowchart of a 2nd generation MOEA based on genetics algorithms [59]

assigning to individuals fitness values based on the Pareto dominance combined with some other criteria usually referred to as density and distribution of the population. That is called environmental selection [79].

It is convenient to realize that in practice, both reproductive and environmental selection (the mechanism of adaptation assignment in terms of dominance and density) constitute the main difference between one MOEA and another. The rest of the algorithm remains the same with a traditional EA.

Nondominated Sorting Genetic Algorithm II

The Nondominated Sorting Genetic Algorithm (NSGA-II) is a well known and extensively used algorithm based on its predecessor NSGA. It was formulated by Deb et al [14] as a fast and very efficient MOEA, which incorporates the features mentioned earlier, i.e., elitism and adaptation assignment in terms of dominance and density. Elitism is possible in NSGA-II due to the use of one steady population and a temporary one. On the other side, adaptation is assessed first by means of a non-dominated sorting procedure followed by a crowding distance measure. The former step correspond to Goldberg's non dominated ranking procedure [20], while the latter consist of the assignment of a density index based on the Manhattan distance between the two closest neighbours of a given solutions into the same rank.

Constraint handling techniques

One of the main issues in evolutionary computation is how to guide the search towards the feasible region in the presence of constraints. The existing approaches can be classified in the following groups: 1) Penalization techniques, 2) Repairing techniques, 3) Separation techniques, and 4) Hybrid techniques. For a review see [5].

The experiments reported in this chapter are based on the approach proposed by Deb et al [14], which could be implemented in the NSGA-II as follows:

- Calculate the normalized sum of constraint violations for all the individuals belonging to $P^t \bigcup P_A^t$.
- Classify the individuals according to the overall constraints violation: when comparing two individuals; if the overall violation of both of them is zero, apply the ordinary ranking assignment, otherwise the individual with the lower (or null) overall violation dominates the other one.
- The rest of the algorithm remains equal.

The integration of this technique to any MOEA promotes feasibility over optimality. Thus the search is guided toward the feasible region. Once it is reached, feasible individuals are sorted according to the particular fashion established by implemented MOEA. Moreover, as the reader may notice, the absence of penalization parameters saves the effort of tuning.

24.6.3 Implementation

NSGA-II is implemented following the pseudo-code presented in the algorithm described in Table 24.4. 1. Notice that we moved from the original formulation allowing different sizes for the two populations involved. Additionally, our implementation is suitable both for integer chromosomes, real chromosomes and for mixed chromosomes (for more details see [59]). For the particular problem studied here, only real variables were needed.

Likewise, the recombination mechanism is one-point crossover, adapted for a mixed chromosome. For real variables, the crossover is performed as a linear combination while the mutation operation is performed like Gaussian mutation of type $R^{new} = R^{old} + N(0, \sigma^2)$.

24.6.4 Computational Example

The example presented in Section 5 is analysed, by comparing both SOP and MOP optimization results. The MO formulation is obtained converting the cost constraint into an objective function. Thus, the following problem is a bi-objective problem with centre specified ($C_i = 0.90$). In those cases where the original SOP considers several constraints, higher dimensional MO formulations are possible. Nevertheless, the bi-objective formulation is the only case suitable for an easy visual inspection; therefore it is appropriate as an introductory example.

Fig. 24.5 shows the Pareto front determined by NSGA-II for two experiments us-ing 12550 evaluations of the objective function. Note that the approximation Pareto fronts obtained are practically equals; for this reason, in the subsequent figures only one experiment is presented. From this figure, it is also clearly visible that the solution achieved during the SO optimization, as expected, shows the same approximation quality.

Fig. 24.6 shows a complete picture of the situation. The minimum and maximum cost as a function of the MIB volume is presented for the solutions obtained earlier. As expected, as long as the range of variation on reliability components is wider, and therefore the MIB is bigger, the range for cost becomes more uncertain. In consequence, DM should define how much uncertainty can be allowed for the system. In other words, the robustness of the system has a worth, which is directly related with the maximal cost of the chosen MIB.

Table 24.4. Pseudo-code for NSGA-II

Input: • M (Population size) • N (Archive size) • t_{max} (Max. number of generations)
Begin: • Randomly initiate P_A^0, set $P^0 = \emptyset$, t = 0. • **While t** $< t_{max}$**:** ∘ $P^t = P^t + P_A^t$ ∘ Assign adaptation to P^t according to non-dominated sorting and crowding distance. ∘ P_A^{t+1} = {N best individuals from P^t} ∘ MP (mating pool) = {M individuals randomly selected from P_A^{t+1} using a binary tournament} ∘ P^{t+1} = {M new individuals generated by applying recombination operators on MP} ∘ t = t+1 • **End loop**
Output: • Non dominated solutions from P_A^t

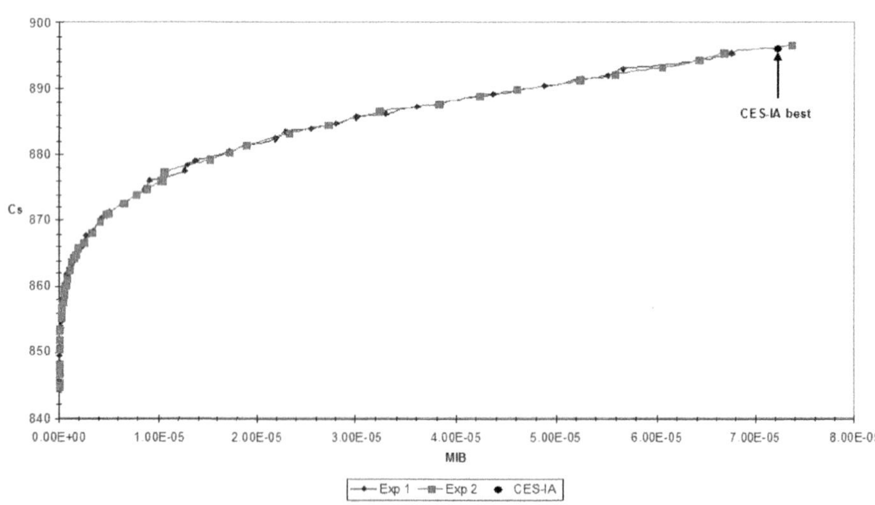

Fig. 24.5. Two sets of trade-off approximations between MIB and C_S^{max}

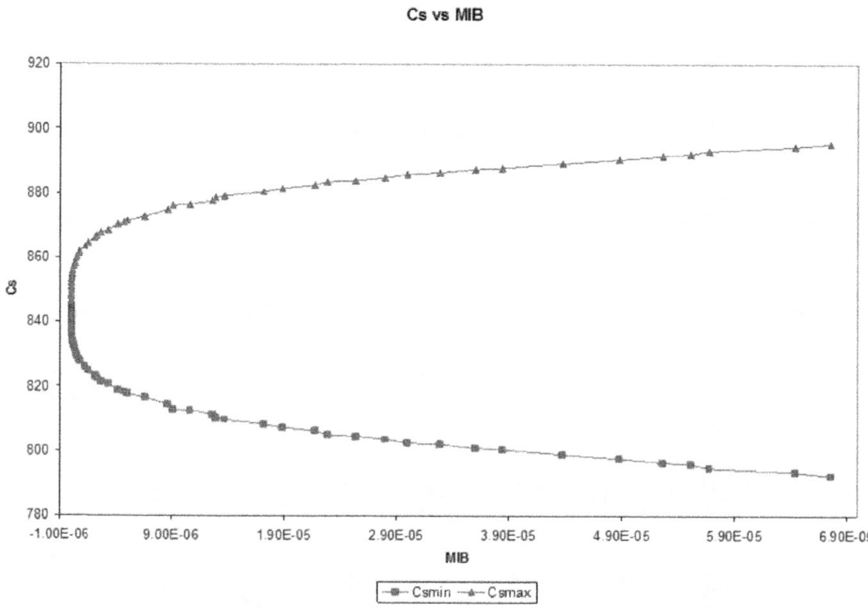

Fig. 24.6. Minimum and Maximum Cost vs MIB

In that sense, Fig. 24.7 shows the average component tolerance as a function of the MIB. For instance, if the DM defines a maximum allowed cost of 870 units, then the maximum average component tolerance would be in the range from 0 to almost 10%.

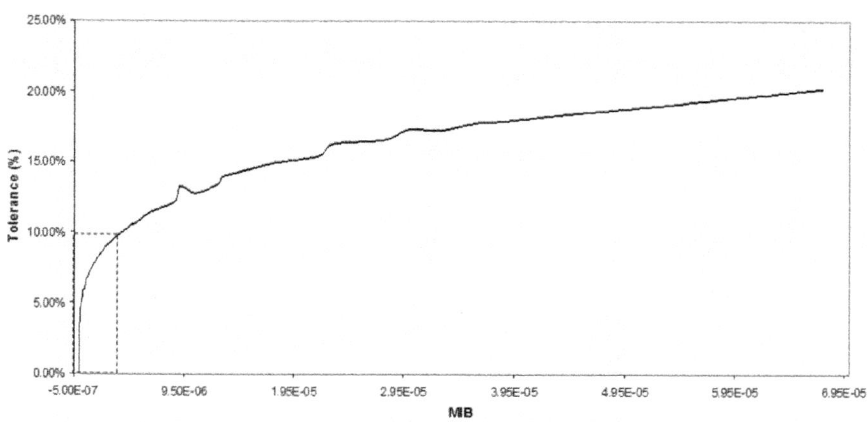

Fig. 24.7. Average component tolerance

The above results show the advantages of the MO formulation for the centre specified formulation.

The MOEAs approach can be also used to analyse the centre-unspecified case. Given the dependency between the vertex and the centroid of each solution, the non-dominated front cannot be directly assessed, instead of that, the determination of the approximation Pareto front requires an iterative procedure like the double-loop strategy used in [36]. First a centroid is selected and then the set of non-dominated solutions for that centroid is investigated. After that, a new centroid is selected and the process continues iteratively. To illustrate the proposed approach, five sets of solutions generated with predefined centroids are shown in Fig. 24.8. It is interesting to note that the two upper curves (centroids $C_i = 0.91$ and $C_i = 0.92$) are completely dominated by the curve obtained with $C_i = 0.90$. This fact shows that a high value for centroid is not necessarily the best option, since the nearer distance regarding the constraint $R_i \leq 1$ limits the range of variation of the components reliability values and consequently reduces the MIB. Likewise, a similar mechanism reduces the extension of the lower curves (centroids $C_i = 0.86$, $C_i = 0.88$ and $C_i = 0.89$), since lower centroids are nearer the system constraint ($R_s \geq 0.99$).

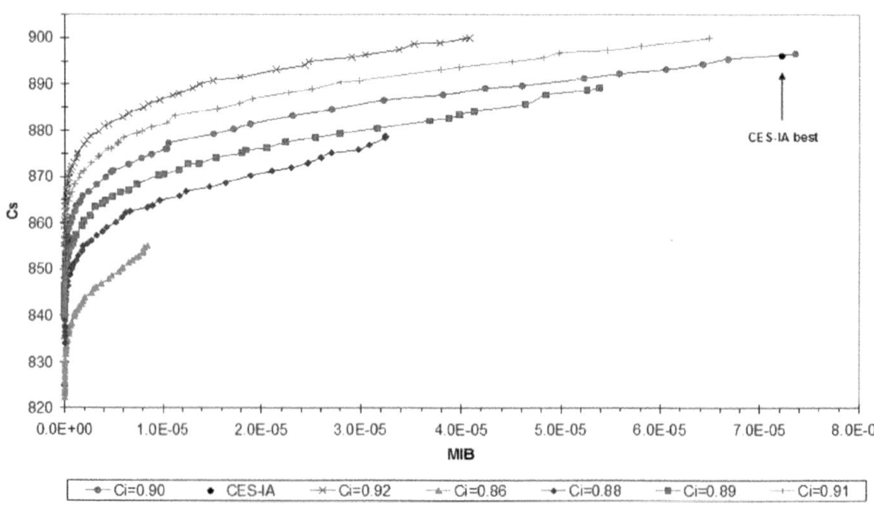

Fig. 24.8. Trade-off between MIB and C_S^{max} for selected centroids

Fig. 24.9 shows the non-dominated front for the set of obtained solutions earlier. The dotted lines correspond to the dominated sectors of the curves, whereas the solid lines are the non-dominated sections. The non-dominated front in this case is formed for efficient solutions with different centroids.

This formulation is evidently better than the one used for the specified-centre case since it allows finding even better solutions. However, this formulation requires more computational work to handle the problem of dependency between vertexes and centroids. Note that this formulation consists on a time-consuming search.

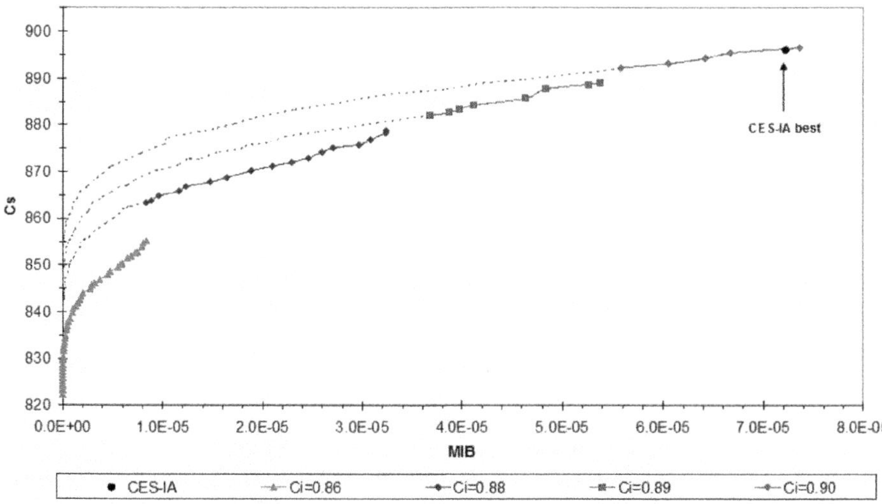

Fig. 24.9. Non-dominated front for selected centroids

A tentative solution to overcome this problem is the use of the "percentage representation" introduced in [60], where centre co-ordinates could be conveniently evolved. However its implementation must be investigated in further researches.

24.7 Conclusions

This chapter proposes an approach based on the use of Evolutionary Algorithms and Interval Arithmetic as an alternative technique to obtain two robust system designs. First, Cellular Evolutionary Strategies are used as the optimization technique while Interval Arithmetic is used as a checking technique that guarantees the feasibility of the design.

The excellent results obtained suggest that the CES-IA approach has a great potential in dealing with difficult system design problems. It is interesting to note that even if Interval Arithmetic can overestimate the size of the hyper box, due to the dependency problem, the box obtained is a valid robust solution, which satisfies all the defined constraints. As mentioned, the overestimation can be treated using special techniques. The added burden to the procedure CES-IA for determining the feasibility verification of the generated box is far outweighed by the flexibility provided by such technique (only one "interval" evaluation and guaranteed ranges) in contrast to multiple vertices evaluation.

The approach initially formulated to consider simultaneously several constraints has been successfully extended to cope with multiple objectives. This new formulation extends the possibilities of the robust design approach providing the DM with a wider horizon of non-dominated alternatives.

The Multiple Objective Evolutionary Algorithms (MOEAs) employed to solve the MOP formulations provide an excellent ways to approximate the Pareto frontier, for both the centre-specified and the centre-unspecified cases. Even so, additional researches are required to improve the efficiency of the MO centre-unspecified solution technique.

References

1. Armengol J, Vehì J, de la Rosa JL, Travé-Massuyés L (1998) On Modal Interval Analysis for Envelope Determination within Ca-En Qualitative Simulator. http://ima.udg.es/SIGLA/X
2. Averbakh I (2000) Minmax regret solutions for minimax optimization problems with uncertainty. Operations Research Letters, 27:57–65
3. Averbakh I, Lebedev V (2002) Interval data minmax regret network optimization problems. Discrete Applied Mathematics 138:289–301
4. Coello C (1999) A Comprehensive Survey of Evolutionary-Based Multiobjective Optimization Techniques. Knowl Inf. Syst 1(3):129–156
5. Coello C (2002) Theoretical and Numerical Constraint-Handling Techniques used with Evolutionary Algorithms: A Survey of the State of the Art. Computer Methods in Applied Mech and Engnng 8(2):1245–1287
6. Coello C. http://www.cs.cinvestav.mx/ẼVOCINV/download/tutorial-moea.pdf
7. Coit DW, Smith AE (1996) Reliability Optimization of Series-Parallel systems using a Genetic Algorithm. IEEE Trans Reliab 45(2):254–260
8. Coit DW, Tongdan J; Wattanapongsakorn N (2004) System optimization with component reliability estimation uncertainty: a multicriteria approach. IEEE Trans on Reliab 53(3):369–380
9. Constantinides (1994) Basic Reliability. In: Annual Reliability and Maintainability Symposium, Anaheim, California, USA
10. Corne DW, Knowles JD (2000) The Pareto-Envelope based Selection Algorithm for Multiobjective Optimization. In: Proceedings of the Sixth International Conference on Parallel Problem Solving from Nature (PPSN VI). Springer, Berlin:839–848
11. Corne DW, Jerram NR, Knowles JD, Oates MJ (2001) PESA-II: Region-based Selection in Evo-lutionary Multiobjective Optimization. In: Proceedings of the Genetic and Evolutionary Computation Conference (GECCO-2001), Morgan Kaufmann Publishers:283–290.
12. Davenport AJ, Beck JC (Unpublished manuscript) A Survey of Techniques for Scheduling with Uncertainty. In: http://www.mie.utoronto.ca/staff/profiles/beck/publications.html
13. Deb K, Gupta H (2005) Searching for Robust Pareto-Optimal Solutions in Multi-objective Optimization. Evolutionary Multi-Criterion Optimization, LNCS 3410. Berlin, Germany: Springer-Verlag:150–164
14. Deb K, Pratap A, Agarwal S, Meyarivan T (2002) A Fast and Elitist MultiObjective Genetic Algorithm: NSGA-II. IEEE Trans Evol. Comput. 6(2):182–197
15. Dhingra A (1992) Optimal Apportionment of Reliability & Redundancy in Series Systems Under Multiple Objectives. IEEE Trans Reliab 41(4):576–582

16. Elegbede C, Adjallah K (2003) Availability allocation to repairable systems with genetic algorithms: a multi-objective formulation. Reliab Engnng Sys Safety 82(3):319–330
17. Ferson S, Long T. Deconvolution can reduce Uncertainty in Risk Analysis. http://ramas.com
18. Giuggioli P, Marseguerra M, Zio E (2001) Multiobjective optimization by genetic algorithms: application to safety systems. Reliab Engnng Sys Safety 72:59–74
19. Granger M, Henrion M (1993) Uncertainty: A Guide to Dealing with Uncertainty in Quantitative Risk and Policy Analysis. Cambridge University Press, UK
20. Golberg D (1989) Genetics algorithms in search, optimization & machine learning. Addison-Wesley Publishing Company, Inc. USA
21. Hadjihassan S, Walter E, Pronzato L (1996) Quality Improvement via Optimisation of Tolerance Intervals During the Design Stage. In: Kearfott RB, Kreinovich V (Eds.) Applications of Interval Computations. Kluwer Academic Publishers, Dordrecht, The Netherlands
22. Hansen E (1992) Global Optimization Using Interval Analysis. Marcel Dekker, Inc., New York
23. Hendrix EMT, Mecking CJ, Hendriks ThHB (1996) Finding Robust Solutions for Product Design Problems. EJOR 92:28–36
24. Hikita M, Nakagawa Y, Nakashima K, Narihisa H (1992) Reliability Optimization of Systems by a Surrogate-Constraints Algorithm. IEEE Trans Reliab 41:473–480
25. Horn J, Nafpliotis N, (1993) Multiobjective optimization using the Niched Pareto Genetic Algorithm. IlliGAL Report 93005, Illinois Genetic Algorithms Laboratory, University of Illinois, USA
26. Jin Y, Branke J (2005) Evolutionary Optimization in Uncertain Environments - A survey. IEEE trans Evol Comput. 9(3):303–317
27. Kim JH, Yum BJ (1993) A Heuristic Method for Solving Redundancy Optimization Problems in Complex Systems IEEE Trans Reliab 42:572–578
28. Knowles JD Corne DW (2000) M-PAES: A Memetic Algorithm for Multiobjective Optimization. In: Proc. of the Congress on Evolutionary Computation (CEC00). IEEE Press, Piscataway, NJ:325–332
29. Kulturel-Konak, S, Smith AE, Coit DW (2003) Efficiently solving the redundancy allocation problem using tabu search. IIE Trans 35(6):515-26
30. Kuo W, Hwang CL, Tillman FA (1978) A note on Heuristic Methods in Optimal System Reliability. IEEE Trans Reliab 27:320–324
31. Kuo W, Lin H, Xu Z, Zhang W (1987). Reliability Optimization with the Lagrange multiplier and branch-and-bound technique. IEEE Trans Reliab 36:1090–1095
32. Kursawe F (1992) Towards Self-Adapting Evolution Strategies. In: Tzeng G, Yu P (Eds.) Proc. of the Tenth International Conference on Multiple Criteria Decision Making, Taipei
33. Kursawe F (1993) Evolution Strategies- Simple "Models" of Natural Process?. Revue Internationale de Systemique 7(5)
34. Kouvelis P, Yu G, (1997) Robust Discrete Optimization and Its Applications. Non Convex Optimization and Its Applications. Kluwer Academis Publishers.
35. Lin HH, Kuo W (1987) A Comparison of Heuristic Reliability Optimization Methods. In: Proc. of the World Productivity Forum & 1987 Int'l Industrial Engineering Conf. Institute of Industrial Engineering:583–589

36. Martorell S, Carlos S, Villanueva JF, Snchez AI, Galvn B, Salazar D, Cepin M (2006) Use of Multiple Objective Evolutionary Algorithms in Optimizing Surveillance Requirements. Reliab Engnng Sys Safety 91:1027–1038
37. Martorell S, Snchez A, Carlos S, Serradell V (2004) Alternatives and challenges in optimizing industrial safety using genetic algorithms. Reliab Engnng Sys Safety 86(1):25–38
38. Medina M, Carrasquero N, Moreno J (1998) Estrategias Evolutivas Celulares para la Optimización de Funciones. In: IBERAMIA'98, 6 Congreso Iberoamericano de Inteligencia Artificial. Lisboa, Portugal
39. Michalewicz Z (1992) Genetic Algorithms + Data Structure = Evolution Programs, Springer-Verlag
40. Milanese M, Norton J, J. Piet-Lahanier J (Eds.) (1998) Bounding Approaches to System Identification. Plenum Press, New York, USA
41. Misra KB, Ljubojevic MD (1973) Optimal Reliability Design of a System: A new look. IEEE Trans Reliab 22:255–258
42. Moore R (1979) Methods and Applications of Interval Analysis. SIAM Studies in Applied Mathematics. Philadelphia, USA
43. Nahas N, Nourelfath M (2005) Ant system for reliability optimization of a series system with multiple-choice and budget constraints. Reliab Engnng Sys Safety 87(1):1–12
44. Nakagawa Y, Nakashima K (1997) A heuristic method for determining optimal reliability allocation. IEEE Trans Reliab 26:156–161
45. Nakashima K, Yamato K (1997) Optimal Design of a Series-Parallel System with time-dependent reliability. IEEE Trans Reliab 26:119–120
46. Neumaier A (1990) Interval Methods for Systems of Equations. Cambridge University Press, UK.
47. Peng-Sheng Y, Ta-Cheng Ch (2005) An efficient heuristic for series-parallel redundant reliability problems. Computers & Operations Research 32(8):2117–2127
48. Ramirez-Marquez JE, Coit DW, Konak A (2004) A Redundancy allocation for series-parallel systems using a max-min approach. IIE Transactions 36(9):891–898
49. Ramírez-Rosado IJ, Bernal-Agustín JL (2001) Reliability and Costs Optimization for Distribution Networks Expansion Using an Evolutionary Algorithm. IEEE Trans on Power Systems 16:111–118
50. Ratschek H, Rokne J. (1984) Computer Methods for the range of functions. Ellis Horwood Limited, UK
51. Ravi V, Murty BSN, Reddy PJ, (1997) Nonequilibrium Simulated Annealing Algorithm Applied to Reliability Optimization of Complex Systems. IEEE Trans. Reliab 46:233–239.
52. Rocco C (2005) A Hybrid Approach based on Evolutionary Strategies and Interval Arithmetic to perform Robust Designs. In: Applications of Evolutionary Computing, Lecture Notes in Computer Science LNCS 3449:623–628, Springer-Verlag
53. Rocco CM, Miller AJ, Moreno JA, Carrasquero N (2000) Reliability Optimisation of Complex Systems. In: The Annual Reliability and Maintainability Symposium. Los Angeles, USA
54. Rocco C, Moreno JA, Carrasquero N (2003) Robust Design using a Hybrid-Cellular-Evolutionary and Interval-Arithmetic Approach: A Reliability

Application. In: Tarantola S, Saltelli A (Eds.) Special Issue: SAMO 2001: Methodological advances and useful applications of sensitivity analysis. Reliab Engnng Sys Safety 79(2):149–159
55. Rosenhead (1989) Rational analysis for a problematic world. Wiley, New York
56. Roy, B. (1998) A missing link in operational research decision aiding: robustness analysis. Foundations of Computing and Decision Sciences 23(3):141–160
57. Roy B. (2002) Robustesse de quoi, vis-à-vis de quoi, mais aussi robustesse pourquoi en aide à la décision? In Newsletters of the European Working Group "Multicriteria Aid for Decisions" 6(3):1–6
58. Sakawa M (1978) Multiobjective reliability and redundancy optimization of a series-parallel system by the Surrogate Worth Trade-off method. Microelectronics and Reliability 17(4):465–467
59. Salazar D, Rocco C, Galvàn B (2006) Optimization of Constrained Multiple-Objective Reliability Problems using Evolutionary Algorithms. Reliab Engnng Sys Safety 91:1057–1070
60. Salazar DE, Martorell SS, Galvn BJ (2005) Analysis of Representation Alternatives for a Multi-ple-Objective Floating Bound Scheduling Problem of a Nuclear Power Plant Safety System. In: Evolutionary and Deterministic Methods for Design, Optimisation and Control with Applications to Industrial and Societal Problems (Eurogen 2005). September 12-14. Munich, Germany
61. Saltelli a, Scott M (1997) Guest editorial: The role of sensitivity analysis in the corroboration of models and its link to model structural and parametric uncertainty. Reliab Enginng Syst Safety 57:1–4
62. Schaffer JD (1984) Multiple Optimization with Vector Evaluated Genetic Algorithms. Ph. D. Thesis. Vanderbilt University. (Unpublished)
63. Schwefel HP, Back Th (1995) Evolution Strategies I: Variants and their computational implementation. In: Periaux J, Winter G (Eds.) Genetic Algorithm in Engineering and Computer Science. John Wiley & Sons
64. Shelokar PS, Jayaraman VK, Kulkarni BD (2002) Ant algorithm for single and multiobjective reliability optimization problems. Quality and Reliability Engineering International 18(6):497–514
65. Srense K (2003) A famework for robust and flexible optimisation using metaheuristics with applications in supply chain design. PhD Thesis. University of Antwerp, Belgium
66. Sevaux M, Srensen K (2004) Robustness Analysis: Optimisation. In Newsletter of the European Working Group "Multiple Criteria Decision Aiding" 3(10):3–5
67. Srinivas N, Deb K (1994) Multiobjective optimization Using Nondominated Sorting in Genetic Algorithms. Evol Comput 2(3):221–248
68. Tillman FA, Hwang CL, Kuo W (1985) Optimization of System Reliability. Marcel Dekker
69. Tsutsui S, Gosh A, Fujimoto Y (1996) A robust solution searching scheme in genetic search. In Parallel Problem Solving from Nature. Berlin, Germany: Springer-Verlag:543–552
70. Tsutsui S, Gosh A (1997) Genetic algorithms with a robust solutions searching scheme. IEEE Trans. Evol. Comput. 1(3):201–208
71. Vehí J (1998) Análisi i disseny de controladors robustos. Ph.D. Thesis, Universitat de Girona
72. Vincke P (2003) About Robustness Analysis. In: Newsletter of the European Working Group "Multicriteria Aid for Decisions" 3(8):7–9

73. Vincke, P (1999) Robust solutions and methods in decision aid. Journal of Multicriteria Decision Analysis 8:181–187
74. Wolfram (1984) Cellular automata as models of complexity. Nature 3H
75. Wong YY, Jong WK (2004) Multi-level redundancy optimization in series systems. Computers & Industrial Engineering 46(2):337–346
76. Yun-Chia Liang, Smith, AE (2004) An ant colony optimization algorithm for the redundancy allocation problem (RAP). IEEE Trans Reliab 53(3):417–423
77. Zitzler E, Thiele L (1999) Multiobjective evolutionary algorithms: a comparative case study and the strength Pareto approach. IEEE Trans Evol Comput 3(4):257–271
78. Zitzler E, Laummans M, Thiele L (2001) SPEA2: Improving the Stregth Pareto Evolutionary Algorithm. TIK Report No. 103. Swiss Federal Institute of Technology (ETH). Computer Engineering and Networks Laboratory (TIK)
79. Zitzler E, Laumanns M, Bleuler S (2004) A Tutorial on Evolutionary Multiobjective Optimization. In: Workshop on Multiple Objective Metaheuristics (MOMH 2002). Springer-Verlag. Berlin, Germany

25

An Evolutionary Approach for Assessing the Degree of Robustness of Solutions to Multi-Objective Models

Carlos Barrico[1] and Carlos Henggeler Antunes[2]

[1] Department of Informatics, University of Beira Interior
6201-001 Covilhã, Portugal and
INESC Coimbra, Rua Antero de Quental 199, 3000-033 Coimbra, Portugal
cbarrico@inescc.pt

[2] Department of Electrical Engineering and Computers
University of Coimbra, Portugal and
INESC Coimbra, Rua Antero de Quental 199, 3000-033 Coimbra, Portugal
ch@deec.uc.pt

Summary. This chapter presents an approach to robustness analysis in multi-objective optimization problems, in which the decision variable space is subject to small perturbations . The concept of degree of robustness is incorporated into the evolutionary algorithm, being operationalized in the computation of the fitness value assigned to solutions, in the selection of the elite, and in the application of the sharing mechanism. Non-dominated solutions are classified according to their degree of robustness. The information on the degree of robustness of solutions is provided to support the decision maker in the selection of a robust compromise solution.

25.1 Introduction

Most complex problems arising in modern technologically developed societies inherently involve multiple, conflicting, and incommensurate evaluation aspects to assess the merits of alternative courses of action. Therefore, mathematical models for decision support become more representative of the actual decision context if those distinct axes of evaluation are explicitly taken into account rather than aggregated into a single indicator, generally of economic nature such as cost or benefit. In multi-objective programming models those axes of evaluation are operationalized by means of objective functions to be optimized subject to a set of constraints. Multi-objective models enable to grasp the conflicting nature of the objectives and the tradeoffs to be made in order to identify satisfactory compromise solutions by providing a basis to rationalize the comparison between non-dominated solutions. A non-dominated (efficient, Pareto optimal) solution is a feasible solution for which no improvement in all objective functions is simultaneously possible; that is, an improvement in an objective function can only be achieved by degrading, at least, another objective

C. Barrico and C.H. Antunes: *An Evolutionary Approach for Assessing the Degree of Robustness of Solutions to Multi-Objective Models*, Studies in Computational Intelligence (SCI) **51**, 565–582 (2007)
www.springerlink.com

function value. Besides contributing to make the model more realistic vis-a-vis actual problems, a multi-objective approach intrinsically possesses a value-added role in the modeling process and in model analysis, supporting reflection and creativity of decision makers in face of a larger universe of potential solutions, having in mind their practical implementation, since a prominent solution no longer exists.

Uncertainty is an intrinsic characteristic of real-world problems arising from multiple sources of distinct nature. Uncertainty emerges from the ever-increasing complexity of interactions within social, economic and technical systems, characterized by a fast pace of technological evolution, changes in market structures, and new societal concerns. It is generally impracticable that decision aid models could capture all the relevant inter-related phenomena at stake, get through all the necessary information, and also account for the changes and/or hesitations associated with the expression of the stakeholders' preferences. Besides structural uncertainty associated with the global knowledge about the system being modeled, input data may also suffer from imprecision, incompleteness or be subject to changes. In this context, it is important to provide decision makers with robust conclusions. The concept of robust solution is generally linked to a certain degree of "immunity" to data uncertainty, to an adaptive capability (or flexibility) regarding an uncertain future or ill-specified preferences, guaranteeing an acceptable performance even under changing conditions (such as model coefficients drifting from "nominal data").

The ability to work in each generation with a population of potential solutions makes evolutionary approaches well suited for multi-objective optimization problems, particularly complex problems of combinatorial nature, in which a set of non-dominated solutions must be identified rather than a single optimal solution (see Coello et al. [7], Deb [8] and Fonseca and Fleming [12]).

The study of a multi-objective optimization problem generally involves the characterization of the set of non-dominated solutions, by performing either an exhaustive computation of these solutions or by computing a representative sample. However, some of these solutions, which could be of interest for a decision maker (DM) as acceptable compromise solutions, in the sense they present a satisfactory balance between the axes of evaluation operationalized through objective functions, can be very sensitive to perturbations. That is, when a given non-dominated solution, selected by using any approach, is implemented in practice, small variations in the values of the decision variables may lead to an important degrading of the objective function performances. Therefore, algorithms must strive for robust solutions, that is, solutions that are relatively insensitive to perturbations in the decision variable space.

Some studies have been devoted to compute robust solutions both in single-objective as well as in multi-objective evolutionary optimization.

The works dealing with robust single-objective evolutionary optimization include the following approaches. Branke [4] suggested some heuristics for computing robust solutions. Branke [5] pointed out key differences between searching for optimal solutions in a noisy environment and searching for robust solutions. Branke and Schmidh [6] suggested a number of methods for alternate fitness estimation. Jin and Sendhoff [16] considered an approach for finding robust solutions in single-objective optimization as a bi-objective optimization problem, in which the objectives to be maximized are the robustness and the performance related with the original function. Tsutsui and Ghosh [22] presented a mathematical model for obtaining robust solutions using the schema theorem for single-objective genetic algorithms. Parmee [20] suggested a hierarchical strategy for searching in several regions with high performance and in the fitness landscape simultaneously. Lim et al. [18] presented an evolutionary design optimization approach that handles uncertainty with respect to the desired robust performance (the so-called inverse robust design, from which they search for solutions that guarantee a certain degree of maximum uncertainty and, at the same time, satisfy the desired nominal performance of the final design solution. Ong et al. [19] presented an evolutionary algorithm based on the combination of a max-min optimization strategy with a Baldwinian trust-region framework, employing local surrogate models for reducing the computational cost associated with robust design problems (focusing on combining evolutionary algorithms with function approximation techniques for robust design).

As far as robust multi-objective optimization is concerned few studies are reported in the literature. Hughes [14] computed the expected error to be used in the deterministic Pareto dominance that depends on the noise in the objective functions. Teich [21] extended existing techniques for space exploration design based on the Pareto-dominance criterion to the case where one or more of the objective functions are subject to uncertainties given by property intervals. Li et al. [17] presented a Robust Multi-Objective Genetic Algorithm (RMOGA) to investigate the trade-off between the performance and the robustness of solutions, considering two objective functions, a fitness value and a robustness index. Deb and Gupta [9, 10] extended an existing approach for finding robust solutions in single-objective optimization problems to a multi-objective setting, by considering the mean effective objective functions computed by averaging a representative set of neighboring solutions instead of the original objective functions. Barrico and Antunes [1–3] presented approaches that use the concept of degree of robustness, which are based on the solution behavior in its neighborhood in the decision variable space. The concept of degree of robustness is also used to assess the solution behavior in the neighborhood of the reference scenario in the space of the objective function coefficients [1].

For more details about both robust single-objective and robust multi-objective evolutionary optimization see Jin and Branke [15].

In this chapter, a multi-objective evolutionary approach to robustness analysis is presented, for assessing the degree of robustness of non-dominated solutions, which is based on the solution behavior in its neighborhood. This concept of degree of robustness (see also the concept of robust solution of type II in Deb and Gupta [9]) permits the user to exert a control on the desired/acceptable level of robustness of the solutions obtained. Users can specify the size of the solution neighborhood, both in the decision variable space and in the objective function space (generally the space the DM is more familiar with).

The concept of degree of robustness is imbedded in the evolutionary process, particularly in the fitness assessment of each individual. The underlying rationale is to bias the evolutionary process towards more robust solutions , that is solutions for which the objective function performances are more immune to perturbations in the decision variable values.

In section 2 the concept of degree of robustness is presented. The main features of the evolutionary algorithm are described in section 3. Illustrative results are presented in section 4, which have been obtained with the evolutionary approach applied to a bi-objective test problem. Finally, some conclusions are drawn in section 5.

25.2 The Degree of Robustness

The definition of robustness (either robust solution or robust method) is not uniform in the literature. However, a common view is shared: a robust solution shall behave well in (slightly) different conditions, meaning that it is as much as possible immune to small changes in the conditions it was designed for.

It is assumed that perturbations of solution x may occur in any dimension $(x_1, x_2, ..., x_n)$. The assessment of the degree of robustness of a solution is based on its behavior within a neighborhood around its "nominal" point. The underlying idea is to determine a set of neighborhoods $k\delta$ around solution x, such that the images of solutions within these neighborhoods are all better than $f(x)$ or belong to a pre-specified neighborhood η around $f(x)$ in the objective function space. The process begins by analyzing randomly generated solutions inside a hyperbox of radius δ around x. This neighborhood (hyperbox) is then progressively enlarged, in multiples of δ $(\delta, 2\delta, ...)$, until the percentage of solutions whose images in the objective space are all better than $f(x)$ or belong to the neighborhood of $f(x)$ is not greater than a pre-defined threshold. This enables to assign a degree of robustness to solutions according to the number of hyperbox enlargements for which that condition is fulfilled (see Fig. 25.1).

The degree of robustness depends on the size of a δ-neighborhood of solution x and the percentage of the h neighboring points whose objective function values are all better than $f(x)$ or belong to the η-neighborhood of $f(x)$. Those h neighboring points are randomly generated around solution x (see also Deb and Gupta [9]).

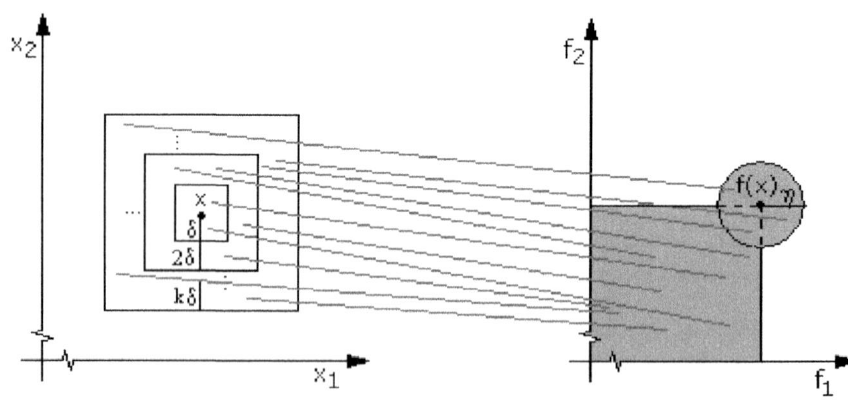

Fig. 25.1. Definition of neighborhoods in the decision variable space and objective function space (for 2-dimension spaces and all functions to be minimized)

The degree of robustness of solution x is a value k, such that (see Fig. 25.1):

a) the percentage of solutions in the $k\delta$-neighborhood of x, whose objective function values are all better than $f(x)$ or belong to the η-neighborhood of $f(x)$, is greater than or equal to a pre-specified threshold p;
b) the percentage of solutions in the $(k+1)\delta$-neighborhood of x, whose objective function values are all better than $f(x)$ or belong to the η-neighborhood of $f(x)$, is lower than p.

The threshold p may be understood as a measure of exigency of the degree of robustness.

The degree of robustness k of a solution x is gradually computed as k increases (neighborhoods $\delta, 2\delta, ..., k\delta$), as well as the number of neighboring points of x $(h, h+qh, ..., h+(k-1)qh)$, such that $h+(t-1)qh$ neighboring points ($t \in \{1, ..., k\}$) are analyzed in the $t\delta$-neighborhood of x.

However, the p, η and q parameter values can be different depending on the solution type (if solutions are non-dominated, or solutions are dominated or non-feasible). Due to the considerable run time of this algorithm, we recommend that the p parameter value should be higher for the dominated and infeasible solutions, and η and q parameters values should be higher for the non-dominated solutions. The underlying rationale is that non-dominated solutions are more important than dominated and infeasible solutions, and therefore a more exhaustive analysis is necessary for the non-dominated solutions.

To analyze whether, for a solution y belonging to δ-neighborhood of solution x, $f(y)$ belongs to the η-neighborhood of $f(x)$, it is necessary to calculate the distance between the images of solutions x and y in the objective space, $f(x)$ and $f(y)$. The normalized distance between $f(x)$ and $f(y)$ is computed by the expressions $||f(y) - f(x)|| \ / \ ||f(x)||$ (relative) or $||f(y) - f(x)||$ (absolute), where the operator $||.||$ can be any suitable metric, such as the city block, Euclidean or Chebycheff metrics. The choice of the metric can also have a role to play: in the city block metric all differences have the same importance, whereas in the Chebycheff metric only the greatest difference in all dimensions matters. The non-normalized distance between $f(x)$ and $f(y)$ can be used too (in this case, $\eta = (\eta_1, ..., \eta_R)$, where R is the number of objective functions of the problem).

The degree of robustness contributes for the evaluation of a solution (individual) of a population and enables to classify the solutions according to their level of robustness with respect to changes in the decision variable values.

25.3 The Evolutionary Algorithm

An evolutionary algorithm encompassing the assessment of the degree of robustness associated with each solution has been implemented, which uses an elitist strategy with a secondary population (with feasible non-dominated solutions only) of constant maximum size. The elitist strategy is aimed at increasing the performance, both accelerating the convergence speed towards the non-dominated frontier and ensuring the solutions attained are indeed non-dominated ones and well spread over the frontier. This is an important issue in real-world problems (see Gomes et al. [13]) since it is necessary to provide the DM with well-distributed and diverse solutions for a well-informed final decision to be made upon.

In order to bias the evolutionary process towards more robust solutions, this concept of degree of robustness is imbedded into the fitness assessment of each individual jointly with the non-dominance test. In each non-dominance level, more robust solutions are more likely to contribute for the next generation.

The main steps of this algorithm are the following:

- The fitness of the individuals composing the main population is computed;

- From the main population (consisting of POP individuals) $POP - E$ individuals are selected by using a tournament technique (E is the size of the elite set);
- A new population is formed by the $POP - E$ offspring generated by crossover and mutation, and E individuals (elite) that are the more robust in the secondary population;
- The evaluation of individuals is carried out by using a dominance test and their degree of robustness, defining an approximation to the non-dominated frontier;
- The non-dominated solutions are computed and they are processed to update the secondary population using a sharing technique, if necessary.

The population used in the algorithm implementation consists of individuals represented by an array of $M = M_1 + ... + M_n$ binary values, where n is the number of decision variables and M_i is the binary size of the ith decision variable representation.

25.3.1 Fitness Assessment

The fitness value of a solution depends on its degree of robustness and the dominance test. For each solution, the fitness computation uses a "non-dominated sorting" technique as in "NSGA-II" ([8], [11]), and involves determining various solution fronts in the following way:

- The first front consists of all non-dominated solutions, a minimum fitness value equal to $POP \times (MaxDegree + 1)$ being assigned to them;
- This fitness value of each one of these solutions is incremented by its degree of robustness;
- The solutions in the first front are temporarily ignored, and the remaining feasible solutions (the dominated solutions) are processed by applying them a dominance test (the non-dominated solutions will belong to the second front);
- The minimum fitness value of the current front is obtained by subtracting $MaxDegree + 1$ to the minimum fitness value of the previous front, which is assigned to the solutions of the current front;
- For each solution in the current front, the fitness value is incremented by its degree of robustness;
- This process continues until all feasible solutions are assigned a fitness value;
- The same process is repeated for the non-feasible solutions until all non-feasible solutions are assigned a fitness value.

If two solutions have the same fitness value, then the best solution is the one with fewer solutions in its neighborhood, according to a defined radius. A niche is defined by a radius *df* around a solution, where *df* is the maximum distance between solutions necessary to obtain a well-spread front and is equal to $MaxDist/POP$. $MaxDist$ is the distance between the pseudo-solutions obtained by considering the best and the worst values for each objective function in the main population.

This procedure is aimed at ensuring that the solutions of the k dominance level have a fitness value greater than the solutions of the $k + 1$ level, and in the same level the solutions with the higher degree of robustness have a higher fitness value. Finally, in the group of solutions with the same dominance level and equal degree of robustness, the best solutions in the group are the ones with fewer neighbors in the population.

25.3.2 The Sharing Mechanism

The sharing mechanism for updating the secondary population uses a niche scheme with a radius of dynamic value. This mechanism is applied after computing all non-dominated solutions candidate for the secondary population. These are all the non-dominated solutions in the set formed by the secondary population and the main population. This mechanism is only applied when the number of solutions candidate for the secondary population ($NCPS$) is greater than the size of this population (NPS).

The sharing mechanism consists in the following steps (adapted for bi-objective problems):

1) Insert the extreme solutions (those with the best values for each objective function);
2) Compute the first niche radius (ds) as the ratio: normalized distance between extreme solutions / NPS (that is, $\sqrt{2}/NPS$);
3) Insert solutions located at a distance greater than ds from the ones already belonging to the secondary population, using the degree of robustness as the priority criterion;
4) Update the value of the niche radius, ds, by reducing it by 10%;

25.3.3 The Initial Population and Genetic Operators

The strategy used to determine the initial population consists in randomly generating feasible non-dominated solutions only. This process consists in the following steps:

1) Randomly generate a feasible solution;
2) If this solution is non-dominated with respect to the initial population then insert this solution into the initial population and apply the dominance test to update this population;
3) If the initial population is not complete then return to step 1.

Uniform crossover and binary mutation have been used, with probability pc and pm.

25.3.4 The Algorithm

The evolutionary algorithm consists in the following steps:

1) Initialization: randomly generate the initial population with POP non-dominated solutions;
2) Determine the degree of robustness of each individual in the initial population;
3) Evaluation: compute the fitness value of each individual in the initial population;
4) Determine the (initial) secondary population of size NPS from the initial population: if $NPS \geq POP$ then copy all solutions from the initial population to the secondary population; else apply the sharing mechanism to the initial population to select NPS solutions;
5) Current population $\leftarrow$ initial population;
   Repeat

6) Build up the (main) population associated with the next generation of size POP:
   a) Introduce E individuals from the secondary population (elite) into the main population;
   b) Select 2 individuals of the current population by tournament (in each tournament, 10% of the individuals in this population are used to generate the selected one);
   c) Apply genetic crossover and mutation operators to the 2 individuals selected;
   d) Insert the new individuals into the main population;
   e) If the main population does not yet contain POP individuals then return to step b);
7) Determine the degree of robustness of each individual in the main population;
8) Evaluation: apply the dominance test and compute the fitness value of each individual in the main population;
9) Determine the $NCPS$ solutions candidate to becoming part of the secondary population;
10) Update the secondary population: if $NPS \geq NCPS$ then copy all candidate non-dominated solutions to the secondary population; else apply the sharing mechanism to all solutions found in step 9 to select NPS solutions;
11) Current population $\leftarrow$ main population;
    Until the pre-specified number of iterations is attained.

25.4 Illustrative Results

This approach incorporating robustness analysis to characterize the non-dominated front has been applied to a bi-objective test problem. The aim is to study the influence of parameters p and η in the determination of the Pareto front of robust non-dominated solutions. These parameters are associated with the exigency level of robustness and the indifference threshold between the objective function values specified by the DM/analyst.

The parameter p may be perceived as a indicator of the robustness exigency. If $p = 100\%$ then the objective function values of all neighboring solutions of x belong to the predefined η-neighborhood of $f(x)$. If $p = 90\%$ then the number of neighboring solutions of x, for which the objective function values belong to the η-neighborhood of $f(x)$ is at least 90%. So, it is more probable that a solution x with degree of robustness k in the first case ($p = 100\%$) has actually this degree of robustness than in the second case ($p = 90\%$) when p is relaxed.

The parameter η is used as a upper bound of the distance between two solutions (of any type) in the objective space. This parameter reflects the indifference thresholds regarding the values of each objective function. As the value of η increases, meaning that the DM is more tolerant with respect to differences in objective function values, the trend is that the number of more robust solutions also increases. The location of the more robust solutions in the non-dominated front is a relevant insight to aid the DM in the selection of a compromise solution.

25.4.1 Test Problem

The test problem is commonly used in the engineering design optimization literature, also studied in Deb [8] and Li et al. [17]. This problem is formulated as follows:

$$\begin{aligned}
&\textit{Minimize } f_1(x) = x_1\sqrt{16 + x_3^2} + x_2\sqrt{1 + x_3^2} \\
&\textit{Minimize } f_2(x) = 20\sqrt{16 + x_3^2}/(x_1 x_3) \\
&\textit{Subject to} \\
&\qquad 20\sqrt{16 + x_3^2} - 100 x_3 x_1 \leq 0, \\
&\qquad 80\sqrt{1 + x_3^2} - 100 x_3 x_2 \leq 0, \\
&\qquad x_1 > 0, \\
&\qquad x_2 \geq 0, \\
&\qquad 1 \leq x_3 \leq 3
\end{aligned}$$

In the binary representation of the decision variables $M_1 = M_2 = 17$ and $M_3 = 6$, and $\Delta x = (0.0001, 0.0001, 0.05)$ where Δx is the variation in the design decision variables. Thus, by construction M_i zeros correspond to the lower bound of x_1 (Δx_1), x_2 (0) and x_3 (1), and M_i ones correspond to the maximum value for x_1 (13.1071), x_2 (13.1071) and x_3 (4.15) in the evolutionary algorithm. These are the satisfactory maximum values for an accurate analysis.

The evolutionary algorithm for determining the non-dominated front uses the following parameter values: $POP = 120$; $NPS = 60$; $E = 0.1NPS$; $pc = 0.95$; $pm = 0.1$; and *number of iterations* $= 20$.

25.4.2 Non-Dominated Front

Fig. 25.2 displays the non-dominated front obtained without considering robustness analysis.

Fig. 25.3 displays the non-dominated front obtained with the following parameter values associated with the robustness analysis: $p = 100\%$; $h = 100$; $\delta = (\delta_1, \delta_2, \delta_3) = (0.1, 0.02, 0.05)$; $\eta = 0.75$, which is used as a upper bound of the absolute normalized distance between two solutions (of any type) in the objective space; and $q = h$.

In this front (Fig. 25.3), non-dominated solutions are categorized into 2 degrees of robustness (0 and 1). This information is relevant from a decision support point of view, since a DM would prefer a non-dominated compromise solution displaying a higher degree of robustness. In this case, with the maximum level of exigency ($p = 100\%$) the most robust solutions are located in the "central" region of the non-dominated front (well-balanced within the range of non-dominated solutions), slightly extending along the non-dominated frontier towards a slight improvement of f_2 at an expense of degrading the value of f_1.

25.4.3 Analysis of Parameter p

In this section the robustness parameter p will be the object of the analysis. All the other robustness parameter values are the same used to obtain the front displayed in Fig. 25.3 (in which $p = 100\%$).

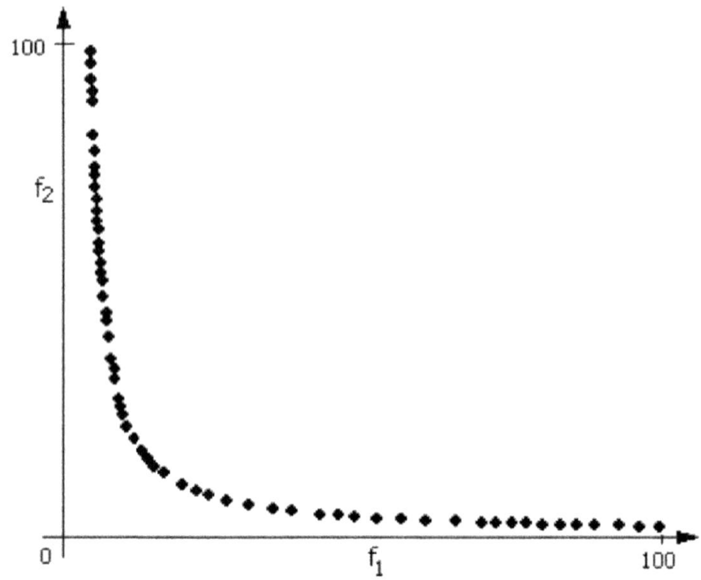

Fig. 25.2. Non-dominated frontier (without robustness analysis)

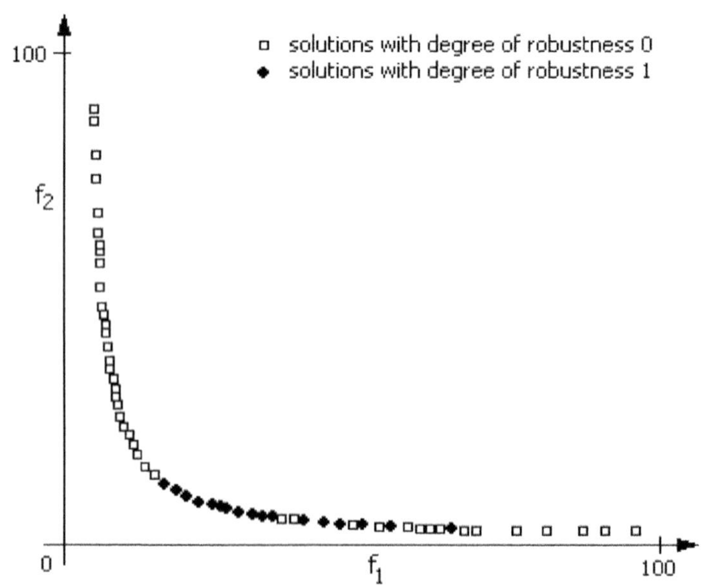

Fig. 25.3. Non-dominated frontier (with robustness analysis, $p = 100\%$, and $\eta = 0.75$)

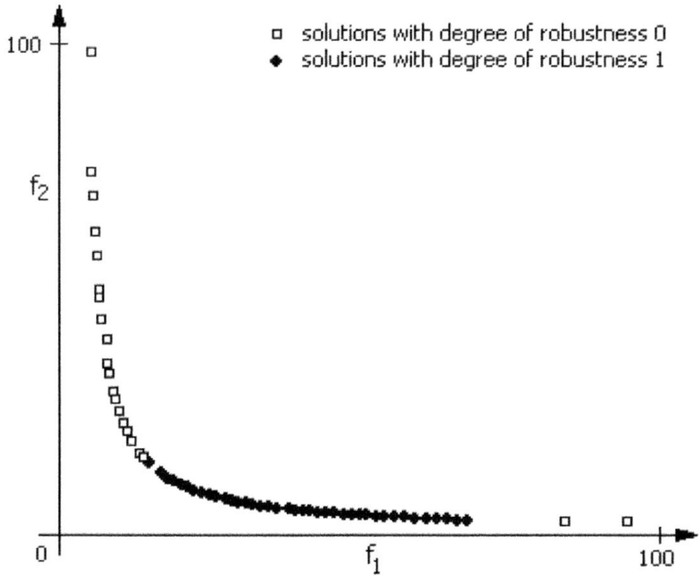

Fig. 25.4. Non-dominated frontier ($p = 90\%$)

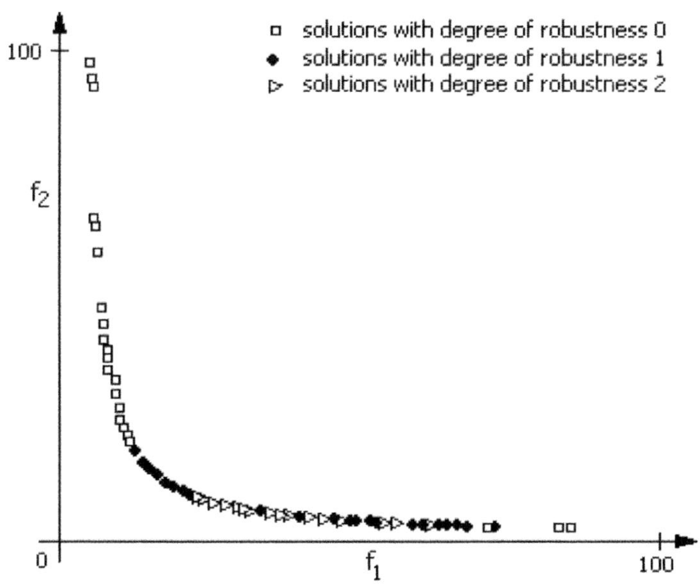

Fig. 25.5. Non-dominated frontier ($p = 70\%$)

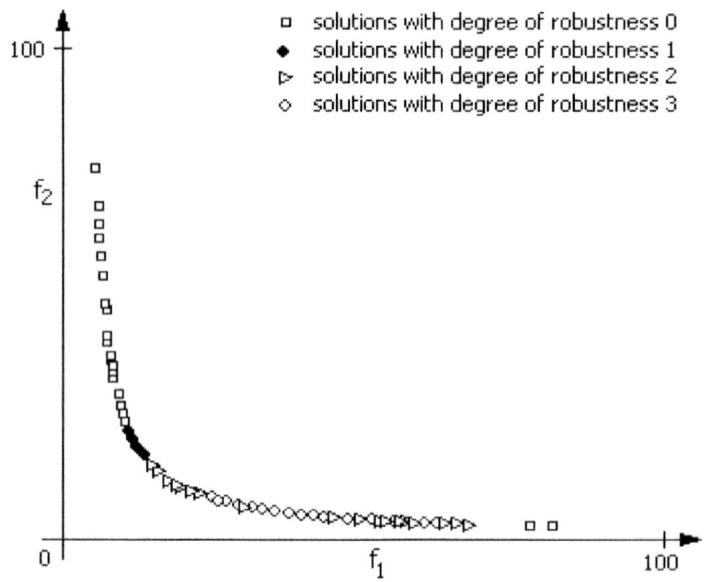

Fig. 25.6. Non-dominated frontier ($p = 50\%$)

Fig. 25.4 - Fig. 25.6 display the non-dominated solutions obtained with different p values for non-dominated solutions: 90% (Fig. 25.4), 70% (Fig. 25.5) and 50% (Fig. 25.6).

In these examples, non-dominated solutions are categorized into 2 degrees of robustness with $p = 90\%$ (such as with $p = 100\%$), 3 different degrees of robustness with $p = 70\%$, and 4 different degrees of robustness with $p = 50\%$ (Fig. 25.4 - Fig. 25.6).

As p decreases, from 100% to 90%, some solutions with a degree of robustness 0 become of degree of robustness 1, reinforcing the trend of extending along the non-dominated frontier towards a slight improvement of f_2 (in a region where f_2 values are not far from its optimum) in exchange of degrading the value of f_1 (Fig. 25.4).

With $p = 70\%$ (Fig. 25.5), non-dominated solutions with the higher degree of robustness are located closer to the optimum of f_2. In this case there are non-dominated solutions possessing a degree of robustness 2, which did not appear in the transition of p from 100% to 90%.

Finally, in the transition of p from 70% to 50% (Fig. 25.6), there is an increase of the number of solutions with degree of robustness 2 also closer to the optimum of f_2. In this front some solutions with degree of robustness 3 appear, which did not happen in the previous examples with higher values for the p parameter. These solutions are broadly located in the same regions where the solutions with degree of robustness 2 are located in the front obtained in the previous example (with $p = 70\%$), that is closer to the optimum of f_2.

This study provides the DM thorough information regarding the selection of a compromise solution (between the objective function values) which also presents a high level of robustness. Non-dominated solutions which privilege f_1 possess a low degree of robustness. A higher degree of robustness is achieved in a region of well-balanced solutions, but closer to the optimum of f_2. This information is relevant for a DM when assessing the merits of competing non-dominated solutions, not just examining the trade-off between the objective function values but also their degree of robustness.

25.4.4 Analysis of Parameter η

In this section the influence of parameter η on the degree of robustness assigned to solutions is analyzed. η contains the upper limits of the absolute normalized distance between two solutions (of any type) in the objective space. All the other robustness parameter values are the same used to obtain the front displayed in Fig. 25.3 (in which $\eta = 0.75$).

Fig. 25.7 - Fig. 25.9 display the non-dominated fronts obtained with different η values: 1.0 (Fig. 25.7), 1.5 (Fig. 25.8) and 0.5 (Fig. 25.9).

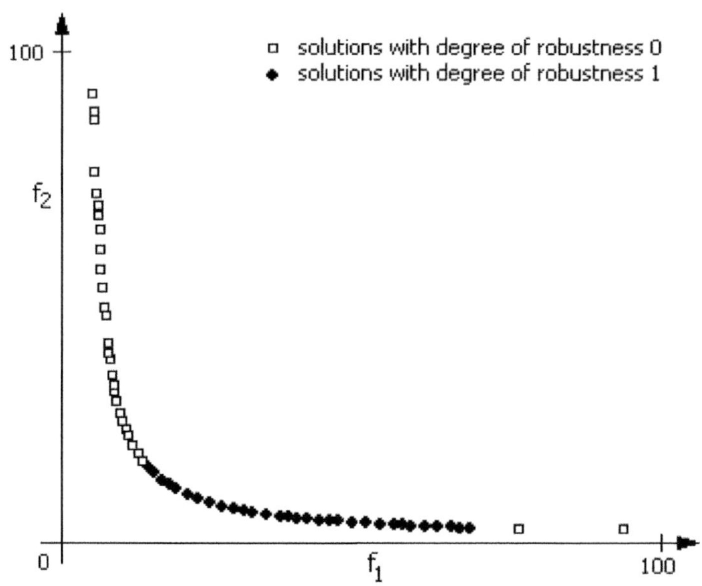

Fig. 25.7. Non-dominated frontier ($\eta = 1.0$)

In the transition of η from 0.75 to 1.0, the non-dominated front obtained is composed by solutions with 2 degrees of robustness (0 and 1) (Fig. 25.7), such as in the front obtained with $\eta = 0.75$ (Fig. 25.3). However, there is an increase of the number of solutions with a higher degree of robustness (1). This happens because the increase of the η value implies relaxing the importance assigned to differences between the objective function values (up to a threshold) and thus enlarging the neighborhood in which the DM is indifferent between non-dominated solutions located therein.

In the transition of η from 1.0 to 1.5, some solutions of degree of robustness 1 become of degree of robustness 2 (Fig. 25.8).

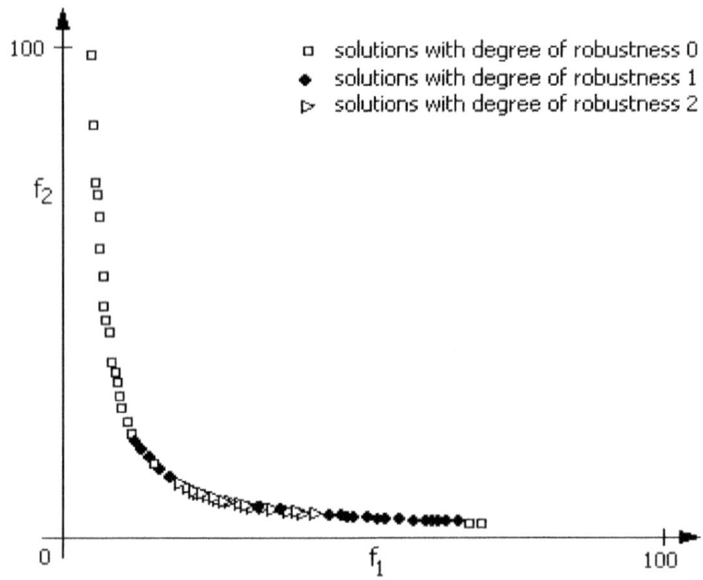

Fig. 25.8. Non-dominated frontier ($\eta = 1.5$)

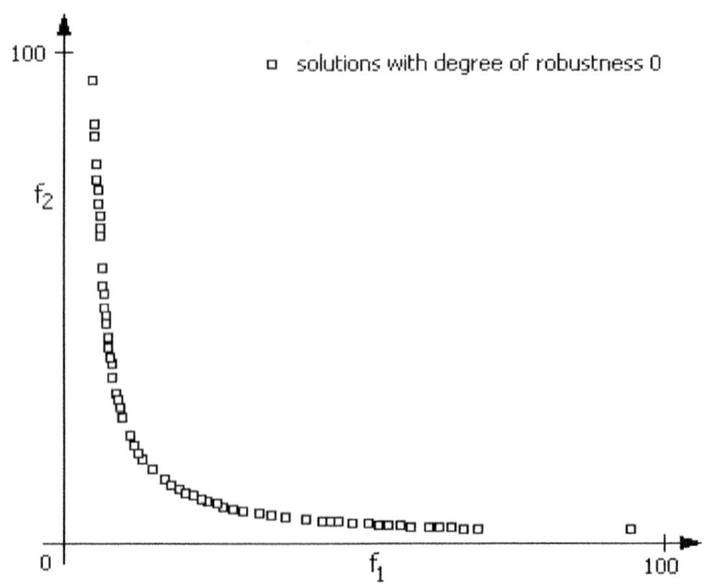

Fig. 25.9. Non-dominated frontier ($\eta = 0.5$)

In the opposite sense, with the decrease of η from 0.75 to 0.5, all non-dominated solutions of the front obtained have the same degree of robustness (0) (Fig. 25.9). In this case, differences in the objective function values become more important for the DM (that is, his/her indifference threshold for considering irrelevant for discriminating purposes the difference between the objective functions values decreases). This means that the amplitude of the neighboring solutions in the objective space, with respect to the solution under analysis, also decreases.

Therefore, solutions with a higher degree of robustness are mostly located in a region of the non-dominated front presenting values for f_2 not far from its optimal value and values for f_1 in the mid of the range of the values f_1 attains within the non-dominated front. This type of insights provided by the assignment of a degree of robustness to the non-dominated solutions is a relevant information for a DM to assess their merits in the selection of a satisfactory compromise solution.

25.4.5 Using Other Metrics

Also, other type of metrics can be used in the robustness analysis, such as a relative normalized and absolute non-normalized distance between two solutions in the objective space.

In Fig. 25.10, a relative normalized distance has been used with the same parameter values used to obtain the front displayed in the Fig. 25.3, except the η parameter value ($\eta = 0.05$, this mean 5% of the reference value). In this case, the most robust solutions are still located towards the best values for f_2, but without sacrificing f_1 too much.

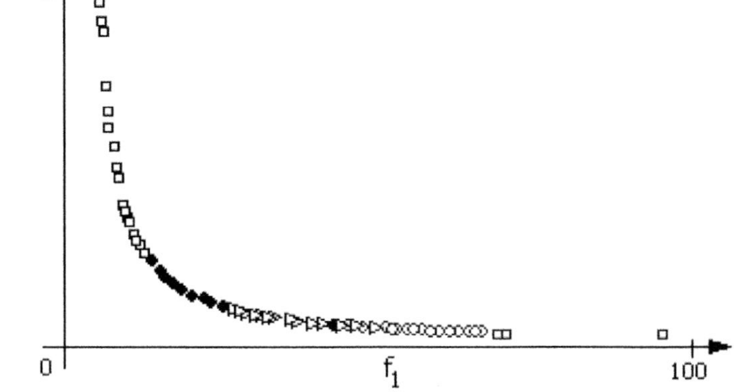

Fig. 25.10. Non-dominated frontier obtained by using the relative normalized distance

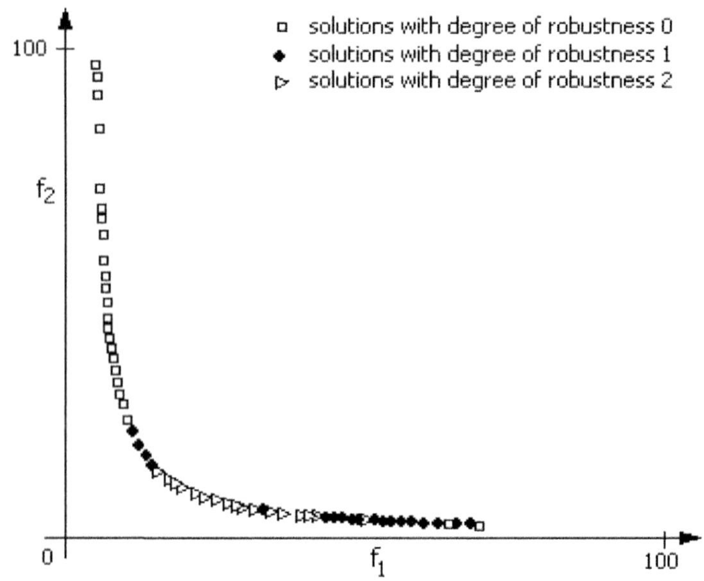

Fig. 25.11. Non-dominated frontier obtained by using the absolute non-normalized distance

In Fig. 25.11, an absolute non-normalized distance has been used with the same parameter values used to obtain the front displayed in the Fig. 25.3, except the η parameter value: $\eta = (\eta_1, \eta_2) = (1.5, 1.5)$. The most robust solutions possess the same characteristics mentioned before: they present very good values for f_2, and values for f_1 in the mid of its range within the non-dominated front.

The specification of parameter η does not impose an excessive burden on the DM since he/she is familiar with the objective function space. Therefore, he/she is able to provide information on the thresholds below which the difference between solutions is not relevant for decision purposes.

25.5 Conclusions

This chapter presented an approach to robustness analysis in evolutionary multi-objective optimization, in which the values of the decision variable are subject to small perturbations. The concept of degree of robustness is incorporated into the evolutionary algorithm, particularly in the computation of the fitness value of the solutions. This approach also enables to classify the solutions of the Pareto-front according to the degree of robustness (solutions with different degrees of robustness can belong to the same Pareto-front), not just classifying solutions as robust or not robust.

Information on the robustness of solutions, and not just on the objective function values, is relevant for assisting a decision maker in assessing the merit of non-dominated solutions and selecting a satisfactory compromise solution that exhibits a higher degree of stability in face of perturbations.

References

1. Barrico C, Antunes CH (2006) "Robustness Analysis in Multi-Objective Optimization". Research Report N3/2006, INESC Coimbra, Portugal.
2. Barrico C, Antunes CH (2006) "A New Approach to Robustness Analysis in Multi-Objective Optimization". Proceedings of the 7th International Conference on Multi-Objective Programming and Goal Programming (MOPGP 2006). Loire Valley (Tours), France.
3. Barrico C, Antunes CH (2006) "Robustness Analysis in Multi-Objective Optimization Using a Degree of Robustness Concept". Proceedings of the 2006 IEEE World Congress on Computational Intelligence (WCCI 2006): 6778-6783.
4. Branke J. (1998) "Creating Robust Solutions by means of an Evolutionary Algorithm". Parallel Problem Solving from Nature, Lecture Notes in Computer Science 1498, Springer: 119-128.
5. Branke J. (2000) "Efficient Evolutionary Algorithms for Searching Robust Solutions". Adaptive Computing in Design and Manufacture (ACDM 2000), Springer: 275-286.
6. Branke J, Schmidh C (2005) "Faster Convergence by means of Fitness Estimation". Soft Computing 9(1): 13-20.
7. Coello C, Veldhuizen D, Lamont G (2002) "Evolutionary Algorithms for Solving Multi-Objective Problems". Kluwer Academic Publishers.
8. Deb K (2001) "Multi-Objective Optimization Using Evolutionary Algorithms". John Wiley and Sons, New York.
9. Deb K, Gupta H (2004) "Introducing Robustness in Multiple-Objective Optimization". KanGAL Report Number 2004016, Kanpur Genetic Algorithms Laboratory, Indian Institute of Technology Kanpur, India.
10. Deb K, Gupta H (2005) "Searching for Robust Pareto-Optimal Solutions in Multi-Objective Optimization". Proceedings of the Third International Conference of Evolutionary Multi-Criteria Optimization (EMO-2005): 150-164.
11. Deb K, Pratap A, Agarwal S, Meyarivan T (2002) "A fast and elitist multi-objective genetic algorithm: NSGA-II". IEEE Transactions on Evolutionary Computation 6(2): 181-197.
12. Fonseca CM, Fleming PJ (1995) "An Overview of Evolutionary Algorithms in Multiobjective Optimization". Evolutionary Computation 3(1): 1-16.
13. Gomes A, Antunes CH, Martins A (2004) "A multiple objective evolutionary approach for the design and selection of load control strategies". IEEE Transactions on Power Systems 19(2): 1173-1180.
14. Hughes EJ (2001) "Evolutionary Multi-Objective Ranking with Uncertainty and Noise". Proceedings of the First International Conference on Evolutionary Multi-Criterion Optimization (EMO-2001): 329-343.
15. Jin Y, Branke J (2005) "Evolutionary Optimization in Uncertain Environments - A Survey". IEEE Transactions on Evolutionary Computation 9(3): 1-15.
16. Jin Y, Sendhoff B (2003) "Trade-Off between Performance and Robustness: An Evolutionary Multiobjective Approach". Proceedings of the Second International Conference on Evolutionary Multi-Criterion Optimization (EMO-2003): 237-251.
17. Li M, Azarm S, Aute V (2005) "A Multi-Objective Genetic Algorithm for Robust Design Optimization". Proceedings of Genetic and Evolutionary Computation Conference (GECCO'05): 771-778.

18. Lim D, Ong YS, Lee BS (2005) "Inverse Multi-Objective Robust Evolutionary Design Optimization in the Presence of Uncertainty". Proceedings of Genetic and Evolutionary Computation Conference (GECCO'05): 55-62.
19. Ong YS, Nair PB, Lum KY (2005) "Max-Min Surrogate-Assisted Evolutionary Algorithm for Robust Aerodynamic Design". IEEE Transactions on Evolutionary Computation 10(4): 392-404.
20. Parmee IC (1996) "The Maintenance of Search Diversity for Effective Design Space Decomposition using Cluster-Oriented Genetic Algorithms (Cogas) and Multi-Agent Strategies (Gaant)". Proceedings of the Second International Conference of Adaptive Computing in Engineering and Control: 128-138.
21. Teich J (2001) "Pareto-Front Exploration with Uncertain Objectives". Proceedings of the First International Conference on Evolutionary Multi-Criterion Optimization (EMO-2001): 314-328.
22. Tsutsui S, Ghosh A (1997) "Genetic Algorithm with a Robust Solution Searching Scheme". IEEE Transactions on Evolutionary Computation 1(3): 201-219.

26

Deterministic Robust Optimal Design Based on Standard Crowding Genetic Algorithm

Qing Ling[1], Gang Wu[2], Qiuping Wang[3]

[1] Department of Automation, University of Science and Technology of China, Hefei 230026, P. R. China
qingling@mail.ustc.edu.cn

[2] Department of Automation, University of Science and Technology of China, Hefei 230026, P. R. China
wug@ustc.edu.cn

[3] National Synchrotron Radiation Laboratory, University of Science and Technology of China, Hefei 230026, P. R. China
qiuping@ustc.edu.cn

Summary. In practical optimization problems, it is often desired that a solution is not only of high performance, but also of high robustness, for example, robust to fabrication tolerances. General forms of robust optimal design models are formulated in this chapter, including deterministic robustness and probabilistic robustness, objective robustness and feasibility robustness. We mainly focus on the deterministic objective robustness model for the single-objective robust optimal design problem. A framework of standard crowding genetic algorithm (SCGA) is used to solve the problem, with a Monte-Carlo simulation (MCS) method to calculate the approximate robust objective function (ROF). Effect of the sampling number of MCS is discussed in the numerical experiments. Design example of the recording optics of varied-line-spacing holographic grating (VLSHG) is provided. This typical deterministic robust optimal design problem is solved by the proposed algorithm successfully.

26.1 Introduction

In practical engineering optimization problems, tolerances of design variables are inevitable. Therefore, it has to be considered whether the system performance will meet the design requirement under the perturbation of optimal solutions. Though designers used to deal with tolerances after the optimization process conventionally, it is better to consider it in the optimization process, i.e. to search for robust solutions. This kind of optimization problems is categorized as robust optimal design in [9].

Q. Ling et al.: *Deterministic Robust Optimal Design Based on Standard Crowding Genetic Algorithm*, Studies in Computational Intelligence (SCI) **51**, 583–598 (2007)
www.springerlink.com

26.1.1 Overview of Robust Optimal Design Model

The robustness of a solution is embodied in the two aspects of optimal design problem: effect of perturbation to objective function and effect of perturbation to constraints. Therefore, there are two main research interests in robust optimal design: 1) selecting proper design variables such that the objective function is not sensitive to the fabrication tolerances, i.e. objective robustness; 2) assuring that the design variables are able to satisfy the constraints under the existence of fabrication tolerances, i.e. feasibility robustness.

The most important problem in robust optimal design is how to define the robustness of solutions, i.e. to build the mathematical model of robust optimal design. Basically, according to the emphasis of the model, it can be classified as objective robustness and feasibility robustness. In objective robustness model, original objective function (OOF) is converted into robust objective function (ROF), under certain definition of robustness. In feasibility robust model, original constraints (OCs) are converted into robust constraints (RCs), too. According to the stochastic properties of the model, it can be classified as deterministic and probabilistic. Furthermore, according to the structure of the model, it can be classified as single-objective and multi-objective.

Probabilistic robustness model introduces the probabilistic properties of tolerances and considers their influence on objective function and constraints. Deterministic robustness model directly considers the worst case of objective function and constraints after the realization of design variables. Probabilistic robustness model is fit for the design problem where the main concern is the average quality of production, while the deterministic robustness model is fit for the design problem where the main concern is the specific case of realization. Various kinds of probabilistic objective robustness models have been discussed in [8]. Generally, they are constructed as multi-objective optimization problem where each objective represents a probabilistic property of OOF. Gunawan [6] studied a kind of deterministic objective robustness model, named as sensitivity region method. Sensitivity region method can be transformed into a multi-objective optimization problem too [13].

Du [5] analyzed the modeling of feasibility robustness problem. Das [3] classified the constraints as hard constraints and soft constraints, equality constraints and inequality constraints, and studied the constraint-handling techniques respectively. Mattson [16] discussed how to handle equality constraints in robust optimal design.

Chen [2] combined objective robustness and feasibility robustness with the quality utility function, and constructed the multi-objective optimization problem. Lee [12] designed the quality utility function to objective robustness and feasibility robustness respectively. Another class of combination method is termed as reliability based design optimization (RBDO), which has been applied to the hyper-elastic structure design [19], air-bearing surface [26], and so on.

Up to now, we mainly focus on the single-objective robust optimal design problem. But the models discussed above can be easily migrated to the multi-objective robust optimal design problem. Deb [4] studied their probabilistic model and deterministic model, and illustrated the relationship between original Pareto front and robust Pareto front. Messac [17] considered the probabilistic model, while Gunawan [7] extended the sensitivity region method to the deterministic model of multi-objective robust optimal design.

26.1.2 Overview of Robust Optimal Design Method

The optimization methods in robust optimal design include local search methods and global search methods. Sorensen [20] discussed how to find robust solutions using local search. Su [21] proposed an automatic differentiation algorithm to solve the robust optimal design problem. For multi-objective robust optimal design, a physical programming algorithm has been developed by Messac [17].

Typical global search methods in robust optimal design are evolutionary algorithms (EAs). Evolutionary algorithms are powerful tools for global search and have been widely applied to the optimization of objective robustness models. For probabilistic objective robustness, Tsutsui [23] introduced genetic algorithms to the robust solution searching scheme. Branke [1] considered improving the efficiency of robust solution searching. Jin [8] proposed a multi-objective evolutionary algorithm to balance performance and robustness. As a natural expansion, Deb [4] applied multi-objective evolutionary algorithms to the multi-objective robust optimal design problem. For the optimization of probabilistic objective robustness model, evolutionary algorithms have achieved successful application in practical engineering problems, including design of multilayer optical coatings [25], design of automobile valve train [10], and stochastic finite element problem [11].

But in deterministic robust optimization problem, researches in optimization methods are still limited. In the sensitivity region method, deterministic robust optimal design is expressed as a max-min problem [6], where the key is how to optimize the embedded minimization problem, i.e. to calculate the sensitivity region. Gunawan used a genetic algorithm (GA) to realize the global search of the sensitivity region. The general form of deterministic robust optimal design discussed in this chapter is expressed as a min-max problem, where the key is how to optimize the embedded maximization problem, i.e. to calculate the robust objective function. Traditional method in approximately calculating ROF is the Monte-Carlo simulation (MCS) method [24]. The most important issue of MCS method in practical application is how to achieve the trade-off between accuracy of solution and consumption of computation. Small sampling number of MCS results in significant evaluation error which will mislead the optimization process, while large sampling number will increase the computation consumption, especially in the high-dimensional problems.

In this chapter, we mainly study the deterministic objective robustness model for the single-objective robust optimal design problem. A general form of deterministic objective robustness model is established and a standard crowding genetic algorithm (SCGA) is used to solve the problem. The effect of the sampling number of MCS method is discussed under the framework of SCGA.

26.1.3 Arrangement of the Chapter

This chapter is organized as follows. Section 26.2 formulates the mathematical models of robust optimal design, including objective robustness and feasibility robustness, probabilistic robustness and deterministic robustness. Section 26.3 provides the robust optimal design framework of SCGA for the general form of deterministic objective robustness model. The MCS method is introduced to calculate approximate ROF. Optimization results in test functions are shown in Section 26.4. In Section 26.5, the design of recording optics of varied-line-spacing holographic grating (VLSHG) is described. This typical deterministic robust optimal design problem is solved by the proposed algorithm successfully. Discussion of future work and conclusion are provided in Section 26.6.

26.2 Models of Robust Optimal Design

In this section, several robust optimal design models are introduced to the single-objective robust optimal design problem.

26.2.1 Formulation of Robust Optimal Design Problem

Considering the following constrained single-objective optimal design problem:

$$\begin{aligned} \min_{\boldsymbol{X}} \quad & f(\boldsymbol{X}) \\ s.t. \quad & \boldsymbol{H}(\boldsymbol{X}) \leq \boldsymbol{0} \end{aligned}$$

Where $f(\boldsymbol{X})$ is the objective function and $\boldsymbol{H}(\boldsymbol{X}) = \{\boldsymbol{H}_i(\boldsymbol{X}), i = 1, 2, ..., N_{con}\}$ are the constraints. $\boldsymbol{X} = \{\boldsymbol{X}_i, i = 1, 2, ..., N_{var}\}$ are the N_{var}-dimensional design variables. In general optimal design problems, a set of design variables $\boldsymbol{X}_D = \{\boldsymbol{X}_{Di}, i = 1, 2, ..., N_{var}\}$ is solved through minimizing objective function $f(\boldsymbol{X})$ under the constraints $\boldsymbol{H}(\boldsymbol{X})$. But in the practical fabrication process, $\boldsymbol{X}_D$ will be affected by the random tolerances. Therefore its realization is a random vector $\boldsymbol{X}_R$:

$$\boldsymbol{X}_R = \boldsymbol{X}_D + \boldsymbol{T}_R \tag{26.1}$$

Where $\boldsymbol{T}_R = \boldsymbol{T}_{Ri}, i = 1, 2, ..., N_{var}$ is the random tolerance vector. Generally speaking, the random noise will result in the degradation of objective function, or even the infeasibility of optimized solutions.

26.2.2 Objective Robustness

Objective robustness describes the sensitivities of objective function to the perturbation of design variables. According to the stochastic properties, it can be classified as deterministic model and probabilistic model. Probabilistic objective robustness is usually formulated as the multi-objective structure.

Probabilistic Objective Robustness

Probabilistic objective robustness model focuses the probabilistic properties of random tolerance vector $\boldsymbol{T}_R \in \boldsymbol{B}_R$. Define tolerance space $\boldsymbol{B}_R = \{L_{Bi} \leq \boldsymbol{T}_{Ri} \leq U_{Bi}\}$, with $i = 1, 2, ..., N_{var}$, where L_{Bi} and U_{Bi} are the lower bound and upper bound of random tolerance element $\boldsymbol{T}_{Ri}$. In general, probabilistic objective robustness model assumes that random tolerance vector $\boldsymbol{T}_R$ are under a known joint distribution $P_R(\boldsymbol{T}_R)$. Therefore, the robustness of a solution can be described by some specific statistical values of OOF, for example, expectation or variance. The ROFs are expressed as f_E and f_V:

$$\min_{\boldsymbol{X}_D} \quad f_E(\boldsymbol{X}_D) = \int_{\boldsymbol{T}_R \in \boldsymbol{B}_R} f(\boldsymbol{X}_D + \boldsymbol{T}_R) P_R(\boldsymbol{T}_R) \mathrm{d}\boldsymbol{T}_R \tag{26.2}$$

$$\min_{\boldsymbol{X}_D} \quad f_V(\boldsymbol{X}_D) = \int\limits_{\boldsymbol{T}_R \in \boldsymbol{B}_R} (f(\boldsymbol{X}_D + \boldsymbol{T}_R) - f(\boldsymbol{X}_D))^2 P_R(\boldsymbol{T}_R) \mathrm{d}\boldsymbol{T}_R \qquad (26.3)$$

In robust optimization problem, design variables must achieve the trade-off between optimality and robustness. ROF f_E reflects this trade-off partially, while OOF f and ROF f_V reflect the optimality and robustness respectively. Therefore, we generally transform the probabilistic objective robustness model to a multi-objective optimization problem, for example, minimizing f and f_E, f and f_V, f_E and f_V, or f, f_E and f_V simultaneously.

Deterministic Objective Robustness

Deterministic objective robustness model focuses the worst case of design variables caused by the random tolerance vector $\boldsymbol{T}_R$. In this meaning, deterministic objective robustness is also termed as worst case optimization. The general form of deterministic robust objective function can be expressed as f_W:

$$\min_{\boldsymbol{X}_D} \quad f_W(\boldsymbol{X}_D) = \max_{\boldsymbol{T}_R \in \boldsymbol{B}_R} \quad f(\boldsymbol{X}_D + \boldsymbol{T}_R) \qquad (26.4)$$

Deterministic objective robustness model does not need the probabilistic information of random tolerance vector. Furthermore, ROF f_W reflects the trade-off between optimality and robustness directly.

The sensitivity region method proposed by Gunawan [6] is another form of deterministic objective robustness model. In (5), robustness is related to the maximal objective function value in the tolerance space. While in the sensitivity region method, robustness is described by the minimal radius of perturbation which breaks a given threshold of objective function value, named as sensitivity region. Thus it is the dual form of (5).

26.2.3 Feasibility Robustness

Feasibility robustness describes the sensitivities of constraints to the perturbation of design variables. It can be divided into probabilistic and deterministic model, too.

Probabilistic Feasibility Robustness

For the constraints in (1), the probabilistic feasibility model is generally expressed as:

$$s.t. \quad P_{con}\{\boldsymbol{H}(\boldsymbol{X}_R) \leq \boldsymbol{0}\} = P_{con}\{\boldsymbol{H}(\boldsymbol{X}_D + \boldsymbol{T}_R) \leq \boldsymbol{0}\} \geq P_E \qquad (26.5)$$

Where P_{con} represents the probability of that all constraints are satisfied. P_E is the expected probability to satisfy the constraints. When the joint distribution of tolerance vector is known as $P_R(\boldsymbol{T}_R)$, (6) is transformed into:

$$s.t. \quad P_{con}\{\boldsymbol{H}(\boldsymbol{X}_D + \boldsymbol{T}_R) \leq \boldsymbol{0}\} = \int\limits_{\boldsymbol{T}_R \in \boldsymbol{B}_R,\, \boldsymbol{H}(\boldsymbol{X}_D + \boldsymbol{T}_R) \leq \boldsymbol{0}} P_R(\boldsymbol{T}_R) \mathrm{d}\boldsymbol{T}_R \geq P_E \quad (26.6)$$

Deterministic Feasibility Robustness

Similarly to the deterministic objective robustness model, deterministic feasibility robustness is also named as worst case analysis. It means that for design variable $\boldsymbol{X}_D$, and for each tolerance vector $\boldsymbol{T}_R$ in the tolerance space $\boldsymbol{B}_R$, the following constraints should be satisfied simultaneously:

$$s.t. \quad \boldsymbol{H}(\boldsymbol{X}_D + \boldsymbol{T}_R) \leq 0. \tag{26.7}$$

Apparently, it is the special case of probabilistic feasibility robustness model of (6), with P_E equal to 1. Deterministic feasibility robustness model does not need the probabilistic information of random tolerance vector too.

26.3 Robust Optimal Design Framework

This section provides the standard crowding genetic algorithm (SCGA) framework to optimize the general form of deterministic objective robustness model. The Monte-Carlo simulation (MCS) method is introduced to calculate the approximate deterministic ROF f_W in (5).

26.3.1 Standard Crowding Genetic Algorithm (SCGA) Framework

In this chapter, we extend genetic algorithm with standard crowding operator to attain the diversity of population. Standard crowding model proposed by De Jong updates population through replacing similar parents [22]. For each child C, select CF (crowding factor) individuals in parents, choose the nearest parent P under some distance metric. For a minimization problem, if the objective function value of P is larger than that of C, use C to replace P, else preserve P. Here we set $CF = PopNum$ (population number of generation) to eliminate the selection error. The procedures of the SCGA for robust optimal design are:

1) Initialization: initialize $PopNum$ uniformly distributed random population in feasible solution space.
2) Crossover: divide population into $PopNum/2$ groups randomly, with two individuals P_1 and P_2 in each group. Generate children C_1 and C_2: $C_1 = A \times P_1 + (1-A) \times P_2$, $C_2 = (1-A) \times P_1 + A \times P_2$ for each group, where A is a random number uniformly distributed in $[-0.25, 1.25]$.
3) Mutation: add a random variation to each child, with the mutation operator in the breeder's genetic algorithm (BGA) [18].
4) Competition: evaluate ROF of parents and children with MCS method, and select the best $PopNum$ individuals into next generation using standard crowding operator with $CF = PopNum$.
5) Stopping criterion: repeat step 2–4, until the maximum generation number $MGGA$ reaches.

To reduce the random error in the previous calculation of approximate ROF of parents, both parents and children must be evaluated in step 4.

26.3.2 Monte-Carlo Simulation (MCS) Method

Computation of ROF is the most important part of robust optimal design for both probabilistic and deterministic objective robustness model. Firstly, the approximation error of ROF will mislead the optimization process. Secondly, the computation of ROF is time-consuming. A popular method to evaluate ROF is the MCS method [24]:

$$f_W(\boldsymbol{X}_D) \approx \max_{1 \le i \le k} \quad f(\boldsymbol{X}_D + (\boldsymbol{T}_R)_i). \tag{26.8}$$

Here k is the sampling number and $(\boldsymbol{T}_R)_i$ represents the i-th realization. Generally speaking, the k sampling points are selected with uniform distribution, because each point in tolerance space is considered as identical in deterministic model. This is different from MCS method used in the probabilistic model.

In most practical optimization problems, main computational consumption comes from evaluation of OOFs. Therefore we use the evaluation time $EvaTim$ as the measurement of computational complexity. In the SCGA framework with MCS method, $EvaTim = 2 \times PopNum \times MGGA \times k$.

26.4 Numerical Experiments

In this section, two test functions are introduced to the robust optimal design framework of SCGA, as shown in Section 26.3.

26.4.1 Test Functions

The first numerical test function comes from [8]:

$$\min_{x} \quad f_1(x) = 2 \times \sin(10 \times x \times \mathrm{e}^{-0.008x}) \times \mathrm{e}^{-0.25x}$$
$$s.t. \quad 0 \le x \le 10$$

Tolerance space of x is $[-Tole, Tole]$, $Tole = 0.2$. Set $k = 1, 10, 20, 50, 100$ in MCS, the OOF, and approximate ROF from MCS with different k, are shown in Fig. 26.1. By increasing the simulation time k, the shape of approximate ROF becomes smoother.

The second test function is the two-dimensional expansion of f_1. Tolerance space of x_1 and x_2 are both $[-Tole, Tole], Tole = 0.2$, too.

$$\min_{x_i,\ i=1,2} \quad f_2(x) = \sum_{i=1}^{2} f_1(x_i)$$
$$s.t. \quad 0 \le x_i \le 10, \quad i = 1, 2$$

For the optimization of test functions, algorithm parameters are constant through the optimization process: $MGGA = 100$, $PopNum = 20$, $CF = 20$.

Typical records of the optimization of original objective function f_1 are shown in Fig. 26.2 (a), noted as non-robust optimal design. SCGA converges to the global non-robust minimum quickly.

Typical records of best solutions and corresponding approximate ROF of f_1 are shown in Fig. 26.2 (b)–(f), in the robust optimal design framework of SCGA, with

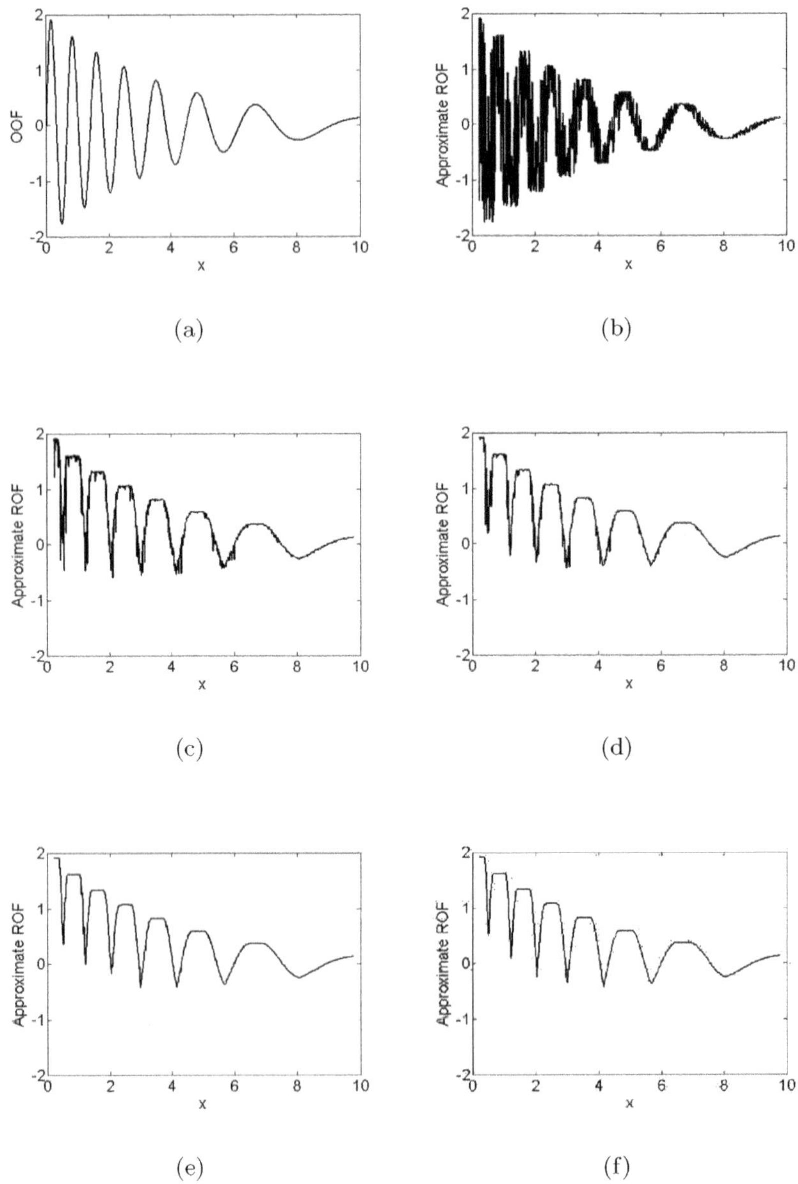

Fig. 26.1. Here (a) is the original objective function (OOF) of f_1, (b)–(f) are the approximate robust objective function (ROF) from MCS, with k = 1, 10, 20, 50, 100, respectively

$k = 1, 10, 20, 50, 100$ in MCS respectively. When $k = 1, 10, 20$, algorithm fails to convergence to the global robust minimum. Convergence is achieved when $k = 50$, but with more fluctuation than that of $k = 100$. It is apparent that SCGA achieves better convergence rate and better stability of solutions when increasing sampling number k in MCS.

To show the accuracy of robust solutions under the different setting of k, run the optimization program 100 times for f_1, select the best one as the solution for each run, and calculate the number of solutions located within [3.5, 4.5], denoted as N. Compute mean value (MV) and standard deviation (STD) of these solutions and their corresponding approximate ROFs, listed in Table 26.1. The exact robust minimum is 4.1611 and the accurate ROF equals to -0.4064.

When $k = 10$ and $k = 20$, algorithm fails in finding the global robust minimum in nearly half of the experiments. When $k = 50$ and $k = 100$, algorithm finds the global robust minimum in almost all experiments. It is the randomness in calculating approximate ROF that leads to the oscillation and inaccuracy of solutions. By increasing k, robust search performance is improved, with the cost of larger computation consumption.

26.4.2 Robust Optimal Design of Test Functions

In test function f_2, optimization program is executed for 100 times and select the best one as the solution for each run. N is denoted as the number of solutions located within $[3.5, 4.5]$ for each variable. MV and STD of these solutions and their corresponding approximate ROFs are listed in Table 26.2. With the expansion of dimension, the robust optimal design results are degraded, due to the degradation of global search ability of SCGA. Therefore, larger population number $PopNum$ and crowding factor CF should be used in high-dimensional practical robust optimal design problems.

Table 26.1. Optimization program is executed for 100 times for f_1. N is denoted as the number of solutions located within [3.5, 4.5]. Mean value (MV) and standard deviation (STD) of these solutions and their corresponding approximate ROFs are also shown for different k in MCS.

sampling number k	solution number N	MV of optimized solutions	STD of optimized solutions	MV of approximate ROF	STD of approximate ROF
1	0	NaN	NaN	NaN	NaN
10	47	4.1554	0.0313	−0.4772	0.0554
20	57	4.1611	0.0153	−0.4368	0.0263
50	94	4.1608	0.0079	−0.4152	0.0141
100	98	4.1615	0.0048	−0.4085	0.0093

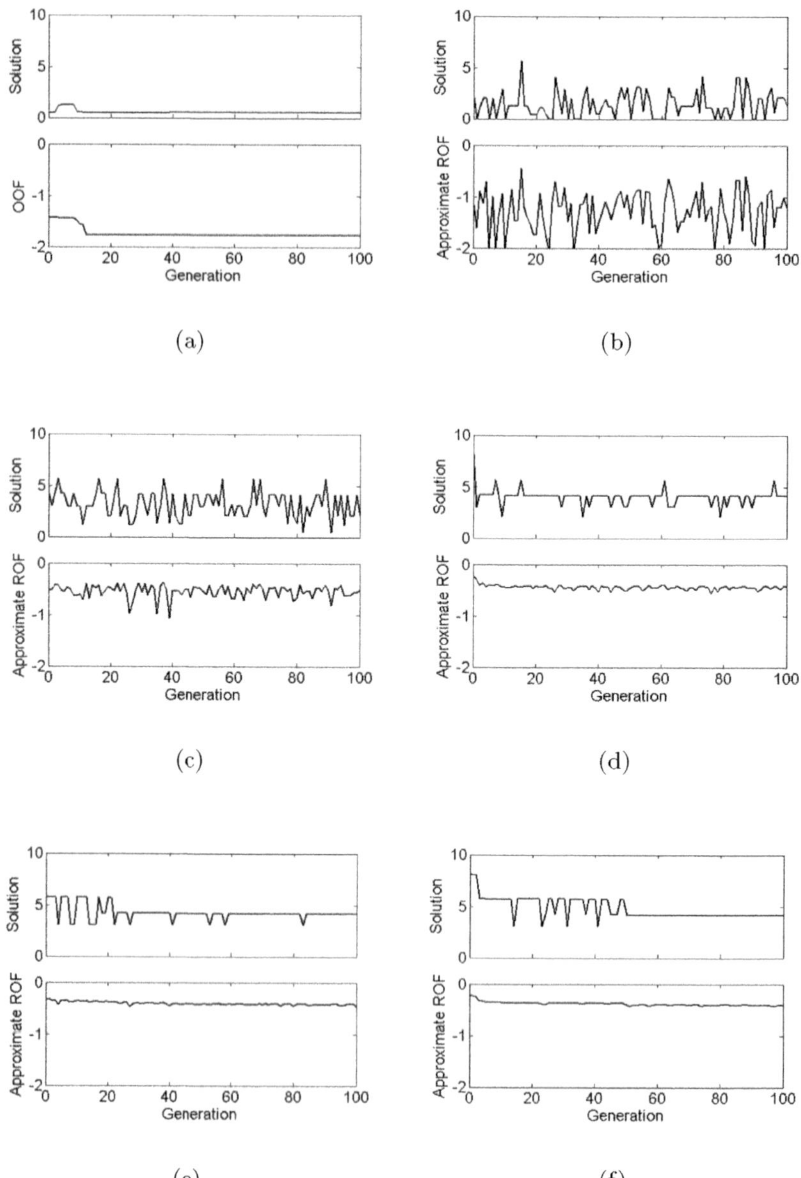

Fig. 26.2. Here (a) is the typical non-robust optimal design of f_1 under the SCGA framework. Typical robust optimal design of f_1 under SCGA framework is demonstrated in (b)–(f), with $k = 1, 10, 20, 50, 100$ in MCS respectively.

Table 26.2. Optimization program is executed for 100 times for f_2. N is denoted as the number of solutions located within [3.5, 4.5] for each variable. Mean value (MV) and standard deviation (STD) of these solutions and their corresponding approximate ROFs are also shown for different k in MCS.

sampling number k	solution number N	MV of optimized solutions	STD of optimized solutions	MV of approximate ROF	STD of approximate ROF
1	0	NaN	NaN	NaN	NaN
10	15	4.1596	0.0396	−1.0400	0.0772
		4.1795	0.0293		
20	18	4.1624	0.0311	−0.9863	0.0940
		4.1662	0.0180		
50	28	4.1594	0.0255	−0.8973	0.0492
		4.1566	0.0236		
100	32	4.1627	0.0253	−0.8592	0.0472
		4.1601	0.0177		

26.5 Practical Engineering Optimization

In this section, the standard crowding genetic algorithm (SCGA) with Monte-Carlo simulation (MCS) method is applied to the practical deterministic robust optimal design of recording optics of varied-line-spacing holographic grating (VLSHG) in National Synchrotron Radiation Laboratory (NSRL).

26.5.1 Recording Optics of Holographic Grating

Holographic gratings are fabricated by recording the interference fringes of two coherent sources in the photoresist coated on grating blanks. Given the shape of blanks, their focal properties can be adjusted by altering the properties of the two recording light sources. Ling [15] introduced auxiliary uniform-line-spacing gratings to generate aspherical wave-fronts to fabricate the varied-line-spacing holographic gratings (VLSHGs). VLSHGs are able to correct high order aberrations in diffractive optical systems and are widely used in high resolution spectrometers and monochromators.

Recording optical system is shown in Fig. 26.3. It consists of two coherent point light sources, C and D, and two uniform-line-spacing gratings, G_1 and G_2. Aim of optimal design is to format the distribution of the groove density as desired. The groove density is the function of four distance parameters and four angle parameters: p_C, q_C, p_D, q_D, γ, η_C, δ, η_D, known as recording parameters [15]. Objective function for minimization is constructed in [14].

26.5.2 Robust Optimal Design of Holographic Grating

One of the main challenges in VLSHG design is to find recording parameters which are insensitive to the fabrication tolerances. Conventional method deals with tolerances after the optimization process. Since the landscape of objective function

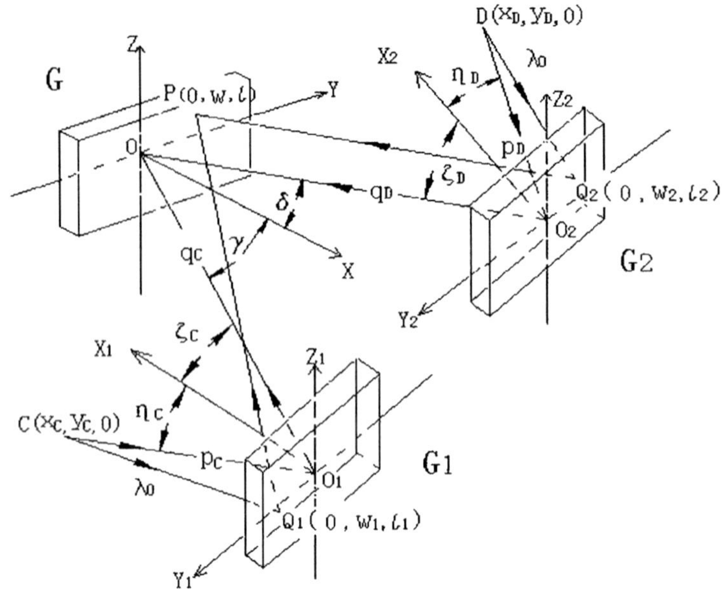

Fig. 26.3. Schematic diagram of recording system consists of two coherent point sources, C and D, two auxiliary uniform-line-spacing gratings, G_1 and G_2, and a plane grating blank G. Recording parameters are four distance parameters and four angle parameters.

Table 26.3. SCGA is applied to the robust optimal design of VLSHG. The best three sets of recording parameters are listed.

Group	$\gamma(rad)$	$\eta_C(rad)$	$\delta(rad)$	$\eta_D(rad)$	$p_C(mm)$	$q_C(mm)$	$p_D(mm)$	$q_D(mm)$
1	0.254	−0.254	0.978	1.130	515	1076	443	937
2	0.190	−0.190	0.875	1.108	434	1105	485	878
3	0.343	−0.343	1.155	1.213	264	1427	425	924

Table 26.4. Corresponding groove parameters are calculated for the optimized groove parameters. Sensitivities caused by fabrication tolerances are considered for groove parameters.

Group	$n_0(groove/mm)$	$b_2(mm^{-1})$	$b_3(mm^{-2})$	$b_4(mm^{-3})$
1	1.4001×10^3	8.2455×10^{-4}	3.0017×10^{-7}	-0.0002×10^{-10}
	$\pm 0.0037 \times 10^3$	$\pm 0.0254 \times 10^{-4}$	$\pm 0.0312 \times 10^{-7}$	$\pm 0.0260 \times 10^{-10}$
2	1.4002×10^3	8.2449×10^{-4}	3.0007×10^{-7}	-0.0005×10^{-10}
	$\pm 0.0038 \times 10^3$	$\pm 0.0304 \times 10^{-4}$	$\pm 0.0368 \times 10^{-7}$	$\pm 0.0314 \times 10^{-10}$
3	1.4004×10^3	8.2463×10^{-4}	3.0017×10^{-7}	-0.2830×10^{-10}
	$\pm 0.0033 \times 10^3$	$\pm 0.0189 \times 10^{-4}$	$\pm 0.0221 \times 10^{-7}$	$\pm 0.0143 \times 10^{-10}$

is highly rugged, conventional method is difficult to generate satisfactory robust solutions.

Optimization of VLSHG is a typical deterministic robust optimal design problem where the main concern is only the special realization of recording parameters. Here we use the proposed SCGA framework with MCS method to design a VLSHG in National Synchrotron Radiation Laboratory (NSRL) with groove density:

$$n = n_0(1 + b_2 w + b_3 w^2 + b_4 w^3), \quad -w_0 \leq w \leq w_0 \tag{26.9}$$

Groove parameters are: $n_0 = 1400 groove/mm$, $b_2 = 8.2453 \times 10^{-4} mm^{-1}$, $b_3 = 3.0015 \times 10^{-7} mm^{-2}$, $b_4 = 0.0000 \times 10^{-10} mm^{-3}$. Half width of grating is $w_0 = 90mm$. Auxiliary grating G_2 is with groove density $n_2 = 1000 groove/mm$, and auxiliary grating G_1 is with groove density $n_1 = 0 groove/mm$, i.e. a plane mirror. Upper bounds, lower bounds and the tolerances of recording parameters are from the physical constraints of recording table. For distance parameters, upper bounds are $2000mm$, lower bounds are $100mm$, and tolerances are $1mm$. For angle parameters, upper bounds are $\pi/2$, lower bounds are $-\pi/2$, and tolerances are $0.001rad$.

In SCGA and MCS, parameters are: $MGGA = 1000$, $PopNum = 100$, $CF = 100$, $k = 100$. The optimization program is executed and the best three sets of recording parameters are shown in Table 26.3. Corresponding groove parameters and the sensitivity of solutions are shown in Table 26.4. The sensitivities of solutions are defined as the error bounds under the fabrication tolerances.

Design results indicate that the SCGA algorithm with MCS method is able to obtain solutions insensitive to fabrication tolerances successfully. Compared to the conventional non-robust optimization methods which ignore the tolerances of design variables [14], the sensitivities of solutions decrease 2–10 times when using the proposed algorithm.

We finally select the recording parameters in group 1 of Table 26.3 as the actual recording parameters. The recording optics is shown in Fig. 26.4.

26.6 Conclusions

In this chapter, we study the robust optimal design problem, and establish the mathematical models, including objective robustness and feasibility robustness, probabilistic robustness and deterministic robustness. The focus is the modeling and optimization of the general form of deterministic objective robustness. The robust optimal design problem is solved in the SCGA framework. MCS method is applied to the calculation of approximated ROF. Numerical experiments indicate that SCGA will achieve better convergence rate and better stability of solutions when increasing simulation time k in MCS. The trade-off between the optimization results and the computation consumption is discussed. The proposed method is applied to the practical optimal design of VLSHG in NSRL, and obtains satisfactory robust solutions.

There are two future directions of the current work in deterministic robust optimal design. The first one is how to decrease the computation consumption effectively. The sampling strategies in [1] are the useful guidance, which are originally applied in the probabilistic robust optimization problems. Other choices include various sampling methods, such as Latin Hypercube Sampling (LHS) [11], and so on. The second direction is how to introduce local search to improve the fine optimization

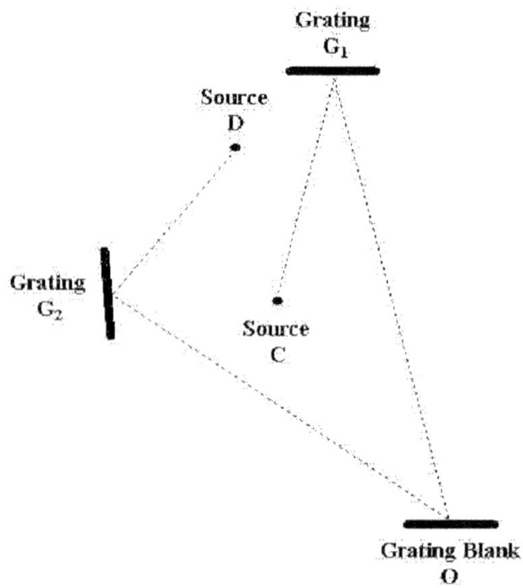

Fig. 26.4. Final recording optics adopts the optimized recording parameters in group 1 of Table 26.3. Sensitivities to the fabrication tolerances are acceptable.

of robust optimal design [20]. The development of hybrid robust optimization algorithm is of great importance in the practical design problems, especially when objective function landscapes are very complicated.

References

1. Branke J (2000): Efficient evolutionary algorithms for searching robust solutions. In: Proceedings of International Conference on Adaptive Computing in Design and Manufacture. 275–286
2. Chen W, Wiecek M, Zhang J (1999): Quality utility: a compromise programming approach to robust design. ASME Journal of Mechanical Design. 121:179–187
3. Das I (2000): Robustness optimization for constrained nonlinear programming problems. Engineering Optimization. 32:585–618
4. Deb K, Gupta H (2004): Introducing robustness in multi-objective optimization. KanGAL Report Number 2004016
5. Du X, Chen W (2000): Towards a better understanding of modeling feasibility robustness in engineering design. ASME Journal of Mechanical Design. 122:357–383
6. Gunawan S, Azarm S (2004): Non-gradient based parameter sensitivity estimation for single objective robust design optimization. ASME Journal of Mechanical Design. 395–402

7. Gunawan S, Azarm S (2005): Multi-objective robust optimization using a sensitivity region concept. Structural and Multidisciplinary Optimization. 29:50–60
8. Jin Y, Sendhoff B (2003): Trade-off between performance and robustness: an evolutionary multiobjective approach. In: Proceedings of International Conference on Evolutionary Multi-criterion Optimization. 237–251
9. Jin Y, Branke J (2005): Evolutionary optimization in uncertain environments: a survey. IEEE Transactions on Evolutionary Computation. 9:303–317
10. Kazancioglu E, Wu G, Ko J, Bohac S, Filipi Z, Hu S, Assanis D, Saitou K (2003): Robust optimization of an automobile valvetrain using a multiobjective genetic algorithm. In: Proceedings of the Design Engineering Technical Conference. 1–12
11. Lagaros N, Plevris V, Papadrakakis M (2005): Multi-objective design optimization using cascade evolutionary computations. Computer Methods in Applied Mechanics and Engineering. 194:3496–3515
12. Lee K, Park G (2001): Robust Optimization considering tolerances of design variables. Computers and Structures. 79:77–86
13. Li M, Azarm S, Aute V (2005): A multi-objective genetic algorithm for robust design optimization. In: Proceedings of the Genetic and Evolutionary Computation Conference. 771–778
14. Ling Q, Wu G, Wang Q (2005): Restricted evolution based multimodal function optimization in holographic grating design. In: Proceedings of IEEE Congress on Evolutionary Computation. 789–794
15. Ling Q, Wu G, Liu B, Wang Q (2006): Varied line spacing plane holographic grating recorded by using uniform line spacing plane gratings. Applied Optics. 45:5059–5065
16. Mattson C, Messac A (2003): Handling equality constraints in robust design optimization. In: Proceedings of the Structures, Structural Dynamics, and Materials Conference. 1–10
17. Messac A, Yahaya A (2000): Multiobjective robust design using physical programming. Structural and Multidisciplinary Optimization. 23:357–371
18. Muhlenbein H, Schomisch M, Born J (1991): The parallel genetic algorithm as a function optimizer. Parallel Computing. 17:619–632
19. Park Y, Kim N, Yim H (2000): Reliability-based design sensitivity analysis and optimization for the hyper-elastic structure using the meshfree method. In: Proceedings of the Pressure Vessels and Piping Conference. 1–11
20. Sorensen K (2004): Finding robust solutions using local search. Journal of Mathematical Modelling and Algorithms. 3:89–103
21. Su J, Renaud J (1997): Automatic differentiation in robust optimization. AAIA Journal. 35:1072–1079
22. Thomsen R (2004): Multimodal optimization using crowding-based differential evolution. In: Proceedings of IEEE Congress on Evolutionary Computation. 1382–1389
23. Tsutsui S, Ghosh A (1997): Genetic algorithms with a robust solution searching scheme. IEEE Transactions on Evolutionary Computation. 1:201–208
24. Tsutsui S (1999): A comparative study on the effects of adding perturbations to phenotypic parameters in genetic algorithms with a robust solution searching scheme. In: Proceedings of IEEE International Conference on Systems, Man and Cybernetics. 585–591

25. Wiesmann D, Hammel U, Back T (1998): Robust design of multilayer optical coatings by means of evolutionary algorithms. IEEE Transactions on Evolutionary Computation. 2:162–167
26. Yoon S, Choi D (2005): Probabilistic designs of air-bearing surface on manufacturing tolerances. ASME Journal of Tribology. 127:149–154

Index

Printed by Printforce, the Netherlands